Modern Cost - Engineering Techniques

Modern
Cost - Engineering
Techniques

An economic-analysis and cost-estimation manual, with comprehensive data on plant and equipment costs in the process industries

Edited by **Herbert Popper**
and the Staff of
Chemical Engineering

McGraw-Hill Book Company

New York · St. Louis · San Francisco
Dusseldorf · London · Mexico · Panama
Sydney · Toronto

Foreword

The data and techniques contained in this eight-section volume are intended for the engineer or technical manager in the process industries, to give him practical help in the economic aspects of his work. While certain parts may be particularly useful to the person who specializes in the cost and economics area, and other parts may be most useful to the nonspecialist, much of the material should be of considerable interest to both.

"Modern Cost Engineering Techniques" is the successor to the pioneering "Cost Engineering in the Process Industries," which appeared in 1960, edited by Cecil H. Chilton, and compiled from articles in *Chemical Engineering*, a McGraw-Hill publication. The current manual's contents have been selected from the broad range of cost and economics articles presented in *Chemical Engineering* during the ensuing decade.

In the area of capital-cost estimation, Sections 2 and 3 include two unified collections of process-equipment costs and related data (pp. 80 to 108 and 111 to 134). This has made it possible to cover the hardware spectrum quite concisely (even after including additional articles on some individual types of equipment), and to give additional space to areas of economic analysis—e.g., risk evaluation—with which technical people are becoming increasingly involved.

All articles that contain cost data include a base-period date or index level, so that the costs can be updated by means of appropriate cost indexes (see pp. 11 to 24). For rough estimates, it can also be assumed that the base period corresponds to the year of original publication; the tabulation on the next page shows the year in which every article in this volume originally appeared.

Herbert Popper

Page	Year	Page	Year	Page	Year
3	1967	209	1967	381	1965
11	1963	212	1967	389	1968
21	1969	218	1967	395	1967
25	1960	220	1968	399	1960
43	1960	225	1966	401	1968
59	1965	226	1968	407	1962
68	1967	232	1966	412	1965
73	1969	237	1966	417	1968
77	1969	240	1968	421	1966
80	1969	243	1964	427	1968
111	1964	245	1961	432	1968
135	1968	251	1962	435	1968
137	1964	253	1964	439	1965
138	1969	256	1966	445	1960
140	1966	263	1967	449	1965
141	1964	269	1967	451	1967
143	1964	273	1965	456	1964
145	1963	274	1964	458	1962
147	1968	277	1964	464	1962
151	1962	285	1961	468	1968
153	1962	305	1968	472	1965
155	1963	307	1967	477	1966
157	1962	316	1967	481	1965
160	1961	322	1968	487	1963
161	1969	329	1968	489	1961
177	1968	333	1967	493	1965
179	1965	342	1969	497	1966
185	1965	350	1967	503	1967
186	1964	356	1963	506	1963
189	1967	362	1966	510	1965
190	1967	367	1960	515	1968
192	1963	373	1960	520	1965
197	1965	377	1961	527	1969

Contents

* The 24-page report on pp. 111–134 (and also the 29-page report on pp. 80–108) contains data on a wide range of equipment and other capital-cost items. Subsequent articles in this section are arranged alphabetically, according to the bold-faced word in the title.

ix

1

Estimating Capital Costs: Principles and Techniques

Rapid Estimation of Plant Costs

It is often necessary to develop a plant-cost estimate on short notice. Which "rapid" estimating technique should be used? Here is an up-to-date discussion of the many methods now available to quickly calculate total installed plant costs, as well as costs of installed process sections, buildings and service systems.

JOHN T. GALLAGHER, *Mobil Oil Corp.*

In deciding on new industrial ventures, quite often it is not possible to wait for the preparation of a definitive capital-cost estimate because of the considerable amount of time and effort required. And yet, very frequently, some kind of cost estimate must be prepared under a pressing deadline.

Fortunately, due to the fast pace of modern business, techniques have been developed and correlations established for several extremely useful shortcut, over-all estimating techniques. Such techniques, in addition to providing quick, preliminary estimates, also serve for order-of-magnitude reviews of definitive estimates.

As with the development of other fields, such as medicine and economics, cost-estimating principles are being developed with the aid of ever-increasing published information. These principles enable one to predict rapidly, and with increasing accuracy, the cost relationships existing in today's chemical industry.

In what follows, we will describe and point out the relative merits and disadvantages of the various rapid cost-estimating methods that exist today. They may be categorized into four major classifications: (1) methods involving major-equipment-to-installed-cost factors, (2) exponential methods of capacity adjustment, (3) curve pricing, and (4) unit-price methods.

Major-Equipment-to-Installed-Cost Factors

The correlations between major-equipment and installed-plant costs developed by H. J. Lang,[1] have long been recognized as the forerunners to the development of rapid cost estimating.

Lang Factors—One of the more popular short-cut estimating techniques for over-all plant cost is the use of Lang factors, which are based on statistical formulas developed from analyses of many projects. The installed capital cost is obtained by multiplying the major-equipment purchase price by a statistical ratio.[2,3] (Typical major-equipment items are pumps, compressors, towers, exchangers, filters, tanks, reactors.) The following relationships apply:

- For solids process plants, $C_i = 3.1\,E$
- For solids-fluids process plants, $C_i = 3.63\,E$
- For fluids process plants, $C_i = 4.74\,E$

Here, C_i stands for installed capital cost, and E for major-equipment purchase cost.

Ratio of Installed Cost to Specific-Equipment Cost— A common error often encountered when using this rapid estimating method is the incorrect selection of ratios for a given material of construction. The installed-capital-cost-to-major-equipment-cost ratio for a predominantly carbon-steel system might be 4:1; if an estimator inadvertently applies this ratio to an all-stainless-steel system, the cost would be highly overestimated.

Table I illustrates the magnitude of the error associated with the misapplication of a carbon-steel cost ratio. The cost of the stainless-steel system is approximately twice that of an equivalent carbon-steel unit. The installed-plant-to-major-equipment purchase-price ratios are 4.0 (100/25) and 2.79 (209/75), respectively, for the carbon and stainless-steel systems. This illustrates the sharp variation in installed-cost ratios among the different primary metals used in chemical plant construction.

A series of installed-cost factors has been compiled[3]

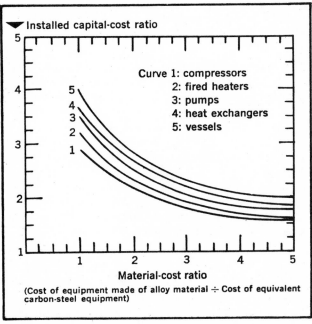

Curve 1: compressors
2: fired heaters
3: pumps
4: heat exchangers
5: vessels

(Cost of equipment made of alloy material ÷ Cost of equivalent carbon-steel equipment)

INSTALLED CAPITAL-COST RATIO for various major-equipment items made of alloy material[4]—Fig. 1

This article is based on a paper presented by the author at the AIChE meeting in New York City, Nov. 26 to 30, 1967.

Installed-cost comparison between plants having carbon-steel systems and those with stainless-steel equipment[7]—Table I

Major-Expense Items	Proportion of Cost of Major Construction Expenses for a Predominantly Carbon-Steel System, %	Relative Cost of an All Stainless-Steel System, as Compared to One of Carbon-Steel (Left Column)
Major-equipment purchase price	25.0	75.0
All other materials	25.5	76.5
Direct labor	21.0	25.0
Testing, field insurance, field equipment and supervision	9.5	9.5
Office cost and overhead	15.0	15.0
Contractor's profit (4% of sum of preceding items)	4.0	8.0
Total cost	100.0	209.0

Ratio of installed cost to equipment costs in a petroleum refinery[3]—Table II

Equipment	Ratio of Installed to Equipment Costs
Pressure vessels	4
Fractionating columns	4
Fired heaters	2
Heat exchangers	3.5
Pumps	4
Compressors	2.5
Instruments	4
Miscellaneous major equipment	2.5

for specific-equipment types relating primarily to the petroleum industry (Table II). These ratios apply to mixtures of materials that consist mainly of carbon steel (with relatively small amounts of alloy material), but are not applicable to all-alloy systems.

A more general approach to installed-cost ratios allows for differences in materials of construction.[4] This calculation is done in two steps: First, the cost ratio of the system material in question is obtained from Table III. Then, locating this cost ratio in Fig. 1, the corresponding installed capital-cost ratio is found. The mathematical formula is as follows:

$$C_i = R_m R_{ci} C_{cs} \qquad (1)$$

where C_i = installed capital cost; R_m = ratio of system material to carbon-steel system; R_{ci} = ratio of installed capital cost to major-equipment cost; and C_{cs} = cost of major equipment made of carbon steel.

The installed capital cost ratio obtained from Fig. 1 provides for major-equipment and commodities procurement, erection costs, various field costs, and home-office costs. This method of calculation is straightforward. For example: find the installed capital cost for a Type 304 all-stainless-steel distillation column, complete with overhead accumulator and recycle pumps, if equipment fob. purchase costs are as follows: tower = $20,000, accumulator = $10,000, pumps ($3,000 each) = $6,000.

Solution: from Table III, R_m for vessels = 2.50; R_m for pumps = 1.8. From Fig. 1, R_{ci} for vessels = 2.5; R_{ci} for pumps = 2.7. Therefore, installed costs = $(30,000 \times 2.5) + (6,000 \times 2.7) = \$91,200$.

Capacity Adjustment by Exponential Method

The exponential capacity-adjustment technique is a classical method, long known to the industry. It yields fairly accurate results, provided that the two plants under consideration are in the same capacity range. For example: if the cost of a 50-ton/day ammonia plant is known, and the cost of a 150-ton/day equivalent unit must be known, the capacity-adjustment formula should yield reasonably good results. However, if one wishes to know the cost of a 1,000-ton/day plant, it is doubtful whether the exponential adjustment of the 50-ton/day plant cost would result in a satisfactory estimate; the adjustment is over too broad a range.

The effect of size on the plant-cost exponent is considerable. In our example, reciprocating compressors would be used in the 50-ton/day plant, whereas centrifugal machines would undoubtedly be used in the 1,000-ton/day plant. This design change alone would distort the results obtained by using a constant exponent over the range of 50 to 1,000 tons/day.

In other situations, a small plant might require only one reactor, while a much larger plant might need two or more operating in parallel. Cost adjustments over the capacity range, where use of a single reactor is practical, can be made by using a constant exponent, but when the plant throughput exceeds the capability of one reactor, the exponent will change abruptly.

The capacity-adjustment exponential formula is as follows:

$$C_A / C_B = (P_A / P_B)^x \qquad (2)$$

where C_A = cost of Plant A; C_B = cost of Plant B; P_A = annual capacity of Plant A; P_B = annual capacity of Plant B; and x = exponential factor.

Selecting the Exponent

The earlier applications of the capacity-adjustment formula used a constant exponent value of 0.6 for most types of processes. Subsequent cost investigations have revealed that the exponent is not constant, but that it varies from process to process. Recent publications indicate that such variations range from 0.38 to 0.9.[5, 6] Use of the wrong exponent can introduce a considerable error into the calculation, which in extreme situations can exceed 25%. For example: with a capacity-adjustment ratio of 2.0, using the 0.6 exponent will

result in a 19% error, if the correct exponent should have been 0.9.

Questions often arise about the variation of cost with capacity of various sections within a process plant. It should be noted that there can be a significant variation in the sectional exponent, depending on the section of the plant under consideration. The over-all ratio for a particular plant may be comprised of widely diverging sectional exponents.

Typical of units where there are exponent sectional variations is the ethylene plant, where the exponent for the pyrolysis (or reaction) section is 0.8 to 0.9 (due to the need for multiple pyrolysis furnaces). The rest of the plant, namely the separation section, has an exponent of approximately 0.55, whereas the over-all plant exponent is 0.58 to 0.60.[7] In scaling up the pyrolysis section, if the over-all average exponent of 0.58 were used, a significant error would be introduced into the estimate.

The foregoing example emphasizes the importance of using extreme care when manipulating capacity adjustments of sections within a process unit. Before using over-all plant exponent values for sectional adjustments, the estimator should satisfy himself that the average value applies to all segments of the plant. A typical application of the exponent formula is presented for illustrative purposes:

If the cost of a 200-ton/day natural-gas-reforming ammonia plant is $4.5 million, what is the approximate cost of a 400-ton/day unit? To solve, we must find the value of C_A in Eq. (2), where $C_B = \$4,500,000$; $P_A = 400$ tons/day; $P_B = 200$ tons/day; and $x = 0.74$. Therefore, $C_A = 4,500,000(400/200)^{0.74} = \$7,500,000$.

Unfortunately, there are some discrepancies among the exponent factors published in recent articles, discrepancies that can be attributed primarily to varia-tions in the definition of the scope and size limitations associated with the respective exponents. In addition, it must be recognized that the exponent may vary among the different processes used to produce a given chemical.

Where exponents are not clearly defined, their use should be restricted to developing order-of-magnitude estimates only. For this type of calculation, the values in Table IV have been found to be satisfactory.

Curve Pricing

The most simplified estimating method consists of comparing the budget under consideration with previous cost data that have been graphically plotted, relating installed capital cost to plant capacity. Curve pricing, however, should be used only with extreme care, as capital costs obtained from plotted curves will at best have an accuracy of ± 15%.

One of the more common errors is the assumption that the cost obtained from a curve is broadly applicable to all plants manufacturing the same chemical, even though quite often published curves do not define clearly the bases for the plotted costs.

Consider the error introduced into an estimate where it is wished to determine the installed battery-limits capital cost for a 1,000-ton/day natural-gas-reforming ammonia plant using captive steam and electric power as the energy source for synthesis-gas compression. From Fig. 2, Curve 1, the approximate cost for this unit would be $11 million. If the estimator used data obtained from a different curve (which had been poorly defined), he might read $13.2 million, which is the approximate cost for a comparable 1,000-ton/day nat-ural-gas-reforming ammonia plant using gas-engine-driven compressors.

Ratio of alloy-material cost to carbon-steel cost for major-equipment items[4] (fob. price)—Table III

Material of Construction	Fabricated Equipment* Clad	Alloy	Heat Exchangers	Centrifugal Pumps
A-285, grade C carbon steel	1.0	1.0	.....	.....
Cast steel	.....	.....	.....	1.0
Stainless steel, Type 410	2.1	2.1	.....	1.5
Stainless steel, Type 405	2.25	2.25	.....	.....
Stainless steel, Type 304	2.75	2.50	.....	1.80
Stainless steel, Type 316	3.0	3.0	.....	2.0
Stainless steel, Type 310	3.25	3.25	.....	.....
Bronze	.....	.....	.....	1.5
Monel	6.5	4.0	.....	2.5
Carbon-steel shell, carbon-steel tubes	.....	.....	1.0	.....
Carbon-steel shell, aluminum-brass tubes	.....	.....	1.25	.....
Carbon-steel shell, Monel tubes	.....	.....	2.25	.....
Monel shell, Monel tubes	.....	.....	3.6	.....
Stainless steel, Type 304, shell and tubes	.....	.....	3.2	.....
Stainless steel, Type 316, shell and tubes	.....	.....	3.0	.....

* Applies to items such as drums, distillation towers and process vessels. Clad vessels have 1/8-in.-thick cladding on 5/8-in. carbon-steel plate
All alloy vessels designed to same pressure and temperature as equivalent carbon-steel vessel.

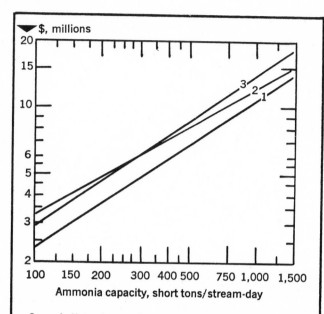

Curve 1: Natural gas reforming using captive steam turbine, electric-motor-driven compressors, and reciprocating compressors up to 400-ton/day capacity. (Above 400 tons/day, centrifugal compressors are used.)

Curve 2: Naphtha reforming. Same equipment as for Curve 1, except that air coolers are used.

Curve 3: Same as Curve 1, except that gas-engine drives are used for the compressors.

Note: Partial-oxidation ammonia plants cost approximately 25% more than above curve values.

INSTALLED CAPITAL COST (battery limits) of ammonia plants (1965 values)[7, 8, 9, 10]—Fig. 2

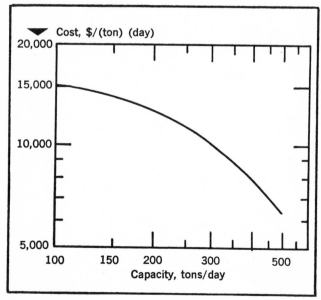

UREA PLANT COST (1964 values[11])—Fig. 3

The above would have introduced an error of 20% that would have to be added to the normal ± 15% accuracy. The reason for the addition would be that the error would have been one of incorrect scope-definition, and would not have been considered in computing the estimating tolerances.

The above example centers around a natural-gas-reforming ammonia plant, but there are also other less-used processes to manufacture ammonia, such as the partial oxidation process. A 1,000-ton/day ammonia plant based on partial oxidation of fuel oil, costs approximately 25% more than the corresponding natural-gas-reforming plant.[7] The higher cost is due primarily to parallel reactor trains, an air separation unit and a more complex synthesis-gas treatment.

The foregoing emphasizes one important aspect of curve pricing, which is that the estimator must be aware of the details of the process. In our example, he would have to know the number of parallel processing lines, the type of feedstock and the basic processing technology.

All aspects of curve pricing are not negative. It is a useful technique that allows rapid approximation of plant costs, the accuracy of which is more than sufficient for preliminary economic studies. Well-defined cost curves serve as a ready check of the order-of-magnitude accuracy of estimates prepared by other more detailed methods. The curve pricing method is an extremely useful management technique, providing the curve is well defined and its limitations are recognized.

Unit-Price Method

This is the most casual and least accurate of the over-all rapid-estimating techniques. It is a rule-of-thumb method, which simply multiplies the annual plant capacity by a unit cost. A typical unit price is expressed as installed capital cost per ton of annual production ($/(ton)(yr.).

Perhaps the most common error committed with the unit-price method is assuming that a particular unit price is essentially constant over a given range of plant capacities; this is definitely not true.

The curve plotted in Fig. 3 illustrates the significant variation in unit installed cost with plant capacity. The curve shows that the unit installed capital cost decreases from $15,300/(ton)(day) to $7,800/(ton)(day), as the plant capacity increases from 100 to 500 tons/day.[11] The chart variation in unit cost is typical of most kinds of chemical processing plants, the major exception being a processing unit composed of many parallel processing trains. As the capacity is increased or decreased, processing trains are either added or eliminated. In such instances, the unit price will remain essentially constant or vary only slightly.

The use of unit-price estimating is not restricted to process battery-limits estimates. Often it is used for estimating utility and other off-site costs, such as that of a closed cooling-water system—which would be represented as $/(gal.)(min.) of circulating water. Here, the unit cost would include provisions for (on an installed-cost basis): the cooling tower; circulating

pumps; tower basin; foundations; yard piping up to the process unit battery-limits; and engineering and contractor costs.

For a cooling-tower system, cooling the circulating water 15 F. and having an 8 to 10 F. approach,* the unit cost will vary from $43/(gal.)(min.) for a 3,000-gal./min. system, to $28/(gal.)(min.) for a 30,000-gal./min. circulating system. The installed cost of a cooling tower system is developed simply by multiplying the total circulating gal./min. by a unit cost. For example, the cost for a 20,000 gal./min. water-circulation system having the same design conditions stated above is (20,000 gal./min.) × [$30/(gal.)(min.)] = $600,000.

Unit-Price for Nonprocess Costs

Unit-cost parameters are also used extensively to develop steam, treated water, refrigeration, tank farm, and a host of other nonprocess system costs. Costs are expressed as $/lb. of steam generated, $/ton of refrigeration produced, $/bbl. of oil stored, etc.

The key to the proper application of the technique lies in the correct definition of a system. The estimator must ascertain that the system to be estimated agrees closely with the design parameters of the historical cost. If the outlet cooling water from the tower in the above example had a 5 F. rather than a 10 F. approach, the estimate would have been 50% too low.

Buildings associated with process plants are often estimated by the unit-price method. Building costs are normally expressed as $/sq.ft. of floor space, or $/cu.ft. of building volume. Building costs for order-of-magnitude estimates are commonly calculated by applying return costs from previous similar buildings. The total cost is obtained by multiplying the floor area by the unit price. This is a reasonably accurate method, provided the historical cost data is obtained from buildings similar in size, type of construction and mechanical complexity. Failure to recognize the importance of design differences or similarities may result in an estimating error of more than 100%.

Typical unit costs also vary considerably within the

---

*Approach=(required cold-water temperature)−(design wet-bulb temperature).

Type of building	Unit Cost, $/Sq. Ft. Floor Space
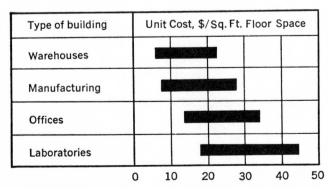

UNIT COSTS for various kinds of buildings (unequipped) associated with chemical plants[7]—Fig. 4

Scaleup factors for chemical plants[7]—Table IV

Type of Plant	Factor
Ethylene oxide	0.79
Ethanol (synthetic)	0.60
Styrene	0.68
Butadiene	0.59
Delayed coking	0.58
Formaldehyde	0.55
Benzene	0.61
Nitric acid	0.56
Oxygen	0.64
Acetylene	0.75
Methanol	0.83
Butyl alcohol	0.55
Isopropyl alcohol	0.60
Caustic	0.35
Phosphoric acid	0.58
Ammonium nitrate	0.54
Urea	0.59
Sulfuric acid (contact)	0.62
Chlorine (electrolytic)	0.35
Hydrogen cyanide	0.71
Ammonia (steam-reformed synthesis gas)	0.74
Ethylene	0.58
Polyethylene (low pressure)	0.67
Polyethylene (high pressure)	0.90

same building classification. The bar chart of Fig. 4 shows the range of costs for the common types of buildings normally associated with a chemical processing plant. Note the considerable overlapping of unit costs. In the extreme case, it is conceivable that on a unit-cost basis, a warehouse might cost more than a laboratory. The wide range of unit prices within a building classification is due to significant variations in unit cost with building size and design.

The careless use of historical unit-cost data can lead to gross errors in building estimates. Consider the development of the budget cost for a 50,000-sq.ft. bulk-products warehouse, where an estimator surmises that a unit cost of $8/sq.ft. would be applicable for pricing the new building (because of a recently built warehouse of similar construction). Thus, for an order-of-magnitude estimate, he assumes that the new building will cost approximately 50,000 × 8.00 = $400,-000, plus a reasonable contingency allowance. However, if the particular product to be stored were hydroscopic (requiring close humidity and temperature control), it would be required that the warehouse be fully air-conditioned. Such a system might cost as much as $6/sq.ft. A much more accurate approximation would have been (8 + 6) × 50,000 = $700,000.

Establishing Estimating Tolerance

Estimation tolerances are based on statistical analyses of return costs from similar projects, yet all too often in industry, the recording and classifying of

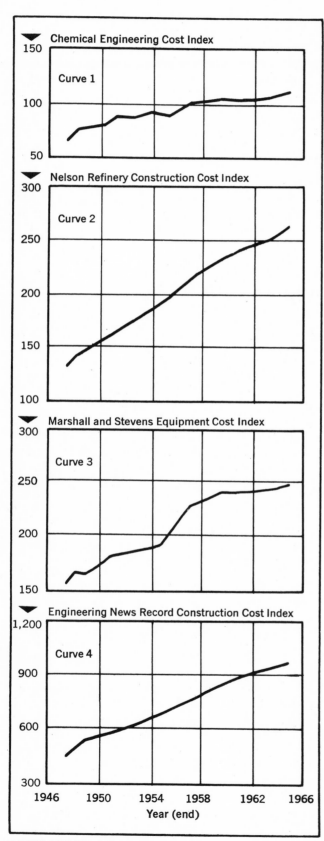

Curve 1
Chemical Engineering Cost Index

Curve 2
Nelson Refinery Construction Cost Index

Curve 3
Marshall and Stevens Equipment Cost Index

Curve 4
Engineering News Record Construction Cost Index

Year (end)

CONSTRUCTION COST INDEXES per year[12, 13]—Fig. 5

return construction costs into statistical data are not given the effort and time deserved. Sufficient and meaningful cost data are required to develop rapid and fairly accurate over-all estimating techniques.

Let us compute a typical estimating tolerance from the return costs of 15 process units, on the assumption that a sound cost-data retrieval system exists. To classify the return cost into statistical data, the cost must be carefully screened to ascertain that all the samples are representative of a homogenenous universe. For example, if 14 of the plants were built with no field labor problems and the 15th plant was delayed for three months because of a strike, the 15th observation should either be adjusted to reflect standard labor conditions, or eliminated from the universe.

All return costs should be compared to a uniform estimating base; in other words, all observations should be referenced to the same estimating method such as definitive, factor, exponential or unit-pricing. Likewise, the universe should not contain samples developed from more than one base, and all assumptions, qualifications and data adjustments should be clearly stated in computation sheets.

In our sample calculation (see Table V), the following assumptions are made: (1) the universe has a normal distribution and is comprised of 15 fluid-processing plants; (2) all observations have been screened and adjusted to reflect a homogeneous universe; (3) all projects were originally estimated by definitive cost estimates; (4) return costs were for inside-battery-limits units only (offsites were excluded).

The statistical formulas for the arithmetic mean and standard deviation apply:

$$\overline{X} = \Sigma fX/n, \text{ and } \sigma = \sqrt{[(\Sigma fX^2)/n] - \overline{X}^2}$$

where $\overline{X}$ = average of observations; f = frequency of occurrence (in our case, $f = 1$); X = value of observation; n = total number of observations; σ = standard deviation of universe.

With a normal universe, the chances are 95.46 out of 100 that the estimate X will lie between -2σ and $+2\sigma$. Since $2\sigma = 12.70$ (see Table V), the range for X is $84 < X < 109.4$.

When referred to the 100% norm, the estimating limits for this type of definitive estimate are -16.0 and $+9.4\%$. The accuracy can thus be stated as ranging from $X + 10\%$ to $X - 16\%$. For budget purposes, it would be recommended that a 10% contingency figure be added to the estimate.

Although the foregoing computation is based on projects originally estimated by definitive estimates, the same procedure can be used for establishing project-contingency values for other estimates.

Escalation of Plant Costs—Price Indexes

A review of the nationally recognized indexes indicates a general upward trend in the installed cost of industrial plants. Although in some years this escalation is negligible, in periods of high economic activity inflation may increase plant capital costs as much as 5% annually. Unfortunately, national indexes are nor-

mally three to six months behind actual conditions. Nevertheless, this published data represents a useful estimating tool. Return cost data used for estimating future projects must be updated by multiplying the historical data by the ratio of the present index and the index value existing during the period from which the return was compiled.

Chemical-plant costs were relatively stable during 1959 through 1964, but 1965 and 1966 were years of rapid escalation because of rising material costs, extreme field-labor shortages and higher wage rates. Present indications are that 1967 will be a period of moderate escalation, due primarily to continually increasing labor costs. In high inflationary periods, use of cost indexes must be tempered with an additional escalation allowance for increased cost not yet reflected in the published data.

The four most popular published price indexes are: *Engineering News Record* construction index (ENR), *Chemical Engineering* plant index (CE), Marshall and Stevens equipment index (M & S), and the Nelson refinery construction-cost index. From an over-all chemical plant-cost viewpoint, the CE and Nelson indexes are generally considered most applicable. The ENR index, which reflects the highest present rate of increase, measures primarily civil-construction cost. Since civil cost is a minor portion of a chemical plant budget, the ENR index is not particularly suited to chemical-plant construction. The following example illustrates the use of the indexes (Fig. 5):

Assume we need to know what would have been the approximate 1965 cost of a certain methanol plant, a duplicate of which cost $5 million in 1960. Using the

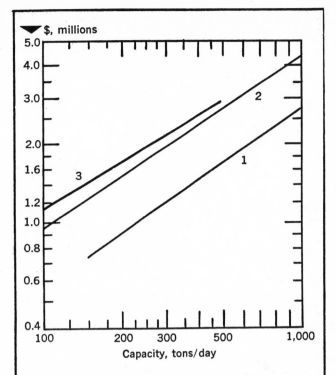

Curve 1: Sulfuric acid (100% basis). All rotating equipment electric motor-driven except main air blower (1966 costs).

Curve 2: Nitric acid (50 to 60% basis). Includes fume elimination, power recovery and waste-heat boiler (1964 costs).

Curve 3: Phosphoric acid (54% basis), wet process. Rock handling and grinding, capacity given as tons/day P_2O_5 (1964 costs).

APPROXIMATE INSTALLED CAPITAL COST (calculated for battery limits only) for acid plants[7, 14, 15]—Fig. 6

Average and standard deviation calculation—Table V

Observation Number	Return Cost Expressed as % of Definitive Estimate (X)	X^2
1	94	8,836
2	103	10,609
3	93	8,649
4	106	11,236
5	94	8,836
6	91	8,281
7	108	11,664
8	106	11,236
9	90	8,100
10	91	8,281
11	101	10,201
12	89	7,921
13	96	9,216
14	102	10,404
15	86	7,396
Total	1,450	140,866

For the above, $\overline{X} = 1,450/15 = 96.7$, and
$$\sigma = \sqrt{(140,866/15) - 96.7^2} = 6.35$$

CE plant index (from Fig. 5), we find that the escalation factors for 1960 and 1965 are 105 and 110, respectively, which makes the approximate 1965 cost $(110/105)(5,000,000) = \$5,230,000$.

Essentials of Over-All Cost-Estimating Methods

Over-all capital cost estimating techniques should: (1) have a systematic approach; (2) be general or all-encompassing; (3) be flexible in their applications; (4) give the required accuracy; (5) require a minimum of preparation time; and (6) be relatively easy to master by engineers and managers not well versed in estimating techniques.

For example: Assume that management has just completed an account-by-account review of the budget for a 1,000-ton/day sulfuric acid plant, the summation of which accounts totaled $2,150,000. Although the individual review of the accounts revealed no significant estimating errors, management was not convinced that this estimate represented an accurate appraisal of the plant capital cost; it was therefore suggested that the estimate be checked by the Lang factor.

Since major-equipment accounts totaled $620,000,

and—as previously stated—the Lang factor (which is the ratio of total installed capital cost to major-equipment cost) is 3.63, the computed cost is $3.63 \times 620,000 = \$2,250,000$. This figure would appear to be a reasonable check of the $2,150,000 detailed-cost estimate.

Assume, however, that someone still challenged the validity of the estimate, and supported the claim with a return over-all cost-capacity curve (Fig. 6), from which the approximate cost was read as $2,700,000. Since this third check revealed a significant deviation from the definitive estimate, it was agreed that the basis for the estimate would be re-evaluated.

Now, assume once more that the estimate-basis re-check revealed that one of the process towers had been omitted from the estimate summary. Since the Lang method is referenced to major-equipment cost, omission of one of the towers would not have been detected by this technique; but pricing curves plotted from return cost data would detect such an omission.

In any systematic approach, the estimator must be aware of the methods used to arrive at the various costs, and their attendant maximum deviations. Quite often, he can detect major errors in the technical design, which serves as the basis for estimating. For instance, the estimate may be quite in order, yet based on a poorly designed plant or method of operation.

While an estimator may not be responsible for the selection of the technology on which the estimate is to be based, he should confirm its correct application.

Most mathematical textbooks include a discussion of the "theory of significant numbers." In broad terms, this means that if certain data are accurate to a given tolerance, any computations performed with it should not be carried to a level more accurate than the least accurate of any of the components. For instance, if two numbers each accurate to the first decimal are multiplied, a meaningful answer is shown with only one decimal figure (i.e., $2.1 \times 1.4 = 2.9$, not 2.94). The same applies to additions: when adding 150,000 and 1,500.51, the total is best expressed as 151,500, not 151,500.51.

The theory of significant numbers has a direct bearing on cost estimating. If a particular cost account represents 1% of the total installed capital cost, and the over-all estimating accuracy is ±15%, there is little value in devoting too much time to estimating the cost of this account. Aside from a brief study to determine the account's approximate value, the estimator's time is best spent on the more significant cost items.

General and Flexible Approach

An estimating method should be selected with as little or as much detail as the particular situation dictates. The cost techniques available give the alternatives of estimating: the cost of each individual unit operation; each identifiable grouping of unit operations forming a process unit; the plant as a whole, as capacity varies. This flexibility is a key characteristic of estimating techniques.

For management to review a capital cost estimate—and to make the decisions dictated by the estimate—it must be aware of the general methods of estimating, as well as of their accuracies and limitations.

Accuracy and Ease of Application of Methods

An estimate is checked simply by using an alternate estimating method. Since it is not logical to spend more time in reviewing than it took to prepare the estimate, the review method normally used is one of the rapid over-all techniques discussed here. In other words, if an estimate is prepared with an accuracy of ±10%, the review method should have an accuracy of ±15 to 20%.

The final prerequisite for a rapid estimating method is that it should be based on uncomplicated techniques that can readily be applied by engineers and managers. The approach must be one based on logic, and statistical historical data and trends. Whenever possible, complex statistical formulas should be left out.

References

1. Lang, H. J., Engineering Approach to Preliminary Cost Estimates, *Chem. Eng.*, Sept. 1947, pp. 130-133.
2. Lang, H. J., Simplified Approach to Preliminary Cost Estimates, *Chem. Eng.*, June 1948, pp. 112-113.
3. Hand, W. E., From Flowsheet to Cost Estimate, *Petrol. Refiner*, Sept. 1958, pp. 331-334.
4. Clerk, Jackson, Multiplying Factors Give Installed Costs of Process Equipment, *Chem. Eng.*, Feb. 18, 1963, pp. 182-184.
5. Frumerman, R., Evaluating the Proposed Plant, *Chem. Eng.*, Oct. 1, 1962, pp. 101-106.
6. Nelson, W. L., What Is Effect of Size on Cost, *Oil Gas J.*, July 22, 1957.
7. Author's personal files.
8. Quartulli, Orlando J., Check List of High Pressure Reforming, *Hydrocarbon Process.*, April, 1965, pp. 151-162.
9. Gignier, J. P., and Quibel, J. H., Cat Reform Naphtha for Ammonia, *Hydrocarbon Process.*, Mar. 1965, pp. 154-156.
10. American Petroleum Institute report, Centrifugals Cut Ammonia Costs, *Hydrocarbon Process.*, May 1966, pp. 179-180.
11. Bellia, Francesco, The Economics of Large Urea Plants, *European Chem. News*, Oct. 16, 1964.
12. Chilton, C. H., Plant Cost Index Points Up to Inflation, *Chem. Eng.*, April 25, 1966, pp. 184-190.
13. Nelson, W. L., 1964 A Year of Challenge, *Oil Gas J.*, Oct. 16, 1964, p. 211.
14. Dark, A. M., Wet Process Phosphoric Acid, *European Chem. News*, Oct. 16, 1964, p. 53.
15. Pinning Down Acid Costs, *Chem. Week*, Jan. 15, 1966, pp. 23-24.

Meet the Author

John T. Gallagher is an associate engineer in Mobil Oil Corp.'s international project engineering department, 150 E. 42nd St., New York City. His earlier experience — obtained with General Aniline and Film Corp., Lummus Co., and Foster Wheeler Corp.—included assignments in project and proposal management in the petroleum and chemical fields. He has M.E. and M.S. degrees in mechanical engineering from Stevens Institute of Technology, and an M.B.A. from the City University of New York.

Key Concepts for This Article

Active (8)	Passive (9)	Purpose of (4)
Speed	Total	Evaluating
Estimating*	Partial	Projects
	Plants*	
	Costs*	

(Words in bold are role indicators; numbers correspond to EJC-AIChE information retrieval system. Asterisks mark key concepts suggested for indexing. Others are added to improve reading as an abstract. Indexing is described in *Chem. Eng.*, Oct. 11, 1965, p. 187.)

New Index Shows Plant Cost Trends

Here is CHEMICAL ENGINEERING's *unique plant construction cost index that is especially tailored to meet the needs of engineers in the process industries.*

THOMAS H. ARNOLD, *New Haven, Conn.*
CECIL H. CHILTON, *Editor-in-Chief*

Although there are literally dozens of published indexes of construction costs, not one has been tailor-made for chemical plant costs. CHEMICAL ENGINEERING is now introducing a chemical plant construction cost index to fill this noticeable void.[8, 18]

Before presenting the index and details about it, we'd like to point out its three specific advantages:

1. All labor and other wage or earning rates are continuously adjusted for trends in worker productivity. Thus, labor wage rates for construction, installation and fabrication, along with engineers' and draftsmen's salaries, are all adjusted by productivity or output-per-manhour factors before being compiled as cost index components.

2. Cost of engineering and supervision manpower for design and construction is included in the index. This component accounts for the cost of engineers, draftsmen, construction supervision, and clerical workers (such as timekeepers, expediters, records clerks, etc.).

3. No extraneous items are priced or weighted into the index. In other words, every item included in the index is actually found or used in chemical process plants—or is closely similar to such items. (It is not generally realized, for example, that a well-known oil refinery cost index includes the price of outboard motors as a component, and at a weight three times that given to centrifugal pumps.)

The four major components of the CE Plant Cost Index, with the percentage weight factor for each, are:

Equipment, machinery and supports............	61%
Erection and installation labor.................	22
Buildings, materials and labor..................	7
Engineering and supervision manpower..........	10
Total..	100%

The major component, by weight percent, is equipment, which has the following subcomponents and percentage weight factors:

Fabricated equipment........................	37%
Process machinery...........................	14
Pipe, valves and fittings.....................	20
Process instruments and controls..............	7
Pumps and compressors......................	7
Electrical equipment and materials............	5
Structural supports, insulation and paint........	10
Total..	100%

The price of pumps and compressors, for example, comprises 4.27% (0.07×61) of the total index. These subcomponents are priced on an unerected, uninstalled basis.

The CHEMICAL ENGINEERING Plant Cost Index, as well as the component indexes and the subcomponents of the equipment component, are given in Tables I and II for the years 1947 through 1961.

Three features of the index require explanation at this point:

1. All index components are based on 1957-59 = 100. This base period has been adopted because it is the period now used for most important official government economic indicators, such as the Federal Reserve System's index of industrial production. Many privately produced economic indexes are changing over to this base period, to provide easier interpretation of several types of data that vary with time. Actually, the choice of a base period is arbitrary; and it is a simple operation to convert to any other period by ratioing to that new base year (see Table III).

2. Our index data go back through 1947, since that is the initial year for which we could obtain valid price information for all components. Since there is rapid technological change in the chemical process industries, necessitating frequent plant modifications, it is unlikely that there is much need today to be concerned with updating costs for facilities built more than 15 years ago.

3. Indexes are reported for the four major components and the seven equipment subcomponents, along with the over-all Plant Cost Index. Therefore, engi-

neers who have component-weight factors specific to their company or operation that are different from our weights may use our component indexes with their own weights to construct a "personalized" composite index. The total or composite index is merely the sum of the products of each individual component index times the weight for that component.

Index Objectives

When one starts out afresh to develop an index that will represent the variation with time of the cost of chemical process plants, it soon becomes apparent that before anything can be accomplished, certain basic terms must be defined, and requirements or goals specified. These definitions and specifications are, of course, interrelated.

By "cost," we mean the most important and largest items of expense incurred by a company *after* it has made a decision to build a plant and has arranged the financing and bought the land (if needed) to build it, and *before* the plant is taken over by the operating personnel for putting it on stream. We have not attempted to include *all* the costs allocated to the construction of a plant between these two points in time. In our index, some of the costs not included are: site clearing and preparation; insurance and taxes during construction; company overhead allocated to the construction project; and contractor's profit.

The definition of "chemical process plants" is more intricate. CHEMICAL ENGINEERING and McGraw-Hill employ the term "chemical process industries" (CPI) to include more than 110 separate four-digit industries as designated by the government's Standard Industrial Classification (SIC) Code of 1957. These industries are conveniently grouped into 18 broad categories, including: chemicals and petrochemicals; fertilizers and agricultural chemicals; lime and cement; man-made fibers; paints, varnishes, pigments and allied products; petroleum refining; soap, glycerin and related products; wood pulp, paper and board.

It is obviously impossible to define a "typical" process plant from the widely varied types of plants suggested by the list above. But we can break down various plants into a common set of constituent or component parts. Thus, a fertilizer plant and an oil refinery, although totally different on the whole, can be characterized by similar parts. These parts form the same group of categories that are included in the component and subcomponent items in our index. Of course, the *weights* given to the different components should be different between a fertilizer plant and an oil refinery, since these represent different types of processing operations. However, as we shall see later, composite indexes calculated with drastically different weights show fairly close agreement.

So far as the term "index" is concerned, we are using it to mean that we have chosen some base period to which we will refer all other variations of costs with time. To get the relative change in costs from time T_1 to time T_2, it is only necessary to divide the value of an index at T_2 by its value at T_1.

We can now appropriately state the fundamental objectives of our indexes:

• To furnish the engineer who does capital cost estimating or economic evaluation with a new and better over-all cost index for complete process plants, as well as to provide indexes for the major components of such plants.

• To include as complete a list of *realistic* components in the over-all index as we reasonably can. Two specific examples are: all wages and salaries are corrected for productivity changes; and the cost of engineering services—one-tenth of our total plant cost—is included as a component.

• To construct an index that accurately and sensitively reflects the true cost trends of plants built in the process industries.

It is almost impossible to test in advance an index of this nature because of (a) the diversity of types of plants built over any meaningful time period, and (b) the rarity—if not impossibility—of finding two process plants built *exactly* alike on a number of occasions at different points in time. But preliminary evaluations of our index indicate that we have met our objectives, and therefore we believe that the CHEMICAL ENGINEERING Plant Cost Index is more accurate than any other published cost index.

Technical Details About the Indexes

It is beyond the scope of this article to discuss technical and statistical details of the various techniques for making index numbers. We used several excellent references, those on index numbers in general,[11, 15, 16, 19, 20, 21] and others particularly concerned with construction cost indexes.[7, 9, 10, 12, 13]

In brief, we decided to average the price changes of a moderate number of commodity items that are the major component parts of all process industry plants, weighting each item by the relative amount of total plant cost that it represents. Since, in using index numbers, the procedure is to find the ratio of indexes for two periods in time, it is evident that users are interested in the *trends* of prices and costs. Then, for every item not accounted for in a generalized index but that does appear in a specific process plant, one must in effect assume that the missing item's price trend is exactly parallel to that of the entire index. To be all-inclusive, therefore, a composite index should have as many price trends (i.e., components) represented in it as possible, approaching the *total* number of items actually found in a plant as the ultimate, most-accurate limit on index sensitivity.

The actual price inputs of our indexes will be government price indexes. In other words, our final index is the weighted average of the price indexes of the component-items. This is a mathematical simplification that reduces by one step the amount of computation in converting actual item-prices to a weighted composite-item index.

We have adopted the usual approach in calling our input data "price indexes" and our output data "cost indexes." By "price" is meant the amount a purchaser

pays a seller for an item, whereas "cost" is the total amount spent by the purchaser to put the item into operating condition in his plant.

Sources of Price Data

The largest source of commodity-price statistics in the U.S. is the Bureau of Labor Statistics (BLS) in the U.S. Dept. of Labor. BLS collects these data for compilation of its Consumer Price Index and Wholesale Price Index. For the Wholesale Price Index, BLS obtains price quotations for over 2,200 separate commodity items monthly and computes price indexes for each of them. In the CHEMICAL ENGINEERING Plant Cost Index, we use 70 separate BLS price indexes for commodities or materials. The largest portion of our Index is made up of 67 BLS price indexes, and because many single commodity indexes are combined into larger group indexes, the prices of 155 individual commodities are included in these 67 BLS indexes. The remaining three BLS indexes we use are large groups of commodity-price compilations that include the prices of over 450 individual items. But together these have a weight in our over-all index only one-ninth the total weight of the other 67 BLS indexes.[1, 2]

BLS also makes monthly surveys of hourly earnings of a large number of employees. We use five of the hourly-earnings data groups from BLS.

Finally, BLS makes annual surveys of salaries of engineers, draftsmen, clerical and administrative workers, from which we obtain trends in earnings for our engineering-services component. For engineers, we combine the BLS data with that from independent salary surveys conducted periodically (generally every two years) by Engineers Joint Council and the National Society of Professional Engineers. We use all three sources to obtain our earnings trends for engineers.

We are fully aware—as are those people concerned with the problem in the BLS—that the prices submitted by manufacturers and suppliers to the BLS are often only list prices, and sometimes are not indicative of the true price for which an item can be bought on the market at that particular time. But since we include a large number of component-items in our index, whose price-discounting factors probably vary widely, the BLS prices are entirely adequate for our use.

Deriving the Weights

The weight by which each price index is multiplied for summation into the total index is the proportional amount of the total cost of process industry plants represented by that particular index component. To arrive at these weights, we conducted a research study among about 60 process industry companies, equipment manufacturers, contractors, and consultants. About one-third were contacted by personal interviews, the rest by a mail survey.

We asked for complete cost breakdowns—into more than 40 subcategories—of process industry plants

built in the past ten years. Data from estimates or projects otherwise uncompleted were excluded. Also, we required that each project be identified by these categories: (a) the type of plant according to its end-products (on the SIC Code four-digit basis); (b) the type of processing done by the plant or unit; i.e., fluids (F), fluids/solids (F/S), or solids (S); (c) the type of project or plant built; i.e., new plant at new site (NP), new unit at an existing site (NU), or expansion of an existing unit (Ex); and (d) monetary size of the project and broad classes; i.e., less than $1 million, $1 million to $10 million, and more than $10 million.

We obtained data on 156 separate projects, broken down by type of processing as: Fluids, 103 projects; F/S, 41; S, 12. On a type-of-project basis, the breakdown was: NP, 35 projects; NU, 97; Ex, 24.

Taking the extremes in distribution of component weights (both of the four major components and the seven subcomponents of equipment) for type of process and project, we calculated composite indexes for each year in the 15-yr. period, 1947-1961. Surprisingly, the agreement between the total index values was extremely close, even though some of the individual weight factors in one index differed greatly from the corresponding weights in the other index. From this analysis, we concluded that for the chemical process industries, the differences between indexes based on the three types of processes and three types of projects were too minor to justify a series of multiple indexes. The CHEMICAL ENGINEERING Plant Cost Index, therefore, is calculated with a weighted average of all factors developed in our analysis of 156 individual projects.*

Description of Equipment Components

Here is a brief description of the seven subcomponents of the equipment component in the CHEMICAL ENGINEERING index:

1. Fabricated equipment. Included in this group are the following representative items of equipment: boilers, furnaces and heaters; columns and towers (with trays); heat exchangers, condensers and reboilers; process drums, reactors, pressure vessels and tanks; storage tanks and spheres; evaporators.

The weights are based upon a detailed study* of the actual amounts of materials and labor going into all the industries producing the above items. Price data for individual items—such as plate steel, structural steel, shapes and forms, sheet and strip steel, and various alloy products—are obtained from the BLS indexes. The manufacturing-labor wage component is from monthly BLS data (SIC Group 34—Fabricated Metal Products) appropriately corrected for productivity changes as described below. There are 16 BLS

---

* Each weight for the four major components was derived from the following number of projects, in the order in which the components appear in the tabulation at the beginning of this article: 156, 156, 145 and 131. The seven subcomponents of the major component (Equipment, machinery, and supports) are each based on the following number of projects, again listed in the order in which they have appeared in tabular form above: 89, 89, 135, 135, 89, 135, and 88.

materials-price indexes and six BLS process-item price indexes (e.g., pressure tanks, liquid storage tanks) combined in this subcomponent.

2. Process machinery. Here we refer to those items of machinery and equipment that are frequently bought off the shelf, rather than custom-fabricated. They usually involve some power source and mechanical-drive equipment, and are thought of as machinery (as opposed to vessels). Typical items of process machinery are: centrifuges; filters; mixing and agitating equipment; rotary kilns and dryers; conveyors and other materials-handling equipment; high-pressure, vacuum or refrigeration-producing equipment; extruders; crushing and grinding equipment; thickeners and settlers; fans and blowers.

This subcomponent index is calculated from: (a) a weighted average of a BLS general industrial machinery and equipment index; (b) an index computed from materials and labor components based on data from the 1958 Census of Manufactures;[3] (c) several process-industry machinery and equipment indexes from the BLS (e.g., conveyors, materials-handling equipment). There are 27 separate BLS price indexes used, plus the BLS average hourly earnings rate for SIC Group 35 (Machinery) corrected for productivity.

3. Pipe, valves and fittings. This subcomponent is made up of 22 separate BLS price indexes for these items.

4. Process instruments and controls. This is computed from a weighted average of materials and labor costs as revealed in the 1958 Census of Manufactures,[4] plus some instrument-items price indexes of BLS. Included in the total index are 20 BLS price indexes plus BLS average hourly earnings data for SIC Group 38 (Instruments and Related Products), corrected for productivity.

5. Pumps and compressors. This is a weighted average of four BLS price indexes for these items.

6. Electrical equipment and materials. There are 11 BLS price indexes weighted and averaged for this subcomponent (e.g., electric motors, transformers, switchgear, wire and cable).

7. Structural supports, insulation and paint. Included in this subcomponent are structural steel, foundation materials (concrete and reinforcing bars), insulation, lumber and paint. It is calculated from a weighted average (weights obtained from our survey) of six BLS price indexes.

Other Index Components

Buildings—This component is composed of a weighted average of a BLS special index of construction materials and the average hourly earnings of general building contractors labor, corrected for productivity. The ratio of materials to labor is 53:47.

Erection and Installation Labor—This is simply the average hourly earnings as determined by the BLS for the contract construction industry, adjusted for productivity changes of labor. Before choosing this source of data, we made an intensive study of alternative sources of construction labor data, both private and governmental. Our conclusion was that although absolute labor rates (e.g., $/hr.) were different from source to source, their trends were almost exactly the same in every case. For consistency of source, and because they represent the type of contractors and construction practices used by the process industries, BLS data were chosen.

Engineering and Supervision Manpower—This is made up of three earnings items: (a) engineers, (b) draftsmen, and (c) supervisory and clerical workers (e.g., construction foremen, expediters, inspectors, estimators, timekeepers, clerks). Earnings data for engineers are obtained from surveys made by the BLS annually,[5] the Engineering Manpower Commission of Engineers Joint Council, and the National Society of Professional Engineers (the last two surveys are made every other year). We use a simple extrapolation technique that assumes that earnings realized since the most recent survey vary at the same rate as they did between the last two surveys. This extrapolation is corrected, of course, as soon as new data from any one of three sources become available.

The average earnings for draftsmen, supervisory and clerical workers are obtained from the same annual BLS survey that includes engineers.

From our survey of process industry projects, we have given the following percentage weights to the three categories: engineers, 33%; draftsmen, 47%; supervisory and clerical workers, 20%. Average earnings are corrected for productivity changes before the component index is calculated.

Correcting for Productivity

Since one of the major advantages built into the CHEMICAL ENGINEERING indexes is a correction of wage and earnings rates for changes in productivity, we will describe in detail our procedure for doing this.

There is a scarcity of data and statistics on the rate of change of productivity or, more explicitly, output of real product per man-hour. The few studies that have been made are of recent origin.[6,14,17] Although productivity itself is a variable quantity—depending on a great number of economic factors—it is possible to correlate change in productivity with time and deduce a trend or average rate of change for any given time period. Expressing productivity trends as "% rate of change per year," the following values have been calculated for the stated time periods.[6,14,17]

	Av. % rate of change per year in output per man-hour		
	1899–1954	1947–1955	1948–1957
Total private economy	2.0	3.6	3.4
Total nonagriculture			2.5
Manufacturing	2.2	3.1	3.4
Chemicals	3.5	6.0	5.9
Durable goods			3.0
Non-manufacturing			2.3
Services		2.8	
Finance and trade	1.2	2.3	2.6
Construction		2.5	

From analyses of these data, along with consultations with experts in the field of productivity measurements and applications, we have chosen an average rate of change of productivity of 2.5%/yr. This rate is applied to all wage or salary figures used throughout our indexes, to correct them for the assumed productivity change.

Our mathematical procedure is to divide the stated earnings figure for any year by the factor $(1.025)^n$, where n is the number of years since an arbitrary base-year, when the factor was taken as 1.0000. Thus, for earnings ten years since the base year, the corrected-earnings figure is the stated or observed earnings rate ($/yr. or $/hr.) divided by $(1.025)^{10} = 1.2801$.

We are aware that productivity varies at different rates from year to year, but nevertheless, as shown in the sample table above, trends are not very different for various time periods, and as repeatedly stated before, cost indexes are built for accurately reflecting *trends* in various component-costs. Moreover, with the growing emphasis on automation and its effect on labor economy, we have received assurance that an ever-increasing amount of productivity data will be forthcoming in the future, with emphasis on even more segments of the economy. Therefore, once we have built a productivity factor into our index, it will be a simple matter to keep correcting it with better data as they appear.

Some mathematical examples should strengthen our conclusion that cost indexes are more accurate when the labor costs are corrected for changes in productivity. Assume that we have a cost index that is composed of 50% labor and other workers' earnings (roughly the case in CHEMICAL ENGINEERING'S total index), and that the true rate of change of productivity is 3.5%/yr. over a ten-year period when we are calculating the total index.

Case (1): Assume we are using a productivity rate of change of 2.5%/yr. Though 3.5% is 40% greater than 2.5%, productivity factors are actually equal to 1.0000 plus the fractional rate of change of productivity. In this example, the error in the labor component of the index in the tenth year due to use of the wrong rate would be $(1.035/1.025)^{10}$ minus 1.000, or 10%. But this means an error in the total index of only (10% × 0.50) or 5%.

Case (2): Assume we make *no* correction for productivity changes at all in our labor rates, which is equivalent to assuming a productivity factor of 1.0000. Then the error in the tenth year in the labor component of the index would be $(1.035/1.000)^{10}$ minus 1.000, or about 41%. In terms of the total index, we find that (41% × 0.50) or about 20% is the error in our computed total cost index.

These examples point out clearly that all cost indexes that have labor rates as components and that *do not* make corrections for labor productivity have built into them what index technicians call an "upward bias." This also shows that *some* correction to account for productivity change—which is based on economic data

Starting with the March 4th, 1963, issue, CHEMICAL ENGINEERING will devote a full page in every issue—just inside the back cover—to the reporting of cost indexes and other economic indicators. In addition to the new Plant Cost Index, this page will be the new location of *CE's* Chemical Consumption Index and the Marshall & Stevens Equipment Cost Index, now appearing regularly in other sections of the magazine. We plan also to list other indexes of value to chemical engineers.

For the Plant Cost Index and its major components, we shall in each issue show figures for three successive months in three stages of finality—preliminary, revised and final. Our "preliminary" figures will be based on BLS preliminary data for all material and labor components except contract-construction and general-building-construction labor; the latter two items will be estimated by *CE.* Our "revised" figures will use BLS final data, except for the two specific construction-labor items; these will be BLS preliminary data (these two figures lag the others, in BLS reporting, by a full month). Our "final" figures will merely reflect any change, a month later, in the BLS final figures for construction labor.

Thus the Plant Cost Indexes for any particular month would have appeared in six or seven consecutive biweekly issues in three different stages. The following figures for October 1962 show the typical variations that might be encountered in progressing from "preliminary" through "revised" to "final":

Cost Indexes October 1962	Preliminary	Revised	Final
Plant Cost Index	102.2	102.1	102.1
Equipment, machinery, Supports	100.7	100.8	100.8
Construction labor	106.1	105.8	105.8
Buildings	101.7	101.6	101.6
Engineering and supervision	102.5	102.5	102.5
Fabricated equipment	101.7	101.7	101.7
Process machinery	101.8	101.7	101.7
Pipe, valves, fittings	100.5	100.8	100.8
Process instruments	105.8	105.7	105.7
Pumps and compressors ...	100.6	100.6	100.6
Electrical equipment	89.5	88.9	88.9
Structural supports, misc...	98.5	98.7	98.7

Annually, we will publish over-all cost indexes for the preceding year, along with a round-up of all the yearly indexes starting with 1947.

We hope that users of the indexes will keep us informed of their experience with them in cost estimating and economic evaluation work. Only through this sort of critical feedback can we hope to increase further the utility of this new estimating tool.

and reasoning—is better than assuming a change of *zero*, at least over any reasonable period of time. It becomes the index maker's responsibility to use the latest and most reliable productivity information to reduce any residual errors in his indexes due to this factor; we will do this with all CHEMICAL ENGINEERING indexes.

Weight Factors and Component Groups

0.61 Equipment machinery & supports

0.37 Fabricated equipment

0.14 Process machinery

0.20 Pipe, valves & fittings

0.25 Typical process machinery

0.50 Process machinery materials & labor

0.25 General industrial machinery

0.74 Pipe & fittings

0.86 Pipe & tubing

BLS Code No.	Weight Factor	Component
10-72-01	0.053	Pressure tank
10-72-03	0.138	Pressure vessel
10-72-11	0.064	Bulk storage tank, 6000 gal.
10-72-12	0.029	Bulk storage tank, 10,000 gal.
10-72-13	0.041	Liquid storage tank, 10,000 bbl.
10-72-15	0.040	Liquid storage tank, 50,000 bbl.
Subtotal	(0.365)	Typical fabricated products
10-14-26	0.216	Plates, carbon steel
10-14-31	0.010	Structural steel shapes
10-14-37	0.007	Bars, alloy steel
10-14-38	0.010	Bars, stainless steel
10-14-39	0.015	Bars, carbon steel
10-14-47	0.047	Sheets, carbon steel
10-14-49	0.011	Sheets, stainless steel
10-14-50	0.008	Sheets, alloy steel
10-14-51	0.006	Strip, carbon steel
10-14-52	0.015	Strip, stainless steel
10-14-56	0.012	Pipe, black, carbon steel
10-14-63	0.010	Mechanical tubing, carbon steel
10-14-65	0.006	Mechanical tubing, stainless
10-15	0.012	Castings & foundry products
10-16	0.006	Ferroalloys & pig iron
10-25	0.022	Non-ferrous mill shapes
Subtotal	(0.413)	Components of fabricated products
SIC 34	0.222	Fabricated products labor
11-44	0.15	Material handling equipment
11-47	0.10	Fans & blowers
11-52-11	0.15	Solids classifier
11-52-12	0.15	Ore flotation machine
11-52-13	0.05	Solids concentrator
11-52-32	0.05	Jaw crusher
11-52-34	0.05	Roll crusher
11-52-41	0.075	Gyratory crusher
11-52-42	0.075	Rod mill
11-52-43	0.075	Ball mill
10-14-26	0.060	Plates, carbon steel
10-14-37	0.010	Bars, alloy steel
10-14-38	0.017	Bars, stainless steel
10-14-39	0.103	Bars, carbon steel
10-14-47	0.033	Sheets, carbon steel
10-14-49	0.019	Sheets, stainless steel
10-14-50	0.010	Sheets, alloy steel
10-14-51	0.014	Strip, carbon steel
10-14-52	0.024	Strip, stainless
10-15	0.120	Castings & foundry products
10-16	0.020	Ferroalloys & pig iron
10-25	0.050	Nonferrous mill shapes
11-73-32	0.032	Electric motor, 3 hp.
11-73-34	0.028	Electric motor, 10 hp.
11-73-35	0.010	Electric motor, 250 hp.
Subtotal	(0.550)	Components of process machinery
SIC 35	0.450	Process machinery labor
Special	1.0729	Total machinery & equipment
11-1	-0.0729	Agricultural machinery (negative)
10-14-56	0.120	Pipe, black, carbon steel
10-14-57	0.060	Pipe, galvanized, carbon steel
10-14-58	0.180	Line pipe, carbon steel
10-14-61	0.040	Pressure tubes, carbon steel
10-14-63	0.100	Mechanical tubing, carbon steel
10-14-65	0.040	Mechanical tubing, stainless steel
10-15-26	0.080	Pressure pipe, cast iron
10-15-15	0.070	Brass tubing
10-25-51	0.020	Copper tubing, ⅜ in. O.D., coils
10-25-52	0.070	Copper tubing, ¾ in. O.D., lengths
10-25-53	0.090	Copper tubing, ⅜ in. O.D., coils

Major weightings:

- 0.22 Erection & installation labor
- 0.07 Buildings, materials & labor
- 0.10 Engineering & supervision

Sub-weightings:

- 0.07 Process instruments & controls
- 0.07 Pumps & compressors
- 0.05 Electrical equipment & materials
- 0.10 Structural supports, insulation & paint
- 0.33 Engineers
- 0.47 Draftsmen
- 0.20 Clerical

- 0.26 Valves
- 0.14 Fittings
- 0.50 Instruments & controls materials & labor
- 0.50 Typical analogous instruments
- 0.50 Engineers (BLS)
- 0.35 Engineers (EJC)
- 0.15 Engineers (NSPE)

Code	Weight	Description
10-74-95	0.130	Fabricated steel pipe & fittings
Special	1.000	Industrial fittings
Special	1.000	Industrial valves
10-14-26	0.024	Plates, carbon steel
10-14-38	0.011	Bars, stainless steel
10-14-39	0.026	Bars, carbon steel
10-14-47	0.044	Sheets, carbon steel
10-14-49	0.013	Sheets, stainless steel
10-14-50	0.010	Sheets, alloy steel
10-14-51	0.006	Strip, carbon steel
10-14-52	0.016	Strip, stainless steel
10-15	0.150	Castings & foundry products
10-25	0.120	Nonferrous mill shapes
11-73-01	0.009	Fractional horsepower motor, 1/6 hp.
11-73-13	0.032	Fractional horsepower motor, 1/4 hp.
11-73-14	0.029	Fractional horsepower motor, 1/2 hp.
Subtotal	(0.500)	Components of instruments
SIC 38	0.500	Instrument manufacturing labor
11-72	1.000	Electrical meters & instruments
11-41-01	0.250	Reciprocating duplex pump
11-41-21	0.400	Centrifugal pump
11-41-31	0.250	Rotary pump
11-41-42	0.100	Stationary compressor
11-73-32	0.160	Electric motor, 3 hp.
11-73-41	0.030	Electric motor, 5 hp.
11-73-34	0.140	Electric motor, 10 hp.
11-73-35	0.020	Electric motor, 250 hp.
11-73-51	0.040	Generator, d.c.
11-73-62	0.040	Generator, a.c.
11-74	0.180	Transformers & power regulators
11-75	0.320	Switchgear, switchboard, etc. equipment
10-26-01	0.030	Copper wire, bare
10-26-11	0.020	Sheathed electrical cable
10-26-21	0.020	Flexible electrical cable
10-14-31	0.280	Structural steel shapes
10-14-41	0.130	Reinforcing bars
13-2	0.190	Concrete
13-72	0.270	Insulation
Special	0.070	Lumber & wood products
06-21	0.060	Paint
Special	1.000	Contract construction labor
Special	0.530	Construction materials
Special	0.470	General building construction labor
BLS	0.170	Engineers, Class II
BLS	0.330	Engineers, Class III
BLS	0.330	Engineers, Class IV
BLS	0.170	Engineers, Class V
EJC	0.200	Engineers, 5 yr.
EJC	0.400	Engineers, 10 yr.
EJC	0.400	Engineers, 15 yr.
NSPE	0.200	Engineers, 5 yr.
NSPE	0.400	Engineers, 10 yr.
NSPE	0.400	Engineers, 15 yr.
BLS	0.340	Draftsman, junior
BLS	0.620	Draftsman, senior
BLS	0.040	Tracer
BLS	0.250	Bookkeeping machine operator I
BLS	0.050	Bookkeeping machine operator II
BLS	0.400	Clerk, accounting, I
BLS	0.300	Clerk, accounting, II

CE's Plant Cost Index
shows the least rate of growth

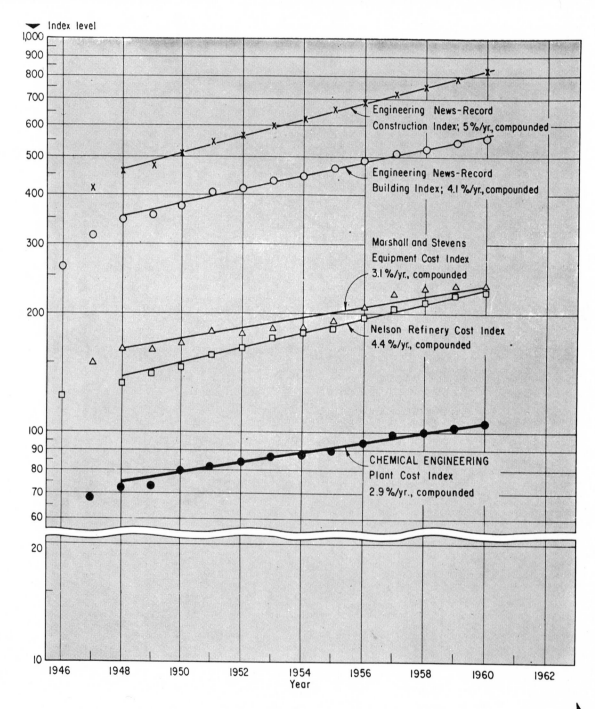

Data on CE Index components since 1947

Trend of plant costs since 1947 – Table I

	Year															
	1947	1948	1949	1950	1951	1952	1953	1954	1955	1956	1957	1958	1959	1960	1961	1962[p]
CHEMICAL ENGINEERING Plant Cost Index	64.8	70.2	71.4	73.9	80.4	81.3	84.7	86.1	88.3	93.9	98.5	99.7	101.8	102.0	101.5	102.1
Equipment, machinery and supports	60.3	65.6	67.2	69.8	77.6	77.8	80.9	82.3	85.1	92.7	98.5	99.6	101.9	101.7	100.2	100.8
Erection and installation labor	71.6	77.7	79.1	80.5	85.1	87.4	91.6	93.5	93.5	95.8	98.6	100.0	101.4	103.7	105.1	105.7
Buildings, materials and labor	71.6	78.9	78.8	82.2	88.1	88.5	91.4	93.1	95.0	98.0	99.1	99.5	101.4	101.5	100.8	101.5
Engineering and supervision manpower	72.5	76.2	74.9	78.2	82.2	84.8	88.3	87.9	92.0	94.2	98.2	99.3	102.5	101.3	101.5	103.0

p – preliminary estimate. Annual BLS price data for 1962 were not available at time of publication.

Trend of equipment costs since 1947 – Table II

	Year															
	1947	1948	1949	1950	1951	1952	1953	1954	1955	1956	1957	1958	1959	1960	1961	1962[p]
Equipment, machinery and supports	60.3	65.6	67.2	69.8	77.6	77.8	80.9	82.3	85.1	92.7	98.5	99.6	101.9	101.7	100.2	100.8
Fabricated equipment	63.2	68.4	69.6	71.5	78.4	79.0	81.3	81.4	84.2	92.5	99.5	99.6	100.9	101.2	100.1	101.5
Process machinery	58.4	63.1	66.2	69.4	76.5	77.5	80.6	82.8	85.3	92.2	98.1	100.1	101.8	101.8	101.1	101.5
Pipes, valves and fittings	53.2	60.1	61.9	64.7	73.1	73.8	78.0	79.5	85.2	94.8	97.9	98.8	103.3	104.1	101.1	100.7
Process instruments and controls	64.2	68.2	69.3	71.9	79.8	80.0	82.9	85.1	86.7	91.2	96.7	100.4	102.9	105.4	105.9	105.8
Pumps and compressors	53.8	58.2	62.3	65.4	73.9	73.4	77.5	79.5	81.7	90.0	97.5	100.0	102.5	101.7	100.8	100.6
Electrical equipment and materials	61.8	64.2	64.2	68.3	79.7	79.3	82.0	83.0	84.3	93.5	98.4	100.6	101.0	95.7	92.3	90.0
Structural supports, insulation and paint	66.6	73.5	75.0	77.6	82.0	83.0	86.0	88.6	90.5	92.5	98.0	100.4	101.6	101.9	99.8	99.0

p – preliminary estimate. Annual BLS price data for 1962 were not available at time of publication.

Converting CE's Plant Cost Index to other base years – Table III

Base year of index	Year														
	1947	1948	1949	1950	1951	1952	1953	1954	1955	1956	1957	1958	1959	1960	1961
1947	100.0	108.3	110.2	114.0	124.1	125.5	130.7	132.9	136.3	144.9	152.0	153.9	157.1	157.4	156.6
1948	92.3	100.0	101.7	105.3	114.5	115.8	120.7	122.6	125.8	133.8	140.3	142.0	145.0	145.3	144.6
1949	90.8	98.3	100.0	103.5	112.6	113.9	118.6	120.6	123.7	131.5	138.0	139.6	142.6	142.9	142.2
1950	87.7	95.0	96.6	100.0	108.8	110.0	114.6	116.5	119.5	127.1	133.3	134.9	137.8	138.0	137.3
1951	80.6	87.3	88.8	91.9	100.0	101.1	105.3	107.1	109.8	116.8	122.5	124.0	126.6	126.9	126.2
1952	79.7	86.3	87.8	90.9	98.9	100.0	104.2	105.9	108.6	115.5	121.2	122.6	125.2	125.5	124.8
1953	76.5	82.9	84.3	87.2	94.9	96.0	100.0	101.7	104.3	110.9	116.3	117.7	120.2	120.4	119.8
1954	75.3	81.5	82.9	85.8	93.4	94.4	98.4	100.0	102.6	109.1	114.4	115.8	118.2	118.5	117.9
1955	73.4	79.5	80.9	83.7	91.1	92.1	95.9	97.5	100.0	106.3	111.6	112.9	115.3	115.5	114.9
1956	69.0	74.8	76.0	78.7	85.6	86.6	90.2	91.7	94.0	100.0	104.9	106.2	108.4	108.6	108.1
1957	65.8	71.3	72.5	75.0	81.6	82.5	86.0	87.4	89.6	95.3	100.0	101.2	103.4	103.6	103.0
1958	65.0	70.4	71.6	74.1	80.6	81.5	85.0	86.4	88.6	94.2	98.8	100.0	102.1	102.3	101.8
1959	63.7	69.0	70.1	72.6	79.0	79.9	83.2	84.6	86.7	92.2	96.8	97.9	100.0	100.2	99.7
1960	63.5	68.8	70.0	72.5	78.8	79.7	83.0	84.4	86.6	92.1	96.6	97.7	99.8	100.0	99.5
1961	63.8	69.2	70.3	72.8	79.2	80.1	83.4	84.9	87.0	92.5	97.0	98.2	100.3	100.5	100.0

Acknowledgements

Two associates on *CE*'s editorial staff furnished highly significant help throughout this project: Theodore R. Olive, who 15 years ago laid important groundwork for this project,[18] and Robert B. Norden, currently responsible for editorial coverage of the cost-engineering area in the process industries.

Finally, here is a partial listing of other individuals and organizations who provided invaluable data and advice throughout this project.

American Cyanamid Co., H. Carl Bauman
Austin Co.
Badger Co.
Bechtel Corp., Donald F. Brosnan
Blaw-Knox Co.
Brown Instruments Div., Minneapolis-Honeywell Regulator Co.
Bureau of the Census, U. S. Dept. of Commerce, Maxwell R. Conklin, Chief, Industrial Div.
Bureau of Labor Statistics, U. S. Dept. of Labor, Harold Goldstein, Chief, Div. of Manpower and Employment Statistics; Leon Greenberg, Chief, Div. of Productivity and Technological Developments; Vincent E. Covins, Chief, Branch of Industrial Prices; Tovio P. Kanninen, Chief, Branch of Occupational Wage and Salary Surveys.
Catalytic Construction Co.
Celanese Chemical Co., F. V. Marsik and J. H. Smithson
Diamond Alkali Co., Blase V. Nemeth
F. W. Dodge Corp.
Dow Chemical Co., Midland, Mich.
Dow Chemical Co., Western Div., William G. Clark
Downingtown Iron Works, Frank Rubin
Engineering News-Record, Elsie Eaves
Engineers Joint Council, Manpower Commission
Federal Reserve System, Div. of Research & Statistics
Foster Wheeler Corp., Irwin Bromberg
Foxboro Co., W. Howe
B. F. Goodrich Chemical Co.
Gulf Oil Corp., J. H. Hirsch
M. W. Kellogg Co., John W. Hackney
Koppers Co., J. F. Rigatti
McGraw-Hill Dept. of Economics
Arthur G. McKee & Co., W. S. Jarvis
Monsanto Chemical Co., Norman Bach
National Bureau of Economic Research
National Industrial Conference Board
National Society of Professional Engineers
Phillips Petroleum Co., Phillips Chemical Co.
Pittsburgh Chemical Co., John F. Lovett
Pittsburgh Plate Glass Co.
Pritchard & Abbott Valuation Engineers, John E. Haselbarth
Procter & Gamble Co., C. R. Hirt and W. H. Patterson
Socony Mobil Oil Co., Wesley J. Dodge
Standard Oil Co. (Ohio), H. F. West

References

1. "A Description of the Wholesale Price Index," *Monthly Labor Review*, Feb. 1952, pp. 180-187.
2. Bureau of Census, U. S. Dept. of Commerce, 1958 Census of Manufactures, Vol. II, part 2, pp. 34C-2; 34C-5; 34C-14; 34C-20; 34C-28-30.
3. Ibid., pp. 35D-5; 35D-23; 35E-27.
4. Ibid., pp. 36A-6; 36A-8; 36A-14; 36A-20; 36A-25; 36A-29.
5. Bureau of Labor Statistics, U. S. Dept. of Labor, National Survey of Professional, Administrative, Technical and Clerical Pay, 1961, BLS Bull. 1310.
6. Bureau of Labor Statistics, U. S. Dept. of Labor, "Trends in Output per Man-Hour in the Private Economy, 1909-1958," BLS Bull. 1249 (Dec. 1959).
7. Chawner, L. J., "Construction Cost Indexes as Influenced by Technological Change and Other Factors," *Jour. of Amer. Statistical Assn.*, 30, no. 191, pp. 561-576 (Sept., 1935).
8. Chilton, Cecil H., "What Today's Cost Engineer Needs," *Chem. Eng.* 66, Jan. 12, 1959, p. 131.
9. Colean, M. L. and Newcomb, R., "Stabilizing Construction: The Record and Potential," McGraw-Hill, New York, 1952.
10. Dennis, S. J., "Recent Progress in Measuring Construction," a paper presented at the Annual Meeting of the Amer. Statistical Assn., New York City, Dec. 28, 1961.
11. Fisher, Irving, "The Making of Index Numbers," 3rd ed. rev., Houghton Mifflin, New York, 1927.
12. Foss, Murray, "How Rigid Are Construction Costs during Recessions?" *The Jour. of Business*, Graduate School of Business, U. of Chicago, 34, no. 3, pp. 374-383 (July, 1961).
13. Kaplan, N. M., "Some Methodological Notes on the Deflation of Construction," Jour. of the Amer. Statistical Assn., 54, no. 287, pp. 535-555, Sept., 1959.
14. Kendrick, John W., "Productivity Trends in the United States," Princeton University Press, Princeton, 1961.
15. Mitchell, W. C., "The Making and Using of Index Numbers," Bureau of Labor Statistics, U. S. Dept. of Labor, BLS Bull. 656.
16. Mudgett, Bruce D., "Index Numbers," John Wiley & Sons, Inc., New York, 1951.
17. National Bureau of Economic Research, "Output, Input, and Productivity Measurement; Studies in Income & Wealth," Vol. 25, Conference on Research in Income and Wealth, Princeton University Press, Princeton, 1961.
18. Olive, T. R., "Indexes in the Plant Cost Picture," *Chem. Eng.*, 54, May 1947, pp. 109-111.
19. Perry, John H., "Chemical Business Handbook," Business Uses of Index Numbers—Sec. 20, pp. 122-155, McGraw-Hill, New York, 1954.
20. Searle, Allan P., "Weight Revisions in the Wholesale Price Index, 1890-1960," *Monthly Labor Rev.*, Feb., 1962, pp. 175-182.
21. Snyder, R. M., "Measuring Business Changes," John Wiley & Sons, Inc., New York, (1954).
22. "Wholesale Price Index," Chap. 10—Techniques of Preparing Major BLS Statistical Indexes—Bureau of Labor Statistics, U. S. Dept. of Labor, BLS Bull. 1168.

Meet the Author

THOMAS H. ARNOLD, JR., *is a chemical engineer with a wealth of experience both in industry and publishing. Educated at Louisiana State University, he has worked for FMC Corp., Humble Oil & Refining Co. and Ethyl Corp., and also put in a hitch with the Army Chemical Corps. From November 1957 to September 1961, Arnold was on the editorial staff of* CHEMICAL ENGINEERING, *his last assignment being that of Southwestern Editor. He is now a student at Yale Law School, preparing to launch into a new career in the legal profession.*

Research for this article was done under the supervision of Editor-in-Chief Cecil H. Chilton, who has long had a special interest in the cost engineering field and is currently president of the American Assn. of Cost Engineers.

Key Concepts for This Article

For indexing details, see Chem. Eng., Jan. 7, 1963, p. 73 (Reprint No. 222). Words in bold are role indicators; numbers correspond to AIChE system.

Active (8)
Estimating
Updating

Organizations
Bureau of Labor
  Statistics
Chemical Engineering

Passive (9)
Costs
Plants
Equipment, process
Construction
Engineering
Labor

Means/Methods (10)
Plant Cost Index, CE

Data
Indexes

CE Cost Indexes:
A Sharp Rise Since 1965

Here are year-by-year
statistics on plant and
equipment costs, with
suggestions on how to use
indexes most effectively
in the updating of
plant-cost data.

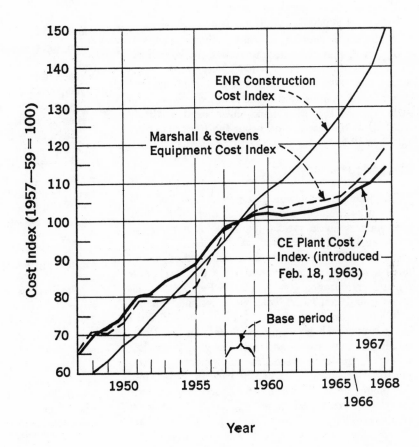

ROBERT B. NORDEN, Chemical Engineering

CHEMICAL ENGINEERING's equipment and plant cost indexes continue to rise, despite efforts by government officials to "take the heat off the economy."

The CE Plant Cost Index is in a sharp upward trend, after a relatively stable 1959-1964 period. During 1959-1964, this index went up a total of 1.5 points, representing a rise of 1.5%, while during the 1964-

1968 period, the index added on 10.3 points, for a 10% rise.

The above chart reflects this trend; the curves trace the 1947-1968 values for the CE index and for two other well-known indexes that have been converted to the same 1957-59 cost base. (More on these other indexes later.)

The *CE* Plant Cost Index is composed of four major components:

	Weight, %
Equipment, machinery and supports	61
Construction labor	22
Buildings, materials and labor	7
Engineering supervision and manpower	10

Of these four, the first three were major contributors to the rise in the index during 1964-1968, particularly during 1968. This is shown in Table I.

The "equipment, machinery and supports" category is made up of the following seven subcomponents:

	Weight, %
Fabricated equipment	37
Process machinery	14
Pipe, valves and fittings	20
Process instruments and controls	7
Pumps and compressors	7
Electrical equipment and materials	5
Structural supports, insulation, paint	10

A statistical summary of these subcomponents is shown in Table II. Note that relatively large rises have taken place in process machinery; pipe, valves and fittings; process instruments and controls; and pumps and compressors. Electrical equipment and materials, on the other hand, have been relatively stable over the last 10 years—but the trend has been up since 1966.

Comparing the Indexes

The Marshall & Stevens Equipment Cost Index (a 47-industry average) is reported in each issue, along with indexes for 12 specific industries. A summary of the annual M&S indexes is shown in Table III.

These M&S indexes are based on detailed equipment appraisals made periodically by the Marshall & Stevens organization (for details see Ref. 1), and include installation labor. They were designed for use in plotting cost trends for *equipment*. The *CE* Plant Cost Index includes equipment, installation labor, buildings, engineering—and was designed to reflect *plant* cost trends. However, the equipment cost component of the plant index can be used to update fob. equipment costs. This component does not include installation labor—such labor appears as a separate component. Thus, if *installed*-equipment costs are desired, the M&S index may be more convenient.

The *CE* Plant Cost Index was introduced six years ago.[2] The editors believed that there was a need for a cost index specifically designed to show chemical plant cost trends. The ENR index was not entirely suitable for this; the M&S index was not designed to be a true plant cost index; and other indexes such as the Nelson Refinery Index were designed for specific industries.

Compiling a Tailor-Made Index

It is important to point out that the *CE* index is a materials-equipment-labor index. It is not based, as is the ENR Index, on "basic" materials such as cement, lumber and steel together with a labor component. The intent was to establish a "typical" chemical plant and develop an index that would chart the changes in costs necessary to reproduce that plant.

A tailor-made chemical plant index is an inherently more difficult undertaking than, say, a tailor-made refinery index.

It is not a difficult task to select a "typical" refinery, since refineries do not differ drastically in their basic makeup. But what is a "typical" chemical plant? A plant producing ammonia bears little physical resemblance to a plant producing electrolytic chlorine.

Fortunately, an ammonia plant and a chlorine plant *do* have certain common characteristics. In fact, most chemical plants have many similar component parts regardless of the particular process used or products turned out. For example, piping, pumps, electric motors, storage tanks, fans and blowers, and instruments, are items found in most chemical processing plants. The differences develop in the effects each of these components has on total cost. Piping in a refinery is a much more significant item in the total cost picture, than it is in, say, a plant for manufacturing cement. As it turns out, however, the overall *CE* index is relatively insensitive to variations in weights assigned to equipment components.

There is no doubt that this is a complicated index to compile. It is made up of 67 Bureau of Labor Statistics' price indexes covering 155 individual commodities, three BLS indexes covering 450 additional items, and BLS hourly earnings data for five groups of employees. The index also includes salary trends for engineers, draftsmen, clerical and administrative workers.

Analysis of Projects

The various components and percentage weights were arrived at after a detailed analysis of data from 156 separate plant construction projects, supplied by a number of process industry companies, contractors and consultants.

The data were analyzed according to three types of processes: fluids handling; fluids/solids handling; and solids handling—and for three types of projects: new plant at new site; new plant at existing site; and expansion at existing site.

The *CE* Plant Cost Index is a calculated weighted average of all factors developed in the analysis of the 156 projects; the analyses showed that the differences between indexes based on the three types of projects were too minor to justify separate reporting.

Correction for Productivity

One of the major advantages, and perhaps the most controversial aspect of the *CE* Cost Index, is the correction of wages and earnings rates for changes in productivity. There is no doubt that productivity of manpower has increased over the years—e.g., because

Trend of plant costs since 1950—Table I

(Base period: 1957-1959 = 100)

	Year																		
	1950	1951	1952	1953	1954	1955	1956	1957	1958	1959	1960	1961	1962	1963	1964	1965	1966	1967	1968
CHEMICAL ENGINEERING Plant Cost Index	73.9	80.4	81.3	84.7	86.1	88.3	93.9	98.5	99.7	101.8	102.0	101.5	102.0	102.4	103.3	104.2	107.2	109.7	113.6
Equipment, machinery and supports	69.8	77.6	77.8	80.9	82.3	85.1	92.7	98.5	99.6	101.9	101.7	100.2	100.6	100.5	101.2	102.1	105.3	107.7	111.5
Erection and installation labor	80.5	85.1	87.4	91.6	93.5	93.5	95.8	98.6	100.0	101.4	103.7	105.1	105.6	107.2	108.5	109.5	112.5	115.8	120.9
Buildings, materials and labor........	82.2	88.1	88.5	91.4	93.1	95.0	98.0	99.1	99.5	101.4	101.5	100.8	101.4	102.1	103.3	104.5	107.9	110.3	115.7
Engineering and supervision manpower ..	78.2	82.2	84.8	88.3	87.9	92.0	94.2	98.2	99.3	102.5	101.3	101.7	102.6	103.4	104.2	105.6	106.9	107.9	108.6

Trend of equipment costs since 1950—Table II

(History of the equipment component and major subcomponents of CE Plant Cost Index. Base period: 1957-1959 = 100)

	Year																		
	1950	1951	1952	1953	1954	1955	1956	1957	1958	1959	1960	1961	1962	1963	1964	1965	1966	1967	1968
Equipment, machinery and supports	69.8	77.6	77.8	80.9	82.3	85.1	92.7	98.5	99.6	101.9	101.7	100.2	100.6	100.5	101.2	102.1	105.3	107.7	111.5
Fabricated equipment..............	71.5	78.4	79.0	81.3	81.4	84.2	92.5	99.5	99.6	100.9	101.2	100.1	101.0	101.7	102.7	103.4	104.8	106.2	109.9
Process machinery...............	69.4	76.5	77.5	80.6	82.8	85.3	92.2	98.1	100.1	101.8	101.8	101.1	101.9	102.0	102.5	103.6	106.1	108.7	112.1
Pipe, valves and fittings	64.7	73.1	73.8	78.0	79.5	85.2	94.8	97.9	98.8	103.3	104.1	101.1	100.6	100.7	101.6	103.0	109.6	113.0	117.4
Process instruments and controls	71.9	79.8	80.0	82.9	85.1	86.7	91.2	96.7	100.4	102.9	105.4	105.9	105.9	105.7	105.8	106.5	110.0	115.2	120.9
Pumps and compressors	65.4	73.9	73.4	77.5	79.5	81.7	90.0	97.5	100.0	102.5	101.7	100.8	101.1	100.1	101.0	103.4	107.7	111.3	115.2
Electrical equipment and materials.....	68.3	79.7	79.3	82.0	83.0	84.3	93.5	98.4	100.6	101.0	95.7	92.3	89.4	87.6	85.5	84.1	86.4	90.1	91.4
Structural supports, insulation and paint .	77.6	82.0	83.0	86.0	88.6	90.5	92.5	98.0	100.4	101.6	101.9	99.8	99.2	97.3	98.3	98.8	101.0	102.1	105.7

Marshall & Stevens annual indexes of comparative equipment costs, 1950 to 1968—Table III

(Base period: 1926 = 100)

	Year																		
	1950	1951	1952	1953	1954	1955	1956	1957	1958	1959	1960	1961	1962	1963	1964	1965	1966	1967	1968
Average of all	167.9	180.3	180.5	182.5	184.6	190.6	208.8	225.1	229.2	234.5	237.7	237.2	238.5	239.2	241.8	244.9	252.5	262.9	273.1
Process Industries																			
Cement.....................	161.6	172.7	172.8	174.6	177.6	182.6	199.4	216.4	222.8	228.7	232.1	231.1	231.8	232.5	235.9	239.3	249.6	258.1	268.3
Chemical....................	169.6	180.7	181.1	183.1	186.2	191.5	209.1	226.5	232.3	236.5	239.2	237.7	238.0	238.7	241.1	243.8	246.1	261.8	271.6
Clay products	156.6	167.7	167.8	169.5	172.4	177.3	193.8	210.2	216.8	222.2	225.7	224.6	225.5	225.8	229.2	232.6	239.0	250.7	260.7
Glass......................	159.7	170.8	171.0	173.0	176.0	180.9	197.5	213.8	219.3	223.2	225.3	224.4	224.7	225.4	227.6	230.1	240.6	271.1	256.3
Paint......................	162.9	174.0	174.4	176.3	179.3	184.3	201.2	217.6	223.2	226.9	229.5	230.0	231.5	232.1	235.0	238.1	243.9	255.7	267.4
Paper......................	163.2	174.3	174.7	176.6	179.6	184.6	201.5	218.2	223.8	227.8	229.9	229.0	229.3	229.9	232.3	234.8	247.5	252.1	261.6
Petroleum products	166.0	177.1	177.6	179.7	182.8	188.0	205.4	222.2	228.0	231.8	234.3	235.0	238.2	238.8	241.8	244.9	253.9	263.4	275.2
Rubber	168.4	179.5	180.0	182.1	185.2	190.5	207.9	224.9	230.8	234.6	237.3	237.9	239.2	240.0	243.0	246.2	251.2	264.7	276.5
Related Industries																			
Elec. power equip.	171.2	182.3	182.8	185.0	188.0	193.3	211.0	229.2	235.2	239.0	241.0	236.3	235.6	234.7	236.8	239.4	246.5	257.3	264.9
Mining, milling................	170.7	181.4	181.9	184.1	187.1	192.6	210.4	227.9	233.8	237.1	240.6	239.2	239.5	240.1	242.6	245.3	253.0	263.5	273.2
Refrigerating	185.2	200.1	200.7	202.8	204.8	211.6	234.3	254.2	260.8	265.1	268.2	268.8	270.4	271.2	274.6	278.2	287.1	299.1	312.5
Steam power	158.4	169.9	170.5	172.6	175.5	180.4	197.0	213.0	218.6	222.9	224.7	225.3	226.6	227.2	230.1	233.0	240.4	250.6	261.8

of technological improvements in construction and engineering.

This is, of course, a complicated subject. Productivity of labor is a variable quantity. It changes from year to year, it is different in various sections of the country, it is seasonal, etc. Certainly, construction-labor trends on the U.S. Gulf Coast in the last few years have shown the effect a local building boom, labor shortage and other factors can have in lowering productivity.

Nevertheless, over the long haul, it can be shown that nationwide productivity of labor has increased at a fairly steady rate. Based on BLS data, a figure of 2.5%/yr. increase in productivity was selected for the *CE* Plant Cost Index. This rate is applied to all wage and salary figures. The correction tends to somewhat lower the cost index compared to indexes having no productivity correction. For more details on this subject, see Ref. 3, pp. 146-147.

Labor Costs: Too Low?

The *CE* Plant Cost Index uses a BLS labor index for contract construction based on monthly average hourly earnings. Some readers have commented on the slow rise of *CE*'s construction labor component. Over the last 10 years, it has gone up about 18%, while other labor indexes show a rise of over 40%. The "conservative" rise of 18% is due primarily to the 2.5% productivity correction—the BLS earnings data are divided by the factor $(1.025)^n$. Here n is the number of years since 1947 when the factor was established as 1.000. This component was then adjusted to establish a construction labor index of 100 in 1958. Thus, the labor earnings in 1968 are divided by a factor of $(1.025)^{21}$ or 1.68 to correct for productivity.

Limitations of Index

There is no perfect index, and while a considerable amount of analysis went into the development of the *CE* Plant Cost Index, there are these limitations:

1. A cost index is a statistical average and suffers from all the disadvantages of averages, as pointed out in regard to productivity.

2. Most indexes represent reproduction costs. They do not account for any radical technological changes in process technology. (They can reflect a technological change in techniques for manufacturing the components that go into the index.) The cost of producing ethylene oxide by direct oxidation is much lower than by the chlorhydrin process. Published cost indexes will not show this reduction.

3. Many plants are improved over the years due to process refinements. Experience often shows that a process step can be entirely eliminated by changing some conditions. These cost reductions, while not as profound perhaps as a technological change, will show up in future plants using a similar process. This is the learning-curve effect. It is doubtful that published cost indexes can reflect such changes.

Using and Adjusting the Index

The *CE* Plant Cost Index is a valuable tool in cost estimation and economic evaluation. However, it is only a guide that primarily indicates inflationary cost trends in the reproduction of a "typical" chemical plant.

The reader should be cautious in his use of *any* cost index, particularly if he is extrapolating plant costs involving a process undergoing radical technological improvements. Of course, "raw" extrapolations over a long span of time—say, more than 10 years—can be dangerous even for a relatively stable process.

A cost index should be used by engineers who know its basic makeup and limitations and can apply the necessary corrections to adjust the index so it reflects actual conditions. Previous articles[2, 3] have pointed out how such adjustments can be made—for instance, how the building component can be given more weight in a pharmaceutical-plant extrapolation, or how the electrical-equipment subcomponent can be given more weight in an electrochemical-plant project.

Similarly, if the geographic region has shown a zero increase in construction-labor productivity over the period in question, the construction-labor component can be adjusted upward by 2.5% per year. (This would still leave an allowance for labor-productivity gains in the wage and salary figures used to compile some parts of the other components.)

Thus, a variety of adjustments is possible; this do-it-yourself aspect was specifically built into the index. ∎

References

1. Stevens, R. W., Equipment Cost Indexes for Process Industries, *Chem. Eng.*, Nov. 1947, pp. 124-126.
2. Arnold, I. H., Jr., Chilton, C. H., New Index Shows Plant Cost Trends, *Chem. Eng.*, Feb. 18, 1963, pp. 143-152.
3. Chilton, C. H., Plant Cost Index Points Up Inflation, *Chem. Eng.*, Apr. 25, 1966, p. 190.

Capital Cost Estimates For Process Industries

JOHN W. HACKNEY, Mobil Oil Corp.

COST estimating for process industries is in the state of transition between art and science. This two-part report is an attempt to collect and integrate established and newly developed estimating methods and procedures. Our objective is to make capital cost estimating more methodical and less dependent on individual judgment and experience. This report is also intended as an aid to engineers who occasionally must make capital cost estimates for process plants but who cannot devote a great deal of time to the review and analysis of current estimating literature.

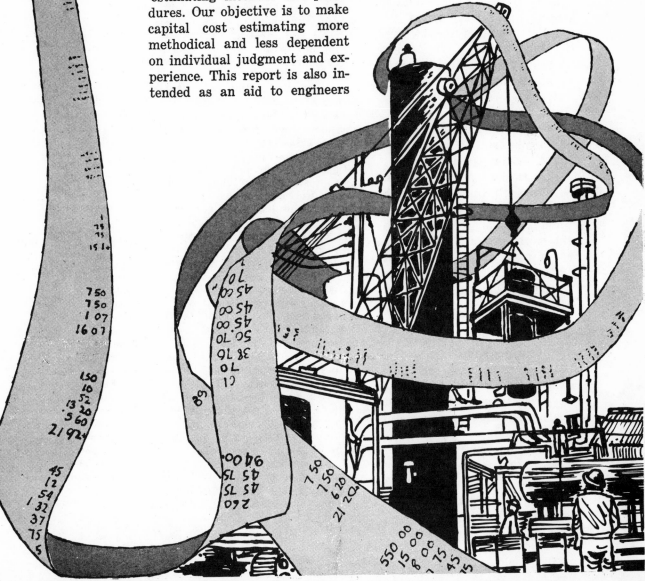

In general, estimating is based on:

- Definition of cost elements so that cost records and estimates have a common basis.
- Collection of cost records based on these definitions.
- Classification and grouping of cost records.
- Analysis of the relationships among cost records.
- Utilization of procedures based on these relationships, and the collected and classified cost record to make cost estimates.

Emphasis in this report is on definitions and procedures. When uniform definitions and procedures can be used, every project record contributes to the common store of estimating data. Hence, every engineer can confidently utilize these data in making his own estimates.

Part I of this report covers the more or less independent procedures for estimating individual portions of project costs. The procedures are:

1. General procedures.
2. Equipment estimating.
3. Material estimating.
4. Labor estimating.
5. Design and field expense.

These procedures can be combined for use in making estimates by one of several different methods. Each method is a co-ordinate grouping of appropriate procedures. Four such methods will be discussed in Part II which will appear on page 43 of this volume.

Also, Part II will include contingencies and appraisal of estimating methods, with a discussion of project information required, time required for estimating and relative accuracy of each method.

This report discusses basic estimating procedures and their application, but does not provide the data necessary for the preparation of a cost estimate.

Provide Adequate Estimating Data

Considerable data must be available and organized before estimating procedures become useful to the cost estimator. Preparatory work for cost estimates includes organization of estimating data, methods of adjusting cost records to current prices, methods of adjusting costs of complete process units for capacity and developing land and site-clearing costs.

Organizing Cost Data

Cost estimates for a modern process plant involve so many features that it is absolutely essential to have some means by which the vast welter of detail can be organized. Costs of past projects must be collected and identified. These costs must be stored in the department records, and they must be easily available from storage for use in making new estimates.

Availability of these data requires a standardized numerical system of classifying cost data. Such a standardized system is indispensable if field accountants, estimators and designers are to have a common basis of understanding for allocation of project costs.

In establishing this account subdivision system the following objectives were considered.

- Account list for a cost estimate should present a logical picture of the scope of the project, with major operating divisions identified and listed in operating sequence.
- Individual account subdivisions should be work units which can be easily segregated and identified in the field.
- Standard accounts should be complete enough to reduce the possibility of overlooking any major element of cost in the estimate.
- Account system should be suitable for use with tabulating machines so as to reduce the cost of obtaining information on the larger projects.
- The system should be as simple as possible, consistent with the great number of units to be classified and the many purposes which must be served.

An account subdivision system must also satisfy requirements with respect to its use for property accounting and for construction planning and control.

Cost estimates should be broken down into primary divisions, each consisting of a major element of operation. These primary divisions state the purpose to be accomplished by the installations in that division and not the method or type of equipment to be used. Where possible, the divisions should coincide with operating departments or sub-departments. They should be listed as nearly as possible in operating sequence. When distribution systems, collection systems, utilities or buildings serve more than one division, they should be listed in a separate division.

As an example, an estimate for chlorine capacity increase might have the following primary divisions.

- Salt unloading.
- D.C. power conversion.
- Chlorine storage.

Work in each primary division should be broken down into construction account subdivisions in

Account Subdivision List

Preparatory	01 to 09
Structures	11 to 19
Storage	21 to 29
Piping and duct work	31 to 39
Equipment	41 to 49
Equipment	51 to 59
Electrical	61 to 69
Site improvements	71 to 79
Engineering expense	
Field	81 to 86
Design	87 to 89
Construction expense	
Procurement	91
Equipment, tools, supplies	92
Temporary facilities	93
Services	94
Administration, fees	95

accordance with the work descriptions given in the table below.

Each subdivision is assigned a four-digit number. The first two digits from the left are from the account subdivision list. The first digit classifies the work as to general type of construction involved, such as structures, piping, electrical.

The second digit classifies the type of construction work. For example, if the first digit is 6, the work is of an electrical nature. If the first digit is 6 and the second digit is 1, the work involved is lighting (61-).

A third digit gives the number of the primary division, and indicates the operating division of which this work is a part. It is separated by a dash from the second digit.

The fourth digit is a zero except when there is more than one item of a like classification in a primary division as shown by 31-30 and 31-31 in the table. Where the project is made up of more than ten primary divisions, ranges are assigned to primary divisions such as —10 to —14 might cover chlorine supply.

Adjust Costs to Current Price

Construction cost estimates based on records of former projects must be corrected to the current value of the dollar. This correction is usually substantial, since construction prices have doubled in the 13 years between 1947 and 1960.

Many indexes are available which purport to measure the change in value of the construction dollar. The most widely used in process industries are: *

Marshall & Stevens (Published in *Chemical Engineering*).

Nelson Refinery Index (Published in *Oil and Gas Journal*).

Engineering News-Record Construction Index.

Another study comparing the merits of 16 of the available indexes came to the conclusion that for process plant estimating purposes there was little basis for a choice among them.

Based on length of publication record and promptness of availability, the *Engineering News-Record* Construction Index is one of the best, and has been used for these studies. A standard value of 600 has been adopted, although almost any other value, including the 1913 base index of 100, would be as good. Using 100 as a base would in some cases reduce the indexing procedure by one arithmetic operation. However, the 600 base, at least for the near future, prevents cataclysmic errors when indexing is overlooked.

*Editor's Note: Mr. Hackney's article was written before the introduction of the CE Plant Cost Index, which is described on pp. 11 to 24.

Two examples indicate the general method of using cost indexes: Find the cost in July 1958 of a filter wheel similar to one purchased in May 1946.

Filter wheel cost, May 1946 .. $19,204
ENR Index, May 1946....... 340
ENR Index, July 1958....... 760
Cost of similar filter, July 1958
$19,200 × 760/340....... $43,000

Find the cost in July 1958 of filter wheel with 600 sq. ft. of filter area.

From capacity-cost curve, based
on ENR 400, filter cost $21,000
ENR Construction Index, July
1958.................. 760
Cost of filter, July 1958
$21,000 × 760/400....... $40,000

Although the *Engineering News-Record* Construction Index measures the change in value of the construction dollar, fluctuation of prices of individual pieces of equipment, buildings, roads, railroads, pipelines and other major items of construction will vary in a somewhat different manner. There are special indexes available to measure the fluctuations of many of these items.

In general, the increase in accuracy obtained by using individual indexes for each item of a cost estimate is outweighed by the resulting unwieldy complexity. In cases where the greater part of the estimate is dominated by a single item, a power generator for example, the accuracy of the estimate can be improved by using the special index for such an

Arrange by Primary Divisions

Salt Unloading —10
03-10 Remove two wood-stave tanks
13-10 Conveyor foundations
16-10 Conveyor supporting frame
57-10 Belt conveyor, 2 x 217 ft. with motor and drive
61-10 Unloading area lighting
66-10 Power supply to conveyor

D. C. Power Conversion —20
63-20 Install new arc limiters on five rectifiers
66-20 Power supply for arc limiters on five rectifiers

Chlorine Storage —30
03-30 Remove scrap from storage tank area
13-30 Chlorine storage tank foundations
21-30 New chlorine storage tank, 55 tons
31-30 Liquid chlorine piping for storage tank
31-31 Snift-gas piping for storage tank

Rearrange by Construction Account

Preparatory
03-10 Remove two wood-stave tanks
03-30 Remove scrap from storage tank area
Structures
13-10 Conveyor foundations
13-30 Chlorine storage tank foundations
16-10 Conveyor supporting frame
Storage
21-30 New chlorine storage tank, 55 tons
Piping
31-30 Liquid chlorine piping from storage tank
31-31 Snift-gas piping from storage tank
Process Equipment
57-10 Belt conveyor, 2 x 217 ft. with motor and drive
Electrical Equipment
61-10 Unloading area lighting
63-20 Install new arc limiters on five rectifiers
66-10 Power supply to conveyor
66-20 Power supply for arc limiters on five rectifiers

Construction Cost Index

1960*	813.2
1959	797.4
1958	759.2
1957	723.9
1956	692.4
1955	659.7
1954	628.0
1953	600.0
1952	569.4
1951	542.7
1950	509.6
1949	477.0
1948	460.7
1947	413.2

*Annual averages, except for 1960 as of Jan. 21, 1960, from *Engineering-News Record.*

Process Unit Capacity-Cost Exponents

$$\frac{\text{Cost of unit A}}{\text{Cost of unit B}} = \left(\frac{\text{Capacity of unit A}}{\text{Capacity of unit B}}\right)^y$$

	y
Refrigeration plants includes auxiliaries [8]	0.85–0.96
Refrigeration plants no auxiliaries [8]	0.80–0.82
Complete refinery, catalytic cracking [1]	0.85
Coke-oven gas separation [2]	0.82
Water-gas plant [2]	0.81
Water-treating plant [3]	0.91
Vacuum distillation [4]	0.80
Catalytic cracking-only [1]	0.78
Solvent dewaxing [5]	0.76
CO and CO_2 removal from hydrogen [2]	0.74
Peterson sulfuric acid towers [2]	0.73
Coking only [1]	0.72
Catalytic polymerization [1]	0.70
Soybean extraction [2]	0.70
Water-gas shift conversion [2]	0.69
Contact sulfuric acid [2,6]	0.64–0.66
Solvent extraction or treating [5]	0.67
Vacuum flash [4]	0.64
Chamber sulfuric acid [6]	0.60
Oxygen plant [2]	0.59
Natural gasoline plants [5]	0.51
Thermal cracking only [5]	0.51
Small oxygen plant [2]	0.47
Hypersorption [5]	0.43

1. Nelson, W. L., "Over-all Plant Costs—Cracking," *Oil & Gas J.,* Nov. 24, 1949, p. 149
2. Sherwood, P. W., "Effect of Plant Process Size on Capital Costs," *Oil & Gas J.,* Mar. 9, 1950, p. 81.
3. Chilton, C. H., "Cost Data Correlated," *Chem. Eng.* June 1949 p. 49.
4. Nelson, W. L. "Cost of Vacuum Plants," *Oil & Gas, J.,* Sept. 18, 1950.
5. Nelson, W. L., "Over-all Plant Costs—Distillation, etc.," *Oil & Gas J.,* Dec. 1, 1949, p. 93.
6. Barr, J. A., "H_2SO_4 Buy or Build," *Chem. Eng.* Apr. 1950, p. 106.
7. "Platforming Process," *Oil & Gas. J.,* Apr. 6, 1950, p. 89.
8. Nelson, W. L., "Refrigeration Plants," *Oil & Gas J.,* July 7, 1949, p. 95.
9. Nelson, W. L., "Cost versus Size," *Oil & Gas J.,* Oct. 5, 1959, p. 263.
10. Williams, Jr., R., "0.6 Factor Aids in Approximating Costs," *Chem. Eng.,* Dec. 1947.
11. Chilton, C. H., "0.6 Factor Applies to Complete Plant Costs," *Chem. Eng.,* Apr. 1950.
12. Nelson, W. L., "What Price Complete Refinery Installations," *Oil & Gas J.,* Nov. 23, 1953.

item. In most cases, a more accurate and almost equally prompt preliminary cost estimate can be obtained from a vendor.

Cost of Process Units

For complete process plant units, making identical products by identical processes with identical site conditions, the following relation gives costs of similar plants at different capacities:

$$\frac{\text{Cost of unit A}}{\text{Cost of unit B}} = \left(\frac{\text{Capacity of unit A}}{\text{Capacity of unit B}}\right)^y$$

where y is the process unit capacity-cost exponent whose value is less than one. The value of y approaches closer to one when unit capacity is increased by adding identical production lines instead of by increasing the size of process equipment.

A process plant unit is defined as a complete installation capable of performing a unit process. It does not include the utilities, raw material supply, service buildings, product shipping yard improvements and other service facilities required to transform a group of process units into an operating plant.

The table* on this page gives capacity-cost exponents for various process units.

An example showing application of the method follows:

Assume recorded cost of a soybean plant, built in 1949 with a capacity of 100 tons/day as $600,000. Find the cost of a similar 300 ton/day plant in July 1958.

ENR Construction Index for 1949	477
ENR Construction Index for July 1958	760
Cost of 100 ton/day plant in July 1958 $600,000 × 760/477	$950,000
Capacity ratio: 300/100	3.0
Cost ratio: (3.0)$^{0.7}$	2.16
Estimated cost, 300 ton/day plant in July 1958 $950,000 × 2.16	$2,100,000

An attempt to check this capacity-cost relationship for four jobs of record yielded only fair results. These were four very similar production units, one put in as a complete new plant, one as an addition with raw material and product storage, one as an addition with product storage and one as an addition with no storage. All units were housed in new buildings. They were built over a 6-yr. period.

Considering the total project costs, there was little trace of exponential relationship between the projects. If the exponential method with ENR index correction had been used for estimating the cost of only the process units common to the four plants, estimating errors would be as follows assuming exponent y is 0.7:

Basis of Estimate	Estimating Errors for Plant B	C	D
Plant A	4%	41%	31%
Plant B		23%	8%
Plant C			−19%

*Data obtained from W. L. Nelson, *Oil & Gas. J.,* Oct. 5, 1950.

Reasons for these discrepancies were many and varied. Local wage rate differences accounted for some. Foundation conditions differed. Technical changes were made in furnace design. Construction supervision and contract arrangements varied. Materials of building construction were not the same. Also the ENR index is not a perfect indicator of price changes. All of these combined to produce the variations shown.

A more legitimate use of this estimating procedure is to approximate the costs of homologous process units at various capacities when the cost at a single capacity has been carefully estimated by other methods.

Assume the capacity-cost exponent to be 0.70. If the estimated cost of the 300 ton/day unit is $1,600,000, then the cost at other capacities may be approximated as shown in the following table.

A Capacity Tons/Day	B Capacity Ratio A/300	C Cost Ratio $B^{0.7}$	D Estimated Cost of Unit $1,600,000 \times C$
100	0.333	0.463	742,000
200	0.667	0.752	1,206,000
300	1.000	1.000	1,600,000
400	1.333	1.223	1,960,000

The exponential relationship between capacity and cost has some general validity, but cannot be relied upon for estimates of a high degree of accuracy.

Estimating Land Costs

Estimates of land cost usually have a low accuracy. Fortunately land cost is a small part of total project cost and usually it is a matter of record before the engineering has advanced to a point where accurate estimates must be made.

Site clearing costs are also difficult to determine with any degree of accuracy, especially in the early stages of a project. Even for a preliminary estimate, the estimator should make every effort to visit the plant site personally. During this visit, he should carefully observe and note every obstacle to the construction of the plant. The cost of removing obstructions which often includes putting service lines in some other location can run into a considerable sum of money.

Some of this work may be estimated by picturing the size of crew and amount of time required to do the work. In other cases, site clearing can be estimated by conventional detailed methods.

Sources of equipment prices, methods of adjusting equipment prices for capacity and methods of estimating auxiliary process equipment are essential to the estimator in making reliable costs estimates.

Equipment Prices

The most accurate method of determining process equipment values is to obtain firm bids from fabricators and suppliers. Often, vendors can supply quick estimating figures which will be very close to bid prices but not involve much of the vendor's time.

Second best in reliability are values from the file of past purchase orders. These prices are indexed for rapid finding in accordance with the standard account subdivision system. When used for pricing new equipment, purchase order prices must be corrected to the current cost index.

Published Sources

Much information has been published on detailed equipment cost estimating. The following have been found to be especially useful:
1. "Chemical Engineering Costs," O. T. Zimmerman and I. Lavine, Industrial Research Service, Dover, N. H.
2. "Cost-imating," W. L. Nelson, *Oil & Gas J.*
3. "Cost File," *Chem. Eng.*

Other sources, such as Chilton's "Cost Data Correlated," *Chem. Eng.*, June 1949; and Aries and Newton's "Chemical Engineering Cost Estimation," Chemonomics, Inc.; give log-log plots of capacity versus installed cost of various types of process equipment. These are difficult to use because installation costs are quite variable and the amount of installation cost included in these graphs is not known.

Capacity vs. Cost

For common types of process equipment such as tanks and heat-exchangers, the information contained in the purchase order file can be summarized on charts of capacity versus cost, corrected to a standard construction cost index of 600. On log-log plotting paper, process equipment cost data often show straight line characteristics which indicate an exponential relationship between capacity and cost.

Capacity and cost of homologous items of equipment are often related in the form:

$$\left(\frac{\text{Cost of A}}{\text{Cost of B}}\right) = \left(\frac{\text{Capacity of A}}{\text{Capacity of B}}\right)^x$$

Here x is the capacity-cost exponent, usually less than one. For equipment in general, an average capacity-cost exponent of six-tenths is suggested by Nelson.*

The following examples illustrate the use of capacity-cost exponents. One example shows how to use the capacity-cost exponent to estimate equipment cost at current prices if capacity and cost are known at some other capacity. The other shows how to apply the ENR cost index to capacity-cost relations.

Find the cost of a 40,000 bbl. cone-roof tank. A recent bid price for a 20,000-bbl. cone-roof tank was $35,000.

Capacity ratio: 40,000/20,000 = 2.0.

Capacity-cost exponent: 0.6 (assumed).

Cost ratio: $(2.0)^{0.6} = 1.565$.

Estimated cost of 40,000 bbl. tank: $35,000 \times 1.565 = $55,000.

Find the cost in July 1953 of a 3,500 lb./hr. steam boiler. A comparable boiler, having a capacity of 5,500 lb./hr. cost $10,560 in July 1946.

ENR construction cost index for July 1953 was 604 while for July 1946 was 354.

Hence, cost of 5,500 lb./hr. boiler in July 1953 becomes: $(604/354) \times $10,560 = $18,000.

Capacity ratio: 3,500/5,500 = 0.636.

Capacity cost exponent = 0.6 (assumed).

Cost ratio: $(0.636)^{0.6} = 0.762$.

Estimated cost in July 1953 of 3,500 lb./hr. steam boiler is $0.762 \times $18,000$ or $13,700.

---

* For typical capacity-cost exponents of process equipment, see W. L. Nelson, "Cost-imating Series," *Oil & Gas J.*, Oct. 5, 1950, p. 263.

Estimate Auxiliary Process Units

In the early stages of a project's development, major items of process equipment are often well established as to type, number, size and materials of construction. This major process equipment can be readily priced by methods previously discussed.

However, there is considerable auxiliary process equipment which cannot be sized or priced without a great deal of additional engineering work. If the process equipment required for a plant is divided into two well-defined classifications, then projects of record can be analyzed to determine cost ratios which have existed in the past between the two classifications. The classifications are: major process equipment and auxiliary process equipment.

Knowing the value of the major process equipment for a new project, and the probable auxiliary process equipment ratio, we can determine the approximate cost of auxiliary process equipment without expending the time required for detailed engineering and pricing of the auxiliary equipment.

To use this ratio method successfully, strict definition of the component parts is essential. Recommended definitions are given in the accompanying table.

Define Process Equipment Carefully

Major Process Equipment E_m

Equipment in which process material undergoes a change of state, condition or composition; or equipment in which process material is stored.

Does include major process equipment complete with drives, agitators, heaters and all factory attached or supplied equipment elements.

Examples:

Bag packers
Circulating pumps
Container fabricating equipment
Furnaces, complete with brick
Installed spare major process equipment
Rectifiers and transformers
Silos and storage tanks
Water treatment equipment

Does not include, except when they are an integral, factory-assembled or supplied part of a major process equipment unit.

Examples:

Chutes and conveyors
Electrical control, instruments
Insulation and piping
Transfer pumps
Vibrating feeders

Auxiliary Process Equipment E_a

Equipment associated with major process equipment for material movement, instrumentation, process control and electrical control.

For example: does include when in process service

Blowers
Control valves
Conveyors, elevators
Instruments and controls
Motor control centers
Over-head cranes
Pumps, vibrating feeders
Weighing devices

Does not include:

Air or gas ducts
Piping
Wiring
Insulation
Power distribution substations
Piping and wiring for instruments

Process Equipment E_p

$E_p = E_m + E_a$

Auxiliary Equipment Ratio

If we have a tabulation of auxiliary process equipment ratios ($e_a = E_a/E_m$) for similar projects, we can obtain an estimate of the value of the auxiliary process equipment. Multiplying the value of the major process equipment which is determined early in the project by the probable auxiliary equipment ratio gives an estimate of the value of the auxiliary process equipment.

Total range of e_a is from 0.00 to about 2.00 with usual values in the range of 0.25 to 0.60. The breadth of these ranges indicates the necessity for selecting the ratios with great care. Tabulation of ratios of auxiliary process equipment to major process equipment for past projects is essential.

Moreover, use of ratios for computing the value of auxiliary process equipment can never be as accurate as detailed design, take-off and pricing of this equipment. As the design of a project advances and the possibility of a more accurate estimate develops, this auxiliary equipment should be listed and priced by the means described for major process equipment.

Freight Costs

Freight cost for equipment, except for remote locations, is a relatively minor item. Average freight for equipment for a chemical plant ranges from less than 1% to about 2% of the equipment cost.

Rates for important shipments should be computed by the company's traffic department. If such help is not available, the railroad serving the plant can provide an estimate of the probable cost of freight.

How to Estimate Other Material

Having discussed the estimating of major and auxiliary process equipment, we can turn our attention to the problem of making preliminary estimates of the additional material which is required to make the process equipment operable. This includes not only material directly associated with the process equipment, but also the material required for site improvement, utilities and other service facilities of a plant.

Equipment ratio method of estimating this material is based on the general observation that the material cost of a process plant tends to be directly proportional to the value of its process equipment.

For detailed analysis of this relationship, the total equipment and material cost of a plant can be broken down into the groups shown in the table. Capital letters with subscripts indicate dollar values while small letters with corresponding subscripts indicate ratios of the dollar value of the item to the dollar value of the process equipment. Since these ratios are typical for similar processes, they can be used to make preliminary estimates of material cost.

Usually the size, type and value of major process equipment can be determined early in the development of the project. Value of auxiliary

Define and Specify Material Groups for Use With Equipment Ratios

Process Equipment Material M_e, $M_e = M_i + M_p + M_w$

Equipment foundations, insulation, supports, paint
For example:
Equipment foundations, insulation, supports, paint
Instrument cases, panel boards
Piping, wiring for process instruments and controls
Platforms, stairs and walkways
Does not include
Building grade slabs or stairways
Pipe insulation, power wiring, process piping

Process Piping Material M_p

For example:
Air and gas ducts
Pipe, fittings, hangers, insulation and supports
Process sewers and stacks
Valves except power-actuated control valves
Does not include
Piping and duct work outside process unit area (M_d)
Sanitary plumbing (M_{sb}) or storm sewers (M_{ys})
Piping and wiring for instruments and controls (M_i)

Process Wiring Material M_w

For example:
Bus bars, clamps, duct and wire
Cable, conduit and fittings, insulators
Does not include
Instruments, controls or motor control centers (E_a)
Electrical equipment outside process unit area (M_d)
Lighting or building service wiring (M_b)

Total Building Material M_b, $M_b = M_{pb} + M_{sb}$
Process Building Material M_{pb}

For example:
Air conditioning, heating, lighting, ventilation
Building space for process units
Building space for in-process storage
Process control rooms, pump houses
Does not include
Space in process building for offices, maintenance, warehousing, lockers, change rooms
Plumbing, heating or lighting for service spaces
Utilities or yard improvements serving process buildings

Service Building Material and Equipment M_{sb}

For example:
Air conditioning, heating, plumbing, ventilation
First aid equipment, lockers, partitions
Lighting, overhead cranes, sprinklers
Shop and storeroom equipment
Buildings for cafeterias, garages, offices, toolrooms and warehouses
Does not include
Pump houses, building space for in-process storage
Utilities and sewers serving the buildings

Distribution Material M_d

Includes all materials and equipment required to distribute to or from process units.
For example:
Electrical power M_{de}
Steam M_{ds}
Water M_{dw}
Raw materials and finishes products M_{dp}
Fuel, air, refrigeration M_{do}
Pipe bridges M_{db}
Buildings or space for utility control rooms
Product lines to customers
Tank cars and barges
Does not include
Items considered to be process equipment such as:
Electrical generators, rectifiers, refrigeration equipment
Sewage and water treatment equipment
Sewers (M_y)
Material distribution in process unit area

Yard Improvement Material M_y

For example:
Railroads M_{yt}
Roads, area paving, sidewalks M_{yr}
Grading, fencing, planting, lighting M_{ym}
Sewers M_{ys}
Waterfront structures and other yard improvements M_{yo}

Total Plant Material M_t
$$M_t = M_e + M_b + M_d + M_y$$

Includes all material for the complete plant except process equipment E_p

Typical Ranges of Material Ratios

Process Equipment Material Ratios

Equipment installation	m_i	$= M_i/E_p$	0.178–0.281
Piping	m_p	$= M_p/E_p$	0.083–0.177
Wiring	m_w	$= M_w/E_p$	0.032–0.080
Total process equipment..................	m_e	$= M_e/E_p$	0.322–0.535

Building Material Ratios

Process building	m_{pb}	$= M_{pb}/E_p$	0.253–0.326
Service Building	m_{sb}	$= M_{sb}/E_p$	0.040
Total building material	m_b	$= M_b/E_p$	0.293–0.366

Distribution Material

Electrical power	m_{de}	$= M_{de}/E_p$	0.085
Steam distribution	m_{ds}	$= M_{ds}/E_p$	0.025
Water distribution	m_{dw}	$= M_{dw}/E_p$	0.041
Raw material and product	m_{dp}	$= M_{dp}/E_p$	.....
Pipe bridge	m_{db}	$= M_{db}/E_p$	0.014
Other	m_{do}	$= M_{do}/E_p$	.....
Total distribution material	m_d	$= M_d/E_p$	0.165

Yard Material

Trackage material	m_{yt}	$= M_{yt}/E_p$	0.013
Road, parking, surfacing	m_{yr}	$= M_{yr}/E_p$	0.017
Sewer	m_{ys}	$= M_{ys}/E_p$	0.020
Fence, lighting, grading	m_{ym}	$= M_{ym}/E_p$	0.006
Other yard material	m_{yo}	$= M_{yo}/E_p$	.....
Total yard material	m_y	$= M_y/E_p$	0.056

Total plant material
(exclusive of process equipment) m_t $= M_t/E_p$ 0.836–1.122

such tables, but the results are usually of doubtful validity because of variations in accounting practices.

Even within a company, estimated costs and recorded costs must be based on identical accounting practices and procedures if satisfactory results are to be obtained. For example: equipment rental can be listed as labor or material. Ratios computed either way are valid. However, if the estimate is made on one basis from costs recorded on another basis, there will be trouble.

Get Unit Material Prices

Unit material pricing for buildings, utilities and yard improvements are often subject to analysis in terms of their material cost per unit of size or capacity. Unit prices can be obtained for items such as buildings, pipelines, sewers, wells, electrical light and power.

Material cost tabulations are also necessary for site improvements such as cost per unit size for finished grading, roads, parking lots, railroad tracks, fences and docks.

All of these unit costs are tabulated for material only. Because of the extreme variations in labor cost per man-hour and variations in job conditions, it is necessary to compute labor on a man-hour basis.

process equipment can be determined by ratio. Adding auxiliary process equipment E_a to major process equipment E_m, we get the value of the process equipment E_p.

Equipment ratios for each of the material groups can be selected from a similar job-of-record. Applied to the value of process equipment for the new project, they will provide an approximation of the cost of materials for the new project. Since each material group can be considered separately, considerable judgment can be exercised in selecting, omitting and combining the equipment ratios to include all features of the new project.

Early material estimates may be entirely on an equipment ratio basis, but as the engineering design advances additional information becomes available. Hence, some material groups may be estimated by more exact methods. However, the more exact figures should be reviewed if they depart too widely from the equipment ratio figures.

When computing or using equipment ratios, the current market value of new process equipment rather than the book value must be used. For example, the same amount of foundation concrete must be bought whether the process equipment is purchased new from the factory at current prices, transferred from some other project or purchased second-hand at a low price. When entering the cost of process equipment in the final estimate, the actual book cost of the item is used. For a general picture of the relative magnitude of equipment ratios, see the table on this page. Projects included in this tabulation were of many types. Hence, the range of values is wide.

Selecting Ratios

Extreme care is necessary in selecting records to be used as sources so that installations selected are comparable to the new project.

Basic guide in selecting equipment ratios is a complete tabulation of experience-ratios for all company jobs-of-record. Published information can sometimes be worked into

Use Appropriation Ledger

With the aid of the appropriation ledger, the material for a completely engineered project can be priced in detail at little expense beyond that which is required for other engineering, procurement and accounting purposes.

The appropriation ledger serves several purposes other than the collection and organization of material cost information.

1. It is a means of finding requisition number, purchase order number and vendor when the general type and use of the material is known.

2. Each subdivision sheet can be checked to see that the material listed on it is adequate to complete the work.

3. Promised delivery can be compared to delivery required by the construction schedule and corrective measures taken if necessary.

4. Prints of appropriation record sheets can be included in final design report as a permanent reference record of material.

How to Estimate Labor Costs

Construction projects to be estimated for an industrial corporation take place under a bewildering variety of conditions. One project may be built in Cleveland by a lump sum contractor who is tops in the business and bidding tight to get the work. A second job may be handled by an inexperienced but hard-working Arkansas contractor. A third job may be in New Jersey with the work to be performed by local plant personnel.

Considering the wide range of possible project conditions, the estimator has little chance of successfully estimating by means of fixed unit prices. Consider two identical wall sections, fifty dollars will buy more or less concrete, depending not only upon the wage rates for the area but also upon the effectiveness of the workmen. This effectiveness is determined by job conditions.

Examine Labor Effectiveness

A complete list of conditions which influence construction-labor effectiveness would be very long. However, the checklist in the table gives the dominant job conditions that may be encountered.

Unfortunately, these job conditions are intangibles and hard to describe. It is difficult to determine quantitatively the change in man-hours caused by a change in one certain job condition.

To attack the problem, we will consider individually each of these dominant job conditions. We will discuss, in a general way, its effect on labor efficiency and try to develop the range of percentage increase which may be experienced as this job condition varies from perfection through good and poor to bad.

Here, the estimator's judgment must be relied upon to assign values in a particular case. This throws a heavy burden on the estimator and assumes a wide background of field experience. It does provide a check list of items to be considered and should aid to focus attention on each. The sum of these individual decisions will provide a reasonably good approximation of the total effect of probable job conditions.

Field Supervision

First on the list of job conditions and first in importance is supervision. Supervisors must not only be competent but also available in sufficient force to plan the work and to dissolve field difficulties as these arise. The number of supervisors required for adequate coverage depends to some extent on the other job conditions. Supervisors must be available in sufficient numbers to give the difficult portions of the job special attention while carrying on efficiently with routine work.

Proper numerical ratios between supervisors and workmen will vary from about one to four up to one to fifteen. In stating these ratios, we are including foremen but not pushers as construction supervisors. We define pushers as any supervisors who work with their hands more than half of the time.

A common ratio for new construction work of ordinary difficulty is one to eight. Higher worker ratios reduce output unless the work from day to day is of an un-

Checklist Gives Range of Job Condition Factors

	Expected Condition	Relative Man-Hour Increase Range	Project
Field supervision			
Number of supervisors		0–0.14	
Competence		0–0.14	
Organization, leadership		0–0.14	
Field labor			
Competence		0–0.35	
Attiude		0–0.35	
Type of contract		0–0.10	
Construction equipment and tools			
Major equipment		0–0.40	
Tools and minor equipment		0–0.35	
Maintenance		0–0.20	
Construction facilities and services			
Storerooms and yards		0–0.05	
Shops and offices		0–0.05	
Personnel facilities		0–0.05	
Temporary utilities		0–0.05	
Physical working conditions			
Weather		0–0.25	
Mud, dust, altitude		0–0.15	
Fumes, heat, hazards		0–0.40	
Access to work			
Roads		0–0.10	
Height		0–0.40	
Other		?	
Construction program			
Buildup rate		0–0.10	
Area loading		0–0.10	
Interruptions		0–0.15	
Material coordination		0–0.12	
Engineering coordination		0–0.08	
Engineering quality		0–0.12	
Other special conditions		?	
Basic job condition factor, perfect conditions		1.00	1.00
Total job condition factor			

usually uniform nature. Lower worker ratios are required for difficult work, and for very technical work such as instrumentation and complicated electrical installations.

Lower worker ratios warn of over-supervising. Over-supervision is quickly apparent in dollars and cents. It is much less likely to be overlooked than the subtle loss of greater sums through decreased work efficiency caused by lack of adequate supervision.

Field supervisors must be provided with the information required to do their jobs efficiently. They must be organized as a team with a minimum of overlapping and gaps in responsibility, upheld in their positions and given a sense of participation in the job.

Moving further up in the project organization, the contractor's supervisors, the contractor's home office, field construction group for the owner and all affected departments of the owner's organization must also be coordinated and must operate as a team for best efficiency. Failure to do this will result in increased field labor costs.

Experience during World War II, with a series of very similar plants of large size, indicated that variations in the ability and performance of the top construction supervisor alone can produce a variation in labor cost of about 14%. A very similar example is quoted by C. Tyler in *Chem. Eng.*, Jan. 1953. He stated that two plants of the same type were built by the same contractors at about the same time but 800 mi. apart. On one, there was an underrun of about 4% and on the other, an overrun of 10%. Tyler attributed the difference in the abilities of the superintendents.

Deficiencies in quality and quantity of subordinate field supervisors might also be expected to increase labor costs by somewhere near 14% each. The top of the ordinary range of the supervision element of the job condition factor might be in the order of three times 14% or 42%.

Field Labor

Unfavorable conditions which may arise on a construction job often influence the labor hours for the project. The estimator must be alert to the following factors:

Sometimes plant production cut-backs make it desirable to use operating personnel for construction work. These men are good at their own jobs, but inexperienced in construction work. Hence man-hour requirements increase.

In some plant construction forces, plant regulations permit promotions to mechanics jobs only through the labor gang on a seniority basis. Few men capable of becoming expert craftsmen will choose this route during good times. Hence, a persistent down-grading of competence in the construction force results, unless counteracted by a good apprentice training program.

Variation in output may be expected in different sections of the country. Construction worker output in larger cities, having over 500,000 population, tends to be low. Some variations are due to a current temporary tightness of the local labor market. During boom times, inexperienced workers are hired because nothing better is available.

Bonus and piecework arrangements have produced remarkable results in reducing man-hour requirements for some types of construction work. Field erection of tanks is outstandingly efficient by major tank builders due to their use of piece rates and the policy of heavy bonuses to field supervisors for low cost. Applying incentive arrangements to this work is possible because the quantity and quality of output can be accurately measured.

Other effective incentives are of a less material nature. Competition between crews, between shifts or between similar projects can produce high output and low costs. High efficiencies are sometimes obtained when a positive deadline must be met—provided that job conditions, especially adequate supervision and planning, are not allowed to deteriorate.

Manufactured emergencies, daily emergencies and unrealistic deadlines have a bad effect. Construction men are keen observers and know the difference between actual, unavoidable emergencies and those resulting from poor management.

Much could be said concerning the willingness of labor as affected by job traditions and union rules. Here, the experience of the estimator is indispensable.

In summary, we find the ordinary range of construction worker com-petence factor is about 0 to 0.35. The willingness element has about the same range, making the total range of the two items 0 to 0.70.

Type of Contract

Lump sum work is driven toward efficiency by the immediate cost to the contractor of inefficiency. Working on a cost plus basis will usually increase labor costs about 5% because of elimination of this driving force. For percentage-fee contractors, the increase may be 10% or more.

Either the percentage-fee or fixed-fee contractor is inclined to shrug his shoulders and accept poor job conditions, especially when he considers the job conditions are due to some act or omission of the owner or engineer. On lump sum work, he will insist on correction of these same conditions.

With or without the contractor's knowledge, these attitudes are almost invariably reflected in the field. However, there are some contractors who handle lump sum and fee work with exactly the same conscientious care and whose employees reflect this policy by working in exactly the same way on both.

On some jobs, lack of complete drawings or the unpredictable nature of the work makes fee arrangements necessary. In these cases, incentive fees can sometimes be arranged which will produce contractor performance equivalent to lump-sum work. These incentive fee contracts are the reverse of the usual percentage-fee arrangements. The more the costs go up, the less the contractor gets. If the costs drop, his fee increases. Usually maximums and minimums are set to prevent unfairness to owner or contractor.

Construction Equipment

Historically, the greatest single factor in reducing the manpower and cost for performing a given amount of work has been the use of construction equipment. Earthmoving is the prime example. Even in the recent period of rapidly rising costs, the increased use of earthmoving equipment, combined with improvements in the size and performance of this equipment has kept costs per unit practically constant. Labor costs will be lowest on projects taking full advantage of

construction equipment such as front-end loaders, fork-lift trucks, cranes and back hoes.

Tools and other minor equipment items have shown similar improvement in recent years. For example: power hand saws, power tampers, steel scaffolding, spray guns, electric drills and power winches contribute to improved efficiency if they can be and are applied to a project.

Tools and equipment must be maintained in good condition if full advantage of their use is to be obtained. Equipment which is out of use because of poor maintenance runs up job costs. Maintenance of small tools and minor equipment is important from an efficiency and safety standpoint.

Loss of efficiency due to lack of major equipment may be 0 to 40%, although special cases can run higher. Lack of adequate small tools and minor equipment can produce losses having a range of 0 to 35%. The tool and equipment maintenance element probably has a normal range of 0 to 20%.

Construction Facilities

Construction facilities, in good order and well located, materially increase the efficiency of a job. If a workman has to walk a half mile each way when he needs supplies from the storeroom, he will probably spend more time walking than working. Efficient service at the storeroom is just as important. Men lined up waiting to be served are building up man-hours and drawing pay.

Location on the site of change houses, construction offices, maintenance shops and craft shops should be carefully established. Some part of the labor cost of a job is built into it when locations for these facilities are chosen. Portable buildings and trailers can be used to meet shifting location requirements as the work progresses.

General clean-up of the working area is a service which will not only reduce accidents but will increase efficiency. Other services and facilities which can contribute to lower costs are: temporary electrical service, compressed air, heat, telephones, drinking water, messenger service, first aid, fire protection, employment and safety inspection.

Although little factual data are

Orderly Construction Crew Buildup Gives Well-Run Job

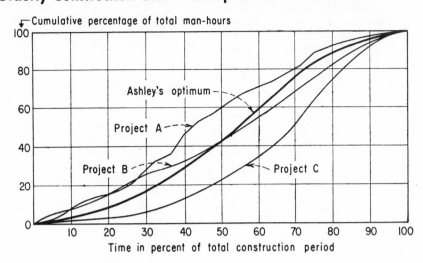

available, it is suggested that lack of effective services and facilities can increase job costs by about 5% for each of the following categories: storerooms and yards, shops and offices, personnel facilities and clean-up, and temporary utilities. The over-all effect for these conditions ranges from 0 to 20%.

Physical Working Conditions

Job efficiency is decreased by climatic conditions, heat, gases, fumes and proximity to hazardous materials. Loss of man-hours due to weather can be approximated by estimating from weather records the number of lost-time days which will be encountered during the construction period of the project. Assume that about a third of that time will be pay time not effectively worked. Usual ranges will be 0 to 25%. Losses due to process fumes, heat and hazards can run as high as 300 to 400% but usually are in the range of 0 to 40%. Often these affect only a limited portion of the project. Mud, dust and altitude effects are probably in the range of 0 to 15%.

Access to Work

Job efficiency decreases with height because of the vertical travel time and the additional care and equipment required to work safely. Extra costs can be minimized by good planning and the use of good equipment such as adequate steel scaffolding and high speed hoists.

Work in deep pits produces a similar loss of efficiency. As a rough rule, the cost of work such as pipe-fitting and wiring can be considered to increase about 1% for every foot above ordinary reach. This applies up to heights of 40 ft. or so.

Good roads are even more important in these days of high-priced labor and wheeled equipment. All-weather roads, laid down as early as possible, get a project off to a good start. For concentrated industrial plant construction, these should be concrete or some other hard surface. Gravel and other low-order surfaces must be constantly maintained, but are usually the economical solution for dispersed projects. Poor roads increase labor costs from 0 to 10% and up.

Construction Program

A construction rate which is faster than normal may produce a decrease in efficiency due to longer hours, crowding of crews, over-rapid buildup and increased difficulty of supervision.

Abnormally fast buildup of crews causes losses. Jobs where employment is soaring seem to develop indigestion. A substantial number of people are wandering around without enough to do, and some who look busy aren't. The graph, shown above, gives actual buildup curves plotted as cumulative percent of total manhours against time in percent of construction period. These are for complete process installations, including foundations, struc-

tural steel, building enclosure, piping equipment and electrical work.

Also plotted is Ashley's* recommended buildup curve for electrical projects costing about $500,000. Curves plotted in this way for well-run jobs should have few sudden breaks and steep sections. Exceptions will be found on the smaller projects where a sub-contractor brings in a good-sized well-trained crew to begin some portion of the project such as steel erection. This will produce a sharp rise in the man-hour curve with little loss of efficiency.

Area loading can be analyzed in terms of the number of field men per 1,000 sq. ft. of work area. A usual rate for jobs costing less than $500,000 is somewhere near five men per 1,000 sq. ft. during the peak weeks of the construction program. Larger jobs have lower apparent rates since not all of the project area is simultaneously active. High concentrations of men result in decreased efficiency; although there is no definite information as to numerical values. Except for emergencies, schedules which require more than seven or eight men per 1,000 sq. ft. are probably not realistic and will almost certainly increase labor costs.

Construction jobs progressing at less than the normal rate have high total costs. Part of this is due to stretching out the period when overhead costs must be maintained, but part of it is due to laxness and bad work habits which develop on a slow job.

Interruptions due to shifting of crews from one part of a project to another, or to emergency repair work can seriously affect job efficiency. When the crews return, they lose considerable time getting workwise. Long interruptions involve the replacement of material, tools and scaffolding which have wandered away.

Waiting time during tie-ins is another source of loss of efficiency. Careful programming and coordination of all the organizations involved will reduce but not eliminate this loss.

Faulty buildup rates can cause losses in the 0 to 10% range. Usual over-crowding can probably cause losses of the same magnitude. Interruptions in the program can cause losses in the 0 to 15% range.

_____
* R. Ashley, "Electrical Estimating," McGraw-Hill, New York, 1956.

Material Coordination

All necessary material should be on hand as required and must be of the correct kind. Otherwise delays result with loss of time and money as crews are shifted from one part of the job to another to work around items which have not yet been delivered or which are not usable. Extreme cases necessitate laying-off crews with the resulting loss of the investment in job education. When such layoffs are necessary, the morale of the remaining men suffers.

Another problem which occasionally arises is the too-early delivery of material. This can cause extra costs due to special unloading arrangements, deterioration in storage, extra warehousing costs, extra weather protection, misuse of materials and loss of identification.

Effect of this job condition may range between 0% to about 12%. If the material situation gets too bad and cannot be corrected, the job should be shut down.

Engineering Coordination

If drawings are available, competent construction superintendents and foremen will study them in advance and layout a program by which the work can be efficiently done. If drawings arrive just before they are needed, this planning cannot be done and job efficiency suffers.

If drawings come in later than required by the construction schedule, there will be times when crews are floating until the required information is obtained. When drawing issue drops too far behind the construction schedule, crews must be laid off with the subsequent costs of loss of morale, re-hiring and re-training.

A check list of questions which can act as a guide in evaluating this element might include:

1. Has a design schedule based on availability of engineering manpower been prepared?

2. Have the design schedule and the construction schedule been coordinated to provide drawings in the order required for construction?

3. Will the scheduled construction starting date give sufficient lead time to the design schedule?

4. Is design completion scheduled far enough ahead of scheduled construction completion?

5. Have procedures been established to assure release of drawings and other types of design information to the field at the optimum timing ahead of actual construction?

Design information availability element of the job condition factor may exhibit variations from 0 to 30%, with usual values ranging from 0 to 8%. The higher values are reached only when faulty planning or considerations other than efficiency force construction progress at rates well beyond the optimum with respect to design progress.

Engineering Quality

Closely associated with the timely receipt of design information is the necessity for accuracy and completeness. Actual errors in engineering information are not as frequent as are ambiguities and omissions. With fully detailed, easily read drawings, the field work goes ahead without holding up crews for field interpretation or detailing by job supervisors, foremen or engineers.

When actual errors exist and are not caught by the design checkers or field supervisors, the cost of removing and replacing faulty work may produce a substantial increase in the labor and materials costs of the job.

Some types of design information which, while not indispensable, can contribute to the efficiency of field work are:

1. Installation instructions from equipment manufacturers.

2. Installation instructions prepared by design engineers.

3. Cut-over and tie-in procedures established by design engineers.

4. Special construction procedures developed during design.

5. Borings and other sub-surface surveys.

6. Surface contour and general area maps.

7. Specifications, especially painting and insulation.

8. Flow diagrams.

9. Single line electrical diagrams.

10. Equipment and instrument lists.

11. Descriptive sections of design reports.

For the quality of engineering information received from the design departments of most corporations and consultants, this element of the

Variables Control Labor Costs

Craft of workmen
Local pay scale
Amount of overtime
Premium pay for overtime
Travel pay agreements
Number of foremen and pushers
Federal and state payroll taxes
Compensation insurance premiums
State revenue payroll tax
Union benefit fund and vacations
Construction equipment cost
Contractor's overhead*
Contractor's profit*
Construction tool and supply cost*

---

*On large jobs these items are included in field expense.

job condition factor will probably vary from about 0 to 12%. Attempts at partial engineering can raise this to 100% or more.

Other Special Conditions

After considering the listed job conditions, the estimator should study his project for any special conditions which may have an adverse or a beneficial effect on labor performance. As an example: on some electrical work there must always be a man on the ground to watch the movements of men on hot equipment.

How to Use These Factors

Total job condition factor is the sum of perfect job condition factor of 1.00 plus the probable decimal increase in labor hours produced by each job condition. Relative range of all the job conditions affecting labor hours is given in the table on p. 121. A convenient form can be prepared from the table to itemize the job factors for each project. On this form, space is provided to list the probable decimal increase in labor hours produced by each job condition as well as a few notes on the conditions expected for the project. Sometimes job conditions are different enough for different parts of the project that it may be necessary to compute more than one job condition factor.

These check lists should be kept until the project is completed. By comparing actual and estimated conditions and man-hours, the accuracy of future appraisals of job conditions and their effects can be improved.

Establish Man-Hour Costs

Effective pay rate is defined by the following equation:

$$\text{Effective pay rate} = \frac{\text{Total labor cost}}{\text{Man hours}}$$

Pay and time of field superintendents, time-keepers and general foremen are excluded from this calculation.

Since there are at least 14 variables affecting the cost of a man-hour, it is easier to determine an effective pay rate by analysis from past similar jobs. Great care must be taken, either to be sure that the job at hand is identical with the job of record as far as pay conditions are concerned, or that necessary corrections are made to allow for the new conditions.

As an aid in establishing effective pay rates, and in correcting recorded effective pay rates to changed conditions, each of the variables will be individually discussed.

Craft of Workmen

Construction work has traditionally been performed by craftsman specializing in one particular type of work. With the passage of time the lines of demarcation between work to be performed by the several crafts have been formalized to a high degree.

The working rules not only specify what crafts shall be employed on certain work, but also what grades of mechanics shall be employed. The rules specify ratios between foremen, mechanics, helpers and apprentices. Working rules may also make necessary such procedures as using electricians in equal numbers with riggers when moving large transformers, or having an operating engineer for every air compressor on the job. Relationships between the unions and the contractors on the job have a tremendous influence on the severity or looseness with which the rules are interpreted.

The estimator with considerable field experience usually knows pretty well what trades will be needed to accomplish a given project. The less experienced estimator has recourse to records or advice from field men.

Local Pay Scale

For a general idea of wage variations across the country consult *Engineering News-Record*. However, it is best in estimating for a new project to get a wage list for the exact location where the work is to be performed. This can be obtained from the local Trade Union Council, the State Employment Office or the local U. S. Department of Labor Field Office.

For large projects it is necessary to check to be sure that the local unions for which the rates are quoted can furnish enough men. If plant maintenance or construction workers are used, the plant supervisors can furnish necessary craft and wage information. Here again it is necessary to check to be sure that sufficient manpower is available in excess of that needed for current plant work. Otherwise it may be necessary to use seasonal contractor's men at higher average pay rates.

Amount of Overtime Work

Average effective pay must be corrected to show the affect of workweeks extended to attract labor to the job, or to expedite the work when the working area is so restricted that additional men cannot be utilized to advantage.

For operations involving heavy construction equipment, longer working hours are often economical, even with the resultant increase in average pay and decrease in worker efficiency.

Construction away from industrial areas sometimes involves travel pay. In cases where equipment manufacturers provide field erection crews, their agreements usually stipulate that the men be paid transportation, travel expenses, travel time and living expenses. These are part of the labor cost of the job.

Foremen and pushers are paid more than the craftsmen they supervise, increasing the average effective pay rate by the wage differential divided by the average number of men in a crew.

Federal and State Payroll Taxes

Federal Old Age Benefit Tax paid by the employer is 3% on all wages and salaries. This tax money pays for the various Social Security

benefits to employees. Unemployment compensation insurance differs from state to state. Some have no tax for this purpose, others have a tax as high as 3%.

Some states as a means of obtaining revenue have established a tax on construction payrolls. This is usually about 0.5%.

Workmen's Compensation

Workmen's compensation insurance covers job accidents to employees. It varies from about $\frac{1}{2}$% to over 30%, depending on the type of work, the state where it is being done and the accident record of the contractor. The *Engineering News-Record* publishes in the annual "Construction Costs" number a complete tabulation of compensation rates by craft and by state.

Union Benefit Fund

In addition to the provisions for accident, death, retirement and unemployment compensation established by state and federal laws, many unions have their own benefit funds and payments. Some New York unions currently have a 3% welfare and a 20¢/hr. pension levy. Rates elsewhere are usually less.

Contractor's Overhead

For small projects, it is usually convenient to consider the contractor's overhead as being a percentage of the labor cost of a man-hour, and to include it in the effective pay rate. On these smaller jobs the contractor cannot possibly keep accurate individual job overhead records, so he charges the job a percentage based on his annual totals for all the work he performs. Contractor's overhead on small projects varies from 10% to 20%.

Although overhead percentages are usually computed and billed on total job costs, the actual overhead expenditure required for all-labor contracts is greater than for a similar dollar volume of labor-and-material contracts. For large projects the contractor's overhead should be budgeted in detail as part of field expense and not considered as part of the average effective pay rate.

Contractor's Profit

Most contractors on small to medium work try to get a 10%

profit on labor and material. Profit margins on the larger jobs are less. Actual average profits for contractors across the country are in the 4% range during normal times.

For small jobs the contractor's profit can be included as a part of the effective pay rate. For the large jobs the contractor's profit is estimated as a lump sum to be included in the field expense budget.

Construction Equipment Cost

Construction equipment cost usually appears in the labor column of the construction ledger. This is based on the consideration that construction equipment is a replacement for human labor. For preliminary estimates, the cost of supplying workmen with equipment such as welding machines, trucks, cranes and bulldozers can be included in the average effective pay rate. For more detailed estimates, machine-hours and machine-hour costs are determined very much as if each machine represented an hourly-paid worker.

Machine-hour costs include rental, or depreciation, as well as operating costs. On large projects a field expense budget is established for the control of equipment operating costs. Individual sub-accounts of the project are charged for the use of machines on an hourly rental basis. Appropriate field expense accounts are credited.

On small jobs construction tools and supplies may be considered part of the contractor's overhead and as such included in the effective pay

rate. Larger jobs should have lump sum items in the construction expense budget to cover tools and supplies.

Accounting Practices

Company accounting practices must be conformed to with respect to charging items to operations, to labor, to material or to overhead. The estimator's problem is to forecast the cost figures as they will be entered on the company's books. Therefore, effective pay rates must be computed on a corresponding basis.

Usually these accounting practices are so tightly built into company procedures that the estimator must conform. Occasionally he may be able to point out changes which will improve the realism with which the construction cost records reflect actual cost to the company.

The sample computation shows how an effective pay rate can be built up from information concerning local pay conditions.

Average craft base rate		$3.09
Foremen and pushers differential	2½%	0.08
Social security	3%	0.09
Unemployment insurance	3%	0.09
Workmen's compensation, avg.	3%	0.09
Expendable tools	½%	0.02
Construction equipment cost	5%	0.15
Overhead (computed on base pay plus equipment)	10%	0.33
Profit (computed on base pay plus equipment plus overhead)	10%	0.38
Average effective pay rate		$4.32

Use Labor-Material Ratios to Find Labor Costs

New project information:

Estimated cost of maintenance shop lighting material, current prices	$10,000
Expected ENR construction cost index	650
Probable job condition factor	1.20

Similar project information:

Standard labor-material ratio at ENR of 600, JCF of 1.50 from lighting installation in similar shop built in Chicago in 1944	0.40 MH/$M

Man-hour computation:

Job labor-material ratio, new project

0.40 × (1.20/1.50) (600/650) = 0.296, say 0.30	0.30 MH/$M
Man-hours to install lighting 0.30 × $10,000	3,000 MH

Labor cost computation:

Average effective pay rate	$6.00/MH
Estimated labor cost is 3,000 × $6	$18,000

For our purpose, a labor-material ratio can be defined as the number of field man-hours required to install a dollar's worth of material or equipment. Any construction labor performed on the plant site is included. Work done off the site is considered to be part of the material cost of the job.

A standard labor-material ratio is the number of field man-hours required under standard job conditions to install the material that could be purchased for one dollar if the ENR Construction Cost Index were at 600.

A job condition factor of 1.50 has been arbitrarily established as standard. It represents conditions about like those encountered when making new installations in operating industrial plants. Construction work on new sites, with good working conditions, will have job condition factors closer to 1.20. Work under poor job conditions will have ratings above 1.50.

Assume, for the purpose of illustration, that a lighting system has been installed in a building during the initial construction of a plant and that the records show that 0.32 man-hours were required for every dollar's worth of material installed. If the job condition factor for this initial work was 1.20, and assuming for simplicity that the ENR Construction Cost Index was 600, then the standard labor-material ratio for this work would be—

$$0.32 \times 1.50/1.20 = 0.40 \text{ MH/\$M}$$

If a new bay is to be added to this same building on a rush basis, with other unfavorable job conditions, the job condition factor for the new work might be 2.00. Assuming the cost index is constant at 600, the job labor-material ratio for the addition would be:

$$0.40 \times 2.00/1.50 = 0.53 \text{ MH/\$M}$$

Standard labor-material ratios must be corrected to a common ENR Index base of 600 because of the variations in the amount of material which can be purchased for a dollar. Assume for the purpose of illustration that the lighting had been installed in the building men-

tioned above at a time when the ENR Index was 400 under job conditions represented by JCF of 1.50, and that under those conditions it took 0.60 field man-hours per dollar's worth of lighting material to do the installation. The standard labor-material ratio for the job would then be:

$$0.60 \times 400/600 = 0.40 \text{ MH/\$M}$$

This standard ratio is less than the record ratio because at ENR 600 a dollar buys less material than at ENR 400. Therefore, less labor will be required to install that dollar's worth of material. No job condition correction is required in this illustrative case, since the actual job conditions have been assumed to be about equal to standard.

Standard ratios should be computed for each subdivision of each job of record. They should be tabulated in accordance with the standard account classifications. Each entry should have a brief description of the type of work done and the type of project.

Correcting ratios to a standard base and grouping them in this fashion improves their usefulness by permitting intelligent interpolation, generalization and averaging. When a sufficient background of such information has been built up, it is not too difficult to select the type of work and job of record most similar to the one at hand and to correct the standard ratios to the job conditions and cost conditions of the current work.

Labor man-hours are computed by multiplying job labor-material ratios by the applicable material dollars. Labor dollars are computed by multiplying labor man-hours by the applicable effective pay rate.

Detailed Unit Labor Pricing

Detailed unit labor estimates with an accuracy of better than plus or minus 3% are a necessity for contractors bidding on lump sum work. They are also needed by engineers for comparisons of alternate designs. Unfortunately these estimates are:
- Expensive—their cost is on

the order of 1 or 2% of the labor cost of the project.
- Slow—all drawings and bills must be complete before the estimate can be completed. With large jobs and the usual shortage of estimators, several weeks may be required to complete the estimate after drawings are made.
- Difficult—construction work has such infinite variety that an estimator must have special experience and voluminous records to be expert in a single portion of the entire labor estimating field.

Usually the owner's engineers, estimators and management need capital cost estimates for (a) determining probable percentage return of a project, (b) checking bid prices and (c) arranging for financing.

For these purposes, an estimate of total labor and material with an accuracy on the order of plus or minus 10 to 15% is often adequate. Labor-material ratio methods may be used to save project completion time and estimating expense in such cases.

Detailed unit labor estimates will usually be required for the smaller jobs and for comparing alternate designs.

Sources of detailed labor estimating information are scattered through many publications. These can be located by means of the references at end of this report. However, the bulk of all labor cost information is in contractor's and estimator's files and is generally inaccessible.

In some cases where a detailed estimate is required, the drawings can be sent out for lump-sum bids. Bids on alternate designs can also be requested. Results will be satisfactory, provided contractors need the work to bid right. In fairness to contractors and because preparing estimates is expensive, bids should be requested only on work which has a better-than-even chance of being authorized for construction. Alternates should be kept to a minimum and alternate bids requested only when it is impossible for the engineer to appraise the relative cost of alternates in some other way.

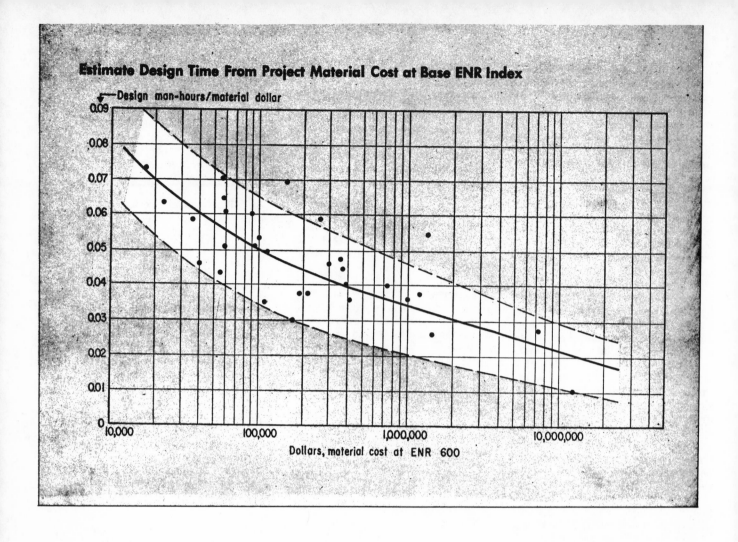

Estimate Design Time From Project Material Cost at Base ENR Index

Design man-hours/material dollar (y-axis)

Dollars, material cost at ENR 600 (x-axis)

Include Design and Field Expenses

Design expense includes all office engineering and drafting required to prepare and issue construction drawings, specifications and bills of material. It does not include preparation of the first issue of the design report nor any of the preliminary engineering studies preceding the design report. Neither does it include field engineering work such as:

- Property survey
- Topographic survey
- Hydrographic survey
- Subsurface survey
- Inspection and testing
- Field layout
- Field engineering liaison

Design expense includes engineering performed by either company or associate engineers. Work performed by company engineers for small projects is part of the general expense of operating the company.

Estimates of Design Expense

Amount of design time required for a project is a function of a number of variables such as size of project, type of project and productivity of engineers and draftsmen.

Size of project is the most dominant variable. There are several possible measures of project size. The graph of design man-hours per material dollar against material has slightly narrower deviations than plots against labor or against labor and material.

The graph at top of this page of design man-hours per material dollar against material cost at ENR 600 for design projects of record shows a distinct tendency toward lower values for the larger projects. There are wide deviations of individual points due to variations in type of project and productivity. Design man-hours for a new project can be approximated from this diagram. However, the estimator must exercise considerable judgment in classifying the project design as easy, standard or difficult.

When the average cost of a design man-hour including the engineering overhead and engineering fee is known, design cost can be estimated from the design man-hour curve. The following analysis illustrates the method:

Estimated material cost of project is $1 million. For estimate basis, use ENR index of 820. Hence, material cost at ENR 600 is $1 million × 600/820 or $733,000.

Assuming standard difficulty, we find design man-hours per material dollar equal 0.036 from the graph. Design man-hours for the project equal 0.036 × 733,000 or 26,300.

Cost per design man-hour is $7.00 which represents design time at

Compare Representative Field Expense Charges of Cost-Plus Contracts

	Total Labor and Material	Labor, Taxes and Insurance	Field Overhead	Profit	Overhead	Profit
		Labor Percentage*			Material Percentage*	
Concrete	$8,000	15%	10%	9%	10%	9%
Concrete	—	10.5%	0%	10%	—	—
Concrete block building	93,700	7%	10%	9%	10%	10%
Structural steel	83,800	12%	12%	10%	—	—
Structural steel	—	11%	15%	10%	—	—
Office partitions	5,000	Actual	10%	9%	10%	9%
Roofing	10,000	8%	17.5%	10%	17.5%	10%
Piping	3,500	Actual	10%	10%	—	—
Piping, gas and water	12,000	13%	10%	10%	10%	10%
Piping	12,000	8%	15%	6%	—	—
Piping	18,000	10%	15%	10%	—	—
Instrumentation	13,000	10%	10%	10%	—	—

* Percentages are applied progressively to the respective labor and material costs.

$3.50/hr. plus 100% overhead and engineering fee. Therefore, estimated design cost equals 26,300 × $7.00 or $184,000.

Since the estimated material cost of the project includes an allowance for unlisted items, the estimated design cost automatically includes the same percentage allowance for engineering of extras. Because of the wide divergence of actual project design man-hours from mean values, the preceding method should be used only for preliminary estimates.

For a layout estimate, more accurate estimates of design cost can be made after the project engineer completes the design report and its list of drawings required for the project. Drawings should be classified as to type of work and man-hours per drawing assigned by the project engineer. When estimated in this way, an unlisted items percentage should be added to the design man-hours to take care of the unforeseen.

When a preliminary bill estimate can be made, much of the design engineering is complete. The actual design cost-to-date, plus the estimated cost of completing the drawings in progress and the drawings not begun, plus a contingency percentage based on the current uncertainty rating, gives the probable cost of the whole. Overoptimism with regard to percent completion should be avoided.

When a design has advanced to the point where a detailed cost esti-

mate can be prepared, almost all of the actual design cost is known. The unlisted items allowance which is relatively small at this stage is added to take care of revisions during construction.

Define Field Expense

Field expense is made up of construction overhead and engineering costs connected with field work and not specifically chargeable to any

single part of a project. These cost items and corresponding account subdivisions are:

Field Engineering
  Field engineering liaison
  Layout survey
  Property, topographic and hydrographic surveys
  Subsurface surveys
  Inspection and testing

Procurement
  Purchasing salaries, expenses and fees

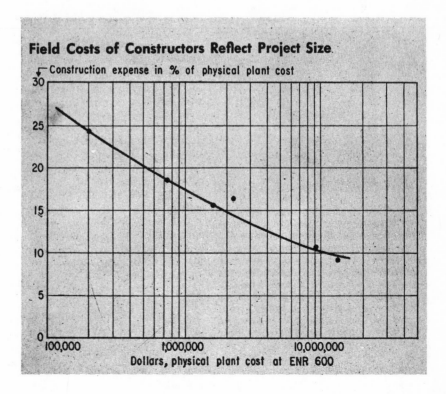

Field Costs of Constructors Reflect Project Size

Transportation of materials, supplies and equipment when unassignable to specific subaccounts

Receipt, storage and issue of materials

Expediting salaries and expenses

Material coordination salaries and expenses

Stores when not chargeable to specific subaccounts

Construction Equipment
Major construction equipment purchase and sale

Major construction equipment rental expense

Tool and minor construction equipment purchase and rental

Supply purchase including office supplies

Repair and maintenance of major construction equipment

Receipt, storage, repair and issue of tools, supplies and minor construction equipment

Construction equipment rental credits

Temporary Facilities
Buildings
Stores, equipment and enclosures
Water supply and distribution
Sanitary facilities
Heating
Electrical supply and distribution
Roads, walks and parking
Compressed air supply and distribution
Office equipment

Services
Field supervision
Cost and timekeeping

Stenographic and messenger service
Salvage
Telephone, telegraph, prints, photographs and publications
Plant protection
First aid and safety
Employment
Clean-up when unassignable to specific subaccounts

Administration and Fees
Contractor's fee and bond
Legal fees
Taxes except payroll taxes
Insurance except workmen's compensation and other payroll-tied insurance
Permits
Employee relations
Public relations

Invisible Field Expense

Estimating field engineering and construction expense is especially difficult because certain items may or may not be considered part of a job's costs. These items depend on circumstances and company policy.

For this reason, construction programs may involve booby-traps for the estimator. Plant construction programs go along for years getting free compressed air, electricity, guard service and other overhead expense items for construction work. Then a really big job comes along. The plant is to be doubled in size. As the construction force builds up, the plant operators start to complain.

How to Estimate Field Expense

Preliminary estimates of field expense for projects under $1 million can be made by applying appropriate percentages to material and labor costs.

For small projects where detailed field expense records are impractical and the owner is supplying all materials, the labor contractor's profit and overhead can be computed as a percentage of the field labor payroll and included in the effective pay rate.

If the contractor is purchasing part of the materials, he will charge a percentage of these items to cover his overhead and profit. If the amount is significant, allowance can be made for it in estimating material cost.

The table on previous page shows some percentages which have been charged by cost-plus contractors on industrial work. Some of these contractors use identical percentages

for labor overhead and material overhead. Actually the overhead expense items chargeable to labor will amount to about 3.5 times those chargeable to the same number of dollars of material.

Even when labor overhead and profit percentages are included in the effective pay rate, the list of account subdivisions for field expenses should be scanned for major items which may later appear in the construction ledger as charges to the project. For example: jobs in the $100,000 to $500,000 range will often require the full-time services of an owner's field liaison engineer and sometimes a full-time accountant. Other field expenses will appear as the jobs increase in size. Such items can be estimated from data filed by account number.

When projects cost much over $1 million most of the field expense items begin to assume such proportions that they can no longer be obtained on a free-loader basis and fewer are hidden in direct labor cost. To handle these jobs properly and to get the best performance from the work force requires a more complete field organization. Its cost will be more than repaid by decreased labor cost and early completion.

In terms of percentage of total physical plant cost, the graph on p. 129 gives the field expense for a completely self-supporting field construction organization. The figures include contractor's home office overhead and profit. Physical plant is defined as the labor and material cost of the project at ENR 600 excluding design and field expense.

When time and available information will permit, field expense should be estimated by drawing up a detailed field expense budget. This is also a necessary step in planning the project. The completed budget is used as a management tool to control actual field expenditures.

John W. Hackney is Manager of Cost Engineering for Mobil Research and Development Corp. His experience has included cost engineering and project management assignments with M. W. Kellogg Co., Beaco Ltd. of Montreal, Diamond Shamrock Co., and Aluminum Co. of America. Mr. Hackney has B.S., M.S., and C.E. degrees from Carnegie-Mellon University, and is a past-president of AACE. His book "Control and Management of Capital Projects" was published by Wiley in 1956.

REFERENCES
"Construction Cost Control," ASCE.
"Indexes in the Plant Cost Picture," *Chem. Eng.*, May 1947, pp. 109-111.
Nelson, W. L. "Construction Cost Indexes," *Oil & Gas J.*, Oct. 21, 1948, p. 141.
Stevens, R. W., "Equipment Cost Indexes for Process Industries," *Chem. Eng.*, Nov. 1947, pp. 124-126.
Ashley, R., "Electrical Estimating," McGraw-Hill, New York, 1949.
Means, R. S., "Building Construction Cost Data," Duxbury, Mass.
Walker, F. R., "The Building Estimator's Reference Book," F. R. Walker, Chicago.
Lang, H. J., *Chem. Eng.*, Sept. 1947, pp. 130-33; Oct. 1947, pp. 117-121; June 1948, pp. 112-113.
Nichols, W. T., "Capital Cost Estimating," *Ind. Eng. Chem.*, 43, 2295 (1951).

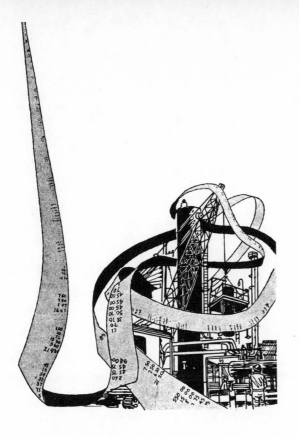

Estimating Methods

For Process Industry Capital Costs

JOHN W. HACKNEY, Mobil Oil Corp.

GOOD estimating requires prior preparation of reliable cost data and development of satisfactory procedures. By selecting suitable combinations of these data and procedures, we can prepare estimates having the optimum degree of accuracy for a specific stage of project development.

In Part I (see p. 25 of this volume) of this two-part report, we developed methods of establishing estimating data sources for equipment, labor and material and provided an analysis of design and field expenses.

In this part of the report, we'll complete the basic data by showing why contingency allowances are necessary and how to prepare and make uncertainty ratings for process-plant projects.

Having established all of the basic estimating procedures, we can apply them to make estimates by the following methods:

- Equipment ratio method
- Layout method
- Preliminary bill method
- Detailed unit cost method

Each of these methods gives estimates having different ranges of accuracy when compared to actual installed cost of a project. On projects under construction, periodic re-estimates based on these methods improve the accuracy of final estimated cost. Re-estimates also provide control over expenditures for project costs including labor and materials.

Each estimating method will require project and cost data commensurate with its range of accuracy. Of course, time and cost of making the estimate will depend on the estimating method chosen.

Provide Contingency Allowances

Contingency allowances can be included in cost estimates to compensate for accidental events, changes in general economic conditions, errors, imperfections of estimating methods and oversights.

Accidental Events

Most accidental events can be neutralized by insurance. For example: fires, construction accidents, floods, explosions, wind-damage can be insured against and the premium made part of the job cost.

Some accidental events such as those encountered in underwater work can be covered by a contingency item. Where possible, it is better to allow for accidental events by paying insurance premiums than by contingency allowances.

Economic Conditions

Forecasting changes in general economic conditions is very difficult. In spite of years of study by statisticians and the expenditure of tremendous sums of money by governmental agencies, there is still no reliable means for predicting major changes in either the national economy or the construction industry.

These problems are so important that we cannot avoid them. If we could forecast construction volume, we might also be able to forecast some of the effects of volume which are not indicated by the usual construction cost indexes. Indexes are usually based on published prices and wage rates and do not include the premiums which are required to obtain labor and material in boom times. In depression periods, the indexes do not reflect special price concessions, prompt delivery, improved worker attitude and similar items which tend to reduce costs.

The *Engineering News-Record* publishes at intervals a forecast of expected trends in the ENR index which deserves careful study.

Another forecasting method, based on a probability study of the ENR index since 1918, suggests the assumption of a continued change of index at the same rate as for the previous 6-mo. period within limits of +13% and −7%. This assumption gives a low spread of results as calculated from the mean probable error.

For comparison, a contractor might assume that the index will be unchanged in the succeeding 12 months. Over a long period of time this contractor, who is assuming no change in prices, will be continually bucking-the-odds to the extent of 4 or 5% between bid time and performance on jobs lasting over two years.

Methods such as the above are not perfect, but in making an estimate for a firm bid all possible information must be considered because we cannot avoid forecasting prices.

Errors

Arithmetic errors in estimates are frequent. Errors can and should be avoided by means of an arithmetical check by someone other than the original estimator. A re-estimate by section, using one of the rapid estimating methods, can be used as a general check. It often uncovers gross errors, if they exist.

Imperfections in Estimates

In estimating the cost of a series of process plants, we might have complete drawings, exactly representing the final physical makeup. We might even have a complete forecast of all conditions which will exist during construction. But even under these impossibly favorable conditions, the individual pre-construction cost estimates of the various plants would fall over and short of the actual costs. Flaws, imperfections and approximations inherent in every estimating method are responsible for this variation.

Assuming that completed design information is available and that a careful detailed estimate is made by an experienced estimator, the effect of inaccuracies in method for major projects is probably in the range of ±3%. Small projects, up to $100,-000, give more erratic results because there is less chance for averaging out.

When design work is incomplete, preliminary estimating methods must be used. These necessarily have a wider accuracy range than detailed estimates. We made a study of nine completed projects varying in size from $100 thousand to $14 million. We assumed perfect knowledge only of the cost of major process equipment. All other costs such as piping, foundations, buildings, electrical work and instrumentation were estimated by ratios. Results were in the range of ±10% and averaged an underrun of 1%.

However, these results should be accepted with caution because of the small size of the sample, the assumption as to full knowledge of major process equipment cost and the difficulty of maintaining strict mental honesty with respect to hindsight vs. foresight. A range of ±25% is probably more realistic.

Oversights

In making an estimate it is easy to overlook items of construction labor and materials which will be required to perform the work described by the drawings and design reports at hand. Moreover, in the preparation of these drawings and reports, it is equally easy to overlook features which will later be required for the proper functioning of the project. This general tendency is accentuated by any uncertainties in project design at its current stage of development.

Consider Project Uncertainty

Accurate appraisal of the current degree of project development is difficult. A chemical project often begins with reaction computations and proceeds through laboratory tests, pilot-planting, process design, site inspection, engineering design and detailed design. It culminates in actual construction.

Estimates to check the desirability of continuing the project will be required at various stages. Of course, the early estimates will be based on incomplete information and therefore will be less accurate. Later estimates will include omitted items uncovered as development proceeds and will be based on a firmer definition of the required installations.

But where are we in the project development? Often pilot-planting, process design, site inspection, en-

gineering, drafting and sometimes construction are proceeding simultaneously. Knowing that a high degree of project uncertainty means a greater number of probable omissions, how can we rate the uncertainty of the project at the time of estimating?

As an approach to this problem, we have divided project development into:

A. General project basis
B. Process design status
C. Site information
D. Engineering design
E. Detailed design
F. Field performance

The six sections have been expanded into a check list, shown on this page, which is made up of individual items of project development. Each item on the list has been given an arbitrary weight which indicates the degree of uncertainty produced in the over-all project if the item is unknown.

In weighting the items on the list, highest values have been given to items such as process background and raw material quantity because uncertainty with respect to such items can have the most serious effects on project cost. Other items such as sanitary sewer design have been given low ratings since they usually have a relatively minor effect on project cost. While the assignment of weights is grossly arbitrary, they represent the sort of percentage overrun which may be expected if the item in question were completely misjudged.

When information with respect to an item has not yet been established, the full uncertainty rating of the item is entered on the check list. When design or other work with respect to an item on the list is partially complete, its uncertainty rating is proportionately reduced from the maximum value.

If project development is complete and reliable except for the items listed under detailed design and field performance, the uncertainty rating will be 120. If the design of this same project is advanced by completion and checking of the detailed drawings and bills of material, the uncertainty rating is reduced to 50. When the field work is 40% complete (60% incomplete), the uncertainty rating is reduced to 30.

By experience, it has been found that uncertainty with respect to items in the general project basis

Uncertainty Rating Check List

	Maximum Rating	This Project
A. General project basis		
Product and byproducts	100	
Raw materials	100	
Process background	200	
Subtotal A	(400)	
Project basis multiplier (One plus 1% of subtotal A)	5	
B. Process design status		
Flow balances (material, heat, power)	70	
Major equipment, type and size	80	
Materials of construction	30	
Review with research, development and operations	70	
Subtotal B	(250)	
C. Site information		
Surveys, including obstruction and subsoil	45	
Re-usable equipment	25	
Re-usable supports, piping and electrical	25	
Buildings available	30	
Utilities available	25	
Yard improvements available	25	
Climatological information	20	
Local ordinances and regulations	20	
Review with operations	25	
Subtotal C	(240)	
D. Engineering design status		
Layouts	35	
Line diagrams, including utilities	50	
Auxiliary equipment, type and size	45	
Buildings, type and size	35	
Yard improvements, type and size	25	
Hazard control	25	
Coating specifications	10	
Review with research, development and operations	70	
Subtotal D	(295)	
E. Detailed design		
Drawings and bills of material	45	
Drawing checks	25	
Subtotal E	(70)	
F. Field performance	(50)	
Subtotal, Sections B through F	(905)	
Uncertainty rating (Multiply subtotal of sections B through F by project basis multiplier), 905 × 5	(4,525)	
Recommended contingency		
Unlisted Items (Percent of base cost)		
Restricted reserve (Percent of base cost plus unlisted items)		

Establish Detailed Uncertainty Rating for Each Step

	Maximum Rating	This Project*
A. General project basis		
Products and byproducts		
Quantity	50	
Physical form	15	
Chemical composition	15	
Allowable impurities	10	
Wastes	10	
Subtotal	(100)	
Raw materials		
Quantity	50	
Physical form	15	
Chemical composition	15	
Allowable impurities	10	
Source	10	
Subtotal	(100)	
Process background		
(Suggested values for certain backgrounds are as follows:†)		
Computation only	(150–200)	
Laboratory data only	(110–160)	
Pilot plant data	(60–110)	
Semi-commercial unit data	(20–60)	
Full-scale experience	(0–20)	
Subtotal	(200)	

* Use dash when item is not involved in project. Use zero when item is completely known.
† When necessary divide equipment into groups of similar background and weight on a percentage basis.

B. Process Design Status

	Maximum Rating
Flow balances	
Material	30
Heat	20
Electricity	20
Sub total	(70)
Major equipment, type and size	
Process	50
Storage, raw material	10
Storage, in-process	10
Storage, products	10
Sub total	(80)
Materials of construction	
Records of durability,	
in identical service	0
in similar service	5
Short-run trials,	
in identical service	5
Laboratory tests, well-simulated conditions	10
Laboratory tests, preliminary	20
No investigation	30
Sub total	(30)
Review of process design	
With research, development	50
With operations	20
Sub total	(70)

C. Site Information

	Maximum Rating
Surveys	
Obstructions, interferences	20
Property	5
Topographic	5
Subsoil	15
Sub total	(45)
Re-usable equipment	
Type	5
Size	5
Materials of construction	5
Condition	10
Sub total	(25)
Re-usable supports	
(Piping and electrical)	
Type	5
Size	5
Materials of construction	5
Condition	10
Sub total	(25)
Buildings Available	
Process space	10
Office, control room and laboratory space	3
Toilet, lunch, change rooms	4
Maintenance, stores space	3
Condition	10
Sub total	(30)
Utilities Available	
Power	10
Water	5
Steam and condensate	5
Fuel, air, refrigeration	5
Sub total	(25)
Yard Improvements Available	
Railroads and docks	5
Roads	5
Parking, sidewalks	3
Fencing, planting, light	2
Sewers, storm	5
Sewers, sanitary	2
Sewers, process	3
Sub total	(25)
Climatological Information	
Rainfall, flooding	10
Wind, earthquake	5
Temperature, dust	5
Sub total	(20)
Local Ordinances, Regulations	
Air and water pollution	10
Zoning and building	10
Sub total	(20)
Review with operations	(25)

D. Engineering Design Status

	Maximum Rating
Layouts	
General plant	15
Process, plan	15
Process, elevations	5
Sub total	(35)
Line Diagrams	
Process pipelines, ducts	15
Conveyors, chutes	10
Distribution and supply	
Power	10
Steam, condensate	7
Water	5
Fuel, air, refrigeration	3
Sub total	(50)
Auxiliary Equipment, Type and Size	
Dry and packaged material movement	10
Pumps, compressors, blowers and fans	10
Instruments, controls	10
Power conversion, control	10
Major spares	5
Sub total	(45)
Buildings, Type and Size	
Process	20
Office, control room and laboratory	5
Toilet, lunch, change rooms	5
Maintenance and stores	5
Sub total	(35)
Yard Improvements, Type and Size	
Railroads and docks	5
Roads	5
Parking, sidewalks	3
Fencing, planting, lighting	2
Sewers, storm	5
Sewers, sanitary	2
Sewers, process	3
Sub total	(25)
Hazard Control	
Ventilation, dust, fumes	8
Explosion, fire prevention	7
Air and stream pollution	7
Preliminary insurance review	3
Sub total	(25)

Coating Specifications
  Insulation   5
  Painting   5
    Sub total   (10)
Review of Engineering Design
  With research,
    development   20
  With operations   50
    Sub total   (70)

E. Detailed Design

                        Maximum Rating
Drawings and Bills of Material
  Structural   5
  Storage   3
  Piping   13
  Equipment   4
  Instrumentation   2
  Insulation (bills only)   2
  Electrical   11
  Site improvements   5
    Sub total   (45)
Drawing Checks
  Safety department   2
  Insurance company   3
  Operations   20
    Sub total   (25)

F. Field Performance
    Sub total   (50)
    Reduce from full value to zero in
    accordance with percent of field
    labor remaining to complete

section will produce uncertainties with respect to all subsequent design work. Therefore, the uncertainty rating of this section is used, not as an added item, but in establishing a multiplier for the sum of the ratings of the other five sections.

Considerable judgment is in-volved in rating the items of the check list, and opinions as to rating individual items will differ. Taking project ratings as a whole, we invariably find agreement within 10% when different estimators rate the same job.

Establish Contingency Allowances

The most dependable guide in establishing proper contingency allowances is past experience. In order to analyze this past experience, actual costs versus estimated costs should be tabulated for all company projects of record. Eliminate from the estimated costs any contingency allowances which may have originally been included. To bring projects of various sizes to a common basis, the data are presented in terms of percentage overruns. Corrections should also be made to compensate for any capacity changes made after the estimate date.

Variations of these overrun percentages are functions of many variables, including among others:

1. Uncertainty status of project at time of estimate
2. General economic condition changes
3. Estimating method used
4. Quantity and quality of cost records available to the estimator
5. Estimator's skill and accuracy
6. Purchasing efficiency
7. Field efficiency
8. Cost efficiency of design if design is completed after the estimate.

When percentage overruns are plotted against uncertainty rating at time of estimate, the pattern shown in the chart on this page develops. While the points have considerable scatter because of the variables previously listed, dominance of the relationship between uncertainty ratings and overruns is apparent.

As additional data become available, it will be possible to classify further the points on the diagram with respect to other variables.

Meanwhile, the overrun pattern of the chart represents an integration of the effect of all of the past practices with respect to estimating, design, purchasing and construction. As such it gives an indication of the contingency allowances which should be applied to future estimates. It also presents a probability picture indicating the probable range of variations between estimated and actual costs as established by a company's historic practice and experience.

Contingency for Unlisted Items

On the chart given here, overruns for only four of the 25 estimates fall below the line marked $U \times 0.06$. In other words, for 84% of the estimates a percentage contingency allowance of at least 0.06 times the uncertainty rating should have been included. Allowing for some improvement in estimating procedures, a percentage contingency allowance called unlisted items in the amount of 0.03 times the uncertainty rating should be included in each estimate. Very seldom will an amount less than this be required.

Restricted Reserve

As a basis for management decision as to whether or not to proceed with a project, it is also desirable to establish the probable maximum overrun which a project may have. This probable maximum can be approximated by the line marked $U \times 0.180$ on the chart. Three points lie above this line so that in 22 out of the 25 cases, or 88% of the time, the overrun percentage has not exceeded 0.180 U. Therefore, the probable cost of a project can be presented as follows:

• Estimated cost includes a percentage contingency allowance for unlisted items of 0.03U to take care of normal contingencies. This

Uncertainty Rating Governs Overrun Allowances

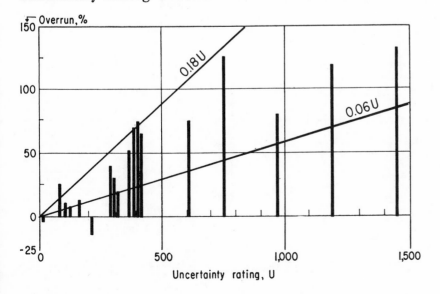

amount or more will be required for about 84% of the projects built.

• Restricted reserve is an additional amount some part of which will be needed for about 72% of the projects. For about 12% of the projects, all of this restricted reserve and more will be needed.

In terms of the estimated cost which already includes the allowance for unlisted items and again allowing for improvement in estimating methods, we can compute restricted reserve from the formula:

$$\frac{R}{E} = \frac{(0.10U - 0.03U)}{(100 + 0.03U)}$$

where R is restricted reserve and E is estimated cost.

How to Uses Restricted Reserve

The amount of the restricted reserve necessary at a given stage of project development, when considered along with the income and return estimate, is a guide to management in deciding whether to:

1. Authorize funds for project completion.
2. Drop the project.
3. Authorize additional study, pilot-planting, site investigation or design work before taking final action.

If the project will be a profitable one, even though all of the restricted reserve should be required, management may decide to release it for all-out design, procurement and construction. Thereby, returns are possible at the earliest possible date. If the project cannot make money at the estimated cost, it usually is dropped. Occasionally, further development or design is justified.

In some cases, the project will be a profitable one if it can be built for the estimated cost but unprofitable if all of the restricted reserve should be required. In such cases, further project development is usually authorized so as to more firmly establish the estimated cost and to reduce the amount of the restricted reserve.

When a project does get the go-ahead, management authorizes only the estimated cost, withholding the restricted reserve. Therefore, the estimated cost is established as a goal to be met by the project team. Restricted reserve funds are released only by approval of supplementary requests supported by detailed justification.

Fit Estimating Method to Needs

Estimating methods require varying degrees of project information for their application. Equipment ratio estimates can be made with a bare minimum of information, while detailed estimates require complete drawings and bills of material. Layout estimates and preliminary bill estimates require less information than detailed estimates but more than equipment ratio estimates. Each of these estimating methods is applicable to a different set of circumstances in project work as shown in the table below.

Equipment Ratio Estimates

When little design information is available and a preliminary estimate of plant cost is required on short notice, the equipment ratio method is recommended. For this estimate, we need to know only:

1. Major process equipment type, size and materials of construction.
2. General extent to which buildings, utilities and yard improvements must be provided.

Accuracy of this information and, therefore, accuracy of the estimate will be affected by the degree

Select Estimating Method Appropriate to Job to Give Best

Information Required [1]	Probable Accuracy [2]	Approximate Cost per $1 million [3]	Process Equipment,[4] Including Instruments
Detailed Unit Cost Method			
General project basis Process design Site information Engineering design Detailed design	±3% at U.R. of 50 or less.	$11,000 to $33,000	Based on firm bids, includes freight, corrected for probable escalation.
Preliminary Bill Method			
General project basis Process design Site information Engineering design Partial detailed design	±6% at U.R. of 100 or less	$3,200	Based on quotes or recent purchase, includes freight, corrected to probable ENR index.
Layout Method			
General project basis Process design Site information Engineering design	±12% at U.R. of 200 or less.	$320	Based on quotes or recent purchase, includes freight, corrected to probable ENR index
Equipment Ratio Method			
General project basis Process design General site information	±25% at U.R. of 400 or less.	$80	Major process equipment based on capacity-cost graphs corrected to probable ENR index. Auxiliary equipment and instruments by ratio to value of major process equipment.

1. See uncertainty rating form (p. 45) for detail of items included in each group.
2. For projects costing more than $100,000. For smaller projects, accuracy ranges will be wide

of uncertainty of knowledge with respect to the following groups of items on the uncertainty rating check list: products and byproducts, raw materials, process background, flow balances, materials of construction background, review of process design, site information and re-usable equipment data.

To organize the estimate, use the equipment ratio estimating form shown on p. 53. Detailed steps in filling out this form are outlined on pp. 51 to 55.

Layout Estimates

When additional site and engineering design information is available, use the equipment ratio method in a modified form to make estimates prior to the completion of drawings and bills of material.

A layout estimating form is shown on p. 130.

Since a well-established plant layout is an essential requirement in making an estimate by this method, it is called the layout method. We require the following information for this type:

1. Major process equipment type, size and materials of construction.
2. Auxiliary equipment type, size and materials of construction; including instruments.
3. General layouts and line diagrams.
4. Field inspection of the site.

Accuracy of the estimate will be affected by the degree of uncertainty of knowledge with respect to the groups of items listed for the equipment ratio estimate plus layouts; line diagrams; auxiliary

equipment type, size and materials of construction; building type and size and yard improvements.

Layouts should include a layout of the process in plan and in elevation, and a general plant layout including the entire area affected by the project.

For small projects this general plant layout may be identical with the process layout provided no parts of the project are outside the area shown by the drawing. However, it is often desirable to have an inset indicating the general location of the project on the plant site.

Line diagrams should show instrumentation and control. Where applicable, line diagrams should be made for all utility supply and distribution systems and material handling equipment. On these line diagrams should be indicated all

Accuracy With Available Information at Minimum Expense and Time

Installation Material for Process Equipment	Building, Distribution, Yard Material	Labor [5]	Design Expense	Field Expense
Based on completed and priced bills of material, or lump sum bids.	Based on completed and priced bills of material, or lump sum bids.	Based on detailed man-hour estimate and expected labor rates corrected for job conditions, or lump sum bids.	Actual plus estimated design cost of changes.	Detailed budget
Based on preliminary priced bills of material, plus allowances for stores material.	Based on preliminary priced bills of material, plus allowances for stores material.	Based on standard labor-material ratios for detailed work classifications, corrected for job conditions and using expected labor rate for each work classification.	Actual plus estimated design cost to complete.	Preliminary budget.
By ratio to value of process equipment.	From layouts and line diagrams on a unit price basis.	Based on standard labor-material ratios for similar work, corrected for job conditions and using expected average labor rates.	Based on complete drawing list.	Percent of physical plant cost from experience records.
By ratio to value of process equipment.	By ratio to value of process equipment; ratio built to fit expected needs.	Based on standard labor experience ratios for similar work, corrected for job conditions and using expected average effective labor rates.	Man-hr./$ of material cost from experience records at average cost per design man-hour.	Percent of physical plant cost from experience records.

3. For project costing about $1 million. See p. 57 for projects of other sizes.
4. All estimates include an allowance for unlisted items based on project's current uncertainty rating.
5. Standard labor-material ratios are for job condition factor of 1.50 and ENR index of 600.

major items of equipment, with notations of size, type and materials of construction. Materials to be used for pipelines and ducts should also be indicated.

Auxiliary equipment and instruments should be listed in tabular form showing type, size, number and materials of construction.

Buildings should be indicated as to type, size and location on the general plant layout. Abnormal ceiling heights and special needs such as air conditioning should be indicated. Structural materials should be noted such as brick with steel frame or corrugated asbestos on steel frame.

Yard improvements should be indicated as to type and size on the general layout drawing. These include railroads and docks, parking, sidewalks, surfacing and sewers.

Hazard control should be indicated on the process line diagrams. These will include arrangements for ventilation, dust and fume control, explosion and fire prevention and control, and air and stream pollution.

Preliminary Bill Estimates

Preliminary bill estimates can be made when the design of a project has advanced to the point where a preliminary material take-off can be made with reasonable accuracy. The following information is required for this estimate:

1. Major process equipment specifications.
2. Auxiliary equipment and instrument specifications.
3. General layouts and line diagrams.
4. Field inspection of site.
5. Preliminary bills of material coded to standard account subdivisions.

Accuracy of this type will be affected by the degree of uncertainty of all items listed on the uncertainty rating check list.

Detailed Estimates

For a detailed unit cost estimate, all material is priced in detail and all labor is priced on the basis of detailed unit cost records. Part or all of the estimate may be derived from firm bids from vendors and contractors. Accuracy of the estimate will be affected by the degree of uncertainty of all of the items on the uncertainty rating check list.

Follow This List to Prepare Cost Estimates

Step in Estimating Procedure		Equipment Ratio Method	Layout Method	Prelim. Bill Method	Detailed Unit Cost Method
Preliminary examination	(1)	X	X	X	X
Field inspection	(2)		X	X	X
Drawing examination	(3)			X	X
Products	(4)	X	X	X	X
Project divisions	(5)	X	X	X	X
Process	(6)	X	X	X	X
Capacity	(7)	X	X	X	X
Location	(8)	X	X	X	X
Estimate basis	(9)	X	X	X	X
Type of estimate	(10)	X	X	X	X
Project number	(11)	X	X	X	X
Estimating form	(12)			X	X
Uncertainty rating	(13)	X	X	X	X
Material take-off	(14)			X	X
Material pricing	(15)			X	X
Unlisted items	(16)	X	X	X	X
ENR cost index	(17)	X	X	X	X
Job condition factor	(18)	X	X	X	X
Effective pay rate	(19)	X	X	X	X
Site clearing; land costs	(20)	X	X	X	X
Reconditioning	(21)	X	X	X	X
Major process equipment	(22)	X	X		
Re-used equipment	(23)	X	X		
Total major process equip.	(24)	X	X		
Auxiliary equipment	(25)	X	X		
Total process equipment	(26)	X	X		
Installation material	(27)	X	X		
Other equipment ratios	(28)	X	X		
Std. labor-material ratios	(29)	X	X	X	X
Job man-hours	(30)	X	X	X	X
Labor take-off	(31)				X
Labor man-hours	(32)	X	X	X	X
Labor cost	(33)	X	X	X	X
Total column	(34)	X	X	X	X
Subtotals	(35)	X	X	X	X
Summary estimate form	(36)			X	X
Division totals	(37)	X	X	X	X
Design expense	(38)	X	X	X	X
Field expense	(39)	X	X	X	X
Estimated cost	(40)	X	X	X	X
Restricted reserve	(41)	X	X	X	X
Estimated cost plus restricted reserve	(42)	X	X	X	X
Signature	(43)	X	X	X	X
Copies	(44)	X	X	X	X
Review	(45)	X	X	X	X
Use of bid pricing	(46)				X

Use Applicable Detailed Procedures in Making Actual Cost Estimates

1. Preliminary examination. Examine all available drawings, sketches and reports for the project.

2. Field inspection. Make field inspection of the project site; preferably by the estimator in the company of the project engineer.

For detailed unit cost method where part or all of the work is to be put out for bids, a company representative thoroughly familiar with the project should accompany the contractor or the contractor's estimator during field inspection. It is important that all contractors be given a complete review of the project and site. To prevent later complications, all adverse as well as favorable conditions should be pointed out to each of the contractors during the field inspection.

3. Drawing examination. Available drawings, sketches and equipment lists should be re-examined in detail after the field inspection. Provide bidders with all available drawings required for intelligent bidding including general layouts and subsurface surveys, if part of the work is to be bid.

4. Products. Determine the products, byproducts and any major wastes which are to be produced by the unit. List them on the products line of the summary form.

5. Project division. Study the flow-sheet for the process. If the project is a large one and more than one major process is involved, then the estimate should be broken down into process divisions. Use a separate estimate sheet for each division. Assign account subdivision numbers in accordance with the principles given in Part I of this report. (Discussed on p. 26 of this volume).

6. Process. On the process line for each summary should be a brief but inclusive description of the unit process performed by the plant unit. This is especially important because it is quite common to make alternate estimates, assuming different processes but the same products.

7. Capacity. Consult the process flowsheet to determine production capacity of the unit for each of the products, byproducts and major wastes. Enter these values on the capacity line. Capacity values serve to identify the estimate further because more than one estimate may be needed to determine the cost of the plant at several capacities.

8. Location. Give geographical location of proposed plant on summary sheet. For study estimates, it may be necessary to make an arbitrary assumption as to plant location in order to have a reasonable and consistent basis for establishing the various elements of plant cost which are affected by geographical location. For layout

estimates, it is preferable that geographical location be established before an estimate is made.

9. Estimate basis. Indicate the general sources of information used by the estimator on the estimate basis line. Include brief notations as to the design information available and notes as to how process equipment was sized, typed and priced. Also give the source of the equipment ratios used in the estimate.

10. Type of estimate. Use suitable code to indicate type of estimate. For example, estimate type numbers are: detailed unit cost method, Type I; preliminary bill method, Type II; layout method, Type III; equipment ratio method, Type IV.

11. Project number. Assign a study number which is the departmental number under which the work on the project is going forward for equipment ratio estimates.

For all other estimates use an appropriation number which is the corporate number assigned to the appropriation request being prepared.

12. Estimating form. A separate detailed estimating form is set up for each sub-account. On each sheet enter the appropriation number and subdivision title and number. List material on form and provide columns for estimating material at unit cost, net cost and gross cost at applicable ENR index. Estimate labor in standard and job man-hours per material dollar at appropriate job condition factor and ENR index and compute labor cost at effective pay rate.

13. Uncertainty rating. With the help of the design engineer, make out and attach to the estimate one of the uncertainty rating forms. This should be done with great care, not only to make sure that the rating is correct but also to use the form as a final check list.

Uncertainty ratings are usually low for detailed unit cost estimates because all drawings for the project must be complete. However, it is entirely possible to have complete detailed drawings on a project where the basic design information is in an incomplete stage. Hence, the uncertainty rating would be correspondingly high.

14. Material take-off. For each drawing, material quantities should be taken off and these quantities and their units entered on preliminary bill estimate sheets opposite the appropriate account subdivision numbers. Since the drawings at this stage of a project are usually incomplete, some of these quantities must necessarily be approximated. For example: piping may not be completely detailed. However, it may be possible to list the approximate lineal feet of pipe of various

sizes and to list valves and major fittings.

For detailed estimates, bills of material are usually available from the design group.

15. Material pricing. For preliminary bill estimates, obtain unit prices from the sources given in Part I (Discussed on p. 29 of this volume). Bills of material are usually available for the detailed unit cost method. Bills are priced and the value of material is entered in the net material cost column. For both methods, show the ENR cost index at which the material is priced at the top of the column.

Net material cost is the product of the quantity and the unit cost of each item. Gross cost of material for an account subdivision is the cost at the probable ENR index of the project including the unlisted items allowance. It is obtained by applying the unlisted items percentage to the net cost.

16. Unlisted items. The unlisted items percentage is determined by multiplying the uncertainty rating of the project by 0.03. This percentage is added to the estimated cost of each of the pieces of process equipment and to all other material items. It is an allowance for the bare minimum of cost items which will inevitably be overlooked in an estimate.

17. ENR cost index. Use the value of the ENR construction cost index that is assumed to be applicable to the period during which the project is under construction. Usually the 12-month forecast of the estimating department can be used for this purpose; or the estimate can be based on a no-price-change assumption and is presented with this qualification.

18. Job condition factor. Determine the job condition factor by filling out a check list of job conditions as described in Part I. (Discussed on p. 33 of this volume). This factor has a very powerful effect on the final estimate amount. However, if time is pressing and assuming that a good contractor will perform the work on a new site under ordinary construction conditions, use a job condition factor of 1.20. Contract work in an operating plant calls for a factor of about 1.50. Spare-time construction by maintenance groups may rate about 2.10. In case of doubt, complete job condition factor sheet and attach to project file copy of estimate. In some cases, job conditions may vary enough between divisions to require separate job condition factors.

19. Effective pay rate. Determine effective pay rates as outlined in Part I of this report. (Discussed on p. 38 of this volume).

20. Site clearing and land. Site clearing and land costs can be estimated according to the principles

Applicable Procedures

given in Part I. (Discussed on p. 29 of this volume). Field inspection of the site is very desirable and an investigation of local land prices may be required. For detailed unit cost estimates, separate subaccounts are established, as required, for these items and a detailed estimate of their cost made.

21. Reconditioning. Equipment to be re-used for the project must be given a field inspection to determine its cost of reconditioning. The necessary work should also be discussed with the design engineer. Materials and man-hours are approximated and the effective pay rate and unlisted items percentage applied to get the total reconditioning cost.

22. Major process equipment. List type, size and number of each of the items of major process equipment. Price the equipment, add the unlisted items percentage and enter the dollar values in the material column. The great influence of major process equipment prices on the estimate total justifies the pricing of the expensive and doubtful items by means of preliminary quotations from vendors.

23. Re-used equipment. The new value of re-used equipment must be considered in using equipment ratios. Any equipment which is to be re-used for the project should be priced at its present market value new; and this value entered in the brackets in the material column. The unlisted items allowance computed on the cost new basis should be entered just below in the material column as shown in the estimate form on p. 53.

24. Total major process equipment. Sum of new major process equipment as listed in the material column plus the new value of re-used equipment. There are no entries in the other columns on this line.

25. Auxiliary equipment. For equipment ratio estimates, use the auxiliary equipment ratio from records of past jobs. The ratio selected as appropriate for the process division is entered in the space marked e_a. The product of e_a and total major process equipment value E_m gives the estimated value of auxiliary equipment for the project.

For layout estimates, list and price auxiliary process equipment, including instruments and motor control equipment, in the same manner as major process equipment. Use a separate sheet and attach to project file copy of the estimate summary. Only the more expensive pieces of auxiliary equipment need be priced by preliminary quotations from vendors. Enter total value of auxiliary process equipment including allowance for unlisted items on the estimate summary form shown on p. 54.

26. Total process equipment. Total process equipment value is the sum of major process equipment value and auxiliary process equipment purchased new plus the new value of re-used equipment. The unlisted items allowance is also included.

27. Installation material. Select the installation material ratio m_i from records of past projects. This ratio is defined and developed in Part I (Discussed on p. 32 of this volume). The product of installation material ratio m_i and total process equipment E_p gives the value of installation material for the project.

28. Other equipment ratios. Ratios for piping material, wiring material, building material and distribution and yard material are selected and applied in accordance with Part I (Discussed on p. 32 of this volume).

For layout estimates, building material and distribution and yard facilities are estimated as follows:

Area, height and construction materials for the process buildings are determined from the layout drawings. The buildings are priced on the basis of costs per unit area or per unit volume plus the unlisted items allowance. Service buildings are priced in a similar manner. Buildings serving more than one process division are listed and estimated in a separate division which also includes distribution and yard facilities.

Essential physical dimensions and distribution facilities such as pipelines for products, steam, water or gas service, sewers and power facilities can be obtained from the layout drawings. Use unit cost data as discussed in Part I to price these facilities and add unlisted items allowance.

29. Standard labor-material ratios. Standard man-hours per material dollar are selected for each material or equipment item from past jobs of record. The heading of the estimate sheet should indicate the projects of record used as the source of these ratios. See Part I for a discussion on the use of standard labor-material ratios.

For detailed unit cost estimates, standard unit man-hours are obtained from the sources indicated in Part I. However, it is sometimes difficult to locate unit labor cost records for a particular type of work.

30. Job man-hours. It is necessary to convert standard labor-material ratios to the conditions of a particular job. This is done by multiplying the standard man-hours per material dollar by the ratio of the job condition factor for the project to the standard job condition factor of 1.50; then multiplying by the ratio of the standard ENR cost index of 600 to the assumed ENR cost index for the project. If more than one job condition factor is required, a separate computation is made for each.

31. Labor take-off. List each item of work required for the project subdivision in the description column. Indicate drawing number which best shows the work to be done on estimate form.

32. Labor man-hours. Labor man-hours are the product of material dollars and job man-hours per material dollar. For used equipment which is to be re-installed, use value based on market value new rather than book value.

33. Labor cost. Labor cost is the product of labor man-hours and the effective pay rate. Labor dollars are rounded so as to avoid an unrealistic number of significant figures in the totals columns of the estimate.

34. Total column. The figures in the total column are the total of material and labor dollars for each item.

35. Subtotals. Columns should be subtotaled and the subtotals cross-checked where possible. If the project has more than one division, subtotals are carried over to the final sheet.

36. Summary estimate form. As a summary of the project estimate, prepare estimate summary sheets. Reference should be made to the detailed sheets for cases of variation in uncertainty rating, ENR cost index, job condition factor, effect pay rate and unlisted items allowance.

37. Division totals. A separate estimate summary sheet is used for each division, and the subtotals from these sheets carried over to a final division summary sheet. On this sheet are listed the divisions of the project with total estimated material, man-hours, and labor cost for each.

38. Design expense. For equipment ratio estimates, standard design man-hours per material dollar are obtained from the graph in Part I (Discussed on p. 40 of this volume). Standard values are adjusted to the job conditions by noting whether the engineering fits an easy, difficult or standard classification. Probable value for the job is entered in the estimate. Multiplying job design man-hours per material dollar by the value of project material at ENR 600 gives design man-hours for project.

A complete drawing list is often available for layout estimates. With the assistance of the project engineer the probable man-hours per drawing can be estimated so as to determine the man-hours required for the project. Unlisted items percentage is added to this figure. As a check, the job design man-hours per material dollar at ENR 600 should be computed and compared with the standard design man-hours per material dollar obtained from the graph in Part I. Effective cost of a design man-hour for both estimates is determined from current department records.

For preliminary bill and detailed unit cost estimates, most of the engineering has already been completed.

(Continued on page 55)

Prepared Estimating Form Saves Time for Equipment Ratio Estimates

EQUIPMENT RATIO ESTIMATE

PRODUCTS	LOCATION	DATE	STUDY NO.
ORTHOSILICATE & METASILICATE	CINCINNATI	6-27-56	OF-00-1

PROCESS	
CONTINUOUS	EST. NUMBER, THIS APPROP. 1

	ESTIMATE TYPE IV

CAPACITY	
INCREASE OF 1,500 # FLAKED PRODUCT PER HOUR	UNCERTAINTY RATING 330
	UNLISTED ITEMS 10 %

ESTIMATE BASIS	
	ENR. COST INDEX 640
RATIOS FROM JOB NUMBER F-6706	JOB CONDITION FACTOR 1.20
MAJOR PROCESS EQUIPMENT DATA BY N.A.	600/640 x 1.20 /1.50 = 0.75
	EFF. PAY RATE $3.00

DESCRIPTION & QUANTITY	MATERIAL	M.H./$M STD.	M.H./$M JOB	LABOR MAN-HRS.	LABOR DOLLARS	TOTAL
SITE CLEARING, RECONDITIONING, AND LAND	1,400	----	---	1,125	3,300	4,780
MAJOR PROCESS EQUIP., INCL. UNLISTED ITEMS						
Caustic Storage Tank & Coil (Ni-clad 15' D. x 20')	14,200	0.137	0.103	1,462	4,350	18,550
Dry Material Storage	500	0.100	0.075	38	110	610
Evaporator (Ni tubes & Ni-clad)	5,800	0.026	0.020	116	350	6,160
Separator	900	0.026	0.020	18	50	950
Ortho Mixer	900	0.083	0.062	56	170	1,070
NEW VALUE OF RE-USED EQUIP.	(200)	0.026	0.020	12	40	240
UNLISTED ITEMS FOR RE-USED EQUIP.	----	----	----	----	----	----
TOTAL, MAJOR PROCESS EQUIP. E_m 22,900	xxxx	xx	xx	xxx	xxxx	xxxx
AUXILIARY EQUIPMENT, INCL. UNLISTED ITEMS e_a 0.358	8,200	0.091	0.068	557	1,670	9,870
TOTAL PROCESS EQUIPMENT E_p 31,100	xxxx	xx	xx	xxx	xxxx	xxxx
INSTALLATION MATERIAL m_i 0.138	4,300	0.280	0.210	904	2,710	7,010
PIPING MATERIAL m_p 0.190	5,900	0.225	0.169	997	2,990	8,890
WIRING MATERIAL m_w 0.048	1,500	0.283	0.212	318	960	2,460
BUILDING MATERIAL m_b 0.068	2,100	0.214	0.161	338	1,020	3,120
DISTRIBUTION & YARD MATERIAL m_{dy} ---	----	----	----	----	----	----

COPIES TO: R.V.N.	SUB TOTAL, NEW MATERIAL	$ 45,700	AV.	AV.		$	$
	SUB TOTAL, INCL. RE-USED MATERIAL	(45,900)	0.17	0.13	5,940	17,800	63,700
	DESIGN EXPENSE @ 5.50 $/M.H.	(43,000) @ ENR. 600	0.060	0.066	(2,840)	15,600	$ 15,600
	FIELD EXPENSE 25 % OF SUB TOTAL.	6,400	0.67	0.50	3,170	9,500	$ 15,900
	TOTAL ESTIMATED COST	$ 52,300	xxx	xxx	FIELD 9,110	$ 42,900	$ 95,200
	RESTRICTED RESERVE 21 % ------------------------------------						$ 20,000
	TOTAL ESTIMATED COST PLUS RESTRICTED RESERVE ----------------						$ 115,200

CONST. FILE	ESTIMATOR	DATA CHECK	ARITH. CHECK	APPROVED	EST. TIME-M.H.
X	J.P.S.	J.L.W.	P.C.H.	E.F.G.	2

Use Summary Form to Consolidate Cost Data for Other Estimates

ESTIMATE SUMMARY

PRODUCTS: TREATED BRINE	LOCATION: PITTSBURGH	DATE: 2-10-56	APPROPRIATION NO. ZF 00 331

PROCESS: DOUBLE CIRCULATION		EST. NUMBER, THIS APPROP.	2
CAPACITY: 200 T/DAY, SOLID BASIS		ESTIMATE TYPE	III
		UNCERTAINTY RATING	366
		UNLISTED ITEMS	11 %
ESTIMATE BASIS: MAJOR PROCESS EQUIPMENT FROM QUOTES. RATIOS FROM ZF-0671 AND G-1112. LAYOUT DRAWING NO. 160-D; FLOW DIAGRAM NO. 159-C.		ENR COST INDEX	650
		JOB CONDITION FACTOR	1.35
		600/ 650 x 1.35 /1.50=	0.83
		EFFECTIVE PAY RATE	$3.20

APPROP. SUB-ACCT. NO.	DESCRIPTION & QUANTITY	MATERIAL	MH/$M STD.	MH/$M JOB	LABOR MAN-HRS.	LABOR DOLLARS	TOTAL
	Site Clearing						
	Remove Brick Building	---	---	---	200	660	660
	Relocate 500' of 2'' Water Line	200	---	---	350	1,100	1,300
	Major Process Equipment						
	Heat Exchangers, 6	76,300	0.020	0.017	1,300	4,160	80,460
	Treat Tanks, 4	21,200	0.120	0.100	2,120	6,780	27,980
	Auxiliary Process Equipment						
	Pumps, 12	9,000	0.070	0.058	522	1,670	10,670
	Controllers, 2	3,100	0.083	0.069	214	690	3,790
	pH Recorder	2,500	0.035	0.029	73	230	2,730
	Power Control Center	5,300	0.160	0.133	705	2,260	7,560
	Total Process Equipment, Ep = $117,400	---	---	---	---	---	---
	Installation Material, Ep x 0.138	16,200	0.280	0.232	3,760	12,040	28,240
	Piping Material, Ep x 0.190	22,300	0.230	0.191	4,260	13,620	35,920
	Wiring Material, Ep x 0.048	5,600	0.300	0.249	1,395	4,460	10,060
	Control House, 20' x 10' @ $4M/sq. ft.	800	0.345	0.286	229	730	1,530
	Access Road, 1400'	4,200	0.754	0.625	2,630	8,400	12,600

COPIES TO:		$	AV.	AV.		$	$
J.A.B.	SUBTOTAL	166,700			17,760	56,800	223,500
F.H.Q.	DESIGN EXPENSE @ 5.50 $/MH	(154,000) ENR 600	0.048	0.050	(7,700)	42,300	42,300
	FIELD EXPENSE 23 % OF SUBTOTAL SPLIT % MAT. % LABOR	20,500	XXX	XXX	9,650	30,900	51,400
	TOTAL ESTIMATED COST	187,200	XXX	XXX	FIELD 27,410	130,000	317,200
	RESTRICTED RESERVE 21 % ---------------------						66,800
	TOTAL ESTIMATED COST PLUS RESTRICTED RESERVE -----------------						384,000
CONSTRUCTION FILE G.V.B.	ESTIMATOR J.A.W.		CHECKED E. F. G.	APPROVED W. J. H.			EST. TIME. M.H. 18

Applicable Procedures

Hence, a major part of the design expense is determined from department records. An allowance for changes and additions to drawings is established by multiplying the unlisted items percentage by the total of the actual design cost to date.

39. Field expense. For small projects, field expense is usually included in the effective pay rate. For the larger projects, it can be approximated from the percentage graph in Part I (Discussed on p. 41 of this volume). For intermediate projects the field expense can be included in the effective pay rate, with special consideration for particular items such as a resident field engineer.

For preliminary bill estimates, field expense is computed on the basis of a preliminary construction expense budget. This can be done on a separate detailed estimate sheet and the total only carried over to the estimate summary sheet.

For detailed unit cost estimates, account subdivisions for field expense are established and the costs connected with each computed on a separate detailed estimate sheet. Totals for these subdivisions are individually listed on the estimate summary sheet and the grand total entered on the final sheet.

40. Estimated cost. This is the total of field expense, design expense and the subtotals of the individual account divisions.

41. Restricted reserve. Restricted reserve is established from the formula

$$\frac{R}{E} = \frac{0.07U}{100 + 0.03U}$$

where R is restricted reserve, E is estimated cost and U is uncertainty rating. The resulting percentage is entered in the space indicated and applied to the estimated cost to determine the amount of the restricted reserve. Values of the constants in the equation may be varied to produce the desired relationship between overruns and underruns.

42. Estimated cost and restricted reserve. This is the probable maximum cost of the project and is entered in the total column.

43. Signature and checking. The estimator should note his initials on the line indicated and turn the estimate over to another member of the department for checking of arithmetic. The estimate then goes to the senior estimator for general checking. Both the arithmetic checker and senior estimator initial the estimate. When checking work is completed, the total man-hours required to prepare and check the estimate are noted.

44. Copies. Copies of the estimate can be prepared by suitable duplicating methods. Copies are distributed to the individual requesting the estimate, to the estimating section's file, to the chief project engineer and any other interested engineering personnel. The detailed estimate sheets are kept in the estimating section's project file. If necessary, duplicates of these sheets can be made for the use of the design engineer and others.

45. Review. Estimates of major importance may be reviewed by a committee made up of:

Staff engineer who made original study

Design engineer for project

Field construction engineer

Estimator

46. Bid prices. If parts of the project have been bid by outside contractors, bid prices are used instead of estimated prices to make up the estimate summary. Source of bids should be indicated.

It is particularly important in contracting part or all of the work to make sure that there are no overlapping contracts, and no gaps between contracts which must be filled by plant labor or other contractors.

Re-Estimates Improve Accuracy and Control Final Cost

Re-estimates are made during the construction of a project to improve the accuracy of the estimate of final cost. They should be made in such a way so as to take advantage of all actual cost information which has become available up to the time of the re-estimate.

A re-estimate form as shown on p. 56 can be used to make labor or material re-estimates. Both forms could be combined but separate forms are used for labor and material re-estimates so as to get concurrent help from the field engineer and the design engineer.

Material Re-Estimates

On re-estimate form for materials show number, title and appropriation estimate for each subaccount of the project. For convenience, each subdivision should be listed on a separate sheet or group of sheets. Design commitment to date totals are obtained from the appropriation ledger and entered in the columns shown.

With the help of the project engineer and the field supervisor, the estimator can approximate the amount of material yet to be requisitioned for each subdivision. This is entered in the required-to-complete column. For some subdivisions, there may as yet be no design commitment. In such cases, the full appropriated amount will presumably be needed to buy material for the subdivision.

The sum of design-commitment to date and the required-to-complete is the revised estimate for each subdivision.

A new uncertainty rating should be made. Separate ratings should be made, if necessary, for the various project divisions. An unlisted items percentage is determined by multiplying the uncertainty rating by 0.03. It is entered in the blank space provided at the bottom of the revised material estimate sheet.

The unlisted items allowance is determined by multiplying the revised material estimate total for the division by the unlisted items percentage. Allowance for unlisted items should be added even though it gets down to a very small percentage. The unlisted items percentage applies to the entire material estimate, but the resulting dollar values should be distributed to open accounts only.

In the latter stages of the project, the appropriation ledger should be checked against the construction ledger which is maintained in the field by the project accountant. This check usually uncovers additional charges such as freight and escalation which must be included in the re-estimate. On occasion, invoiced amounts will be greater or less than the order amounts.

After completion of the material re-estimate, a field inspection should be made to check this re-estimate and to obtain information with respect to labor progress.

Labor Re-Estimates

Number, title and appropriation estimate amount for each account subdivision should be entered on a re-estimate form similar to the one shown on p. 56. Separate sheets or groups of sheets should be used for each of the project's divisions. If more convenient, the form can be made out in terms of man-hours rather than dollars.

Labor expenditures to date are

LABOR PROGRESS REPORT AND RE-ESTIMATE, As of _____							
Title _____						No. _____	
Reported by					Sheet No.		
Account Number	Account Title	Appr'n. Estimate	Expended to Date	% Complete	Req'd. to Complete	Revised Estimate	

obtained from the field accounting group and entered in the column as shown. Percent labor completion of each subdivision is estimated by the field engineer or by the estimator after a detailed field inspection. It is important that these completion figures be based on conditions as they existed at the closing date of the expended-to-date amounts.

In some cases, especially when work on a subdivision is approaching completion, it is easier and more accurate to estimate manhours and dollars required to complete the work rather than estimating the percent completion.

It is necessary to check the current revised material estimate when estimating percent completion or required to complete. Proposed additions or changes are often known to the design engineer and included in the material re-estimate well in advance of revision and issue of drawings to the field engineer.

Revised labor estimate is either the sum of the expended-to-date and the required-to-complete figures or the expended-to-date figure divided by percent complete. Whichever method seems likely to produce the more accurate result should be used. If work for a subdivision is contracted on a lump-sum basis, the bid amount plus change orders to date plus estimated work by others is totaled to get the revised labor estimate.

Unlisted items are computed for each division in the same manner as for material. In connection with the monthly analysis report, the unlisted items allowance can be distributed to appropriation divisions in proportion to the required-to-complete dollars.

Labor re-estimate forms are also useful in construction-labor control. If the field supervisor finds that the re-estimated cost of a subdivision departs substantially from the appropriation estimate, he should take immediate steps to find out why and initiate any action necessary.

As a summary of the project re-estimate, estimate summary sheets of the form shown on p. 54 should be prepared. The type of estimate will be marked Re-. Appropriation number, current uncertainty rating, current ENR index and the unlisted items allowance should be entered. Job condition factor and effective pay rate are re-estimated if necessary.

Division totals only from the detailed estimate sheets are transferred to the estimate summary. Each division entry includes the title of the division, the gross cost of material, labor dollars or manhours and total re-estimated cost for the division.

Design expense is the recorded cost of design to date plus the current unlisted items percentage to take into account changes which will be required during completion of the project.

Field expense is re-estimated on a detailed subaccount basis in a manner similar to that used for other divisions.

Required Estimating Time

In scheduling estimating work and in appraising and selecting estimating methods, it is desirable to have a means of predicting estimating time and money which will be required to make an estimate by a given method for a project of a certain size.

On the next page is a graph of estimated total project cost versus estimator man-hours for some 90 projects. Estimates were made by three different methods as shown. Man-hours include checking-time but not stenographic time.

The points of the graph show considerable scatter. Part of this is due to the fact that eight estimators with varying degrees of experience were at one time or another employed on the work. There was also a considerable variation in the degree of development of the basic project information supplied to the estimators. In some cases part or all of the equipment and material pricing was done by design or staff engineers whose time was not recorded with the estimating time.

Some of the scatter, especially for jobs involving only a few man-hours, is due to the difficulty of determining just exactly how many hours should be assigned to a particular job. This is particularly true since all of the individuals concerned had concurrent assignments

Find Total Engineering Man-Hours

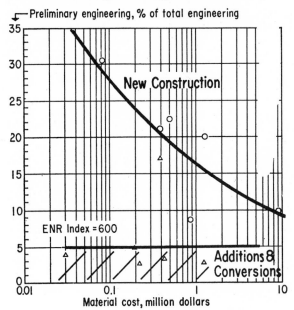

Preliminary engineering, % of total engineering

New Construction

ENR Index = 600

Additions & Conversions

Material cost, million dollars

Compare Fees to Project Costs

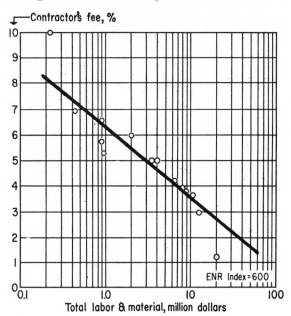

Contractor's fee, %

ENR Index = 600

Total labor & material, million dollars

Estimating Time Depends on Project Cost and Estimating Method

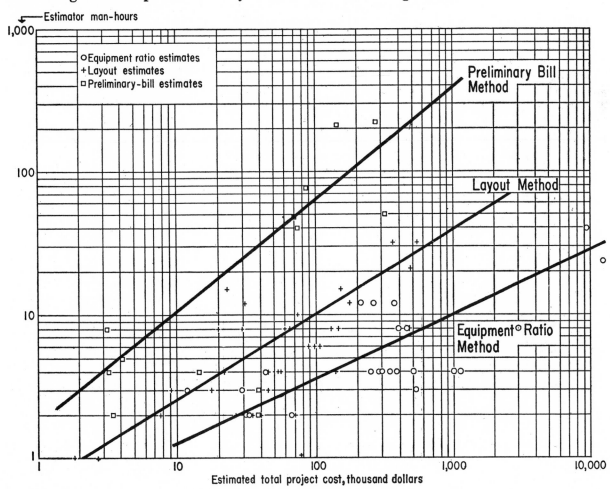

Estimator man-hours

○ Equipment ratio estimates
+ Layout estimates
□ Preliminary-bill estimates

Preliminary Bill Method

Layout Method

Equipment Ratio Method

Estimated total project cost, thousand dollars

in cost analysis work and field construction service.

One further point of difficulty is the classification of estimates into the three groups. Very few estimates are entirely true to type. Estimates shade from one type into the next making the boundaries between the groups indefinite.

Time and cost of making estimates by the several different methods must be considered along with their probable accuracy in selecting the estimating method currently most suitable for a project.

Accuracy of Methods

In the section on establishment of contingency allowances, we listed the variables which affect the divergence of actual from estimated costs. Since estimating method is only one of the list of variables affecting the accuracy of an estimate, it is very difficult to segregate its effects.

Quite often, successive cost estimates are made by different methods by the same estimator for a single project. In this case, the cost is common to the several successive estimates. Hence, the effects of cost efficiency of the design group, purchasing efficiency, field efficiency and general economic condition changes are nullified. There remains the possibility that skill in executing one type of estimate may be greater than for another, or the cost records available for one may be better than those for another type.

As yet not enough of this comparative information is available to establish reliable figures as to the accuracy which may be expected for the four different types of estimates. Available figures indicate that the percentage accuracy range for small projects will be much wider than for large projects. The accuracy band is also wider for projects having a high uncertainty rating than for projects estimated by the same method with a low uncertainty rating.

Detailed unit cost estimates are made by contractors with an accuracy range of $\pm 3\%$ disregarding design changes. This assumes complete drawings and specifications with an uncertainty rating in the 50 range.

Preliminary bill estimates should also have a relatively high accuracy. The material portion of an estimate made by this method is firmly established. Labor which is computed by labor-ratios, however, cannot be as accurate as a good unit man-hour estimate. Although complete information is not available, it is believed that preliminary bill estimates can be made with an accuracy of $\pm 6\%$ provided the project is in the over $100,000 range and the uncertainty rating at the time of the estimate is less than 100.

Layout estimates have a wider accuracy range than preliminary bill estimates since much less information is used. Accuracy of the equipment portion of layout estimates should be good because all of these items are listed and priced. Installation material and labor which are based on ratios will have a lower degree of accuracy. Although no actual information is available, it has been tentatively assumed that for projects costing more than about $100,000, with uncertainty ratings at the time of estimate of 200 or less, an accuracy of $\pm 12\%$ can be obtained.

Equipment ratio estimates are less accurate than layout estimates, mostly because of the reduced degree of definition with respect to auxiliary equipment and instruments and the sketchy definition of utilities, buildings and yard improvements. It has been tentatively assumed that for projects costing over $100,000 with uncertainty ratings of 400 or less at the time of estimate, an accuracy of $\pm 25\%$ may be attained by equipment ratio methods.

Additional Cost Data

Relationship between preliminary engineering manhours and total engineering man-hours is shown in the graph on p. 133 for process projects. This information is sometimes valuable in the early stages of project planning. Preliminary engineering for process projects is arbitrarily defined as including:

Mechanical flowsheets

General arrangement drawings and block models

Preliminary equipment, drawing, account subdivision and material lists

Specifications for major equipment

Site information

Preliminary project schedule

Preliminary estimate of capital cost, manufacturing costs and income and return.

Another chart on p. 133 provides some information with respect to construction fees which have been charged in the past by contractors on major projects. The contractor's fee in this case includes his home office overhead and profit but does not include any of the other items of field expense previously listed in Part I (pp. 25–42).

ACKNOWLEDGEMENTS

Many of the procedures in this report are variations of practice described in the literature referred to in Part I. These have been combined with new material to form a complete, inclusive estimating system.

The graphical data represent the combined efforts of the cost engineers of the Diamond Alkali Co., especially Blase Nemeth, chief construction engineer.

The restricted reserve principle is based on suggestions made by C. A. Butler, Jr.

REFERENCES

U. S. Bureau of Mines, "Bibliography of Investment and Operating Costs for Chemical and Petroleum Plants," Information Circular 7516, 1930–Sept. 1948; Information Circular 7705, Sept. 1948–June 1952; Information Circular 7751, July 1952–June 1954; Information Circular 7847, July 1954–Dec. 1956; Information Circular 7916, Jan. 1957–Dec. 1958. Series is continuing.

Weaver, J. B., "Annual Cost and Profitability Estimation Review," for 1926–1953, Chem. Eng., Oct. 1954, pp. 185–193; for 1948–1954, Chem. Eng., June 1955, pp. 247–252; for 1955, Ind. Eng. Chem., May 1956, pp. 934–942; for 1956, Ind. Eng. Chem., June 1957, pp. 936–946; for 1957, Ind. Eng. Chem., May 1958, pp. 753–762; for 1958, Ind. Eng. Chem., May 1959, pp. 689–696.

Aries, R. S., R. D. Newton, "Chemical Engineering Cost Estimation," McGraw-Hill, New York, 1955.

Vilbrandt, F. C., C. E. Dryden, "Chemical Engineering Plant Design," 4th Ed., McGraw-Hill, New York, 1959.

New Cost Factors Give Quick, Accurate Estimates

This novel method, incorporating new and refined ratio cost factors, is claimed to be accurate enough for the appropriation of funds for the construction of process plants.

C. A. MILLER
Canadian Industries, Ltd.

Never before has industry been so aggressive in looking ahead, assessing markets, investigating new products, and evaluating profitability. In a highly competitive market, the company that can get answers quickly will be the one out in front. The old adage of "the early bird gets the worm" was never more true than it is today.

Along with other factors, the ap-

Based on a paper given by the author at the 9th annual meeting of the American Assn. of Cost Engineers, June 29, 1965, in Los Angeles.

propriation of funds is largely dependent on markets, selling prices, and capital costs. The quicker one can provide answers in these three areas, the quicker a company can act. It is a well-known fact that market research is the weakest link in this chain. Market estimates are frequently in error by over 50%.

In contrast, estimates for the appropriation of funds must be within ±10%. This 10% accuracy is a magic number that appears to be a common requirement in a great many companies.

In a system where one component may have no better accuracy than

±50%, it is my contention that too much stress has been placed on the 10% accuracy of another component. Further, despite this requirement, the 10% accuracy is not being obtained in practice.

At the time the appropriation estimates are made, the scope of the work is not known in detail. As a result, a great many assumptions are made. The appropriation estimate is prepared, based on these assumptions. If the assumptions are correct, there is a good chance that the estimate will be within 10%. If the assumptions are not correct, the accuracy may be anything.

Comparing conventional and factor estimates—Table I

Project	Estimated Cost of Equipment	Factor Estimate	Conventional Estimate	Actual Cost	Remarks
A	$3,397,100	$8,370,000	$6,476,500	$8,014,000	
B	99,000	263,000	288,000	——	Built with scope changed
C	122,600	357,800	354,000	——	Built with scope changed
D	202,780	456,400	433,720	——	Built with scope changed
E	1,360,000	4,400,000	4,592,000	4,229,479	
F	893,000	1,894,000	1,716,000	——	Not built
G	474,000	985,000	922,300	——	Not built
H		6,624,000	6,709,000	6,421,500	
I	257,000	407,000 to 497,000	450,000	——	Not built
J	1,869,000	4,796,000	4,310,000	4,610,100	

Of course, if the scope of the work is reasonably complete, the estimate may be made in considerable detail with a substantial expenditure of man-hours, and accuracy will have a high probability of being within 10%. If the scope of work is poorly defined, the estimate will not require the same expenditure of man-hours and the chances of being within 10% are greatly reduced.

Accuracy of Cost Estimates

Let us look at the actual performance of a number of projects. A study of 53 projects ranging from $10,000 to $9 million revealed that twenty projects overran the initial amount authorized by an average of 19%, three projects broke even, and thirty projects underran by an average of 14%. No other criterion was used other than to compare the "initial" request for funds with the final cost. Thus the comparison may not be entirely fair since no attempt was made to investigate legitimate reasons for the differences, such as a later decision to alter the capacity of the plant, and other approved changes in scope.

Nevertheless, it does serve to demonstrate that the 10% accuracy claim has little validity when comparing final cost with initial request. Furthermore, the importance of having the 10% accuracy is being overly emphasized since the accountants still seem to do a fair job of budgeting for the funds, and the profits are such that the company remains in business.

The executive who approves the appropriation, on the premise that the final cost will be within 10% of the estimate, is either naive or appreciative of the true situation and willing to accept it.

Most often, I believe, the latter case is the correct one. In other words, management is satisfied that a system calling for ±10% with an actual performance of about ±15% is good enough to make major decisions.

Advantages of Factor Estimates

Actually the factor estimate in its present state of development is capable of meeting these requirements. This is contrary to the general opinion that factor cost estimating is only suitable for order-of-magnitude estimates.

There are several reasons for the interest in the factor estimates:

• They can be made quickly.

• They require fewer number of man-hours.

• When the scope of work is vague, their statistical averaging techniques produce a higher overall accuracy than a detailed estimating technique predicated on a great many assumptions.

• Statistics show that they get the same accuracy performance as many conventional appropriation estimates.

Let us look at an actual performance of the two approaches.

Table I illustrates a number of projects in recent years for which funds were requested in the conventional way—using the conventional appropriation type of estimate that was supposed to be within 10%. In each case, a factor estimate was also made. In some cases, the project failed to get approval and was not built. In several cases, the projects were approved but involved major changes in scope so that comparison with the final cost is of no value. The remaining cases are complete with the final achieved costs.

On actual performance, we can conclude from Table I that the factor estimate is as good or better than the more detailed estimate in these cases.

Refined Factor Method

The factor system used on the above mentioned projects has been further refined, and it is this refined system that will now be described.

The factor system depends on some relationship such as: total cost = $f(x)$. In one form using the Lang factor: Cost of a plant = a factor × total equipment cost.

A different factor is used for plants handling solids, liquids, or liquids and solids. This is the broadest and least reliable type of factor estimate. Various adaptations and refinements of this theory have been presented in recent years. For example, methods have been developed that separate the costs into major components such as piping, electrical, buildings, etc., and provide separate factors for each component.

The most recent method is the "module technique," which applies the same theory of $\$ = f(x)$, but in this case, x is the cost of a single piece of equipment.

The Lang method is fast but not accurate enough and does not account for certain variables that often are known and hence should be taken into account to improve the accuracy.

The module method is much more precise and accurate but requires more time and data. A specific piece of equipment is assessed and the cost of each auxiliary item attributable to that piece of equipment is obtained by factors to get the overall total cost. A significant advance in this technique is the recognition and allowance for the effect of variation in size of equipment on the factors.

As a system becomes more precise, however, its application becomes more specific, and thus a substantial fund of data is required to make it applicable to a wide range of different plants. We believe that the module method is excellent in concept and principle, but until this substantial fund of data is available, its use will be somewhat limited.

Combination Technique

The system detailed here was developed quite independent of the module system and is probably best described as a compromise or blend of the module method and the other systems that have been published. It makes allowance for variation in size but takes advantage of the effects of over-all averaging to give a wide coverage of different types of plants.

There are a number of items that cause discrepancies in factor estimates, and these must be taken into account if greater accuracy and reliability are to be obtained. In addition to the nature of the process, such as liquids, solids, etc., there are three significant items that

Unit-cost concept compensates for plant complexity

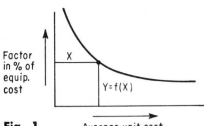

Fig. 1

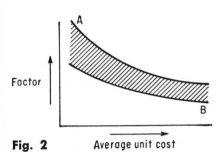

Fig. 2

cause variation in factor estimates. These are:

- Size of equipment.
- Materials of construction.
- Operating pressures.

If the size of the equipment gets larger, the over-all factor becomes smaller.

If the equipment is made from more-expensive materials such as stainless steel, glass-lined vessels, etc., the factors again become smaller.

If the operating pressures increase, it is also known that the over-all factor will decrease.

To a considerable degree, all these items can be taken into account by one number, the "average unit cost" of the process equipment.

Average unit cost of the equipment is defined as:

$$\frac{\text{Total cost of process equipment}}{\text{No. of equipment items}}$$

Fig. 1 illustrates how this relationship can be used:

1. If the size of a plant is increased, the equipment becomes larger and the average cost per item increases. Thus the point on the curve will be farther to the right,

and the corresponding factor will be lower.

2. Alternatively, if the equipment has been carbon steel and is changed to stainless steel, the average cost of each item will increase, and again the point on the curve will be farther to the right and the factor will be reduced.

3. Similarly, if the operating pressure is increased from atmospheric to a high-pressure operation, the average cost per item will again increase and the resulting factor will go down.

It follows that regardless of what issues are causing variations in the factors, the principle of basing them on average unit cost of the equipment will have a narrowing effect on the differences.

In practice, the curve becomes a band, as indicated in Fig. 2. Thus for any average unit cost the problem is to select the right position in the width of the band at that point.

In overly simplified terms, point "A" indicates the factor for a very small plant operating at atmospheric pressure with mild steel equipment, while point "B" represents the factor for a large high-pressure plant with stainless-steel equipment.

New Factors

Following this principle, we have analyzed the feedback from a great many plants and have developed factors for all components of the battery limit, i.e., for foundations, piping, insulation, electrical, etc. The study embraced a wide range of plants including ethylene, fertilizer, caustic-chlorine, polyethylene, and TNT. The resulting factors are provided in Table II.

Each column represents a value for average unit cost of process equipment, and all factors are given as a range. The precise selection must be based on knowledge of the project and experience of the estimator. In practice, we have used a three-number selection (as in PERT) picking a high, probable and low, and we would expect an experienced estimator with a detailed appreciation of the system to select suitable ranges. Subsequently,

treatment of highs and lows will provide an upper and lower figure for the estimate in which we have a high degree of confidence.

These data provide a complete set of factors for estimating a wide variety of battery-limit plants.

If the process-equipment estimate is order-of-magnitude, the total cost will be order-of-magnitude. However, if a good estimate has been made for the process equipment, and scope of work is fairly well defined, then the total estimate should be adequate for the appropriation of funds.

Area Concept

So far, our comments have applied strictly to the battery-limit plant. Actually there are four areas involved in a plant estimate:

Battery limit	(B/L)
Storage and handling	(S&H)
Utilities	(U)
Services	(S)

For the sake of clarity, precise definitions of all four areas are provided on p. 236.

A great deal has already been published on estimating the cost of auxiliaries and we will not cover this item in detail. Where practical, significant items are estimated by large component techniques such as steam plants in $/1,000 lb. of steam, warehouses in $/sq. ft. of floor, etc.

However, when inadequate information is available for this type of estimate, we have found again that factors can produce a very acceptable estimate.

Tables III, IV, V provide suitable cost factors for storage and handling, utilities, and services.

Note that services have been found to have greater correlation when expressed in % of (B/L + S&H + U), rather than a direct function of the B/L.

With the cost factors provided in the various appendices, there are now sufficient data to make a complete estimate.

Four-Step Procedure

The following procedure will be followed:

1. Estimate cost of B/L from process equipment.

Ratio factors for estimating battery-limit costs—Table II

Range of Factors as Percent of Basic Equipment

		AVERAGE UNIT COST OF M P I IN 1958 DOLLARS						
		UNDER $3,000	3,000 to 5,000	5,000 to 7,000	7,000 to 10,000	10,000 to 13,000	13,000 to 17,000	OVER 17,000
BASIC EQUIPMENT Delivered to site, excluding sales taxes & catalyst	M P I (Main plant items)	X	X	X	X	X	X	X
	M U E (Miscellaneous unlisted items) Early flowsheet stage	20 to 10% of M P I 's in all categories						
	Scope of work well defined	10 to 1% of M P I 's in all categories						
	NOTE: Top of ranges: Complicated processes Many process steps							
	Bottom of ranges: Simple processes Few process steps							
	BASIC EQUIPMENT = M P I + M U E	100	100	100	100	100	100	100
FIELD ERECTION OF BASIC EQUIPMENT	High percentage of equipment involving high field labor	23/18	21/17	19.5/16	18.5/15	17.5/14.2	16.5/13.5	15.5/13
	AVERAGE (Mild steel equipment)	18/12.5	17/11.5	16/10.8	15/10	14.2/9.2	13.5/8.5	13/8
	High percentage of corrosion materials and other high unit-cost equipment involving little field erection	12.5/7.5	11.5/6.7	10.8/6	10/5.5	9.2/5.2	8.5/5	8/4.8
EQUIPMENT FOUNDATIONS AND STRUCTURAL SUPPORTS	HIGH - Predominance of compressors or mild steel equipment requiring heavy fdns.			17/12	15/10	14/9	12/8	10.5/6
	AVERAGE - For mild steel fabricated equipment			12.5/7	11/6	9.5/5	8/4	7/3
	AVERAGE - For predominance of alloy and other high unit-price fabricated equipment	7/3	8/3	8.5/3	7.5/3	6.5/2.5	5.5/2	4.5/1.5
	LOW - Equipment more or less sitting on floor	5/0	4/0	3/0	2.5/0	2/0	1.5/0	1/0
	PILING OR ROCK EXCAVATION	Increase above values by 25 to 100%						
PIPING includes ductwork excludes insulation	HIGH - Gases and liquids, petrochemicals, plants with substantial ductwork	105/65	90/58	80/48	70/40	58/34	50/30	42/25
	AVERAGE FOR CHEMICAL PLANTS Liquids, electrolytic plants	65/33	58/27	48/22	40/16	34/12	30/10	25/9
	LIQUIDS AND SOLIDS	33/13	27/10	22/8	16/6	12/5	10/4	9/3
	LOW - Solids	13/5	10/4	8/3	6/2	5/1	4/0	3/0
INSULATION OF EQUIPMENT ONLY	VERY HIGH - Substantial mild steel equipment requiring lagging and very low temperatures	13/10	11.5/8.5	10/7.4	9/6.2	7.8/5.3	6.8/4.5	5.8/3.5
	HIGH - Substantial equipment requiring lagging and high temperatures (petrochemicals)	10.3/7.5	9/6.3	7.8/5.2	6.7/4.2	5.7/3.4	4.7/2.8	4.8/2.5
	AVERAGE FOR CHEMICAL PLANTS	7.8/3.4	6.5/2.6	5.5/2.1	4.5/1.7	3.6/1.4	2.9/1.1	2.2/0.8
	LOW	3.5/0	2.7/0	2.2/0	1.8/0	1.5/0	1.2/0	1/0
INSULATION OF PIPING ONLY	VERY HIGH - Substantial mild steel piping requiring lagging and very low temperatures	22/16	19/13	16/11	14/9	12/7	9/5	6/3.5
	HIGH - Substantial piping requiring lagging and high temperatures (petrochemicals)	18/14	15/12	13/10	11/8	9/6	7/4	4.5/2.5
	AVERAGE FOR CHEMICAL PLANTS	16/12	14/10	12/8	10/6	8/4	6/2	4/2
	LOW	14/8	12/6	10/5	8/4	6/3	4/2	2/1
ALL ELECTRICAL except building lighting and instrumentation	ELECTROLYTIC PLANTS (includes rectification equipment)		55/42	50/38	45/33	40/30	35/26	
	Plants with mild steel equipment, heavy drives, solids	26/17	22.5/15	19.5/12.5	17/10	14/8.5	12/7	10/6
	Plants with alloy or high unit-cost equipment, chemical and petrochemical plants	18/9.5	15.5/8.5	13/6.5	11/5.5	9/4.5	7.3/3.5	6/2.5
	NOTE: Above figures include 1 to 3% for B/L outside lighting which is not covered in Building Services							

INSTRUMENTATION

		AVERAGE UNIT COST OF M.P.I. IN 1958 DOLLARS						
		UNDER $3,000	3,000 to 5,000	5,000 to 7,000	7,000 to 10,000	10,000 to 13,000	13,000 to 17,000	OVER 17,000
INSTRUMENT-ATION	Substantial instrumentation, central control panels, petrochemicals		58/31	46/24	37/18	29/13	23/10	18/7
	MISCELLANEOUS CHEMICAL PLANTS		32/13	26/10	20/7	15/5	11/3	8/2
	Little instrumentation, solids		21/9	17/7	13/5	10/3	7/2	5/1

NOTE: Total instrumentation cost does not vary a great deal with size and hence is not readily calculated as a percent of Basic Equipment. This is particularly true for distillation systems. If in doubt, detailed estimates should be made.

MISCELLANEOUS

MISCELLANEOUS Includes site preparation, painting & other items not accounted for above

Top of range: large complicated processes

Bottom of range: smaller, simple processes

RANGE FOR ALL VALUES OF BASIC EQUIPMENT 6 to 1%

BUILDINGS - ARCHITECTUAL & STRUCTURAL (excludes bldg. services)

NOTE: When building specifications and dimensions are known, a high-speed building cost estimate is recommended especially if buildings are a significant item of cost. If a separate estimate is not possible, evaluate the buildings as follows before selecting the factors.

BUILDING EVALUATION when most of process units are located inside buildings					
	HIGH Brick and Steel	MEDIUM		LOW Economical	EVALUATION
QUALITY OF CONST.	+4	+2		0	
	VERY HIGH UNIT COST EQUIPMENT	MOSTLY ALLOY STEEL	MIXED MATERIALS	MOSTLY CARBON STEEL	
TYPE OF EQUIPMENT	−3	−2	−1	0	
OPERATING PRESSURES	VERY HIGH	INTERMED.		ATMOS.	
	−2	−1		0	

BUILDING CLASS = ALGEBRAIC SUM =

	BLDG. CLASS	AVERAGE UNIT COST OF M.P.I. IN 1958 DOLLARS						
		UNDER $3,000	3,000 to 5,000	5,000 to 7,000	7,000 to 10,000	10,000 to 13,000	13,000 to 17,000	OVER 17,000
MOST OF PROCESS UNITS INSIDE BUILDINGS	+2	92/68	82/61	74/56	67/49	59/44	52/39	46/33
	+1 to −1	72/49	62/43	56/38	51/33	45/29	41/26	36/21
	−2	50/37	44/33	40/29	35/25	30/21	27/18	23/15
OPEN-AIR PLANTS WITH MINOR BUILDINGS		37/16	32/13	28/11	24/8	20/6	17/4	14/2

BUILDING SERVICES

NOTE: The following factors are for Battery Limit (process) buildings only and are expressed in percent of the Building-Architectual & Structural cost. They are not related to the Basic Equipment cost.

	HIGH	NORMAL	LOW
Compressed air for general service only	4	1½	5
Electric lighting	18	9	5
Sprinklers	10	6	3
Plumbing	20	12	3
Heating	25	16	8
Ventilation: without air conditioning	18	8	0
with air conditioning	45	35	25
TOTAL OVER-ALL AVERAGE*	85	55	20

The above factors apply to those items normally classified as building services. They do not include:

1. Services located outside the building such as sub-stations, outside sewers, outside water lines, etc., all of which are considered to be outside the Battery Limit, as well as outside the building.

2. Process services.

* The totals provide the ranges for the type of building involved and are useful when the individual service requirements are not known. Note that the over-all averages are not the sum of the individual columns.

2. Estimate S&H by factors from the B/L cost.

3. Estimate utilities as a function of B/L.

4. Estimate services as a function of (B/L + S&H + U).

It should be noted that the above procedure applies to grass-roots plants or to battery-limit additions. It is not intended that the data in this article be applicable to any project that is less than a battery-limit addition.

Estimating the Battery Limit

It is necessary to break B/L down into its major components so that all items are included and the summation adds up to the total cost of the entire B/L. Strict definition and understanding of process equipment is particularly important since this is the base for all the factors.

Process equipment is called "basic equipment" in this factor technique. It is defined as the cost of all process equipment, delivered to the site. It does not include foundations or structural supports, insulation, painting, or erection.

For practical purposes:

Basic equipment = Main plant items + miscellaneous unlisted equipment.

Where: Main plant items (MPI) represent all the usual major items of equipment that would be indicated on a flowsheet down to and including pumps. And miscellaneous unlisted equipment (MUE) represents the minor miscellaneous equipment that turns up on all projects but rarely gains a position on the flowsheet or on the early equipment lists.

The cost of these two together, delivered to the site, represents the "basic equipment" cost that is the basis of this factor estimating system.

The average unit cost is based on the MPI's only, and does not include the MUE, i.e. the total number of MPI's are counted and divided into the total delivered cost of the MPI's.

It is the basic equipment, however, that represents 100 when applying the factors.

To compensate for inflationary effects, the average unit cost of the

Storage and handling in percent of battery-limit cost—Table III

	Grass Roots Plant	B/L Addition on Existing Site
	% of B/L Cost	
Low: Raw material by pipeline. Little warehouse space	2	0
Avg.: Average raw-material storage & finish-product warehousing	15–25	2–6
High: Tank farm for raw-material. Substantial warehousing for finished product	70	20

Utilities in percent of battery-limit cost—Table IV

		Range for Grass Roots
Utilities buildings		3–10
Arch'l. & struct'l.	2–7	
Mechanical services	0.5–4	
Compressed air system		0.1–4
Electrical systems		1.5–6
Substation	0.5–3.5	
Distribution	0.5–3	
Outside lighting	0.15–1.5	
Gas system		0–0.6
Sewers & drainage system		1.3–3.5
Steam system		1.5–11
Generation	1–9	
Distribution	0.5–3	
Water system		1–10
Pumphouse	1–8	
Cooling towers & recirc.	0.5–5	
Distribution	0.15–3	
Fire protection	0.2–1	
Water treatment	0.2–1.5	
Miscellaneous		0.5–3

	Grass Roots			B/L Additions		
	Low	Avg.	High	Low	Avg.	High
Over-all averages for all utilities	10	20–30	50	3	6–14	30

Services in percent of (B/L + S&H + U)—Table V

	Range for Grass Roots
Main office	1–5
Laboratories	0–2.8
Shops & stores	1–8
Lunch rooms	0–2.2
Change houses	0–2.2
Personnel & gatehouses	0–1
Roads, railroads & fences	1.3–5.5
Service equipment	0.5–4.5
Miscellaneous	0.5–2

	Grass Roots			B/L Additions		
	Low	Avg.	High	Low	Avg.	High
Over-all averages for total services	5	10–16	20	0	2–6	15

Calculation sheet for factor estimating—Table VI

Division & Location	Proj. or Study No.	TITLE Chlorine Plant			DATE June '65
	Requested by	CAPACITY			

NO. of M.P.I.'s	COST INDEXES		FACTOR OR ACCURACY	LOW	PROBABLE	HIGH
	1958	CURRENT				
100	100	112				

	FACTOR OR ACCURACY	LOW	PROBABLE	HIGH
AVERAGE UNIT COST OF M.P.I.'s IN 1958 DOLLARS ____ 9000				
M.P.I. (Main plant items)	Estimated		1,000,000	
M.U.E. (Miscellaneous unlisted equipment)	7%		70,000	
BASIC EQUIPMENT (M.P.I. + M.U.E.) (Excluding sales taxes and catalyst)	100 +10-10	963,000	1,070,000	1,177,000

	REMARKS	FACTOR			
Field erection of Basic Equipment	Slightly under Average	8/10/12			
Equipment Foundations & Structural supports	Average	5/7/9			
PIPING	Aver. for Chem Plants	20/28/35			
INSULATION Equipment Piping	Relatively low " "	0.5/1/2 4/5/6			
ELECTRICAL	Electrolytic Plant	35/39/43			
INSTRUMENTATION	Lower range of average for Chem Plants	6/10/14			
MISCELLANEOUS		3/4/5			
BUILDINGS Architectural & Structural	Evaluation – 1 to 2	30/35/40			
BUILDING SERVICES: (in % of Arch'l. & Struct.)					
Compressed air Electrical lighting Sprinklers Plumbing Heating Vent. & Air Conditioning	1.5 9 10 8 16	45% =	14/16/18		
TOTAL SERVICES	45				
SUB TOTAL–FACTORED ITEMS		125.5/155/184			
ADJUSTMENTS: LOWS + 10 HIGHS –10					
TOTAL FACTORED ITEMS ADJUSTED		138/155/166	1,330,000	1,660,000	1,955,000
DIRECT COST OF B/L (Excluding taxes and catalyst)		+15% -16%	2,293,000	2,730,000	3,132,000

MPI's should be in constant dollars, and for our purposes we have selected 1958 as the base year.

Sample Estimate

Table VI illustrates a sample B/L factor estimate.

We will assume that the estimate is for a chlorine plant, also that there are 100 main plant items that are estimated to cost $1 million. With a current cost index of 112, this gives an average unit cost in 1958 dollars of $9,000.

The factors are selected from the data in Table II and the estimate can be completed easily by following the form. The following trouble spots, however, are worthy of mention:

Sales taxes are treated as a separate item later on in the estimate and should not be included in cost of the basic equipment. The same for catalyst. Based on specific knowledge of the project, an accuracy evaluation is placed on the basic equipment. Subsequently, when factoring the various components, the low factor is applied against the low basic equipment cost, and the high factor is applied against the high basic equipment cost.

Instrumentation and buildings has the least correlation with basic equipment, and if scope is adequately defined other quick methods of assessing them are recommended.

Building services are estimated as a function of the architectural and structural cost of the buildings. Once established, it can then be converted to a factor of the basic equipment so that all factors placed in the factor column refer to basic equipment.

When we arrive at the total of all the factored items, the range is very wide since the extreme factors have been applied against the extremes of the basic equipment. In practice, it is obvious that there is no chance of all the lows or all the highs occurring at the same time. Thus an adjustment is required and in practice we have found that it can be reasonably made by adding 10% to the lows and deducting 10% from the highs.

The method of making this adjustment offers an interesting area for some mathematical studies. We have done some work in this direction and are currently pursuing it further.

We now have an estimate of the direct cost of the B/L. Now auxiliaries have to be estimated.

Estimating the Auxiliaries

Table VII provides the estimate for the auxiliaries. If scope of the work is well defined, then these can be estimated quickly, using large component techniques. For purposes of this illustration, however, we are estimating all auxiliaries by factors.

Use of high and low factors again poses a problem since all the lows will not occur at the same time nor all the highs at the same time. In this case, because the B/L estimate is significantly larger than the auxiliaries and is generally more accurate, we have been applying the three factors for each auxiliary against the probable cost of the B/L. The values so obtained are then added without any other adjustment.

At this point, we have the total direct cost of the B/L + auxiliaries.

Handling Contingencies

Most companies that appropriate funds for capital projects provide for contingencies: 10% is a common allowance.

If we consider our estimate, then we can say with a fair degree of confidence that the high figure has already provided for all the contingencies. That is why it is high. However, the low figure has no allowance for contingencies and hence there is good reason to provide one. Thus, when determining how much money should be appropriated, it seems reasonable to say that it should be in the range of the: (Low +10% to the High).

This has the effect of narrowing the range and appearing to make the estimate more accurate. Nevertheless, it seems to be consistent with normal company thinking and practice.

Summary sheet for factor estimating—Table VII

Division & Location / Proj. or Study No. / Requested by	TITLE: CHLORINE PLANT / CAPACITY			DATE
	FACTOR OR ACCURACY	LOW	PROBABLE	HIGH
DIRECT COST OF B/L		2,293,000	2,730,000	3,132,000
STORAGE AND HANDLING *R.M. warehouse only + related handling*	6/8/10	164,000	218,000	273,000
UTILITIES *Rectifiers incl. in B/L Water supply available Little steam required*	7/10/15	191,000	273,000	410,000
SERVICES *Somewhat less than average* (in per cent of (B/L + S&H + U))	7/10/13	226,000	322,000	418,000
TOTAL B/L + AUXILIARIES	± 19%	2,874,000	3,543,000	4,233,000
CATALYST				
TAXES				
TOTAL DIRECT COST				
INDIRECT COSTS				
CONSTRUCTION FIELD, O.H. & PROFIT				
ROYALTIES, LICENCES AND PATENTS				
ENGINEERING				
TOTAL INDIRECT COSTS				
TOTAL DIRECT AND INDIRECT				
CONTINGENCIES				
TOTAL APPROPRIATION				

Comparing accuracy of estimates—Table VIII

Accuracy on Equipment	Accuracy on Estimate		Accuracy With Contingency Allowance	
	B/L Plant	Grass Roots	B/L Plant	Grass Roots
±10%	±15%	±19%	±11%	±14%
± 5%	±10%	±15%	± 6%	±10%

Accuracies of Factor Estimates

It is interesting to examine the accuracies of the estimate at various points.

The starting point was a basic equipment estimate of ±10%.

For this situation: The B/L estimate is within ±15%, and the B/L + auxiliaries are within ±19%. If a 10% contingency is added to the lows, then the B/L estimate has an accuracy of ±11%, while the B/L + auxiliaries has an accuracy of ±14%. Such accuracies are typical of the ranges that are generally obtained with factor estimates.

Table VIII shows these potential accuracies along with what they might be if the basic equipment estimate were within ±5%.

Refinement of the significant items in the B/L can also improve the accuracy.

Acknowledgement

We wish to acknowledge the contribution of W. G. Cowie of Imperial Chemical Industries, Billingham, England, who conducted the initial studies on the average unit cost technique and provided considerable reference information.

Definitions of Four Areas

Battery Limit (B/L)—This area represents all process operations. It can be defined as the boundaries enclosing a plant or process unit so as to include those facilities directly involved in the conversion of raw material to finished product. It applies to all buildings, equipment, piping, instruments, etc., that specifically involve the process or manufacturing operation. It includes that portion of the compressed air, electrical, refrigeration, steam, water, plumbing, fire protection, process-waste disposal, and air-conditioning systems, etc., that are inside the process area, but does not include the outside lines, etc., that convey such utilities or services to or from the battery-limit buildings.

Storage and Handling (S&H)—Consists of all warehouses, storage tanks, loading, unloading, and handling facilities, etc., required for raw materials and finished products associated directly with the product being made. It includes the necessary pipelines from the point of storage to the walls or boundaries of the battery limit. It does not include storage and handling of raw materials for utilities, such as coal, fuel oil, etc., which are included with the cost of the utility. Similarly it does not include in-process storage, which is normally charged to the battery limit, unless it is a large intermediate storage station.

Meet the Author

C. A. Miller is Supervisor of Cost Studies at Canadian Industries, Ltd., Montreal, Quebec, Can. Most of his experience has been with CIL, following a brief teaching career. His experience covers design of chemical plants, field work on construction of munitions and atomic energy plants, production and management of process plants. In 1957, he was appointed to his present position. Mr. Miller holds a B. A. Sc. in mechanical engineering from the U. of Toronto. He is a past president of AACE, a member of the Engineering Institute of Canada, and a registered professional engineer in Ontario and Quebec.

Utilities (U)—Utilities refer, in general, to the production of energy and its transportation to and from the battery limit as well as to other buildings on the site. It consists of: Compressed-air plant if located outside the battery limit and outside air lines; electric power supply consisting of substation, outside lines, and yard and fence lighting; refrigeration system if located outside the battery limit consisting of refrigeration machines, and outside refrigerant lines; steam plant and outside steam lines; water supply, pumphouses, main cooling tower, and outside water lines; drains and sewers including normal sewerage treatment systems (process waste-treatment systems are part of battery limit up to a point where the discharge is safe to enter a main effluent sewer); storage and handling facilities for raw materials used in the production of utilities.

Services (S)—Represents all the remaining items of investment that are necessary to round out the plant into a fully operating unit. It includes items such as offices, laboratories, shops and lunchrooms, change houses, gatehouses, roads, ditches, railways, fences, communication system, service equipment, track scales, etc.

The last three areas are often referred to as chemical plant auxiliaries or off-site facilities.

Key Concepts for This Article

Active (8)
Estimating

Means/Methods (10)
Factors
Ratios

Passive (9)
Costs
Costs, plant
Storage
Utilities
Insulation
Piping
Instrumentation
Electrical
Buildings

Updated Investment Costs for 60 Types of Chemical Plants

Here is a timely and useful summary of investment-cost data, together with Lang factors, for a large variety of processing facilities.

JOHN E. HASELBARTH, *Pritchard & Abbott*

In making any sort of plant-investment cost study, one must remember that there are three categories for capital projects: grass roots, new unit on established site, or enlargement of existing unit.[1] The cost tabulation on the opposite page is based on the second (and most common) category—i.e., production units that are constructed on a previously developed plant site. If the project fell into the so-called grass-roots category, the total cost would have to include all off-site facilities such as utility buildings, general services, etc., and the final investment figures would be roughly 30% to 40% more. The third major division, which is simply the enlargement of an existing plant, is roughly 20% to 30% less than the category considered.

For plant sizes other than those given in the table, the estimator can use size factors (popularly known as Lang factors); these are shown for each individual product, and apply within roughly a two-fold ratio, extending either way from the plant size as given. Separate data should be obtained for any size beyond this limitation.

All costs are as of the first quarter of 1967. Sources of information include published data, along with company files that contain cost studies of more than 100 plants producing several hundred products.

Trends in Plant and Operating Costs

In the 18 years since CHEMICAL ENGINEERING published C. H. Chilton's pioneering studies of capital and operating costs for various types of plants,[2] the ENR Construction Cost Index has increased from 400 to 1,000, which would indicate a vast general increase in chemical-plant construction costs. However, there were many important technical advances, coupled with the construction of multiproduct plants and large single-product plants, that countered this long-range inflationary trend. The manufacture of sulfuric acid by the contact process is a dramatic example of the way in which design improvements can more than off-set inflation; one can now make twice as much acid for half as much investment as was true 20 years ago.

For most products, it is quite difficult to compare process-labor costs today with the costs that applied in 1950, because the manpower deployment system has changed. In 1950, it was quite reasonable to expect a given processing unit to employ a specific number of operators whose task was confined to the production of the one material being manufactured in the unit to which they were assigned. Now, with the addition of computerized control and long-distance control-panel arrangements, we find that the operators can spread their work load so that they may be able to control the processing of several products simultaneously. Because of this, it is sometimes almost impossible to relate man-hour requirements directly to production quantities for a given product.

The nature of the operation often dictates the personnel placement. For plants primarily engaged in solids processing, a greater percentage of materials handling is required than in plants engaged in solid-fluid or fluid processing, and more people will thus be needed for shipping, receiving and storage.

For a fluids processing plant (e.g., ethylene, oxygen, ammonia), the labor requirements today may range from $\frac{1}{3}$ man-hour to 1 man-hour per ton of product. For solid-fluid plants (e.g., polyethylene, caustic/chlorine, sulfuric acid), the labor requirements may range from 2 to 4 man-hours per ton.

References

1. Haselbarth, J. E., Berk, J. M., Analysis of Ethylene Plant Costs, *Chem. Eng.*, Apr. 18, 1960.
2. Chilton, C. H., Cost Data Correlated, *Chem. Eng.*, June 1949; "Six Tenths Factor" Applies to Complete Plant Costs, *Chem. Eng.*, Apr. 1950.

Meet the Author

John E. Haselbarth has spent the last ten years with Pritchard & Abbott, 2530 W. Holcombe Blvd., Houston, where he has been concerned with investment costs of large industrial plants, as well as economic evaluation studies. His earlier experience includes nine years with Olin Mathieson Chemical Corp. at Pasadena, Tex., where he was engaged primarily in process design, economic evaluation, and plant-investment cost studies. A chemical engineering graduate of Ohio State University, he is a registered professional engineer in Texas, and maintains membership in AIChE and the Amer. Assn. of Cost Engineers.

1967 Capital-Cost Data for Processing Plants

Compound	Source or Route	Typical Plant Size, Tons/Yr.	Investment Cost, $	Investment, $ per Annual Ton	Size Factor (Lang),* L	Remarks
Acetaldehyde.................	Ethylene	50,000	3,500,000	70	0.70	Metallic catalyst required
Acetylene.....................	Natural gas	75,000	9,500,000	127	0.70	High purity
Alumina.......................	Bauxite	100,000	9,000,000	90		
Aluminum sulfate.............		75,000	2,000,000	27		
Ammonia......................		500,000	16,000,000	32	0.70	
Ammonium phosphate........		250,000	2,500,000	10	0.68	Fertilizer grade
Ammonium sulfate............		140,000	1,200,000	9	0.68	
Carbon black.................		30,000	3,000,000	100		
Carbon dioxide...............		200,000	2,400,000	12		
Carbon tetrachloride..........		30,000	2,500,000	85		
Butadiene....................	Butane	100,000	50,000,000	500	0.70	
Butadiene....................	Butylenes	200,000	70,000,000	350	0.70	
Chlorine/caustic.............. Cl₂:		70,000	13,000,000			
NaOH:		78,000			0.69	
Cyclohexane.................		100,000	750,000	8	0.70	Does not include hydrogen plant
Diphenylamine...............		10,000	2,400,000	240		
Ethanolamine................		25,000	1,750,000	70		
Ethyl alcohol.................	From ethylene by direct hydration or via ethyl sulfuric acid	75,000	3,750,000	50	0.72	Manufacturing costs are lower in the direct hydration process
Ethylbenzene ⎱		20,000	1,800,000			These chemicals are produced simultaneously
Paraxylene ⎰		8,500	1,100,000			
Ethyl chloride................		15,000	3,000,000	200		
Ethyl ether...................		35,000	1,200,000	35		
Ethylene.....................	Refinery gases or hydrocarbons	300,000	15,000,000	50	0.71	
Ethylene dichloride...........		25,000	3,200,000	127	0.71	
Ethylene oxide...............	Direct oxidation of ethylene	100,000	9,000,000	90	0.67	Cost also includes conversion to ethylene glycol as needed
37% Formaldehyde...........	Hydrocarbons	100,000	13,000,000	130		
Glycerin (synthetic)...........		35,000	5,500,000	157	0.67	
Hydrofluoric acid.............		15,000	2,600,000	175		
Hydrogen....................		60,000	6,500,000	108	0.80	
Isopropyl alcohol.............		150,000	7,500,000	50		
Maleic anhydride.............		50,000	18,000,000	360		
Melamine....................		70,000	11,500,000	164		
Methanol....................	Natural gases	210,000	9,000,000	43	0.71	
Methyl chloride..............	Methanol	10,000	500,000	50	0.72	
Methyl ethyl ketone...........		35,000	3,750,000	107		
Methyl isobutyl ketone ⎱		25,000	1,250,000	50		
Methyl isobutyl carbonal ⎰		10,000	750,000	75		
Nitric acid...................		50,000	5,000,000	100		
Oxygen plants................		150,000	2,250,000	15	0.71	
Phenol......................		45,000	9,000,000	200		
Phosphoric acid (as P₂O₅).....		100,000	2,400,000	24	0.66	Wet process—contains 30% P₂O₅
Cis-polybutadiene.............		50,000	12,000,000	240	0.67	
Polyisoprene (includes manufacture of the monomer)		30,000	5,000,000	320	0.74	
Soda ash....................	Natural brine	400,000	34,000,000	85		No synthetics plants built since 1934
Sodium metal.................		20,000	7,000,000	350		
Styrene......................		20,000	8,500,000	425		
Sulfuric acid.................	Contact process	280,000	2,100,000	8	0.67	
Sulfur recovery...............	Refinery gases	15,000	1,500,000	100		
Toluene diisocyanate..........		12,500	7,500,000	600		
Urea........................		140,000	4,300,000	31		
Vinyl acetate.................		40,000	7,000,000	175		
Vinyl chloride monomer........		100,000	2,000,000	20		
Refinery Products		**(Bbl./Day)**				
Alkylation units (H₂SO₄ or HF)..		10,000	7,750,000	775		
BTX extraction................	From reformer streams; e.g. Udex	10,000	3,400,000	340	0.70	
Cat. cracker (fluid)...........	Cost based on fresh feed	35,000	14,000,000	400		Includes vapor recovery and CO boiler
Cat. reformer.................		23,000	7,500,000	375		
Crude distillation units........		100,000	4,700,000	47		
Delayed coker................		14,000	5,000,000	357		
Hydrocracker.................		28,000	21,000,000	750		
Wax plants...................		7,500	900,000	120		
Gas absorption and dehydration plants		50 MM cfd.	2,000,000			

* Where no size factor appears, assume a value of 0.70. Use: To obtain investment for a capacity other than the one shown, multiply the stated investment cost by the ratio of the desired capacity to the stated capacity, raised to the power L.

2

The Module Approach to Capital Cost Estimating

If you want to establish the cost of process modules from
estimated or quoted equipment costs, you can do so quickly by using these . . .

"Rapid Calc" Charts

KENNETH M. GUTHRIE, W. R. Grace & Co.

Published data, personal files and vendor quotations can result in a reliable estimate of the cost of this or that type of processing equipment. However, this fob. cost is seldom all that is needed, because each individual piece of equipment must be installed on foundations and connected into a circuit that includes piping, supports, instruments, electrical wiring, insulation, platforms and other items necessary to properly operate a given process. This complex of installed equipment is established as a chemical process module.

Not much has been published on how to quickly and accurately convert equipment dollars into the variable costs associated with chemical process modules. The data in this Cost File provide a short-cut technique that will generate field-installation, indirect or total costs associated with a "norm" chemical process module.* Adjustment factors for alloy relationships and dollar magnitude are provided, and their use is illustrated in worked-out examples representing projects in two different stages of development.

The Chemical Process Module

Consider this module to represent the total cost of process equipment (such as heat exchangers, pressure vessels, pumps, compressors, etc.), together with field materials, field labor and indirect costs necessary to install the equipment on a prepared jobsite. Cost-element relationships are complex, and adjustment from the "norm" is necessary to accurately

reflect individual installations. It is necessary, therefore, to assemble primary cost elements in a consistent format:

1. Fob. equipment cost	(E)	0000		
Field materials*	(M)	000		
2. Direct material cost	(E+M)	0000		
3. Direct labor cost	(L)	00		
4. Direct cost C_{M+L}	(E+M+L)	00000		
5. Indirect costs	(% of 4)	0000		
6. Bare-module cost	(4+5)	00000		
Contingency†	(% of 6)	000		
7. Total module cost		$00000		

(bracketed at right: Field Installation, Bare-Module Cost, Total Module Cost)

*Field materials is a secondary cost element that includes piping; instrumentation; foundations and concrete work (except piling); local steel such as ladders, platforms and minor supports (not steel structures); electrical materials; insulation, and paint.

†Contingency in this case is partly an accuracy statistic and partly an allowance for unlisted items. Assume 4% for chart reading, 6% for unlisted items—and an amount of 5 to 10% for equipment cost accuracy, depending on the method used to generate these costs.

The dollar value of each primary cost element (except indirect costs) is obtained from the equipment cost (E) by using factor relationships that have been established from feedback data. On the charts facing the next page, these relationships are expressed in a series of cost equations that can be adjusted for alloy materials (F_r) and dollar magnitude (F_m) by factors read directly from the "rapid calc" charts appropriate to each equation.

Alloy Ratio Factor (F_r)—The "norm" module factors are based on carbon steel (CS) relationships; it is often necessary, however, to specify the circuit in alloy materials, and this directly affects the cost of equipment and piping. The relative value of other module costs (such as non-piping field materials, field labor,

---

* The "norm" chemical process module reflects mid-1968 costs on a prepared jobsite in the U.S. Gulf Coast area. Process module costs do not include site development, piling, buildings, major storage or other offsite facilities.

indirect costs) should, however, be maintained on the carbon steel basis. It is therefore necessary to convert the alloy value of total equipment to carbon steel equivalents by using the parameter:

$$\text{Alloy ratio} = \frac{\text{Total equipment cost (in alloy) \$}}{\text{Total equipment cost (CS basis) \$}}$$

to obtain the adjustment factor (F_r) for each primary element or module cost. This factor can be read directly from the charts and substituted in the appropriate cost equations.

Magnitude Factor (F_m)—The basic module was set up on a carbon-steel equipment value of $2.2 million. Correction for dollar magnitude (F_m) above or below this "norm" can be read directly from the magnitude-factor charts for each primary cost element, and substituted in the appropriate equations.

Indirect Costs

This indirect-cost account includes the following secondary cost elements:

Construction Overhead: Field-labor fringe benefits and statutory burdens, field supervision, temporary field facilities, construction equipment rental, small tools, etc.

Engineering and Contractor Fee: Direct engineering labor, indirect office costs, burden and overhead, contractor fee.

These costs are more properly related to total direct material and labor (C_{M+L}) than to equipment dollars, and are particularly sensitive to alloy ratios, labor/material ratios, job location, dollar magnitude. A "norm" 42% of direct costs has been included in the chart data, which can be used for early process-module evaluations. Adjust when necessary to reflect individual project characteristics.

Total Module Cost

Process modules are usually integrated with other module costs (such as site preparation, buildings, off-site facilities) at the direct-cost level (C_{M+L}); it may be necessary, however, to know the value of the total process module for alternate evaluations, such as replacing equipment and other revamp considerations. In this case indirect costs are included, and one can obtain the bare-module cost from Eq. (6),* correcting for alloy or magnitude if necessary. One can then add a contingency (the amount reflecting the quality of the equipment cost data) to obtain total module cost.

Module Adjustment

The terms "Index" in the equations is an adjustment factor representing a numerical evaluation of material escalation, labor escalation, labor productivity, area wage rates, indirect cost ratios, (e.g., more or less field supervision), or an area factor combination of

---

* Equation numbers refer to the "rapid-calc" charts—i.e., Eq. (6) is the equation above Chart 6 on the opposite page.

all these variables. It is assumed that some data are presently available to the estimator to be able to adjust the appropriate equations as required; this is outside the scope of this Cost File.

Consider the accuracy of this "rapid calc" method to be approximately ±3% compared with actual module integrations. It is, therefore, valid for early evaluations, conceptual estimating and some definitive analysis when considered with the other modules that make up the total plant investment.

The author's report on the cost of various kinds of process equipment, which is scheduled to be published in the Mar. 24 issue, will contain other applications of the module concept.

Let us now consider two illustrative cases:

Example 1: Embryo Estimate

The total fob. equipment cost in a process circuit is estimated to be $1,500,000 in carbon steel. Use the "rapid calc" charts to predict the total module cost for (1) carbon steel; (2) 50% of the equipment having an alloy ratio of 3.0; and (4) 100% alloy ratio 5.0. Assume that a 20% contingency is necessary.

Carbon Steel

Alloy ratio	= 1.0

Bare-module cost:

E	= 1,500,000
From Chart 6A, F_r	= 3.22
$(E)(F_r)$	= 4,830,000
From Chart 6B, F_m	= 1.03

Eq. (6): (1,500,000(3.22)(1.03)	=	4,970,000
Contingency @ 20%	=	994,000
Total module cost	=	$ 5,964,000

50% Alloy Ratio 3.0

CS magnitude: (1,500,000) (0.5)	=	750,000
Alloy magnitude: (1,500,000)(0.5) (3.0)	=	2,250,000
Equipment fob. magnitude, CS/alloy, (E):	=	3,000,000
Alloy ratio, 3,000/1,500 = 2.0		

Bare-module cost:

E	= 3,000,000
From Chart 6A, F_r	= 2.60
$(E)(F_r)$	= 7,800,000
From Chart 6B, F_m	= 0.99

Eq. (6): (3,000,000) (2.60) (0.99)	=	7,720,000
Contingency @ 20%	=	1,544,000
Total module cost	=	$ 9,264,000

100% Alloy Ratio 5.0

CS magnitude	=	1,500,000
Alloy magnitude (1,500,000)(5)	=	7,500,000
Equipment fob. magnitude, alloy, (E)	=	7,500,000
Alloy ratio, 7,500/1,500	=	5.0

Cost Equations and "Rapid-Calc" Adjustment Charts

1. Total equipment: $E = (\text{Source}) \text{X(Index)}$

    Source = Published data, personal files or vendor quotations

2. Direct material: $(E+M) = (E)\,(F_r)\,(F_m)\text{X(Index)}$

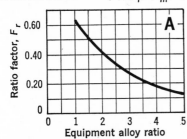 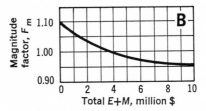

3. Direct field labor: $L = (E)\,(F_r)\,(F_m)\text{X(Index)}$

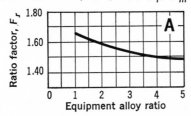 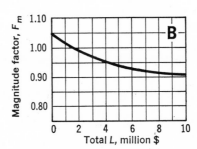

4. Direct material-and-labor cost: $C_{M+L} = (E)\,(F_r)\,(F_m)\text{X(Index)}$

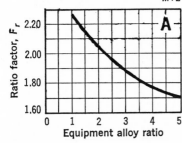 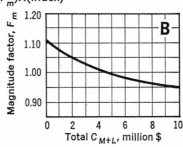

5. Indirect costs $= (C_{M+L})\,(F_r)\,(F_m)\text{X(Index)}$

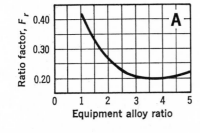 

6. Total bare-module cost $= (E)\,(F_r)\,(F_m)\text{X(Index)}$

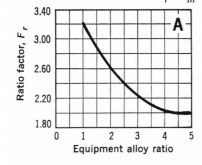 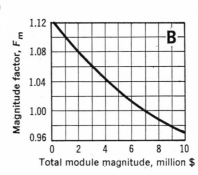

Bare-module cost:

E	$= 7,500,000$
From Chart 6A, F_r	$= 2.0$
$(E)(F_r)$	$= 15,000,000$
From Chart 6B, F_m	$= 0.96$
Eq. (6): $(7,500,000)(2.0)(0.96)$	$= 14,400,000$
Contingency @ 20%	$= 2,880,000$
Total module cost	$= \$17,280,000$

Example 2: The More Advanced Project

The following multiple equipment-classifications have been estimated for a carbon-steel process circuit:

Shell-and-tube (S/T) exchangers	$=$	398,500
Vessels (vert.)	$=$	510,400
(horiz.)	$=$	255,600
Pumps and drivers	$=$	143,500
Fob. magnitude, CS	(E)	\$1,308,000

From the above evaluation, compute the dollar value of primary elements in the process module, using the "rapid calc" charts. Also evaluate the same module in stainless steel (SS) given the following alloy factors: exchangers (SS/SS) = 4.50, vertical process vessels = 3.5 (solid), horizontal process vessels = 3.5 (solid), pumps = 1.91.

Note that in this case the dollar value of each equipment classification has been established, the scope of the project having advanced beyond the embryo stage. A comparable degree of definition and accuracy is now expected of the installation cost. This means that we must evaluate *all* the primary cost elements either to total direct cost, C_{M+L}, or total module cost, as required. Adjustments can then be made for time, location and circuit complexity as required (i.e., via the Index).

Carbon Steel

1. Equipment fob. magnitude, CS,

(E)	$=$	1,308,000
Alloy ratio	$= 1.0$	

2. Direct materials, $(E + M)$:

E	$= 1,308,800$
From Chart 2A, $F_r = 1.65$	
$(E)(F_r)$	$= 2,158,000$
From Chart 2B,	
F_m	$= 1.04$
From Eq. (2):	
$(1,308,000)(1.65)(1.04)$	$= 2,244,000$

3. Direct field labor (L):

E	$= 1,308,000$
From Chart 3A, $F_r = 0.62$	
$(E)(F_r)$	$= 811,000$
From Chart 3B,	
F_m	$= 1.02$
From Eq. (3):	
$(1,308,000)(0.62)(1.02)$	$= 827,000$

4. $C_{M+L} = 2,244,000 + 827,000 = 3,071,000$

5. Indirect costs:

C_{M+L}	$= 3,071,000$
From Chart 5A, $F_r = 0.42$	
$(C_{M+L})(F_r)$	$= 1,290,000$
From Chart 5B, $F_m = 1.03$	
From Eq. (5):	
$(3,071,000)(0.42)(1.03)$	$= 1,328,000$

6. Total bare-module cost $= \$ 4,399,000$
   Contingency @ 15% $= 659,000$

7. Total module cost $= \$ 4,458,000$

Stainless Steel

	CS	Factor	SS
Exchangers	$398,500 \times 4.50 =$		1,793,300
Vessels, vert.	$510,600 \times 3.50 =$		1,787,100
Vessels, horiz.	$255,400 \times 3.50 =$		893,900
Pump and drivers	$143,500 \times 1.91 =$		274,100
	\$1,308,000		\$4,784,400

1. Equipment fob. magnitude,

stainless (E)	$=$	4,748,400
Alloy ratio,		
4,784/1,308	$= 3.63$	

2. Direct materials $(E + M)$:

E	$= 4,748,000$
From Chart 2A, $F_r = 1.52$	
$(E)(F_r)$	$= 7,217,000$
From Chart 2B, $F_r = 0.96$	
From Eq. (2):	
$(4,748,400)(1.52)(0.96)$	$= 6,928,000$

3. Direct field labor

(L): E	$= 4,748,000$
From Chart 3A, $F_r = 0.21$	
$(E)(F_r)$	$= 997,000$
From Chart 3B, $F_m = 1.20$	
From Eq. (3):	
$(4,748,400)(0.21)(1.20)$	$= 1,196,000$

4. $C_{M+L} = 6,928,000 + 1,196,000 = \$ 8,124,000$

5. Indirect costs:

C_{M+L}	$= 8,124,000$
From Chart 5A, $F_r = 0.21$	
$(C_{M+L})(F_r)$	$= 1,706,000$
F_m	$= 1.02$
From Eq. (5):	
$(8,124,000)(0.21)(1.02)$	$= 1,740,000$

6. Total bare-module cost $= \$ 9,864,000$
   Contingency @ 15% $= 1,480,000$

   Total module cost $= \$11,344,000$

Meet the Author

Kenneth M. Guthrie is chief estimator and cost analyst in the Engineering Div., Technical Group, of W. R. Grace & Co. (7 Hanover Sq., New York City). For details of his professional background, see p. 79.

Field-Labor Predictions
For Conceptual Projects

With these techniques, you can not only analyze field-labor requirements
but also visualize the construction schedule for capital
projects in embryonic or conceptual phases of development.

KENNETH M. GUTHRIE, Fluor Corp.

When the budget or timetable for a capital project has to be drastically revised because of field-labor developments, the question arises whether such developments could have been foreseen at the conceptual stage. Often, the answer is yes.

The Feature Report on conceptual capital-cost estimating on page 80 shows how module techniques can be used to predict dollar figures for direct field labor, within well-defined limits—i.e., on a "norm" basis. The norm represents average mid-1968 wage rates,* site conditions and productivity in the U.S. Gulf Coast area. (Correction factors can be used, if necessary, to adjust for other project locations, local wage rates, productivity, etc.)

Further appraisal is necessary to test the realism of the labor manhours in relation to material cost, manpower density, craft availability and other limitations. The data in this Cost File have been developed to assist such analysis, and provide a basis from which to predict approximate manpower requirements and field durations prior to full definition or capital appropriation. An illustrative example will be provided.

Data Grouped in Five Figures

Fig. 1 relates direct labor manhours (equipment setting, field erection, etc.) to total direct material cost (process equipment, piping, field materials, etc.) Adjustments have been made to represent average wage rates and productivity in the norm location. Use this chart to check the field manhours generated by conceptual estimating techniques.

Relative efficiencies are usually expressed in percentages of base 100 (U.S. Gulf Coast). Fig. 2 provides easy reference to convert percent values into the linear productivity factors that are handy for labor analysis work. For instance, 60% efficiency in relation to base 100 is converted to a productivity factor 1.67 (or 167 manhours are required to perform a norm task of 100 manhours).

Fig. 3 shows "norm" manpower density as a function of direct labor dollars, and indicates a maximum number of men acceptable on an average jobsite without loss of "norm" productivity or without requiring an abnormal amount of field supervision. The data are based on feedback that suggests a minimum working area of 50 or 60 sq. ft. per man. Use this chart to check labor availability, plot plan areas, etc.

Fig. 4 correlates field duration (in months) with direct labor dollars, and enables you to predict linear time for "norm" field construction, assuming average Gulf Coast productivity and mid-1968 wage rates. Adjust for other locations and labor efficiencies as required.

Using peak manpower density and field duration from Fig. 3 and 4, a series of graphic man-month integrations can be made; these correspond to the typical manning profiles shown in Fig. 5. The area under each profile represents the amount of work to be accomplished in terms of man-months; this is a constant for each particular project unless the basic scope is changed. (Fig. 5 is based on a composite

---

* A composite wage rate of $4.80 per hour has been derived from an average assortment of crafts required for process-plant construction. Indirect costs such as fringe benefits, statutory burden, and field supervision are not included at this stage.

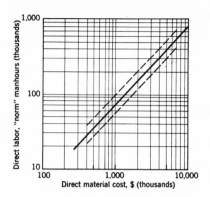

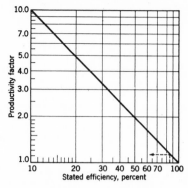

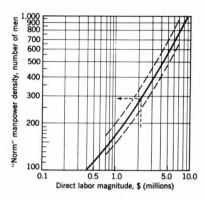

DIRECT LABOR MANHOURS vs. material cost—Fig. 1

PRODUCTIVITY: Linear factors vs. percentages—Fig. 2

MANPOWER DENSITY vs. estimated labor costs—Fig. 3

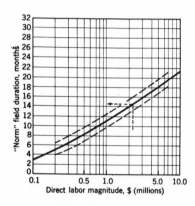

FIELD DURATION vs. estimated labor costs—Fig. 4

MANNING PROFILES: Norm man–month integrations for four projects of increasing field-labor cost—Fig. 5 ▶

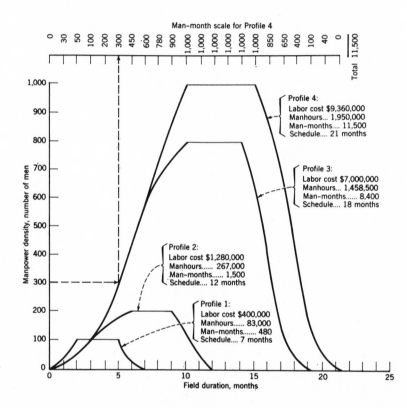

wage rate of $4.80 per hour, and a labor output of 170 manhours per month.)

A word about the contours of the profile. Field construction activity tends to peak out at approximately 60% along the linear time scale, when most of the major equipment is onsite or set in place. At that point, piping and material erection are fully activated.

Data from job manning profiles indicate that both sharp- and flat-top contours have evolved, and resemble skewed distribution curves found in statistics. The rate of adding or terminating craftsmen from a project, however, can vary considerably according to equipment and material deliveries, craft availability, and so on.

The contours indicated in Fig. 5 represent typical "smoothed averages" obtained from a series of observed projects. The mean peak is 60% (e.g., during the 13th month of an expected field duration of 22 months) with a flat top of approximately 23 to 25% of field duration. In this form, the data can be successfully used to analyse and predict during the conceptual phase.

Profiles Lend Realism to Duration Estimate

Graphic integrations are achieved by analyzing a series of trial-and-error profiles. For each profile, the number of men required in each month is projected

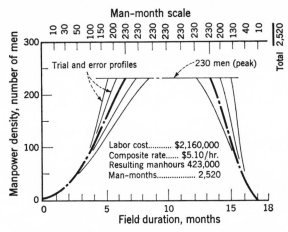

Man-month scale

10 30 50 100 150 200 230 230 230 230 230 230 200 130 40 10   Total 2,520

Trial and error profiles ——→   - - - 230 men (peak)

Labor cost........... $2,160,000
Composite rate...... $5.10/hr.
Resulting manhours 423,000
Man-months................ 2,520

FIELD LABOR man–month integration of example—Fig. 6

on a man-month scale; the summation of these values equals the area under the profile in terms of man-months. This, in turn, is compared with the estimated manhours divided by the productivity rate per man. Both should agree.

Reference to Fig. 5, for example, shows that a total labor cost of $9.36 million will result in approximately 1.95 million manhours if the norm composite rate of $4.80 per hour is considered valid. This, in turn, equals 11,500 man-months at the productivity rate of 170 manhours per month. The area under this particular profile, therefore, represents 11,500 man-months under the restraint of peak manpower density of 1,000 men for 5 months, with a minimum field duration of 21 months necessary to complete the work.

It is clear that the variables "manpower density" and "field duration" are interrelated. If the peak number of men indicated by Fig. 3 are not available, the field duration will be longer than anticipated by Fig. 4, assuming the same amount of work has to be accomplished.

The data presented are based on personal exposure to the complexities of the field-labor cost account. The development of a clearly defined norm reference is essential to predict labor costs of process plant projects that are in conceptual phases of development. Although estimators are encouraged to establish similar references to best describe individual experience and requirements, the data and techniques in this Cost File can be used with confidence.

Field-Labor Analysis: Illustrative Example

A conceptual estimate is being prepared for a process plant complex at Location A. Direct material costs are predicted as $5.0 million.

Local labor productivity is 85% of the norm reference, the area craft rate composite for heavy construction is $5.10 per hour. The limit of craft availability is expected to be 230 men during peak construction activity.

Predict the field labor cost for this project and analyze the result for peak manpower relationships and construction schedule. Indicate the earliest field start to achieve a commissioning date of Jan. 30, 1971, allowing two months for startup activities.

Direct material cost = $5,000,000

From Fig. 1:

Norm direct labor manhours = 370,000

Adjust for productivity and labor rate:

From Fig. 2, the productivity factor = 1.15

Adjusted manhours = (370,000) (1.15) = 423,000

Area labor rate, $/hr. = $5.10

Direct labor cost = (423,000) (5.10) = $2,150,000

Establish norm relationships:

Direct labor magnitude = $2,150,000

From Fig. 3:

Manpower density (peak) = 280 men

From Fig. 4:

Norm field duration = 14.5 months

Analysis: From the above data, the norm peak manpower density is 50 men above the expected labor availability at the jobsite during construction. This means that the indicated field duration of 14.5 months will prove to be inadequate to accomplish all the work (premium time is not considered at this stage). A man-month integration is, therefore, necessary to establish the linear schedule required by 230 men with a relative productivity of 85% as follows:

From Fig. 1, norm manhours = 370,000

Relative output per man-month at 85% efficiency
= (170)(0.85) = 144

Adjusted man-months = 370,000/144 = 2,520

The 2,520 man-months is the area under the manpower profile in Fig. 6.

The integration now indicates that about 17 months will be required by the local field forces to complete the work, provided productivity can be maintained and 230 men are available each day between the 7th and 13th months on the linear schedule.

If the commissioning date of Jan. 30, 1971 is mandatory and two months are allowed for startup, then field construction should commence not later than June 1969. Equipment and material deliveries would have to be geared to this date.

This kind of analysis is often ignored in conceptual work. Even at this stage, however, management should be presented not only with an estimate of dollars, but also with a realistic estimate of time. ∎

Meet the Author

Kenneth M. Guthrie has recently joined Fluor Corp. (2500 S. Atlantic Blvd., Los Angeles). Previously, he was chief estimator and cost analyst in the Engineering Div., Technical Group, of W. R. Grace & Co. Guthrie is a member of AIChE and the American Assn. of Cost Engineers. In addition to authoring the cost-estimating report in the Mar. 24 issue, he has written two Cost Files, including: "Estimating the Cost of Process Modules via Rapid-Calc Charts," which appeared in the issue of Jan. 13, 1969.

Data and techniques for preliminary . . .
Capital Cost Estimating

Here is a compilation of costs for a large variety of
plant equipment—from air coolers to weigh scales—and the introduction
of a "module" technique for making fast, accurate and consistent cost estimates.

K. M. GUTHRIE, W. R. Grace & Co.

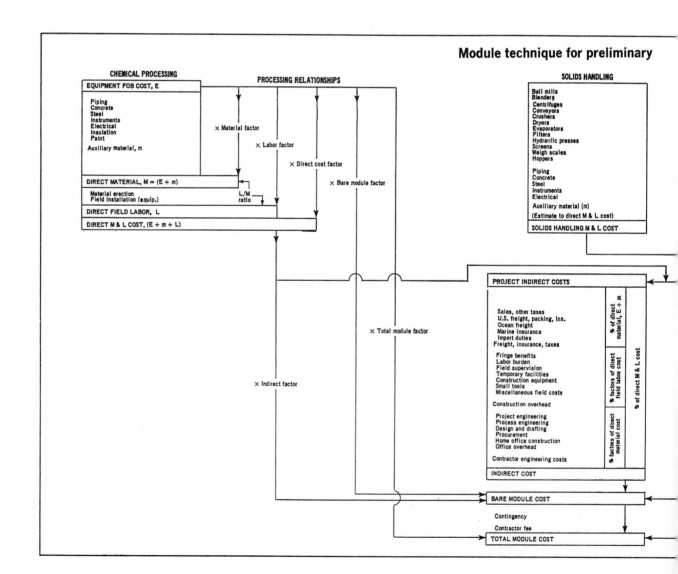

Module technique for preliminary

One of the critical phases in the early, conceptual stage of a process-plant project occurs when an economic evaluation must be prepared.

Often many cost studies are generated and analyzed for process alternatives before a good picture of the economic structure of a project emerges.

It is important, at this stage, to have capital cost estimates that are as accurate as possible. And it is also important to use consistent estimating techniques so that alternatives can be compared on the same basis, and comparisons can also be made between projects.

The information in this Report is directed at these two objectives. Charts and tables are presented for the capital cost of a great variety of major equipment components used in process plants. These data are based on recent information obtained from equipment vendors.

Also, a "module" technique is presented for estimating the cost of an installed unit or an installed plant. This method was developed based on feedback data from 42 process plant projects.

We hope that use of the data and techniques presented here will enable engineers to rapidly assemble a wide variety of relatively accurate cost estimates—recognizing that there are fluctuations and uncertainties in conceptual work—and inspire some professional consistency in estimating techniques, terminology and cost-data presentation.

Module Concept

The application of the module concept to process plant estimation is shown in Table I. All major cost elements are grouped into six distinct modules, five direct and one indirect, as follows:

- Chemical processing.
- Solids handling.
- Site development.
- Industrial buildings.
- Offsite facilities.
- Project indirects.

An estimating module represents a group of cost elements having similar characteristics and relationships. Each module can be integrated or combined with other modules at the material and labor (M&L)

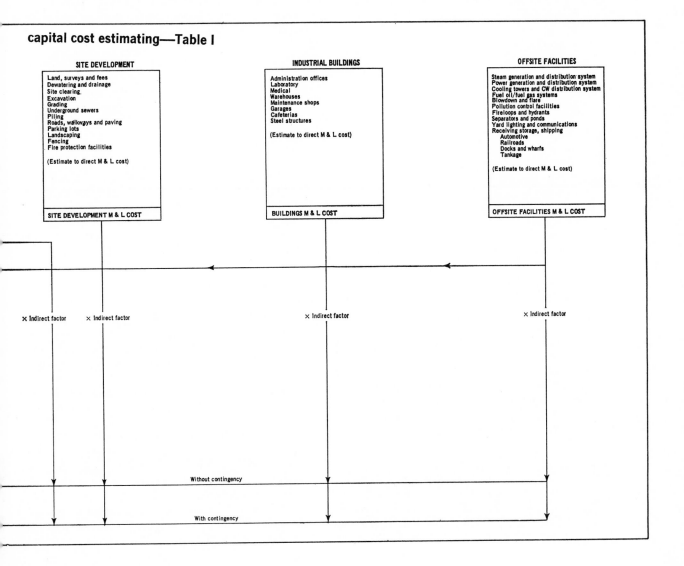

capital cost estimating—Table I

SITE DEVELOPMENT

Land, surveys and fees
Dewatering and drainage
Site clearing.
Excavation
Grading
Underground sewers
Piling
Roads, walkways and paving
Parking lots
Landscaping
Fencing
Fire protection facilities

(Estimate to direct M & L cost)

SITE DEVELOPMENT M & L COST

INDUSTRIAL BUILDINGS

Administration offices
Laboratory
Medical
Warehouses
Maintenance shops
Garages
Cafeterias
Steel structures

(Estimate to direct M & L cost)

BUILDINGS M & L COST

OFFSITE FACILITIES

Steam generation and distribution system
Power generation and distribution system
Cooling towers and CW distribution system
Fuel oil/fuel gas systems
Blowdown and flare
Pollution control facilities
Fireloops and hydrants
Separators and ponds
Yard lighting and communications
Receiving storage, shipping
    Automotive
    Railroads
    Docks and wharfs
    Tankage

(Estimate to direct M & L cost)

OFFSITE FACILITIES M & L COST

× Indirect factor   × Indirect factor   × Indirect factor   × Indirect factor

Without contingency

With contingency

cost level, provided data are in consistent cost terms.

Each direct-cost module contains such items as equipment costs, erection labor, etc., which are developed by using estimating techniques based on cost data presented in this Report. And once total indirect costs are established, all direct costs can be extended to become "bare module costs," as shown in Table I, by using the indirect factor (considered here as a constant for a particular project). This permits alternates to be quickly weighed without having to adjust all the indirect cost elements for each evaluation.

Equipment Modules

All chemical process modules are assembled from combinations of seven primary cost elements:
- Equipment fob. cost.
- Direct material.
- Direct field labor.
- Direct M&L cost.
- Indirect costs.
- Bare module cost.
- Total module cost.

And fourteen secondary cost elements, the field materials direct costs:
- Piping.
- Concrete.
- Steel.
- Instruments.
- Electrical.
- Insulation.
- Paint.

The installation costs:
- Material erection.
- Equipment setting.

And the indirects:
- Freight, insurance, taxes.
- Construction overhead.
- Engineering.
- Contingency.*
- Contractor fee.**

The primary elements establish key relationships and the dollar framework of the estimate: secondary elements "fill in" the details when necessary as the project develops.

Piping factors in the installation modules were developed from an analysis of some 42 refinery, petrochemical and chemical plants. Each factor is based on an evaluation of sample modules of each major equipment item—this "factor" includes associated process piping, flanges, fittings, valves, control valves (all above 2 in.), together with auxiliary piping

(drains, etc.) under 2 in., local underground piping, and a relative portion of underground piping, yard piping and supports (including fuel oil, fuel gas, steam, cooling water, feed and product lines, etc., and all other services within the process battery limits). Most of this piping is not shown on process flowsheets but must be covered in monetary predictions. These factors have been successfully used to *assist* conceptual work and generate "norm" cost estimates. Of course, adjustments are usually necessary to reflect individual characteristics as a particular project develops and more information becomes available.

Thus, a process equipment module represents the cost of a major piece of process equipment (such as heat exchanger, pressure vessel, pump, compressor, etc.) together with dollars for field materials, field labor, and indirects necessary to install the equipment in a chemical process "circuit." The dollar value of each cost element (except indirects) in a particular module is obtained from equipment fob. cost by using factor relationships established from feedback data, and "normalized" for carbon steel (CS) and dollar magnitude.

These factors are located in the cost-data section (Figs. 1-18) adjacent to each process equipment cost chart. Identification in each case is by multiple dollar magnitude.

Module	Dollar Magnitude, $100,000
A	Up to 2
B	2 to 4
C	4 to 6
D	6 to 8
E	8 to 10

The term "multiple" here applies to the total dollar value of a group of equipment, such as all the exchangers, all the pumps, etc., in a process circuit. If one piece of equipment only is being considered, use the data for singular moludes as presented in Table III.

Chemical Process Module

A chemical process module is a compilation of multiple process equipment modules at the M&L level and represents the direct cost of a process circuit. The cost includes equipment together with piping and instrumentation, minor steelwork such as ladder platforms, supports (not major structures), concrete foundations and substructures (no piling), insulation, and paint.

Because of the interrelationships of many variables within this module (such as equipment mix, alloy ratios, labor/material ratios, dollar magnitude), it is useful to assemble a standard or "norm" process module for reference.

The "norm" process module assembled here also serves to illustrate how to use the information in the cost-data charts. It is based on an average equipment

---

* Contingency is not a statistical relationship as used here. It is essentially an allowance to cover items not estimated directly, but that are known to exist in a process plant. Conceptual estimating techniques apply to approximately 80-90% of the total expected cost of the project: contingency monies cover the additional costs to bring the prediction up to 100%. These costs are expected to be spent. Contingency, in this sense, is not a measure of accuracy.
** Contractor fee is usually based on "total monies handled" and should be the last calculation before total module cost or "selling price" is established.

mix and dollar magnitude as obtained from observed projects:

	Average % Total Equipment	Average Dollar Magnitude, $
Furnace...................	14.0	300,000
Exchangers.................	18.0	396,000
Process vessels (V).........	15.0	340,000
Process vessels (H).........	8.0	176,000
Pumps and drivers..........	7.0	154,000
Compression...............	30.0	660,000
Onsite tankage.............	8.0	174,000
Total equipment..........	100.0%	$2,200,000

Using the equipment dollar magnitude and the multiple installation modules in the cost-data sections, primary element cost integration can be made in terms of % total equipment cost (E) as shown in Table II.

Using these direct-cost data and an indirect cost factor of 1.34, the following typical "norm" chemical process module can be assembled based on percent equipment cost. These factors can be used for early conceptual work, provided adjustments are made for known deviations. Of course, where data are available, it is preferable to build your own module costs.

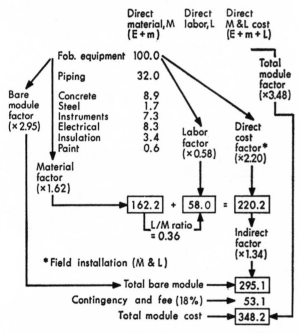

The chemical process module represents mid-1968 costs and labor productivity on a U.S. (Gulf Coast) jobsite. Factor relationships are based on carbon steel. All factors are % total equipment.

Key Relationships

Certain key relationships between the primary cost elements develop at this stage: these are essential for rapid assembly of estimates and provide important checkpoints for definitive or bid analysis.

Material factor indicates the relationship between total equipment costs (E) and total field materials

associated with the equipment (m). This factor includes all direct material costs and should range from 1.42 to 1.75 depending on the circuit complexity. The "norm" module indicates 1.62.

Labor factor includes all field labor required to install the equipment and erect the field materials. It represents direct field labor (L) that should generally range from 0.54 to 0.66 of total equipment cost, depending mostly on the dollar relationship between labor and material (L/M ratio). The "norm" module indicates 0.58. The labor rate as used here is based on a composite rate of $4.80/hr. (Gulf Coast). This is derived from an average mix of crafts for process construction. This rate does not include fringe benefits or labor burdens, which are included in the construction overhead account.

L/M ratio relates direct labor dollars to direct material dollars. The L/M ratio, which varies with each activity (see cost data), is an important measure of productivity, and should range from 0.32 to 0.40 for construction within the U.S. The "norm" module indicates 0.36. (This factor is particularly sensitive to overseas locations where wage rates, labor efficiencies and site conditions vary extensively).

Direct cost factor ($M\&L$) relates fob. equipment dollars to the cost of the equipment together with cost of field materials and field labor necessary to install the equipment on a prepared jobsite.

It represents field installation costs for the process circuit and should range from 1.8 to 2.6. The "norm" module indicates 2.20.

Indirect cost factor includes all the indirect cost elements associated with the module or project as listed in Table I. The total dollar value of these elements is particularly sensitive to alloy ratios, L/M ratios, labor productivity, field supervision, site location and dollar magnitude, and range from 32 to 45% of direct costs. A "norm" factor of 1.34 has been established to represent the process module indirects for the U.S. (Gulf Coast) in mid-1968.

Bare module factor includes all the direct and indirect cost elements in the process module; and, used as a multiplier on equipment cost, it is a measure of the dollars required to integrate single or multiple pieces of equipment into a particular process circuit. This factor can vary between 2.38 and 3.64: the "norm" module indicates 2.95.

Total module factor represents all the estimated costs in the bare module plus the contingencies considered necessary to adjust for unlisted items or insufficient scope definition (10-20%) and contractor fee (usually 3-5%). The "norm" module indicates 3.48 and this includes factors of 15% contingency and 3% fee.

Solids Handling Module

The solids handling module involves heavy mechanical equipment items (see Table I). Approximately 60% of direct labor manhours is consumed by handling and setting the equipment, compared with 10-

Developing a process module using equipment cost data—Table II

$ Magnitude	300,000 Furnace	396,000 Exchangers	340,000 Vessels (V)	176,000 Vessels (H)	154,000 Pumps & Drives	660,000 Compression	174,000 Tanks	2,200,000 Total Module
Equip. fob. cost (E)	14.0%	18.0%	15.0%	8.0%	7.0%	30.0%	8.0%	100%
Piping / Concrete / Steel / Instruments / Electrical / Insulation / Paint / Field materials (m)	×1.34	×1.71	×2.03	×1.63	×1.72	×1.58	×1.20	×1.62
Direct material, $M=(E+m)$	18.8	30.78	30.45	13.04	12.04	47.40	9.60	162.
Material erection / Material setting	×0.22	×0.37	×0.47	×0.36	×0.41	×0.37	×0.11	×0.36
Direct field labor (L)	4.14	11.38	14.31	4.69	4.93	17.53	1.06	58.
Direct $M\&L$ cost	22.94	42.16	44.76	17.73	16.97	64.93	10.66	220.

Only the primary cost elements need be evaluated at this stage by the above over-all factors. The secondary elements can be filled in when necessary from the individual module factors:

Equipment fob. cost, E	14.0%	18.0%	15.0%	8.0%	7.0%	30.0%	8.0%
Piping	2.52	8.12	8.92	3.20	2.13	6.13	0.88
Concrete	1.40	0.99	1.47	0.49	0.28	3.64	0.64
Steel	.....	0.54	1.18	.....	.....	.....	.....
Instruments	0.56	1.82	1.75	0.48	0.22	2.44	.....
Electrical	0.28	0.36	0.74	0.41	2.17	4.28	.....
Insulation	.....	0.86	1.20	0.41	0.18	0.76	.....
Paint	.....	0.09	0.19	0.05	0.06	0.15	0.08
Field materials, m	4.8	12.78	15.45	5.04	5.04	17.40	1.60
Material erection	4.14	9.85	12.21	4.08	4.25	14.40	1.00
Equipment setting	Incl.	1.53	2.10	0.61	0.68	3.13	0.06

Note: All data are based on % total equipment, E

15% in the process module. Foundations and electrical work are predominant in field materials while piping is a relatively minor element.

The following is a "norm" mechanical equipment module assembled on principles similar to the process module. These factors can be used for early conceptual work provided adjustments are made for known deviations:

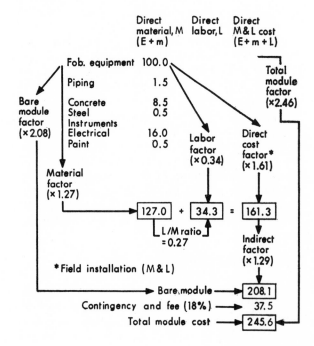

This module represents mid-1968 productivity and costs on a U.S. (Gulf Coast) jobsite. Factor relationships are on a carbon-steel basis and all factors are % total equipment.

Project indirects can be assembled on the same basis as the chemical process module. Use 10.6% of direct material cost for engineering, 52% of direct labor cost for construction overhead, and 3% for contractor fee.

It should be noted that the above modules are typical "norms" based on data available to the author. They have been used successfully to assist early capital cost predictions and bid-analysis work.

Project Indirect Costs

Once the direct material and labor costs of the project have been calculated, it is necessary to predict the indirect costs to cover work by a contractor for project management, engineering, procurement, field supervision and general overhead expenses. These costs can be related to direct $M\&L$ costs during early phases of development by indirect factors, but this account soon becomes sensitive to alloy ratios, labor/material ratios, plant location (labor productivity, field supervision, etc.) and magnitude. The indirect factor can vary considerably with each individual jobsite.

Freight, Taxes, Duties: These costs are a function of direct material dollars. For conceptual work, use 5% of direct material value for domestic freight and

3% for sales and use tax within the U.S. Overseas freight can amount to 10-12%, and import duties up to 10-15%, of the material value shipped, depending on distance and port of entry.

Construction Overhead: These costs are properly related to direct-field-labor dollars and can range from 60-80% of direct labor for construction job sites within the U.S.—up to 100% for overseas locations where additional field supervision is required and construction equipment has to be transported to grass roots areas. A "norm" value for conceptual work is established at 67.4% of direct-field-labor costs. This results in the following cost-element format and % values of equipment costs:

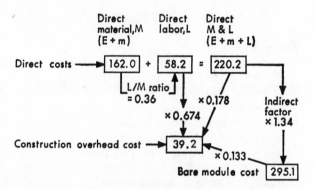

From the previous tabulation, construction overhead costs can be related to direct *M&L* dollars. This results in a 0.178 factor between construction overhead and *M&L* costs and a 0.133 factor with bare module cost. This is the "norm" case and is tabulated based on the following:

	Base Percent	% Direct Labor Cost*	% Direct M&L Cost**	% Bare Module Cost
Fringe benefits	14.8	10.0	2.6	2.0
Labor burden	22.4	15.0	4.0	2.9
Field supervision	17.8	12.0	3.2	2.4
Temporary facilities	8.9	6.0	1.6	1.2
Construction equipment	14.8	10.0	2.6	2.0
Small tools	3.6	2.4	0.6	0.5
Miscellaneous	17.7	12.0	3.2	2.3
Total construction overhead	100.0%	67.4%	17.8%	13.3%

The elements in overhead include:

Fringe Benefits: Employer's contribution to union funds for health and welfare, vacations, holidays, sick leave, retirement. Add monies for travel and subsistence when required.

Labor Burden: Employer's mandatory contributions for Federal Social Security (FICA), Federal Unemployment Insurance (FUI), State Unemployment Insurance (SUI), Workmen's Compensation.

Field Supervision: Salaries, fringe benefits and bur-

* At $1.1 million direct labor magnitude.
** On L/M ratio of 0.36.

den items for supervisory and field personnel. Add travel and subsistence when necessary.

Temporary Facilities: Buildings, roadways, parking yards, work areas, scaffolding, rigging, utilities, fencing, and miscellaneous.

Construction Equipment: Equipment rental and handling, plus freight to and from the jobsite.

Small Tools: Expendable construction tools, usually valued at $200 or under.

Miscellaneous Field Costs: Job cleanup, watchmen, minor equipment repairs, medical services, welder tests, consumable supplies, warehousing, vendor services, job insurance such as public liability (PL), public damage (PD), automobile, and all-risk.

Correction factors are necessary to adjust for other L/M ratios and $M\&L$ magnitudes. Use these charts for such corrections:

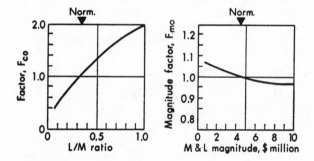

Total Construction Overhead

From the above data, total construction overhead costs can be obtained from total direct ($M\&L$) dollars using the cost equation:

Construction overhead costs, $\$ = (M\&L)\ (0.178)\ \times (F_{co})\ \times (F_{mo})$ Index

Engineering Costs

The amount of engineering manhours can vary considerably between projects and is particularly sensitive to unusual design problems, process complexities, job duplication, schedule requirements and job magnitude. These costs are basically related to the value of direct materials engineered. A "norm" value for conceptual work is about 13.6% or direct material. This results in the following cost-element format and % of equipment costs:

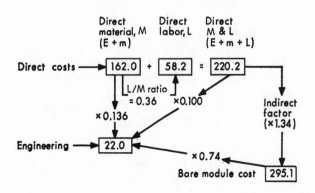

Engineering costs are related to direct *M&L* dollars for early conceptual work. Using the base factor of 0.136 on direct material will result in a 0.10 relationship between the cost of engineering and the *M&L* costs, and 0.074 relationship with bare module cost.

This is further tabulated below:

	Base Percent	% Direct Material Cost*	% Direct M&L Cost**	% Bare Module Cost
Project engineering	14.3	1.9	1.4	1.0
Process engineering	4.5	0.6	0.4	0.3
Design and drafting	26.8	3.6	2.6	1.8
Procurement	2.7	0.4	0.3	0.2
Home office construction	0.8	0.2	0.1	0.2
Engineering direct labor	49.1	6.7	4.8	3.5
Office indirects and overhead†	50.9	6.9	5.2	3.9
Total engineering cost	100.0	13.6	10.0	7.4

Correction factors are necessary to adjust for other *L/M* ratios and *M&L* magnitudes. Use these charts for such corrections:

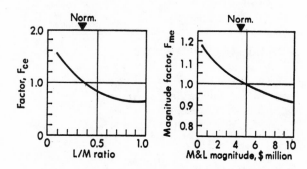

Adjusting for Other Characteristics

Adjust for other project characteristics if necessary by using the factor F_{pt} from:

Project Type	F_{pt}
Chemical complex............	1.4
Chemical processing plant......	1.0
Solids/fluid processing........	0.8
Solids handling................	0.6
Buildings only................	0.4

Total Engineering Costs

From the above data, calculate total engineering cost from direct (*M&L*) dollars by using the cost equation:

Engineering cost, $= (M\&L) \times (0.10)\ (F_{ce})\ (F_{me})\ (F_{pt}) \times$ Index

Cost Data

The charts in this section have been prepared from recent pricing data and adjusted, when necessary, to the base year (mid-1968) and base material (carbon steel). Range of the observed data is indicated by the dotted lines.

All cost data are in a constant state of change, and no dollar value can be considered absolute. Experience, judgment and analysis must be constantly applied to the use of these or any other cost data, and adjustments made to more accurately reflect the project on hand, particularly when scope and definition become available.

PROCESS EQUIPMENT

Process Furnaces

Process furnaces (Fig. 1) are used for high-capacity heat transfer to a process stream.

Structural profiles are usually "box-type" or "A-frame" with multiple tube banks and integral stacks. Location is within the process area and unit costs can vary considerably with individual designs.

The chart is based on field-erected costs for furnaces designed to elevate hydrocarbon streams to 700 F. at 500 psi. (max.) with absorbed heat duties in excess of 10 million Btu./hr. All tube banks are carbon steel except pyrolysis or reformer furnace include stainless tubes; over-all efficiency is rated at 75% with oil/gas combination burners.

Factors are provided to adjust for pyrolysis or reformer design types, alloy tube materials, and design pressures up to 3,000 psi.

Furnace costs are on a subcontract basis and include all direct materials, field labor, subcontractor overhead and profit necessary to construct the furnace on the jobsite. Add dollars for foundations, external piping, and other field costs from the installation cost modules.

Example: Predict the cost in mid-1969 of a catalytic reformer furnace designed to liberate 95 million Btu./hr. at 450 psi. Stainless-steel tubes are required in the radiant section; catalyst is expected to cost $80,000. Assume escalation to be 8%/yr. from mid-1968 and evaluate field installation and module costs for the furnace.

Base cost (from Fig. 1)	= $250,000
Adjust for material, pressure and escalation	
$F_d = 1.35$; $\quad F_p = 0.00$; Index = 1.08	
Expected cost in 1969	= (250,000) (1.35 + 0.00) (1.08)
	= $364,500
From module 1B:	
Direct cost factor	= 1.64
Direct *M&L* (CS basis)	= (250,000) (1.64)
	= 410,000
Field installation (*M&L*)	= 410,000 + (364,500 − 250,000)
	= $524,500
Bare module (CS basis)	= (250,000) (2.19)
	= 547,000

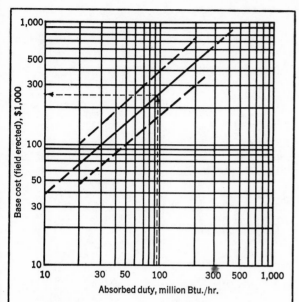

Required
Furnace type
Absorbed heat duty, Btu./hr.
Design pressure, psig.
Radiant tube material

Basis of chart
Process heater
Box or "A-frame" construction
Carbon steel tubes
Design pressure, 500 psi.
Field erected

Time base
Mid-1968

Exponent
Size exponent 0.85
$Cost_2 = cost_1 (size_2/size_1)^*$

Included
Complete field erection
Subcontractor indirects

Process Furnace Cost, $ = [Base cost$(F_d + F_m + F_p)$]Index
Pyrolysis or Reformer Furnace Cost, $ = [Base cost $(F_d + F_p)$] Index
Adjustment factors

Design Type	F_d	Radiant Tube Material	F_m*	Design Pressure, Psi.	F_p*
Process heater	1.00	Carbon steel	0.00	Up to 500	0.00
Pyrolysis	1.10	Chrome/moly	0.35	1,000	0.10
Reformer (without catalyst)	1.35	Stainless	0.75	1,500	0.15
				2,000	0.25
				2,500	0.40
				3,000	0.60

*If these factors are used individually, add 1.00 to the above values.

Field installation modules

Module	1A	1B	1C	1D	1E
Base dollar magnitude, $100,000	Up to 2	2 to 4	4 to 6	6 to 8	8 to 10
Heater erected cost, E	100.0	100.0	100.0	100.0	100.0
Piping	18.5	18.0	17.7	17.5	17.5
Concrete	10.3·	10.0	9.9	9.8	9.7
Steel	–	–	–	–	–
Instruments	4.1	4.0	3.9	3.9	3.9
Electrical	2.1	2.0	2.0	2.0	1.9
Insulation	–	–	–	–	–
Paint	–	–	–	–	–
Field materials, m	35.0	34.0	33.5	33.2	33.0
Direct material, $E + m = M$	135.0	134.0	133.5	133.2	133.0
Material erection	30.5	29.6	29.2	28.9	28.7
Equipment setting	–	–	–	–	–
Direct field labor, L	30.5	29.6	29.2	28.9	28.7
Direct M & L cost	165.5	163.6	162.7	162.1	161.7
Freight, insurance, taxes	–	–	–	–	–
Indirect cost	61.2	55.6	53.7	53.5	51.7
Bare module cost	226.7	219.2	216.4	215.6	213.4
L/M ratios	0.23	0.22	0.22	0.21	0.21
Material factor, $E + m$	1.35	1.34	1.34	1.33	1.33
Direct cost factor, M & L	1.65	1.64	1.63	1.62	1.62
Indirect factor	0.37	0.34	0.33	0.33	0.32
Module factor (norm)	2.27	2.19	2.16	2.15	2.13

Note: All data are based on 100 for equipment, E.
Dollar magnitudes are based on carbon steel.

Process furnaces—Fig. 1

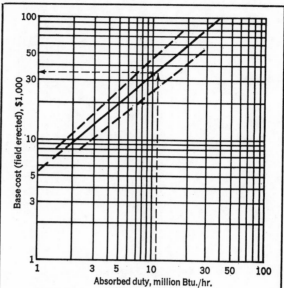

Required
Absorbed heat duty, Btu./hr.
Design pressure, psig.
Radiant tube material

Basis of chart
Process heater type
Cylindrical construction
Carbon steel tubes
Design pressure, 500 psi.

Time base
Mid-1968

Exponent
Size exponent 0.85

Included
Complete field erection
Subcontractor indirects

Fired Heater Cost, $ = [Base cost$(F_d + F_m + F_p)$]Index

Adjustment factors

Design Type	F_d	Radiant Tube Material	F_m*	Design Pressure, Psi.	F_p*
Cylindrical	1.00	Carbon steel	0.00	Up to 500	0.00
Dowtherm	1.33	Chrome/moly	0.45	1,000	0.15
		Stainless	0.50	1,500	0.20

*If these factors are used individually, add 1.00 to the above values.

Field installation modules

Module	2A	2B	2C	2D	·2E
Base dollar magnitude, $100,000	Up to 2	2 to 4	4 to 6	6 to 8	8 to 10
Heater erected cost, E	100.0	100.0	100.0	100.0	·100.0
Piping	15.5	15.0	14.8	14.6	14.6
Concrete	10.3	10.0	9.9	9.8	9.7
Steel	–	–	–	–	–
Instruments	5.1	5.0	4.8	4.8	4.7
Electrical	2.1	2.0	2.0	2.0	1.9
Insulation	–	–	–	–	–
Paint	–	–	–	–	–
Field materials, m	33.0	32.0	31.5	31.2	30.9
Direct material, $E + m = M$	133.0	132.0	131.5	131.2	130.9
Material erection	29.9	29.0	28.6	28.3	28.1
Equipment setting	–	–	–	–	–
Direct field labor, L	29.9	29.0	28.6	28.3	28.1
Direct M & L cost	162.9	161.0	160.1	159.5	159.0
Freight, insurance, taxes	–	–	–	–	–
Indirect cost	60.3	54.7	52.8	52.6	50.9
Bare module cost	223.2	215.7	212.9	212.1	209.9
L/M ratios	0.22	0.22	0.22	0.21	0.21
Material factor, $E + m$	1.33	1.32	1.31	1.31	1.31
Direct cost factor, M & L	1.63	1.61	1.60	1.59	1.59
Indirect factor	0.37	0.34	0.33	0.33	0.32
Module factor (norm)	2.23	2.16	2.13	2.12	2.10

Note: All data are based on 100 for equipment, E.
Dollar magnitudes are based on carbon steel.

Directed fired heaters—Fig. 2

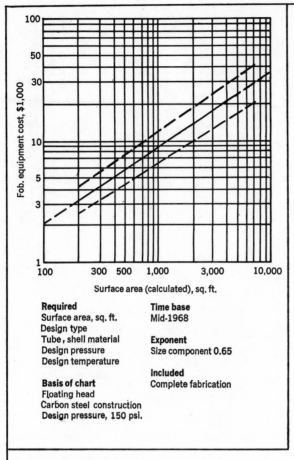

Required

Surface area, sq. ft.
Design type
Tube, shell material
Design pressure
Design temperature

Time base

Mid-1968

Exponent

Size component 0.65

Included

Complete fabrication

Basis of chart

Floating head
Carbon steel construction
Design pressure, 150 psi.

Exchanger Cost, $ = [Base cost $(F_d + F_p) \times F_m$] Index

Adjustment factors

Design Type	F_d	Design Pressure, Psi.	F_p*	
Kettle, reboiler	1.35	Up to 150	0.00	*If these factors are
Floating head	1.00	300	0.10	used individually,
U tube	0.85	400	0.25	add 1.00 to these
Fixed tube sheet	0.80	800	0.52	values.
		1,000	0.55	

Shell/Tube Materials, F_m

Surface Area, Sq. Ft.	CS/ CS	CS/ Brass	CS/ Mo	CS/ SS	SS/ SS	CS/ Monel	Monel/ Monel	CS/ Ti	Ti/ Ti
Up to 100	1.00	1.05	1.60	1.54	2.50	2.00	3.20	4.10	10.28
100 to 500	1.00	1.10	1.75	1.78	3.10	2.30	3.50	5.20	10.60
500 to 1,000	1.00	1.15	1.82	2.25	3.26	2.50	3.65	6.15	10.75
1,000 to 5,000	1.00	1.30	2.15	2.81	3.75	3.10	4.25	8.95	13.05
5,000 to 10,000	1.00	1.52	2.50	3.52	4.50	3.75	4.95	11.10	16.60

Field installation modules

Module	3A	3B	3C	3D	3E
Base dollar magnitude, $100,000	Up to 2	2 to 4	4 to 6	6 to 8	8 to 10
Equipment fob. cost, E	100.0	100.0	100.0	100.0	100.0
Piping	45.6	45.1	44.7	44.4	44.3
Concrete	5.1	5.0	5.0	5.0	5.0
Steel	3.1	3.0	3.0	3.0	3.0
Instruments	10.2	10.1	10.0	9.9	9.8
Electrical	2.0	2.0	2.0	2.0	2.0
Insulation	4.9	4.8	4.7	4.7	4.7
Paint	0.5	0.5	0.5	0.5	0.5
Field materials, m	71.4	70.5	69.9	69.5	69.3
Direct material, $E + m = M$	171.4	170.5	169.9	169.5	169.3
Material erection	55.4	54.7	54.2	53.9	53.8
Equipment setting	7.6	6.5	5.9	5.5	5.2
Direct field labor, L	63.0	61.2	60.1	59.4	59.0
Direct M & L cost	234.4	231.7	230.0	228.9	228.3
Freight, insurance, taxes	8.0	8.0	8.0	8.0	8.0
Indirect cost	86.7	78.8	75.9	75.5	73.0
Bare module cost	329.1	318.5	313.9	312.4	309.5
L/M ratios	0.37	0.36	0.35	0.35	0.35
Material factor, $E + M$	1.71	1.70	1.70	1.69	1.69
Direct cost factor, M & L	2.34	2.32	2.30	2.29	2.28
Indirect factor	0.37	0.34	0.33	0.33	0.32
Module factor (norm)	3.29	3.18	3.14	3.12	3.09

Note: All data are based on 100 for equipment, E.
Dollar magnitudes are based on carbon steel.

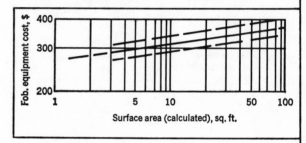

Double-pipe exchanger costs (for process requirements less than 100 sq. ft. Specify double pipe units).

Adjustment factors

Material: CS/CS = 1.0, CS/SS = 1.85
Pressure: up to 600 psi. 1.00
900 1.10
1000 1.25

Module factors

Field installation 1.35
Module factor (norm) 1.83

Shell-and-tube exchangers—Fig. 3

Bare module cost (mid-1969)	= 547,000 + (364,500 − 250,000) = $661,500
Contingency (15%)	= 99,200
Total module cost (mid-1969)	= $760,700

Direct Fired Heaters

Fired heaters (Fig. 2) are designed to elevate the temperature and entropy of process streams usually without change to the molecular structure. Designs are usually cylindrical, with vertical radiant tube banks fired by oil/gas combination burners.

The chart is based on field-erected costs for direct fired heaters with carbon-steel tubes; such cost includes all direct materials, field labor, subcontractor overhead and profit. Add dollars for foundations, external piping and other field costs from the installation modules.

Shop fabricated units can usually be purchased up to 25 million Btu./hr. Beyond this, field fabrication is necessary. Factors are provided on the chart to adjust for design type, tube materials and design pressures up to 1,500 psi.

Example: Find the current cost of a vertical direct-fired heater designed to liberate 12 million Btu./hr.

Tubes are chrome/moly and design pressure 1,000 psi. Evaluate the field installation cost for *M&L* integration of the heater.

Base cost (from Fig. 2) = $35,000
Adjust for material and pressure:
$F_d = 1.00$; $F_m = 0.45$; $F_p = 0.15$;
Expected cost in 1968 = (35,000)(1.00+0.45+0.15)
 = $56,000
Direct cost factor = 1.63
Direct *M&L* cost = (35,000) (1.63)
 (CS basis)
 = $57,000
Field installation (M&L) = 57,000+(56,000−35,000)
 = $78,000

Shell-and-Tube Exchangers

Heat transfer is essential in most process operations. The most versatile equipment for this purpose is the shell-and-tube (S/T) exchanger (Fig. 3).

Classification is defined by construction characteristics such as floating head, U-tube, fixed tubesheet or kettle reboiler. Alloys and high pressures are usually involved on the tubeside.

The chart is based on floating-head construction with carbon-steel shell and tubes. Dollars are related to surface area up to 10,000 sq. ft. Factors are included to adjust for design type, materials of construction, and design pressures up to 1,000 psi. Dollars for foundations, field materials, field labor, indirect costs are obtained from the appropriate installation modules.

Example: Establish the current cost of multiple U-tube exchanger units designed for a total of 9,000 sq. ft. of surface in a 400-psi. pressure circuit. Both shell and tube are stainless steel. Assume escalation to be 6%/yr. to mid-1969.

Base cost (Fig. 3) = $33,000
Adjust for type, pressure, material and escalation:
$F_d = 0.85$; $F_p = 0.25$; $F_m = 4.50$; Index = 1.06
Expected cost (mid-1969) = (33,000)(0.85+0.25)(4.50)(1.06)
 = $173,000

Air Cooler Costs

Air coolers are used extensively where heat rejection from a process is possible, using ambient air as a coolant. Fin-tube bundles provide maximum exchange surface within a minimum size casing—for conceptual estimating, consider the "air cooler" area to represent the calculated area divided by 15.5 (the fin-tube factor).

Fig. 4 is based on field-erected costs for individual air-cooler units* designed for 150 psi. with "aircooler areas" from 400 to 10,000 sq. ft. (6,200 to 155,000

(Con't. p. 92)

* Conceptual pricing early in project development is based on individual units; structural integration is, however, possible as project definition develops, and this can reduce costs up to 15%.

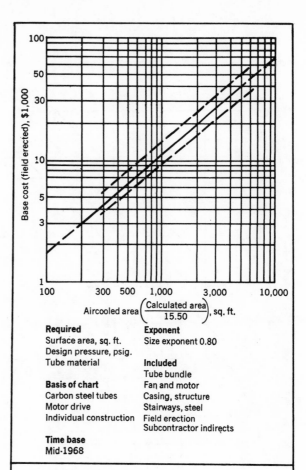

Aircooler area $\left(\dfrac{\text{Calculated area}}{15.50}\right)$, sq. ft.

Required	Exponent
Surface area, sq. ft.	Size exponent 0.80
Design pressure, psig.	
Tube material	**Included**
	Tube bundle
Basis of chart	Fan and motor
Carbon steel tubes	Casing, structure
Motor drive	Stairways, steel
Individual construction	Field erection
	Subcontractor indirects
Time base	
Mid-1968	

Air Cooler Cost, $ = [Base cost$(F_p + F_t + F_m)$]Index

Adjustment factors

Pressure Rating, Psi.	F_p	Tube Length, Ft.	F_t*	Tube Material	F_m*
150	1.00	16	0.00	Carbon steel	0.00
250	1.05	20	0.05	Aluminum	0.50
500	1.10	24	0.10	Stainless	1.85
1,000	1.15	30	0.15	Monel	2.20

*If these factors are used individually, add 1.00 to the above values.

Field installation modules

Module	4A	4B	4C	4D	4E
Base dollar magnitude, $100,000	Up to 2	2 to 4	4 to 6	6 to 8	8 to 10
Exchanger erected cost, E	100.0	100.0	100.0	100.0	100.0
Piping	15.0	14.1	13.9	13.8	13.7
Concrete	1.5	1.4	1.4	1.4	1.4
Steel	–	–	–	–	–
Instruments	4.0	3.8	3.7	3.6	3.6
Electrical	10.0	9.4	9.3	9.2	9.2
Insulation	–	–	–	–	–
Paint	0.6	0.6	0.6	0.6	0.6
Field materials, m	31.1	29.3	28.9	28.6	28.5
Direct material, E + m = M	131.1	129.3	128.9	128.6	128.5
Material erection	25.8	24.3	24.0	23.7	23.6
Equipment setting	6.0	5.2	4.8	4.5	4.4
Direct field labor, L	31.8	29.5	28.8	28.2	28.0
Direct M & L cost	162.9	158.8	157.7	156.8	156.5
Freight, insurance, taxes	8.0	8.0	8.0	8.0	8.0
Indirect cost	60.3	53.9	52.0	51.7	50.1
Bare module cost	231.2	220.7	217.7	216.5	214.6
L/M ratios	0.24	0.23	0.22	0.22	0.21
Material factor, E + m	1.31	1.29	1.29	1.28	1.28
Direct cost factor, M & L	1.63	1.59	1.58	1.57	1.56
Indirect factor	0.37	0.34	0.33	0.33	0.32
Module factor (norm)	2.31	2.20	2.18	2.16	2.14

Note: All data are based on 100 for equipment, E
Dollar magnitudes are based on carbon steel.

Air coolers—Fig. 4

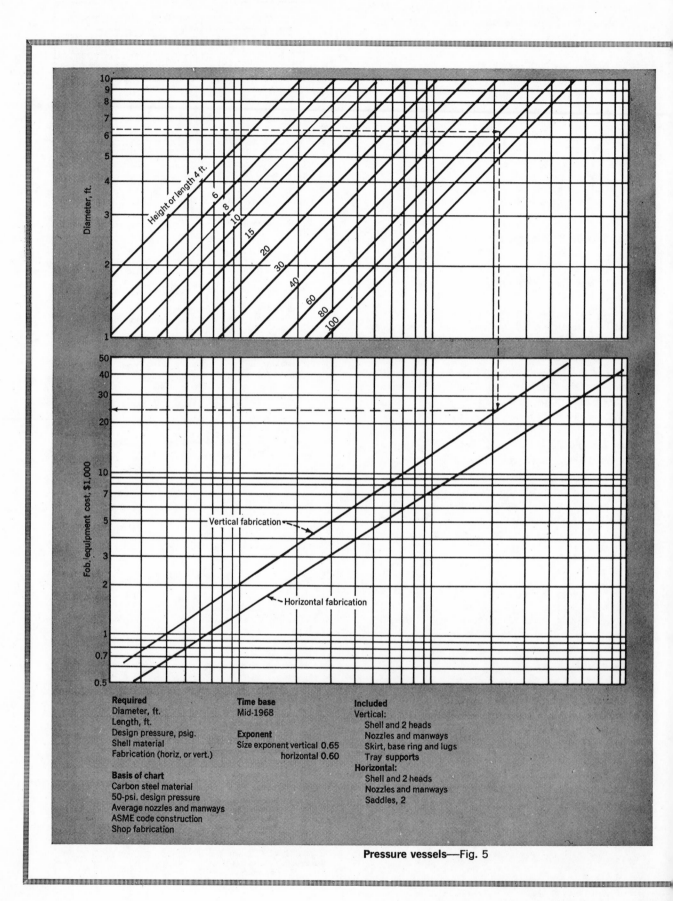

Required
Diameter, ft.
Length, ft.
Design pressure, psig.
Shell material
Fabrication (horiz. or vert.)

Basis of chart
Carbon steel material
50-psi. design pressure
Average nozzles and manways
ASME code construction
Shop fabrication

Time base
Mid-1968

Exponent
Size exponent vertical 0.65
horizontal 0.60

Included
Vertical:
Shell and 2 heads
Nozzles and manways
Skirt, base ring and lugs
Tray supports
Horizontal:
Shell and 2 heads
Nozzles and manways
Saddles, 2

Pressure vessels—Fig. 5

Process Vessel Cost, \$ $= [\text{Base cost} \times F_m \times F_p]\text{Index}$

Adjustment factors

Shell Material	F_m Clad	F_m Solid		Pressure Factor Psi.	F_p
Carbon steel	1.00	1.00	Up to	50	1.00
Stainless 316	2.25	3.67		100	1.05
Monel	3.89	6.34		200	1.15
Titanium	4.23	7.89		300	1.20
				400	1.35
				500	1.45
				600	1.60
				700	1.80
				800	1.90
				900	2.30
				1,000	2.50

Field installation modules

Vertical fabrication

Module	5A (V)	5B (V)	5C (V)	5D (V)	5E (V)
Base dollar magnitude, $100,000	Up to 2	2 to 4	4 to 6	6 to 8	8 to 10
Equipment fob. cost, E	100.0	100.0	100.0	100.0	100.0
Piping	60.0	59.6	59.5	59.4	59.3
Concrete	10.0	9.9	9.8	9.8	9.8
Steel	8.0	7.9	7.8	7.8	7.8
Instruments	11.5	11.5	11.4	11.3	11.3
Electrical	5.0	4.9	4.9	4.9	4.9
Insulation	8.0	8.0	8.0	8.0	8.0
Paint	1.3	1.3	1.3	1.3	1.3
Field materials, m	103.8	103.1	102.7	102.5	102.4
Direct material, $E + m = M$	203.8	203.1	202.7	202.5	202.4
Material erection	84.0	83.5	83.2	83.0	82.9
Equipment setting	15.2	14.9	14.0	13.5	13.2
Direct field labor, L	99.2	98.5	97.2	96.5	96.1
Direct $M \& L$ cost	303.0	301.6	299.9	299.0	298.5
Freight, insurance taxes	8.0	8.0	8.0	8.0	8.0
Indirect cost	112.0	102.5	98.9	98.7	95.5
Bare module cost	423.0	412.1	406.8	405.7	402.0
L/M ratios	0.48	0.47	0.47	0.47	0.46
Material factor, $E + m$	2.04	2.03	2.03	2.02	2.02
Direct cost factor, $M \& L$	3.03	3.02	3.00	2.99	2.98
Indirect factor	0.37	0.34	0.33	0.33	0.32
Module factor (norm)	4.23	4.12	4.07	4.06	4.02

Note: All data are based on 100 for equipment, E.
Dollar magnitudes are based on carbon steel.

Field installation modules

Horizontal fabrication

Module	5A (H)	5B (H)	5C (H)	5D (H)	5E (H)
Base dollar magnitude, $100,000	Up to 2	2 to 4	4 to 6	6 to 8	8 to 10
Equipment fob. cost, E	100.0	100.0	100.0	100.0	100.0
Piping	41.1	40.1	39.7	39.4	39.2
Concrete	6.2	6.1	6.0	5.9	5.9
Steel	—	—	—	—	—
Instruments	6.2	6.1	6.0	5.9	5.9
Electrical	5.2	5.1	5.0	5.0	5.0
Insulation	5.2	5.1	5.0	5.0	5.0
Paint	0.5	0.5	0.5	0.5	0.5
Field materials, m	64.5	63.0	62.2	61.7	61.5
Direct material, $E + m = M$	164.5	163.0	162.2	161.7	161.5
Material erection	52.2	51.0	50.4	50.0	49.8
Equipment setting	9.3	8.3	7.7	7.2	7.0
Direct field labor, L	61.5	59.3	58.1	57.2	56.8
Direct $M \& L$ cost	226.0	222.3	220.3	219.0	218.3
Freight, insurance, taxes	8.0	8.0	8.0	8.0	8.0
Indirect cost	83.6	75.6	72.7	72.3	69.8
Bare module cost	317.6	305.9	301.0	299.3	296.1
L/M ratios	0.37	0.36	0.35	0.35	0.35
Material factor, $E + m$	1.64	1.63	1.62	1.62	1.61
Direct cost factor, $M \& L$	2.26	2.22	2.20	2.19	2.18
Indirect factor	0.37	0.34	0.33	0.33	0.32
Module factor (norm)	3.18	3.06	3.01	2.99	2.96

Note: All data are based on 100 for equipment, E.
Dollar magnitudes are based on carbon steel.

Trays, packings, and linings

Packings

Raschig Rings		Size, In.		
	1	1½	2	3
Materials & Labor, \$/Cu. Ft.				
Stoneware	5.2	4.3	3.5	2.9
Porcelain	7.0	5.8	4.7	3.9
Stainless	70.2	45.8	32.5	22.8
Berl saddles	3/4	1	1½	
Stoneware	18.8	14.5	7.8	
Porcelain	20.7	15.9	8.7	

	M & L, \$/Cu. Ft.
Activated carbon	14.2
Alumina	12.6
Coke	3.5
Crushed limestone	5.8
Silica gel	27.2

Linings

	In. Thick	M & L, \$/Sq. Ft.
Acid brick	3	3.80
	4	5.50
	6	8.25
Firebrick	4½	7.16
	9	10.79
Rubber	3/16	4.37
	1/4	4.75
Refractory	2	7.50
	4	10.52
Gunite	2	3.20
	4	4.55
Chemical lead	5 lb.	6.25
	10	7.13
	15	8.86

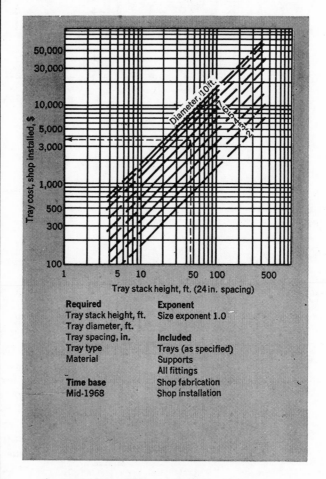

Required	Exponent
Tray stack height, ft.	Size exponent 1.0
Tray diameter, ft.	
Tray spacing, in.	**Included**
Tray type	Trays (as specified)
Material	Supports
	All fittings
Time base	Shop fabrication
Mid-1968	Shop installation

Tray Cost, \$ $= [\text{Base cost}(F_s + F_t + F_m)]\text{Index}$

Adjustment factors

Tray Spacing, In.	F_s	Tray Type	F_t*	Tray Material	F_m*
24	1.0	Grid		Carbon steel	0.0
18	1.4	(no downcomer)	0.0	Stainless	1.7
12	2.2	Plate	0.0	Monel	8.9
		Sieve	0.0		
		Trough or valve	0.4		
		Bubble cap	1.8		
		Koch Kascade	3.9		

*If these factors are used individually, add 1.00 to the above values.

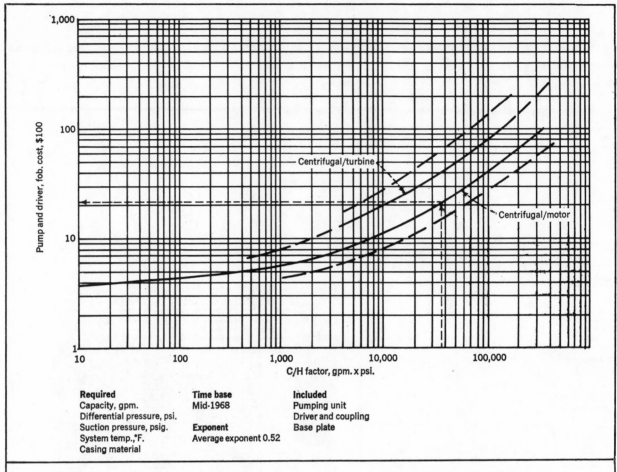

Required
Capacity, gpm.
Differential pressure, psi.
Suction pressure, psig.
System temp., °F.
Casing material

Time base
Mid-1968

Exponent
Average exponent 0.52

Included
Pumping unit
Driver and coupling
Base plate

Centrifugal Pump Cost, \$ = [Base cost × F_m × F_o]Index

Centrifugal pumps and drivers—Fig. 6

Adjustment factors

Material	F_m
Cast iron	1.00
Bronze	1.28
Cast steel	1.32
Stainless	1.93
Carpenter 20	2.10
Worthite	2.44
Hastelloy C	2.89
Monel	3.23
Nickel	3.48
Titanium	8.98

Operating Limits		Max. value	
Suction pressure, psig.	150	500	1,000
System temperature, °F.	250	550	850
Factor F_o	1.0	1.5	1.9

Field installation modules

Module	6A	6B	6C	6D	6E
Base dollar magnitude, $100,000	Up to 2	2 to 4	4 to 6	6 to 8	8 to 10
Equipment fob. cost, E	100.0	100.0	100.0	100.0	100.0
Piping	30.2	29.8	29.6	29.5	29.4
Concrete	4.0	3.9	3.9	3.9	3.9
Steel	–	–	–	–	–
Instruments	3.0	2.9	2.9	2.9	2.9
Electrical	31.0	30.5	30.3	30.3	30.2
Insulation	2.5	2.5	2.5	2.4	2.4
Paint	0.8	0.8	0.8	0.8	0.8
Field materials, m	71.5	70.4	70.0	69.8	69.6
Direct material, $E + m + M$	171.5	170.4	170.0	169.8	169.6
Material erection	60.0	59.2	59.0	58.6	58.5
Equipment setting	9.7	9.2	8.9	8.7	8.6
Direct field labor, L	69.7	68.4	67.9	67.3	67.1
Direct M & L cost	241.2	238.8	237.9	237.1	236.7
Freight, insurance, taxes	8.0	8.0	8.0	8.0	8.0
Indirect cost	89.2	81.2	78.5	78.2	75.7
Bare module cost	338.4	328.0	324.4	323.3	320.4
L/M ratios	0.41	0.40	0.40	0.40	0.40
Material factor, $E + m$	1.72	1.70	1.70	1.70	1.69
Direct cost factor, M & L	2.41	2.39	2.38	2.37	2.36
Indirect factor	0.37	0.34	0.33	0.33	0.32
Module factor (norm)	3.38	3.28	3.24	3.23	3.20

Note: All data are based on 100 for equipment, E.
Dollar magnitudes are based on carbon steel.

calculated area). Tubes and casing are carbon steel; efficiency is rated at 70%. Factors can adjust for other tube materials such as stainless and pressures up to 1,000 psi.

Cooler costs are on a subcontract basis and include all direct materials, field labor, subcontractors' overhead and profit. Add dollars for foundations, external piping and other prime contractor costs from the installation modules.

Due to the relatively rough control achieved by air coolers, it is usually necessary to install a shell-and-tube trim cooler downstream of the air cooler to achieve close temperature control. There trim coolers can cost approximately 15-20% of the air cooler dollars.

Example: Due to limited water supplies, air coolers will remove most of the heat from an exothermic process plant operating at 240 psi. Some 150,000 sq. ft. of calculated surface is required for this purpose, with an additional 40% of this area for trim control. The plant is expected to be built in late 1970. Stainless tubes, 16 ft. long, are specified, and escalation is expected to be 4%/yr. after mid-1968.

Predict the anticipated capital cost required for the air cooler facilities (end-1970).

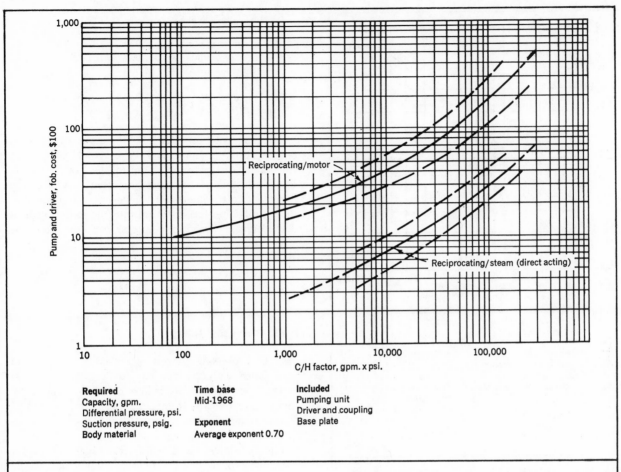

Reciprocating pumps and drivers—Fig. 7

Required | Time base | Included
Capacity, gpm. | Mid-1968 | Pumping unit
Differential pressure, psi. | | Driver and coupling
Suction pressure, psig. | **Exponent** | Base plate
Body material | Average exponent 0.70

Reciprocating Pump Cost, $ = [Base cost $\times F_m \times F_o$]Index

Adjustment factors

Material	F_m	Operating Limits			
Cast iron	1.00	Suction pressure, psi.	150	500	1,000
Bronze	1.25	System temperature, °F.	250	550	850
Cast steel	1.55				
Stainless	2.10	Factor F_o	1.0	1.2	1.4

Calculated surface area = 150,000 sq. ft.
Air cooler area = 9,700 sq. ft.
  (150,000 ÷ 15.50)
From Fig. 4:
  Base cost (mid-1968) = $70,000
Adjust for pressure, material and escalation:
  $F_p = 1.05$; $F_t = 0.00$; $F_m = 1.85$; Index = $(1.04)^2(1.02)$
Expected cost (end-1970) = $(70,000)(1.05 + 1.85)(1.04)^2(1.02)$
  = $224,000
From module 4A:
  Module factor = 2.31
  Bare module (CS basis) = (70,000)(2.31)
  = 162,000
  Bare module cost = 162,000 + (224,000 − 70,000)
  (end-1970) = $316,000
Trim cooler cost (stainless tubes)
  (20% of 40% of air = (224,000)(0.20)(0.40)
  cooler cost) = $18,000
From data Fig. 3,
  F_m(CS/SS) = 1.78
Carbon steel equivalent = 18,000/1.78
  = 10,000
From module 3A:
  Bare module factor = 3.29
  Bare module (CS basis) = (10,000)(3.29)
  = 32,900
  Bare module cost = 32,900 + (18,000 − 10,000)
  (end-1970) = $41,000

Field installation modules

Module	7A	7B	7C	7D	7E
Base dollar magnitude, $100,000	Up to 2	2 to 4	4 to 6	6 to 8	8 to 10
Equipment fob. cost, E	100.0	100.0	100.0	100.0	100.0
Piping	30.2	29.8	29.6	29.5	29.4
Concrete	4.0	3.9	3.9	3.9	3.9
Steel	–	–	–	–	–
Instruments	3.0	2.9	2.9	2.9	2.9
Electrical	31.0	30.5	30.3	30.3	30.2
Insulation	2.5	2.5	2.5	2.4	2.4
Paint	0.8	0.8	0.8	0.8	0.8
Field materials, m	71.5	70.4	70.0	69.8	69.6
Direct material, $E + m = M$	171.5	170.4	170.0	169.8	169.6
Material erection	60.0	59.2	59.0	58.6	58.5
Equipment setting	9.7	9.2	8.9	8.7	8.6
Direct field labor, L	69.7	68.4	67.9	67.3	67.1
Direct M & L cost	241.2	238.8	237.9	237.1	236.7
Freight, insurance, taxes	8.0	8.0	8.0	8.0	8.0
Indirect cost	89.2	81.2	78.5	78.2	75.7
Bare module cost	338.4	328.0	324.4	323.3	320.4
L/M ratios	0.41	0.40	0.40	0.40	0.40
Material factor, $E + m$	1.71	1.70	1.70	1.70	1.69
Direct cost factor, M & L	2.41	2.39	2.38	2.37	2.36
Indirect factor	0.37	0.34	0.33	0.33	0.32
Module factor (norm)	3.38	3.28	3.24	3.23	3.20

Note: All data are based on 100 for equipment, E.
  Dollar magnitudes are based on carbon steel.

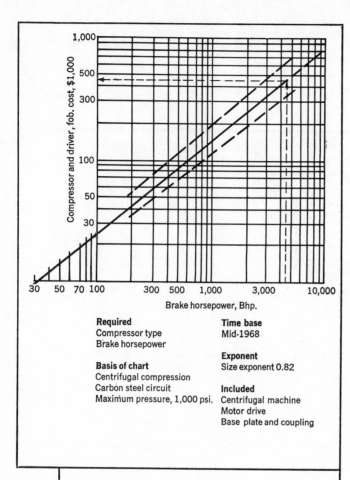

Required
Compressor type
Brake horsepower

Basis of chart
Centrifugal compression
Carbon steel circuit
Maximum pressure, 1,000 psi.

Time base
Mid-1968

Exponent
Size exponent 0.82

Included
Centrifugal machine
Motor drive
Base plate and coupling

Compressor Cost, $ = [Base cost × F_d]Index

Design Type	F_d
Centrifugal/motor	1.00
Reciprocating/steam*	1.07
Centrifugal/turbine*	1.15
Reciprocating/motor*	1.29
Reciprocating/gas engine*	1.82

*Includes interstage pots and coolers but not snubbers or other flow dampening devices.

Field installation modules

Module	8A	8B	8C	8D	8E
Base dollar magnitude, $100,000	Up to 2	2 to 4	4 to 6	6 to 8	8 to 10
Equipment fob. cost, E	100.0	100.0	100.0	100.0	100.0
Piping	20.6	20.2	20.1	20.0	19.9
Concrete	12.3	12.1	12.0	11.9	11.9
Steel	–	–	–	–	–
Instruments	8.2	8.0	8.0	8.0	8.0
Electrical	15.4	15.2	15.0	14.9	14.8
Insulation	2.6	2.6	2.5	2.5	2.5
Paint	0.5	0.5	0.5	0.5	0.5
Field materials, m	59.6	58.6	58.1	57.8	57.6
Direct material, $E + m = M$	159.6	158.6	158.1	157.8	157.6
Material erection	49.8	49.0	48.5	48.0	47.8
Equipment setting	11.6	10.9	10.7	10.5	10.4
Direct field labor, L	61.4	59.9	59.7	58.5	58.2
Direct $M \& L$ cost	221.0	218.5	217.3	216.3	215.8
Freight, insurance, taxes	8.0	8.0	8.0	8.0	8.0
Indirect cost	81.8	74.3	71.7	71.4	69.1
Bare module cost	310.8	300.8	297.0	295.7	292.9
L/M ratios	0.38	0.38	0.37	0.37	0.37
Material factor, $E + m$	1.59	1.58	1.58	1.58	1.57
Direct cost factor, $M \& L$	2.21	2.18	2.17	2.16	2.16
Indirect factor	0.37	0.34	0.33	0.33	0.32
Module factor (norm)	3.11	3.01	2.97	2.96	2.93

Note: All data are based on 100 for equipment, E.
Dollar magnitudes are based on carbon steel.

Process gas compressors and drivers—Fig. 8

Total air cooler facilities
Air cooler module	= 316,000
Trim cooler module	= 41,000
Total bare module	= 357,000
Contingency 15%	= 54,000

Total module cost = $411,000
(end 1970)

The air cooler facilities are, therefore, expected to cost $411,000 at the end of 1970, or $2.74/sq. ft. of calculated surface. According to the base data on Fig. 4, this analysis can vary between ±30% of the expected cost. There are, however, some potential savings of 15% for structural integration during final engineering. The maximum cost can, therefore, be stated as:

Maximum module cost = (411,000) (1.30 − 0.15)
= $473,000

Pressure Vessel Costs

Process pressure vessels are usually designed in accordance with the current ASME pressure vessel codes. They are usually cylindrical shells capped by two elliptical heads. Installation can be either vertical or horizontal, according to particular process requirements.

Fig. 5 relates vessel diameter, tangent height or length to fob. cost. Wall thickness is included in the pressure factors.

The base cost represents pressure vessels fabricated in carbon steel to resist 50-psi. internal pressure with average nozzles, manways and supports. Factors can be used to adjust for other shell materials, and pressures up to 1,000 psi. Add dollars for foundations, platform steel, field labor and indirect costs from the field installation modules.

Tray assemblies, packed beds, linings and other internals are priced separately and added to the shell cost as required.

Example: The physical dimensions of a distillation tower are 6-ft. 3-in. dia., and 78-ft. tangent height. Shell material is specified as stainless 316 (clad); the operating pressure is 100 psi., and 49 stainless valve trays at 12-in. spacing are required. Predict the field installation and module costs in mid-1969, assuming 4%/yr. escalation from mid-1968:

From Fig. 5:
Base cost = $24,000
Adjust for material, pressure and escalation:
F_m (clad) = 2.25; F_p = 1.05; Index = 1.04
Expected cost, mid-1969 = (24,000) (2.25) (1.05) (1.04)
= $59,000

From Table III:
Direct cost factor = 3.04
Direct $M\&L$ (CS bases) = (24,000) (3.04)
= 73,000
Field installation ($M\&L$) = 73,000 + (59,000 − 24,000)
= $108,000

Bare module factor $= 4.34$
  Bare module (CS basis) $= (24,000) (4.34) = 104,000$
Bare module cost (mid-1969) $= 104,000 + (59,000 - 24,000)$
  $= \$139,000$
Tray assembly cost:
  Stack height $= 49$ ft.
  Tray diameter $= 6$ ft., 3 in.
  Tray spacing $= 12$ in.
  Tray type $=$ Valve
  Tray material $=$ Stainless
From tray cost chart:
  Shop installed cost (base) $= \$3,800$
Adjust for spacing, type, material and escalation:
  $F_s = 2.2$; $F_t = 0.4$; $F_m = 1.7$; Index $= 1.04$
  Expected cost, mid-1969 $= (3,800) (2.2 + 0.4 + 1.7) (1.04)$
  $= \$17,000$

Centrifugal Pump and Driver

The chart on centrifugal pumps (Fig. 6) is based on fob. pump and drives suitable for general-process applications (up to 100 gpm. at 1,000 psi., or 5,000 gpm. at 20 psi.). The base cost limits suction conditions to 150 psi. at 250F. Factors are included to adjust for other materials and other operating conditions.

The capacity-head (C/H) factor provides a close correlation of the raw data. This factor is obtained by multiplying the capacity (gpm.) by total dynamic head (psi).

Add dollars for foundations, field materials, field labor and contractor indirect costs as required from the installation modules.

Example: Additional motor driven pumping facilities are being considered as part of a process plant expansion scheduled for late 1971. Process flowrate is 300 gpm., total head 250 psi., suction pressure 500 psi. Service is corrosive, requiring stainless steel materials. Two centrifugal pumps are specified with one common spare. Predict the module cost of this facility installed within the process area. Assume escalation at 6%/yr. after mid-1968.

Total pumping rate $= 300$ gpm.
  For one pump:
    Pumping rate $= 150$ gpm.
    Total head $= 250$ psi.
    C/H factor $= (150) (250)$
    $= 37,500$
From Fig. 6:
  Base cost $= \$2,100$
    For three pumps $= (2,100) (3)$
    $= 6,300$
Adjust for material, suction pressure, escalation
  $F_m = 1.93$; $F_o = 1.50$; Index $= (1.03) (1.06)^3$
For three pumping units:
Expected cost (end-1971) $= (6,300)(1.93)(1.50)(1.03)(1.06)^3$
  $= \$22,000$
From module 6A:
  Bare module factor $= 3.38$
  Bare module (CS basis) $= (6,300) (3.38)$
  $= 21,300$
  Bare module cost $= 21,300 + (22,000 - 6,300)$
    (end-1971) $= \$37,000$

Contingency 15% $= 5,600$
Total module cost $= \$42,600$
  (end-1971)

According to the base data on Fig. 6, the above costs can vary between 35% of the expected value.

Reciprocating Pump and Driver

Reciprocating pumps are usually specified for low capacities, high pressure flows and some metering applications. Direct-acting steam drives are most commonly used, although motor and turbine drives can also be specified.

Fig. 7 is based on fob. costs for direct-acting reciprocating equipment suitable for specific process applications up to 100 gpm. at 1,000 psi., or 5,000 gpm. at 20 psi. Base costs limit suction conditions to 150 psi. and 250 F.; factors are included to adjust for other materials and operating conditions.

The C/H factor should be used on a similar basis as centrifugal pumps. Add dollars for foundations, field materials, field labor and indirect costs from the installation modules.

Process Gas Compression

Process gases are required in a wide range of capacities, pressures and temperatures; process gas compression is a complex unit operation involving either centrifugal or reciprocating machines. Commercial equipment has the following general characteristics:

Centrifugal: High-capacity, low-discharge pressures usually within a range of 400 to 12,500 cu. ft./min. up to 3,000 psi.

Reciprocating: Lower capacity, high-discharge pressures usually within a range of 100 to 25,000 cu. ft./min. up to 4,000 psi.

Custom-designed equipment is usually necessary for applications exceeding the above capacities and pressures. Motor, turbine or gas engine drives can also be used.

For cost estimating purposes, consider the many variables such as volume or weight capacity (cu. ft./min. or lb./hr.), molecular weight, k-values, compression ratios as represented by the Bhp. parameter. These calculations are required before Fig. 8 can be used.

This chart is based on field-assembled costs for centrifugal machines with motor drive ranging from 200 to 5,000 Bhp. Factors are included to adjust for other compressor/driver combinations; carbon steel is assumed in all applications. Add dollars for foundations, field materials, field labor, and indirects from the installation modules as required.

Example: Two turbine-driven centrifugal compressors will circulate synthesis gas in a process plant; total horsepower requirements are calculated to be 9,000 Bhp. Equipment will be ordered in mid-1969

and installed in late 1970. For study purposes, predict the field installation and module costs of this compression facility using a project indirect factor of 1.42. Assume 6%/yr. for material escalation and 4% for labor escalation.

For one machine: Bhp. $= 4,500$
Base cost (Fig. 8) $= \$440,000$
Adjust for design type and escalation:
$F_d = 1 15;$ Index $= 1.06$
Expected cost, mid-1969 $= (440,000) (1.15) (1.06)$
$= \$536,000$
For two machines $= (536,000) (2)$
$= \$1,072,000$
From module 8E:
$= 1,072,000$

Material factor $= 1.57$
Material index $= 1.06$
Direct material cost $(E+m) = (1,072,000) (1.57) (1.06)$
$= \$1,795,000$
Labor index $= (1.02) (1.04)^2$
Direct field labor cost $(L) = (1,072,000)(0.58)(1.02)(1.04)^2$
$= \$690,000$
Field installation cost $= (1,795,000) + (690,000)$
$(E+m+L)$
$= \$2,485,000$
Indirects @ 42% $= 1,050,000$
Bare module $= 3,535,000$
Contingency @ 15% $= 530,000$
Total module cost $= \$4,065,000$

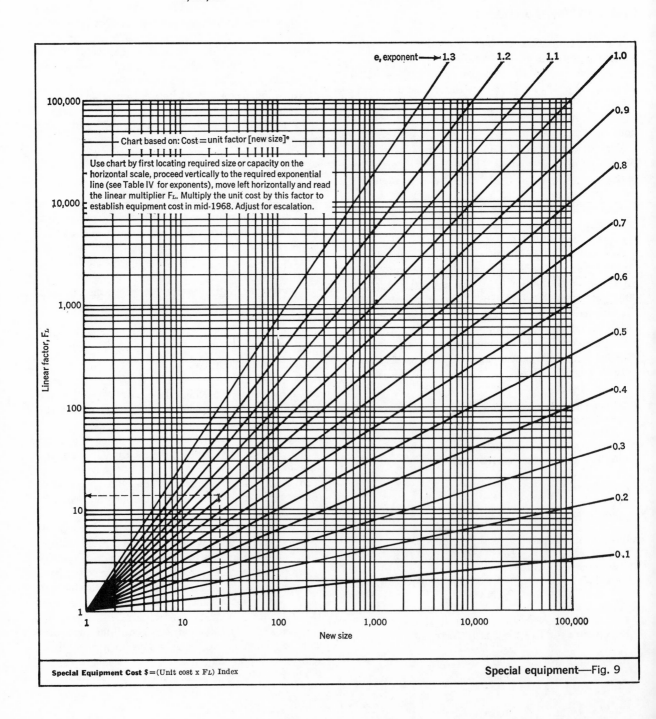

Chart based on: Cost $=$ unit factor [new size]^e

Use chart by first locating required size or capacity on the horizontal scale, proceed vertically to the required exponential line (see Table IV for exponents), move left horizontally and read the linear multiplier F_L. Multiply the unit cost by this factor to establish equipment cost in mid-1968. Adjust for escalation.

Special Equipment Cost $ $= $(Unit cost x F_L) Index

Special equipment—Fig. 9

Installation modules for single units—Table III

Module	Shell-and-Tube	Air Coolers	Process Vessel (V)	Process Vessel (H)	Pump & Driver	Compressor & Driver
Module	IIIA	IIIB	IIIC	IIID	IIIE	IIIF
Equipment fob. cost (E)	100.0	100.0	100.0	100.0	100.0	100.0
Piping	46.1	18.2	60.6	42.0	30.5	20.9
Concrete	5.1	1.9	10.0	6.3	4.0	12.4
Steel	3.1	......	8.0	......	......	......
Instruments	10.2	4.8	11.5	6.3	3.0	8.3
Electrical	2.0	12.0	5.0	5.2	31.0	15.8
Insulation	4.9	......	8.0	5.2	2.5	2.6
Paint	0.5	0.6	1.3	0.5	0.8	0.5
Field materials (m)	71.9	37.5	104.4	65.5	71.8	60.5
Direct material ($E+m$)	171.9	137.5	204.4	165.5	171.8	160.5
Material erection	55.8	31.5	84.5	53.3	60.5	50.4
Equipment setting	8.5	6.4	15.5	10.5	10.4	12.5
Direct field labor (L)	64.3	37.9	100.0	63.8	70.9	62.9
Direct $M\&L$ cost	236.2	175.4	304.4	229.3	242.7	223.4
Freight, insurance, taxes	8.0	8.0	8.0	8.0	8.0	8.0
Indirect cost	94.4	70.2	121.6	91.7	97.1	89.4
Bare module cost	338.6	253.6	434.0	329.0	347.8	320.8
L/M ratios	0.37	0.27	0.49	0.39	0.42	0.39
Material factor ($E+m$)	1.72	1.38	2.04	1.65	1.72	1.61
Direct cost factor ($M\&L$)	2.36	1.75	3.04	2.29	2.43	2.23
Indirect factor	0.40	0.40	0.40	0.40	0.40	0.40
Module factor	3.39	2.54	4.34	3.29	3.48	3.21

Note: All data are based on 100 for equipment, E. Unit is within battery limits.

SPECIAL EQUIPMENT

Data in this section apply mostly to heavy mechanical equipment associated with the solids handling module. Much of this equipment is custom designed and represents a "block of money" involving high initial costs and high labor cost for field setting. Material factors are, however, minimal.

The exponential chart technique is used here (Fig. 9) to generate all special equipment costs from the tabulation of unit costs* and size exponents (Table IV). Field installation factors (which include the cost of equipment plus foundations, electrical, paint, field labor) and labor/material (L/M) ratios are also included for evaluating installation costs.

Various process module "special equipment" items (such as agitators, ejectors, tank heaters) are also included, together with a selection of miscellaneous items found in chemical plants.

Due to some diversity in the correlations, the data in this section should be regarded as "first approximations." If the project proves to be economically sound, obtain vendor quotations as soon as appropriate design data can be developed.

Example: Find the cost of installing a bucket elevator in a fertilizer plant capable of raising 75 tons/ hr. of solid material from grade to a belt conveyor system located 25 ft. above grade. Assume this equipment will be required in mid-1970, and that escalation is predicted to be 6%/yr. Do not include revamp work.

From tabulation:

Unit cost (75 tons/hr.)	= $400/ft.
Exponent	= 0.83

From Fig. 9:

Linear factor (25 units @ 0.83)	= 14.0
Expected equipment cost (mid-1970)	= (400) (14.0) (1.06)^2
	= $6,300
Field installation ($M\&L$)	= 11,600 = 1.84
	(6,300)
Norm indirects @ 29%	= 3,400
Base module	= 15,000
Contingency (15%)	$\doteq$ 2,300
Total module cost (mid-1970)	= 17,300

The Piping Account

Piping systems are an integral part of chemical plants. They usually include a complexity of linear pipe, flanges, fittings, valves supports, etc., which interconnect the associated process equipment into an integral circiut. Installation costs ($M\&L$) are high and can vary from 50% to 75% of total equipment dollars (CS basis) depending on circuit complexity, plot layout and equipment spacing. Predicting piping

---

* This value is not a generalized unit cost. To be able to use the exponential chart, the base cost of each item is reduced to "unit cost at unity"—that is, the cost of one unit of size or capacity at unity on the size/capacity scale.

	Unit	Unit Cost,[a] $	Size Exponent	Field Installation Factor,[b] M & L	L/M Ratio
Agitators					
Propellers	Hp.	350	0.50	1.62	0.27
Turbine	Hp.	750	0.30	1.62	0.27
Air compressors (cap.)					
125 psig. (cap.)	Cu. ft./min.	2,900	0.28	1.60	0.27
Air conditioners					
Window vent	Ea.	300	.....	1.12	0.12
Floor-mounted	Ea.	200	.....	1.12	0.12
Rooftop 10 ton	Ea.	3,800	.....	1.20	0.20
20	Ea.	6,500	.....	1.20	0.20
30	Ea.	8,100	.....	1.20	0.20
Air dryers (cap.)	Cu. ft./min.	200	0.56	1.74	0.37
Bagging machines (cap.)					
Weight	Bags/min.	3,300	0.80	1.45	0.11
Volume	Bags/min.	1,000	0.80	1.45	0.11
Blenders (cap.)	Cu./ft.	850	0.52	1.61	0.27
Blowers & fans (cap.)	Cu./ft.	7	0.68	1.59	0.25
Boilers (industrial)					
15 psig	Lb./hr.	400	0.50	1.50	0.26
150	Lb./hr.	440	0.50	1.50	0.26
300	Lb./hr.	500	0.50	1.50	0.26
600	Lb./hr.	560	0.50	1.50	0.26
Centrifuges					
Horizontal basket	Dia., in.	140	1.25	1.57	0.23
Vertical basket	Dia., in.	310	1.00	1.57	0.23
Solid bowl (SS)	Hp.	1,900	0.73	1.60	0.27
Sharples (SS)	Hp.	5,200	0.68	1.60	0.27
Conveyors (length)					
Belt,† 18 in. wide	Ft.	450	0.65	1.69	0.33
24	Ft.	540	0.65	1.69	0.33
36	Ft.	620	0.65	1.64	0.28
42	Ft.	700	0.65	1.64	0.28
48	Ft.	750	0.65	1.64	0.28
Bucket (height)					
30 tons/hr. (8 in. × 5 in.)	Ft.	220	0.65	1.84	0.44
75 tons/hr. (14 in. × 7 in.)	Ft.	400	0.83	1.84	0.44
120 tons/hr. (15 in. × 8 in.)	Ft.	500	0.83	1.84	0.44
Roller, 12 in. wide	Ft.	7	0.90	1.69	0.33
15	Ft.	8	0.90	1.69	0.33
18	Ft.	9	0.90	1.65	0.29
20	Ft.	10	0.90	1.65	0.29
Screw, 6 in. dia	Ft.	230	0.90	1.59	0.25
12	Ft.	270	0.80	1.59	0.25
14	Ft.	290	0.75	1.59	0.25
16	Ft.	300	0.60	1.59	0.25
Vibrating, 12 in. wide	Ft.	80	0.80	1.64	0.28
18	Ft.	110	0.80	1.64	0.28
24	Ft.	120	0.90	1.60	0.26
36	Ft.	150	0.90	1.60	0.26
Cranes (cap.)					
Span 10 ft.	Tons	1,800	0.60		
20	Tons	2,400	0.60		
30	Tons	3,800	0.60	Field erected costs	
40	Tons	4,800	0.60		
50	Tons	6,300	0.60		
100	Tons	8,500	0.60		
Crushers (cap.)					
Cone	Tons/hr.	750	0.85	1.57	0.23
Gyratory	Tons/hr.	55	1.20	1.57	0.23
Jaw	Tons/hr.	85	1.20	1.57	0.25
Pulverizers	Lb./hr.	520	0.35	1.59	0.25
Crystallizers (cap.)					
Growth	Tons/day	5,500	0.65	1.75	0.38
Forced circulation	Tons/day	7,900	0.55	1.75	0.38
Batch	Gal.	170	0.70	1.60	0.26
Dryers (area)					
Drum	Sq. ft.	3,000	0.45	1.74	0.36
Pan	Sq. ft.	1,900	0.38	1.74	0.36
Rotary vacuum	Sq. ft.	3,100	0.45	1.74	0.36
Ductwork					
(Shop fabricated and field erected)					
Aluminum	Lin./ft.	5.42	0.55	Incl.	0.87
Galvanized	Lin./ft.	8.00	0.55	Incl.	0.84
Stainless	Lin./ft.	15.12	0.55	Incl.	0.44

plant equipment—Table IV

	Unit	Unit Cost,[a] $	Size Exponent	Field Installation Factor,[b] M & L	L/M Ratio
Dust collectors (cap.)					
Cyclones	Cu. ft./min.	3	0.80	1.69	0.32
Cloth filter	Cu. ft./min.	25	0.68	1.69	0.32
Precipitators	Cu. ft./min.	390	0.75	1.69	0.32
Ejectors (cap.)					
4 in. Hg suction	Lb./hr.	2,000	0.79	1.10	0.10
6	Lb./hr.	200	0.67	1.10	0.10
10	Lb./hr.	200	0.55	1.10	0.10
4-stage barometric					
2.5 mm. Hg suction	Lb./hr.	2,500	0.45	1.12	0.12
5.0	Lb./hr.	1,400	0.48	1.12	0.12
10.0	Lb./hr.	900	0.53	1.12	0.12
20.0	Lb./hr.	700	0.54	1.12	0.12
5-stage barometric					
0.5-mm. Hg suction	Lb./hr.	4,200	0.50	1.15	0.15
0.8	Lb./hr.	3,200	0.50	1.15	0.15
1.0	Lb./hr.	2,800	0.48	1.15	0.15
1.4	Lb./hr.	2,500	0.49	1.15	0.15
Elevators (height)					
Freight 3,000 lb	Ft.	3,600	0.32		
5,000	Ft.	4,000	0.32	Field erected costs	
10,000	Ft.	5,400	0.32		
Passenger 3,500 lb	Ft.	3,900	0.48		
Evaporators					
Forced circulation	Sq. ft.	6,000	0.70	1.90	0.35
Vertical tube	Sq. ft.	1,200	0.53	1.90	0.35
Horizontal tube	Sq. ft.	800	0.53	1.90	0.35
Jacketed vessel (glasslined)	Gal.	1,000	0.50	1.74	0.37
Filters (effective area)					
Plates and press	Sq. ft.	330	0.58	1.79	0.42
Pressure leaf-wet	Sq. ft.	410	0.58	1.79	0.42
dry	Sq. ft.	1,500	0.53	1.79	0.42
Rotary drum	Sq. ft.	1,400	0.63†	1.60	0.27
Rotary disk	Sq. ft.	1,000	0.78	1.60	0.27
Flakers (effective area)					
Drum	Sq. ft.	1,300	0.64	1.59	0.25
Generator sets (portable)					
10 kw	Ea.	1,500	.....	.....	.....
15 kw	Ea.	2,000	.....	.....	.....
25 kw	Ea.	3,000	.....	.....	.....
50 kw	Ea.	5,000	.....	.....	.....
100 kw	Ea.	7,000	.....	.....	.....
Hoppers (cap.)					
Conical	Cu./ft.	1.0	0.68	1.04	0.04
Silos	Cu./ft.	0.9	0.90	1.10	0.10
Hydraulic presses (plate area)					
100 psig	Sq. ft.	2,500	0.95	1.74	0.36
300	Sq. ft.	3,600	0.95	1.74	0.36
500	Sq. ft.	5,000	0.95	1.74	0.36
1,000	Sq. ft.	6,200	0.95	1.74	0.36
Mills (cap.)					
Ball	Tons/hr.	550	0.65	1.70	0.34
Roller	Tons/hr.	5,000	0.65	1.70	0.34
Hammer	Tons/hr.	500	0.85	1.70	0.34
Screens (surface)					
vibrating single	Sq. ft.	900	0.58	1.32	0.18
double	Sq. ft.	1,100	0.58	1.32	0.18
Stacks (height)					
24 in. (CS)	Lin./ft	25.83	1.00	1.24	0.16
36 in. (CS)	Lin./ft.	58.20	1.00	1.24	0.16
48 in. (CS)	Lin./ft.	78.25	1.00	1.24	0.16
Tank heaters (area)					
Steam coil*	Sq. ft.	94.12	0.32	1.25	0.25
Immersion	Kw.	18.75	0.85	1.20	0.20
Weigh Scales					
Portable beam	Ea.	250	.....	.....	.....
dial	Ea.	1,500	.....	.....	.....
Truck 20 ton	Ea.	4,000	.....	1.08	0.08
50	Ea.	7,200	.....	1.08	0.08
75	Ea.	8,500	.....	1.08	0.08

[a] All unit costs are based on mid-1968. These are not general unit costs.

[b] Field installation included equipment foundations, electrical, paint and field labor. (No indirects) *Stainless factor 2.4

† For enclosed conveyors walkway multiply by 2.10

costs is difficult, even when design engineering is complete and material takeoff is possible.

Circuit complexity is a measure of the "tightness" of the layout, equipment spacing or plant location, specified as follows:

Tight. Limited plot area, minimum equipment spacing, few pipe bends, maximum fittings, indoor location or limited work area (rel. factor = 1.08).

Normal. Average plot layout with central pipe rack, standard equipment spacing, maximum pipe bends and shop fabricated spools, liberal work areas (rel. factor = 1.00).

Loose. Mostly straight pipe runs, (yard and some offsites), minimum bends, fittings, and hangers, few valves, open layout (rel. factor = 0.85).

The data in Table V are useful in conceptual work to evaluate isolated piping runs that are not included in the modules (see p. 116 for discussion of piping in modules). Only lengths of pipe, installation characteristics, and valve count are necessary. Use the exponential chart and Table V to generate individual piping costs.

Example: Product storage from a process facility is located some 1,500 ft. from the battery-limit plant. Pumping rate is 500 gpm. against a friction/head resistance of 50 psi.; a 4-in.-dia. transfer line is specified. Predict the cost of the total transfer facility (carbon steel) in mid-1970 if two centrifugal pumping units are located 100 ft. within the process area. Include two line-size shutoff valves (gate) outside battery limits. Assume escalation to be 4%/yr. from mid-1968, and the contractor indirect factor of 1.40.

Within battery limits:

Pump capacity	=500 gpm.
Total head	=50 psi.
C/H factor	=(500) (50)
	=25,000
Index	=(1.04)²

From Fig. 6:

Pumping cost (E)	= (2) (1,800) (1.04)²
	=$3,900

From module 6A:

Direct cost factor	=2.41
Pump installation cost (M&L)	= (3,900) (2.41)
	=$9,400

From piping data and exponential chart:

Unit cost process piping (CS)	=1.56
Size exponent	=0.93
Linear feet	=100
Chart linear factor (for 4 in. at 0.93)	=3.75
Installed piping cost (M&L)	= (1.56) (100) (3.75) (1.04)²
	=$6,000
Total battery-limit installation cost (M&L)	= (9,400 + 6,000)
	=$15,400

Outside battery limits (offsite):

From piping data and exponential chart:

Unit cost offsite piping (CS)	=0.82
Size exponent	=1.05
Linear feet	=1,500
Linear factor (for 4 in.)	=4.25
Index	=(1.04)²
Installed offsite piping (M&L)	=$5,700

Add two 4-in. gate valves:
From piping data and exponential chart:

Unit cost (each) CS	=60
Size exponent	=1.35
Number	=2
Index	= (1.04)²
Chart linear factor (each)	=8.5
Installed valve cost (M&L)	= (60) (2) (8.5) (1.04)²
	=$1,100

Total piping installation cost

Outside battery limits (M&L)	= (5,700 + 1,100)
	=$6,800

Total transfer facility:

Within battery limits	=15,400
Outside battery limits	=6,800
Total field installation (M&L)	=22,200
Indirects @ 40%	=8,900
Bare module	=31,100
Contingency @ 15%	=4,600

Total expected transfer module =$35,700
mid-1970

SITE DEVELOPMENT

Most jobsites require extensive preparation before foundation work and equipment setting can start. The extent and cost of this preparation depends on the amount of surface clearing, rock blasting, grading, excavation piling, etc., specified.

A normal jobsite assumes good soil conditions, average grading, excavation, and minimum piling requirements. The checklist on Table I and the items included in Table VI indicate the most likely activities to be accounted for in most conceptual cost estimates.

Quantity data are not always available in early conceptual work; in this case, assign reasonable M&L dollar allowances for each appropriate item. This work varies extensively with each jobsite; survey

(Con't. p. 103)

Piping system unit costs—Table V

	Unit	Unit Cost,[a] $	Size Exponent	Field Installation Factor	L/M[b] Ratio
Linear pipe					
Carbon steel	Lin. ft.	0.26	1.20	....	.....
Chrome/moly	Lin. ft.	1.02	1.20	....	.....
Stainless	Lin. ft.	1.51	1.20	....	.....
Offsite piping					
Carbon steel	Lin. ft.	0.82	1.05		0.64
Chrome/moly	Lin. ft.	2.29	1.05		0.18
Stainless	Lin. ft.	3.26	1.05		0.12
Process piping					
Carbon steel	Lin. ft.	1.56	0.93		0.78
Chrome/moly	Lin. ft.	4.18	0.93	Included	0.22
Stainless	Lin. ft.	5.85	0.93		0.15
Valves[c]					
Check	Ea.	50	1.35		0.03
Gate	Ea.	60	1.35		0.03
Plug	Ea.	70	1.35		0.03
Globe	Ea.	74	1.35		0.03
Control	Ea.	240	1.05		0.05

Table includes direct material and field labor (M&L). Multipliers (if required): average insulation 1.05, steam tracing and insulation 1.10.

[a] At unity only, on the horizontal scale. Not a general unit cost.

[b] "Norm" circuit complexity.

[c] Multipliers: CS = 1.00; chrome/moly = 1.95; stainless = 3.69.

Costs for site development—Table VI

	Unit	Min.	Norm.	Max.	M&L L/M Ratio
Dewatering and drainage					
Pumping system (rented)..	Day	25	32	40	.....
Wellpoint dewatering system..........	Month	6,500	7,500	8,500	.....
Drainage trench........	Lin. ft.	0.75	0.85	0.95	.....
Fencing					
Complete fence (light)...	Lin. ft.	1.34	1.88	2.42	0.32
Complete fence (heavy).	Lin. ft.	1.51	2.13	2.75	0.32
Chain link.....	Lin. ft.	5.48	5.93	6.38	0.48
Gates 6 ft. (light)........	Ea.	55.50	67.50	79.50	0.15
(heavy).......	Ea.	69.25	82.50	95.75	0.15
(chain)........	Ea.	105.65	128.00	150.65	0.15
Corner posts..	Ea.	31.50	32.00	32.50	0.19
Fire protection					
Pumps Firehouse Firetrucks (2)	Allowance	100,000	150,000	200,000	.....
Land surveys and fees					
General surveys and fees.	% total cost	4.0	9.0	14.0	.....
Soil tests......	Ea.	300	400	500	.....
Landscaping					
General.......	Sq. yd.	1.50	1.70	1.90	0.90
Piling					
Wood (untreated)...	Lin. ft.	1.70	2.15	2.60	2.20
Wood (creosoted)...	Lin. ft.	2.15	2.60	2.90	2.50
Concrete: precast......	Lin. ft.	6.75	7.00	7.25	0.50
cast in place..	Lin. ft.	4.75	6.62	8.50	1.25
Steel pipe (concrete filled)........	Lin. ft.	7.50	9.50	11.50	0.25
Steel section..	Lin. ft.	7.40	8.50	9.50	0.27
Sheet piling, steel.........	Sq. ft.	1.45	2.60	3.75	0.20
wood.........	Sq. ft.	1.25	1.75	2.25	0.35
Pile driver setup........	Ea.	6,800	7,500	8,200	.....
Roads, walkways, paving					
aving 4-in.-thk. reinf., 6-in. subbase....	Sq. yd.	6.35	7.87	8.39	1.75
6-in.-thk. reinf., 6-in. subbase....	Sq. yd.	7.61	9.37	10.13	1.75
2-in.-thk. asphalt top, existing base	Sq. yd.	2.37	3.12	3.87	0.22
2-in.-thk. asphalt top, 4-in. subbase.......	Sq. yd.	3.58	4.68	5.78	0.22
3-in.-thk. asphalt top, 12-in. subbase.......	Sq. yd.	6.37	7.62	8.87	0.22
Gravel surface 2-in.-thk. gravel.......	Sq. yd.	0.33	0.58	0.83	0.52
4-in.-thk. gravel.......	Sq. yd.	0.55	0.87	1.19	0.52
6-in.-thk. gravel.......	Sq. yd.	1.00	1.38	1.76	0.52
Parking lots					
Black-top surface.....	Sq. yd.	5.30	6.25	7.54	0.45
Sewer facilities					
Asbestos Cement pipe (general)......	Lin. ft.	4.55	4.85	5.15	0.38
Concrete pipe (reinforced) 18 in. dia.....	Lin. ft.	5.65	5.80	5.95	0.38
36 in..........	Lin. ft.	14.75	15.96	17.23	0.39
72 in..........	Lin. ft.	50.33	52.18	54.03	0.40
Vitrified clay piping 18 in. dia.....	Lin. ft.	7.55	7.80	8.45	0.86
24 in..........	Lin. ft.	15.15	16.10	17.05	0.88
36 in..........	Lin. ft.	33.95	36.20	38.95	0.89
Septic tank (45,000 gal.)..	Ea.	......	7,500	......	0.05
Site clearing, excavation and grading					
Site preparation Machine cuts.........	Cu. yd.	0.50	0.56	0.63	0.30
Clearing and grubbing....	Sq. yd.	0.13	0.15	0.18	.....
General grading.....	Sq. yd.	0.63	0.44	0.48	.....
Final leveling	Sq. yd.	0.25	0.31	0.38	.....
Foundation excavation					
Machine excavation..	Cu. yd.	1.50	1.63	1.75	0.58
Machine plus hand trim........	Cu. yd.	2.50	3.44	3.75	0.90
Hand work..	Cu. Yd.	7.56	10.00	12.50	.....
Trench excavation					
Machine 3½ ft. deep × 2 ft. wide.....	Lin. ft.	0.38	0.44	0.50	0.50
4 ft. × 3 ft...	Lin. ft.	0.56	0.63	0.68	0.50
Machine 4½ × 4 ft.	Lin. ft.	1.12	1.13	1.25	0.50
Machine 5 ft. × 5 ft......	Lin. ft.	1.38	1.50	1.63	0.50
Hand labor..	Cu. yd.	8.75	10.12	13.75	.....
Trench shoring					
Sheeting.....	Sq. ft.	1.25	1.52	1.75	1.12
Trench and foundation backfill					
Machine plus hand trim........	Cu. yd.	1.44	1.56	1.68	0.90
Hand labor only........	Cu. yd.	5.79	6.25	6.75	.....
Miscellaneous materials					
Sand........	Cu. yd.	3.05	4.80	5.55	.....
Gravel......	Cu. yd.	1.50	2.25	3.00	.....
Dirt fill......	Cul yd.	1.30	2.15	3.00	.....
Crushed stone.......	Cu. yd.	2.55	4.37	5.19	.....

Unit costs for offsite facilities—Table VII

	Unit	Field Installation, $ M&L Min.	Norm.	Max.	L/M Ratio
Air systems					
Instrument air					
Compression facilities, air dryer, air receiver, and distribution......	$M	18.75	43.75	62.50	0.80
Plant air					
Compression facilities, air receiver and distribution...............	$M	12.50	31.25	50.85	0.80
Blowdown and flare					
For general purposes (including flare lines, blowdown drum and disposal pit)..............	$M	81.75	102.58	187.52	0.45
Cooling tower and CW distribution					
Use 1.15 design factor on estimated throughput.					
Cooling tower costs....		See cost data section			
Distribution systems					
For general purposes	Gpm.	12.58	36.50	43.25	0.85
River intake installation					
For general purposes	Gpm.	8.22	16.25	24.37	1.12
Fireloop and hydrants					
For general purposes..	$M	12.58	22.50	40.24	0.80
Fuel systems					
Fuel oil (includes pumps, storage, piping, controls and distribution)..................	$M	6.25	25.12	43.75	0.85
Fuel gas (includes receiver, piping, controls and distribution).......	$M	12.50	37.52	62.58	0.85
General water systems					
Treated water					
Filtered and softened	Gal.	0.15	0.23	0.30	
Distilled.............	Gal.	0.65	0.92	1.20	
Drinking & service water					
General facilities.....	$M	2.50	5.40	7.58	
Power generation and distribution					
Use 1.10 design factor on estimated consumption.					
Generating facilities....	Kw.	See cost data Section			
Electrical distribution					
For general purposes	Kw.	87.5	93.75	98.75	0.75
Main transformer stations					
Three phase, 60 cycle					
Capacity 3,000 kva.	Kva.	33.0	37.0	44.0	
5,000	Kva.	20.0	23.0	26.0	
10,000	Kva.	13.0	14.0	16.0	
20,000	Kva.	10.0	12.0	13.0	
Secondary transformer stations					
4,200/575 v.					
600 kva...........	Kva.	30.1	33.8	42.3	
1,000................	Kva.	20.1	25.2	31.5	
1,500................	Kva.	15.6	19.5	24.3	
2,000................	Kva.	14.8	18.5	23.2	
13,200/575 v.					
600 kva...........	Kva.	28.2	35.3	44.2	
1,000................	Kva.	21.2	26.5	33.1	
1,500................	Kva.	16.6	20.8	26.2	
2,000................	Kva.	15.4	19.3	24.1	
Receiving, shipping, storage					
Automotive					
Forklift trucks					
3,000 lb..........	Ea.		7,800		
5,000 lb..........	Ea.		11,000		
10,000 lb..........	Ea.		16,200		

	Unit	Field Installation, $ M&L Min.	Norm.	Max.	L/M Ratio
Pallet truck					
Hydraulic 4,000 lb.	Ea.		930		
Electric 4,000 lb...	Ea.		3,600		
Payloaders					
2 cu. yd. (gas).....	Ea.		21,000		
4 cu. yd. (gas).....	Ea.		33,700		
2 cu. yd. (diesel)...	Ea.		22,900		
4 cu. yd. (diesel)...	Ea.		36,500		
Tank trailers					
Carbon steel.......	Ea.		14,800		
Aluminum.........	Ea.		21,600		
Stainless..........	Ea.		36,500		
Tractors					
Gasoline..........	Ea.		12,500		
Diesel.............	Ea.		27,500		
Tractor shovel					
2-cu. yd. bucket...	Ea.		26,300		
3-cu. yd. bucket...	Ea.		28,700		
4-cu. yd. bucket...	Ea.		35,600		
Automotive shipping facilities					
One outlet per 2,000 bbl.			9,800		0.85
Docks and wharves					
Light construction					
2-in. deck..........	Sq. ft.	5.15	5.63	6.25	0.45
3-in................	Sq. ft.	6.25	6.87	7.50	0.45
Medium construction					
3-in................	Sq. ft.	8.75	9.38	10.12	0.45
4-in................	Sq. ft.	10.15	11.25	12.50	0.45
Heavy construction					
4-in................	Sq. ft.	12.50	15.62	18.75	0.45
concrete...........	Sq. ft.	17.50	21.25	25.25	0.45
Dredging					
General operations...	Cu. yd.	4.32	10.81	17.28	
Tankage					
General.............		See cost data section			
Railroad					
Straight track (railroad siding)......	Lin. ft.		26.25		0.58
Turnout.............	Ea.		2,800		0.10
Bumper.............	Ea.		790		0.12
Blinker and gate.....	Ea.		9,300		0.15
Grade and ballast....	Lin. ft.		6.25		0.95
Locomotives (battery)					
9 ton............	Ea.		35,000		
12 ton............	Ea.		41,800		
Locomotives (diesel)					
1½ tons...........	Ea.		11,000		
3 tons.............	Ea.		14,000		
Tank car (10,000 gal.)	Ea.		10,800		
Railroad shipping facilities					
One outlet per 2,000 bbl........	Ea.		4,800		0.85
Steam generation and distribution					
Use 1.10 design factor on estimated consumption.					
Package boilers (up to 150,000 lb./hr.).....	Lb./hr.	See cost data section			
Field erected (above 150,000 lb./hr.).....	Lb./hr.	See cost data section			
Steam distribution					
For general purposes	Lb./hr.	0.94	1.52	1.68	0.85
Yard lighting and communications					
For general purposes..	$M	18.75	52.25	93.75	0.75
Yard transfer lines and pumps					
For general purposes..	$M	17.25	31.25	56.25	0.65

reports should be consulted as soon as a particular jobsite is chosen.

The normal unit costs in Table VI represent total direct material and labor costs for each item. Use the *L/M* ratio to break out the material and labor component values for *M&L* integration if necessary.

INDUSTRIAL BUILDINGS

Typical process plant buildings are listed in Table I; unit costs vary considerably due to the wide diversity of architectural forms, materials of construction, and functional layouts. It is usual for a local subcontractor to erect these buildings since he is familiar with area practices and labor conditions for this class of work.

The data in this section present range of *M&L* unit costs from which normal building modules can be assembled during the conceptual phase. Use 30% of *M&L* to represent subcontractor indirects (including overhead and profit); otherwise the prime contractor indirects will apply to an *M&L* integration.

To provide some flexibility, certain standardized specification characteristics have been established:

Low Cost. Prefabricated light-steel frame and roof, transite or metal sheet walls, concrete floor, minimum furnishing and accommodations inside.

Average. Prefabricated medium steel frame and trusses, good brick or concrete walls, concrete floor and roof, average furnishing and accommodations inside.

High Cost. Custom-designed heavy-steel frame, heavy sidings, face brick on concrete block walls, concrete floor and roof, expensive finishing and accommodations inside.

Base Shell Cost

The following data represent average unit cost ranges within, but not confined to, the above specifications for *single story* building shells. Foundations and concrete floor slab are included:

| | Base Height, Ft. | Shell M&L Cost,* $/Sq. Ft. Floor Area | | |
		Low	Average	High
Administration offices	10	4.26	6.88	9.51
Cafeterias	12	2.21	5.85	8.48
Compressor houses (with bridge crane)	20	3.02	4.73	5.66
Control house (equipped)	10	3.55	5.85	9.36
Garages	15	1.82	2.89	4.43
Maintenance shops	20	2.49	4.09	4.93
Laboratories and medical	10	5.32	7.67	10.15
Process buildings	20	2.58	4.09	7.34
Warehouses	20	2.29	3.57	4.83

* Included: Foundations, concrete floor slabs, structural frame, outside walls or sidings, interior partitions, finishing and paint. No indirects or piling.

Adjustment Factors

Use the following factors to adjust the base shell cost:

Number of Stories (Multiple of Base Height)	F_n
1	1.0
2	1.4
3	1.9
4	2.5

Foundations	F_f*
Concrete floors, spread footings, min. depth (norm)	0.00
Concrete floor, 10-ft.-deep piers to footings	0.10
Concrete floor, supporting structure on piles (piles not included)	0.15
Concrete floor and structure on piles (piles not included)	0.35

Roof structure	F_{ro}*
Flat roof (norm)	0.00
Steel truss	0.10
Flat roof and monitor	0.15
Sawtooth roof construction	0.30

Industrial Building Services

Add dollars for services and equipment as indicated in the following data:

| Building Services, F_s | Cost Range, $/Sq. Ft. Effective Floor Area | | |
	Low	Average	High
Air conditioning	3.75	5.30	7.00
Lighting and electrical:			
Process buildings, cafeterias	1.50	1.75	2.00
Offices, laboratories, medical, control, compressor houses	2.25	2.50	2.75
Warehouses, maintenance shops	0.70	0.90	1.10
Heating and ventilating	1.00	1.50	2.00
Plumbing (general)	1.21	1.70	2.18
Fire prevention equipment (includes alarms, extinguishers and sprinkling systems)	0.90	1.10	1.30

| Furniture and equipment, F_e | Cost Range, $/Sq. Ft. Effective Floor Area | | |
	Low	Average	High
Laboratory equipment	8.00	16.00	25.00
Office equipment	3.00	5.00	7.00
Shop equipment	4.00	6.00	8.00
Cafeteria equipment	3.50	4.50	5.50

The data in this section generate total direct *M&L* costs; integration with other modules at this level requires a material and labor breakdown. This is done by using *L/M* ratios:

	Low	Average	High
L/M Ratio	0.45	0.34	0.32

* If these factors are used individually, add 1.00 to the above values.

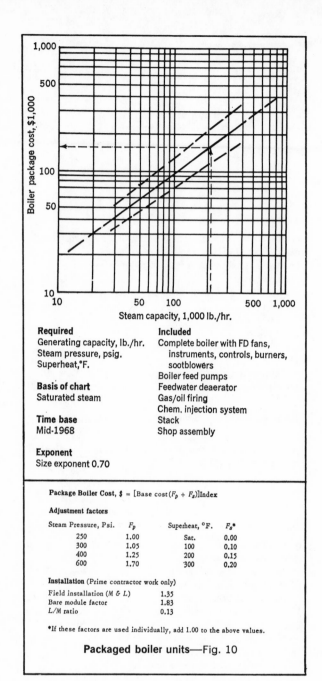

Required
Generating capacity, lb./hr.
Steam pressure, psig.
Superheat, °F.

Included
Complete boiler with FD fans,
    instruments, controls, burners,
    sootblowers
Boiler feed pumps

Basis of chart
Saturated steam

Feedwater deaerator
Gas/oil firing
Chem. injection system

Time base
Mid-1968

Stack
Shop assembly

Exponent
Size exponent 0.70

Package Boiler Cost, $ = [Base cost $(F_p + F_s)$]Index

Adjustment factors

Steam Pressure, Psi.	F_p	Superheat, °F.	F_s*
250	1.00	Sat.	0.00
300	1.05	100	0.10
400	1.25	200	0.15
600	1.70	300	0.20

Installation (Prime contractor work only)

Field installation ($M \& L$)	1.35
Bare module factor	1.83
L/M ratio	0.13

*If these factors are used individually, add 1.00 to the above values.

Packaged boiler units—Fig. 10

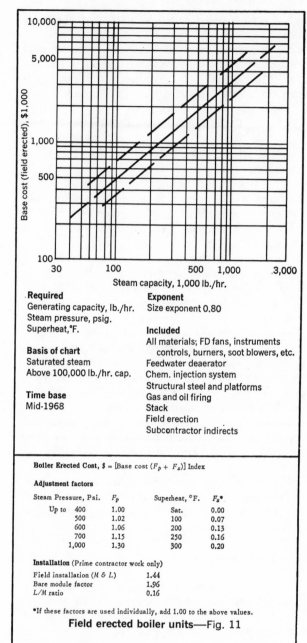

Required
Generating capacity, lb./hr.
Steam pressure, psig.
Superheat, °F.

Exponent
Size exponent 0.80

Basis of chart
Saturated steam
Above 100,000 lb./hr. cap.

Included
All materials; FD fans, instruments
    controls, burners, soot blowers, etc.
Feedwater deaerator
Chem. injection system
Structural steel and platforms

Time base
Mid-1968

Gas and oil firing
Stack
Field erection
Subcontractor indirects

Boiler Erected Cost, $ = [Base cost $(F_p + F_s)$] Index

Adjustment factors

Steam Pressure, Psi.	F_p	Superheat, °F.	F_s*
Up to 400	1.00	Sat.	0.00
500	1.02	100	0.07
600	1.06	200	0.13
700	1.15	250	0.16
1,000	1.30	300	0.20

Installation (Prime contractor work only)

Field installation ($M \& L$)	1.44
Bare module factor	1.96
L/M ratio	0.16

*If these factors are used individually, add 1.00 to the above values.

Field erected boiler units—Fig. 11

Building Indirect Costs

If complete module costs are required for sub-contracted buildings, use indirect factor of 1.30 on direct $M\&L$ value; otherwise assume project indirect factor for prime contractor operations. In each case, integrate at the bare-module cost level.

Total Building Costs

The above data can be assembled into a variety of building modules, using the cost equation:

Building cost =
[(Base shell) $\underbrace{(F_n + F_f + F_{ro}) + (F_s + F_e)]}_{\text{Direct } M\&L \text{ cost}} \times$ [Indirect cost]

Use 1.30 for indirect cost if building subcontract is anticipated.

Example: A grassroots process-plant complex requires the following building facilities, equipped and fully serviced:

	No. of Stories	Specification	Ground Floor Area, Sq. Ft.
Control house	1	Average	600

Establish the "norm" cost of the above facilities (on a subcontract basis) in mid-1970, assuming building escalation of 4%/yr. from mid-1968. No air conditioning is required. Assume normal foundations and flat roof.

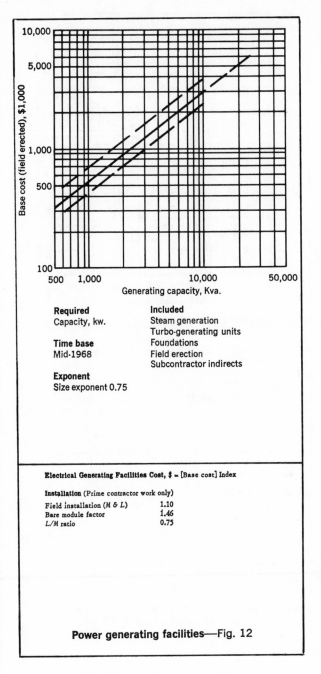

Required
Capacity, kw.

Time base
Mid-1968

Exponent
Size exponent 0.75

Included
Steam generation
Turbo-generating units
Foundations
Field erection
Subcontractor indirects

Electrical Generating Facilities Cost, \$ = [Base cost] Index

Installation (Prime contractor work only)

Field installation (M & L)	1.10
Bare module factor	1.46
L/M ratio	0.75

Power generating facilities—Fig. 12

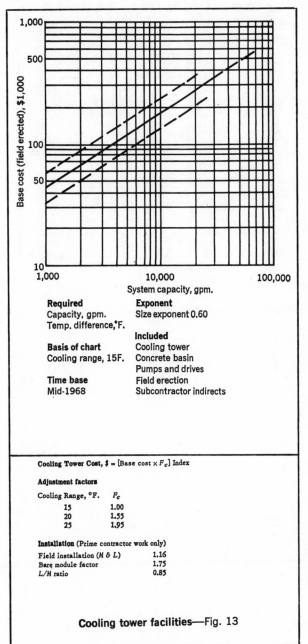

Required
Capacity, gpm.
Temp. difference,°F.

Basis of chart
Cooling range, 15F.

Time base
Mid-1968

Exponent
Size exponent 0.60

Included
Cooling tower
Concrete basin
Pumps and drives
Field erection
Subcontractor indirects

Cooling Tower Cost, \$ = [Base cost × F_c] Index

Adjustment factors

Cooling Range, °F.	F_c
15	1.00
20	1.55
25	1.95

Installation (Prime contractor work only)

Field installation (M & L)	1.16
Bare module factor	1.75
L/M ratio	0.85

Cooling tower facilities—Fig. 13

$F_n = 1.00$; $F_f = 0.00$; $F_{ro} = 0.00$; Index $= (1.04)^2$

Shell cost	$= (600)(5.85)(1.04)^2$	$= \$3,800$
Lighting and electrical	$= (600)(2.50)(1.04)^2$	$= 1,600$
Heating	$= (600)(1.50)(1.04)^2$	$= 1,000$
Total services		$= 2,600$
Total direct $M\&L$		$= 6,400$
Indirect @ 30%		$= 1,900$
Bare module		$= 8,300$
Contingency @ 15%		$= 1,200$
Subcontractor module cost, mid-1970		$= \$9,500$

Chemical Plant Structures

Steel structures are necessary in most chemical plants to support elevated equipment and to provide general access for operation and maintenance. Cost of these structures varies considerably according to layout, magnitude of deadload, process considerations, etc. During early conceptual work, elevation requirements can usually be determined from a quick analysis or knowledge of the process characteristics; if this is not available use a nominal 5-10% of total equipment cost (CS basis).

As the project develops, sketch layouts can be made to indicate plant area and height; at this stage use the following data to evaluate these structures within three major specification groups:

Light. Not more than 50 ft. high, mostly access and platform areas, minor equipment support and stairways, open construction.

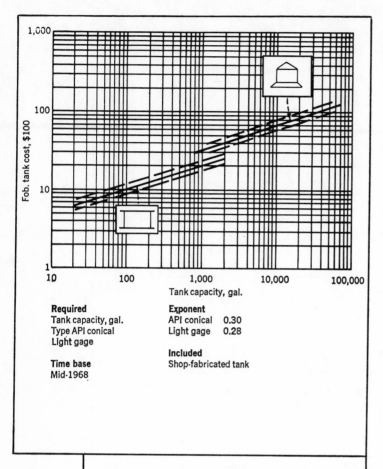

Required
Tank capacity, gal.
Type API conical
Light gage

Time base
Mid-1968

Exponent
API conical 0.30
Light gage   0.28

Included
Shop-fabricated tank

Storage Tank Cost, $ = [Base cost × F_m]Index

Adjustment factors

Material	F_m	Installation (Prime contractor work only)	
Carbon steel	1.00	Field installation (M & L)	1.47
Aluminum	1.40	Bare module factor	1.96
Rubber lined	1.48	L/M ratio	0.23
Lead lined	1.55		
Stainless	3.20		
Glass lined	4.25		

Note: Consider glass lining limited to 10,000 gal. (max.).

Storage tanks up to 40,000 gal.—Fig. 14

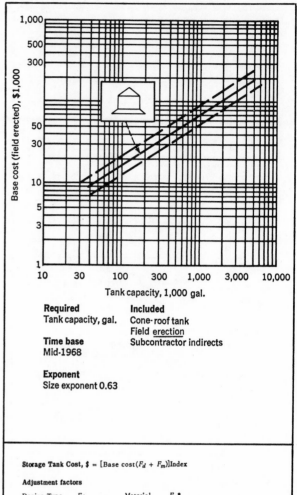

Required
Tank capacity, gal.

Time base
Mid-1968

Exponent
Size exponent 0.63

Included
Cone-roof tank
Field erection
Subcontractor indirects

Storage Tank Cost, $ = [Base cost(F_d + F_m)]Index

Adjustment factors

Design Type	F_d	Material	F_m*
Cone roof	1.00	Carbon steel	0.00
Floating roof	1.06	Rubber lined	0.48
Lifter roof	1.25	Lead lined	0.55
		Stainless	2.20

Installation (Prime contractor work only)

Field installation (M & L)	1.85
Bare module factor	2.52
L/M ratio	0.21

*If these factors are used individually, add 1.00 to the above values.

Storage tanks above 40,000 gal.—Fig. 15

Medium. Structures up to 200 ft. high, designed for average elevated equipment supports, platform areas and stairways, open construction used for most of the structure.

Heavy. Structures up to 300 ft., designed for heavy equipment supports, crane or catalyst loading structures, some covered construction (metal sheeting with potential wind loads), outside freight elevator structure. (Elevator not included).

Steel Structure Cost Factors
Field Installation Factors,
F_s, $/Cu. Ft.

Height, Ft.	Light	Medium	Heavy
Up to 10	0.05	0.20	0.40
10 to 20	0.10	0.29	0.52
20 to 30	0.15	0.35	0.62
30 to 40	0.18	0.40	0.70
40 to 50	0.20	0.44	0.76
50 to 60		0.48	0.80
60 to 70		0.52	0.86
70 to 80		0.54	0.90
80 to 90		0.58	0.93
90 to 100		0.60	0.95
100 to 200		0.80	1.42
200 to 300			1.60

Steel structure cost (M&L),

$$\$ = \{(\text{Area} \times \text{height}) \times F_s\} \times \text{Index}$$

Installation: Field installation (M&L) included in F_s; Indirect factor ("norm") = 1.34; L/M ratio = 0.60

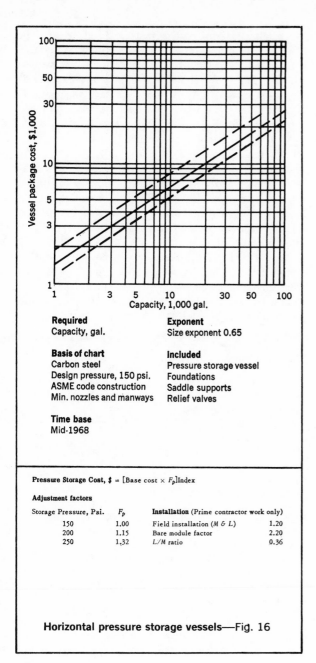

Required
Capacity, gal.

Exponent
Size exponent 0.65

Basis of chart
Carbon steel
Design pressure, 150 psi.
ASME code construction
Min. nozzles and manways

Included
Pressure storage vessel
Foundations
Saddle supports
Relief valves

Time base
Mid-1968

Pressure Storage Cost, $ = [Base cost × F_p]Index

Adjustment factors

Storage Pressure, Psi.	F_p	Installation (Prime contractor work only)	
150	1.00	Field installation (M & L)	1.20
200	1.15	Bare module factor	2.20
250	1.32	L/M ratio	0.36

Horizontal pressure storage vessels—Fig. 16

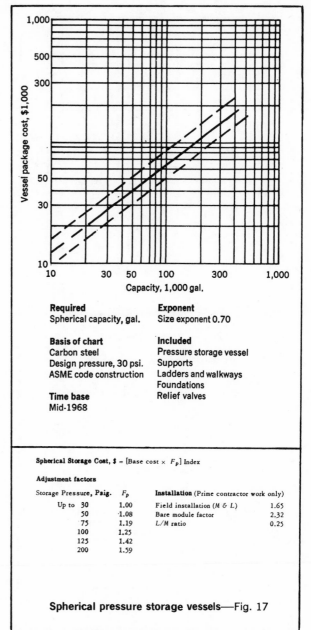

Required
Spherical capacity, gal.

Exponent
Size exponent 0.70

Basis of chart
Carbon steel
Design pressure, 30 psi.
ASME code construction

Included
Pressure storage vessel
Supports
Ladders and walkways
Foundations
Relief valves

Time base
Mid-1968

Spherical Storage Cost, $ = [Base cost × F_p] Index

Adjustment factors

Storage Pressure, Psig.		F_p	Installation (Prime contractor work only)	
Up to	30	1.00	Field installation (M & L)	1.65
	50	1.08	Bare module factor	2.32
	75	1.19	L/M ratio	0.25
	100	1.25		
	125	1.42		
	200	1.59		

Spherical pressure storage vessels—Fig. 17

Included: steel structure, equipment supports, checker plate, stairways, handrails, concrete footings (no piling or concrete substructures), shop paint, field labor, no indirects. Not included: local pipe supports, tower ladders and platforms.

Use of the data will generate average field installation cost (M&L) for steel structures within the above specifications. Integrate at the direct cost (M&L) level or add 27% for subcontractor indirects to obtain bare module cost.

Example: An ethylene plant is being considered for construction in mid-1970. It will require a "cascade" refrigeration process with a series of evaporators and let-down drums in a gravity flow circuit. An open steel structure will support the equipment and provide operating and access areas from a height of 60 ft. above grade. If the plan area is 100 ft. x 50 ft., predict module cost of this structure for heavy loading assuming escalation at 4%/yr. from mid-1968.

Area	= (100) (50) sq. ft
Height	= 60 ft.
Index	= $(1.04)^2$
From tabulated data:	
F_s (60 ft. heavy)	= 0.80
Field installed structure (M&L)	= (5,000) (60) (0.8) $(1.04)^2$
	≃ 260,000
Indirects @ 27%	= 70,000
Bare module	= 330,000
Contingency, 10%	= 33,000
	————
Subcontractor module cost, mid-1970	= $363,000

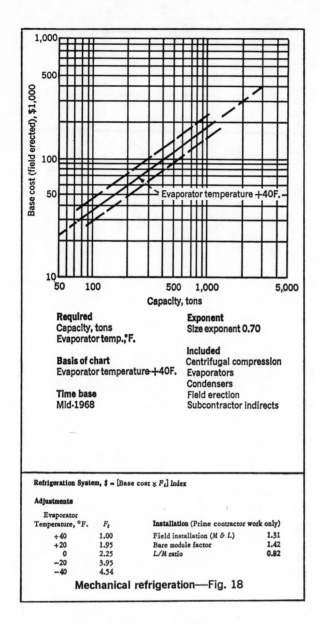

Required

Capacity, tons
Evaporator temp., °F.

Basis of chart
Evaporator temperature +40F.

Time base
Mid-1968

Exponent
Size exponent 0.70

Included
Centrifugal compression
Evaporators
Condensers
Field erection
Subcontractor indirects

Refrigeration System, $ = [Base cost × F_t] Index

Adjustments

Evaporator Temperature, °F.	F_t	Installation (Prime contractor work only)	
+40	1.00	Field installation (M & L)	1.31
+20	1.95	Bare module factor	1.42
0	2.25	L/M ratio	0.82
−20	3.95		
−40	4.54		

Mechanical refrigeration—Fig. 18

Average system costs (including offsite piping, electrical, instrumentation, and other bulk materials) can be generated from the Table VII unit cost ranges. This is suitable for conceptual normal estimates until more-precise data are available for a particular project. Use the L/M ratios to obtain direct material and labor components for M&L integration.

Figs. 10-18 present cost/capacity charts for some off-site facilities.

Example: A process complex requires an additional boiler plant capable of generating 200,000 lb./hr. of saturated steam at 400 psi. The new facilities will be located in an existing boiler house outside the process area. Predict the cost of this installation (1) as part of a current revamp program with 42% indirects, (2) as an individual installation with new distribution system late in 1969. Assume escalation to be 4%/yr. from mid-1968.

Base cost (Fig. 10) = $150,000. Adjust for pressure and escalation to end-1969.

$F_p = 1.25$; $F_s = 0.00$; Index = 1.02

Estimated boiler cost (end-1968)	= (150,000) (1.25) (1.02)
	= $191,000
Field installation (M&L)	= (191,000) (1.35)
	= 258,000
Revamp indirects @ 42%	= 108,000
Bare module cost	= $366,000
Contingency 15%	= 55,000
Total module cost (revamp)	= $421,000

$F_p = 1.25$; $F_s = 0.00$; Index = 1.04

Estimated boiler cost (end-1969)	
	= 202,000
Field installation (M&L)	= (202,000) (1.35)
	= 273,000
Distribution system (M&L); Table VII	
(200,000) (0.94) (1.02) (1.04)	= 199,000
Direct costs (M&L)	= 472,000
"Norm" indirects @ 34%	= 160,000
Bare module installation	= 632,000
Contingency 15%	= 95,000
Total module cost (end-1969)	= $727,000

OFFSITE FACILITIES

Offsite facilities are major equipment items and utility systems generally located outside the process battery limits. These items are listed on Table I and the checklist in Table VII. Major elements such as steam generators, power generators, cooling towers, tankage, etc., are presented as subcontractor base costs on cost/capacity charts, together with prime-contractor installation modules. The use of this data will generate direct material and labor costs that can be integrated at the M&L level.

Meet the Author

Kenneth M. Guthrie is chief estimator and cost analyst in the Engineering Div., Technical Group, of W. R. Grace & Co. (7 Hanover Sq., New York City). His responsibilities include capital cost estimating, cost analysis and cost control. Previously associated with Mobil Oil Corp. and M. W. Kellogg Co., Mr. Guthrie has over ten years of experience in cost engineering. He is a member of AIChE and the American Assn. of Cost Engineers.

3

Costs of Process Equipment and Other Capital Items

COSTS OF PROCESS EQUIPMENT

. . . For More-Accurate Chemical Plant Estimation

Based on the author's estimating notebook, here is a comprehensive compilation of equipment cost data for estimating expenditures for process plants. All information comes from recent vendor quotes, covering a wide variety of equipment used in chemical plants.

H. E. Mills, General Chemical Div., Allied Chemical Corp.

Chemical engineers, whether in research and development, design, or production, are no strangers to the problems of estimating the cost of process equipment or chemical plants.

Many engineers maintain their own private collection of cost data, and use this information for quick, preliminary estimates of capital investments. However, unless the engineer is devoting most of his time to cost estimating, his files tend to become a somewhat disorganized collection of published data, company information and quotes from vendors.

With this in mind, a comprehensive study of a wide variety of chemical plant items has been made. Pertinent information concerning each item, obtained from an average of vendors' quotations are presented here in charts and tables. They form a usable basis or "norm" for estimating chemical plant costs. This collection of cost data includes many items commonly used to make order-of-magnitude or preliminary chemical plant cost estimates. Although this list is not to be considered exhaustive, many items of equipment are included that can be used for preparing more-detailed and specific estimates.

Included in this collection are items relating to manufacturing equipment,

fabrication, piping, instrumentation, power and electrical, building and construction, and yard improvements. The items are defined as to purchase cost, capacity, horsepower, size, materials of construction, to avoid chance of error in interpretation and application. Reasonable installation time is given to avoid guesswork in determining total estimated cost.

It should be stressed that the data represent a "norm." There may be some condition or circumstance, especially in the area of labor, that deviates from this norm; in such a case, an empirical factor should be applied to the base cost to take care of the deviation.

For instance, consider the man-hr. shown on the cost chart for installing a 4-ft. x 4-ft. ball mill (all "hours to install" on the charts are in man-hr.). As for all items, these hours are for equipment setting only, and not for concrete foundations or supporting steel. They represent the time to bring the equipment into the plant, place it on the foundation, align and secure, align the drive, fit the liner, and connect the feed end. To these hours, a factor must be applied to cover any unusual condition. For example, a percentage could be added if this mill were placed in an out-of-the-way corner of a third floor, or a percentage deducted if it were placed outside on clear, level ground.

Location or topography of the job site, season of the year, etc., may at times cause a deviation from the norm for labor or material or both. Again, a factor should be applied to the basic cost.

In view of this, it should be noted that no set of cost data should be accepted per se. This collection is intended and should be considered as a firm base adjustable to any particular deviation.

All data are based on a Marshall and Stevens index of 238.8.

PROCESSING

AGITATORS (Open tanks)

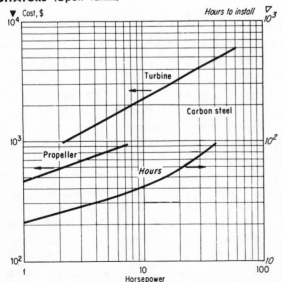

Multiply cost by following for:

Type 304 stainless steel	1.12	Rubber covered	1.36
Type 316 stainless steel	1.17	Monel alloy	1.37

Motor included. Factors apply for open and closed tanks.

AGITATORS (Closed tanks)

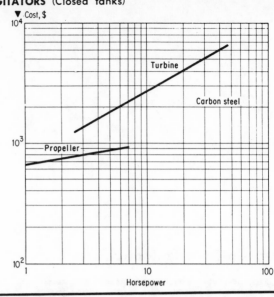

BLOWERS (heavy duty, industrial type)

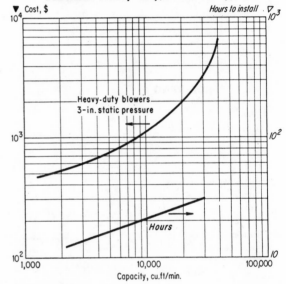

Cu. Ft. Air/Min.	Hp.
2,000	3
20,000	20
40,000	30

Motor and drive not included.

CRUSHING AND GRINDING: BALL MILLS

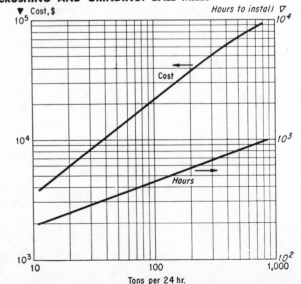

(Ball Mills)

Size, Ft.	Tons/24-Hr.*	Hp.	Weight Mill, Lb.	Weight, Ball Charge, Lb.
3x3	13	10	8,500	2,800
4x4	36	25	20,000	6,500
5x5	71	50	29,000	12,500
6x6	152	125	50,000	21,600
7x6	230	200	80,000	29,400
8x8	450	300	125,000	51,000
9x7	626	400	150,000	60,000

* Crushing material to ½-in. to 48 mesh.

Ball charge: $190/ton.
Liner, motor, drive and guard, included.

CRUSHING AND GRINDING: CONE CRUSHERS

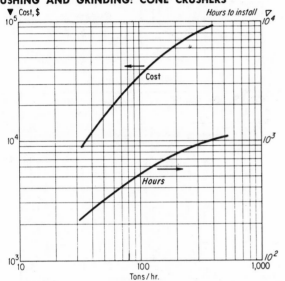

Size, Ft.	Tons/ Hr.*	Hp.	Weight, Lb.	Size, Ft.	Tons/ Hr.*	Hp.	Weight Lb.
2	40	30	12,000	5½	230	200	90,000
3	75	75	25,000	7	400	300	175,000
4	125	150	56,000				

* Approx. 1-in. product.

Motor, drive and guard, included.

CRUSHING AND GRINDING: GYRATORY CRUSHERS

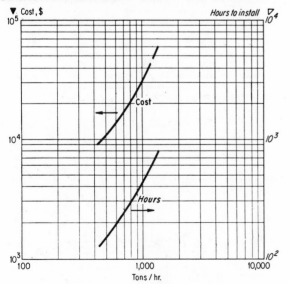

Size, Ft.	Tons/Hr.	Hp.
2 1/3	600	75
3	900	125
4	1,200	200

Motor and drive included.

CRUSHING AND GRINDING: JAW CRUSHERS

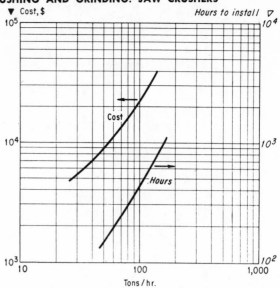

Jaw Opening, In.	Tons/Hr. (Approx.)	Hp.	Weight, Lb.	Size, In.	Tons/Hr. (Approx.)	Hp.	Weight, Lb.
15x24	50	50	14,500	21x36	80	75	25,000
18x24	50	50	15,000	25x40	90	100	36,500
18x36	80	60	21,000	32x40	100	125	47,000

Motor and drive included.

CRUSHING AND GRINDING: PULVERIZERS

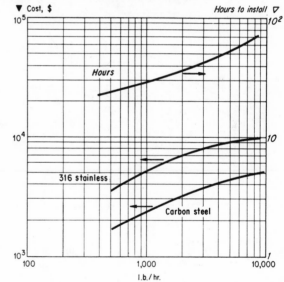

Size	Lb./Hr.	Hp.	Size	Lb./Hr.	Hp.
1	300	3	6	2,100	25
2	600	5	7	3,600	30
3	800	7½	8	4,800	40
4	1,200	10	9	6,000	50
5	1,800	15	10	7,200	60
			11	8,500	75

Legs for units 1 thru 5: $50
Stands for units 6 thru 11: $150
Add 15% for explosion-proof construction
Flexible coupling: $200

Motor and drive not included.

CRUSHING AND GRINDING: ROLL CRUSHERS

Size, In.	Tons/ Hr.*	Hp.	Weight, Lb.	Size, In.	Tons/ Hr.*	Hp.	Weight, Lb.
24x14	20	25	8,500	42x16	50	45	30,000
30x14	30	30	14,000	62x24	120	65	88,000
36x16	40	40	18,800				

* Approx. 1¼-in. product.

Motor and drive included.

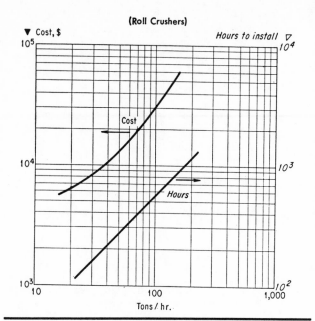

(Roll Crushers)

▼ Cost, $ Hours to install ▽

Cost

Hours

Tons / hr.

DRYING EQUIPMENT: PAN DRYERS

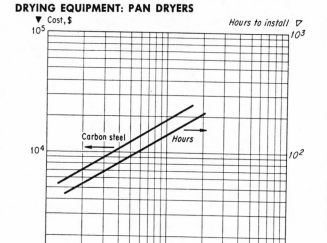

▼ Cost, $ Hours to install ▽

Carbon steel   Hours

Drying surface, sq.ft.

— Size —		Drying		— Size —		Drying	
Dia., Ft.	Depth, In.	Surface, Sq. Ft.	Hp.	Dia., Ft.	Depth, In.	Surface, Sq. Ft.	Hp.
3	18	15	3	8	36	90	25
4	24	25	5	10	36	130	40
6	24	50	10				

Cost Multipliers

20% 304 stainless-steel clad, 1.25; 20% 316 stainless clad, 1.60

Motor and drive included.

DRYING EQUIPMENT: ROTARY VACUUM DRYERS (Horizontal, indirect, jacketed)

— Size —		Drying Area,		— Size —		Drying Area,	
Dia. In.	Length Ft.	Sq. Ft.	Hp.	Dia. In.	Length Ft.	Sq. Ft.	Hp.
18	6	50	5	24	24	140	15
24	12	70	7½	30	24	190	25
18	12	110	10	44	15	230	30

Cost Multipliers

20% 304 stainless-steel clad, 1.25; 20% 316 stainless clad, 1.60

Motor and drive included.

(Rotary Vacuum)

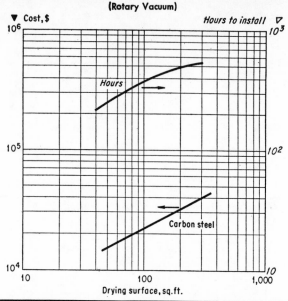

▼ Cost, $ Hours to install ▽

Hours

Carbon steel

Drying surface, sq.ft.

DUST COLLECTOR (Cloth bay with electric motor shaker)

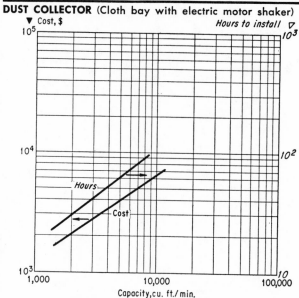

▼ Cost, $ Hours to install ▽

Hours

Cost

Capacity, cu. ft./ min.

Capacity, Cu. Ft./Min.	Hp.	Capacity, Cu. Ft./Min.	Hp.
1,000	1	4,000	10
1,500	2	5,000	10
2,000	3	6,000	15
2,500	5	8,000	20
3,000	7½	10,000	20

Motor and drive included. Add 40% for continuous operation.

FILTERS: CENTRIFUGALS (Continuous solid bowl)

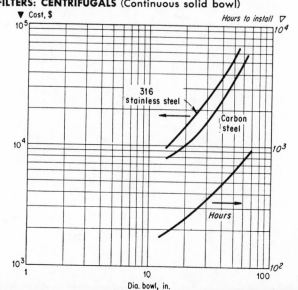

▼ Cost, $ Hours to install ▽

316 stainless steel

Carbon steel

Hours

Dia. bowl, in.

(Centrifugals) Bowl Dia., In.	Solid Holding Cap., Cu. Ft./Min.	Hp.	Weight, Lb.
18	0.6	15	3,000
24	1.3	25	5,500
36	3.0	60	12,000
40	10.0	75	17,000
54	40.0	150	32,000

Motor and drive not included.

HEAT EXCHANGERS (Steel shell)

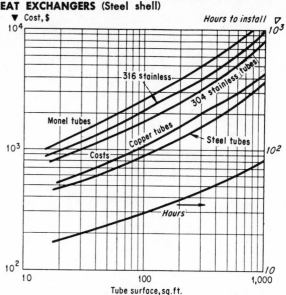

FILTERS: VERTICAL LEAF PRESSURE (Wet cake)

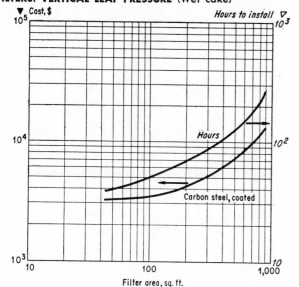

Dia., In.	No. Leaves	Spacing of Leaves, In.	Filter Area, Sq. Ft.	Dia., In.	No. Leaves	Spacing of Leaves, In.	Filter Area, Sq. Ft.
18	7	1¾	25	36	17	2½	200
24	11	1⅞	50	42	20	2½	270
30	13	1⅞	130	48	23	2½	350

Cost Multipliers

304 stainless steel, 1.15; 316 stainless, 1.25; dry cake filter, 1.20

HEAT EXCHANGERS (Stainless-steel shell and tubes)

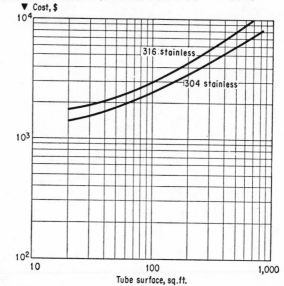

Straight tube, fixed tubesheet (Two passes on tube side, one pass on shell side Exchangers are 150 psi. For 300 psi, multiply costs from above charts by 1.3.

PUMPS: GENERAL PURPOSE CENTRIFUGAL (Single and two stage, single suction)

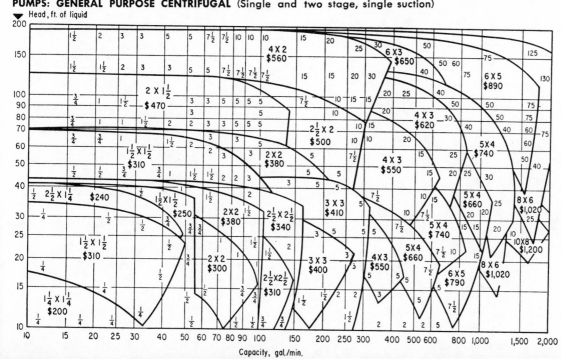

Cost Multipliers

Cast iron, bronze fitted, 1.00; bronze, 1.26; cast steel, 316 stainless fittings, 1.50; 316 stainless, 1.80; Worthite, 1.90; Hastelloy alloy C, 2.80.

Small numbers within selection blocks indicate approx. hp.

Price includes pump, steel base, and coupling. No motor.

PUMPS: INSTALLATION TIME (Man-hr. to install pump and motor)

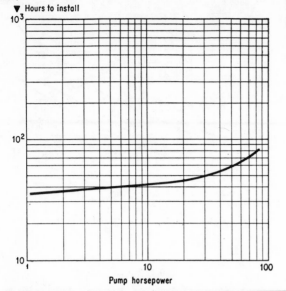

▼ Hours to install

Pump horsepower

(Screens)

Size, Ft.	No. Decks	Hp.	Weight, Lb.	Size, Ft.	No. Decks	Hp.	Weight, Lb.
1x3	1	½	400	3x6	1	2	1,300
	2	¾	550		2	3	1,700
1½x3	1	½	600	4x8	1	3	2,400
	2	¾	675		2	5	3,400
2x4	1	¾	800	5x10	1	5	3,600
	2	1	1,100		2	7½	5,200

REACTORS: GLASS-LINED, JACKETED

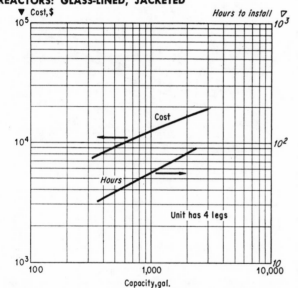

▼ Cost,$ Hours to install ▽

Cost

Hours

Unit has 4 legs

Capacity, gal.

Capacity, Gal.	Agitator Hp.	Speed, Rpm.	Capacity, Gal.	Agitator Hp.	Speed, Rpm.
500	5	120	1,000	10	120
750	7½	100	2,000	15	125

STACKS

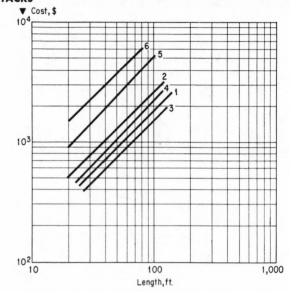

▼ Cost, $

Length, ft.

Curve No.	
1	24-in. dia. Redwood, with hoops, shipped, set up.
2	36-in. Redwood, with hoops, shipped, set up.
3	24-in. carbon steel, delivered.
4	36-in. carbon steel, delivered.
5	24-in. 304 stainless steel, delivered.
6	36-in. 304 stainless steel, delivered.

SCREENING EQUIPMENT: VIBRATING SCREENS

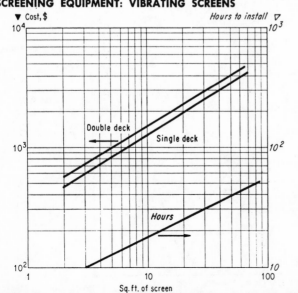

▼ Cost, $ Hours to install ▽

Double deck

Single deck

Hours

Sq. ft. of screen

STORAGE TANKS: VERTICAL

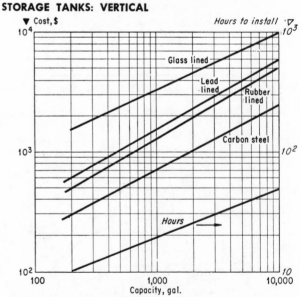

▼ Cost, $ Hours to install ▽

Glass lined

Lead lined

Rubber lined

Carbon steel

Hours

Capacity, gal.

Description: Cone roof, 18-in. manhole, one 4-in. nozzle, one 6-in. nozzle four 2-in. nozzles.

Cost Multipliers (times 304 stainless cost, next page)

Monel, 1.60; Inconel, 1.80; Nickel, 1.85.

(Storage tanks: vertical cont'd. on next page.)

(Vertical Tanks)

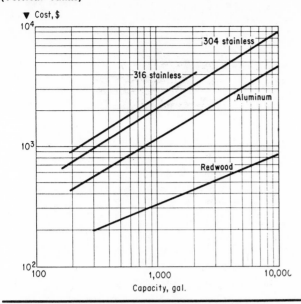

STORAGE TANKS: LARGE VERTICAL

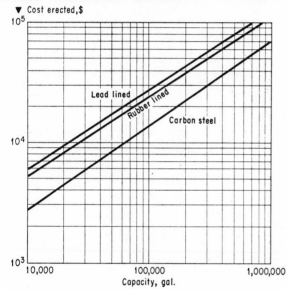

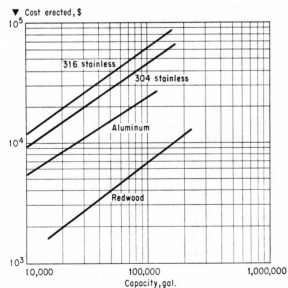

Description: Cone roof, 18-in. manhole, one 4-in. nozzle, one 6-in. nozzle four 2-in. nozzles.

Cost Multipliers (times 304 stainless cost)

Monel, 1.60; Inconel, 1.80; Nickel, 1.85.
Prices are for erected tanks not including foundations.

STORAGE TANKS: HORIZONTAL

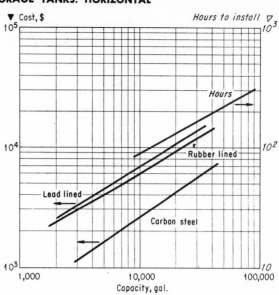

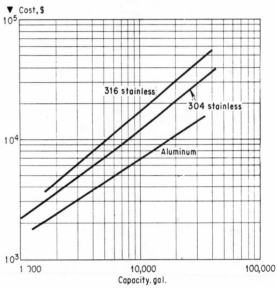

Description: 18-in. manhole, one 4-in. nozzle, one 6-in. nozzle, four nozzles, flanged and dished heads.

Cost Multipliers (times 304 stainless cost)

Monel, 1.5; Inconel, 1.8

FABRICATION

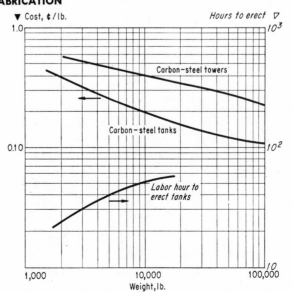

117

Steel, 1.0; 304 stainless, 4.0; 347 stainless, 4.8; 316 stainless, 5.2; 310 stainless, 5.8; Monel, 7.0; Inconel and nickel, 8.2.

Bubble Cap Trays (Carbon Steel)

Dia., In.	Weight, Lb.	Cost, $	Dia., In.	Weight, Lb.	Cost, $
12	100	90	48	400	220
24	200	110	60	500	330
36	300	140			

Multiply by 2.5 for 304 stainless trays.

Weight of Steel Plate

Thickness, In.	Weight Lb./Sq. Ft.	Thickness, In.	Weight Lb./Sq. Ft.
3/16	7.65	1/2	20.40
1/4	10.20	9/16	22.95
5/16	12.75	5/8	25.50
3/8	15.30	3/4	30.60
7/16	17.85	7/8	35.70
		1	40.80

Weight of Welded Steel Nozzles

Dia., In.	Weight, Lb. 150 Psi.	Weight, Lb. 300 Psi.	Dia., In.	Weight, Lb. 150 Psi.	Weight, Lb. 300 Psi.
1	3	4	10	75	110
1½	5	8	12	105	155
2	7	9	14	135	220
2½	10	13	16	155	282
3	12	17	18	190	350
3½	16	20	20	233	484
4	19	25	24	304	585
4½	21	28	26	350	710
6	33	50	28	380	840
8	48	75	30	415	920

Work Example

Estimate cost of a vessel, 10½-ft. dia., 8-ft. high; no covers; ⅜-in. steel plate; four 3-in., 150-lb. nozzles; two 6-in., 150-lb. nozzles; one 18-in. manhole.

Under fabrication, the area of the shell (p. 141) is:

32.99 × 8	= 263.92 sq. ft.
And the flat head	= 86.59
Total sq. ft.	**= 350.51**

Weight of steel plate is:
⅜-in., 15.3 lb./sq. ft. × 350.5 sq. ft. = 5,355 lb.

Weight of nozzles:

4 × 12 lb.	=	48
2 × 33 lb.	=	66
1 × 190 lb.	=	190
Total	**=**	**5,659 lb.**
Add for misc.	=	541 lb.
Total	**=**	**6,200 lb.**

From fabrication chart for steel tanks, $0.25/lb. × 6,200 lb. = $1,550

PACKING FOR TOWERS

Rashig Rings, Stoneware Dia., In.	$/Cu. Ft.	Berl Saddles, Stoneware Dia., In.	$/Cu. Ft.
1	4.60	½	24
1½	3.70	¾	16
2	3.20	1	9.60
3	2.70	1½	7.20

Cost Multipliers

Porcelain, 1.2; steel, 2.0; stainless steel, 25.0; Karbate, 10.0.

Misc. Packing Materials	$/Cu. Ft.	Misc. Packing Materials	$/Cu. Ft.
Activated carbon	12	Crushed limestone	3
Alumina	11	Resin	68
Calcium chloride	3	Silica gel	25
Coke	1.50		

LININGS FOR TANKS AND TOWERS

Acid Brick Thickness, In.	$/Sq. Ft. Installed	Fire Brick Thickness, In.	$/Sq. Ft. Installed
2½	2.75	4½	5.50
4½	4.70	9	9.00
9	10.00		

Chemical Lead Weight, Lb.	$/Sq. Ft. Installed	Rubber Thickness, In.	$/Sq. Ft. Installed	
			Field	Shop
8	4.20	3/16	4.20	3.30
10	5.00	1/4	4.70	3.80
12	6.00	Manways $30 ea.		
16	8.00	Nozzles $10 ea.		

Miscellaneous Linings

	$/Sq. Ft. Installed		$/Sq. Ft. Installed
Metalizing	3	Refractory (2-in.)	6.30
Heresite	2.50	Gunite	
		2-in. reinforced	2.50
		4-in. reinforced	3.50

INSULATION AND PAINTING

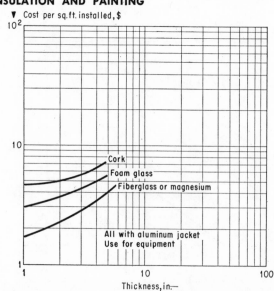

▼ Cost per sq. ft. installed, $

Painting: Vessels, $0.40/sq. ft.; structural steel, $0.25/sq. ft. (assume 400 sq. ft./ton.)

TABLES OF AREAS

Area of Shell, Sq. Ft./Lineal Ft. of Length

Diameter, Ft.*

In.*	1	2	3	4	5	6	7	8	9	10	11	12	13	14	15
0	3.14	6.28	9.42	12.57	15.70	18.80	21.99	25.13	28.27	31.42	34.56	37.70	40.84	43.98	47.12
1	3.39	6.53	9.68	12.82	15.96	19.10	22.24	25.38	28.53	31.67	34.81	37.95	41.09	44.23	47.38
2	3.68	6.82	9.96	13.10	16.24	19.38	22.53	25.67	28.81	31.95	35.09	38.23	41.37	44.52	47.66
3	3.93	7.07	10.21	13.35	16.49	19.64	22.78	25.92	29.06	32.20	35.34	38.48	41.63	44.77	47.91
4	4.18	7.32	10.46	13.60	16.74	19.89	23.03	26.17	29.31	32.45	35.59	38.74	41.88	45.02	48.16
5	4.46	7.60	10.74	13.89	17.03	20.17	23.31	26.45	29.59	32.74	35.88	39.02	42.16	45.30	48.44
6	4.71	7.85	11.00	14.14	17.28	20.42	23.56	26.70	29.85	32.99	36.13	39.27	42.41	45.55	48.69
7	4.96	8.11	11.25	14.39	17.53	20.67	23.81	26.95	30.10	33.24	36.38	39.52	42.66	45.80	48.95
8	5.25	8.39	11.53	14.67	17.81	20.95	24.10	27.24	30.38	33.52	36.66	38.80	42.95	46.09	49.23
9	5.50	8.64	11.78	14.92	18.06	21.21	24.35	27.49	30.63	33.77	36.91	40.06	43.20	46.34	49.48
10	5.75	8.89	12.03	15.17	18.32	21.46	24.60	27.74	30.88	34.02	37.17	40.31	43.45	46.59	49.75
11	6.03	9.17	12.32	15.46	18.60	21.74	24.88	28.02	31.16	34.31	37.45	40.59	43.73	46.87	50.01

Area of One Flat Head, Sq. Ft.

Diameter, Ft.*

In.*	1	2	3	4	5	6	7	8	9	10	11	12	13	14	15
0	0.785	3.14	7.07	12.57	19.64	28.09	38.37	50.27	63.62	78.54	95.03	113.1	132.7	153.9	176.8
1	0.916	3.40	7.45	13.07	20.27	29.03	39.37	51.28	64.75	79.85	96.47	114.7	134.4	155.7	178.6
2	1.08	3.70	7.89	13.66	20.99	29.90	40.38	52.42	66.04	81.18	97.93	116.3	136.1	157.6	180.6
3	1.23	3.98	8.30	14.19	21.65	30.68	41.28	53.46	67.20	82.52	99.40	117.9	137.9	159.5	182.7
4	1.39	4.26	8.71	14.73	22.31	31.47	42.20	54.50	68.37	83.86	100.1	119.5	139.6	161.3	184.6
5	1.58	4.60	9.19	15.34	23.07	32.32	43.24	55.68	69.69	85.22	102.4	121.1	141.3	163.2	186.6
6	1.77	4.91	9.62	15.90	23.76	33.18	44.18	56.75	70.88	86.59	103.9	122.7	143.1	165.1	188.7
7	1.96	5.23	10.07	16.48	24.45	34.00	45.13	57.82	72.08	87.97	105.4	124.4	144.9	167.0	190.7
8	2.19	5.60	10.58	17.13	25.25	34.94	46.20	59.04	73.44	89.36	106.9	126.0	146.6	168.9	192.7
9	2.40	5.94	11.04	17.72	25.97	35.78	47.17	60.13	74.66	90.76	108.4	127.7	148.5	170.9	194.8
10	2.63	6.29	11.52	18.32	26.69	36.64	48.15	61.24	76.36	92.17	110.0	129.4	150.2	172.8	196.8
11	2.90	6.70	12.07	19.01	27.53	37.61	49.27	62.49	77.29	93.59	111.5	131.0	152.1	174.7	198.9

Area of One Flanged and Dished ASME Head, Sq. Ft.

Diameter, Ft.*

In.*	1	2	3	4	5	6	7	8	9	10	11	12	13	14	15
0	1.33	4.52	9.62	16.62	25.07	35.26	47.78	62.21	77.24	95.03	114.7	136.4	159.4	184.7	211.6
1	1.54	4.91	10.18	17.35	25.52	36.32	49.02	63.62	78.54	96.47	116.3	137.8	161.3	186.7	213.8
2	1.77	5.31	10.75	17.72	26.42	36.85	49.64	64.33	79.85	97.93	117.9	139.6	163.2	188.6	215.9
3	2.01	5.73	11.34	18.10	26.88	37.39	50.27	65.04	81.18	99.40	119.5	141.3	165.9	190.7	218.1
4	2.27	5.94	11.64	18.86	27.34	38.48	51.53	66.48	83.86	102.4	121.1	144.9	168.9	192.7	220.3
5	2.54	6.16	11.95	19.64	28.27	40.72	54.11	67.93	85.22	103.9	124.4	146.6	170.8	194.8	222.5
6	2.66	6.61	12.57	20.43	30.19	41.85	55.42	69.40	86.59	105.4	126.0	148.4	172.8	196.8	224.8
7	2.90	7.07	13.20	21.24	31.17	43.01	56.75	70.88	87.97	106.9	127.7	150.2	174.7	198.9	227.0
8	3.14	7.55	13.85	22.06	32.17	44.18	57.41	72.38	89.36	108.4	129.4	152.1	176.7	201.0	229.2
9	3.17	8.04	14.52	22.90	33.18	44.77	58.09	73.90	90.76	110.0	131.0	153.9	178.6	205.2	231.4
10	3.87	8.55	14.86	23.76	33.70	45.36	59.45	75.43	92.17	111.5	132.7	155.7	180.6	207.3	233.7
11	4.15	8.81	15.21	24.63	34.21	46.57	60.82	76.98	93.59	113.1	134.4	157.6	182.6	209.5	235.9

Area of One Elliptical Dished Head, Sq. Ft.

Diameter, Ft.*

In.*	1	2	3	4	5	6	7	8	9	10	11	12	13	14	15
0	2.14	5.94	12.57	21.24	34.21	46.57	63.62	81.18	103.9	126.0	152.1	180.6	211.6	245.1	283.5
1	2.41	6.38	13.20	22.06	35.26	47.78	64.33	82.51	105.4	127.7	153.9	182.6	213.8	249.7	286.0
2	2.63	6.61	14.19	22.90	36.32	49.02	65.04	85.22	108.4	131.0	155.7	184.7	215.9	252.1	288.5
3	2.90	7.07	14.52	23.76	36.85	50.27	66.48	86.59	110.0	132.7	157.6	186.6	220.3	254.4	291.0
4	3.14	7.55	15.21	24.63	37.39	51.53	69.40	89.36	111.5	134.4	161.3	188.6	222.5	256.8	298.6
5	3.46	8.55	15.90	25.52	38.48	52.81	70.88	90.76	113.1	137.8	163.2	190.7	227.0	259.2	301.2
6	3.80	8.81	16.62	26.42	39.59	55.42	72.38	92.17	114.7	139.6	165.1	192.7	229.2	263.9	303.7
7	3.98	9.08	17.35	28.27	40.72	56.75	73.90	93.59	116.3	141.3	167.0	194.8	231.4	266.3	308.9
8	4.34	9.62	18.10	29.23	41.85	57.41	75.43	95.03	117.9	143.1	168.9	196.8	235.9	271.2	311.5
9	4.60	10.18	18.86	43.01	58.09	76.98	97.93	121.1	146.6	172.8	201.0	238.2	273.6	316.7	
10	5.31	10.75	19.64	32.17	44.18	59.45	78.40	100.1	122.7	148.4	176.7	203.1	240.5	278.5	319.4
11	5.73	11.95	20.43	33.18	45.36	62.21	79.85	102.4	124.4	150.2	178.6	207.3	242.8	281.0	322.0

* Combine ft. and in. to get area.   For example, a 7ft.–6 in. shell would have an area of 23.56 sq. ft./ft.

PIPING

HEAT-RESISTANT GLASS PIPE & FITTINGS†

Size, In.→		1 Cost, $	1 Hr. to Install	1½ Cost, $	1½ Hr. to Install	2 Cost, $	2 Hr. to Install	3 Cost, $	3 Hr. to Install	4 Cost, $	4 Hr. to Install	6 Cost, $	6 Hr. to Install
Pipe, lineal ft.*		2	0.06	2.90	0.06	3	0.08	6.10	0.15	8.70	0.2	12.75	0.3
Ells, ea.	45°	7.15	0.6	10.30	0.6	13.80	1.0	21.65	1.1	30.90	2.1	45.30	4.0
	60°	7.15	0.6	10.30	0.6	13.80	1.0	21.65	1.1	30.90	2.1	45.30	4.0
	90°	7.15	0.6	10.30	0.6	13.80	1.0	21.65	1.1	30.90	2.1	45.30	4.0
Tees, ea.		8.70	0.7	12.50	0.7	16.75	0.9	26.25	1.7	37.50	3.0	55	5.0
Tees reducing, ea.		—	—	12.50	0.5	16.75	0.6	26.25	1.5	37.50	2.5	55	5.3
Reducers, ea.		—	—	8	0.5	10.70	0.6	16.80	1.0	24	1.5	35.20	2.5
Crosses, ea.		—	—	16	0.8	21.45	1.2	33.60	2.0	48	3.5	70.40	5.5
Cast-iron flange sets, ea.		2.60	0.5	3.20	0.5	4.30	0.5	5.40	0.75	16.30	1.0	31	2.0
Gasket & bolt set,** ea.		0.13	0.13	0.75	0.14	0.85	0.15	1.60	0.15	2.75	0.15	4	0.16
Hangers, ea. (Spacing), ft.		1.20 (6)	0.5	1.40 (6)	0.75	2 (8)	0.75	2.50 (12)	1.0	3.50 (16)	1.25	7 (18)	1.25
Caps, ea.		2.45	0.5	3.50	0.5	4.70	0.5	7.35	0.75	10.50	1.0	15.40	2.0

* Based on 5-ft. lengths.   ** Teflon gasket.   † Pyrex brand.

STEEL SCREWED PIPE & FITTINGS

Size, In.→		½ Cost, $	½ Hr. to Install	¾ Cost, $	¾ Hr. to Install	1 Cost, $	1 Hr. to Install	1¼ Cost, $	1¼ Hr. to Install	1½ Cost, $	1½ Hr. to Install	2 Cost, $	2 Hr. to Install	3 Cost, $	3 Hr. to Install
Pipe, lineal ft., Sch. 40	Welded	0.11	0.15	0.14	0.20	0.20	0.22	0.30	0.24	0.35	0.26	0.45	0.31	0.90	0.45
	Seamless	0.43	0.15	0.48	0.20	0.60	0.22	0.80	0.24	0.90	0.26	0.60	0.31	1.20	0.45
Pipe, lineal ft., Sch. 80	Welded	0.15	0.20	0.20	0.27	0.27	0.28	0.36	0.29	0.50	0.31	0.65	0.33	1.30	0.60
	Seamless	0.50	0.20	0.55	0.27	0.70	0.28	0.95	0.29	1.10	0.31	0.95	0.33	1.70	0.60
90° ells, ea.	150 lb. Malleable iron	0.20	0.60	0.25	0.70	0.32	0.80	0.50	0.90	0.68	1.00	0.97	1.40	2.95	2.50
	300 lb. Malleable iron	0.78	0.65	0.90	0.75	1.10	0.85	1.50	0.95	1.90	1.10	2.65	1.60	5.75	2.70
	2,000 lb. Forged steel	0.75	0.65	0.91	0.75	1.21	0.85	1.72	0.95	2.49	1.10	3.15	1.60	10.47	2.70
	3,000 lb. Forged steel	0.90	0.70	1.10	0.80	1.55	0.90	2.30	1.00	3.30	1.20	4.52	1.70	13.60	2.80
45° ells, ea.	150 lb. Malleable iron	0.21	0.60	0.29	0.70	0.35	0.80	0.60	0.90	0.75	1.00	1.05	1.40	3.30	2.50
	300 lb. Malleable iron	0.80	0.65	0.95	0.75	1.00	0.85	1.45	0.95	1.85	1.10	2.70	1.60	5.70	2.70
	2,000 lb. Forged steel	0.87	0.65	1.07	0.75	1.43	0.85	1.94	0.95	2.69	1.10	3.54	1.60	11.47	2.70
	3,000 lb. Forged steel	1.20	0.70	1.38	0.80	1.86	0.90	2.15	1.00	3.61	1.20	4.87	1.70	14.05	2.80
Tees, ea.	150 lb. Malleable iron	0.28	0.80	0.37	1.00	0.47	1.20	0.78	1.50	0.96	1.70	1.40	2.00	3.30	4.00
	300 lb. Malleable iron	0.90	0.85	1.00	1.10	1.20	1.30	1.60	1.60	1.95	1.80	2.86	2.20	8.10	4.30
	2,000 lb. Forged steel	1.01	0.85	1.33	1.10	1.76	1.30	2.40	1.60	3.82	1.80	4.80	2.20	15.73	4.30
	3,000 lb. Forged steel	1.27	0.90	1.63	1.20	2.22	1.40	3.20	1.65	4.32	1.90	5.94	2.30	19.14	4.40
Unions, ea.	150 lb. Malleable iron	0.35	0.50	0.45	0.70	0.55	0.90	0.75	1.00	0.95	1.20	1.20	1.40	3.95	2.00
	300 lb. Malleable iron	0.75	0.55	0.85	0.75	1.15	0.95	1.50	1.10	1.85	1.30	2.30	1.50	7.70	2.30
	2,000 lb. Forged steel	2.60	0.55	3.12	0.75	3.77	0.95	4.64	1.10	5.96	1.30	8.34	1.50	24.00	2.30
	3,000 lb. Forged steel	2.60	0.60	3.12	0.80	3.77	1.00	4.64	1.20	5.96	1.40	8.34	1.60	24.00	2.50
Reducers, ea.	150 lb. Malleable iron	0.16	0.55	0.19	0.65	0.28	0.85	0.37	0.90	0.45	1.10	0.70	1.40	2.00	2.50
	300 lb. Malleable iron	0.70	0.60	0.75	0.70	0.90	0.90	1.00	0.95	1.35	1.20	1.90	1.50	4.55	2.60
	2,000 lb. Forged steel	0.73	0.60	1.04	0.70	1.40	0.90	1.94	0.95	2.07	1.20	2.90	1.50	8.24	2.60
	3,000 lb. Forged steel	0.73	0.70	1.04	0.80	1.40	1.00	1.94	1.10	2.07	1.20	2.90	1.60	8.24	2.70
Half couplings, ea.	3,000 lb. Forged steel	0.24	0.30	0.30	0.35	0.40	0.50	0.60	0.70	0.81	0.90	1.25	1.40	4.52	1.70
Couplings, ea.	3,000 lb. Forged steel	0.44	0.60	0.50	0.75	0.67	0.90	1.10	1.10	1.50	1.20	2.20	2.20	5.95	2.80
Strainers, ea.		2.70	0.80	3.50	1.00	4.00	1.30	6.00	1.50	7.10	1.70	10.50	1.90	25.00	2.50
Traps, ea.		15.00	0.60	15.00	0.90	23.00	1.20	30.00	1.40	40.00	1.50	—	1.90	—	2.70
Hangers, ea. (Spacing,) ft.		1.00 (6)	0.50	1.00 (6)	0.50	1.20 (8)	0.50	1.20 (8)	0.75	1.40 (10)	0.75	2.00 (12)	0.75	2.50 (18)	1.00

STEEL WELDED PIPE & FITTINGS

Size, In. →		¾ Cost $	¾ Hr. to Install	1 Cost $	1 Hr. to Install	1½ Cost $	1½ Hr. to Install	2 Cost $	2 Hr. to Install	3 Cost $	3 Hr. to Install	4 Cost $	4 Hr. to Install	6 Cost $	6 Hr. to Install	8 Cost $	8 Hr. to Install	10 Cost $	10 Hr. to Install	12 Cost $	12 Hr. to Install
Welds, ea.	Sch. 40	—	0.7	—	0.9	—	1.1	—	1.2	—	1.8	—	2.0	—	3.0	—	4.0	—	5.0	—	6.0
	Sch. 80	—	0.8	—	1.0	—	1.2	—	1.5	—	2.1	—	2.5	—	3.5	—	4.5	—	6.0	—	8.0
Pipe, lineal ft., Sch. 40	Welded	0.15	0.06	0.22	0.07	0.34	0.09	0.45	0.1	0.91	0.14	1.45	0.16	—	0.25	—	0.3	—	0.5	—	0.6
	Seamless	0.56	0.06	0.56	0.07	0.86	0.09	0.60	0.1	1.11	0.14	1.52	0.16	2.61	0.25	3.80	0.3	5.40	0.5	6.80	0.6
Pipe, lineal ft., Sch. 80	Welded	0.28	0.07	0.33	0.08	0.50	0.1	0.65	0.12	1.30	0.18	2	0.22	—	0.35	—	0.45	—	0.6	—	0.65
	Seamless	0.65	0.07	0.70	0.08	1.10	0.1	0.85	0.12	1.55	0.18	2.15	0.22	4	0.35	5.85	0.45	7.40	0.6	9.10	0.65
90° ells, ea.	Std. wt.	0.80	1.4	0.90	1.8	1.35	2.2	1.80	2.4	3.65	3.6	5.65	4.0	13.25	6.0	24.85	8.0	52.60	10.0	82.35	12.0
	Ex. Hvy.	1.20	1.6	1.20	2.0	1.75	2.4	2.30	3.0	4.70	4.2	7.35	5.0	17.90	7.0	32.05	9.0	68.35	12.0	107	16.0
45° ells, ea.	Std. wt.	0.80	1.4	0.90	1.8	1.35	2.2	1.60	2.4	3.30	3.6	4.25	4.0	10.35	6.0	18.50	8.0	39.45	10.0	61.75	12.0
	Ex. Hvy.	1.20	1.6	1.20	2.0	1.75	2.4	2.10	3.0	4.25	4.2	5.50	5.0	13.40	7.0	24.05	9.0	51.25	12.0	80.25	16.0
Tees, ea.	Std. wt.	2.50	2.1	2.50	2.7	4.20	3.3	6.10	3.6	11.10	5.4	15.75	6.0	32.94	9.0	61.70	12.0	105	15.0	150	18.0
	Ex. Hvy.	2.50	2.4	2.50	3.0	4.20	3.6	6.10	4.5	14.40	6.3	20.50	7.5	42.85	10.5	80.20	13.5	136	18.0	195	24.0
Reducers, ea.	Std. wt.	1.00	1.0	1.20	1.0	1.95	1.5	2.15	2.0	2.52	3.0	3.51	3.5	7.85	5.0	14.16	7.0	21.50	8.0	33.10	10.0
	Ex. Hvy.	1.85	1.5	2.15	1.5	3.51	2.0	3.70	2.5	4.50	4.0	6.30	5.0	13.80	6.0	27.60	8.0	39.70	10.0	57.90	13.0
Slip-on flanges, ea.	150 lb.	1.90	1.0	1.90	1.0	2.42	1.5	2.73	2.0	3.71	2.5	4.90	3.0	7.75	4.5	11.76	6.0	16.12	7.0	27.30	8.0
	300 lb.	2.75	1.0	2.75	1.0	3.50	1.5	4.00	2.0	5.23	3.0	7.10	4.0	13.05	6.0	20.60	8.0	29.05	9.0	48.40	11.0
	600 lb.	4.80	1.0	5.07	1.0	6.03	1.5	6.60	2.0	9.45	4.0	16.40	5.0	31.40	7.0	46.25	9.0	72	10.0	98.80	12.0
Weld-neck flanges, ea.	150 lb.	3.65	0.7	3.65	0.7	3.65	0.8	3.65	1.0	4.78	1.8	6.50	2.0	10.15	3.0	18.35	3.5	26.50	4.5	43.27	6.0
	300 lb.	4.65	0.7	4.65	0.7	4.65	0.9	4.65	1.0	6.15	2.0	9.95	2.5	17.30	3.5	26.90	4.5	37.60	6.0	66.30	7.0
	600 lb.	11.20	0.8	11.20	0.8	11.20	1.0	11.20	1.5	15.40	2.5	26.30	3.0	48.52	4.0	69.10	5.0	106	6.5	144	8.0
Bolt & gasket sets*, ea.		0.40	0.04	0.40	0.04	0.45	0.05	0.80	0.07	0.90	0.08	1.60	0.09	2.20	0.15	2.40	0.2	5	0.25	5.50	0.3
Hangers, ea.		1	0.50	1.20	0.50	1.40	0.75	2	0.75	2.50	1.0	3.50	1.25	7	1.25	9	1.25	11	1.50	13	1.50
(Spacing), ft.		(6)		(8)		(10)		(12)		(18)		(20)		(25)		(30)		(35)		(40)	

* Red-rubber gasket.

CHEMICAL-LEAD PIPE

Size, In. →		1½ Cost $	1½ Hr. to Install	2 Cost $	2 Hr. to Install	3 Cost $	3 Hr. to Install	4 Cost $	4 Hr. to Install	6 Cost $	6 Hr. to Install	8 Cost $	8 Hr. to Install	10 Cost $	10 Hr. to Install	12 Cost $	12 Hr. to Install
Pipe, lineal ft.	¼-in. wall	1.50	0.4	1.90	0.5	2.70	0.9	3.50	0.9	5	1.3	7	1.8	8	2.0	9	3.0
	⅜-in. wall	2.30	0.6	3	0.8	4.20	1.6	5.50	1.6	8.13	2.0	9	2.8	12	3.5	14	4.0
	½-in. wall	—	—	—	—	5.50	2.0	7.50	2.0	10.50	3.0	13	3.5	16	4.5	18	5.5
Burning neck flanges, ea.		7.50	1.2	8.50	1.5	12	2.0	16	2.5	19	3.0	30	3.5	40	4.5	50	6.0
Supports, lineal ft.	Vert.	0.30	0.1	0.30	0.1	0.30	0.1	0.30	0.1	1.50	0.5	1.70	0.5	2	0.5	2.50	0.5
	Horiz.	0.80	0.2	1	0.2	1.30	0.2	1.50	0.2	4.50	0.6	5	0.6	5.50	0.6	6	0.6

CAST IRON, MECHANICAL JOINT PIPE & FITTINGS

Size, In.→	3 Cost $	3 Hr. to Install	4 Cost $	4 Hr. to Install	6 Cost $	6 Hr. to Install	8 Cost $	8 Hr. to Install	10 Cost $	10 Hr. to Install	12 Cost $	12 Hr. to Install	14 Cost $	14 Hr. to Install	16 Cost $	16 Hr. to Install	18 Cost $	18 Hr. to Install	20 Cost $	20 Hr. to Install	24 Cost $	24 Hr. to Install
Pipe, lineal ft. 150 lb.	1.40	0.11	1.50	0.15	2.20	0.25	3.20	0.35	4.10	0.45	5.20	0.6	6.50	0.75	7.60	0.85	9	0.95	10.50	0.95	14.50	1.3
250 lb.	1.40	0.12	1.50	0.16	2.20	0.25	3.20	0.45	4.10	0.55	5.40	0.7	7	0.8	8.20	0.95	10	1.05	11.70	1.05	15.10	1.5
Bends, ea. ⅛	13.30	2.5	14.20	3.0	19.05	4.0	26.90	5.0	36.80	6.0	49.35	6.5	68.55	7.5	84.15	10.0	103	11.0	131	12.0	179	14.0
¼	14.30		15.20		20.80		29.50		42.85		56.25		82		102		127		161		231	
Tees, ea.	20.40	3.0	22.30	3.5	30.75	5.0	43.80	6.0	68.60	8.0	89.10	9.0	140	12.0	172	16.0	219	18.0	274	20.0	409	25.0
Reducers, ea.	—	—	15.40	2.0	16.50	3.0	23	4.0	32	5.0	40	6.0	—	—	68	8.0	—	—	105	10.0	156	12.0
Crosses, ea.	—	—	30	5.0	40.30	5.5	43	8.0	87	10.0	111.30	12.0	—	—	213	18.0	—	—	334	26.0	488	30.0
Line joints, ea.	—	1.5	—	1.5	—	2.0	—	2.0	—	2.5	—	2.5	—	2.5	—	2.5	—	3.0	—	3.0	—	3.5

CAST IRON, BELL & SPIGOT PIPE & FITTINGS

Size, In.→	3 Cost $	3 Hr. to Install	4 Cost $	4 Hr. to Install	6 Cost $	6 Hr. to Install	8 Cost $	8 Hr. to Install	10 Cost $	10 Hr. to Install	12 Cost $	12 Hr. to Install	14 Cost $	14 Hr. to Install	16 Cost $	16 Hr. to Install	18 Cost $	18 Hr. to Install	20 Cost $	20 Hr. to Install	24 Cost $	24 Hr. to Install
Pipe, lineal ft. 150 lb.	1.35	0.11	1.60	0.15	2.20	0.25	3.10	0.35	4	0.45	5.00	0.60	6.30	0.75	7.50	0.85	9	0.95	10.30	1.1	14.00	1.3
250 lb.	1.35	0.12	1.60	0.16	2.20	0.3	3.10	0.45	4	0.55	5.30	0.70	6.90	0.8	8.20	0.95	10	1.05	11.20	1.2	14.70	1.5
Bends, ea. Class D	10.50	3.2	12.20	4.5	16.50	6.5	25.60	9.0	47.10	11.0	47.70	14.0	88.50	15.0	132	18.0	159	22.0	189	25.0	293	33.0
Tees, ea. Class D	16.60	5.0	20.30	6.0	26.30	9.0	40	13.0	69	17.0	90	20.0	132	23.0	170	27.0	209	31.0	255	36.0	355	50.0
Line joints, ea.	—	1.0	—	1.4	—	2.0	—	3.0	—	4.0	—	4.5	—	5.0	—	6.0	—	7.0	—	8.0	—	9.0

CAST IRON, FLANGED PIPE & FITTINGS

Size, In.→	3 Cost $	3 Hr. to Install	4 Cost $	4 Hr. to Install	6 Cost $	6 Hr. to Install	8 Cost $	8 Hr. to Install	10 Cost $	10 Hr. to Install	12 Cost $	12 Hr. to Install	14 Cost $	14 Hr. to Install	16 Cost $	16 Hr. to Install	18 Cost $	18 Hr. to Install	20 Cost $	20 Hr. to Install	24 Cost $	24 Hr. to Install
Pipe, lineal ft. 250 lb.	7.50	0.9	10.50	1	16	1.1	22	1.3	31	1.7	40	1.9	50	2.0	—	—	—	—	—	—	—	—
90° ells, ea. 250 lb.	12	2.2	18	3.5	30	4.5	43	5.5	67	6.5	95	8.5	140	10.5	—	—	—	—	—	—	—	—
Tees, ea. 250 lb.	17	3.5	27	4.5	47	6.0	65	8.0	99	10.0	143	12.0	215	14.0	—	—	—	—	—	—	—	—
Crosses, ea. 250 lb.	—	—	36	5.4	60	8.0	82	11.0	126	13.0	178	16.0	269	19.0	—	—	—	—	—	—	—	—
Reducers, ea. 250 lb.	—	—	14	4	23	4.5	34	5.5	50	6.5	77	7.5	115	8.5	—	—	—	—	—	—	—	—
Line joints, ea.	—	1.5	—	2.3	—	2.3	—	2.8	—	2.8	—	3.0	—	3.0	—	—	—	—	—	—	—	—
Trenches 5 ft. deep, ea.	0.60	0.05	0.60	0.05	0.70	0.05	0.70	0.05	0.70	0.06	0.80	0.06	0.90	0.06	0.90	0.08	0.90	0.08	1	0.09	1.20	0.1

Note: Cost of pipe includes jointing material.

STAINLESS-STEEL WELDED PIPE & FITTINGS

Size, In.→	½ Cost $	½ Hr. to Install	¾ Cost $	¾ Hr. to Install	1 Cost $	1 Hr. to Install	1½ Cost $	1½ Hr. to Install	2 Cost $	2 Hr. to Install	2½ Cost $	2½ Hr. to Install	3 Cost $	3 Hr. to Install	3½ Cost $	3½ Hr. to Install	4 Cost $	4 Hr. to Install	6 Cost $	6 Hr. to Install	8 Cost $	8 Hr. to Install
Welds, ea. Sch. 10	—	1.3	—	1.4	—	1.5	—	1.75	—	2.0	—	2.3	—	2.5	—	3.0	—	3.4	—	5.0	—	6.0
Sch. 40	—	1.4	—	1.45	—	1.6	—	1.8	—	2.2	—	2.5	—	2.9	—	3.2	—	3.5	—	5.1	—	6.6

STAINLESS PIPE (cont'd.)

Item		½ Cost $	½ Hr.	¾ Cost $	¾ Hr.	1 Cost $	1 Hr.	1½ Cost $	1½ Hr.	2 Cost $	2 Hr.	2½ Cost $	2½ Hr.	3 Cost $	3 Hr.	3½ Cost $	3½ Hr.	4 Cost $	4 Hr.	6 Cost $	6 Hr.	8 Cost $	8 Hr.
Pipe, lineal ft., Sch. 10	Welded	1.60	0.09	1.80	0.12	2.80	0.13	3.25	0.17	3.80	0.18	4.35	0.22	5.75	0.26	5.40	0.28	6.05	0.3	8.30	0.4	11.40	0.5
	Seamless	1.90	0.09	2.60	0.12	4.05	0.13	4.70	0.17	5.45	0.18	5.60	0.22	6.30	0.26	7.10	0.28	8.25	0.3	—	0.4	—	0.5
Pipe, lineal ft., Sch. 40	Welded	2.05	0.1	2.40	0.13	2.85	0.14	4.30	0.18	5.00	0.2	—	0.24	—	0.28	—	0.3	—	0.32	—	0.45	—	0.55
	Seamless	2.85	0.1	3.70	0.13	4.05	0.14	6.10	0.18	6.15	0.2	8.25	0.24	9.90	0.28	11.95	0.3	12.90	0.32	17.45	0.45	38.90	0.55
90° ells, ea.	Sch. 10	3.15	2.6	3.15	2.8	3.80	3.0	4.70	3.5	6.55	4.0	12.95	4.6	12.95	5.0	21.40	6.0	21.40	6.8	56.90	10.0	106.00	12.0
	Sch. 40	3.60	2.8	3.60	2.9	4.70	3.2	7.25	3.6	8.90	4.4	24	5.0	24.00	5.8	40.50	6.4	40.50	7.0	85.00	10.2	188.00	13.2
45° ells, ea.	Sch. 10	2.85	2.5	2.85	2.8	3.40	3.0	4.25	3.5	5.90	4.0	10.35	4.6	10.35	5.0	17.10	6.0	17.10	6.8	39.50	10.0	74.20	12.0
	Sch. 40	3.25	2.8	3.24	2.9	4.25	3.2	6.55	3.6	8	4.4	19.20	5.0	19.20	5.8	32.40	6.4	32.40	7.0	59.50	10.2	132	13.2
Tees, ea.	Sch. 10	6.90	3.9	6.90	4.2	6.90	4.5	9.80	5.3	13	6.0	27.50	6.9	27.50	7.5	48.75	9.0	48.75	10.2	94.10	15.0	184	18.0
	Sch. 40	8	4.2	8	4.4	9.90	4.8	18	5.4	22.30	6.6	46.10	7.5	46.10	8.7	62.00	9.6	62.00	10.5	134	15.3	230	19.8
Reducers, ea.	Sch. 10	—	1.5	4	2.0	6	2.0	6.40	3.0	6	4.0	12.05	5.0	13.50	6.0	16.90	6.5	16.90	7.0	39.65	10.0	46.90	14.0
	Sch. 40	—	2.0	5	3.0	5.75	3.0	7.65	4.0	9.70	5.0	13.70	6.0	14.20	8.0	19.80	7.0	19.80	10.0	42	12.0	66.25	16.0
Slip-on flanges, ea.	150 lb.	4.95	1.5	5.65	2.0	6.65	2.0	9.55	3.0	12.90	4.0	18.60	4.5	19.90	5.0	22.40	5.5	25.50	6.0	40.50	9.0	63.50	12.0
	300 lb.	6.95	1.5	7.90	2.0	9.60	2.0	15.25	3.0	18.85	4.0	26.05	5.0	27.85	6.0	36.00	7.0	40.15	8.0	90	12.0	138.50	16.0
	600 lb.	13.85	1.5	15	2.0	19.75	2.0	26.90	3.0	29.85	4.0	39	6.0	45.75	8.0	54.00	9.0	80.10	10.0	142.75	14.0	220.40	18.0
Weld-neck flanges, ea.	150 lb.	6.55	1.4	6.80	1.4	8.50	1.4	12	1.6	14.75	2.0	21.80	2.5	24	3.6	27.70	4.0	30.80	4.0	55.40	6.0	98.05	7.0
	300 lb.	9.15	1.4	9.85	1.4	11.90	1.4	18.45	1.8	22.15	3.0	29.10	3.5	33.60	4.0	42.95	4.5	47.80	5.0	98.35	7.0	157.90	9.0
	600 lb.	14.95	1.6	18.65	1.6	20.45	1.6	32	2.0	35.80	3.0	43.65	3.5	50.10	5.0	60.95	5.5	83.10	6.0	161.50	8.0	231.35	10.0
Caps, ea.	Sch. 10	4.60	1.3	4.60	1.4	4.60	1.5	4.60	1.8	5.20	2.0	8.30	2.3	8.30	2.5	9.70	3.0	9.70	3.4	16.50	5.0	26.40	6.0
	Sch. 40	5.05	1.4	5.05	1.5	5.05	1.6	5.05	1.9	5.80	2.2	10.10	2.5	10.10	2.9	14.50	3.2	14.50	3.5	28.10	5.1	52.50	6.6
Gasket & bolt sets, ea.*		0.70	0.05	0.80	0.05	0.90	0.05	1	0.06	1.30	0.05	2.30	0.08	2.50	0.09	3.00	0.12	3.50	0.12	6	0.15	9	0.2
Hangers, ea.		1	0.5	1	0.5	1.20	0.5	1.40	0.75	2	0.75	2	1.0	2.50	1.0	3	1.0	3.50	1.0	7	1.25	9	1.25
(Spacing), ft.		(6)		(6)		(8)		(10)		(12)		(14)		(18)		(18)		(20)		(25)		(30)	

*Teflon gasket.
Note: Prices are for type 304 & 304L. For 316 & 347 multiply by 1.25; for 316L multiply by 1.45.

ALUMINUM WELDED PIPE & FITTINGS

Size, In.→		1 Cost $	1 Hr. to Install	1½ Cost $	1½ Hr. to Install	2 Cost $	2 Hr. to Install	2½ Cost $	2½ Hr. to Install	3 Cost $	3 Hr. to Install	3½ Cost $	3½ Hr. to Install	4 Cost $	4 Hr. to Install	6 Cost $	6 Hr. to Install	8 Cost $	8 Hr. to Install	10 Cost $	10 Hr. to Install
Pipe, lineal ft.	Sch. 40	0.64	0.09	1.04	0.1	1.25	0.12	1.91	0.24	2.50	0.27	2.90	0.3	3.44	0.34	6.17	0.5	9.29	0.8	14.33	1.0
90° ells, ea.		3.75	2.8	4.50	3.2	5.50	4.0	13	4.6	13	5.2	20	5.6	2	6.2	40	9.0	75	12.0	150	14.0
45° ells, ea.		3.40	2.8	4.05	3.2	4.95	4.0	10.40	4.6	10.40	5.2	16	5.6	16	6.2	32	9.0	53	12.0	105	14.0
Tees, ea.		7.55	4.2	12.80	4.8	16.75	6.0	33.50	6.9	33.50	7.8	45	8.4	45	9.3	85	13.5	150	18.0	320	21.0
Reducers, ea.		5	2.5	6.40	3.0	9.50	3.8	13.50	4.5	13.50	5.0	17.10	5.4	17.10	6.0	43.70	8.6	55.20	11.5	97.10	13.5
Slip-on flanges, ea.	150 lb.	5.60	1.5	6.40	2.4	7.60	3.0	11.20	3.5	11.20	3.6	15.60	4.0	15.60	4.5	22.40	6.0	40	7.5	124	9.0
	300 lb.	6.60	1.6	8.30	2.5	9.10	3.1	10.80	3.6	13.60	3.8	18.10	4.2	18.10	4.7	34	6.5	121	8.0	154	10.0

ALUMINUM PIPE (cont'd)

Item		1 Cost $	1 Hr. to Install	1½ Cost $	1½ Hr. to Install	2 Cost $	2 Hr. to Install	2½ Cost $	2½ Hr. to Install	3 Cost $	3 Hr. to Install	3½ Cost $	3½ Hr. to Install	4 Cost $	4 Hr. to Install	6 Cost $	6 Hr. to Install	8 Cost $	8 Hr. to Install	10 Cost $	10 Hr. to Install
Weld-neck flanges, ea.	150 lb.	7	1.4	8	2.2	9.50	2.8	14	3.2	14	3.9	19.50	4.2	19.50	4.2	28	6.0	50	7.0	155	8.0
	300 lb.	11.75	1.5	11.75	2.0	11.80	2.9	13.90	3.4	17.50	3.7	27.90	4.6	27.90	4.9	38	6.5	142	7.5	202	9.0
Gasket & bolt sets, ea.		0.45	0.05	0.80	0.07	0.90	0.08	0.90	0.08	0.90	0.08	1.20	0.09	1.60	0.09	2.20	0.15	2.40	0.2	5	0.25
Hangers, ea. (Spacing), ft.		1.20 (8)	0.5	1.40 (10)	0.75	2 (12)	0.75	2.50 (18)	1.0	2.50 (18)	1.0	2.50 (18)	1.0	3.50 (20)	1.25	7 (25)	1.25	9 (30)	1.25	11 (35)	1.5
Sch. 40 welds, ea.		—	1.4	—	1.6	—	2.0	—	2.3	—	2.6	—	2.8	—	3.1	—	4.5	—	6.0	—	7.0

PVC PLASTIC PIPE & FITTINGS

Item		½ Cost $	½ Hr. to Install	¾ Cost $	¾ Hr. to Install	1 Cost $	1 Hr. to Install	1½ Cost $	1½ Hr. to Install	2 Cost $	2 Hr. to Install	2½ Cost $	2½ Hr. to Install	3 Cost $	3 Hr. to Install	3½ Cost $	3½ Hr. to Install	4 Cost $	4 Hr. to Install	6 Cost $	6 Hr. to Install
Pipe, lineal ft., Sch. 40	Plain	0.22	0.09	0.27	0.09	0.41	0.09	0.65	0.12	0.86	0.14	1.35	0.14	1.75	0.14	—	—	2.50	0.19	4.30	0.32
	Threaded	0.22	0.07	0.27	0.07	0.41	0.07	0.65	0.09	0.86	0.09	1.35	0.1	1.75	0.1	—	—	2.50	0.13	4.30	0.25
Pipe, lineal ft., Sch. 80	Plain	0.32	0.09	0.42	0.09	0.60	0.09	0.90	0.12	1.30	0.14	2	0.14	2.60	0.14	—	—	3.70	0.19	7.10	0.32
	Threaded	0.32	0.07	0.42	0.07	0.60	0.07	0.90	0.09	1.30	0.09	2	0.1	2.60	0.1	—	—	3.70	0.13	7.10	0.25
90° ells, ea.	Socket	1	0.11	1.25	0.13	1.60	0.19	2.15	0.35	3.00	0.45	6.10	0.65	10.20	0.9	—	—	15.20	1.2	37	1.3
	Screwed	0.90	0.26	1.20	0.35	1.45	0.45	1.95	0.6	2.70	0.8	5.60	0.8	9.50	2.0	—	—	13.80	2.3	—	2.5
45° ells, ea.	Socket	1	0.11	1.25	0.13	1.60	0.19	2.15	0.35	3.00	0.45	6.10	0.65	9.00	0.9	—	—	13.00	1.2	37	1.3
	Screwed	0.90	0.26	1.20	0.35	1.45	0.45	1.95	0.6	2.70	0.8	5.60	0.8	8.00	2.0	—	—	11.80	2.3	—	2.5
Tees, ea.	Socket	1.20	0.17	1.50	0.20	1.70	0.29	2.50	0.55	3.50	0.68	8.40	0.98	16.00	1.4	—	—	24.00	1.8	53	2.0
	Screwed	1.10	0.39	1.30	0.53	1.60	0.67	2.30	0.9	3.10	1.2	8.00	1.25	14.70	3.0	—	—	21.80	3.5	—	3.8
Adapters, ea.	Socket	0.90	0.1	1.10	0.14	1.30	0.16	1.75	0.35	2.40	0.4	5.10	0.6	4.80	0.75	—	—	10	0.9	—	1.1
	Screwed	0.90	0.25	1.10	0.3	1.30	0.4	1.75	0.5	2.40	0.6	5.10	0.7	4.80	1.2	—	—	10	2.0	—	2.5
Unions, ea.	Socket	4.40	0.1	5.10	0.14	6	0.76	9.00	0.35	13.20	0.4	40	0.6	43.00	0.75	—	—	—	—	—	—
	Screwed	4	0.25	4.70	0.3	5.45	0.46	8.30	0.5	12.10	0.6	37	0.7	39.00	1.2	—	—	—	—	—	—
Valves (globe), ea.	Socket	19.50	0.15	19.50	0.18	23	0.25	36	0.4	49	0.5	—	0.9	110	1.0	—	—	—	1.1	—	1.5
	Screwed	19.50	0.35	19.50	0.4	23	0.5	36	0.6	49	0.8	—	1.0	—	2.0	—	—	—	2.6	—	3.5
Caps, ea.	Socket	0.90	0.05	1.10	0.04	1.30	0.1	1.70	0.15	2.50	0.2	5.10	0.25	4.90	0.35	—	—	10	0.4	—	0.7
	Screwed	0.85	0.1	1.02	0.1	1.20	0.15	1.60	0.2	2.10	0.3	5.60	0.3	5.10	0.7	—	—	8	0.8	—	1.0
Flanges, ea.	Screwed	1.50	0.32	1.70	0.4	1.80	0.45	2.70	0.5	3.30	0.6	10.10	0.6	12.00	1.1	—	—	16	1.2	16	1.2
Gasket & bolt sets*, ea.		0.80	0.05	0.95	0.05	1.10	0.05	1.40	0.06	1.90	0.07	2.30	0.08	2.30	0.09	—	—	4	0.1	4	0.15
Hangers, ea. (Spacing,) ft.		1 (6)	0.5	1 (6)	0.5	1.20 (8)	0.5	1.40 (10)	0.75	2.00 (12)	0.75	2.30 (18)	0.75	2.50 (18)	1.0	—	—	3.50 (20)	1.25	7 (25)	1.25

*Teflon gasket.

INSTALLATION TIME: PLASTIC-LINED PIPE & FITTINGS

Size, In.→		½ Hr. to Install	¾ Hr. to Install	1 Hr. to Install	1½ Hr. to Install	2 Hr. to Install	2½ Hr. to Install	3 Hr. to Install	3½ Hr. to Install	4 Hr. to Install	6 Hr. to Install
Plastic-lined steel pipe, lineal ft.		—	—	0.3	0.4	0.5	0.6	0.7	0.8	0.9	1.4
Ells flgd., ea.	90°	—	—	2.5	3.5	5.0	6.5	8.0	9.0	12.0	14.0
	45°	—	—	2.5	3.5	5.0	6.5	8.0	9.0	12.0	14.0
Tees, flgd., ea.		—	—	3.5	5.5	6.5	8.5	10.0	13.0	16.0	20.0
Reducers, flgd., ea.		—	—	2.5	3.5	5.0	6.5	8.0	9.0	12.0	14.0

INSTALLATION TIME: SWEAT FITTINGS

		½	¾	1	1½	2	2½	3
Tubing, lineal ft.	Hard	0.2	0.22	0.26	0.27	0.3	0.32	0.34
	Soft	0.2	0.22	0.26	0.27	0.3	0.32	0.34
Ells, ea.	45°	0.3	0.38	0.58	1.2	1.5	2.2	2.6
	90°	0.3	0.38	0.58	1.2	1.5	2.2	2.6
Adapters, ea.		0.28	0.36	0.53	1.1	1.5	2.0	2.6
Plugs, ea.		0.12	0.12	0.25	0.5	0.6	0.8	1.2
Reducers, ea		0.3	0.4	0.6	1.2	1.5	2.2	2.7
Unions, ea.		0.32	0.41	0.61	1.3	1.7	2.5	3.0
Couplings, ea.		0.29	0.4	0.6	1.2	1.4	2.2	2.8
Flanges, ea.		0.29	0.29	0.45	0.9	1.5	1.7	2.0
Valves, ea.		0.4	0.5	0.7	1.5	1.8	2.6	3.0
Tees, ea.		0.41	0.55	0.8	1.6	2.3	3.0	3.5

INSULATION AND PAINTING

Pipe Size, In.→ Fiberglass Insulation		½	¾	1	1½	2	2½	3	3½	4	5	6	8	10	12
						Cost, $/Lineal Ft. Pipe									
Standard, thickness	Std.	1.30	1.30	1.30	1.30	1.30	1.30	1.30	1.40	1.70	1.80	2.10	2.30	3.50	4.50
	W.P.	1.80	1.80	1.80	1.80	1.80	2	2.20	2.50	2.80	3.50	3.80	5.	6.50	—
1-½-in. thick	Std.	1.20	1.20	1.40	1.70	1.90	2.10	2.30	2.40	2.60	2.80	3.30	4.10	5	5.50
	W.P.	1.80	1.80	2.10	2.60	2.70	3.10	3.30	3.50	3.90	4.30	4.70	6	7	9
2-in. thick	Std.	2.20	2.30	2.40	2.60	2.80	3.10	3.30	3.70	3.90	4.30	4.80	5.80	6.80	7.70
	W.P.	3	3.20	3.40	3.80	4	4.20	4.50	5	5.30	6	6.80	8	9.50	10

Foamglass Insulation

		½	¾	1	1½	2	2½	3	3½	4	5	6	8	10	12
1-in. thick	Std.	1.30	1.30	1.30	1.30	1.40	1.40	1.50	1.80	1.90	2.30	2.50	3.60	4.30	—
	W.P.	2	2	2	2	2.10	2.10	2.30	2.70	2.90	3.60	3.80	5.50	6.70	—
2-in thick	Std.	2.10	2.30	2.30	2.50	2.70	3	3.20	3.60	3.80	4.20	4.70	5.80	6.60	7.60
	W.P.	3.10	3.30	3.50	3.90	4.10	4.30	4.70	5.10	5.50	6.30	7	8.30	9.60	11
3-in. thick	Std.	3.20	3.40	3.40	3.70	4.20	4.50	4.80	5.40	5.80	6.30	7	8.50	9.70	11.50
	W.P.	4.70	5	5	5.50	6	6.60	7	7.80	8.50	9.10	10	12	14	17
Painting		0.15	0.15	0.19	0.19	0.19	0.25	0.30	0.35	0.40	0.45	0.58	0.70	0.85	1
Sq. Ft outside surface per L.F. pipe		0.220	0.275	0.344	0.497	0.622	0.753	0.916	1.047	1.178	1.456	1.735	2.26	2.81	3.34

Note: Cost includes material & labor, no overheads. Use double the L. F. price for each fitting. W.P.=Waterproof cover; Std.=Standard cover.

VALVES

Size, In.→ Plug Valves		½	¾	1	1¼	1½	2	2½	3	3½	4	5	6	8
						Cost, $ for each								
316 stainless	Screwed	31	45	50	—	84	92	—	—	—	—	—	—	—
	Flanged	40	52	57	—	90	102	—	175	—	340	—	580	—
Monel	Screwed	36	54	66	—	105	129	—	—	—	—	—	—	—
	Flanged	42	61	75	—	115	140	—	242	—	420	—	—	—

Diaphragm Valves

Body	Diaphram	½	¾	1	1¼	1½	2	2½	3	3½	4	5	6	8
Iron	Neoprene	9	10.50	12.50	16.80	22	30	42	51	—	—	—	—	—
	Teflon	21	22.50	25	29.80	40	51	74	90	—	—	—	—	—
S.S.	Neoprene	33	37	46	55	68	99	145	185	—	—	—	—	—
	Teflon	44	49	59	68	76	120	177	224	—	—	—	—	—
Aluminum	Neoprene	19	21	25	31	39	49	72	91	—	—	—	—	—
	Teflon	31	33	38	44	57	70	104	130	—	—	—	—	—

Valves cont'd. next page

VALVES

Diaphragm Valves

Size, In. →		½	¾	1	1¼	1½	2	2½	3	4	5	6	8
						Cost, $ for each							
Cast steel	Neoprene	23	25.50	33	39.50	50.50	63.50	95.	121	—	—	—	—
	Teflon	35	37.50	46	53	69	85	128	150	—	—	—	—
Glass lined	Neoprene	47	47	48	52	59	67	90	105	—	—	—	—
	Teflon	60	60	62	65	77	88	123	145	—	—	—	—
Hours to install		0.6	0.7	0.8	1.0	1.2	1.3	1.8	2	—	—	—	—
Pinch Valves Rubber	50 lb.	—	—	—	—	—	—	—	100	133	170	180	360
	150 lb.	—	—	—	—	—	—	—	110	160	210	230	420
Neoprene	50 lb.	—	—	—	—	—	—	—	105	140	172	185	355
	150 lb.	—	—	—	—	—	—	—	—	—	213	238	422
Hours to install		—	—	—	—	—	—	—	2	3.0	3.8	4.5	5.5

VALVES — For Water, Oil, Gas

Size, In. →		¼	⅜	½	¾	1	1¼	1½	2	2½	3	3½	4	5	6	8	10	12	14	16
										Cost, $ for each										
Gate 125 lb.	Screwed	3.20	3.20	3.60	4.10	4.95	6.50	7.90	11.45	19.60	29	—	—	—	—	—	—	—	—	—
	Flgd.	—	—	—	—	—	—	—	32.70	37.40	44	54.50	54.50	92.95	92.95	155.65	266.20	376.20	520.30	739.20
Gate 150 lb.	Screwed	3.50	3.50	4.10	4.80	5.70	7.70	9.40	13.50	22.70	34	—	—	—	—	—	—	—	—	—
	Flgd.	—	—	—	—	—	—	—	81.60	103.40	112.20	143	143	225.50	225.50	355.20	512.60	665.50	995.50	1,359
Gate 300 lb.	Screwed	7	7	8.20	10	13.50	20.30	26	36.10	58.30	78	—	—	—	—	—	—	—	—	—
	Flgd.	—	—	—	—	—	—	—	115.50	138.60	161.70	—	209	374	374	519.20	770.90	1,070	1,606	2,200
Globe 125 lb.	Screwed	2	2	2.40	3.50	4.60	6.10	8	12.30	18.60	28.70	—	—	—	—	—	—	—	—	—
	Flgd.	—	—	—	—	—	—	—	40.60	45.40	51.80	79.40	79.40	126.60	126.60	234.70	273.60	578	—	—
Globe 150 lb.	Screwed	5.30	5.30	6.10	7.40	10	13.80	18.60	28.60	44.50	66.40	—	—	—	—	—	—	—	—	—
	Flgd.	—	—	—	—	—	—	—	102.30	139.70	156.70	198	198	304.65	304.65	540	—	—	—	—
Globe 300 lb.	Screwed	6.50	6.50	7.30	9.60	13.10	18.10	22.70	35.80	62	91.30	—	—	—	—	—	—	—	—	—
	Flgd.	—	—	—	—	—	—	—	127.80	147.40	173.80	247.50	247.50	376.70	376.70	583	—	—	—	—
Check 125 lb.	Screwed	2	2	3	3.90	4.80	6.20	7.60	11.10	18.80	27.50	—	—	—	—	—	—	—	—	—
	Flgd.	—	—	—	—	—	—	—	21.15	25.35	30.20	45.87	45.87	72.75	72.75	154	258.40	363.10	612.70	860
Check 150 lb.	Screwed	3.10	3.10	3.60	4.80	6.30	8.60	10.80	16.30	29.50	45	—	—	—	—	—	—	—	—	—
	Flgd.	—	—	—	—	—	—	—	79.20	94.60	111.10	144.10	144.10	220	220	360.50	502.70	660.40	—	—
Check 300 lb.	Screwed	5	5	5.50	7	8.50	10.20	13	19.20	36	64	—	—	—	—	—	—	—	—	—
	Flgd.	—	—	—	—	—	—	—	95.20	106.70	122.10	162.80	162.80	269.30	269.30	440.25	632.50	832.70	—	—
Hours to install	Screwed	0.3	0.3	0.4	0.7	0.8	1.0	1.2	1.4	1.6	2.0	2.5	2.8	3.6	4.2	5.0	7.0	9.0	11.0	14.0
	Flgd.	—	—	—	—	—	—	1.2	1.4	1.8	2.0	—	—	—	—	—	—	—	—	—

Stainless

		¼	⅜	½	¾	1	1¼	1½	2	2½	3	3½	4
Gate	Screwed	—	—	43.20	55.20	61.80	81.60	96	115.80	168	192	—	281
	Flgd.	—	—	60.60	66	79.20	100.20	114.60	136.20	175.80	247.80	—	331.30
Globe	Screwed	—	—	—	—	67.20	—	100.20	123.60	168.40	192.20	—	317.20
	Flgd.	—	—	—	—	—	—	—	—	—	—	—	—
Check	Screwed	—	—	66	—	—	—	96	116	160	184	—	272
	Flgd.	—	—	—	—	—	—	—	—	—	—	—	—

INSTRUMENTATION

INSTRUMENTATION: PRESSURE INSTRUMENTS

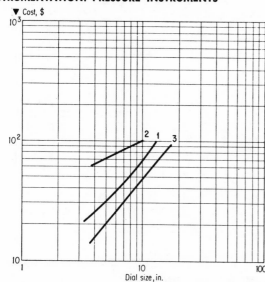

Curve No.	Description
1	Pressure gage (process). Dial type, flanged phenolic case, 316 stainless Bourdon and socket. Bottom connected with safety blowout.
2	Low-pressure gage with 316 stainless bellows.
3	Receiver pressure gage, 3-15 psi., pneumatic, flush mounted, phenolic case, bronze Bourdon. Bottom connection.

Pressure Switches

Diaphram type for instrument air service, 3-15 psi., 10 amp. Microswitch
Single setting $30
Dial setting $40

Pressure indicating controllers: Field-mounted pressure controller, stainless Bourdon tube, 0-3,000 psi., with pushbutton, $100.

Pressure transmitters, pneumatic: 316 stainless element, 30-1,000 psig., with mounting bracket and output gage. Indicating, $280; blind, $280.

INSTRUMENTATION: LEVEL GAGE GLASSES (REFLEX)

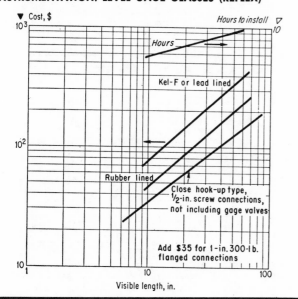

INSTRUMENTATION: LEVEL GAGE GLASSES (TRANSPARENT)

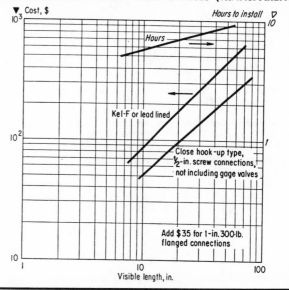

INSTRUMENTATION: THERMOCOUPLES

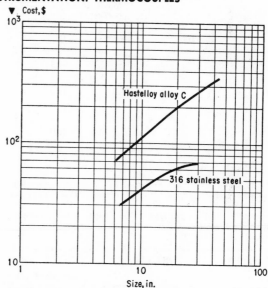

INSTRUMENTATION: POTENTIOMETERS

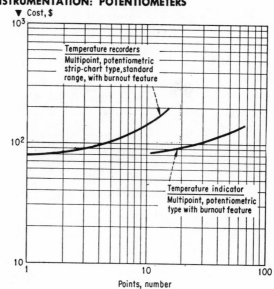

Temperature transmitters: Blind temperature transmitter, potentiometric electro to pneumatic converter, $550. Alarm contact, adjustable, $160. Air purge connection, $10.

Temperature switches: Bimetallic 15-amp. single-pole. single-throw Microswitch, to 400 F., $100.

Temperature indicating controller: Field mounted, 347 stainless bulb-5-ft. capillary, to 1,000 F., with pushbutton, $200.

INSTRUMENTATION: LEVEL CONTROLLERS
(External Displacement)

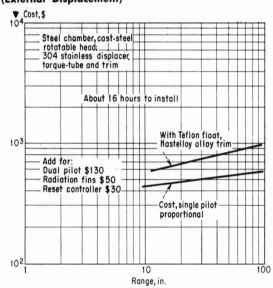

Level Switches

Float operated external steel cage, 1-150 lb. flanges, single-pole, double-throw contact. Weatherproof housing: $230; explosion-proof housing: $260.

Top-mounted displacement type, 3-125 flanges, 2-level tandem switch, 2 single-pole double-throw contacts. Weather-proof housing: $110; explosion-proof housing: $150.

INSTRUMENTATION: ORIFICE PLATES

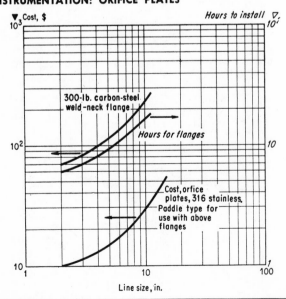

INSTRUMENTATION: ROTAMETERS (ARMORED)

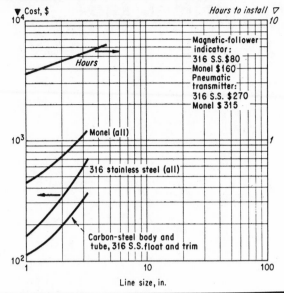

INSTRUMENTATION: PITOT TUBE

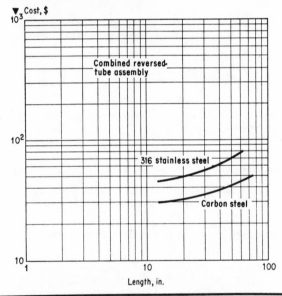

INSTRUMENTATION: FLOW TRANSMITTERS

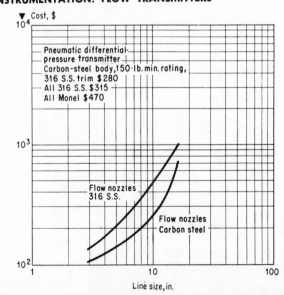

INSTRUMENTATION: VALVES

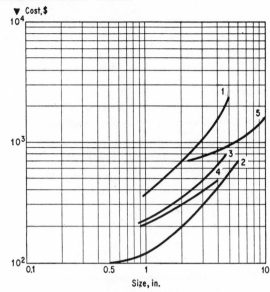

Curve No.	Description
1	Flow control valves. Stainless-steel body and trim.
2	Pressure-reducing valves. Steam, pilot operated, cast-iron body, stainless inner valve and seat, bronze diaphragm
3	Pressure-reducing valves, gas and liquid, spring loaded, steel body, stainless trim.
4	Temperature-reducing valves. Vapor-pressure type, bronze body, stainless trim, 8-ft. of copper capillary, integral thermometer on head.
5	Butterfly valves. Manual-geared-handwheel operated, wafer type, 304 stainless body and disk.

MATERIALS HANDLING

CHUTES AND GATES

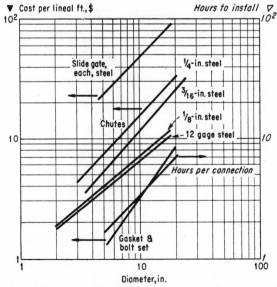

Prices are for carbon steel. For 304 stainless, multiply by 3; for aluminum, multiply by 1.85.

Flexible connections, $20 to $30 ea.

Transition piece, round to square, five times the per ft. price.

CONVEYING EQUIPMENT: APRON CONVEYORS

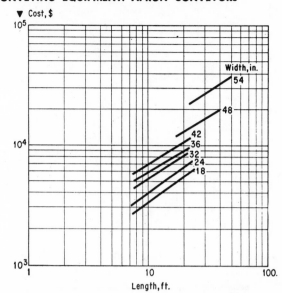

Size		Tons/Hr.	Hp.	Shipping Wt., Lb.	Hours to Install
Width, Ft.	Length, Ft.				
18	10	20	2	7,000	36
24	10	40	2	8,000	42
30	10	60	5	9,000	54
36	10	80	5	10,000	64
42	10	100	5	11,000	70
48	20	150	7½	21,000	132
54	30	200	10	51,000	262

CONVEYING EQUIPMENT: BELT CONVEYORS

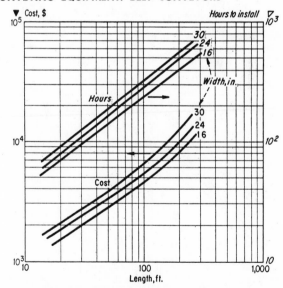

Cost includes steel welded framing, head pulley, tail pulley, 5-in. dia. troughing idlers 5-in. dia. return idlers, roller bearings on head shaft, screw take-up on tail shaft, 4 ply 28-oz. duct belt, cold-rolled steel shafts.
Not included: motor, drive or supports.

Capacities, Tons/Hr. (100 Lb./Cu. Ft. Material)

Belt Width, In.	Speed, Ft./Min.						Shipping Wt. Lb./Lineal Ft.
	150	200	250	300	350	400	
14	50	65	80	100	112	...	60
16	65	84	106	126	147	...	65
18	80	110	135	160	190	215	70
20	100	132	165	197	230	263	80
24	150	200	250	300	350	400	90
30	240	320	400	479	560	641	100

Extras: Belt enclosures, add 10%; tripper, add 12%; asbestos ply in belt, add 12%; deck plate, dust shield, add 10%.

CONVEYING EQUIPMENT: BUCKET ELEVATORS

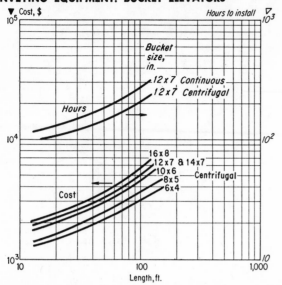

Cost includes casing, chain and sprockets, malleable iron buckets, roller bearings, screw take-up, boot and head.
Not included: motor, drive or guard.

Capacities, Tons/Hr. (100 Lb./Cu. Ft. Material)

Bucket Size, In.	Centrifugal	Continuous	Belt	Shipping Wt., Lb./Lineal Ft.
6x4	14	...	14	100
8x5	31	30	31	150
10x6	48	51	53	200
12x7	85	62	85	210
14x7	100	72	103	230
16x8	156	110	152	250

Extras: Machinery platform, $600; safety cage ladder, $6.50/lineal ft.; Heavy-duty chain and buckets, add 15% to cost.

For continuous elevators, multiply costs by 1.10.

For belt elevators, multiply costs by 0.90.

CONVEYING EQUIPMENT: PNEUMATIC SYSTEMS

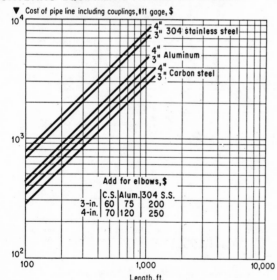

Add for elbows, $	C.S.	Alum.	304 S.S.
3-in.	60	75	200
4-in.	70	120	250

Cap., Tons/ Hr.*	Line Size, In.	Hp.	Cost of Powerpack, $†	Cap., Tons/ Hr.*	Line Size, In.	Hp.	Cost of Powerpack, $†
2	3	10	2,000	7	4	20	2,300
3	3	10	2,000	8	4	20	2,300
4	3	15	2,100	10	4	25	2,400
5	3	20	2,200	15	4	30	2,500
6	3	20	2,200				

* For free-flowing materials, 20 to 60 lb./cu. ft.
† Includes blower, motor (TEFC), base, coupling, checkvalve, pressure gages and filter.

Airlock and Manifold:

Pipe Dia., In.	Hp.	Cost, $*
3	¾	1,200
4	1	1,600

* Includes airlock, manifold, motor, reducer, roller chain drive and guard.

Cyclones

Pipe Dia., In.	Carbon Steel	Cost, $ Aluminum	304 Stainless
16	420	750	1,100
22	500	900	2,300

Bag dumping stations: Air feeder, aluminum, with safety grate and slitting knives, $800 ea.

Bin vent (carbon steel) with weather hood: 3-in., $350; 4-in. $700.

Flexible hose (galvanized steel with connections): 3-in. dia., $50/10 ft.; 4-in. dia., $60/10 ft.

CONVEYING EQUIPMENT: ROLLER CONVEYORS

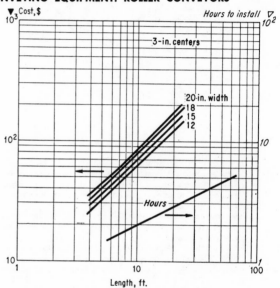

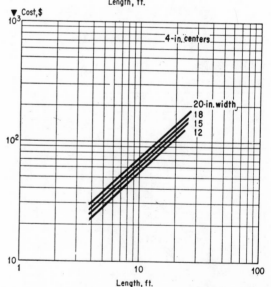

Width, In.	Capacity, Lb. 3-In. Center	4-In. Center	90° Curve, $ 3-in.	4-in.	Tripod Support, $
12	2,000	1,500	65	53	10
15	2,000	1,500	75	60	10
18	2,000	1,500	80	65	15
20	2,000	1,500	90	70	20

CONVEYING EQUIPMENT: ROTARY FEEDERS

Size, In.	Hp.	Shipping Wt., Lb.	Size, In.	Hp.	Shipping Wt., Lb.
4	1/3	300	10	1/3	925
6	1/3	550	12	1/3	1,100
8	1/3	750			

Not included: motor and drive.
Extras: Reversing mechanism: 4 to 10-in. size, $100; 12-in. size, $230.

(Rotary Feeders)

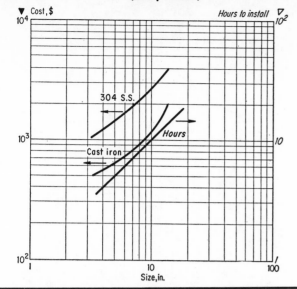

Not included: motor and drive. For 316 stainless, multiply cost by 1.50. If balancing is necessary, add 50%. For 10 or more tons/hr.

Width of Pan, In.	Hp.	Hours to Install Conveyor Per L. Ft.	Cost of Cover, $ Carbon Steel	Stainless
12	2	2	20	30
18	3	2	26	40
24	5	2	30	45
36	7½	3	35	55

HOISTING EQUIPMENT: HAND AND ELECTRIC

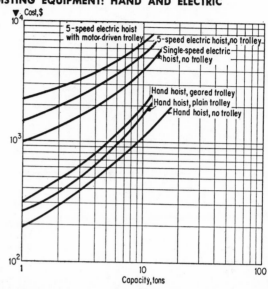

Extras: Acid-resistant construction, $70; dust-tight construction, $30; power reels, $150; chain container, $30.

Capacity, Tons	Hoist Hp.	Trolley Hp.	Lift, Ft. Hand	Elect.	Hours to Install Hoist	Trolley
1	1½	¾	8	16	2	3
2	3	¾	8	16	3	4
3	5	¾	10	18	3	4
4	…	…	10	..	4	5
5	7½	1	12	18	4	5
6	…	…	12	..	4	5
8	7½	1	12	20	5	6
10	7½	1½	12	20	5	7
12	7½	1½	12	18	5	7

CONVEYING EQUIPMENT: SCREW CONVEYORS

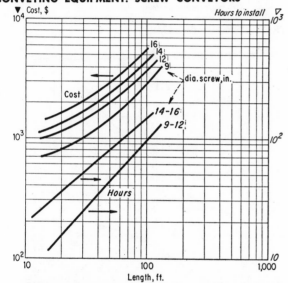

Cost includes trough and cover, screw flight, bearings, inlet and outlet.
Not included: motor, drive or supports.
For 304 stainless, multiply by 3.5.

Dia. of Screw, In.	Cap., Tons/Hr. Grain	Sand	Hp.	Shipping Wt., Lb./L. Ft.
9	22	55	1	30
12	46	100	1.5	50
14	72	160	2	70
16	110	250	3	90

DRIVE EQUIPMENT

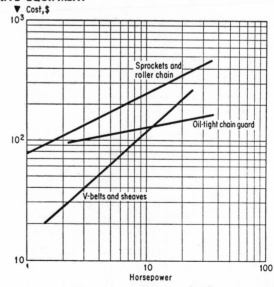

Hp.	Reducers	Hours to Install Guards	Motors	Sprockets & Roller Chain
1	12	8	5	4
1½	12	8	5	4
2	12	8	5	4
3	12	8	6	4
5	16	10	7	6
7½	16	10	9	6
10	20	12	13	8
15	20	12	16	8
20	24	12	19	12
25	24	12	24	12

CONVEYING EQUIPMENT: VIBRATING CONVEYORS

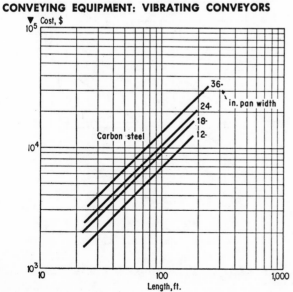

(Drive Equipment)

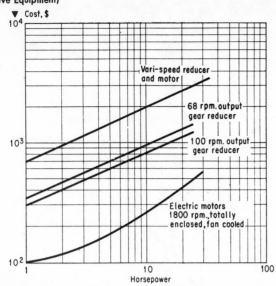

SCALES

	Cost, Each, $
Beam scale	
500 lb. capacity	120 (2)*
1,000 lb. capacity	160 (2)*
Bench, dial & beam, to 2,000 lb.	900 (6)*
Floor, dial & beam, to 6,500 lb.	
Semi-frame	1,300 (18)*
Full-frame	1,500 (20)*
Suspended platform, to 2,600 lb.	1,700 (24)*
Tank scales (complete with weigh bridge and saddles)	

Tank size, ft.	Live load	Dead load	
4½x12	10,000 lb.	15,000 lb.	
(on stand with tape drive cabinet dial)			2,400(48)*
6x20	50,000 lb.	10,000 lb.	
(with 12 ft. extension levers, beam)			2,100(48)*
10x36	150,000 lb.	25,000 lb.	
(with 8ft. extension levers, beam)			4,500(48)*

	Cost, Each, $
Mercury magnetic switch	65
Slave relay	35

* Number in parenthesis is installation man-hr.

POWER

POWER EQUIPMENT: AIR COMPRESSORS

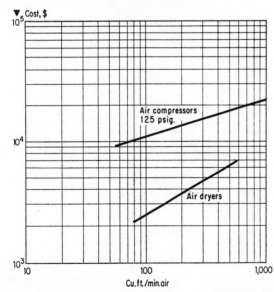

Cost is for single stage, double acting, heavy duty, water cooled, electric driven, 125 psig. Included are compressor, motor, air filter, aftercooler and receiver.

Cu. Ft./Min. Air	Rpm.	Hp.
143	230	50
275	275	75
325	300	100

POWER EQUIPMENT: COOLING TOWERS

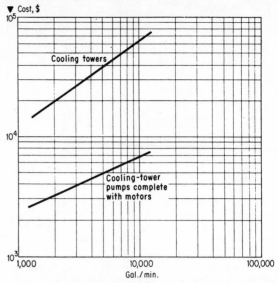

Cost includes wooden tower construction, fans, motors and drives.

POWER EQUIPMENT: BOILERS

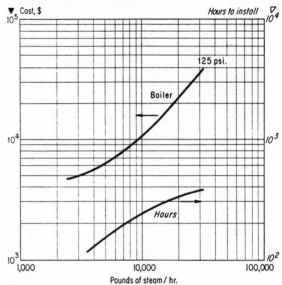

Cost includes fuel-burning equipment, induced or forced draft fan and motor, all controls and integral piping.

POWER EQUIPMENT: REFRIGERATION
(Mechanical Compression)

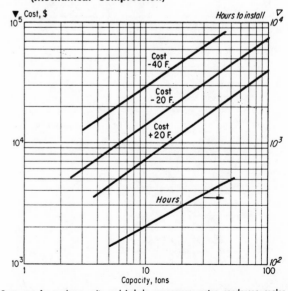

Costs are for package units and include compressor, motor, condenser, cooler and controls.

ELECTRIC WIRE AND CONDUIT

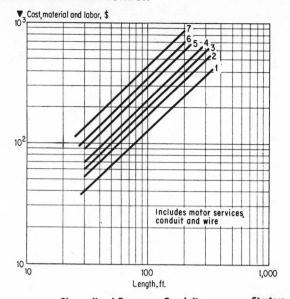

Includes motor services, conduit and wire

Curve No.	No. of Wires	Wire Size	Sherardized Raceway Conduit, 600-Volt Wire On Centers, In.	Max. Hp. 440-V.	Max. Hp. 220-V.	Starters, 440-V., Cost, $ Oil	Starters, 440-V., Cost, $ NEMA-1
1	3	12	¾	....	....	...	...
2	6	12	1	1–10	1–5	200	100
3	3 3	10& 12	1	15	7½	200	110
4	3 3	8& 12	1¼	20–25	10	250	150
5	3 3	2& 12	1½	50–60	25	280	150
6	3 3	0& 12	2	75	30	400	250
7	3 3	000& 12	2	100	50	400	250

Safety switches: 30 amp., $80; 60 amp., $100; 100 amp., $170; 200 amp., $230; 400 amp., $500.

Extras: Pushbuttons are $20 ea.; starter rack, $80/ft. plus 8 hr. installation time per ft.; equipment lighting fixtures, $12 ea.; substation, $25/kva.; overhead feeder, $8/ft.

BUILDINGS, YARD

BUILDINGS AND CONSTRUCTION

Total Cost of Erected Buildings	Unit	Cost, $	Man-Hr. to Install
Laboratory: steel frame, masonry walls, floor and roof; heating, lighting and plumbing	Sq. ft.	30	*
Office: steel frame, masonry walls, floor and roof; heating, lighting and plumbing	Sq. ft.	25	*
Process building: multi-level, 12-ft. clearance, steel platforms, heating, lighting and plumbing.			
Masonry construction	Sq, ft.	22	*
Aluminum on steel	Sq. ft.	20	*
Transite on steel	Sq. ft.	18	*
Open structure: 3-level, steel, with lighting & plumbing	Sq. ft.	16	*
Warehouse: single story, 15-ft. clearance. Steel frame, masonry walls, floor and roof; heating, lighting and plumbing.	Sq. ft.	15	*
Doors			
Metal: Steel frame, 8x8 ft., automatic	Ea.	850	30
Steel rolling, 12x12 ft. manual	Ea.	600	24
Single slide, 4x8 ft.	Ea.	100	5
Wood: Sectional, overhead, 12x10 ft.	Ea.	200	6
Swing exterior, 3x7 ft.	Ea.	60	4
Excavation			
Machine	Cu. yd.	2 to 4	*
Hand	Cu. yd.	3 to 6	*

	Unit	Cost, $	Man-Hr. to Install
Floors			
Asphalt tile	Sq. ft.	0.60	*
Concrete, reinforced, 4-in. thick	Sq. ft.	2.00	*
Steel grating	Sq. ft.	3.65	*
Wood deck, 2-in. thick	Sq. ft.	1.00	*
Foundations, included excavation, backfill and forming: flat slab, 1 cu. yd. concrete, 5.3 sq. ft. forms, 100 lb. reinforcing steel†	Cu. yd.	25	6
Pits & basins: 1 cu. yd. concrete, 17.5 sq. ft. forms, 115 lb. reinforcing steel†	Cu. yd.	28	8
Walls and piers: 1 cu. yd. concrete, 57.0 sq. ft. forms, 160 lb. reinforcing steel†	Cu. yd.	60	16
Lumber			
Structural, plain	MBM	170	30
Structural, creosoted	MBM	235	30
Plywood: ⅜-in.	Sq. ft.	0.12	1
½-in.	Sq. ft.	0.16	1
⅝-in.	Sq. ft.	0.17	1
Piling (20–25 ton load, 60-ft. long)			
Wood treated	Ea.	140	*
Wood plain	Ea.	110	*
Composite	Ea.	340	*
Concrete	Ea.	420	*
Test pile	Ea.	1,600	*
Load test	Ea.	2,000	*
Equipment "on and off site"	Per Job	3,000	*
Roofs			
Aluminum	Sq. ft.	0.62	*
Built-up 5-ply	Sq. ft.	0.35	*
Reinforced concrete	Sq. ft.	1.30	*
Steel coated	Sq. ft.	0.52	*
Transite	Sq. ft.	0.70	*
Sprinkler Systems			
Wet system	Sq. ft.	0.70	*
Dry system	Sq. ft.	0.90	*
4-in. alarm valve, wet system	Ea.	400	40
4-in. alarm valve, dry system	Ea.	600	50
6-in. alarm valve, wet system	Ea.	450	40
6-in. alarm valve, dry system	Ea.	650	50
Compressor to operate 500 heads	Ea.	250	16
(Valves and compressor not included in sq. ft. price)			
Structural Steel			
Grating, 1¼-in. standard	Sq. ft.	2.00	0.33
Grating, expanded metal	Sq. ft.	1.50	0.17
Grating, checker plate, 5/16-in.	Sq. ft.	1.80	0.13
Handrail, standard pipe	Lineal ft.	3.60	0.67
Handrail, bar type	Lineal ft.	1.50	0.70
Ladder, with safety cage	Lineal ft.	7.00	0.50
Ladder, without safety cage	Lineal ft.	2.50	0.35
Operating platforms, including stairs	Sq. ft.	12.00	1.8
Stairway	Vert. ft.	50.00	....
Stair treads	Ea.	8.00	1
Building steel, shop fab	Lb.	0.20	0.01
Platform and support steel, shop fab	Lb.	0.25	0.02
Toeplate steel, standard	Lineal ft.	0.75	0.10
Walls			
Siding			
Transite	Sq. ft.	0.68	*
Aluminum	Sq. ft.	0.65	*
Steel-coated	Sq. ft.	0.55	*
Brick			
4-in.	Sq. ft.	1.15	*
8-in.	Sq. ft.	2.30	*
12-in.	Sq. ft.	3.40	*
Concrete block			
6-in.	Sq. ft.	1.10	*
8-in.	Sq. ft.	1.30	*
12-in.	Sq. ft.	1.70	*
Windows, Industrial			
Steel	Sq. ft.	4.50	*
Aluminum	Sq. ft.	6.00	*
Wood	Sq. ft.	1.80	*

* Labor cost is included in material cost.
† To adjust: forms, $1/sq.ft.; reinforcing steel, 15¢/lb.

YARD IMPROVEMENTS

		Unit	Cost, $	Man-Hr. to Install
Car spotters & accessories				
Stationary				
3 full cars, 5 hp		Ea.	890.00	20
5 full cars, 10 hp		Ea.	1,650.00	30
Portable				
3 full cars, 5 hp		Ea.	1,510.00	24
6 full cars, 10 hp		Ea.	2,500.00	32
Drum type car pullers				
Hp.	Wt. lb.			
10	1,900	Ea.	2,470.00	24

		Unit	Cost, $	Man-Hr. to Install
15	3,200............	Ea.	3,050.00	36
20	4,300............	Ea.	4,100.00	48
	(Motors are included).			

Rope

Dia., In.		Unit	Cost, $	
½	wire............	L.f.	$0.40	—
¾	wire............	L.f.	0.50	—
⅞	wire............	L.f.	0.63	—
1	wire............	L.f.	0.80	—
1¼	manila............	L.f.	0.40	—
1½	manila............	L.f.	0.57	—
1⅝	manila............	L.f.	0.70	—
1¾	manila............	L.f.	0.84	—
2	manila............	L.f.	1.00	—

Docks & wharfs

	Unit	Cost, $	
Concrete & steel, 100 lineal ft. wide	L.f.	5,000	*
Wood, 100 lineal ft. wide........	L.f.	1,500	*

Dredging

	Unit	Cost, $	
....	Cu. yd.	1.20	*

Pipe bridges (includes structural steel and foundations)

	Unit	Cost, $	Man-Hr.
Heavy duty.....................	L.f.	19.00	2.5
Light duty.....................	L.f.	12.00	1.8
Pipe column			
16ft.—4in.................	Ea.	60.00	9
16ft.—6 in.................	Ea.	75.00	10
22ft.—6in.................	Ea.	100.00	20

Plant fence

	Unit	Cost, $	
7ft. Chain link (3-strand B.W.)....	L.f.	5.00	*
3ft. Wide man gate............	Ea.	30.00	*
10ft. Equipment gate...........	Ea.	100.00	*
20ft. Truck or R.R. gate (manual)...	Ea.	400.00	*
20ft. Truck or R.R. gate (automatic).	Ea.	1,800.00	*
Relocate plant fence............	L.f.	2.60	*

Railroads

	Unit	Cost, $	
Track (standard)...............	L.f.	$21.50	*
(tram)...................	L.f.	13.50	*
Turnout.....................	Ea.	2,500.00	*
Bumper.....................	Ea.	350.00	*
Blinker & gate................	Ea.	7,500.00	*
Ties, creosoted (6"x8"x8').......	Ea.	7.00	*
Grade & ballast...............	L.f.	5.00	*

Cars, ore, 24-in. track

Capacity, Cu. Ft.	Wt., Lb.	Unit	Cost, $	
12	660.......	Ea.	320.00	*
16	700.......	Ea.	340.00	*
20	930.......	Ea.	380.00	*

Locomotives, mine, battery type

Size, Tons	Wt., Lb.	Unit	Cost, $	
9	19,000........	Ea.	28,000.00	—
12	26,000........	Ea.	32,000.00	—

Locomotive, mine, Diesel type

1½	3,000	2 cyl.....	Ea.	8,600.00	—
3	7,000	4 cyl.....	Ea.	10,600.00	—

Roads & walkways

Concrete, slab, mesh reinforced

Thickness, in.	Sub-base, in....	Unit	Cost, $	
4	6.........	Sq. yd.	6.50	*
6	6.........	Sq. yd.	7.50	*
8	6.........	Sq. yd.	8.50	*
Paving asphalt				
3	12.........	Sq. yd.	6.00	*

Sewers

Reinforced Concrete, Thickness, in.

	Unit	Cost, $	
12	L.f.	4.40	*
15	L.f.	5.50	*
18	L.f.	6.50	*
24	L.f.	10.00	*
30	L.f.	12.00	*

Vitrified Clay, Thickness, in.

	Unit	Cost, $	
4	L.f.	3.00	*
6	L.f.	3.50	*
8	L.f.	5.50	*
12	L.f.	8.00	*
15	L.f.	11.00	*
18	L.f.	15.00	*
24	L.f.	20.00	*

Site development

	Unit	Cost, $	Man-Hr. to Install
Clearing & grubbing............	Sq. yd.	0.20	*
Grade out.................	Cu. yd.	0.50	*
Cut, fill & compact.............	Cu. yd.	0.70	*
New fill compacted.............	Cu. yd.	0.60	*
Topsoil.....................	Cu. yd.	0.65	*
Gravel fill.................	Cu. yd.	2.50	*
Crushed stone...............	Cu. yd.	3.50	*
Seeding.................	Sq. yd.	0.25	*

Sluiceway

	Unit	Cost, $	
Open, piled & sheathed.........	L.f.	150.00	*

Wells

	Unit	Cost, $	
200 Gpm. 400ft. deep, 15 hp....	Ea.	16,000.00	*
500 Gpm. 200ft. deep, 40 hp....	Ea.	21,000.00	*
1200 Gpm. 400ft. deep, 75 hp....	Ea.	29,000.00	*

* Labor cost is included with material cost.

AUTOMOTIVE EQUIPMENT

Fork lift trucks

			Cost, $
3,000 lb.	15-in. load center	72-in. lift	6,200
4,000	15	72	6,750
5,000	24	72	7,500
6,000	24	84	8,500
8,000	24	84	9,500
10,000	24	92	11,000

	Cost, $
Hand truck, heavy duty...............................	50
Hydraulic pallet truck, 4,000 lb......................	750
Jack lift electric pallet truck, 4,000 lb.............	2,700
Payloaders	
18 cu. ft., gas........................	5,000
20 cu. ft., gas........................	7,200
1 cu. yd., gas........................	8,200
1.75 cu. yd., gas........................	15,000
1.75 cu. yd., diesel........................	16,700
4.0 cu. yd., gas........................	25,000
4.0 cu. yd., diesel........................	27,000
Railway tank cars (8,000 gal.)	
Steel........................	11,000
Aluminum........................	16,000
Stainless steel........................	27,000
Tank trailers (4,300 gal. unlined)	
Carbon steel........................	10,000
Aluminum........................	13,000
Stainless steel........................	16,000
Tractors	
Gasoline........................	10,000
Diesel........................	20,000
Tractor shovel	
2¼ cu. yd. bucket 105 hp........	21,000
2¼ 125 hp........	23,000
3¼ 150 hp........	28,500
Walkie pallet truck, 4,000 lb........................	1,900
Battery........................	500
Charger........................	400

Meet the Author

Herbert E. Mills is a cost estimator for General Chemical Div., Allied Chemical Corp., Morristown, N. J. He has a background of over 20 years in methods, time and motion study, and cost estimation.

A frequent contributor to this magazine, Mr. Mills was educated in industrial management at Temple University. He has served as treasurer of the Delaware Valley section of the American Assn. of Cost Engineers.

Estimating the Cost of Process Buildings via Volumetric Ratios

When using ratio methods to estimate the cost of a building that will house a processing unit, the "volumetric ratio" concept can lead to greater accuracy and to a better visualization of the building's dimensions.

WILLIAM G. KNOX, *Koppers Co.*

The ratio method of investment evaluation is a widely accepted tool to provide economic guidance for research projects involving new processes; with it, the engineer can use simple multiplication to convert his estimate of bare-equipment cost to an estimate of total plant investment.

During the 20 years since Hans Lang[1] first presented the idea of using ratios, numerous articles have been presented in an effort to improve the accuracy of this technique. These improvements generally involved methods that would permit the influence of total equipment cost, materials of construction, multiplicity of units, extent of field fabrication, location factors, etc., to be applied to the installation ratio, thereby improving the reliability of the calculated investment.

One of the least evaluated components of plant investment is the cost of buildings. This Cost File will outline a procedure to improve building cost estimation.

Every process has a unique building requirement. A highly automated plant constructed entirely out-of-doors will require only a control room for the operator and the control panel. In many instances, the relatively minor cost of such a building could be completely ignored. At the other extreme of the scale is the process that is entirely enclosed in a building or buildings.

The scale of operations can also dictate the building requirements. For example, the distillation columns for a particular plant might be of a size that could be housed in the process building, thereby eliminating the complications that weather may bring to operations and maintenance. On a larger scale, the distillation facilities for the same process might logically be located out-of-doors. Thus the building requirement of a process is distinctive to that process at the selected scale of production.

One available building-cost estimating method is to include the cost of a building in the investment ratio. Tables have been developed and published that show building cost as a percentage of the total plant cost. These tables incorporate the variables of plant size and the general type of processing involved. Generally, the buildings average about 8% of the total plant cost. The range of costs, however, appears to be from 3% to 13%. This means that the use of an average building factor might by itself cause the total plant investment to be in error by ±5%. Since the probable accuracy attainable by the ratio method is ±25% of the total plant investment according to Hackney,[2] the possible error in building cost can be a sizable portion of the total estimating inaccuracy.

An accurate method of obtaining the cost of buildings is to prepare a general equipment-arrangement drawing so that the size and shape of the building can be estimated. This method is generally too time-consuming and sophisticated for use with the ratio-type estimate for research projects.

Proposed: The Volumetric Ratio Method

The proposed estimating method is based on the observation that the processing equipment actually occupies a relatively small but nearly constant fraction of the total volume of a process building.

Nine actual installations were reviewed. Equipment volumes were calculated for major equipment items, such as reactors, tanks, dryers, centrifuges, condensers, heat exchangers, distillation columns, hoppers, compressors, etc. Miscellaneous volumes such as piping and panelboards were not included. Pumps were not included (although they might be if there were an unusually large number or if the pumps were unusually large). The building volumes were obtained from specification drawings and by actual measurement.

135

Warehouses and separate control rooms were not included.

Based on this limited sample, it was calculated that the process equipment occupied only 3.75% of the total building volume. At first glance, this appears to indicate that the volumetric utilization is extremely low. However, the remaining "empty" space must contain all of the piping, the electrical systems, services, structural work, ventilation equipment, and panelboards. In addition, there must be floor area and empty space to provide access to the equipment for operations and maintenance, and to permit movement of personnel.

The equipment volumes used in the volumetric ratio method are those most readily obtained. Reactor volumes are taken as the total internal volume. If this value is not available, it can be calculated by assuming that the reactor operates with a 20% free-board; the internal volume, then, is the nominal volume divided by 0.80. Distillation columns are generally sized by internal cross-sectional area; their total volume is calculated using the internal diameter.

Heat exchangers present more of a problem, since they are available in a wide range of types and tube sizes. Fortunately, the volume of shell-and-tube exchangers tends to be proportional to the heat-transfer surface for a given tube size, almost independent of total exchanger area. The catalogs of a number of manufacturers were reviewed, and it was found that the factor relating exchanger volume to the heat-transfer surface varied from 0.033 cu.ft./sq.ft. for $\frac{3}{8}$-in. tubes, to 0.062 cu.ft./sq.ft. for 1-in. tubes. A value of 0.046 ($\frac{5}{8}$-in. tubes) is a good general factor to use, because tube sizes are not established in a ratio estimate unless the cost curves used relate tubing size to cost.

Vertical vaporizers, of the type used for propane, butane or ammonia, have a volume:area relationship of about 0.2 cu.ft./sq.ft. of transfer surface. Factors of this kind can be established for other heat exchanger types.

In actual practice, the engineer-estimator sizes the processing equipment to obtain an estimate of equipment costs. Equipment items such as reactors, tanks, distillation columns, dryers and hoppers are sized by volume and, therefore, the engineer will have an actual volume for these units. The volumes of other items, such as heat exchangers and compressors, will have to be obtained from manufacturers' catalogs, or by using a volume factor as described above for heat exchangers. Occasionally, because the research project is still in the laboratory stage, there will not be sufficient data for the engineer to even roughly design a specific item of equipment. In this case, the engineer will have to exercise judgment in estimating both equipment cost and volume.

Once the volumes are all available, the engineer decides on the specific items to be included in the building. He then applies the volumetric ratio to the equipment volumes to obtain the estimated building volume.

Sufficient published data are available on the volumetric cost of buildings to permit a building cost to be calculated. Actual inhouse building costs would be more suitable, where available.

Sample Calculation

Let us estimate the size and cost of a building to house a process requiring three reactors, each having a nominal volume of 2,000 gal., three 500-sq.ft. condensers, a 5,000-gal. holding tank, and a 12-in. dia. stripping column 20-ft. high with a 500-gal. reboiler.

Item	Volume in Cu. Ft.
Reactors, 3 x 2,000 x 1/0.8	
= 7,500 gal.	1,000
Condenser, 3 x 500 x 0.046	69
Holding tank, 5,000 gal.	670
Column, $(0.5)^2$ x (π) x (20)	16
Reboiler, 500 gal.	67
Total	1,822 cu. ft.

Volume of building = 1,822/0.0375 = 48,600 cu. ft.
Total building cost at $0.80 per cu. ft. = $39,000

Advantages of This Approach

The engineer obtains a bonus from the use of this volumetric ratio: Not only will the accuracy of the cost estimate be improved, but he obtains an estimate of the actual physical size of the building. In reference to the example above, the reactors will be about 22 ft. high over-all. If a total building height of 30 ft. is assumed, then the foundation area can be calculated to be 1,600 sq. ft. This is a building 40 ft. square or, more probably, a building 60 ft. long by about 27 ft. wide so that the reactors could be placed side-by-side. At this point, then, the engineer can visualize his process building. In spite of all attempts to make estimating a mechanical procedure, judgment on the part of the engineer is still a vital factor, and this visualization of the process in a building should be helpful here.

The volumetric ratio will vary from process to process and from company to company, depending upon the design philosophy. It is recommended that, wherever possible, appropriate volumetric ratios be determined, based on actual experience. The reported value of 3.75% is based on a rather limited sample, but it should be adequate until a larger quantity of data is evaluated.

References

1. Lang, H., Simplified Approach to Preliminary Cost Estimates, *Chem. Eng.*, June 1948, p. 112.
2. Hackney, Estimating Methods for Process Industry Capital Costs, *Chem. Eng.*, April 4, 1960, p. 119.

Meet the Author

William G. Knox is Manager of the Engineering Group in the Research Dept. of Koppers Co. (15 Plum St., Verona, Pa. 15147), where he is responsible for providing engineering and cost analysis for laboratory and pilot-plant development projects. His background includes engineering experience with Monsanto Co. and Commonwealth Engineering. He graduated from the University of Cincinnati in 1949 with a B.S. in Chemical Engineering, and is a member of AIChE.

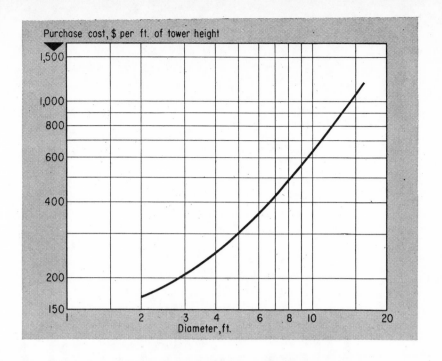

Costs for Rotating Disk Contactors

JACKSON CLERK, *Holland, Mich.*

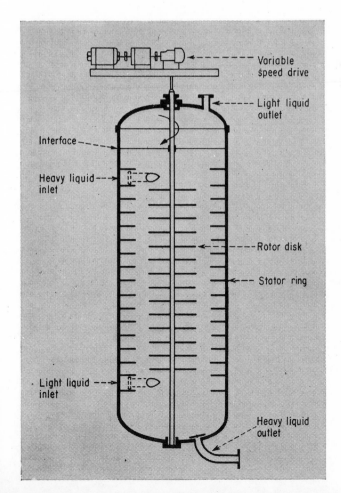

Surprisingly few data have been published on cost of extraction equipment, particularly equipment for liquid-liquid extraction. Bauman has some costs for sieve trays and towers.[1] Aries and Newton presented a cost-curve for continuous centrifugal extractors[2] (but the data, which appear to be for 1950, probably cannot be updated with assurance using a cost index). A 1961 Cost File[3] contained a nomograph for the cost of various packings (Raschig and Berl).

During a recent investigation, data was required on the popular rotating disk extraction column. This is a relatively new continuous contactor, consisting of a tower divided into a number of compartments by a series of stator rings. Rotors are attached to a common vertical shaft and are centered in each compartment. The shaft is driven by an electric motor (see drawing).

Purchase-cost information developed as a result of our investigation is shown on the chart: this is for carbon-steel contactors 25 to 50-ft. high, with a design pressure of 150 psig. Costs include disks, rings, manways and nozzles, supports and skirt, but not motor or drive. Data are for July 1964.

Ratio factors for 304 stainless are: 2.6, 3.3, 3.4, 3.5 for 2, 5, 10, 16-ft. dia. respectively (c.s. is 1.0).

References

1. Bauman, C. H., "Fundamentals of Cost Engineering in the Chemical Industry," pp. 66-67, Reinhold, 1964.
2. Aries, R. S., Newton, R. D., "Chemical Engineering Cost Estimation," p. 42, McGraw-Hill, 1955.
3. Wroth, W. F., *Chem. Eng.,* July 10, 1961, p. 166.

Key Concepts for This Article

Active (8)	Passive (9)	Means/Methods (10
Estimating	Extractors	Charts
	Contactors	Curves
	Rotating disk	Factors
	Continuous	
	Columns (process)	
	Towers	

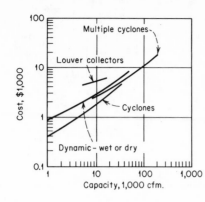
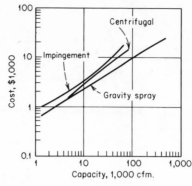
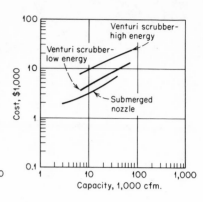

Dust Collection Equipment

GORDON D. SARGENT, Nopco Chemical Div., Diamond Shamrock Chemical Co.

Costs shown on Fig. 1, are manufacturers' prices for dust collectors only. System costs must be developed for the individual case, adding the cost of freight, foundations, supports, erection, and auxiliary equipment. Collector costs may be adjusted for special materials of construction. The cost basis year is 1968. (Marshall and Stevens Index = 273).

For any one type of collector, price may vary widely depending on the required duty. The costs presented are for "typical" applications suitable for

Calculation of collector efficiency——Table I

Particle Size Distribution, Microns	Particle Size Midpoint of Range, Microns	Fraction of Total Dust in Range, %		Cyclone Efficiency at Midpoint, %*		Percent Dust Collected	Dust Emitted, %†		Electrostatic Precipitator Efficiency at Midpoint, %		Dust Collected,%
<2½	1¼	12	×	33.5	=	4.0	50.3	×	77.0	=	38.7
2½–5	3¾	8	×	64.5	=	5.2	17.6	×	90.5	=	16.0
5–7½	6¼	6	×	76.7	=	4.6	8.8	×	94.0	=	8.3
7½–10	8¾	4	×	84.2	=	3.4	3.8	×	95.0	=	3.6
10–15	12½	8	×	89.3	=	7.1	5.7	×	95.5	=	5.4
15–20	17½	7	×	92.0	=	6.4	3.8	×	96.0	=	3.7
20–30	25	10	×	94.3	=	9.4	3.8	×	96.5	=	3.7
30–40	35	10	×	96.0	=	9.6	2.5	×	96.8	=	2.4
40–60	50	15	×	97.3	=	14.6	2.5	×	97.7	=	2.4
60–75	67½	10	×	98.5	=	9.9	0.6	×	98.7	=	0.6
75–104	89½	7	×	99.1	=	6.9	0.6	×	99.2	=	0.6
104–150	127	3	·×	100.0	=	3.0	—		—		—
						84.1%					**85.4%**

Data are those of Stairmand (Ref. 12).

*Efficiencies as read from an enlarged curve.
†Calculating emitted dust from collected:
Percent of 1¼ micron dust emitted = $\frac{12 - 4.0}{100 - 84.1}$ = 50.3%

Calculating combined efficiency:
Cyclone fraction emitted = 1.000 − 0.841 = 0.159
Precipitator fraction emitted = 1.00 − 0.854 = 0.146
Combined fraction emitted = 0.159 × 0.146 = 0.0232
Combined fraction collected = 1 − 0.0232 = 0.977
Combined percent collected = 97.7%

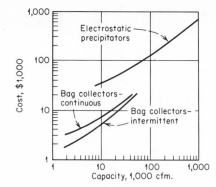

TYPICAL COSTS of various kinds of dust collectors plotted against capacity (costs are for equipment only)—Fig. 1

a preliminary comparison. Where costs of systems are of the same order of magnitude, it is probably worthwhile to obtain quotations for each dust collector.

Over-All Cost of Gas Cleaning

In an economic comparison between different collectors for a particular application, the efficiencies are likely to be different. The higher cost for better collection can be compared with the cheaper installation, as was compared in Table I Costs above those shown tend toward an uneconomical equipment choice, while those below tend toward being a good buy.

The original data for this figure were presented by Stairmand[12] and are adjusted to current U.S. costs in Table II. Stairmand developed the operating costs of the various collector systems cleaning 60,000 cfm. gas at 68 F. with an inlet dust-loading of 5 grains/cu.ft. of the standard test dust given in Table V, with 30% less than 10 μ particle size.

It is interesting to note the high-cost units are the spray chamber (with a high water-rate) and the dynamic scrubber (with a high power-rate). Also, the electrostatic precipitator may have a high first cost but efficiency is high and power costs are reasonable.

Over-all cost of gas cleaning—Table II

Collector Type	Efficiency on Standard Dust, %	Average Pressure Drop, in w.g.	Installed Cost, $	Power Cost, $/yr.	Water Required, Gal/1,000 Cu. Ft.	Water Cost $/Yr.	Maintenance, $/Yr.	Total Annual Cost, $/Yr.	Total Annual Cost ¢/(Yr.)(Cfm.)
Dry									
1. Louver collector	58.6	1.7	34,500	1,560	——	——	300	5,310	8.9
2. Medium efficiency cyclone	65.3	3.7	25,000	3,380	——	——	200	6,080	10.1
3. High efficiency cyclone	84.2	4.9	48,500	4,520	——	——	200	9,570	16.0
4. Multiple cyclone	93.8	4.3	52,500	3,960	——	——	200	9,410	15.7
5. Electrostatic Precipitator	99.0	0.9	233,000	2,000	——	——	1,300	26,600	44.4
6. Fabric filter, shaker*	99.7	2.5	165,000	3,740	——	——	10,000	30,240	50.4
7. Fabric filter, envelope*	99.8	2.0	152,000	3,380	——	——	9,500	28,080	47.0
8. Fabric filter, reverse jet	99.9	3.0	231,000	7,920	——	——	19,000	50,020	83.6
Wet									
9. Submerged Nozzle	93.6	6.1	66,700	5,640	0.7	1,010	700	14,020	23.3
10. Spray Chamber	94.5	1.4	139,000	4,760	21.7	31,250	1,000	50,910	84.8
11. Impingement Scrubber	97.9	6.1	82,200	5,800	3.6	5,190	1,000	20,210	33.7
12. Wet Dynamic Scrubber	98.5	——	136,000	45,400	6.0	8,640	700	68,340	141
13. Low energy venturi	99.7	20.0	107,000	18,820	8.4	12,100	1,000	42,620	71.2
14. High energy venturi	99.9	31.5	117,000	29,740	8.4	12,100	1,000	54,540	91.0

Notes:
The installed cost includes equipment, auxiliaries, such as fans, pumps, and motors (but not including solids disposal equipment), site preparation and installation. Installation charges are 100% of all equipment except electrostatic precipators and dynamic scrubbers which are 50%.
Prices are adjusted to 1968 with a Marshall and Stevens index of 273. The British pound is taken at $2.84 U.S.
The cost of power is taken at $0.01 per kwh. Efficiencies of fans and motors are assumed 60%. Water cost is $0.05 per 1,000 gal. and reported usage converted from Imperial (assumed) to U.S. gallons. (Data are from Stairmand, Ref. 11.)

*Maintenance charges include bag changes, once per year for envelope-type and twice for shaler and reverse-jet filters.

Estimating Costs of Process Dryers

Data for five popular dryers—direct rotary, louver, rotary vacuum, vacuum shelf, and spray—are presented as an aid in making rough cost estimates.

Most processes involving the treatment of liquids and solids include a drying step somewhere on the flowsheet. As an aid in estimating costs for drying a chemical product, we have correlated data from published sources for five widely used process dryers.

One of the simplest and least expensive of commercial dryers is the direct rotary. It consists of a horizontal cylinder in which solids to be dried make direct contact with a hot gas stream. The costs shown represent data for a "standard" rotary dryer, equipped with a rotary drum and lifting flights. There are numerous variations on this basic design, but for rough approximations, the costs given should be suitable.

Usually steel construction is specified—alloys will add 100 to 200% to the steel costs shown. Data plotted include the fan, dust collector, oil burner, feeder and controls.

Louver dryers involve use of overlapping shelves or louvers to support the solids. The feed material is dried as hot gases pass between the shelves and up through the bed of material. The Roto-Louvre dryer outwardly resembles a direct rotary since it has a rotating, horizontal cylinder. Costs for steel construction are shown in the figure and include heater, dust collector, fans, motor, drive and controls.

The rotary vacuum dryer is a relatively high-production-rate unit. It usually consists of a jacketed, horizontal rotating cylinder with internal flights. Again, costs are for steel construction and include motor and drive, vacuum pump, dust filter, condenser and receiver.

The vacuum shelf dryer consists basically of shelves and a cabinet. Heavy-wall construction is used;

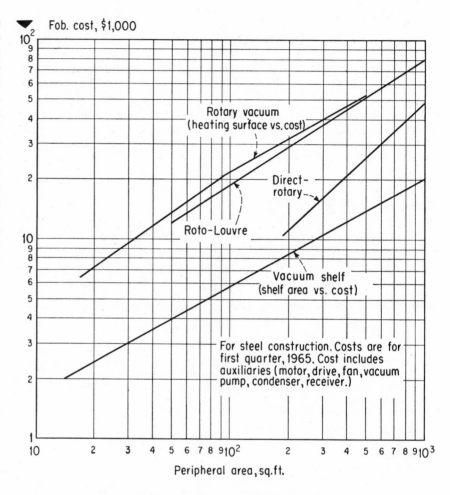

Fob. cost, $1,000

Rotary vacuum (heating surface vs. cost)

Direct-rotary

Roto-Louvre

Vacuum shelf (shelf area vs. cost)

For steel construction. Costs are for first quarter, 1965. Cost includes auxiliaries (motor, drive, fan, vacuum pump, condenser, receiver.)

Peripheral area, sq.ft.

the door swings on heavy-duty hinges and is held in place with eyebolts. This unit comes equipped with condenser, receiver and vacuum pump.

Spray dryers are usually rated on the basis of evaporative capacity of water, since heat-transfer area cannot be correlated with size or costs. For rough approximations of fob. costs, data shown in the table (right) should prove useful in a "quick" estimate:

| | Spray Dryer | |
	Small	Large
Operating temp., °F..	500–1,000	500–1,000
Evap. cap., lb./hr....	400–1,000	6,500–16,000
Cost, fob. $1,000.....	31–33	170–190

These costs are subject to a number of variables, including inlet temperatures and degree of atomization, which can greatly affect equipment size and costs.

Key Concepts for This Article

Active (8)	Passive (9)	Means/Methods (10)
Estimating	Costs*	Charts
	Dryers*	Tables

(Words in bold are role indications, numbers correspond to EJC-AIChE system except for Role 8 modification. Asterisks mark key concepts suggested for indexing. Others are added to improve reading as abstract. Indexing is described in *Chem. Eng.*, Oct. 11, 1965, p. 187.)

Preliminary Estimates for Steam-Jet Air Ejectors

These charts are useful for obtaining order-of-magnitude cost estimates when evaluating steam-jet ejectors, and for detecting gross errors in bids submitted by vendors.

The potential user of steam-jet ejectors frequently wants to obtain a quick, reasonable estimate of the price of an ejector to accomplish a given job. It is also desirable to be able to check quotations from vendors to detect gross errors. The cost charts presented here are an attempt at meeting the requirements for order-of-magnitude estimating.*

In the typical vacuum-process system, an ejector is used to remove air and other noncondensable gases from the vent of a condenser (because air and noncondensables are saturated with process vapor at the vent-gas temperature, the condenser must be designed for some vapor subcooling to prevent excessive vapor carryover to the ejector).

Need Size Variables

Before using the cost charts, two quantities must be determined: equivalent air load, and suction pressure. The total vapor load (including air leakage) should be converted to equivalent dry air at 70 F. This can be done using conversion ratios as described in Heat Exchanger Institute standards† or assuming all the vapor is air.

Design suction pressure is determined by establishing the lowest desired operating pressure in the system and subtracting pressure drops in the line to the ejector (including process condenser and vent line if ejector is attached to the vent of the condenser). In addition, a small safety factor should be used.

* Based on information in paper "Steam-Jet Air Ejectors: Specification, Evaluation, and Operation," by R. B. Power, equipment engineer, Union Carbide, Chemicals Div., South Charleston, W. Va., given at winter annual meeting ASME, Philadelphia, Pa., Nov. 17-22, 1963.
† Heat Exchanger Institute Standards for Steam-Jet Ejectors, 3rd ed., available from Heat Exchanger Institute, 122 E. 42nd St., New York, N. Y.

Equipment costs for steam-jet ejectors—Fig. 1

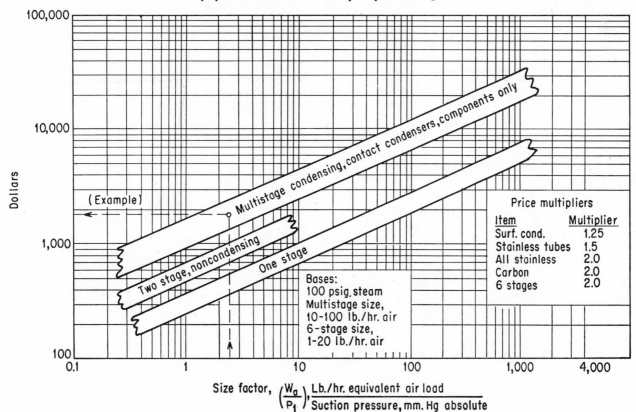

Size factor, $\left(\dfrac{W_a}{P_f}\right)$, $\dfrac{\text{Lb./hr. equivalent air load}}{\text{Suction pressure, mm. Hg absolute}}$

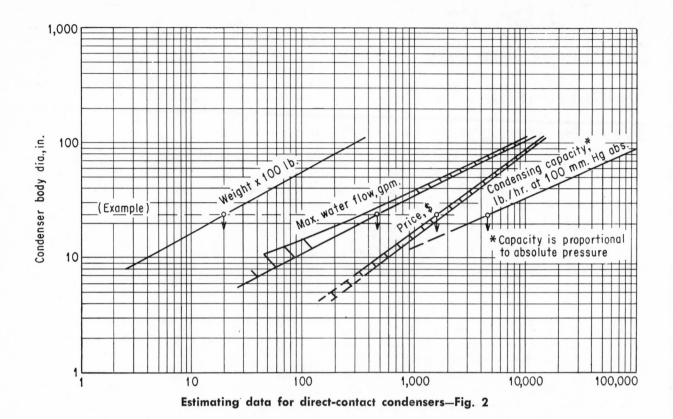

Estimating data for direct-contact condensers—Fig. 2

Prices for condensers in multistage systems (sometimes called intercondensers) are for direct-contact (barometric) condensers. Cooling water flows down as a spray or curtain of water.

Correlating Costs

Fig. 1 can be used for estimating the cost of single and multistage steam-jet ejectors. The size factor correlates well with the price of individual stages but is less accurate for multistage ejectors. And the curves of Fig. 1 fit best for ejectors handling 10 to 100 lb./hr. of equivalent dry air.

Accuracy is adequate for preliminary estimates, but the curves do not show the effects of a number of variables. Large loads, low steam pressure, warm cooling water (over 50 C.), all will tend to increase ejector prices. If the load contains much condensable vapors, the price will tend to be lower.

Rough estimates of price, weight, water capacity, water-vapor condensing capacity of direct-contact condensers may be obtained from Fig. 2. By making a preliminary estimate of ejector design and using Fig. 2, it is possible to prepare a detailed estimate of the price of each ejector stage and contact condenser. This will yield a more accurate estimate of a multi-stage-ejector cost than that obtained from Fig. 1.

When making a comparison of various ejector alternatives, an evaluation must be based on total economic cost. This would include consideration of (1) **costs of steam**, cooling water, electricity; (2) price **including necessary spare parts**, freight, costs of test-

ing; (3) costs of installing and maintaining the ejector.

Additional stages require extensions of steam piping and may require individual supports. Each additional condenser must be supported and each requires an extension of cooling-water piping, and a condensate drain leg plus associated steam tracing where applicable. And qualitative factors such as expected performance of each equipment supplier in offering prompt technical assistance and spare-parts service should not be ignored.

Examples

As an illustration of the use of Fig. 1: What is the price of a stainless-steel condensing multistage ejector with surface condensers, for 25 lb./hr. air at 10 mm. Hg abs.? Chart shows cost will be $1,900 × 2 × 1.25 = $4,750.

A 24-in.-dia. contact condenser weighs 2,000 lb., has a water capacity of 490 gpm., will condense 2,200 lb./hr. steam at 50 mm. Hg abs. Fig. 2 gives a price of $1,600 for such a condenser (carbon steel).

All costs are based on second-quarter, 1963 prices.

Key Concepts for This Article

For indexing details, see *Chem. Eng.*, Jan. 7, 1693, p. 73 (Reprint No. 222). Words in bold are role indicators; numbers correspond to AIChE system.

Active (8)	Passive (9)	Means/Methods (10)
Estimating	Costs	Charts
	Ejectors	Curves
	Aspirators	
	Condensers (process equipment)	
	Sprays	

Capital Cost Data
For
Electrical Systems
In Process Plants

Transformers, distribution feeders, conduit, lighting, motor controllers and motors are all correlated here in charts and tables for quick, preliminary cost estimates.

The costs of electrical items in a typical chemical plant can add up to a sizable percentage of total capital costs. All too often, some over-all "factor" is used in arriving at electrical costs, particularly in a preliminary chemical-plant estimate because of the scarcity of data on electrical costs.

A more accurate way to handle this problem is to separate the electrical items into several categories and total up the costs for these categories. James R. Auld of Du Pont of Canada, in an interesting paper delivered last month at an American Assn. of Cost Engineers meeting* gave a logical breakdown of categories and presented unit costs per kva.

Actually, the data are intended only for preliminary estimates. And the information is good for plants with the usual motors, heaters and lighting—large electrolytic systems or electric furnace loads require another type of analysis.

Nine Categories

Assuming electric power is purchased, the categories of electric items are:
- Incoming high-voltage supply lines.
- Main transformer stations.
- Main distribution feeders.
- Secondary transformer stations.
- Secondary distribution feeders.
- Conduit and wiring for motors.
- Lighting.
- Motor control.
- Motors.

The values shown on the accompanying charts and tables are "typical" costs. How close they may be for any single situation will depend on a great number of variables. Costs are total direct costs and are based on first quarter 1964 data.

Even before thinking about the capital-cost estimate, it is necessary to decide on what type of electrical distribution system is required for the process. For instance, a chemical plant using batch processes can be easily shut down a day or two a year to service the main items of electrical equipment. This plant does not require the same type of system as a large continuous-process plant requiring a minimum of shutdowns. A low-cost radial electrical system, for example, is simple to maintain but will not provide continuous service and a failure of some item of electrical equipment will cause a plant shutdown. The so-called "spot-network system" will give continuous service even when an item in the system fails, but it is relatively expensive. When the system has been decided upon, a simple line diagram can be made to show the system, and an estimate is usually made from this diagram.

In the estimate, assumptions must be made concerning the ratio of peak electric demand load to the total installed load. This demand factor can vary from 0.25 to 0.80. It is also desirable to know whether the chemicals involved in the process are corrosive, flammable or explosive, since special electrical equipment and wiring may be necessary.

(Turn page for charts and tables.)

(Turn page for charts and tables.)

---

* American Assn. of Cost Engineers, 8th Annual Meeting, June 29-July 1, 1964, Astor Hotel, New York. Also see *Chem. in Canada*, May, 1964, p. 29.

Key Concepts for This Article

Active (8)	Passive (9)	Passive (9) (con't.)	Means/Methods (10)
Estimating	Costs	Motors	Charts
	Electrical	Transformers	Tables
	Equipment	Lighting	
	Wires	(illumination)	
	Conduits	Controllers	

(Words in bold are role indicators; numbers correspond to AIChE information retrieval system. Indexing details are described in *Chem. Eng.*, Jan. 7, 1963.)

Typical costs for conduit and wiring for motors—Fig. 1

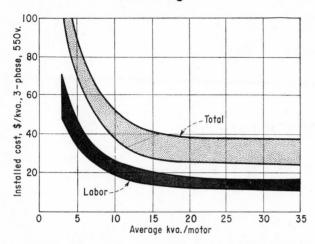

Costs of main transformer stations—Table I

Capacity, Kva.	Primary Voltage 25 Kv.		Primary Voltage 46 Kv.		Primary Voltage 69 Kv.	
	Installed Cost, $/Kva.					
	A*	B†	A*	B†	A*	B†
3,000	36	29	41	34	48	40
5,000	22	19	25	21	28	24
10,000	14	12	15	13	17	15
20,000	11	10	12	11	13	12

* A—Two three-phase transformers, 60 cycles, two main incoming supply lines.
† B—Two three-phase transformers, 60 cycles, one main incoming supply line.

Lighting costs on the basis of watts—Fig. 2

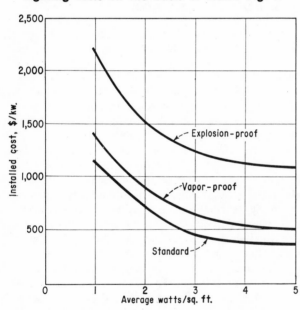

Costs of main distribution feeders—Table II

Capacity, Kva.	Primary Voltage 42 Kv.		Primary Voltage 13.2 Kv.	
	Installed Cost, $/Kva. Main Transformer·Capacit y			
	A*	B†	A*	B†
3,000	15–30	5–10		
5,000	10–20	4–15		
10,000			6–12	2–5
20,000			4–8	1–3

* A—Underground.
† B—Overhead.

Secondary transformer stations—Table III

Capacity, Kva.	4,200/575 Volts	13,200/575 Volts
	Installed Costs, $/Kva.	
600	32	34
1,000	24	25
1,500	18.5	19
2,000	19.5	20

Note: For secondary distribution feeders with average run of 150 ft., use $5/kva. of secondary transformer capacity. For others, use length x $5/150.

Motor controllers can be expensive—Fig. 3

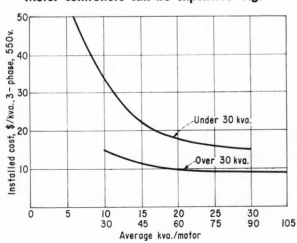

Electric-motor costs for preliminary estimates—Fig. 4

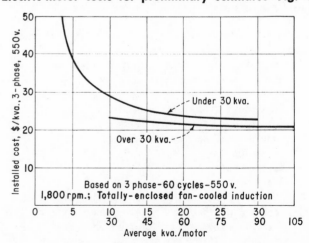

EVAPORATOR ECONOMICS AND CAPITAL COSTS

FERRIS C. STANDIFORD, JR.   W. L. Badger Associates, Inc.

Here are the many factors that must be considered in order to evaluate capital costs, installed costs, operating costs and maintenance costs of complete evaporator systems.

Evaporator costs can be calculated quite closely on the basis of type, heating surface, and materials of construction. The purchase price of the evaporator normally includes interconnecting vapor, liquor and condensate piping; pumps and bases (but not motors); condenser, ejectors, condensate flash system; engineering; a guarantee of steam economy and capacity when clean; and engineering assistance during startup of an evaporator system.

Costs on this basis are correlated in Fig. 1 per square feet of heating surface in each effect of the evaporator. These costs are based on use of steel or cast-iron bodies and admiralty metal tubes, except as

noted. Where other tube materials are used, approximate corrections can be made for cost of the tubes. Correction for other body materials is difficult. In some cases, welded alloy bodies are cheaper than those of cast iron. In this event, it is best to make corrections on the basis of estimated vapor head weights and fabricated material costs.

The fully erected cost of evaporator plants, assuming outdoor construction, includes the selling price plus foundations, supports and walkways, motors, switchgear, wiring, equipment and piping erection, customer engineering, major instrumentation, insulation, condenser water pumps, and service facilities. In the case of *LTV* black-liquor evaporators, this amounts to from 180 to 200% of the purchase price. Thus total erected plant cost would be about twice the purchase price of the evaporator.

For other types of evaporators, the erection costs are similar per square foot of heating surface; so total installed cost can be taken as the sum of the purchase price of the evaporator chosen and the purchase price of an *LTV* evaporator of the same size. Tubing

Uninstalled costs of evaporators based on steel or cast-iron bodies and admiralty metal tubes (except as noted) for various heating surfaces per effect—Fig. 1

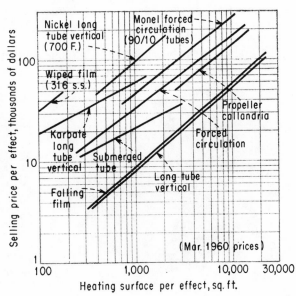

materials usually represent such a large part of evaporator cost that erection cost usually does not go up nearly as rapidly as evaporator cost when expensive tubing alloys are used.

Operating Costs

The cost of steam, power, and condenser water can readily be calculated. Operating labor requirements are generally very small. Usually one man can handle an evaporator of any number of effects—sometimes several such evaporators. Where boilouts are frequent or the evaporator plant is isolated, two men are generally required. Routine maintenance, such as care of pumps, may require an additional man on one shift. Major maintenance is generally so infrequent that plants are usually not specifically staffed for this duty alone. Maintenance labor costs are included in total maintenance costs, which are discussed below.

Maintenance and Life

Aside from care of pumps and interconnecting piping, maintenance of evaporators is required mainly for tube cleaning and tube replacement. Tubes must be thin for good heat conductivity and usually must be replaced several times during the life of the evaporator body, which must be made quite thick to withstand the vacuum, pressure, or structural stresses involved. Except for tubes, evaporators seem to last almost forever. Cast iron or steel bodies and tubesheets of salt evaporators (with copper tubes), which operate under fairly severe corrosive conditions, usually last 40 to 50 years.

There have been a number of replacements since the last war. Most salt evaporators were installed between the turn of the century and World War I, so some of the replacements may well have been only because of technological improvements. One that was erected in 1900 is still in use. Kraft black liquor is quite corrosive, but none of the steel LTV evaporator bodies in this service are known to have required replacement or major repair even though such evaporators came into general use over 30 years ago. Thus evaporator life is generally governed by obsolescence rather than corrosion.

Tube life is highly variable. Kraft black-liquor

evaporators normally have steel tubes (about 11 gage) and life is usually about one year times the number of the effect (i.e., six years in the sixth effect). Only in the first, and usually the second effect, is a more expensive material (in this case, stainless steel) justifiable, and these tubes last "forever." Sometimes tube life is governed by a factor other than actual failure. In one case, it was more economical to replace copper alloy tubes every two years than suffer loss in capacity that resulted from fouling of the tubes under the severe conditions encountered.

Even where the most expensive tube materials are necessary, a very limited tube life may be justified. The Dowtherm-heated caustic high concentrator[17] is a good example. These LTV evaporators have nickel tubes, but the throughput is so high that even though the nickel picked up by the caustic is less than 2 ppm., tube life is less than a year at full production. To keep the nickel pickup this low requires treatment of the feed caustic to reduce any chlorates and dissolved oxygen present. Usually sugar is used for this purpose[18], admitted as a solution into the caustic as it enters the evaporator. Sometimes results are far better than expectations.

Fig. 2 shows the inlet end of a tube from such an evaporator after four years of continuous service. Not only was treatment effective enough to eliminate corrosion completely, it was enough to precipitate the trace of nickel present in the feed. The tubes were so obstructed with nickel "scale" in the first few inches at the inlet end that feed could no longer be gotten into the machine. We hope that this is sufficient evidence that care in design, fabrication and operation are all important in achieving good service life from an evaporator.

Little numerical information is available on actual maintenance costs. In the salt industry, maintenance cost of cast-iron evaporator systems where the evaporators are 20 to 50 years old works out to about 3% per year of the present day installed-replacement cost of the evaporator. In the concentration of diaphragm-cell caustic to 50% solution, which involves crystallization of salt, maintenance of plants with forced-circulation nickel evaporators is about 5%/yr. Tube life in this case is essentially unlimited as long as liquor velocities are not unreasonably high and the steam is free of dissolved CO_2.

Fig. 2

Meet the Author

Ferris C. Standiford, Jr., is president of W. L. Badger Associates, Inc., Ann Arbor, Mich. Joining them in 1946, he has been active in consulting engineering and has specialized in the fields of heat transfer and evaporation, high-temperature heating, and inorganic chemical production. Among his numerous technical publications is the Evaporator Section in the new fourth edition of "Perry's Chemical Engineers' Handbook." He has a B.S.E. and an M.S. in chemical engineering from the University of Michigan, and is a member of ACS and AIChE.

Estimating the Cost of High-Pressure Equipment

These charts provide up-to-date cost information on high-pressure (up to 5,000 psi.) heat exchangers, process vessels, reciprocating pumps and reciprocating compressors.

KENNETH M. GUTHRIE, W. R. Grace & Co.

As brought out in the series of articles that started in the Sept. 23, 1968 issue, high-pressure technology for process plants has advanced considerably in recent years, particularly in ammonia synthesis and other circuits in the 2,000 to 5,000 psi. range. Process equipment required to operate in these high-pressure environments is usually custom designed for specialized fabrication techniques.

The charts in this Cost File have been prepared from recent pricing data. The base cost on each chart is for carbon steel at 5,000 psi.; factors are included to adjust for other materials and pressure levels. The dotted lines represent the range of the data.

Costs for complete installations can be obtained by multiplying the base cost on each chart by a module factor.* The module includes the cost of the equipment itself, field materials (e.g., piping, concrete, steel platforms, instrumentation, paint, insulation), field labor (including equipment handling, erection and foundation work, but not including pilings), and contractor's indirect costs (overhead, profit, etc.) with no contingency.

---

* The module factor is applied to the *base* cost because installation expenses are more directly related to the size of the unit than to its materials of construction or pressure rating. Although the piping used in installing a stainless-steel heat exchanger, for instance, would be more expensive than if the exchanger were of carbon steel, the difference in terms of over-all installation cost is generally not large enough to make much of a difference in the modular factor. The module factor applies to single units. In multiple installations, there is a saving of about 10%.

High-Pressure Heat Exchangers

The cost of heat exchange at high pressure can be quickly obtained from Fig. 1, which is based on equipment designed for 5,000 psi., on both shell and tubeside. Fabrication is for vertical installation with floating tubesheet bundle. The observed data range from 50 to 2,000 sq. ft. of surface area.

Example: Find the current cost of a heat exchanger with 450 sq. ft. of surface designed for 4,000 psi. on

Time base: Mid-1968

Exponent for extrapolating to other sizes: 0.45

Data required: Surface area (sq. ft.)
Design pressure (psi.)

Basis of chart: Carbon steel, 5,000 psi.
High-pressure on both sides
Vertical installation

Range: 70 to 2,000 sq. ft.
3,000 to 5,000 psi.

Installation: Module installation factor = 3.69

Adjusted cost = (Base cost)$(F_m)(F_p)$(Cost Index)

Adjustment Factors

Materials* (Shell/Tube)	F_m	Shell Pressure	F_p
CS/CS	1.00	5,000 psi	1.00
CS/SS	2.69	4,000 psi	0.90
Clad SS/SS (316)	3.49	3,000 psi	0.82

* Carbon steel = CS
Stainless steel = SS

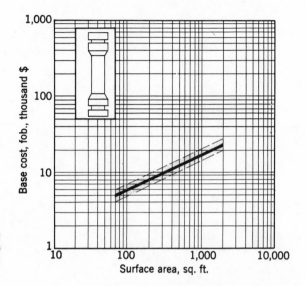

COST OF SHELL-AND-TUBE EXCHANGERS designed to operate in high-pressure circuits from 3,000 to 5,000 psi.—Fig. 1

Time base: Mid-1968

Exponent for extrapolating to other sizes: 0.48

Data required: Dia. (ft.)
Height or length (ft.)
Design pressure (psi.)
Shell material

Basis of chart: Carbon steel
Multilayer or spiral-wound construction
5,000-psi. design pressure
vertical installation; internals not included

Installation: Module installation factor (vertical) = 4.75
Module installation factor (horiz.) = 3.59

Adjusted cost = (Base Cost)$(F_m)(F_p)$(Cost Index)

Adjustment Factors

Shell Material	F_m	Pressure, Psi.	F_p
Carbon steel	1.00	5,000	1.00
Stainless 304 (clad)	2.30	4,000	0.93
Stainless 316 (clad)	2.60	3,000	0.85

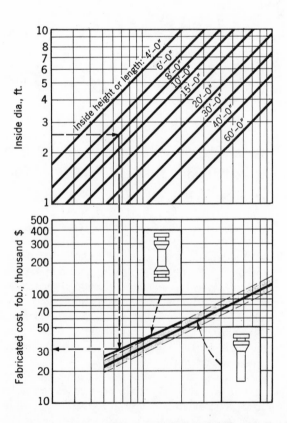

COST OF MULTILAYER OR SPIRAL-WOUND PROCESS VESSELS designed to contain pressures of 3,000 to 5,000 psi—Fig. 2

Time base: Mid-1968

Exponent for extrapolating to other sizes: 0.35

Data required: Capacity (gpm.)
Discharge pressure (psi.)

Basis: Cast iron
Suction pressure = 100 psi.
Discharge pressure = 5,000 psi.

Range: 8 to 500 gpm.
3,000 to 5,000 psi.

Included: Pumping unit
Gear and coupling
Driver (motor)
Baseplate

Installation: Module Installation factor = 3.78

Adjusted cost = (Base Cost)$(F_m)(F_p)$(Cost Index)

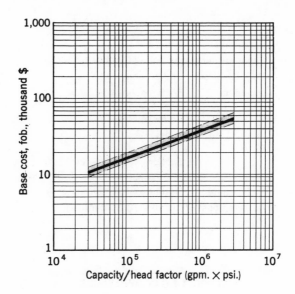

Adjustment Factors

Pump Material	F_m	Discharge Press., Psi.	F_p
Cast iron	1.00	5,000	1.00
Stainless 304	1.69	4,000	0.91
Stainless 316	1.91	3,000	0.83

COST OF RECIPROCATING PUMPS to push liquids into circuits operating at 3,000 to 5,000-psi.—Fig. 3

both shell and tubeside. Shellside material is specified as carbon steel (cs), with stainless tubes. Evaluate the module installation.

Base Cost (from chart) = $12,000

Adjusting for material and pressure, the expected exchanger fob. cost in 1968 is:

($12,000) (2.69) (0.9) = $29,000
Module installation (cs base) = ($12,000) (3.69) = $44,300
Module cost, 1968 = $44,300 + (29,000 − 12,000)
= $61,300

High-Pressure Process Vessels

Process-vessel shells in high-pressure service can be quickly priced from Fig. 2. Internals or special appurtenances are not included. Although based on vertical installation, the data can be used for horizontal installation without adjustment.

Example: Find the current cost of a process reactor shell to be installed vertically and designed to operate at 3,000 psi., with two pressure heads. The inside volume is approximately 40 cu. ft. and stainless steel (304) is required to resist corrosion.

40 cu. ft. = 2′ 6″ inside dia., 8′ 0″ inside length
Base cost (from chart) = $32,000

Adjusting for material and pressure, the expected vessel fob. cost in 1968 is:

($32,000) (2.30) (0.85) = $63,000
Module installation (cs base) = ($32,000) (4.75) = $152,000
Module cost, 1968 = $152,000 + (63,000 − 32,000)
= $183,000

High-Pressure Process Pumps

Reciprocating pumps are used almost exclusively to boost process fluids into high-pressure circuits; data for estimating the cost of such pumps are shown in Fig. 3. The capacity/head (C/H) factor, which has been generated for the purpose of close correlation of the data, is obtained from the product of capacity (gpm.) and total head (psi.).

Example: Predict the cost in 1969 of pumping facilities designed to boost 400 gpm. of liquid from 100-psi. storage into a pressure system at 5,000 psi. No corrosion is anticipated. Escalation is expected to be 5% per yr.

Total head = 5,000 − 100 = 4,900 psi.
C/H Factor = (400) (4,900) = 1,960,000
Base cost (from chart) = $50,000

Adjusting for material, pressure and escalation, the expected pump fob. cost in 1969 is:

($50,000) (1.0) (1.0) (1.05) = $52,500
Module installation (cs base) = ($52,500) (3.78) = $198,450
Module cost, 1969 = $198,450

High-Pressure Compression

Fig. 4 is based on the cost of reciprocating equipment capable of compressing process gases from 200 psi. into 5,000-psi. pressure circuits over a range of 100 to 4,000 bhp. (Gas characteristics, volumetric

Time base: Mid-1968

Exponent for extrapolating to other sizes: 0.80

Data required: Brake horsepower

Included: Reciprocating machine
Gear assembly
Driver (motor)
Baseplate

Basis: Cast iron
Suction pressure = 200 psi.
Discharge pressure = 5,000 psi.

Installation: Module installation factor = 3.49

Adjusted cost = (Base Cost)(Cost Index)

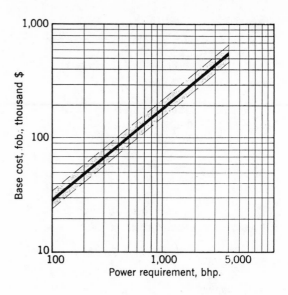

COST OF RECIPROCATING COMPRESSORS to boost gas pressure from 200-psi. suction to 5,000-psi. discharge—Fig. 4

capacity and pressure differentials are calculated into the bhp. parameter.)

Example: Predict the cost in late 1970 of gas compression into a process-plant circuit operating at 5,000 psi. The power requirements are calculated at 900 bhp. Escalation is expected to be 6% per year.

Base cost (from chart) = $170,000
Anticipated cost in 1970 = ($170,000)(1.06)(1.06)
= $191,000
Module installation (cs base) = ($191,000)(3.49)
= $666,600
Module cost, 1970 = $666,600

Meet the Author

Kenneth M. Guthrie is chief estimator and cost analyst in the Engineering Div., Technical Group, of W. R. Grace & Co. (7 Hanover Sq., New York City). His responsibilities include capital cost estimating, cost analysis and cost control. Previously associated with Mobil Oil Corp. and M. W. Kellogg Co., Mr. Guthrie has over ten years of experience in cost engineering. He is a member of AIChE and the American Assn. of Cost Engineers.

Guide to Estimating Costs of Installed Instruments

*Calculating installed costs
for process instruments
is a complicated, tricky business.
Here are some useful data for
making a quick, rough estimate.*

D. K. RIDLEY, *Industrial Products Group
Minneapolis-Honeywell Regulator Co., Philadelphia, Pa.*

There are very few published data on installation costs for process instruments. And for a good reason. The great variety of instruments and the many techniques available for connecting instruments into a chemical process make it difficult to come up with general cost information that might be useful for rough estimating.

The table on the next page is an attempt at correlating some installation-cost data for order-of-magnitude estimates. The time basis is 1961.

However, before using this table, you should know what it does, and does not, cover. In putting together an estimated cost for an instrument installation, we should first start by identifying the various costs involved such as direct, fixed, variable and semi-variable.

For the purposes of this article, I will define direct costs as including instrument, material (tubing, gaskets, unions, etc.), direct labor of installation and commissioning labor (checking, adjusting, etc.). Excluded are fixed costs such as supervision, light, heat, administrative expenses; variable costs such as fringe benefits, taxes, vacations; semi-variable costs such as engineering-design time.

Tabulates Direct Costs

The table is a collection *only* of direct costs according to the above definition. And at that, it does not itemize all direct costs for a complete system. The table does not include cost of control valves (see Cost File 55, Aug. 7, 1961, p. 137). And it does not include cost of panels—panels can be flat face, cubicle, graphic; steel, stainless steel, Formica, slate, dust-proof, console, etc. For example; a 10-ft.-long steel panel with about 1 instrument/ft. of length would cost about $1,500 installed; a more complicated panel with 3.4 miniature instruments/ft. of length might cost $4,000; a graphic panel should run about $6,000 installed.

Also, cost of running long lines of conduit, wiring or tubing is not included. Much depends on how tubing is bracketed (whether it is bundled, concealed, sweated, flared or put together with compression fittings). And tubing can be copper, plastic, stainless steel or aluminum. Much the same is true of conduit and wire. As an example, copper tubing ($\frac{1}{4}$-in. O.D.) costs about $0.10/ft. and requires about 1 hr. of installation time for each 10 ft. Electrical conduit ($\frac{1}{2}$-O.D.) costs about $0.18/ft. and calls for 1 hr. installation time per 10 ft.

Installation labor is listed as man-hr. in the table. Labor rates vary according to geographical location. But for very rough estimating, $5.75/hr. for pipefitters and $6/hr. for electricians can be used. These rates include overhead, benefits, tool amortization.

Also, in any large installation, the costs of installing multitude bundles or multiconductor mains, field junction boxes, bulkhead connections, source of compressed air (for pneumatic systems), all must be included in an over-all instrumentation cost estimate.

Data for estimating installed direct costs for process instrumentation*

Instrument	Instrument Costs, $				Installation Material Cost, $				Installation Labor, Man-Hr.			Commissioning, Man-Hr.		
	Direct-Connected	Transmitter	Receiver Miniature	Receiver Standard	Direct-Connected	Transmitter	Receiver Miniature	Receiver Standard	Direct-Connected	Transmitter	Receiver	Direct-Connected†	Transmitter	Receiver
Alarms														
Flow	400	—	30	30	200	—	20	20	7	—	9	1	—	1
Level	80	—	30	30	50	—	20	20	9	—	9	1	—	1
Pressure	60	—	30	30	30	—	20	20	7	—	9	1	—	1
Temperature	60	—	30	30	30	—	20	20	8	—	9	1	—	1
Flow														
Indicator	300	380	180	170	160	60	10	10	20	40	3	1 to 3	5	5
Controller	680	380	560	550	180	60	40	40	40	40	25	3 to 5	5	5
Indicator-controller	680	380	560	550	180	60	40	40	40	40	25	3 to 5	5	6
Recorder 1 pen	320	380	230	180	160	60	10	10	25	40	4	1 to 3	5	6
2 pen	—	760	320	240										
Recorder-controller	690	380	670	560	180	60	40	40	40	40	25	3 to 5	5	8
Level														
Gage	150	—	—	—	20	—	—	—	20	—	—	1 to 3	—	—
Indicator	300	380	180	170	100	50	10	10	15	40	3	1 to 3	3	5
Controller	680	380	560	550	50	50	40	40	40	40	25	3 to 5	3	5
Indicator-controller	680	380	560	550	50	50	40	40	40	40	25	3 to 5	3	6
Recorder 1 pen	320	380	230	180	30	50	10	10	20	40	4	1 to 3	3	6
2 pen	—	760	320	240										
Recorder-controller	690	380	670	560	50	50	40	40	40	40	25	3 to 5	3	8
Pressure														
Indicator	50	280	180	170	20	50	10	10	5	30	3	1 to 3	3	5
Controller	540	280	560	550	50	50	40	40	30	30	25	3 to 5	3	5
Indicator-controller	540	280	560	550	50	50	40	40	30	30	25	3 to 5	3	6
Recorder 1 pen	160	280	230	180	30	50	10	10	10	30	4	1 to 3	3	6
2 pen	220	560	320	240										
Recorder- 1 pen	550	280	670	560	50	50	40	40	30	30	25	3 to 5	3	8
controller 2 pen	850	560	—	960										

Temperature	Res. Therm.	Thermo-couple	Res. Therm.	Thermo-couple										
Indicator	220	820	340	550	180	170	10	50	10	10	4	30	3	1 to 3	3	5
Controller	610	1,180	340	550	560	550	40	50	40	40	30	30	25	3 to 5	3	5
Indicator-controller	610	1,180	340	550	560	550	40	50	40	40	30	30	25	3 to 5	3	6
Recorder 1 pen	230	800	340	550	230	180	10	50	10	10	10	30	4	1 to 3	3	6
2 pen	350	—	680	1,110	320	240									
Recorder- 1 pen	620	1,180	340	550	670	560	40	50	40	40	30	30	25	3 to 5	3	8
controller 2 pen	1,090	—	680	1,110	—	960									
Multipoint indicator	1,120				10	—	—	—	6+2 for each point	—	—	1 to 3	—	—
Multipoint recorder	1,465				10	—	—	—	6+2 for each point	—	—	1 to 3	—	—
Thermocouple														
Switch panel	30				10	—	—	—	6	—	—	1	—	—
Wells	50				—	—	—	—	1	—	—	1	—	—
Thermocouple	20				—	—	—	—	1	—	—	1	—	—

* **Not included in above table:** long runs for pneumatic tubing and electrical conduit; panels; cutting, patching, foundation; source of compressed air; control valves; setting of valves, venturis, flowmeters, flanges, orifices.
Table can be used for pneumatic or electronic systems.

† Use following for controllers (except on-off): 1 to 4 instruments, 5 man-hr./inst.; 5 to 15 instruments, 4 man-hr./inst.; > 15 instruments, 3 man-hr./inst. For non-controllers and on-off controllers use: 1 to 4 instruments, 3 man-hr./inst.; 5 to 15 instruments, 2 man-hr./inst.; > 15 instruments, 1 man-hr./inst.

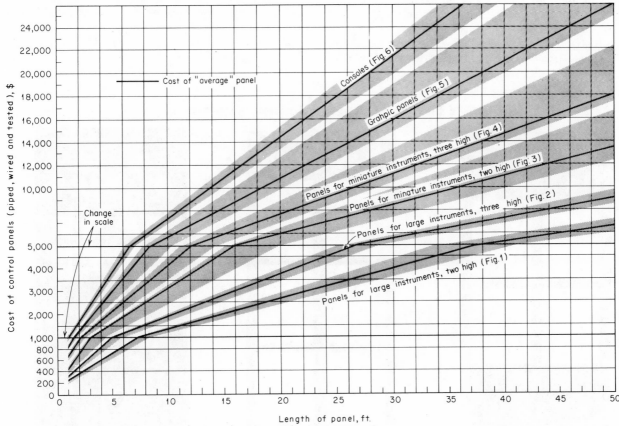

Note: "Average" is a free-standing panel, 3/16-in. steel, angle-iron framework, cadmium-plated copper air tubing.

Costs Correlated For Process Control Panels

Curves for six panel designs show that costs are more sensitive to the number of instruments mounted on a control panel than to panel length.

BELA G. LIPTAK
Principal Instrument Engineer
Crawford & Russell, Inc., Stamford, Conn.

Relatively little has been published on the cost of panels for process control instruments, perhaps because these panels are available in a great variety of designs and materials, making it difficult to correlate costs.

However, for preliminary estimation, we have found that installed panel costs can be correlated with panel length and number of instruments on the panel. Even here, the data yield bands of costs rather than single curves (see chart above, and Figs. 1 to 6 on next page).

The cost chart shows that the cost per unit panel length increases with the density of instruments on the panel. For example, a 20-ft.-long panel with two rows of large case instruments will cost about the same as a 10-ft.-long panel with two rows of miniature instruments, because the density (number of instruments/ft. of panel length) of the miniature instruments is twice the density of the large instruments.

Estimated costs shown include material and labor costs for a completely piped, wired and tested panel. Costs of instruments are not included.

Graphic, cubicle and console panels with the same density of instruments as flat-face panels cost about 15 to 30% more, due to added layout work.

Further, panel cost depends not only on instrument density and panel length but also on complexity. The cost of a panel will increase with the number of sole-

noid valves, pressure switches and other accessories. For this reason, the cost chart shows ranges of costs. The solid line indicates cost of an average panel.

In addition, the data were plotted assuming good design and installation practices such as:

• Instrument grouping follows process sequence.

• Only secondary instruments are located at the top and bottom of the panel.

• At least five feet of servicing space are behind the panel.

• Air piping comes in at the top of the panel, electrical wiring at the bottom, for orderly separation.

• Panel is located in central, pressurized, air-conditioned control room (explosion-proof housings are not required).

All costs are based on March 1962 data.

Six basic panel designs for process instruments

Fig. 1—Large instruments, two high; with 1.15 control instruments/ft. of panel length.

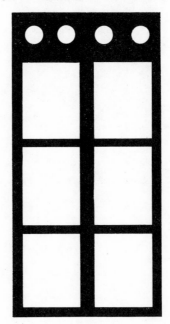

Fig. 2—Large instruments, three high; with 1.7 control instruments/ft. of panel length.

Fig. 3—Miniature instruments, two high; with 2.3 control instruments/ft. of panel length.

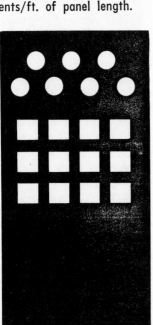

Fig. 4—Miniature instruments, three high; with 3.4 control instruments/ft. of panel length.

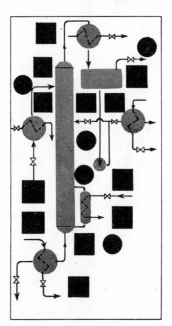

Fig. 5—Miniature instruments, in graphic arrangement; with a maximum of 3.4 control instruments/ft. of panel length.

Fig. 6—Miniature instruments, in console panel; with 3.4 control instruments/ft. of panel length. Maximum density on consoles can be as high as 6 instruments/ft.

Guide to Insulation Costs for Vessels

Tables of material costs and installation labor permit you to come up with a quick answer on how much it will cost to insulate a process vessel.

T. N. DINNING, *Senior Estimator*
*The Dow Chemical Co., Houston, Tex.**

With a few exceptions, the problem of whether a process vessel should or should not be insulated centers around the question "Can we afford heat loss (or gain) better than we can afford the cost of insulation?"

Since this question comes up frequently in prelim-

* Material in this article is based on a paper on costs of insulation given by the author at a Capital Cost Symposium conducted by the American Association of Cost Engineers, Dec. 1962 in Houston, Tex.

inary process evaluations, we have developed some data on the cost of insulating process vessels for high or low-temperature service.

Tables I and II give costs per square foot for hot and cold insulation blocks of varying thicknesses. Costs for accessories for finishing vessel insulation are shown on a "per sq. ft." basis, except that wire and bands are priced per 100 lineal ft. In general, the finishes shown are suitable for outdoor areas and will withstand reasonably severe external corrosive conditions.

Direct field labor man-hours for various vessel insulation operations are shown in Table III.

Table IV is an attempt at combining two factors to adjust "ideal" conditions to actual job conditions for an insulating job. The "job factors" and "management factors" are judgment factors for arriving at a more realistic cost estimate.

All the cost data presented here were developed assuming good design and installation practices. In gen-

---

Cost of insulation materials — Table I

High-temperature insulation (85% magnesia, calcium silicate, etc.):

Thickness, in.	1	1½	2	2½	3	4
Price, $/sq. ft.	0.327	0.49	0.654	0.817	0.981	1.30

Low-temperature insulation:

Polystyrene foam, uncoated, curved segments 12 in. wide × 108 in. long

	Price, $/Sq. Ft.			
Vessel O.D., In.	1½-in. Thick	2-In. Thick	3-In. Thick	4-In. Thick
16–18	1.10	1.17	1.56	1.82
20–24	1.07	1.07	1.37	1.69
26–30	0.91	1.07	1.37	1.53
32–58	0.78	0.91	1.24	1.53
60–62	0.78	0.91	1.07	1.37
64–94	0.65	0.78	1.07	1.37
96 and over (flat sections 6 in. × 108 in.)	0.46	0.69	0.95	1.24

Polyurethane, uncoated
Curved segments for 26-in. dia. and larger vessels

Thickness, in.	¾	1	1½	2	2½	3
Price, $/sq. ft.	0.90	1.20	1.50	1.80	2.10	2.40

For flat block polyurethane

Board ft.	<600	600–3,000	3,000–15,000
Price, $/b. f.	0.26	0.22	0.20

Material costs: accessories to be applied over vessel insulation — Table II

Insulation for heat

1-in. Monel hex mesh (over wired-on insulation)	$0.24/sq. ft.
⅛-in. asphalt mastic (with asbestos) coat	$0.15/sq. ft.
½-in. × 0.020-in. Monel bands	$9.05/100 lineal ft.
½-in. × 0.020-in. stainless bands	$3.02/100 lineal ft.

Insulation for low temperature
Polystyrene

¼-in-thick hard asbestos and Portland cement coat	$0.03/sq. ft.
1-in. galvanized hex mesh	$0.03/sq. ft.
Hot asphalt (for installing lags)	$0.04/sq. ft.
Bands	See above
Weatherproof coating of ⅛-in-thick Insulmatic	$0.17/sq. ft.

Polyurethane

Hot asphalt (for installing lags)	$0.04/sq. ft.
Bands	See above
Weatherproof coating of ⅛-in.-thick Insulmatic	$0.17/sq. ft.
Glass cloth	$0.05/sq. ft.

	$/100 Lineal Ft.		
Tie Wire	16-Gage	14-Gage	12-Gage
Monel alloy	1.45	2.22	3.60
Stainless steel	1.12	1.69	2.76
Galvanized steel	0.21	0.33	0.57

Direct field labor for vessel insulation — Table III

Operation	Man-Hr./Sq. Ft.
One-layer blocks wired on	0.070
One additional layer wired on	0.046
Cement coat ¼-in. thick	0.040
Wire mesh	0.015
Hot asphalt for installing lags	0.015
Glass cloth	0.010
Weatherproof coating ⅛ in.	0.050
Asphalt mastic ⅛ in.	0.040

Job and management factors — Table IV

	Management Factors			
Job Factors	Excellent	Good	Fair	Poor
Excellent	0.84	0.81	0.76	0.70
Good	0.78	0.75	0.71	0.65
Fair	0.72	0.69	0.65	0.60
Poor	0.63	0.61	0.57	0.52

Job factors: Inherent to the job, such as anticipated weather conditions, accessibility to the work, requirements of the specifications governing job procedure, work in hazardous areas, etc.

Management factors: Subject to control by management, such as selection, training and direction of workmen, care and repair of equipment, planning and coordinating job progress and sequence, selection of capable supervision, etc.

How to Use: Multiply direct labor in $ by [(1/factor) — 1] and add to direct labor.

---

eral, for "hot" insulation these practices include single insulation below 800 F., multiple layer for 800 to 1,200 F.; insulation is always applied in sectional or segmented form using the "broken joint" technique; all joints are sealed with asbestos cement. Vessel nozzles are usually insulated with cement (below 2 in. dia.) or blocks (above 2 in.); Monel alloy wires and bands are used for highly corrosive conditions, otherwise stainless steel is suitable (on 9-in. centers).

For "cold" insulation, our experience has been with

Key Concepts for This Article

For indexing details, see Chem. Eng., Jan. 7, 1963, p. 73 (Reprint No. 222). Words in bold are role indicators; numbers correspond to AIChE system.

Active (8)
Estimating

Passive (9)
Costs, installed
Insulation, vessel

Means-Methods (10)
Factors
Labor

Matl. of Constr.
Magnesia, 85%
Calcium silicate
Polystyrene
Polyurethane

polyurethane and polystyrene. Here again the "broken joint" technique should be applied and weatherproof cement used on all joints.

All costs given are based on November 1962 prices.

Work an Example

Assume we want to insulate four storage vessels (8-ft. dia., 35-ft. long) to contain 350 F., and two towers (6-ft. dia., 70-ft. high) operating at —25 F. Eight men and a foreman will work on the job at an average rate of $4.66/man-hr.

The storage vessels have a surface area (heads and shells) of 4,264 sq. ft., and 525 lineal ft. of bands are required (9-in. centers). Using Tables I, II, III and IV:

	Unit Matl. $/Sq. Ft.	Unit Labor $/Sq. Ft.	Total Matl., $	Total Labor, $
1½-in. calcium silicate	0.49	0.33	2,089	1,407
1-in. Monel mesh	0.24	0.07	1,023	298
⅛-in. asphalt mastic	0.15	0.19	640	810
stainless bands (per 100 ft.)	3.02	....	16	.....

The towers have a total surface area of 3,126 sq. ft., and 410 ft. of steel bands are required. Thus:

4-in. Styrofoam	1.37	0.33	4,283	1,032
¼-in. asbestos hard coat	0.03	0.19	94	594
1-in. galvanized mesh	0.03	0.07	94	219
hot asphalt coat	0.04	0.07	125	219
stainless bands (per 100 ft.)	3.02	....	12	.....
⅛-in. weatherproof coat	0.17	0.23	531	719
scaffolding	....	....	100	106

This gives subtotals of $9,007 for material and $5,404 for labor. A correction factor of 0.78 will be used on the labor (1/0.78 = 1.29; 0.29 × 5,404 = 1,567.) to give a total labor charge of $6,970. Contractors overhead and profit will add another $3,640 (23% of material and labor). Insulating the nozzles on the towers and vessels will add $2,350 to the cost (15% of labor and material). Thus, total cost will be $22,000.

How Much for Jacketed Equipment?

Cost curves cover a broad range of jacketed equipment used to keep process materials hot and liquid. This includes jacketed pipe, fittings, valves, sumps and sump pumps.

MICHAEL M. KIRK,
Nitrogen Division
Allied Chemical Corp., Hopewell, Va.

Many process materials used in chemical operations solidify or form slurries when cooled below their "solidification" temperatures. Often it is necessary to keep these materials above such temperatures so they can be handled in process equipment as liquids. Jackets are widely used for this purpose.

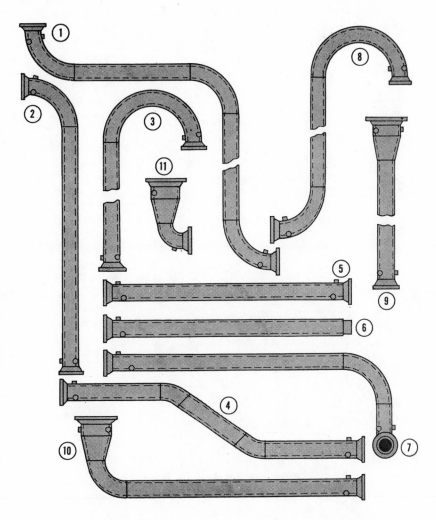

Variety of jacketed-pipe units for process applications—Fig. 1

Jacketed-pipe costs based on 20-ft. of straight sections—Fig. 2

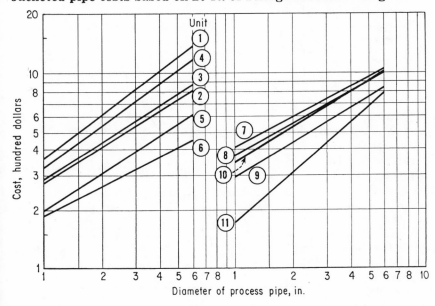

Jacketed Pipe Fittings and Valves

Where uniform heating of a process line is required, jacketing does a better job than steam tracing.

The units of jacketed pipe shown in Fig. 1 have a Schedule 10 inner pipe made of Type 304 stainless steel with 150-lb. flanges. Jackets of Schedule 40 carbon steel are a nominal size larger than the inner pipe or large enough to allow free flow of steam and condensate in the annular area. All units shown are equipped with half couplings for steam and condensate tie-ins. Radius of return bends vary from 6 in. to 24 in.

All jacketed units, with the ex-

Costs of welding, insulation, straight sections, for pipe—Fig. 3

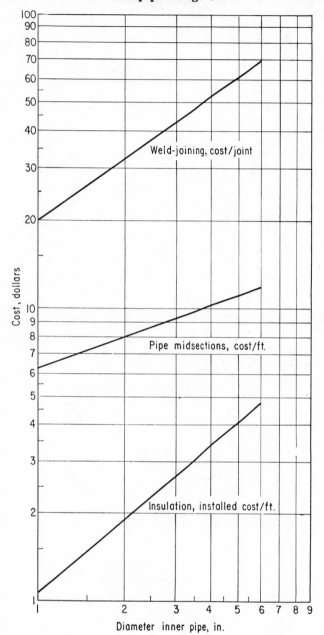

middle-pipe-section cost per ft. is $9.20. Then, $570 − ($9.20 × 8) = $496.40, the cost of a 12-ft. unit.

For units with greater than 20 ft. of straight section, the additional pipe plus joining and welding charges must be added. Assume it is desired to add 10 ft. to the 20-ft. unit. Use basic data, plus joining and welding charges from Fig. 3. Joining and welding charges for a 3-in.-dia. section is $44. Then, 570 + (10 × $9.20) + $44 = $706, cost of 30-ft. unit.

Fig. 4 gives the costs of jacketed ells, tees and gate valves. These units are constructed of 304 stainless steel with carbon-steel jackets, flanged and of 150-lb. design. Standard pipe sizes are 1-in., 1¼-in., 1½-in., 2, 2½-in., 3-in., 4-in., 6-in., etc. Type 304 stainless steel bolts for the above units cost from about $0.60 for the smaller to $1.50.

Magnesia insulation installed costs are also shown in Fig. 3.

Jacketed stainless ells, tees and valves—Fig. 4

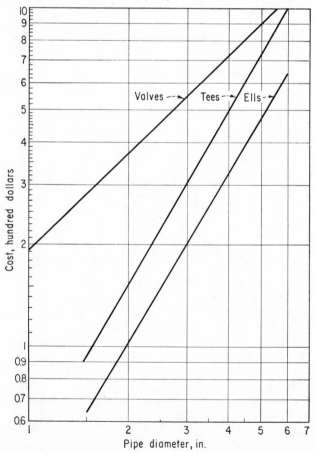

ception of Unit No. 11 include 20 ft. of straight pipe (Fig. 2).

To determine cost of shorter units, determine the straight-pipe-section cost per ft. from the curve in Fig. 3. Multiply this amount by the number of feet removed from the section and subtract this from the cost determined from Fig. 2.

As an example: what is the cost of a 3-in.-dia. Unit No. 3 with 12 ft. of straight pipe? From Fig. 2, using the curve for Unit No. 3, total cost is $570. From Fig. 3, the

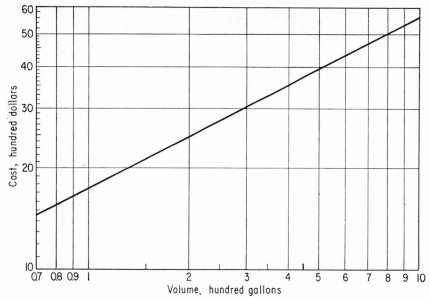

Costs of jacketed sumps with stainless inner shell—Fig. 5

Jacketed Sumps for Process Liquids

In some chemical processes, it is necessary to hold materials at elevated temperatures to prevent solidification. Jacketed sumps are often used for this purpose.

The jacketed sumps (costs shown in Fig. 5) discussed here are constructed with an inner shell of Type 304 stainless (straight sides and flat bottom), a jacket of carbon steel and covers of carbon steel. The inner shell is jacketed on the bottom and to full height on the side.

Sumps are designed for 55-psi. steam in the jackets. Units are mounted on legs and have clips for insulation.

Costs of these units are based on vendor's quotations. For the cost of a sump with an all-stainless-steel jacket, add 15% to the cost from Fig. 5.

Magnesia insulation installed costs for the sumps are indicated in Fig. 6.

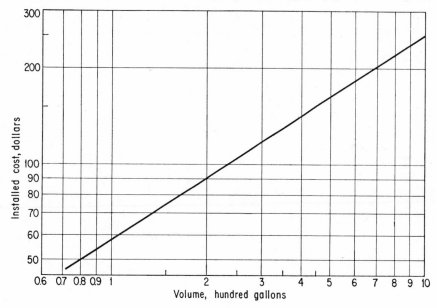

Installed insulation costs for jacketed sumps—Fig. 6

Sump Pumps Prevent Solidification

Although conventional centrifugal pumps may be used external to heated sumps, there is always the possibility of material solidifying in the pump. Submerged pumps overcome this hazard since they are at the same temperature as the product.

Pump costs, shown in the table (right), are for submerged centrifugal, single stage, single suction, open impeller designs. They are of Type 304 stainless steel, and the pump is mounted on plates designed to fit the openings in the sump cover. The pumps can operate with up to 150 ft. of head. Vertical motors, totally enclosed, fan cooled, 440 volt, 1,750 rpm., connected directly to the pumps, are included in the costs. Starters are not included.

Gal./Min.	Cost, $
30	2,950
50	3,000
100	3,150
200	3,400
300	3,680
400	4,000

Except where noted, all costs are purchase costs, adjusted to ENR index of 875.

Packed-Tower Costs

Handy chart quickly and accurately gives cost of packing. This can be added to tower costs, available from a number of sources.

WILLIAM F. WROTH
Dow Chemical Co.
Freeport, Tex.

You can get a quick and accurate estimate of tower packing costs by using the chart at the right. These data, added to cost of a column, will result in total purchased cost for a packed tower.

There are many ways to obtain column costs. For rough estimating I use the following (6 to 24-in. dia., top blind flange, bottom bumped head, skirt, base ring):

Column purchased cost (without packing) = $18.5 \, L_s f_m \, (D/12) + 57 \, L_b \, (D/12)$.

D = Nominal tower dia., in.

L_s = Length of shell (bend line to flange), ft.

f_m = Material factor. (1.0 for steel. For stainless: 6-in. dia., 2.9; 12-in. dia., 3.25; 24-in. dia., 3.5).

L_b = Length of skirt, bottom of steel to bottom bend line, ft. (L_b = 0.2 L_s, for L_s less than 50 ft., L_b = 10 for L_s greater than 50 ft.).

The chart presented here for various packings should be useful when comparing a number of alternate packings, or calculating packing costs for a large number of towers of different sizes.

Charts are based on an approximation of packing height as follows:

Depth of packing = 0.744 × length of shell

In the example shown on the graph, the material cost of packing a 12-in. dia., 30-ft.-high tower with 2-in. Berl saddles is about $135. Uninstalled cost of a carbon-steel 12 x 30 tower would be $900.

Packing data are based on Sept. 1959 costs.

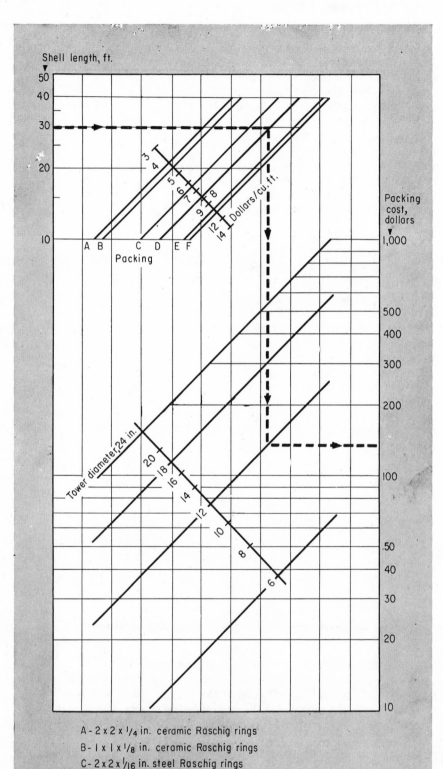

A - 2 x 2 x 1/4 in. ceramic Raschig rings

B - 1 x 1 x 1/8 in. ceramic Raschig rings

C - 2 x 2 x 1/16 in. steel Raschig rings

D - 2 in. Berl saddles

E - 1 in. Berl saddles

F - 1 x 1 x 1/16 in. steel Raschig rings

Piping, Pumps and Valves

After selecting, sizing and designing the liquid handling system comes
the all-important question: How much will it cost?
Data presented here will allow the engineer to estimate
piping circuit costs with accuracy of about 10%

KENNETH GUTHRIE, W. R. Grace & Co.

Liquid handling systems usually include piping circuits, pumping facilities, power drivers, storage facilities and control elements. Cost of these components can be quickly evaluated on a definitive basis by using the data in this article.

Piping circuits are characterized by a complexity of linear pipe, flanges, fittings, valves, etc. The cost of each component contributes only a small percent to total material dollars. However, fabrication and erection charges quickly escalate this percentage into a major element of total installed costs. It is important, therefore, to have a definitive fix on piping dollars.

Many short-cut methods have been devised to evaluate piping system cost. These are valid for conceptual estimating or when the project is not fully defined. The data that follow, however, are applicable only after rigorous circuit analysis and when piping layouts and isometrics are available for quantity take-off (e.g., pipe size length and specification, flange and valve count, etc.).

Piping Costs

Unit costs have been developed for bare pipe material, fittings, flanges and valves. These are arranged on a series of graphs, Fig. 1 through 21, and grouped according to three widely used materials:

    A—Carbon steel (ASTM A-53 Grade A)
    B—Chrome/molybdenum (ASTM A-335 P12)
    C—Stainless steel (ASTM A-312 Type 316)

Costs are average for mid-1968 and do not include any negotiated discounts. Labor is based on standard manhours for handling, aligning and welding or bolting the components together in the field. The productivity measure represents a typical jobsite in the Gulf Coast area. The composite crew rate, including a foreman, for field fabrication and erection is averaged out at $5.75/hr. Typical costs of steam tracing and sewer piping are shown in Fig. 24.

Material and labor (M&L) data are, therefore, on a common base and can be used immediately to estimate the dollar value of the piping account. Adjustments are necessary to reflect productivity and wage rates in locations other than the Gulf Coast or for any time beyond mid-1968. It is assumed that the estimator has such adjustment data available; discussion is outside the scope of this article.

Pumping Equipment

Energy for liquid transportation is generated by electric motors, steam turbines or gas engines and transmitted to the liquid by centrifugal or reciprocating pumps. The cost of general purpose pumps (including gears, couplings and base plate) is shown in Fig. 23.

The data in Fig. 23 are based on the C/H factor (capacity in gpm. multiplied by total dynamic head in psi.), which provides close correlation of raw cost information. Horsepower requirements at 60% efficiency can be found from the scale below the graph. Factors are included to cover installation and materials other than cast iron. To obtain total investment in pumping equipment, the cost of various drivers is indicated in Fig. 25.

Field installation factors include the cost of equipment plus material and labor dollars for excavation, foundation work, handling and setting up prior to pipe erection. These factors are multiplied by equipment costs to obtain total direct material and labor. Indirect costs are not included at this stage.

In Fig. 23 and 25, costs for complete installations can be obtained by multiplying the base cost on each chart by the Base Module Factor shown. This includes the cost of the equipment itself, field materials (piping, concrete, steel platforms, instrumentation, paint, insulation) field labor (including equipment handling, erection and foundation work but not including pilings), and contractor's indirect costs.

Indirect Costs

Once all the direct costs have been assembled it is necessary to evaluate associated indirect costs. These include:

- Sales taxes, freight and packing.
- Construction overhead: field-labor fringe benefits and statutory burdens, field supervision, temporary facilities, equipment and tools.
- Engineering and contractor fee: direct engineering labor (including model work), indirect office costs, burden and overhead and contractor fee.

All of the above items can be included in an average indirect factor of 1.30, which is derived as follows:

Direct Material (M)		Direct Labor (L)		Direct M&L Cost
100.00	+	78.00	=	178.00
$\times 0.148^1$	+	$\times 0.492^2$		
↓		↓		
Engineering		Construction Overhead		
14.80	+	38.37	=	53.17
		Sales taxes 3% of 100	=	0.03
		Freight 5% of 100	=	0.05
		Total Indirects	=	53.25
		Total Bare Cost	=	231.25
		Indirect factor	=	231/178 = 1.30

[1] Direct engineering costs include piping circuit analysis, analytical engineering, plot plans, and isometrics (or scale models). This factor also includes indirect office costs, burden and overhead, and contractor fee.
[2] This factor includes field labor benefits and statutory burdens, field supervision, some average scaffolding, rigging, equipment rental, small tools, etc.

Contingency

The above data and techniques will give cost estimates for 85 to 90% of the elements in a liquid handling circuit. A contingency of 10 to 15% should now be added to cover unlisted items such as hangers, supports, piping under 1 in., etc. Contingency monies bring the total estimated cost to represent 100% of the elements in the system.

Use of the unit cost data in this section together with design and material specifications, quantity take-off from planning drawings, models or isometric sketches, will enable the chemical or mechanical engineer to predict the direct material and labor costs of an infinite variety of piping configurations and systems, either individually or integrated within a plant complex.

Accuracy limits on using these data and techniques depend mostly on the care taken in assembling the estimate and, to a certain extent, the accuracy of the equipment costs. One should allow ± 3% for chart reading errors and 5 to 10% on equipment costs. On this basis over-all accuracy should fall between 8 and 13%.

EXAMPLE: ESTIMATING PIPING SYSTEM COST

A piping system is required to transport liquid feedstock from atmospheric storage to a distillation column. The stream must be preheated to 450 F. by heat exchange before entering the column. Line control is necessary to maintain a flow rate of 400 gpm. Terminal pressure in the column is 470 psi. Line friction, control and head losses amount to 30 psi. Two centrifugal pumping units with motor drive are specified (one spare).

Piping Circuit

Cost of linear pipe materials (Fig. 1a)

6-in. piping:	$40 \times 3.0 =$	120
4-in. piping:	$400 \times 1.8 =$	720
3-in. piping:	$20 \times 1.1 =$	22
3-in. piping:	$20 \times 1.1 =$	22
4-in. piping:	$200 \times 1.8 =$	360
		1,244

Cost of linear pipe erection (Fig. 2a)

6-in. piping:	$40 \times 2.9 =$	116
4-in. piping:	$400 \times 2.2 =$	880
3-in. piping:	$20 \times 1.9 =$	38
3-in. piping:	$20 \times 1.9 =$	38
4-in. piping:	$200 \times 2.2 =$	440
		1,512

Cost of fittings (Fig. 3a, 5a, 8a)

6-in. Tee:	$1 \times 30.0 =$	30.0
6-in. Ell:	$3 \times 17.0 =$	51.0
4-in. Tee:	$2 \times 15.8 =$	31.6
4-in. Ell:	$2 \times 7.5 =$	15.0
4-in. Reducer:	$1 \times 7.2 =$	7.2
3-in. Ell:	$1 \times 5.0 =$	5.0
3-in. Ell:	$1 \times 5.0 =$	5.0
3-in. Reducer:	$1 \times 5.2 =$	5.2
4-in. Tee:	$3 \times 15.8 =$	47.4
4-in. Ell:	$7 \times 7.5 =$	52.5
4-in. Reducer:	$2 \times 7.2 =$	14.4
		264.3

Materials

Pipe Size, in.	Pipe Length, ft.	Line Rating, lb.	Pipe Weight	Tee	90° Ell	Flange	Reducer	Gate Valve	Control Valve	Line Welds
6	40	150	std.	1	3	3	—	2	—	—
4	400	150	std.	2	2	2	1	2	—	20
3	20	150	std.	—	1	2	—	—	—	—
*3	20	300	std.	—	1	6	1	1	1	—
*4	200	300	std.	3	7	5	2	3	—	10

* Insulated to conserve heat

Text continued on page 176

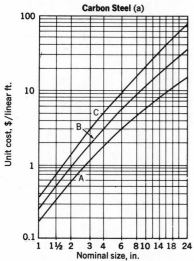

Carbon Steel (a)

Unit cost, $/linear ft. vs Nominal size, in.

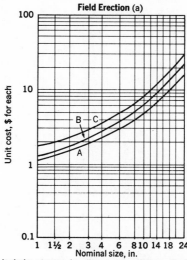

Field Erection (a)

Unit cost, $ for each vs Nominal size, in.

Includes storage, handling, cutting, fitting, aligning on pipe racks. No line welds, stress relief or X-ray.

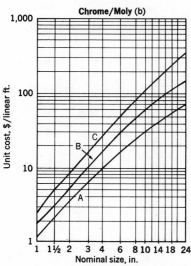

Chrome/Moly (b)

Unit cost, $/linear ft. vs Nominal size, in.

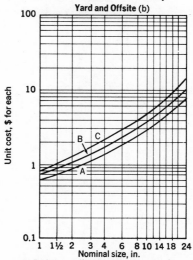

Yard and Offsite (b)

Unit cost, $ for each vs Nominal size, in.

Includes storage, handling, cutting, fitting, aligning on pipe racks. No line welds, stress relief or X-ray.

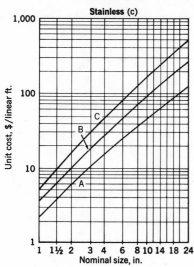

Stainless (c)

Unit cost, $/linear ft. vs Nominal size, in.

A Standard weight, sch. 40
B Extra strong, sch. 80
C Double extra strong, sch. 160

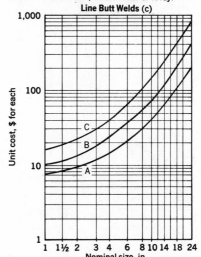

Line Butt Welds (c)

Unit cost, $ for each vs Nominal size, in.

Includes cutting, beveling and line welds. No stress relief or X-ray.

A Standard weight, sch. 40   C Double extra strong, sch. 160
B Extra strong, sch. 80

LINEAR PIPE material cost (seamless)—Fig. 1

AVERAGE PIPE erection cost—Fig. 2

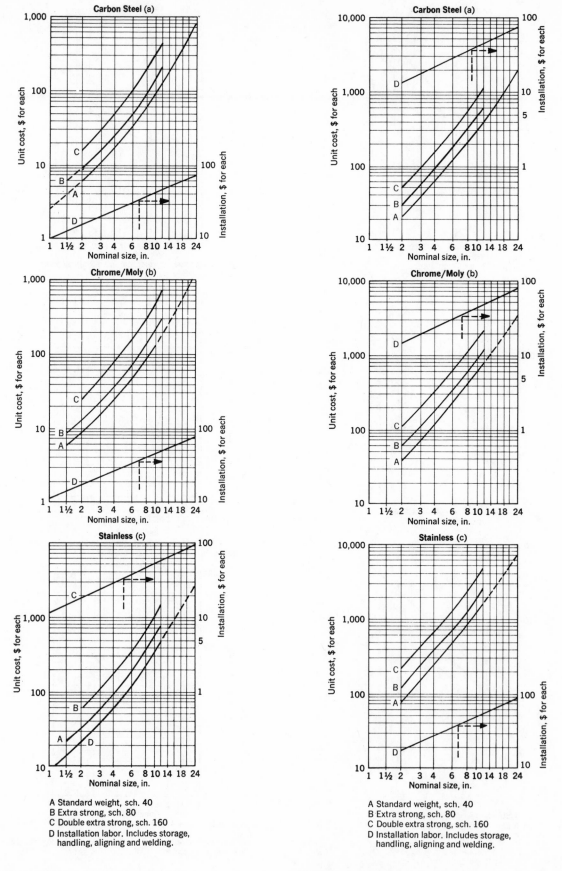

Carbon Steel (a)

Chrome/Moly (b)

Stainless (c)

A Standard weight, sch. 40
B Extra strong, sch. 80
C Double extra strong, sch. 160
D Installation labor. Includes storage,
 handling, aligning and welding.

AVERAGE FITTING costs, equal tees—Fig. 3

Carbon Steel (a)

Chrome/Moly (b)

Stainless (c)

A Standard weight, sch. 40
B Extra strong, sch. 80
C Double extra strong, sch. 160
D Installation labor. Includes storage,
 handling, aligning and welding.

AVERAGE FITTING costs, crosses—Fig. 4

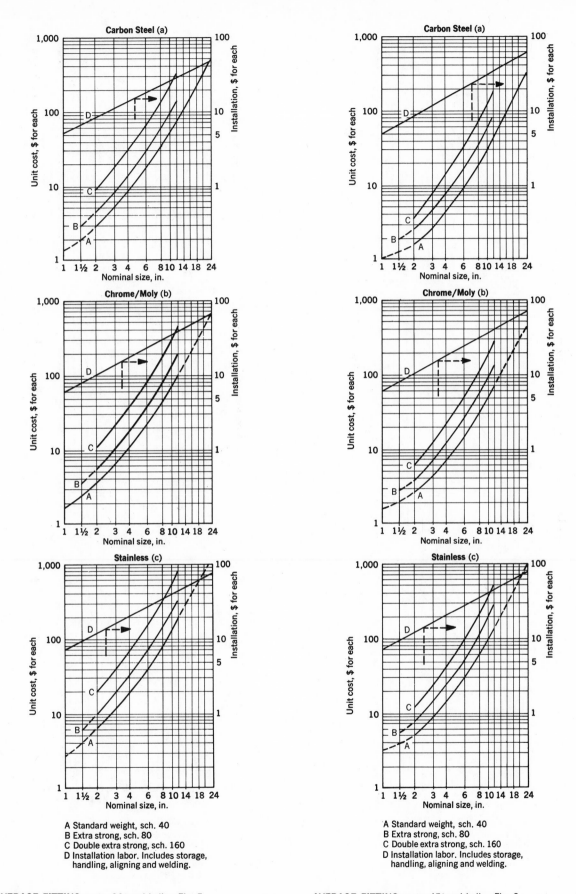

A Standard weight, sch. 40
B Extra strong, sch. 80
C Double extra strong, sch. 160
D Installation labor. Includes storage,
  handling, aligning and welding.

AVERAGE FITTING costs, 90° weld ells—Fig. 5

A Standard weight, sch. 40
B Extra strong, sch. 80
C Double extra strong, sch. 160
D Installation labor. Includes storage,
  handling, aligning and welding.

AVERAGE FITTING costs, 45° weld ells—Fig. 6

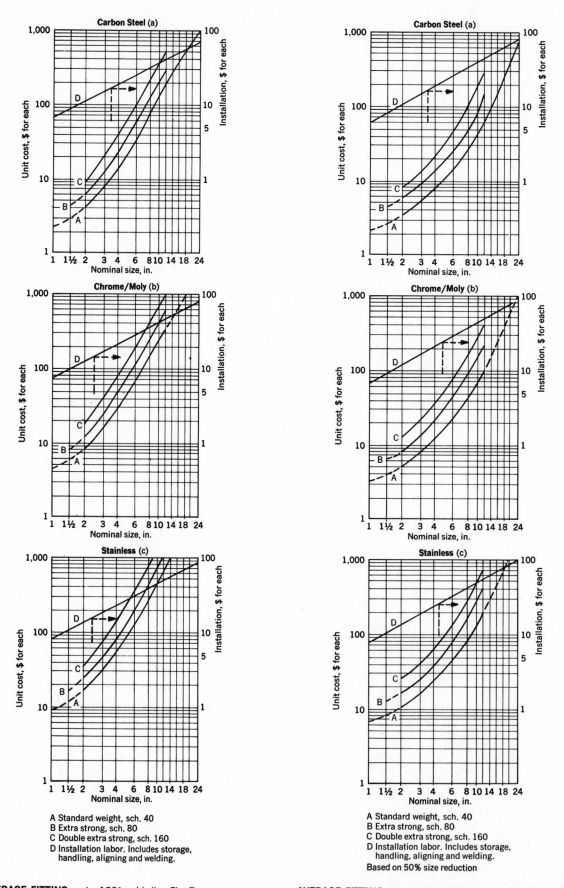

AVERAGE FITTING costs, 180° weld ells—Fig. 7

A Standard weight, sch. 40
B Extra strong, sch. 80
C Double extra strong, sch. 160
D Installation labor. Includes storage,
handling, aligning and welding.

AVERAGE FITTING costs, eccentric reducers—Fig. 8

A Standard weight, sch. 40
B Extra strong, sch. 80
C Double extra strong, sch. 160
D Installation labor. Includes storage,
handling, aligning and welding.
Based on 50% size reduction

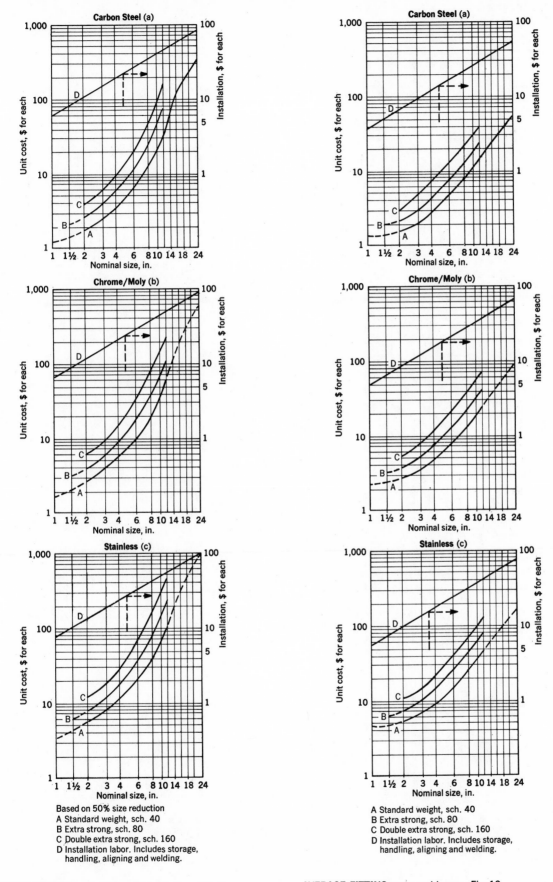

Based on 50% size reduction
A Standard weight, sch. 40
B Extra strong, sch. 80
C Double extra strong, sch. 160
D Installation labor. Includes storage,
  handling, aligning and welding.

A Standard weight, sch. 40
B Extra strong, sch. 80
C Double extra strong, sch. 160
D Installation labor. Includes storage,
  handling, aligning and welding.

AVERAGE FITTING costs, concentric reducers—Fig. 9

AVERAGE FITTING costs, weld caps—Fig. 10

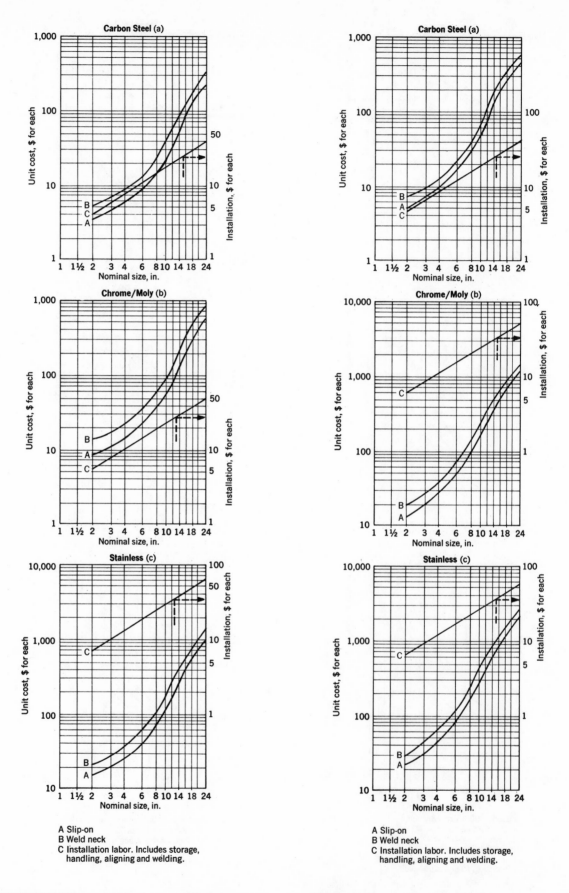

AVERAGE FLANGE costs, 150-lb. raised face—Fig. 11

AVERAGE FLANGE costs, 300-lb. raised face—Fig. 12

A Slip-on
B Weld neck
C Installation labor. Includes storage,
   handling, aligning and welding.

A Slip-on
B Weld neck
C Installation labor. Includes storage,
   handling, aligning and welding.

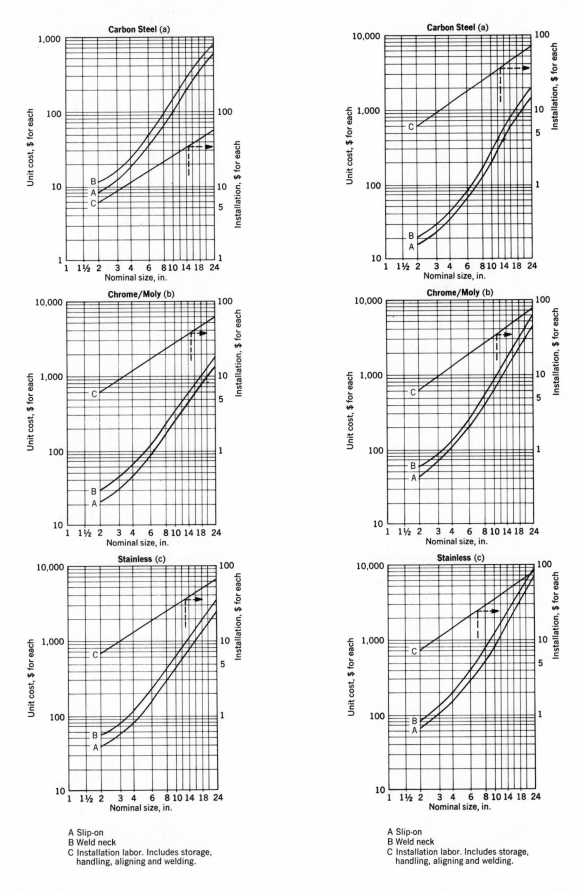

Carbon Steel (a)

Chrome/Moly (b)

Stainless (c)

A Slip-on
B Weld neck
C Installation labor. Includes storage,
  handling, aligning and welding.

AVERAGE FLANGE costs, 600-lb. ring joint—Fig. 13

AVERAGE FLANGE costs, 900-lb. ring joint—Fig. 14

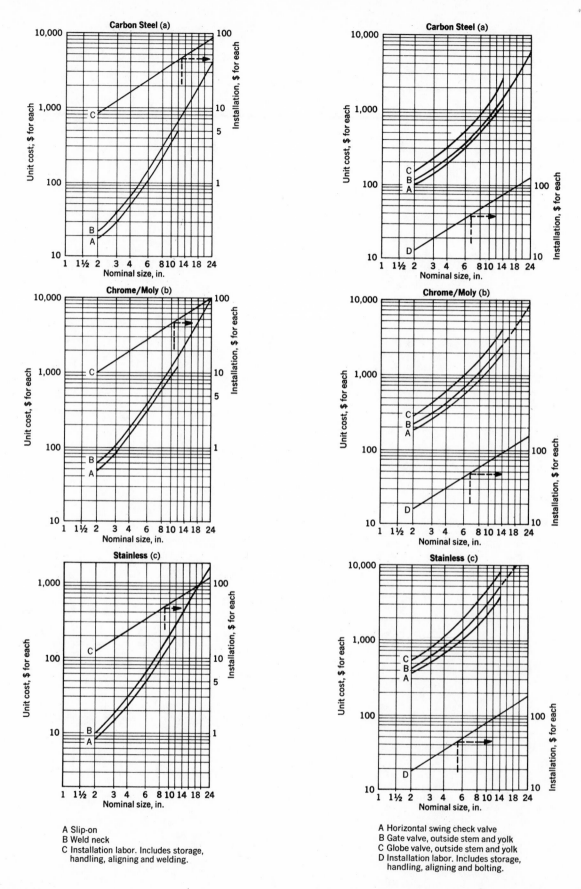

Carbon Steel (a)

Chrome/Moly (b)

Stainless (c)

A Slip-on
B Weld neck
C Installation labor. Includes storage,
   handling, aligning and welding.

AVERAGE FLANGE costs, 1,500-lb. ring joint—Fig. 15

A Horizontal swing check valve
B Gate valve, outside stem and yolk
C Globe valve, outside stem and yolk
D Installation labor. Includes storage,
   handling, aligning and bolting.

AVERAGE VALVE costs, 150-lb. raised face—Fig. 16

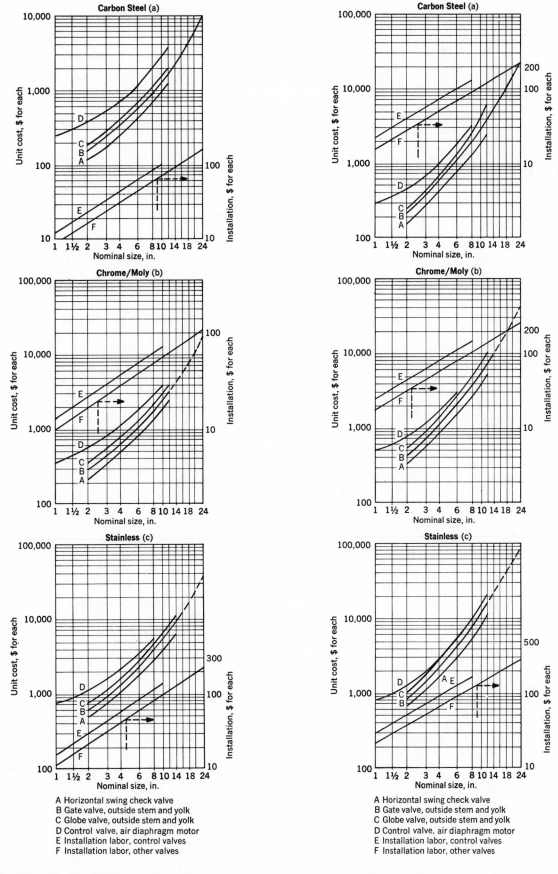

Carbon Steel (a)

Chrome/Moly (b)

Stainless (c)

A Horizontal swing check valve
B Gate valve, outside stem and yolk
C Globe valve, outside stem and yolk
D Control valve, air diaphragm motor
E Installation labor, control valves
F Installation labor, other valves

A Horizontal swing check valve
B Gate valve, outside stem and yolk
C Globe valve, outside stem and yolk
D Control valve, air diaphragm motor
E Installation labor, control valves
F Installation labor, other valves

AVERAGE VALVE costs, 300-lb. raised face—Fig. 17

AVERAGE VALVE costs, 600-lb. ring joint—Fig. 18

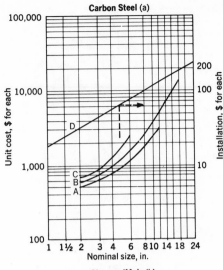

Carbon Steel (a)

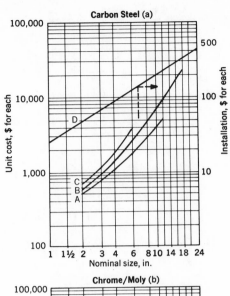

Carbon Steel (a)

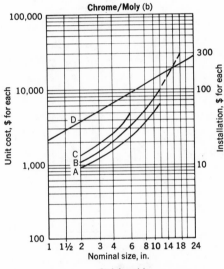

Chrome/Moly (b)

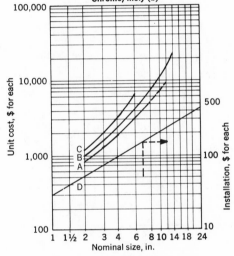

Chrome/Moly (b)

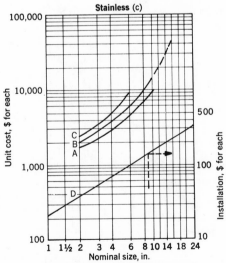

Stainless (c)

A Horizontal swing check valve
B Gate valve, outside stem and yolk
C Globe valve, outside stem and yolk
D Installation labor. Includes storage,
   handling, aligning and bolting.

AVERAGE VALVE costs, 900-lb. ring joint—Fig. 19

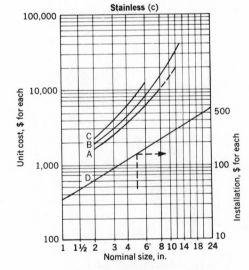

Stainless (c)

A Horizontal swing check valve
B Gate valve, outside stem and yolk
C Globe valve, outside stem and yolk
D Installation labor. Includes storage,
   handling, aligning and bolting.

AVERAGE VALVE costs, 1,500-lb. ring joint—Fig. 20

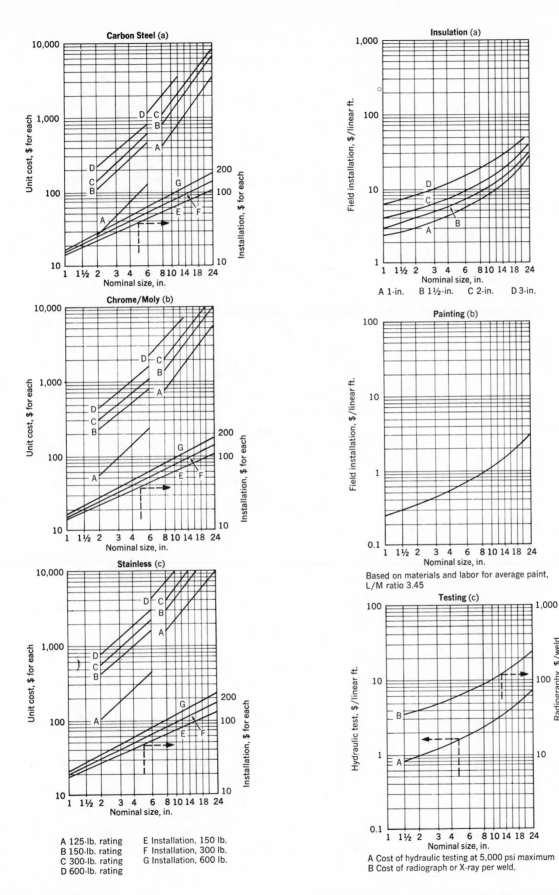

Carbon Steel (a)

Unit cost, $ for each / Installation, $ for each vs. Nominal size, in.

Chrome/Moly (b)

Unit cost, $ for each / Installation, $ for each vs. Nominal size, in.

Stainless (c)

Unit cost, $ for each / Installation, $ for each vs. Nominal size, in.

A 125-lb. rating   E Installation, 150 lb.
B 150-lb. rating   F Installation, 300 lb.
C 300-lb. rating   G Installation, 600 lb.
D 600-lb. rating

AVERAGE VALVE costs, flanged plug valves—Fig. 21

Insulation (a)

Field installation, $/linear ft. vs. Nominal size, in.

A 1-in.   B 1½-in.   C 2-in.   D 3-in.

Painting (b)

Field installation, $/linear ft. vs. Nominal size, in.

Based on materials and labor for average paint, L/M ratio 3.45

Testing (c)

Hydraulic test, $/linear ft. / Radiography, $/weld vs. Nominal size, in.

A Cost of hydraulic testing at 5,000 psi maximum
B Cost of radiograph or X-ray per weld.

AVERAGE PIPE insulation, painting and testing costs—Fig. 22

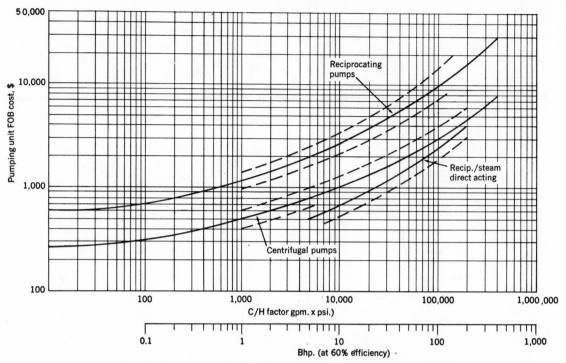

Adjusted pump cost = (base cost) (F_m) (F_p) (Cost index)

Adjustment Factors

Material F_m			Suction Pressure F_p		
	Centrifugal	Reciprocating		Centrifugal	Reciprocating
Cast iron	1.00	1.00	Up to 150 psi	1.00	1.00
Cast steel	1.32	1.55	150 to 500 psi	1.50	1.20
Stainless	1.93	2.10	500 to 1,000 psi	1.90	1.40

Field Installation (M&L) = 1.76   Base Module Cost = 2.35   L/M Ratio = 0.28

PROCESS PUMPING units costs—Fig. 23

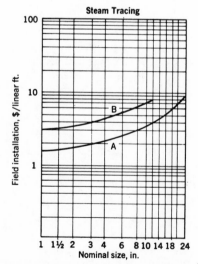

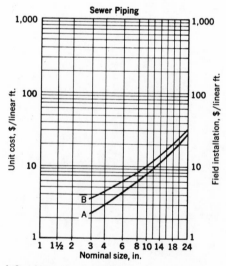

A Field installation includes material and labor for copper tubing. L/M ratio = 6.3. No insulation or indirects.

B Field installation includes material and labor for stainless steel tubing. L/M ratio = 2.5. No insulation or indirects.

A Cast iron sewer pipe with hub and spigot joints, extra heavy wall.

B Installation labor includes storage, handling, aligning, trench and backfilling. No indirects.

COST of steam tracing and sewer piping—Fig. 24

Basis
Standard (enclosed fan cooled)
Speed 1,750 rpm.
230/460 volts
3 phase/60 cycle
Motor Cost = (Base cost) (F_t) (F_s) (Cost index)

Motor type	F_t
Drip proof	0.67
Standard	1.00
Explosion proof	1.13

Motor speed	F_s
1,750	1.00
3,600	1.15

Field Installation (M&L) = 1.84
Base Module Factor = 2.47
L/M Ratio = 0.35

Basis
Steam pressure 300 psi.
Non-condensing
Single stage
Turbine cost = (Base cost) (F_d) (F_p) (Cost index)

Design type	F_d
Non-condensing	1.00
Condensing*	1.75

Stream pressure	F_p
Up to 300 psi	1.00
300 to 600 psi	1.18
600 to 900 psi	1.26

*Includes steam condensing unit, local piping and accessories

Field Installation (M&L) = 1.79
Base Module Factor = 2.41
L/M Ratio = 0.35

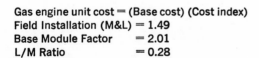

Gas engine unit cost = (Base cost) (Cost index)
Field Installation (M&L) = 1.49
Base Module Factor = 2.01
L/M Ratio = 0.28

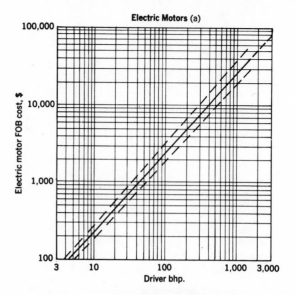

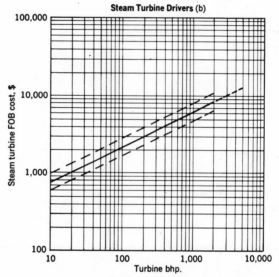

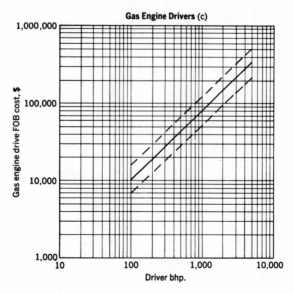

PROCESS PUMPING driver costs—Fig. 25

Cost of fittings installation (Fig. 3a, 5a, 8a)

6-in. Tee:	$1 \times 30 =$	30
6-in. Ell:	$3 \times 19 =$	57
4-in. Tee:	$2 \times 14 =$	28
4-in. Ell:	$2 \times 14 =$	28
4-in. Reducer:	$1 \times 19 =$	19
3-in. Ell:	$1 \times 12 =$	12
3-in. Ell:	$1 \times 12 =$	12
3-in. Reducer:	$1 \times 15 =$	15
4-in. Tee:	$3 \times 14 =$	42
4-in. Ell:	$7 \times 14 =$	98
4-in. Reducer:	$2 \times 19 =$	38
		379

Cost of flanges (Fig. 11a, 12a)

6-in., 150-lb.:	$3 \times 13.0 =$	39.0
4-in., 150-lb.:	$2 \times 8.5 =$	17.0
3-in., 150-lb.:	$2 \times 6.9 =$	13.8
3-in., 300-lb.:	$6 \times 9.5 =$	57.0
4-in., 300-lb.:	$5 \times 13.0 =$	65.0
		191.8

Cost of flanges installation (Fig. 11a, 12a)

6-in., 150-lb.:	$3 \times 11.0 =$	33.0
4-in., 150-lb.:	$2 \times 7.6 =$	15.2
3-in., 150-lb.:	$2 \times 6.9 =$	13.8
3-in., 300-lb.:	$6 \times 9.5 =$	57.0
3-in., 300-lb.:	$5 \times 9.5 =$	47.5
		166.5

Cost of gate valves (Fig. 16a, 17a)

6-in., 150-lb.:	$2 \times 350 =$	700
4-in., 150-lb.:	$2 \times 210 =$	420
3-in., 300-lb.:	$1 \times 240 =$	240
4-in., 300-lb.:	$3 \times 320 =$	960
		2,320

Cost of valves installation (Fig. 16a, 17a)

6-in., 150-lb.:	$2 \times 35 =$	70
4-in., 150-lb.:	$2 \times 25 =$	50
3-in., 300-lb.:	$1 \times 25 =$	25
4-in., 300-lb.:	$3 \times 31 =$	93
		238

Cost of control valves (Fig. 17a)

3-in., 300-lb.:	$1 \times 540 =$	540

Cost of control valves installation (Fig. 17a)

3-in., 300-lb.:	$1 \times 32 =$	32

Line welds labor (Fig. 2c)

4-in.:	$20 \times 15 =$	300
4-in.:	$10 \times 15 =$	150
		450

Insulation materials and labor (Fig. 22a)

3-in., 1½-in. thick:	$20 \times 4.8 =$	96
6-in., 1½-in. thick:	$200 \times 7.0 =$	1,400
		1,496

Cost of pumps and drivers

Capacity (C)	$= 400 \times 1.1^* = 440$ gpm.
Head (H)	$= 470 + 30 = 500$ psi.
C/H factor	$= 440 \times 500 = 220,000$
	bph. $= 200$ (Fig. 23a)
Pump cost	$= 2 \times 5,300 \times 1.76^{**} = 18,600$ (Fig. 23)
Motor cost	$= 2 \times 5,000 \times 1.84^{**} = 18,400$ (Fig. 24a)
Total pumps and drivers	$= 37,000$

* Design factor
** Field installation factor

Summary

	Direct Material	Direct Labor	Total
Linear pipe	1,244.0	1,512.0	2,756.0
Fittings	264.3	379.0	643.3
Flanges	191.8	166.5	358.3
Valves	2,320.0	238.0	2,558.9
Control valves	540.0	32.0	572.0
Line welds	....	450.0	450.0
	4,560.1	2,777.5	7,337.6
Take-off allowance*	456.0	277.5	733.5
Piping investment	5,016.1	3,055.0	8,071.1
Insulation			1,496.0
Pumps and drivers			37,000.0
Total field installation			46,564.1
Indirect costs @ 30%			13,970.1
Total base cost			60,534.2
Contingency @ 15%			9,080.1
Total system cost			$69,614.3

* Statistical adjustment based on comparisons between measured pipe and actual pipe ordered. It varies according to the consistency of take-off techniques and estimating methods. An average of 10% is indicated for general purposes.

Piping Material Discounts

The piping material costs in this article represent average list prices effective during the middle of 1968. It should be pointed out that substantial discounts—up to 30 or 40%—can be negotiated on major projects. Discounts, however, are so volatile that no credit should be applied to estimating work or economic evaluations.

Meet the Author

Kenneth M. Guthrie is now with Fluor Corp., 2500 S. Atlantic Blvd., Los Angeles, Calif. 90022. At the time this article was written he was chief estimator and cost analyst in the Engineering Div., W. R. Grace & Co. His responsibilities include capital cost estimating, cost analysis and cost control. Previously associated with Mobil Oil Corp. and M. W. Kellogg Co., Mr. Guthrie has over 10 years experience in cost engineering. He is a member of AIChE and the American Association of Cost Engineers.

Installed Costs of Outside Piping

These curves can be a major time-saver when estimating the installed cost of aboveground or underground piping; they reflect the varying amount of fittings that may be used for long and short runs.

D. A. BOSWORTH, *Hercules Inc.*

Battery-limit piping, for the order-of-magnitude type of study, is commonly estimated by applying various percentages to total equipment costs, total plant costs, or to some portions of the equipment (e.g., see *Chem Eng.*, Sept. 14, 1964, p. 228).

Frequently, however, there is the problem of applying a dollar figure to that portion of the estimate containing the piping outside the battery limit—i.e., piping between processing units and utility or storage facilities, etc. It is very difficult (and dangerous) to obtain this figure by applying a percentage to any other item in the estimate.

The usual time-consuming procedure consists of listing the lineal feet of pipe involved, together with the assumed number of fittings, flanges, valves, etc., associated with each pipe run. These are then priced for (1) material, (2) labor to fabricate, and (3) labor to erect.

The curves shown in this article obviate the necessity of listing and pricing these items; they present basic unit prices that can be adapted to a variety of conditions, leading to quick nondefinitive estimates.

Aboveground Piping

Fig. 1 and 2 give the installed cost, in $/ft., of short aboveground runs with an average number of fittings: the curves cover pipe of various diameters in standard-weight steel, and in two types of stainless steel (304L and 316L). On the opposite page, Fig. 3 and 4 show the corresponding cost data for longer runs with fewer fittings. The curves are based on these assumptions:

Short Runs	Longer Runs
100 ft. pipe	1,000 ft. pipe
6 ells	6 ells
1 tee	1 tee
12 flanges	104 flanges
1 valve	2 valves

Note that a short run containing few fittings and valves may be estimated with the "Longer Runs" curve. Conversely, a longer run containing many fittings and valves may be estimated using the "Short Runs" curve. Accordingly, values between the two curves, or on either side, may be used depending on the estimator's judgment as to the number of fittings, etc., involved.

Two Types of Underground Piping

Cost curves are also herewith presented for underground piping for two of the more common types of pipe—i.e., vitrified clay pipe and cast-iron mechanical-joint pipe (Fig. 5). The basis for the installed cost ($/ft.) data in these curves is as follows:

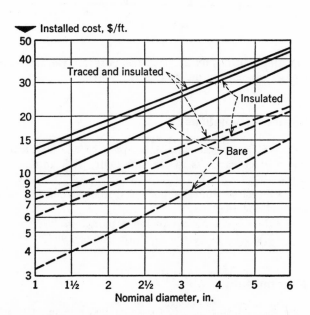

SHORT RUNS of aboveground, external piping: Dashed line shows installed cost in steel (standard-weight); solid line in Type 304L stainless steel (Schedule 10S)—Fig. 1

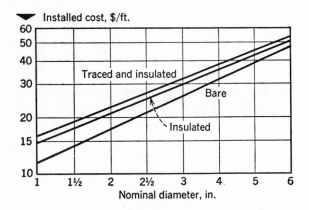

INSTALLED COST for short runs of aboveground piping in Type 316L stainless steel (Schedule 10S)—Fig. 2

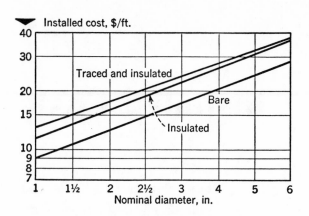

INSTALLED COST for longer runs of aboveground piping in Type 316L stainless steel (Schedule 10S)—Fig. 4

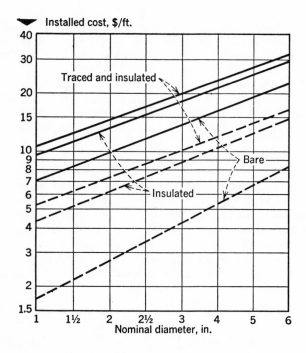

LONGER RUNS of aboveground, external piping: Dashed line shows installed cost in steel (standard weight); solid line in Type 304L stainless steel (Schedule 10S)—Fig. 3

Vitrified Clay	Cast Iron
1,000 ft. pipe	500 ft. pipe
6 quarter bends	3 90° ells
1 tee	1 tee
Excavate and backfill	1 post indicator valve
(5-ft. depth)	5 thrust blocks
	Excavate and backfill
	(5-ft. depth)

As in the case of aboveground piping, variations may be made from the curve values to reflect special conditions such as extra fittings.

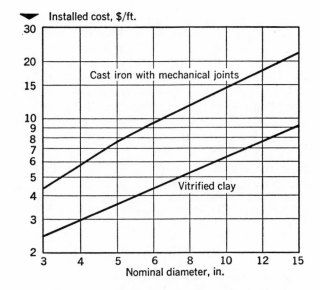

UNDERGROUND PIPING: Installed cost for two common types used outside the battery limits—Fig. 5

General Assumptions

The following assumptions apply to all the curves:
 • Direct labor is $6 per man-hour, with no indirect costs.
 • Supports and hangers for aboveground piping are not included.
 • Bare steel piping includes painting.
 • Material costs are as of December 1967.

Meet the Author

D. A. Bosworth performs capital cost estimation work for Hercules Inc., Wilmington, Del. He joined that firm in 1951, and was a senior engineer with project and maintenance engineering responsibilities before his 1966 appointment as First-Class Estimator. He has a B.Ch.E. degree from Rensselaer Polytechnic Institute, and belongs to the Amer. Assn. of Cost Engineers.

Reinforced-Plastic Pipe: A User's Experiences

Eight years of experience with thousands of feet of reinforced plastic pipe will be summarized in these articles on costs, and installation and maintenance ideas.

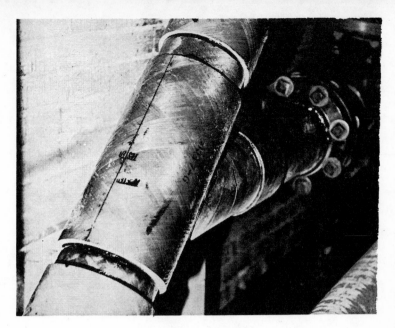

JOHN H. MALLINSON, *FMC Corp.*

Unfortunately, no material yet discovered is a panacea for all the corrosion and maintenance problems besetting the chemical industry today. Nor is it likely that such a material will supersede all others in the immediate future. But, obviously, we cannot wait indefinitely before taking action on new materials of construction.

Because of the severe corrosive environment involved in the manufacture of rayon, American Viscose Div. of FMC Corp. is engaged in a continuing program that comprises both extensive laboratory tests and evaluation of on-stream performance. This series is devoted to a detailed analysis of the performance of glass-reinforced plastic (GRP)* piping. Primary emphasis will be given to test and cost data as well as criteria for selecting GRP pipe.

Depth of Experience

At American Viscose, in Front Royal, Va. (the world's largest rayon plant with 43 acres under roof), our depth of experience with epoxy systems spans eight years, and with polyester systems, five years. During this time, approximately 22,000 ft. of glass-reinforced epoxy systems have been installed and more than 12,000 ft. of polyester piping placed in service. The size and distribution of this piping is shown in Figs. 2 and 3.

Service Conditions

At our plant, GRP piping has been installed in a wide spectrum of tough applications. In the production of viscose rayon, our reinforced-plastic process equipment is subjected to:

• 25% sulfuric acid and salt solutions.

---

* Used as an abbreviation for glass-reinforced plastics throughout this report. It is not intended as a tradename.

• Solutions of zinc sulfate and sodium sulfate.
• Chlorine bleach.
• Weak sodium sulfide solutions.
• Condensate.
• Rayon spin-bath solution, which is a mixture of sulfuric acid, zinc sulfate, sodium sulfate and various additives.
• Derivatives of the rayon spin-bath solution.

Service is rugged, with temperatures to 215F. and pressures to 100 psig. Cleaning these acid transfer systems is accomplished by successive chemical operations with 30% HCl and 10% caustic soda solutions (incorporating wetting agents), flushing for 6 to 8 hr. with each chemical at 150F.

In plant maintenance, reinforced-plastic products are used extensively in plant sewers, spin-bath circulating and cascade systems, condensate lines, soft water piping where iron contamination would en-

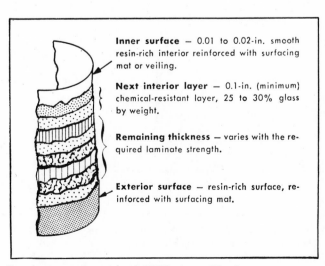

Inner surface — 0.01 to 0.02-in. smooth resin-rich interior reinforced with surfacing mat or veiling.

Next interior layer — 0.1-in. (minimum) chemical-resistant layer, 25 to 30% glass by weight.

Remaining thickness — varies with the required laminate strength.

Exterior surface — resin-rich surface, reinforced with surfacing mat.

Laminate construction of GRP* pipe.—Fig. 1

danger product quality, bleach systems, and many other areas.

Criteria for Selecting Piping

The important elements to be considered in a piping installation where corrosive liquids are being handled are:

- Strength and corrosion resistance.
- Cost of a given type of pipe.
- Cost of fittings.
- Insulation.
- General ease of installation.
- Joining costs.
- Ability of the piping installation to provide uninterrupted service.
- Expected service life of the piping system.
- Ease of modification and repair.
- Delivery and field inventory.

Strength and Corrosion Resistance

We require a pipe of sufficient strength to withstand operating pressures and physical abuse, while providing maximum resistance to corrosion. These vital qualities cannot be separated in glass-reinforced plastic pipe. The component materials used in the formulation of the pipe and the method of fabrication determine both the strength and degree of corrosion resistance that the finished pipe will have.

There are a number of good chemical-resistant polyester resins that lend themselves well to the fabrication of chemical equipment. The choice of the particular resin depends primarily upon the conditions to be encountered. It is our opinion that a bisphenol-A-fumarate resin represents the optimum choice for the type of oxidizing acids that we must contend with. It is used in all our polyester piping systems at the present time.

This does not mean, however, that we have neglected to continually examine other resins in this field. For example, an intensive testing program is under way on the class of polyester resins based on glycols and a phthalic acid that contain a large amount of chlorine in the molecule. These are commonly called chlorinated polyester resins.

We find the preliminary results from this program in exhaust systems, where fire retardancy rates high, this type of resin is the material of choice. Recently, such resin, with 5% Sb_2O_3 added to further enhance its already excellent fire-retardant qualities, was chosen for a GRP duct and fan system in an acid-laden area. The capital cost of this installation, recently completed, is in excess of $200,000. The blend of excellent corrosion resistance, ease of construction and erection, coupled with good fire-retardant properties, make these resins superb materials of construction for this type of work.

Testing programs are also under way with other polyester resins such as the bisphenol-A hydrogenated bisphenol resins and other resins manufactured by reputable vendors.

There has been much discussion on the glass/resin ratio used in pipe construction. Some vendors push this ratio up to 75% to 85% glass with 15% to 25% resin. With this formulation, they are able to build a tremendously strong pipe, reduce the wall thickness, and sell their product at a cost below that of other types of GRP pipe.

We evaluate all grades and types of glass-reinforced epoxy and polyester pipe, from those with extremely high glass/resin ratios to those with the more-moderate ratios. We presently believe that the maximum corrosion/strength characteristics are obtained with glass-to-resin ratios of 25% to 40% glass to 60% to 75% resin. The corrosion resistance of any reinforced-plastic pipe is created by the resin, and the strength by the glass. Pipe made according to these specifications still has a safety factor of 10 to 1.

Tests made by responsible investigators have shown that in extremely corrosive conditions, structures made with a ratio of 75% glass and 25% resin lose a great deal of their strength in a relatively short time —so that they rapidly become weaker than the 25% to

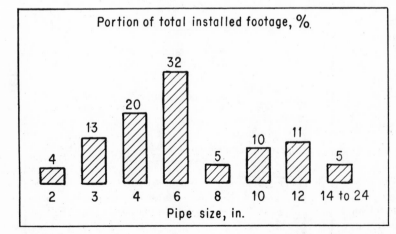

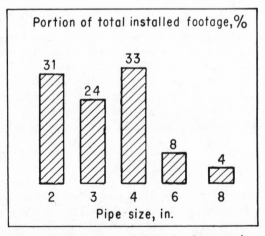

Size distribution of glass-reinforced epoxy pipe, 22,000 ft., all sizes, (left, Fig. 2) and of glass-reinforced polyester pipe, 12,000 ft., all sizes, (right, Fig. 3).

40% glass and 60% to 75% resin laminates. In our field experience, if the "E" grade glass is exposed in any appreciable degree to our spin bath solution, a high-glass-content pipe will rapidly disintegrate.

We have seen high-glass-content pipe fail in two months when the glass fibers became exposed. Glass reinforcement in pipe must always be liberally covered with resin to protect it from reagents. Industry could well use a high-strength chemical-grade glass fiber as the reinforcing material for GRP pipe, and some investigative work is now being done in this area by persons who recognize that this problem exists.

But as in so many other fields, gray areas exist. The degree of attack is quite often related to the solution temperature. In acidic solutions where high temperatures (above 180 F.) exist, a completely different type of problem is encountered. In high-temperature solutions (because both the epoxies and polyesters are excellent insulators), the inside of the pipe wall is trying desperately to expand while the outside of the pipe wall, being relatively cool, is doing its utmost to limit the expansion. In a moderate glass/resin ratio pipe of, let us say 25% glass/75% resin, the physical strains set up within the pipe wall may be sufficient to produce piping failure in the small piping sizes, such as 2, 3 and 4 in.

Under a stereo microscope, an examination of the resin-rich inner surface reveals that failure has occurred through intensive internal stress even though no chemical attack on the resin has taken place. As the pipe sizes are increased to 6 in. and above, this phenomenon (even in glass/resin rations of 25 to 75) no longer occurs. Nor is it observed in the smaller pipe sizes of 2 to 4 in. when the glass/resin ratio is increased to 40% glass/60% resin and higher. This indicates that optimum materials of piping construction in high-temperature systems above 180 F. should be a relatively high glass/resin ratio pipe, especially in the small sizes. This can be extended to tank nozzles where the contents of the tank are extremely hot. It is a simple matter for a fabricator to adjust the glass/resin ratio in this type of construction to obtain the desired results.

Laminate Construction

In our field applications, we have found that sound laminate construction is vital to pipe strength and corrosion resistance. In glass-reinforced polyester pipe, the chemical resistant laminate construction illustrated in Fig. 1 starts with an all-important 0.010-in. resin-rich interior surface, that creates a corrosion-proof barrier, reinforced with chemical-grade glass mat. Successive layers of resin-saturated reinforcement are then applied to produce the desired thickness and strength. Where pipe-design specifications call for glass cloth, woven roving or similar materials, a minimum of two layers of 1½-oz. chopped-strand mat with a high-solubility binder is used between the inside shell and the first layer of major reinforcement. The outer layer is covered with a resin-saturated glass mat that resists fumes, spillage and weathering.

Before we leave the subject of corrosion resistance, a point or two should be made concerning the exterior of GRP pipe. One of the constant hazards in chemical plant operations is the alarming number of failures that occur from the outside in. Thus, to be truly corrosion proof, pipe must be resistant both inside and out; reinforced plastics are.

To provide for the proper cure-out of polyester pip-

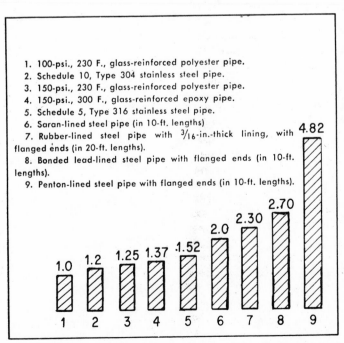

1. 100-psi., 230 F., glass-reinforced polyester pipe.
2. Schedule 10, Type 304 stainless steel pipe.
3. 150-psi., 230 F., glass-reinforced polyester pipe.
4. 150-psi., 300 F., glass-reinforced epoxy pipe.
5. Schedule 5, Type 316 stainless steel pipe.
6. Saran-lined steel pipe (in 10-ft. lengths)
7. Rubber-lined steel pipe with 3/16-in.-thick lining, with flanged ends (in 20-ft. lengths).
8. Bonded lead-lined steel pipe with flanged ends (in 10-ft. lengths).
9. Penton-lined steel pipe with flanged ends (in 10-ft. lengths).

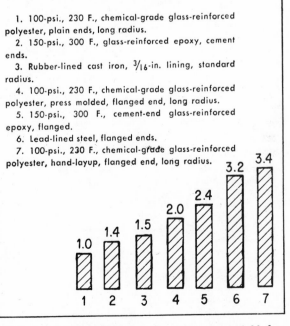

1. 100-psi., 230 F., chemical-grade glass-reinforced polyester, plain ends, long radius.
2. 150-psi., 300 F., glass-reinforced epoxy, cement ends.
3. Rubber-lined cast iron, 3/16-in. lining, standard radius.
4. 100-psi., 230 F., chemical-grade glass-reinforced polyester, press molded, flanged end, long radius.
5. 150-psi., 300 F., cement-end glass-reinforced epoxy, flanged.
6. Lead-lined steel, flanged ends.
7. 100-psi., 230 F., chemical-grade glass-reinforced polyester, hand-layup, flanged end, long radius.

Comparison of average purchase price of uninstalled pipe, 2 to 12 in. (left, Fig. 4) and comparative costs of 90-deg. elbows. (right—Fig. 5). Numbers below the bars refer to the listed materials of construction.

ing (bisphenol-A fumarate resin), many vendors prevent air inhibition by adding a paraffin wax to the final coat. This produces a beautiful glossy appearance but may cause subsequent trouble in developing high wrapped-joint reliability. The removal of this wax is difficult except by using solvents, which all too often are dangerous in themselves. Attempting to sand to provide the proper surface for a wrapped joint only smears the wax and will drive it successively deeper into the pipe. One of the best answers is to eliminate wax in the final coat and provide a finish coat with one of the glycol-phthalic acid-chlorine additive corrosion-resistant resins. Pipe made up exclusively from the chlorinated resins do not suffer from air inhibition problems.

We have subjected several sections of reinforced polyester pipe of standard manufacture to a rigorous test where they were completely submerged in a spin-bath circulating system for two years—testing the degree of corrosion generated from the outside in. This was equal to many years of intermittent dripping, splashing and fume exposure that occur in some of our locations. The pipe appeared to be in good shape from all normal tests. Further, reinforced plastic pipe requires no maintenance painting. This alone amounts to considerable savings in time, materials and manpower.

Comparison of Basic Pipe Costs: 1965

Our purposes will be served best if a first-class, top-quality product is obtained at a competitive price. We buy top-quality material even though it costs a bit more, and have resisted bargain hunting even though we know that some glass-reinforced plastic pipe can be bought for less than we are currently paying. Also, we have constantly examined the quality of pipe and stock received and have called to the vendor's attention any noticeable flaws (for purpose of quality control improvement).

Lead or lead-lined-steel pipe is used extensively throughout our plant. However, lead-lined pipe is largely unsatisfactory because of its high initial cost, high installation cost and high cost of maintenance. While rubber-lined pipe costs less initially than lead pipe, it is still far more expensive than GRP pipe. Rubber-lined steel pipe is used on soft water lines where iron contamination is a problem. Rubber-lined pipe, however, is difficult to repair, and even more difficult to maintain than either lead or GRP pipe. In fairness, the maintenance on rubber-lined pipe has been generally less than on lead or lead-lined-steel pipe.

Cost savings of reinforced plastic pipe, over lead-lined-steel pipe, can be conservatively stated as being at least 40%. These savings are also about the same for reinforced plastic tanks in comparison to lead-lined steel tanks.

A few factors indicate the relative economy of GRP piping versus lead. For example, we can install about $3.40 worth of lead piping for each leadburner man-hour. We can install $6.25 worth of plastic piping for each pipefitter man-hour. The ease of installation is perfectly clear from these figures. Further, note that in relative basic costs, bonded-lead-lined steel pipe

Comparative costs, both relative and in actual dollars/linear ft., for various materials and sizes—Table I

Item	Description		2	4	6	8	10	12	Weighted Average
					Pipe Size, In.				
1.	100 psi., 230F., chemical grade glass-reinforced polyester, plain ends, long radius								
	Relative cost:		1.0	1.0	1.0	1.0	1.0	1.0	1.0
	Actual cost:		15.71	20.05	25.33	35.70	57.48	78.50	
2.	150 psi., 300F., cement end, glass-reinforced epoxy								
	Relative cost:		.65	.84	1.87	2.02	1.8	1.27	1.4
	Actual cost:		10.10	17.15	47.70	72.20	103.00	99.95	
3.	Rubber-lined cast iron, 3/16-in. lining, standard radius								
	Relative cost:		1.54	1.5	1.68	1.5	1.43	1.34	1.5
	Actual cost:		24.24	30.08	42.70	53.29	81.81	105.35	
4.	100 psi., 230F., chemical grade glass-reinforced polyester, press molded, flanged end, long radius								
	Relative cost:		1.8	2.0	2.1	2.3	1.9	2.1	2.0
	Actual cost:		28.19	40.91	53.07	81.36	108.26	166.73	
5.	150 psi., 300F., cement-end, glass-reinforced epoxy, flanged end								
	Relative cost:		1.86	2.18	2.74	2.94	2.80	2.07	2.4
	Actual cost:		29.40	44.70	69.00	105.30	160.00	162.30	
6.	Lead-lined steel, flanged								
	Relative cost:			3.1	3.58	3.1	3.27	3.07	3.2
	Actual cost:			62.45	90.59	110.39	187.85	241.73	
7.	100 psi., 230F. chemical grade glass-reinforced polyester, hand laid-up flanged end, long radius								
	Relative cost:		3.56	4.0	3.8	3.5	2.7	2.5	3.4
	Actual cost:		56.00	82.50	96.50	125.50	155.00	197.00	

costs 2½ times as much as its polyester counterpart. In general, glass-reinforced plastic means at least a 40% reduction in piping costs where it replaces lead, rubber-lined steel, or lead-lined steel.*

Competition in piping in the GRP field is growing progressively intense as better and more-economical methods are developed for producing this product.

This applies to both the reinforced polyesters and the epoxies. In general, the latest piping information shows that GRP polyester pipe may be purchased for 20 to 30% less than GRP epoxy and in some cases even less than this. In any large piping system, it will generally be found that if both the polyester and epoxy are suitable for the conditions indicated, the polyester system can be installed for less money. Most of the high-quality corrosion-resistant polyester pipe sold today is made on mandrels by hand layup process, while the bulk of the GRP epoxy piping is a machine-made pipe. Some machine-made polyester pipe is available, however.

The Society of the Plastics Industry is working aggressively to standardize the industry output. Fittings of standard dimensions, and pipe manufactured to meet standard specifications, should be produced instead of the previous wide nonstandard selection available in the industry. This, of course, is a step forward, one that is most necessary. Fittings and pipe need to be interchangeable regardless of which vendor they are purchased from, and if interchangeability is to be

---
* On a project just bid for an ion-exchange system, the vendor allowed a deduction of $7,050 if GRP epoxy pipe was used instead of rubber-lined steel (a sizeable savings and a better job).

possible with other corrosion-resistant fittings, then all fittings must ultimately meet an ASA standard.

The "Average purchase price comparison graph" (Fig. 4) and associated "Comparative Basic Materials Costs—GRP Piping Table" (Table I) are to be used as general guides and for rough estimating only. Prices shown vary considerably from vendor to vendor, and from one geographic area to another. These prices were obtained from first-class sources and represent those from reputable vendors in early 1965.

While Table I indicates relative purchase prices for different corrosion-resistant materials of construction, these are actually an average that exists in pipe from 2 to 12 in. dia. For some types of material, this ratio will hold substantially constant over the complete range. The economy of other types of pipe improves as the sizes increase. For example, up to 4 in. in size, Schedule 10—Type 304 stainless steel costs about the same as, or less than, 100-psi. polyester. In sizes above 4 in., the average is heavily weighted in favor of polyester.

Although we have included Type 304 stainless steel in our studies, this grade of stainless shows comparatively little corrosion resistance to anything but the mildest corrosive substances.

In some cases, it was difficult to arrive at exact cost comparisons because of variations in vendors' prices, and because of changes made in the manufacturing specifications. In cases where product lines were not complete, it was necessary to use prices from several vendors to establish continuity.

When buying reinforced plastic pipe, we do not buy

Comparative costs, both relative and in dollars/unit, for 90-deg. elbows in various materials and sizes—Table II

Item	Description	2	3	4	6	8	10	12	Weighted Average
1.	100 psi., 230F., glass-reinforced polyester								
	Relative cost:	1.0	1.0	1.0	1.0	1.0	1.0	1.0	1.0
	Actual cost:	2.22	2.57	3.55	4.80	6.79	9.26	12.15	
2.	150 psi., 230F., glass-reinforced polyester								
	Relative cost:	1.0	1.2	1.16	1.35	1.33	1.33	1.38	1.25
	Actual cost:	2.22	3.08	4.10	6.50	9.00	12.30	16.75	
3.	150 psi., 300F., glass-reinforced epoxy								
	Relative cost:	1.18	1.38	1.50	1.47	1.42	1.38	1.24	1.37
	Actual cost:	2.60	3.55	5.35	7.05	9.60	12.20	14.55	
4.	Schedule 10, Type 304 stainless steel								
	Relative cost	.82	1.1	.98	1.36	1.4	1.4	1.32	1.2
	Actual cost	1.83	2.80	3.47	6.51	9.51	12.84	16.08	
5.	Rubber-lined steel, 3/16-in. rubber lining, 20-ft. lengths, flanged								
	Relative cost:	1.93	2.62	2.32	2.42	2.34	2.23	2.26	2.30
	Actual cost:	4.28	6.72	8.26	11.61	15.87	20.60	27.48	
6.	Saran-lined steel, 10-ft. lengths								
	Relative cost:	1.21	1.87	1.84	2.46	2.62	.....	.....	2.0
	Actual cost:	2.69	4.82	6.51	11.83	18.36	.....	.....	
7.	Schedule 5, 316 stainless steel								
	Relative cost:	1.04	1.50	1.58	1.67	1.53	1.63	1.69	1.52
	Actual cost:	2.30	3.85	5.62	8.02	10.37	15.11	20.41	
8.	Lead-lined steel, homogeneously bonded, flanged ends, 10-ft. lengths								
	Relative costs:	2.16	2.60	2.40	2.80	3.10	2.92	2.95	2.70
	Actual cost:	4.81	6.71	8.52	13.41	21.01	27.18	35.91	
9.	Penton-lined steel, 10-ft. lengths								
	Relative cost:	3.10	4.10	4.6	6.1	6.2	.....	.....	4.82
	Actual cost:	6.88	10.47	16.15	29.14	42.36			

the product alone. While it is true that the product of any pipe manufacturer must be competitive in terms of price and quality, the buyer has a right to expect the backup of a top-flight service organization. Many companies pay lip service to backup, and from our side of the desk it is easy to recognize the great differences that exist among reinforced plastic pipe suppliers. Technical depth is important to back up the product. This depth extends throughout the manufacturer's organization right down to the shop personnel who fabricate the pipe and is equally important in the supplier's field construction crews. These crews must have a high degree of competence and a thorough knowledge of their product.

Fittings and Costs

Flanges, reducers, branches, tees and all other fittings are available from the better vendors, in standard sizes, configurations and lengths. Custom-designed fittings are also available to meet special requirements. GRP molded flanges are now being produced by high-compression techniques to meet ASA specifications.

Our field experiences have shown that fittings designed with smooth interior surfaces, and with strength and corrosion resistance qualities comparable to the pipe, are vital to over-all system integrity. These qualities receive primary attention when we make fittings selections. As an example, several of our major suppliers manufacture flanges that have chemical resistance and strength properties comparable to the piping system. These flanges are designed to standard ASA dimension schedules for bolt circles, bolt sizes and orientation, thus assuring proper hookup to pumps, valves, equipment and existing systems.

The comparative basic-material costs for a variety of 90-deg. elbows are shown in Table II. Their relative costs are illustrated in Fig. 5. These same conclusions apply to fittings in general:

1. It is apparent that minimum system-cost can be achieved through using the contact-molded elbow with wrapped joints in sizes 6 through 12 in. Some small price advantage appears to exist in cement-end elbows in the 2, 3, and 4-in. sizes.

2. A good choice on a flanged system is to use the contact-molded elbow with a butt and strap press-molded flange, or separate flange, attached. This produces a system that is generally less expensive than the conventional 150-psi. epoxy flanged elbow.

3. Again, the higher cost of using contact-molded stub ends as a means of providing a system of flanging (whether it involves tees, elbows, reducers, etc.) becomes evident. The use of contact-molded stub ends should be minimized in order to keep system costs down.

4. In fittings, per se, there does not appear to be a great deal of difference in price between a polyester fitting with press-molded flanges and a flanged epoxy fitting in sizes 2 through 4 in. However, in sizes 6 through 10 in., the savings in the press-molded, flanged-end reinforced polyester fittings are considerable. Corrosion resistance characteristics of these two

basic resins, however, need to be evaluated to determine the proper resin system to be used. In severely oxidizing environments, the resin of choice will often be polyester.

5. In an epoxy system, cement-end fittings will provide considerable economies over flanged-end fittings. Couplings will provide a less-expensive installation than flanges for runs of pipe. To achieve a minimum standard cost in epoxies, use cement couplings and cement fittings wherever possible. Use the minimum number of flanges necessary to provide system assembly and maintenance.

6. To provide the minimum cost in the assembly of a polyester system, use wrapped joints and plain-end fittings wherever possible. Use the minimum number of flanges necessary to permit assembly and proper maintenance.

Strict adherence to the guidelines shown above, and particularly Items 5 and 6, can produce real economies in system assembly.

Insulation

Glass-reinforced plastic pipe has unusually low thermal conductivity. In medium-temperature applications such as ours (0 to 200 F.), no insulation is needed for GRP pipe. This represents significant savings on capital costs and maintenance. For example, 1½ in. of insulation applied to 6-in. metal pipe costs approximately $3 per lineal foot. Another important factor related to GRP pipe's low thermal conductivity is condensation. Condensation is held to a minimum or is at least retarded by GRP pipe. Again, in this area, GRP polyester piping is generally somewhat thicker than GRP epoxy piping owing to the method of manufacturing. Under these conditions, GRP polyester piping is more resistant to sweating than GRP epoxy pipe, although both have approximately the same thermal conductivity per inch of thickness.

Meet the Author

John H. Mallinson is a senior staff engineer of the American Viscose Div. of FMC Corp., Front Royal, Va. 22630. He received his degree in chemical engineering from the University of Pittsburgh. Aside from two years with Pennsylvania Salt Mfg. Co., and war service in the U. S. Navy. (Ship Repair and Mine Countermeasures), he has spent his entire career with American Viscose. He received his professional engineering license in 1949.

Key Concepts for This Article

Active (8)	Passive (9)	Ind. Variable (6)
Specifying*	Pipes*	Material*
Selection	Plastics*	
	Fibers*	Dep. Variable (7)
	Glass*	Costs*

(Words in bold are role indicators; numbers correspond to EJC-AIChE information retrieval system. Asterisks mark key concepts suggested for indexing. Others are added to improve reading as an abstract. Indexing is described in *Chem. Eng.*, Oct. 11, 1965, p. 187.)

Chemical Feed Pumps

Costs, capacities and discharge pressures are correlated on a chart useful for estimating the price of reciprocating chemical feed pumps.

MICHAEL M. KIRK,
Allied Chemical Corp., Hopewell, Va.

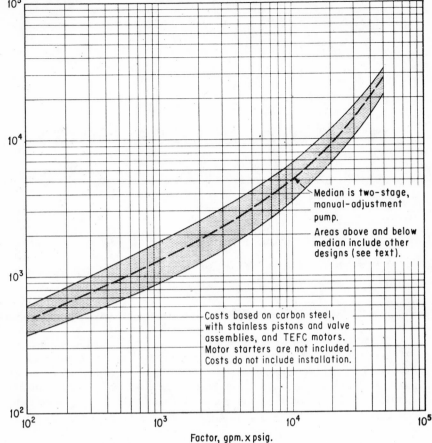

Purchase cost, $

Median is two-stage, manual-adjustment pump.

Areas above and below median include other designs (see text).

Costs based on carbon steel, with stainless pistons and valve assemblies, and TEFC motors. Motor starters are not included. Costs do not include installation.

Factor, gpm. x psig.

The costs of reciprocating chemical feed pumps are influenced by a variety of **factors** such as number of cylinders, materials of construction, valve design, type of drive, type of flow adjustment, control system, fluid viscosity, volume of liquid pumped and pressure.

For preliminary, order-of-magnitude estimating, it is not necessary to go into details on all these variables. Actually we have found that a correlation between volume pumped, discharge pressure and costs is adequate for many purposes. The pump costs shown in the figure (shaded area, see above) were plotted vs. a factor obtained by multiplying gpm. by pump discharge pressure. The curve covers a range of $\frac{1}{4}$ to 60 gpm. and **75** to 4,600 psig.

Some judgment must be used in applying the curve to a particular problem. For instance, reciprocating feed pumps are generally classified as simplex, duplex, triplex (1, 2, 3 cylinders). Single-stage pumps deliver liquids in a surging pattern, multistage pumps level out the surge effect. It should be kept in mind that a single-stage pump for a high-pressure application would be of heavy construction and operated at relatively low speeds. Multistage pumps can be of lighter construction and operated at higher speeds—thus there would not be a great difference in costs between the two types.

Another factor affecting costs involves flow adjustment. Lower-cost pumps are designed to pump at a constant rate. The higher-priced units are usually variable-feed pumps and there are two types: one must be shut down to adjust flow; the other can be adjusted while the pump is running. The latter may be adjusted manually; or air, hydraulic pressure or an electric motor can be used in a remote-control system.

Median Costs

The majority of data used as the basis for the cost curve were for pumps designed for manual adjustment while the unit is operating. Most pumps priced for the curve were of carbon steel with stainless pistons and valve assemblies. Motors (totally enclosed, fan cooled) are included in the costs shown. Costs were converted to CE Plant Cost Index 105.

No installation costs are included, and costs do not include motor starters (see Cost File 77, June 10, 1963 for motor starters).

In using the cost data, assume a two-stage, manual-adjustment-in-motion pump, with carbon-steel construction and stainless trim, as the median (dotted line on figure). Choosing a point above or below the curve should be guided by the following variables:

- Number of cylinders.
- Materials of construction.
- Control design, control-point location.
- Lubrication system and type of drive.

Key Concepts for This Article

Active (8)	Passive (9)	Means/Methods (10)
Estimating	Costs	Charts
	Pumps, reciprocating	Curves
	Liquid	
	Feeders	

(Words in bold are role indicators; numbers correspond to AIChE information retrieval system. Indexing details are described in *Chem. Eng.*, Jan. 7, 1963.)

Low-Temperature Refrigerators: Costs and Operating Power

These cost and performance curves will allow you to make an accurate, preliminary estimate for the new refrigerators operating in the —452 F. range.

R. E. BERNERT, *Principal Research Engineer,*
Avco-Everett Research Laboratory, Everett, Mass.

Recent interest in systems and processes requiring temperatures in the liquid-helium range makes the cost of refrigeration machines generating these temperatures a worthwhile subject. The cost and perform-ance curves presented here will allow reasonable cost and power estimates to be made for refrigeration systems producing either liquid helium (normal boiling point: 4.2 K.) or continuous 4.2 K. refrigeration.

The cost curves (Fig. 1) have been fitted to available data on present machines and to cost estimates of future large machines. The refrigerators generally include all necessary controls and safety devices, requiring only hookup to power and cooling water. The lower curve on the figure corresponds to an open-loop system designed primarily to produce liquid helium. The area between the upper curves is for a closed refrigerant loop. Cycle design will, of course, be affected by the lowest temperatures desired and the coolant

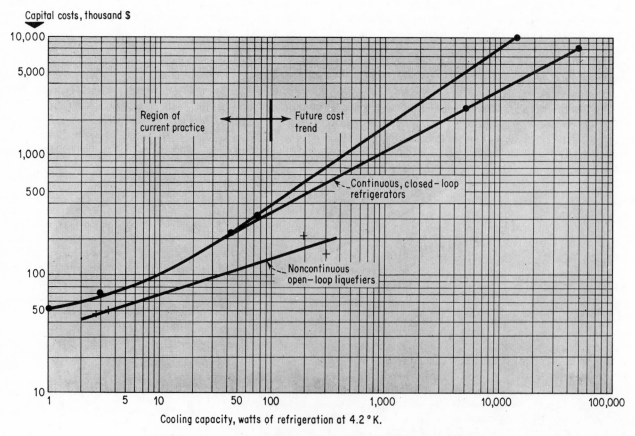

Estimated costs for refrigerators operating at 4.2 K.—Fig. 1

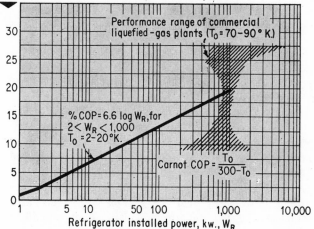

% COP = 6.6 log W_R, for
$2 < W_R < 1,000$
$T_0 = 2$–$20°$ K.

Carnot COP = $\dfrac{T_0}{300 - T_0}$

Efficiency of low-temperature refrigerators—Fig. 2

Nomenclature
(Basis: One Hour)

Q_{refrig}	Heat absorbed by refrigerant, watts
W_{ideal}	Ideal work of refrigeration, watts
T_o	Heat absorption temperature, °K. (4.2K.)
T	Heat rejection temperature, °K. (300K.)
Q_c	Heat leak due to thermal conduction, watts
Q_R	Heat leak due to thermal radiation, watts
Q_r	Refrigerator cooling capacity, watts
k	Avg. thermal conductivity, watt/(ft.) (°K.)
L	Length of conductor, ft.
A_σ	Cross-section of conductor, sq. ft.
h_r	Apparent radiant heat-transfer coefficient between two infinite parallel surfaces, watt/ sq. ft.
A_r	Surface area of radiation, sq. ft.
W_r	Installed power of refrigerator, kw.

return temperature to the refrigerator. Thus user operating requirements will have an effect on the cycle design as well as on the cost.

Another important cost consideration is the anticipated duty cycle of the machine. Generally, continuous processing places a premium on downtime. Thus, the additional expense of spare components may be warranted. The most important item is the compressor: this will add approximately 10% to the system cost.

In Fig. 2, the percent Carnot coefficient of performance (efficiency) of the refrigerator is plotted as a function of the system size. In using this curve, it is suggested that a limiting value of 20% be taken for large systems until operating data become available. This value is based on the comparison with oxygen/ nitrogen plants operating in the 70 to 90 K. range.

Using the data from Fig. 2, we can construct Fig. 3, which shows the increase in installed power vs. cooling capacity of the system. This curve will allow an estimate of the cost of furnishing utilities to the installation. The curve is based on the equation for the Carnot cycle work for refrigeration, defined as the ratio of the heat absorbed by the refrigerant to the work done on the fluid, or:

$$\frac{Q_{refrig.}}{W_{ideal}} = \frac{T_o}{T - T_o} \tag{1}$$

This ratio is called the Carnot coefficient of performance, or Carnot COP. The work calculated is the minimum or ideal and must be multiplied by $(100)/(\%$ COP) to obtain the actual installed power requirement.

For a liquid helium refrigerator having $T_o = 4.2$ K. and rejecting heat to ambient $T = 300$ K., the ideal work requirement is $1/COP$ or 295.8/4.2 which equals 70.3 watts power per watt refrigeration. The actual power requirement will be between 5 and 50 times the ideal, according to Fig. 2.   *(Turn page)*

Actual power required for 4.2 K. refrigerators—Fig. 3

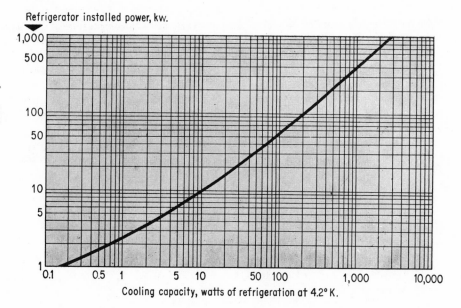

How Refrigeration Cycle Works

In a typical refrigeration system, helium gas first goes to a compressor. It is then cooled to 80 F. and flows to a Dewar cold-box vessel.

In the Dewar, the helium passes through heat exchangers, and is cooled to −319 F. while transferring heat to cold gaseous nitrogen and to liquid nitrogen at −319 F. The helium temperature is lowered to −415 F. in another heat exchange with recycle helium. A portion of this cold helium is cooled to −436 F. by expansion in an engine. This material, combined with recycle helium, is used to cool the main stream to −435 F. Another heat exchange, then flow through a Joule-Thomson valve condenses about 15% of the stream at about −452 F. Gaseous helium is recycled. All equipment in the Dewar is supported by long, slender rods. For more details on this process, see *Chem. Eng.*, Oct. 1, 1962, pp. 76-78.

Estimating Refrigeration Requirement

Components operating at cryogenic temperatures are generally contained in the vacuum-insulated container, or Dewar. Also, liquid-helium temperatures generally imply a radiation barrier operating at 77 K., the boiling point of liquid nitrogen. This barrier may be incorporated as part of the over-all refrigeration system. The "heat leak" to the 4.2 K. environment then consists essentially of conduction Q_C and radiation Q_R. Heat is conducted into the 4.2 K. zone via support members, instrument leads, etc. Radiation results from heat interchange between the 77 K. and 4.2 K. surfaces. If power leads are used, joule heating of the power leads should be included. Thus, the total heat leak Q_T consisting only of conduction and radiation is:

$$Q_T = Q_C + Q_R$$

or

$$Q_{\text{refrig.}} = Q_T = (k/L) A_C (77 - 4.2) + h_r A_r \qquad (2)$$

For stainless steel, k, the average thermal conductivity between 77 K. and 4.2 K. is 1.53 watt/(ft.) (°K.) and for high-purity copper it's 1,500 watt/(ft.) (°K.) The apparent heat-transfer coefficient due to radiation, h_r, between 77 K. and 4.2 K. will be about 5 x 10^{-3} watt/sq. ft. using an emissivity of the reflecting surfaces of 0.05. The area, A_C, is the total cross-sectional area of metal supports and leads passing between the two temperatures; and the area, A_r, is the surface area of radiation. Infinite parallel flat plates at 4.2 K. and 77 K. are assumed. For other geometries, h_r must be modified.

Cost Estimate

We may now estimate the refrigerator cost based upon the working volume to be maintained at 4.2 K., and the amount of instrumentation and structural supports required for the working vessel.

The system data are (closed loop):

Cross-section A_C of copper leads	= 0.02 sq. in.
Cross-section A_C stainless-steel support	= 10 sq. in.
Average length L of leads and supports	= 2 ft.
Vessel area (taken as A_r)	= 150 sq. ft.

From Eq. (2) for copper, we have the following:

$$(1,500) \left(\frac{0.02}{144 \times 2} \right) (77 - 4.2) = 7.59 \text{ watts}$$

For the stainless:

$$(1.53) \left(\frac{10}{144 \times 2} \right) (77 - 4.2) = 3.86 \text{ watts}$$

And for radiation:

$$(5 \times 10^{-3}) (150) = 0.75 \text{ watts}$$

The system's total heat leak will be 12.20 watt/hr. From Fig. 1, refrigerator capital cost will be $112,000, and from Fig. 3, the system will require approximately 12 kw. of installed power. As with any refrigeration system, some capacity in excess of the minimum is required to initially cool the system to the operating temperature. The amount will depend upon available time and system weight.

Many applications require only intermittent operation, and open-loop systems may be used. This arrangement permits a smaller open-loop liquefier combined with helium storage tank and gas recovery system.

Alternatively, the infrequent user may consider purchasing liquid directly from commercial sources. Liquid-helium ranges in price from $4 to $7/liter, depending upon location and use rate. With one watt of refrigeration equivalent to approximately 1.4 liters of liquid helium, the above refrigeration requirement could be satisfied by 12.2 × 1.4 = 17.1 liters/hr. of helium. At say $6/liter, the cost per hour of operation would be $103. To this must be added the helium required to initially cool the system. After precooling with liquid nitrogen (a small cost compared to the helium) to 77 K., approximately ½ liter/lb. of stainless steel will be required.* Used alone, approximately 20 liters/lb. would be necessary.

In comparing the systems, the capital and operating costs of the continuous system would have to be broken down to an hourly basis to determine the most economical approach.

All data presented here are based on 1963-64 costs.

Low-temperature helium refrigerators are produced commercially by several firms, such as: Arthur D. Little, Inc., Cambridge, Mass.; Cryonetics Corp., Burlington, Mass.; Air Products and Chemicals Corp., Allentown, Pa.; Linde Co., Div. of Union Carbide, Tonawanda, N. Y.; North American Phillips Co., Ashton, R. I.; Cryogenics Engineering Corp., Denver, Colo.; Garrett Airesearch Manufacturing Co., Los Angeles, Calif.

*Danzberger, A. H., "Handling Liquid Helium in a Propulsion Test Facility" American Rocket Society preprint 1636-61, **Mar.** 1961.

Key Concepts for This Article

Active (8)	Passive (9)	Means/Methods (10)
Estimating	Costs	Charts
	Power	Curves
	Utilities	
	Refrigerators	
	Helium	

(Words in bold are role indicators; numbers correspond to AIChE information retrieval system. Indexing details are described in *Chem. Eng.*, Jan. 7, 1963.)

Estimating the Cost of Jacketed, Agitated and Baffled Reactors

Here is new cost data on reactors of various sizes and materials of construction.

GEORGE C. DERRICK, *Hooker Chemical Corp., Niagara Falls, N.Y.*

TABULAR SUMMARY of prices quoted on reactors ▶

	Reactor Capacity, Gal.					
	500	1,000	2,000	4,000	5,000	10,000
Fob. price, $:						
Carbon steel	3,550	5,430	7,440	....	13,430	....
Type 316 stainless*	6,880	10,560	15,810	....	23,830	34,840
Nickel*	9,020	13,330	19,950	....	30,800	44,550
Glass-lined	11,630	15,674	20,440	33,140	....	....
Motors:						
Suitable hp.	5	10	20	40	40	75
TEFC list price, $	138	240	393	753	753	1,405

* Process side is solid stainless or nickel; the shell is dimpled Inconel.

BASIC SPECIFICATIONS for jacketed, agitated reactors ▶

	Rating:				Process Nozzles	Type of Seal	No. of Baffles	Corrosion Allowance In.
	Process		Shell					
	Psig.	°F.	Psig.	°F.				
Carbon steel	50	353	125	353	9	packed	4	1/16
Type 316 stainless	50	353	125	353	9	packed	4	1/16
Nickel	50	353	125	353	9	packed	4	1/16
Glass-lined	100	650	100	350	9	mech.	1	......

ESTIMATING CHART, based on a Marshall & Stevens cost index of 260 ▶

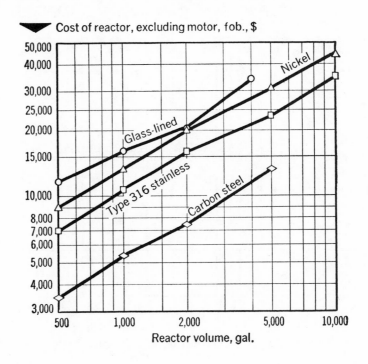

▼ Cost of reactor, excluding motor, fob., $

Reactor volume, gal.

What Do Chemical Reactors Cost In Terms of Volume?

These curves, which should prove particularly useful in process scaleup work, permit the costs of many different types of reactors to be estimated and compared on a volume basis.

T. E. CORRIGAN, W. E. LEWIS and K. N. McKELVEY*

In most of the literature, heat exchanger costs are expressed on the basis of area; pipe and tubing on the basis of linear feet; columns on the basis of height, diameter and tray spacing; and tanks on either a stor-

* Dr. Corrigan is with the research and technical division of Mobil Chemical Co., Metuchen, N.J. 08840. The other authors are from Ohio State University.

age capacity or weight basis. In the case of chemical reactors, the major dimension of interest is the volume. Yet, because so many different types of equipment (including the types just cited) can be used for reactors, one encounters costs that are based on all sorts of variables but only seldom on volume. Our aim here is to fill this gap.

A chemical reactor is defined as any piece of equipment whose sole purpose is to carry out a chemical

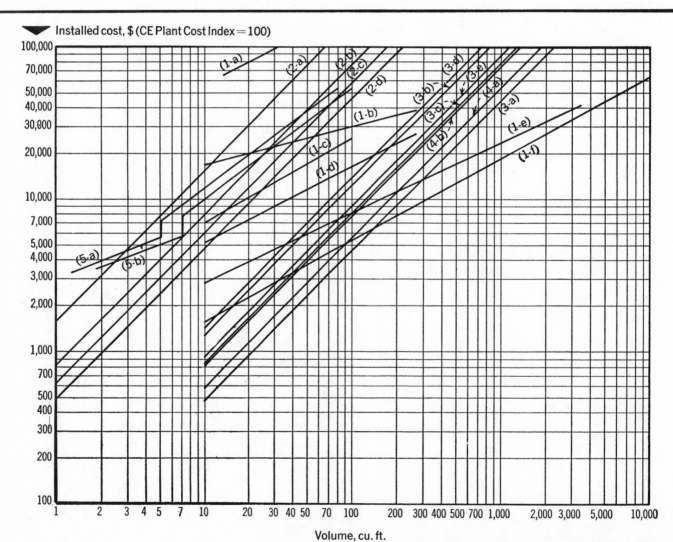

▼ Installed cost, $ (CE Plant Cost Index = 100)

Volume, cu. ft.

INSTALLED COSTS of the various types of reactors described in the key at the right are here shown on a volume basis.

reaction. Therefore, reactors can be classified in many ways, some of the major categories being:

- Batch.
- Semi-batch.
- Tubular.
- Continuous stirred tank.
- Towers (packed or plate).
- Catalytic fixed bed.
- Catalytic fluidized bed.

The selection and design of the proper type of reactor for a given process involves the consideration of temperature control, materials of construction, effect of backmixing and, of utmost importance, cost. The final selection as to type would depend upon the degree of conversion desired, the importance of temperature control, and the effect of backmixing on the reaction. The desired size would be calculated on a volume basis. The final selection should also be influenced by cost. It can be seen from the graph below that there is a wide variation in installed cost for equal-volume reactors.

The cost information was obtained in the standard literature; the references are cited in the "key" to the graph. Some of these references are older than others; they have all been adjusted to a CE Plant Cost Index of 100, a common base that is easy to use for further updating to any time desired. For instance, multiplying the cost figures by 1.09 will bring the results up to January 1967, the CE Plant Cost Index having been at 109.0 that month.

Although the use of indexes on data that is more than a few years old does involve some inherent inaccuracy, this does not negate the value of the graph for preliminary estimates and for process scaleup comparisons. Of course, current bids should be used for definitive estimates, or to help decide cases where the graph shows two equally usable types of reactors to be very close to each other in cost for a given volume.

Cost and Other Selection Considerations

Aside from temperature control and materials of construction, certain other practical considerations arise in selecting the proper type of reactor. For instance, a gas-phase reaction with a several-second holding time would not be carried out in a batch process. On the other hand, the manufacture of a fine organic chemical that involves several synthesis steps having reaction times of several hours each would be very amenable to batch processing.

A tubular-flow reactor or a shell-and-tube heat exchanger would be used for a homogeneous reaction with a large heat of reaction. The liquid-phase reaction could also be carried out in a continuous stirred tank, or a series of tanks with jackets and internal cooling coils.

One example of cost saving is the use of a packed tower reactor. If the process is one that cannot tolerate backmixing, a plug-flow reactor is needed. Although one usually thinks of the tube or pipe reactor as the only plug flow, the packed tower has very near plug-flow behavior. (The values of (D/uL) are as low as 0.02 or 0.01.) On an equal-volume basis, the packed tower costs only about one-tenth as much as a tubular reactor.

From the chart, it can be seen that for 100 cu. ft. of reactor volume, 2-in. welded pipe would cost $75,000, while a packed tower (a 100-in.-dia. column with 3-in. Raschig rings) would cost only $7,500.

We thus believe that reactor costs on a volume basis, as presented by the various curves on the chart, will be helpful to engineers in their evaluation and selection of reactor types.

Key

1—Tanks
  a—Jacketed, pressurized for 1,500 psig. (Ref. 2, p. 62)
  b—Autoclaves (Ref. 1, p. 57)
  c—Jacketed, pressurized for 300 psig. (Ref. 2, p. 62)
  d—Jacketed, pressurized for 50 psig. (Ref. 2, p. 62)
  e—Agitated tank (Ref. 1, p. 57)
  f—Storage tank (Ref. 2, p. 66)

2—Welded pipe (Ref. 1, p. 197)
  a—1 in.
  b—2 in.
  c—4 in.
  d—6 in.

3—Packed column, 10 in. (Ref. 1, p. 51)
  a—Empty
  b—Raschig rings (½ in.), 0.53 void fraction (Ref. 3, p. 366)
  c—Raschig rings (1 in.), 0.68 void fraction (Ref. 3, p. 366)
  d—Berl saddles (½ in.), 0.53 void fraction (Ref. 3, p. 366)
  e—Berl saddles (1 in.), 0.69 void fraction (Ref. 3, p. 366)

4—Packed column, 100 in. (Ref. 1, p. 51)
  a—Empty
  b—Raschig rings (3 in.), 0.74 void fraction (Ref. 3, p. 366)

5—Heat exchangers, Type 316 stainless-steel tubes, triangular pitch (Ref. 1, p. 119)
  a—With ¾-in. tubes
  b—With 1-in. tubes

OTHER EXCHANGERS	COST FACTOR
Floating head	1.00
U-Tube removable	0.85
Fixed tube-sheet	0.80
Lengths: A— 8 ft.	1.3
B—12 ft.	1.1
C—16 ft.	1.0
D—20 ft.	0.95

References

1. Chilton, C. H., (Ed.), "Cost Engineering in the Processing Industries," McGraw-Hill (1960).
2. Aries, R. S., Newton, R. D., "Chemical Engineering Cost Estimation," McGraw-Hill (1955).
3. Zimmerman, O. T., Lavine, I., "Chemical Engineering Cost," Industrial Research Service (1950).

Key Concepts for This Article

Active (8)	Passive (9)	Ind. Variable (6)
Estimating*	Reactors*	Volume*
Comparing	Costs, installed*	Specifications
Selecting		

	Means Methods (10)	Dep. Variable (70)
	Curves	Costs

Comparing Costs Of Materials for Cryogenic Containers

For low-temperature vessels, when is stainless more economical than aluminum? Where should 9% nickel steel be used? These material comparison charts, based on costs of fabricated cryogenic containers, give you the answers.

The problem of materials selection for cryogenic containers is primarily an economic one. There are a large number of different materials of construction available to do an adequate job. But is is possible, for example, to design and build an 80,000 gal. cryogenic vessel for $200,000. It is also possible to pay $400,000 for essentially the same unit, using another material without a significant improvement in low-temperature performance.

Comparisons Based on Equipment Costs

Recognizing the need for cost information, United States Steel Corp., working in conjunction with Arthur D. Little (ADL), has completed a detailed study on the costs of commonly used cryogenic storage containers.* The study gives cost comparisons for available materials, not on a $/lb. basis, but on finished, fabricated vessels, using a standardized cost-analysis technique.

The charts and table (shown on p. 138) are based on the USS-ADL report, and show costs of the inner container. It should be noted that the information presented here does not necessarily represent actual competitive costs or have a relation to selling price. This is because fabricator efficiencies and labor rates vary over the country. And a number of other variables such as cleaning procedures and use of additional container openings were impractical to consider in this

* "USS Low Temperature and Cryogenic Steels Materials Manual" will be available after Jan. 1, 1964 from R. E. Petsinger, U. S. Steel Corp., Market Development Division, 525 William Penn Place, Pittsburgh, Pa. 15230.

study. However, these items do not significantly affect the net results. The data on the charts and tables are intended for use *only* in comparing the relative economics of materials of construction.

Procedure Used in Study

In developing the cost data, USS and ADL assumed a hypothetical fabricating shop with qualified welders capable of semi-automatic welding of stainless steel, the 5000 series aluminum alloys, 9% nickel steel and 3½% nickel steel. It was assumed that the storage vessels were fabricated in accordance with the latest revision of Section VIII of the ASME Unfired Pressure Vessel Code and, in part, with the recommendations of the applicable API code. Items of common cost were excluded (cold testing, outer wall vessel—majority of cryogenic vessels have a carbon-steel jacket, cost of evacuation, supports, etc.). Standard container openings were assumed for all designs (relief valve, inlet, outlet, vacuum outlet, liquid level, and vapor openings).

Cost data were collected based on discussions with a cross-section of container fabricators.

Estimating the Cost

In estimating costs, two construction classifications were considered: shop fabrication and fabrication in the field. Shop fabrication usually involves cylindrical vessels with three possible heads: ASME flanged and dished, elliptical or hemispherical.

The lower pressure shop-fabricated vessel would be representative of large bulk containers; the 50-psia. vessel is typical of containers used in over-the-road distribution; 100-psia. pressure represents some of the lower working pressure distributor stations. Most vessels used at distributor stations approach 250 psia.

For shop work, USS used a typical labor rate of $7/hr., which includes shop overhead, shop efficiency and fringe benefits. Allowable stress for each material was taken as that permitted by the ASME code.

Field-erected tanks are most often spherical or flat-bottomed cylinders with elliptical heads. Only a non-evacuated design was considered for the cylindrical vessel. For the spherical container, length to diameter ratio is fixed and pressure is the important variable (an evacuated design was used in the cost analysis). For example purposes only, the table represents summary results of an estimate on a spherical container of 6,000-bbl. (252,000-gal.) capacity.

As the charts show, costs vary with materials, pressure, vessel configuration and volume. In developing this information, the most economic length/diameter ratio was found for each alloy and this was used in calculating the most economic design for each material.

According to the charts, 3½% nickel steel is the most economical material considered in the study.† Of

† These conclusions, of course, are based on first cost of the vessel and do not take into account cost of operating and maintaining the equipment.

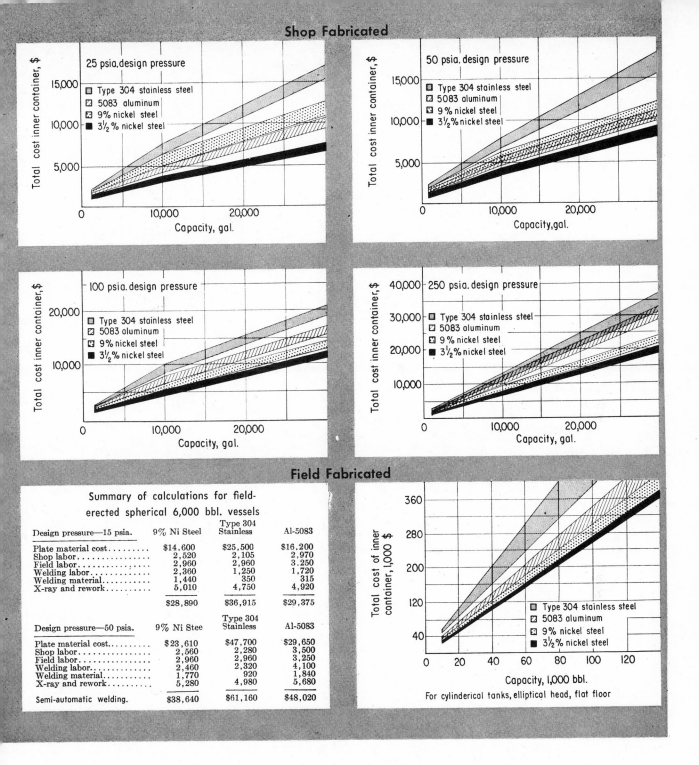

Field Fabricated

Summary of calculations for field-erected spherical 6,000 bbl. vessels

Design pressure—15 psia.	9% Ni Steel	Type 304 Stainless	Al-5083
Plate material cost.........	$14,600	$25,500	$16,200
Shop labor...............	2,520	2,105	2,970
Field labor............; ...	2,960	2,960	3,250
Welding labor...........	2,360	1,250	1,720
Welding material..........	1,440	350	315
X-ray and rework..........	5,010	4,750	4,920
	$28,890	$36,915	$29,375

Design pressure—50 psia.	9% Ni Stee	Type 304 Stainless	Al-5083
Plate material cost.........	$23,610	$47,700	$29,650
Shop labor...............	2,560	2,280	3,500
Field labor...............	2,960	2,960	3,250
Welding labor.............	2,460	2,320	4,100
Welding material..........	1,770	920	1,840
X-ray and rework.........	5,280	4,980	5,680
Semi-automatic welding.	$38,640	$61,160	$48,020

For cylinderical tanks, elliptical head, flat floor

---

course, this steel has a limited temperature range (down to about −175 F.).

For applications below −175 F., aluminum is competitive with 9% nickel in shop-fabricated vessels to about 50 psia. at capacities larger than 3,000 gal. Above 50 psia., 9% nickel seems to be less expensive. Stainless is competitive with aluminum in small containers of 1,000 gal. or less, and in the 150-200 psia. range at all capacities. Above 250 psia., stainless 304 is the economically preferable material over aluminum. Aluminum is less expensive than stainless in the 25 to 100-psia. shop-fabricated containers.

In the cylindrical, field-erected container, aluminum is competitive with 9% nickel up to 20,000 bbl. (840,000 gal.). Above this capacity, 9% nickel seems

less expensive than aluminum for cryogenic containers.

For evacuated, spherical containers of 6,000-bbl. capacity, 9% nickel and aluminum are competitive at low pressures; 9% nickel is less expensive above about 10 psig.

All data refer to July 1963 costs.

Key Concepts for This Article

Active (8)	Passive (9)	Passive (9) cont'd	Means/Methods (10)
Comparing	Costs	Aluminum	Curves
Evaluation	Metals	Nickel steels	Charts
	Alloys	Cryogenics	Tables
	Iron alloys	Vessels	
	Steels	Tanks	
	Stainless steel		

4

Buyer/Seller/Contractor Relation-ships, and the Control of Capital Costs

Negotiating With Engineering Contractors

A broad spectrum of engineers should be interested in these negotiations, which range all the way from the preliminary discussions held when bid specifications are prepared, to the hammering out of the contract itself.

SIDNEY A. BRESLER, *Borden Chemical Co.*
and
MARTIN J. HERTZ, *Attorney at Law*

The stage has been set for the start of the negotiating cycle.

The gleam in the mind's eye of a process engineer or a market analyst has swelled into a report recommending the construction or expansion of a chemical processing plant. Preliminary technical reviews, market surveys, cost estimates and pro-forma income statements have confirmed the attractiveness of the proposed project. Management approval having been obtained, a sequence of actions is now put into motion that will transform the still somewhat generalized concept into a set of detailed specifications, and result in the physical creation of the desired facility.

If an outside company will be involved in designing and/or constructing the plant, the extent of its work and its compensation will be established by a written contract. Each of the parties to the contract will wish to obtain terms most favorable to himself. Neither will wish to exact terms so injurious to the other that the contract may be voided deliberately because the financial burden has become unsupportable. Both parties are aware of the value of defining clearly the key aspects of the client-contractor relationship.

Negotiating such a contract is an art, combining elements of scientific methodology and clever bargaining, tempered by engineering judgment and an appreciation of the law of contracts. Certain principles have been found effective in arriving at mutually acceptable agreements, and we will show how these principles can be applied. They pertain not only to the engineers and attorneys of the operating company, but also to those of the contractor.

It should be noted that an understanding of the salient portions of a contract, and of the factors that must be considered before fruitful negotiations with a contractor can begin, will be of considerable help to engineers other than those directly involved in the discussions. While contract negotiations may be conducted on a fairly high level in the corporate structure, key decisions may be made by the operating company's process engineers, who are charged with finding the right questions to ask during the preparation of the definitive plant specification.

These same engineers, or their counterparts on the contractor's staff, must find the correct answer—the optimum plant design—and then construct the plant

within the limits imposed by the signed contract.

We shall see how important it is for these engineers to proceed in a systematic and informed manner, so that costly design changes, afterthoughts, recriminations and wasted effort can be minimized.

Deciding to Employ a Contractor

Some chemical and petroleum companies have such large capital expansion and improvement budgets that their annual volume of design and construction work exceeds that handled by most plant contractors. Such corporations find it advantageous to maintain complete engineering staffs, including requisite specialists in thermodynamic analysis, corrosion control, computer technology, cost estimation, etc. Generally they find little advantage in going outside their own organiza-

tion if the process involved is one that has been developed within the company, except during relatively short periods of intense activity when work may be subcontracted to avoid undesired fluctuations of permanent staff level. If the process is one that has been intensively studied by engineering contractors who have developed proprietary innovations, even a very large company may find it advantageous to tap this special know-how.

The small or medium-size chemical company can seldom justify a large staff of engineering specialists. Usually, in the small company, a minimum permanent central engineering staff will be maintained, to prepare feasibility studies and maintain liaison with selected plant contractors. Minor construction work, not exceeding a pre-established value, may be assigned to operating plant engineers.

What guarantees can an owner reasonably expect to

Situation I:

The contractor offers a proprietary process, purchases the equipment and erects the plant. He may offer guarantees in these areas:

Plant-Output

• The plant will produce a minimum quantity of the desired product or products.
• Such products will meet prescribed limits of purity.
• Products will be delivered to the plant battery limits at prescribed temperatures and pressures.
• Discharge of noxious liquids or gases will be within prescribed limits and will not subject the owner to suits brought by private parties or governmental authorities.

Note: It may be desirable to specify both the instruments and analytical methods that are to be used for testing product purity as well as the allowable tolerance of instrument measurements.

Also note that the guarantee of plant output and operating requirements will generally be based on plant performance demonstrated during a limited period, not exceeding several days or weeks, under the supervision or direction of a representative of the contractor. It is unusual for either of these conditions to be waived.

Operating Requirements

• Raw-material and utility usage will not exceed prescribed amounts. (This guarantee may be applied to individually specified materials or to a total requirement based upon unit costs and needed amounts of each material. The latter approach is preferable.)

In order for this guarantee to be effective, raw-material and utility quality, temperature and pressure must be specified as fully as are the requirements for the plant product.

Usage of catalysts and other materials that are

not expected to require replacement within relatively short periods of time is generally not guaranteed, as useful life of these materials is greatly influenced by the skill of the plant operators.

Operating labor and maintenance requirements cannot be absolutely guaranteed, since staffing is established by the owner in accordance with his philosophy of plant operations, and perhaps also in accordance with local custom of the area in which the plant is to be located. However, maximum man-hours are sometimes proposed, based on certain assumptions.

Design

• The plant design will follow the accepted proposal specification.
• The plant will be designed in accordance with good engineering principles, by competent personnel.

Patent Liability

• Neither the process nor any of the equipment supplied as part of the plant will infringe any valid patent.

Note: Plants designed and erected in this country usually are given an unlimited guarantee against patent liability. However, the contractor may offer no guarantee at all if the plant is to be constructed in another country, if his attorneys have not made a thorough patent search or are unfamiliar with local patent jurisprudence.

Equipment

• Equipment will meet all applicable industrial codes, such as TEMA, ASME, etc., and all local government regulations.
• All equipment will be new, and its performance will be guaranteed by the manufacturer as well as the contractor.
• Proper materials of construction will be

A company of medium size (or, on occasion, a small company that has developed its own process) may ask its engineering staff to at least prepare the necessary process flowsheets. Furthermore, depending on its skill and knowledge of feasible mechanical design (and on the man-hours that can be allocated to the work), the staff may prepare the engineering flowsheets and plot plans, and purchase the equipment. In this case, only the detailed engineering and construction will be done by an outside general contractor. Usually the competition for such work is greater, and overhead costs and fees smaller, than if the project were handled by a consulting engineering company.

In some cases, the staff of the operating company may be oriented toward mechanical design and construction rather than process work; in this case, the company's engineers may undertake the job of general contractor for all major civil work and for off-site facilities such as roads, railroad sidings, water supply, and administration buildings—particularly if the new plant is to fit into an existing complex where special site requirements are known and relationships with local subcontractors well developed.

Of course, whenever responsibility for a major portion of the plant design or construction is retained by the operating company and the remaining part assigned to a contractor, special care will be required to cope with problems of coordination and ultimate responsibility for plant performance.[1,2]

To summarize, an operating company may retain engineering contractors because it does not have a sufficiently large staff to do all or part of the work itself, because it is willing to pay a premium for the advantage of having assigned undivided responsibility,

obtain from a contractor or a consultant?

selected and the plant will not be subjected to excessive corrosion, erosion or other form of chemical attack.

Note: Usually it is quite difficult to determine the cause of observed equipment inadequacy, whether of capacity, efficiency, or resistance to attack. This is particularly true if the plant has been running for some time and there is a question of whether or not operating limits established by the designer have been exceeded.

Construction

• Work will be performed in an expeditious manner, by qualified personnel, in accordance with accepted practices.

• All reasonable efforts will be made to have the plant completed and in operation on a promised date. (The question of damages for late completion will be discussed later.)

Situation II

The contractor offers a process that he has licensed from a third party, purchases the equipment and erects the plant.

As far as the owner is concerned, it does not matter whether the contractor is offering his own process or one that he has licensed. The owner will expect the contractor to provide all of the guarantees that he would have if the process had been his own.

Situation III

A contractor or consultant supplies basic process information and equipment specifications, and the owner elects to have his own staff, or a general contractor, complete the detailed engineering and construction.

The owner will expect the contractor to offer all of the guarantees with respect to plant output, operating requirements, design (basic equipment design only) and patent liability outlined above. However, he must be prepared to transmit all purchase orders (unpriced, if so desired) and detailed drawings to the contractor for review, and have the contractor's representative at the job site during the installation of important equipment and during the test run.

Situation IV

The owner supplies the basic process know-how that he has developed or licensed and hires a contractor to complete the detailed design, purchase the equipment and construct the plant.

This is the converse of the condition described immediately above. The owner will expect guarantees that the plant will be designed in accordance with accepted engineering principles and that sound judgment will be used in the selection of equipment. All of the guarantees relating to construction, as indicated above, are applicable.

Although there can be no guarantee with respect to process, the owner should expect the contractor to call his attention to any aspect of the basic design that does not appear reasonable and, in general, to help the owner attain a sound and profitable plant.

Situation V

The owner gets complete engineering information from a contractor or a consultant but retains a general contractor to construct the plant.

The only guarantees that the owner may expect are those pertaining to construction. However, even those guarantees may be effectively waived if the owner or his advisor overspecify the procedure that the constructor must use, and relieve him of any discretion in the manner in which he conducts his work.

or because it is convinced that a particular process plant can be designed most efficiently by a contractor whose staff is expert in this area. In many instances, it is the last reason that is decisive.

Negotiate, or Require Competitive Bids?

If it has been decided to retain an engineering and construction firm, a second decision must be made as to whether to select one contractor and negotiate only with him, or to request competitive bids from a number of contractors. Assuming that there is more than one contractor qualified to handle the project, the matter may resolve into a question of time versus money.

If a single contractor is selected and impressed with the necessity of having the plant in operation as rapidly as possible, he will start the engineering design at once and place orders for long-delivery equipment immediately. The design will be based on work done before, and there will be a minimum of process or purchasing optimization. If more time is allowed, process alternates can be studied and the design modified where this appears advisable.

Alternatively, if several bids are solicited for a major project, two or three months may elapse before the bids are received and evaluated. However, if the owner has foreseen this delay and allows the necessary time for the bidding procedure, selects competent bidders, and allows the competitive spirit to prevail, he will probably obtain a lower priced and more efficient plant.

When asked to prepare a bid, the contractor must decide whether or not to spend the relatively large sum—perhaps one percent of the total plant cost—required to prepare the estimate. Some engineering companies will accept a competitive or lump-sum invitation to bid only with the greatest of reluctance, and today many others will refuse to bid if they believe that the number of their competitors is so large (i.e., more than three or four) that the likelihood of obtaining a profitable contract is small.

Initial Discussions

If the owner decides to prepare specifications and ask for competitive bids, he should first review basic criteria with the selected contractors who have expressed an interest in the project. The owner's objective is to learn the latest state of the art, to discover any reasons that might induce him to modify his proposed specifications, and to make sure that the requirements of the specification will be clear enough to result in bids that will be comparable with each other.

The contractor probably will not disclose all the features of his design at this time, because they may not have been completely developed or because he is afraid of the possibility of disclosure of process innovations to competitors. However, he will want to be sure that the final specification, when transmitted, will not preclude his design. Before determining how much effort he will devote toward preparing a proposal, he will want to form an opinion of the seriousness of the project, the likelihood of its coming to fruition and the timing thereof, the possibility of obtaining future work from the same client, and the caliber of his competition.

The initial discussions should center around the following specific subjects:

Plant size and cost—Generally the contractor can immediately submit rough estimates of operating and capital cost requirements. These should agree with the amounts used in the feasibility study. Estimated time for plant construction should be compared with previous assumptions. More important is the variation of unit costs with rated capacity. If possible, the selected plant size should be such that minor variations of capacity do not significantly modify operating costs. If this is not true, the lower costs associated with the larger plant should be fed into the market study to determine whether or not greater sales and a greater financial return can be obtained from this larger, more efficient plant.

At times, it may be found advantageous to purchase a large plant even though it may be necessary to operate it at far below its rated capacity, and at lower efficiency, until the full market develops.[3]

The question is not always simply one of scaleup, which can be estimated using a capacity ratio raised to a 0.6 or 0.7 power, but may involve equipment limitation, such as the possibility of using centrifugal compressors in an ammonia plant. The question of "standard" plant designs may also influence size considerations. If a contractor has already completed a set of detailed design drawings for a plant of a given size, and even has precommitted orders for some of the major equipment, his price for a large "standard" plant may be less than that for a smaller plant designed to the precise capacity selected by the operating company.

Process innovations—There is always a risk in being the first to use a new process or an untried piece of equipment. This risk must be weighed against expected benefits. Some operating companies are more venturesome than others. In deciding whether to consider or reject the innovation, the engineers of the operating company should use the initial discussions to form opinions both as to the general capability of the contractor's staff and its specific capability with regard to the process under study.

Standard specifications—Most contractors have design standards and accounting standards that they are accustomed to using. If the operating company requests that its own standards be used instead, the consequences of such a request should be explored. Also falling within this general category of design standards are questions of safety factors on uncoded vessels, life of fired equipment, extent of usage of stainless steel or other corrosion-resistant alloys, degree of instrumentation or automatic control, etc.

Seriousness of the project—As all contractors are quite busy now, they are extremely reluctant to assign engineers to prepare a detailed estimate unless there is a reasonable likelihood that work on the project will be authorized in the near future. Failing to be con-

Some Guidelines for Personal Negotiations

Do's

1. Be prepared to listen. Have the other party either state his terms or present a draft contract. Terms may be proposed that are more favorable than you expected to settle for, or that protect you from contingencies you had not considered. You will have an opportunity later to pick desirable terms and to voice your objections to undesirable ones.

Permit the other party to explain his desires, and his reasons for having specific terms in the contract. This will help you appreciate the relative importance of those terms to him, and may reveal potentially valuable information to you.

2. Maintain a sense of perspective. If the important points have been settled reasonably, try to cover the remaining points as expeditiously as possible, and aim for a fast windup to negotiations. A lengthy debate over some non-essential item may cause the other party to develop second thoughts about the whole contract.

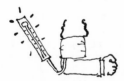

3. Take note of emotional reaction. This is related to the above. Listen to the inflection and tones of the other person and watch the expressions on his face. They may reveal which points are more valuable to him, and when he is losing patience or getting edgy; they may also reveal how difficult he will be to live with during the course of the project. Some points may be crucial enough to pursue no matter what the reaction—for instance, a damage clause that is highly undesirable to one party may serve a needed purpose in giving him a strong incentive to perform properly in all respects, so that the clause will never be invoked. If the point is less justifiable or less important, an emotional reaction may indicate that you have gone too far and that the point should be withdrawn.

To maintain a healthy emotional climate, try to avoid whispering or writing furtive notes to people in your group; also avoid giving the impression that you consider the other party's word as being less trustworthy than your own.

Don'ts

1. Don't make rash decisions. Don't impulsively agree to terms you do not understand. Reserve your verdict on undecided terms. When all other terms of the contract have been concluded, if substantial concessions are made, you may decide to modify your position or waive the questionable terms. Take notes as the talks progress so that you know precisely what has been agreed to.

One way to avoid rash decisions is to know the economic effect of what is being proposed. Some points have greater dollar implications than others. Take the time to work out the dollar effect of a suggested modification. Every clause has some real or contingent calculable economic effect.

2. Don't press the other side to the wall. You should, of course, be thorough, and request every term you wish, being prepared to explain your reasons. However, never press for more than 80 to 90% of the terms you desire if they are not promptly yielded. Don't press for 100% acceptance.

Nor should you expect to go into such great detail as to methods and scheduling of construction that the contractor is presented with an inflexible situation, because this could result in delay rather than expedition of the work.

3. Don't crow. The other party may have weaknesses or special needs that cause him to yield to some demands that you would have been willing to modify. A situation in which you obtain a contract "too good to be true" is to be viewed with caution, but not necessarily with suspicion if entered into freely and considered fair by the other party. The opposite extreme of "crowing"—voicing guilt feelings — should also be avoided. When a point is conceded, do not belabor it or continue to explain why it is needed. Doing so is the mark of an amateur and may create unwarranted suspicions.

neers can be used more profitably working on active projects.

The contractor will probably also explore the ability of the owner to obtain funds for his new plant. Especially if the operating company is small, or has its main office out of the country, the contractor will want to check the credit rating and total assets of the proposed client before proceeding.

The Bid Specification

At this stage of the study, it is relatively easy for the operating company to evaluate the effect of all the changes proposed by the contractors, and either to explicitly define all the important variables or indicate an allowable range when requesting bids. Failure to do so is disadvantageous to both parties. If the operating company later accepts a bid that is very attractive even though not in accordance with its specifications—and many such bids have been accepted because the offering reflected sound engineering and financial reasoning—the operating company has not gained all of the advantages it had hoped to achieve by requesting competitive bids.

Similarly, the contractor who has spent more time and money to comply with all given specifications only to see some of them waived at the last minute is apt to be somewhat less willing to cooperate with the operating company on future studies.

Thus, a unilaterally prepared specification, no matter how conscientiously produced, is not enough to assure that the contractor will be of greatest service to a prospective client. At times, contractors have gained contracts by submitting proposals based on their own evaluation of the client's real needs; at other times, they have seen their fully engineered and documented proposals discarded because these proposals did not conform with the requirements of the specification. The basic problem is one of communication.

The specifications, as transmitted to the contractors, need not be very detailed but should include certain technical and financial data beyond the basic ones of desired plant size (or allowable variation of capacity), product quality, and raw-material description.

The nature and permissible use of any available utilities, such as boiler feedwater, cooling water, steam, electric power, should be indicated.

Although operating companies are understandably reluctant to reveal their exact costs of raw materials and utilities, and their method of evaluating decreased operating cost versus increased capital cost, it is most important that some indication of these values be transmitted to the contractor. One of the operating company's major objectives in allowing the contractors the time to prepare detailed competitive bids is to gain a proposal that will optimize operations within the framework of general corporate policy, including financial policy.

The availability of funds, and the cost of acquiring funds in the market will vary considerably from company to company.[6] The contractor should therefore be

Unfair Procedures

1. Contractor A sends in a bid that is lower than any other bids received by the purchaser. The latter shows the bid and proposal to his personal favorite, Contractor B, who then revises his own bid and is awarded the contract. This type of behavior will result in a refusal of contractors to bid on future contracts. Moreover, while the bid may have a low price on it, the revisions could open the way for extras, changes, and longer time for construction, all adding up to a more costly contract.

2. Here's an actual case: During the course of negotiation for a site, the seller agreed upon the terms, but later declined to sign the contract because he claimed his son-in-law was the actual owner and had not agreed to the price. Eventually, the price was increased 10% and the contracts redrawn. At the contract appointment, the seller asked for a higher interest rate for the portion of the price that was to be financed, again claiming the original rate had not been approved by his son-in-law.

This type of procedure is unfair because it removes the purchaser from the market as far as other sites are concerned, and wastes the time of the negotiating people. Simultaneously, it destroys the effectiveness of the person negotiating for the seller.

The purchasers handled this particular situation by threatening to withdraw unless the son-in-law sent them a signed irrevocable offer containing his terms and attended the next meeting. When the son-in-law complied, the purchasers refused to deal with the original negotiator, and worked out a deal with the son-in-law at the original rate of interest, and 10% lower than the original price.

If the seller had not used an unfair procedure at the start, he would not later have had to reveal how anxious he was to sell.

vinced, the contractor may give the project a low priority, resulting in a delay of months. An offer by the operating company to pay for the cost of preparing the estimate will not always be considered a sufficient inducement, because the cost of making up a detailed proposal frequently exceeds what might appear to be a reasonable price, and also because the design engi-

given a quantitative method to help him decide between recommending larger pipe sizes or larger pumping costs, and between larger reactors or lower conversion rates. Other decisions of this type involve the installation of spare equipment, greater use of alloy or stainless steel, more-extensive instrumentation—as against the expectation of increased maintenance costs.

Of course, the contractor must be told whether his bid is to be for the process plant alone or for a complete facility including such things as water supply and treatment, power supply, storage tanks or buildings, railroad or truck loading facilities, control room or administration buildings, plant communications, fire water system, etc. Responsibility must be established for determining the extent of required plant leveling or fill, the source of fill or dump area, soil bearing value, meteorological data, underground or visible obstructions, taxes, insurance requirements, and local ordinances and codes.

The contractor must be allowed a reasonable amount of time in which to prepare his bid. He will expect to be given an indication of the accuracy required—whether the bid is to be used as the basis of a study, a request for budget authorization, or for final evaluation of proposals. Before preparing his bid, the contractor will want to visit the job site, determine the availability of construction labor and note local practices, judge the availability of construction material such as concrete that is normally purchased locally, and check the equipment unloading facilities.

If the area is entirely unknown to the contractor, and particularly if the plant is to be constructed in a place where local ordinances are subject to varying interpretations, the contractor may find it impossible to submit a lump sum bid. However, if he decides to bid, the contractor has to consider all of the information given to him and, if necessary, modify his process accordingly. For instance a design optimized for plants being erected in the southeastern part of the U. S., where natural gas costs are low, should not be proposed in an area of low-cost hydroelectric power but high-cost gas. (This appears to have been overlooked by several unsuccessful bidders for a major contract recently awarded.)

Evaluating the Bids

After the bids have been received, the staff of the operating company must show some degree of sophistication in evaluating them, making sure that all bids are complete and have been brought to a common base before the final comparison is made. Among the specific items to be noted, in addition to those described above, are the purity, pressure and temperature of the product and coproducts; the guarantee of raw material and utility usage (a careful distinction must be made between "expected" and "guaranteed" quantities); whether or not the initial charges of catalyst, chemicals and lubricants are being supplied; the extent of patent guarantees; and the number and caliber of startup personnel and operator training facilities that will be provided.

Frequently, insufficient attention is given to the type and extent of instrumentation that will be supplied. An instrument listing or preliminary flowsheet and specifications should be requested. Of particular interest are the:

• Number and type of analytical instruments to be furnished.

• Extent to which control valves will be provided with manual bypasses or hand wheels.

• Number of automatically controlled valves as contrasted with manually loaded valves.

• Type of control panel—standard, semigraphic or graphic.

• Proportion of instrument signals fed to indicating or recording instruments in the control room rather than having to be read locally.

• Extent to which safety alarms and shutdown devices, as well as relief valves and rupture disks, have been provided.

A plant that has a minimum of instrumentation may well operate with lower efficiency and probably will be subject to unnecessarily high operating and maintenance costs. Also, if instrumentation has been reduced below a reasonable level, it is quite possible that the same philosophy of minimum cost has been applied to the design of other plant items that cannot be checked as readily.

Of course, if computer control is being contemplated, a much more critical review of instrumentation must be made.

Type of Contract Desired

Bids may be requested in various forms, the most common being lump sum, which is a specific price for the project, or cost plus, which is a specific percentage or fee over the contractor's provable cost. Detailed discussion of these forms can be found in the literature.[4, 5] The contract may incorporate additional terms that would affect the final cost, of which the most common are:

Bonus incentive, based on a factor desired by the owner for which he is willing to pay additional compensation. Such a factor may be completion several months ahead of proposed schedule, or use of less electricity per ton, or reduced unit cost of production.

Guaranteed maximum, whereby the contractor for a cost-plus project guarantees that its total cost will not exceed a specific amount.

Share of cost savings, whereby the contractor is rewarded with a percentage of the amount by which the cost is reduced below a given figure. If the contractor is also required to absorb a portion of the amount by which the cost of the project exceeds the estimated cost, the form of the contract will be similar to the incentive type currently favored by the Pentagon.[7]

Cost Plus vs. Lump Sum

A substantial number, although certainly not all, of the major contractors prefer to obtain contracts based

on lump-sum bids rather than on any of the cost-plus alternates. This is particularly true if the process is one with which they are familiar, if they have accumulated equipment and construction costs from previous jobs and have had adequate time in which to prepare a detailed estimate.

There are two basic advantages that a contractor hopes to gain by lump-sum bidding. The first is that his own costs remain hidden. Therefore, if he has discovered a significant process improvement, or has been able to effect economies by other means such as bulk purchasing or the reuse of engineering or drafting designs, he need only pass along enough of the saving to cause his bid to be preferred over that of his competitors. Also, if he wants to facilitate the acceptance of a new design or his entrance into a process area in which he has not heretofore demonstrated his competence—or if he is short of work and willing to sacrifice profit for some contribution to overhead and the ability to retain his staff—he can "buy" a job by submitting a very low bid.

The second advantage is that, once a specification has been accepted, there is a minimum of participation, review or questioning by the owner. Work tends to proceed steadily, and designers are freed for other profitable jobs as rapidly as possible. Although under a cost-plus contract the owner bears the cost of any delay, when work is plentiful this may not compensate the contractor for the fact that his personnel are still working on an old job that may even have a fixed or a maximum fee.

Other contractors, of equal capability, are reluctant to offer or accept a lump-sum contract. Their attitude is that they are offering professional services and can be of greatest help to the owner by working as closely with him as he may desire on all phases of the project. They may neither wish to obtain an excessive profit if their costs are less than had been anticipated nor suffer a loss resulting from an error in estimating or some other cause beyond their control (such as increases in equipment prices or labor rates).

The owner's position with respect to financial risk is diametrically opposite that of the contractor. A lump-sum contract enables the owner to prepare a reasonably firm estimate of expected plant costs, and therefore of potential profit. Alternatively, the cost-plus contract provides no protection if the total costs far exceed the contractor's estimate—regardless of whether or not the discrepancy can be justified—but such a contract does enable the owner to benefit if costs are reduced by process modifications, astute purchasing, or for any other reason.

The extent to which the owner wishes to participate in the detailed plant-design work is a matter of company policy. However, supervision and control can be overdone. Such action can result in delays that may prove more costly to the operating company than to the contractor. To this extent a lump-sum contract, which tends to minimize owner participation, can be advantageous to both parties.

Of course, if the operating company is not in a position to prepare complete specifications—perhaps because of unfamiliarity with the process—a meaningful comparison of undetailed lump-sum bids becomes rather difficult, and may prove hazardous if some of the bidders are without reputation in the field.

Incidentally, it should not be assumed that specifications permitting a contractor to purchase inexpensive equipment can always be upgraded in favor of more-expensive items merely by offering to pay the difference to the contractor. He may be unwilling to integrate a piece of equipment of another design or standard into his process for many reasons, such as unfamiliarity with the equipment, exposure to patent liabilities, increased cost through delays, or loss of guarantees by owners of the process.

Bonus incentive terms should sometimes only be brought up after the prime contract is signed—i.e., after the operating company has already obtained the best promise under competitive bidding. A bonus for early completion, for example, may be desirable if the product can be best sold in a seasonal period and the purchaser is starting at a late construction date, or if the purchaser is starting on time but wants to assure himself more than adequate margin of time for all contingencies.

Sometimes a bonus may set a bad precedent. However, note that a moved up completion date might result in:

• Gaining an increased share of the market for the product.

• Saving the interest charges on a nonproductive facility.

• Earlier availability of tax advantages, such as depreciation.

• Gaining sales a season earlier.

• Getting the jump on competitors also entering a particular market area.

Letter of Intent

Once the operating company has selected a contractor and reached an understanding with respect to the basic criteria of plant design and payments, it will generally want the work to start as rapidly as possible. However, both parties recognize that many elements of the final agreement are still unresolved—in fact, they may not have been raised at this point—so that several months may be required to negotiate the terms of a contract. To save time, it is usual for the purchaser to give the contractor a letter of intent, authorizing the expenditure of funds on the purchaser's behalf.

In preparing this interim document, it is important that the following salient points be adequately covered:

1. The features of the plant design and equipment specifications, and the extent of the information (basic process design, detailed engineering design, etc.) that will be supplied by the contractor. Usually reference will be made to the contractor's proposal and to a supplementary letter that contains all of the modifications to which both parties have agreed.

2. Contract price. If the total has not been established, it is usually possible to establish a base amount

Fair or Unfair?

1. "Contractor must complete the plant by Oct. 30, 1967, or pay $10,000 per day as liquidated damages." The operator would want this clause so that, assuming a 30-day production cycle, he could promise shipment on Nov. 30, 1967. If he only intended to start shipping on Dec. 31, 1967, he might still legitimately want a 30-day insurance period to provide for delays if something went wrong. Even if the operator's loss of profit is only $5,000 per day, it is not illegal to seek an edge, and the clause would be considered valid.

However, what if the loss of profit is only $500 per day and, in addition, no raw materials will be supplied until Apr. 30, 1968? If the contractor delivers the complete plant on Nov. 10, 1967, and the operator seeks to collect $110,000 in damages for the 11-day delay, how is this regarded?

It can be seen that the damages sought are 20 times those that could be incurred; furthermore, since there is no possibility of production for six months, no damages at all are actually sustainable. The provision is thus clearly a penalty clause (even though the contract implies that it is a damages clause) and the courts will not normally enforce the penalty.

Generally, the rule is that if the imposition of a clause results in a forfeiture that is extensive and has no reference to the expected or actual situation, the clause is illegal and unenforceable.

2. "A full set of blueprints and specifications must be provided when the plant is turned over as complete." The contract for the first of a number of sulfuric acid plants contained this clause. Thereafter, the same clause appeared in separate, consecutive contracts for the construction of six more sulfuric acid plants. For five of these plants, the clause was not observed; the contractor only provided plans and data on those parts of the plants that differed from the first one.

When a dispute arises about the last plant, the operator decides to enforce every provision of the contract. It might cost the contractor $50,000 in fees to produce a complete set of plans and specifications; he may not have included this amount in his bid because the clause had not been observed in the last five instances. Nevertheless, the clause is a valid one, and the operator is entitled to a complete set of plans.

However, if the operator's insistence on a clause of minimal value to himself but of great cost to the contractor were publicized, other contractors might become reluctant to do business with him in the future.

3. "Contractor requires that all but 5% of the price be paid pursuant to a schedule of dates." This is unfair because payments should be geared to progress of the construction and of purchasing commitments. For instance, if a lengthy strike occurred and supplies were delayed, or a subcontractor or supplier went bankrupt, a fixed payment schedule might require the operating company to pay 95% of the total price even though the contract is only 50% complete (unless there were protective clauses for such contingencies).

4. "Purchaser guarantees to contractor that he will not use nor permit any other company to use these designs or processes in any other plant belonging to purchaser or otherwise." This clause is unfair to the purchaser because it is too broad. The contractor has the duty to police his own patents or secret processes.

Normally, however, the purchaser is required to state that he will not knowingly and intentionally disclose secret processes to other companies.

5. "Purchaser shall indemnify and defend contractor against any claims by reason of construction and use of the plant." This clause is unfair. If a contractor is selling a process, he should make sure he has the right to sell it; he should defend all law suits brought against him, and a limited class of suits brought against the purchaser.

(However, in a foreign country where the contractor is not certain of his patent situation, he will not generally represent that he is selling the rights to a secret process that the purchaser may use to the exclusion of any other company in that country.)

6. "Contractor shall be liable for any damages whatsoever accruing to purchaser, whether or not caused by the delay or negligence or improper construction by the contractor, if these damages occur by reason of the failure of contractor to complete the project by a certain date." This sort of "open damage" clause is unfair, because delays caused by such things as a steel-industry strike are beyond the control of the contractor.

Damages shall accrue only if certain exceptions such as wars, strikes, catastrophes, acts of God or like causes have not brought about the delay. Moreover, they shall be limited by type and amount, so that it is possible to ascertain the prospective amount in advance, based upon certain hypothetical conditions or facts.

and define the factors that remain subject to negotiation.

3. Terms of payment. Since the payment schedule and terms are frequently subject to later negotiation, a cost-plus-percentage overhead usually is used during the interim period.

4. Limit of liability. Having authorized expenditures on its behalf, the operating company is liable for all equipment orders placed, as well as for the office costs of the contractor. To reduce its total liability, if satisfactory contractual terms cannot be obtained, the operating company usually will limit the amount of funds that the contractor may commit. There may be one limit on the contractor's out-of-pocket expenses and another on purchase commitments. (In the past, such orders were placed subject to pre-established maximum cancellation charges. At the present time, suppliers' shops are so busy that practically no work will be done on orders that appear to have a fair possibility of being cancelled. Therefore this precautionary expedient is of little use today.) While the contract is being negotiated, the limits may be extended if the talks are proceeding satisfactorily.

5. Guarantee of payment. If the operating company is large and has a good credit rating, no further guarantee of payments will be required. However, if the company does not have a good credit rating, or is a foreign firm with no assets in this country, the contractor may require the establishment of an irrevocable letter of credit. (This is of particular importance to the contractor if equipment orders are being placed in his name.)

6. Cancellation. Both parties stand to lose substantial sums of money if the work is cancelled. The contractor has lost his potential profit, plus the time spent by his engineering and drafting staffs. The operating company, if it still wishes to proceed with the project, has had its plant startup delayed and is committed to either accept the equipment that already has been ordered or pay the cancellation charges. Since both parties thus have a strong inducement to arrive at a satisfactory contract, specific cancellation penalties are not always included in the letter of intent. However, under some circumstances, the contractor may require the acceptance of one or both of two conditions:

The first is the imposition of a very high overhead rate or fee for the work done during the interim period, these additional charges to be credited to the operating company upon signing of the contract.

The second stipulation, designed to protect the contractor from unwarranted use of his know-how, is that, if the talks fail, the operating company cannot proceed with the construction of the project for a specified number of years by using the services of either another or its own staff.

The pitfalls of commencing with a letter of intent rather than a complete and detailed contract are almost the same as the disadvantages of working without a contract. There may be no meeting of the minds on specific points, each party believing a problem has been solved in a different way. Differences of opinion may arise over the handling of contingencies.

Also, as time goes on, the terms agreed to may become hazy in the minds of the parties, and the recollection of what was intended will vary more and more. This makes it increasingly difficult to ascertain whether performance is being properly rendered. It is even more difficult for other officers or divisions of the company to know what rights and responsibilities have accrued from the deal, and thus the orderly proceedings of a corporation are affected.

Furthermore, no standard has been established whereby one individual can be substituted for another in case of job change, illness or death, and whereby rights can be transferred to another unit or company under certain conditions (e.g., merger, sale of company, etc.).

Nevertheless, the advantages of a properly executed letter of intent almost always exceed the disadvantages, and they are widely used today.

Preparing for Negotiation

The suggestions that follow apply in varying degrees to all stages and types of negotiations—from the first contact between an operating company and an engineering contractor (e.g., to find out if the latter is interested in bidding on a project) to the detailed negotiating of contractual terms. The suggestions are pertinent whether you proceed upon a letter of intent or a complete contract. They involve a certain kind of homework, but doing this homework can save endless headaches and avoid misunderstandings.

Prior to any meeting, put the following items into written form:

1. Terms that are essential to you and the transaction.

2. Terms that are desirable to you but that may be waived if necessary.

3. Terms the other party may present that you cannot agree to.

4. Terms the other party may present that you could agree to if necessary, but would not find desirable.

By doing this, you will note that you have decided what the contract should contain, i.e., the terms essential to both parties.

Decide in advance who is to be your spokesman in the talks. Don't "send a boy to do a man's job." Statements and requests made by a more recognized authority in the field will be accorded greater consideration by the other side.

Most people would like to deal with the real principal. A negotiator with little authority may have difficulty getting the party with whom he is dealing to give him a final commitment, thus delaying conclusion of the deal.

A two-man bargaining team, one a "good guy" and the other a "bad guy," permits reversal of position, if necessary, and facilitates toughening or relaxing terms as negotiating proceeds. Both aspects may at times be combined in one skillful person.

Estimate the time necessary for each meeting and find out who the other party will have representing

The Contract: Eight Basic Sections

"Whereas" Section.—Too few persons take the opportunity that the "whereas" part of a contract gives them. This is the place to state the basis for the agreement in prose form, easily understandable. Here you may tell why the contract is important, what it concerns, what results you expect, what sort of conditions the contract is based upon, and why performance by a certain date may be important. If ten days—or even years—later there appears to be an ambiguity in the contract, its solution can be aided by reference to the whereas clause.

The Subject—This section describes what the plant consists of and what is to be done. The specific details are either given in the body of the contract or in an annexed book of specifications, or by reference to another identifiable plant or plant design.

The Price—This section stipulates how the price is to be arrived at, and the method of payment. After the initial payment (which can be made on a letter of intent, at commencement of construction, or at some other stipulated time), subsequent payments are based upon the stage of progress, computed in any of the following ways:
- Generally upon percentage of completion of the entire plant.
- A percent at indicated dates.
- Fixed sums on completion of specified portions of work.
- Fixed sums on indicated dates.
- A percentage of actual completion, leaving an unpaid percentage to await demonstration-of-performance tests when the project is totally finished.
- A combination of the above methods.

Testing or the Standard—A project must measure up to certain yardsticks or independent tests. The nature, extent and time for these tests of actual performance must be set forth. This is different than a comparison of construction with specifications; a project may comply with specifications and yet not be able to perform at expected levels. This section will set forth:
- A timetable of tests and the duration of each.
- Maximum requirements, per unit of manufacture, of raw material, labor, power, maintenance and supervision. Based on assumed costs for each of the above items, a guarantee may be made that the total manufacturing cost per unit will not exceed a stipulated maximum.
- Minimum plant capacity.
- Maximum power and utility cost on a continum with raw-materials consumption.
- Extent of freedom from mechanical defects, excessive wear and tear and maintenance.

Damages—The consequences of a failure to measure up to the standards are then set forth. Opportunities to correct are usually provided. Damages may be:
- Unlimited. (To cover delays in completion, a flat rate per day of delay is sometimes agreed on.
- Unlimited except regarding consequential damages or loss of profits.
- Unlimited respecting representations as to plant capacity, end-product purity and compliance with specifications as to plant design, but pursuant to a formula with regard to usage of raw materials, or utility or power consumption.

- Waived if purchaser waives the tests or fails to operate the plant within a given period of time.

Related to waiver is the problem of establishing the stage at which equipment is turned over to the owner; this involves not only financial responsibility but responsibility for the condition of the equipment and conformance to specifications. This turnover may occur at one of five stages, typically:
- As the order is placed, if the contractor is acting as agent for the operator.
- On arrival at the site.
- After erection and physical check.
- Upon testing in operation.
- Upon complete plant acceptance.

Responsibility for maintenance of equipment, as well as risk of loss, may be based upon time of acceptance. To minimize the possibility of disputes as to condition or performance, all equipment tests should be witnessed and signed at each transfer of custody. Also, where feasible, at least one representative of the contractor should be in the plant at all times—from the moment the first burner is lit until all conditions set forth in the contract have been met.

Specification of Equipment and Personnel—The specifications may refer to particular process and equipment manufactured by a particular company, or this requirement may be separately listed, giving alternatives among enumerated manufacturers. Equally important may be the requirement that specified individuals supervise the job and that they will not be assigned to another job without the purchaser's consent. Clauses with reference to avoidance of delays or work stoppages, working in an expeditious manner and bonus-incentive tie-ins are equally applicable at this point.

Site Representations—The careful contractor will want to base his estimates upon certain assumed conditions, one of which is the nature of the project site. Representations of this sort are normally an escape valve for the contractor. However, properly used, these representations may also protect the purchaser against claims that there are unforeseen conditions rendering the job more expensive or causing work delays. Basic representations relate to:
- Soil content.
- Underground obstructions, natural and man made (piping).
- Bearing capacity (pilings).
- Ground levels and inclines.

Protective Clauses—The clauses set forth above are essential to the contract or highly desirable. To furnish protection, and to tie up the loose ends, some additional clauses are usually included. These involve:
- Insurance coverage.
- Workmen's liens.
- Disputes and arbitration.
- Jurisdiction.
- Indemnifications against claims for patent and process rights.
- Secrecy.
- Re-use of drawings and plans for plant expansion or plants at other locations.
- Extra work, cancellation, substitution and amendment procedures.
- Burden of taxes.

him at the meetings. Prepare adequate time and facilities.

Also, find out as much as you can about the reputation of the company with whom you are dealing, and the reputation and personality of the individual representatives who will negotiate with you. Try to decide in advance what approach is best with this type of company and its representative. Use introductory social chatter to see the reactions of the other party before the formal atmosphere sets in. Ascertain the authority of the negotiator. You may not be able to depend upon the representations of someone who is unauthorized or not knowledgeable.

Techniques of Negotiation

The object of the negotiator is, of course, to obtain the best possible deal for his company, within the limits of fairness. Some do's and dont's that are in keeping with this goal are shown on p. 201.

It should be emphasized that a deal can be "tough" without being unfair. For instance, a contractor anxious to establish his proficiency in a new field may accept a very low profit margin so as to give him an edge over his more-experienced competitors. Or an operating company with a poor credit rating, in dealing with a contractor who is busy enough to be selective, may agree to payment terms that would normally be considered unfavorable, in order to obtain a plant in a hurry.

The final contract in these two cases may reflect the relative economic position of the two parties, but will only be considered "unfair" if one side has exploited its preponderant position to press the other side to the wall in regard to terms. If these terms deny the other party protection to which he is normally entitled, deprive him of the opportunity to remedy defaults, or saddle him with harsh penalties that bear no relation to the likely damages, the contract may be unfair and may, in effect, be bad business. The result of such a harsh contract can be that the other party eventually:

• Cuts corners, and refuses to give more than the bare minimum.

• Abandons the contract, bringing about costly consequences some of which cannot be fully redressed by legal action.

• Refuses to go ahead, or slows down deliberately.

• Publicizes the one-sidedness of the contract. Mere telling of the story, with no deliberate attempt to cause damage, can still create difficulties in subsequent contract negotiations with others.

Some examples of unfair clauses and procedures are given on p. 202 and 205.

The Terms of the Contract

The contract is the formal expression of the meeting of minds of the parties; its terms will be referred to innumerable times during the course of the project by engineers with a variety of functions. The eight basic sections found in most construction contracts are outlined on p. 207; they are the main items that need to be covered during the negotiations referred to in the previous section.

While this outline may serve as a checklist and prototype, each project has different features and problems. The ingenuity and industry of the contracting parties are necessary to insure that these different features be combined and integrated into the contract. No contract will fit every situation. Although the variations may be slight, they should be given proper attention, so that they do not develop into major problems.

References

1. McGregor, L. W., Process Supplied by Third Party, *Hydro. Proc.*, July 1965.
2. Upson, F. A., Process Supplied by Plant Owner, *Hydro. Proc.*, June 21, 1965.
3. Coleman, J. R., Jr., Optimum Plant Design for a Growing Market, AIChE Meeting, Buffalo, N. Y., May 5-8, 1963.
4. Hackney, J. W., Pros and Cons of Construction Contracts, *Chem. Eng.*, June 21, 1965.
5. Steinke, J. J., What Engineers Should Know About Contracts, *Chem. Eng.*, Feb. 17, 1965.
6. Moskovits, P. D., Find What Money Costs Your Company. *Hydro. Proc.*, Aug. 1964.
7. Department of Defense, Incentive Contracting Guide, DSAH 7800.1, Jan. 1965.

Meet the Authors

Sidney A. Bresler is a project manager for Borden Chemical Co., 350 Madison Ave., New York City. He has sat on both sides of the negotiating table, having formerly been associated with Chemical Construction Corp. Mr. Bresler co-authored the article "Questions and Answers on Today's Ammonia Plants" in the June 21, 1965, issue, and has a number of other articles and patents to his credit. He has an M.Ch.E. degree from Brooklyin Polytechnic Institute, and an M.B.A. from Columbia University.

Martin J. Hertz is an attorney in private practice, with offices at 60 E. 42 St., New York City. He specializes in negotiating agreements for construction and purchase of chemical and other industrial plants. Mr. Hertz conducts seminars in negotiating techniques for corporate executives, and is the author of "Making Agreements That Are Most Favorable For You," to be published later this year by the Industrial Education Institute. He acquired his L.L.B. from Harvard Law School, and his L.L.M. from New York University.

Keywords for This Article

Active (8)	Passive (9)	Purpose of (4)
Preparation	Contracts	Construction*
Negotiation*	(Agreements)*	Engineering
Comparing	Specifications*	Design
	Bids*	Plants
	Guarantees*	Projects, capital*

Means/Methods (10)
Contractors
(Vendors)*
Purchaser
Laws*

(Words in bold are role indicators, numbers corresponding to EJC-AIChE system except for Role 8 modification. Asterisks mark key concepts suggested for indexing. Others are added to improve reading as abstract).

For Big Savings—Control Costs While Defining Scope

The time spent defining and optimizing the scope of a project can make the difference between an efficient, low-cost facility and one that falls victim to runaway costs.

R. P. STURGIS, *Mobil Oil Corp.*

Despite the increasingly determined and sophisticated methods that are being used to control the cost of new facilities in face of inflationary and competitive pressures, capital expenditures for similar facilities still show large variations from job to job. Why will one oil refinery cost $40 million and a similar one cost $50 million? Why will chemical plants of a certain capacity vary in cost by the same 20% or more? Why does this pattern of inconsistency follow through many other industries?

Although some of the differences are unavoidable, others can be minimized through increased control measures taken during the period of project "scope" definition.

There are many areas of cost control; listed below are seven of them in the order of time sequence. (This often is the order of potential-savings magnitude as well.) We can think of these areas as successive periods that involve:

1. Plant size selection and determination of the general location.
2. Establishing the number of products and their quality specifications.
3. Examination and selection of the best site.
4. Optimizing the process design.
5. Screening and minimizing the cost of nonprocess scope.
6. Detailing the design.
7. Actually constructing the plant.

This review will be limited to the second through the sixth points listed above. These are areas that often get too little attention.

Too few engineers and managers realize that even through only a small percentage of a project's total cost and time is spent in defining and optimizing the detailed scope, the impact of this activity on the project's final cost is very significant. Also, too few realize that the basic cost of a project is really determined during the first few months of design. This basic cost is the theoretical minimum required for the desired facilities at the time and at the location selected. At the point in time when the owner or customer/client approves the process flowsheet, the plot-plan layout, and process and mechanical specifications (as a basis for detailed design engineering, purchasing and erection), the owner has "cast the die," thereby establishing the base cost of the particular project. The problem from then on becomes one of policing—to hold costs down to the established base cost. We will discuss how to "cast a better die" by exercising sound definition-cost-control. Here are some methods that can be used to control costs in these early definition stages.

Cost Control in Action

First and perhaps foremost, is the necessity to get everyone who has anything to do with the project in a cost conscious frame of mind. Although it may sound like a cliché, cost consciousness must indeed start at the top and extend all the way down. But cost consciousness and proper frames of mind are of little avail unless a yardstick is established on which to base decisions. The unique difference between cost reporting and cost control is that control implies action, and action must involve decisions, and decisions require justification, and justification must be measured—hence the need for a yardstick. During periods of peak work, when many contractors prefer not to bid lump sum, the need for a yardstick is particularly critical. Staff groups and other project contributors continually discover changes and improvements that might be incorporated; knowing which changes to resist is difficult unless there is a measure of the impact on cost and completion-schedule to accommodate an ever-moving target. The yardstick can provide this measure.

The second requirement for good cost control can be implemented when the project organization is established. Each member appointed to the project staff must be made a practicing cost engineer in his specialty. This should be discussed with him at the time of his appointment, and it should be stressed that each decision he makes must be made with cost control firmly in mind. If the decision has an effect of increasing or decreasing project costs, he should pass the word along to the cost engineer assigned to the project. He also should enlist the services of the cost engineer when he needs cost estimates for comparative purposes.

The cost estimating engineer assigned is intended

Based on a paper presented on July 12, 1967 at the National Meeting of the American Assn. of Cost Engineers, held at Cleveland.

to devote full time to cost analysis and control activities. He should have access to a reading file of all correspondence that takes place between the project staff and the contractor. He should also read minutes of meetings to pinpoint any discussion or decisions that will affect project cost.

One of the greatest mistakes that can be made is to use the cost engineer for purposes other than cost work—e.g., for troubleshooting when a crisis arises. Nothing should be permitted to distract him from his assigned duty of keeping track of the financial status.

For a major project, it is important to set up formalized cost-control procedures to be followed throughout. During these early design phases, specific cost cutting conferences should be scheduled for the primary purpose of reviewing matters pertaining to project cost. Each flow diagram, plot plan, major equipment selection or major project decision should be discussed.

During technical reviews, flow diagrams are marked-up to show additional equipment that is required by one or another of the operating people or project staff. How often does a project staff examine flow diagrams in detail to see what can be omitted? All additions should be justified economically, unless they are necessary for the safe operation of the plant. Plot plans can be reviewed from the standpoint of compactness; every foot of increase in plot-plan area represents a corresponding increase in cost.

One of the most important tasks during the early design stages is the preparation or updating of a monthly estimate. It is often felt that such estimates have little value because the job has not yet been defined. Nevertheless, an estimate has been prepared on some basis or other, even though assumptions have been made for many areas. As the job progresses, it must be updated periodically to indicate the effect of the decisions made during the previous period. Not only should improved scope definition be reflected in this updating, but all items that come to the attention of the cost engineer should be adequately covered. For example; if a major increase in wages has been negotiated during the previous months, the effect on final job costs should be reflected at once.

All scope items can be placed in one of three general areas. The first includes the firm requirements of the project. The second area includes intangible items that are not firm requirements, but matters of company policy. The third area includes all items that are subject to economic justification.

Are Firm Requirements Really Firm?

What are some firm requirements that would fall in the first area?

One might be the methods of receiving and shipping products. Even though they have been set up by definition as part of the project scope, they must be examined to see if they are realistic. If, for example, a truck-loading platform is included to load only one truck per week, abandonment of that shipping method should be considered.

Economic Decisions Involving Flexibility, Reliability and Efficiency

Flexibility refers primarily to the processing capacity of the individual operating units comprising a plant. Included is the ability to process raw materials of different characteristics and in varying combinations and amounts, to vary the product yield and specifications, and to add capacity efficiently in the future. The ability to operate or shut down a unit independently of the other units in a complex is also included.

Reliability refers to the capability of a plant to operate as desired in the face of countless known and unknown obstacles. Desired levels of service or on-stream factors of the equipment must be known in order to evaluate the need for dual, alternate or standby facilities. This area involves such items as spare equipment and parts on hand or permanently installed; storage facilities for raw materials and intermediate products; equipment bypasses; corrosion and erosion protection; degree of conservation in design of equipment. Included here are such items as:
- Rate of heat release in furnaces.
- Fouling factors of heat exchangers.
- Storm-resistance factors for wharves.
- Light-duty vs. heavy-duty machinery.
- Hardware quality.
- Planned obsolescence.
- Reliability of purchased services (electric power, water supply, all transport services, maintenance equipment and labor).

Safety—including fire protection control—possibly may be included here, if the added facilities are justified on an economics basis, over and above the minimum required for hazard protection.

Efficiency covers two distinct areas, the yield efficiency in converting lower priced raw or semi-finished materials to higher priced materials, and the economy with which this is done.

Included in the first category is the ability to:
- Extract maximum amounts of high-grade materials from the raw-material supply.
- Minimize losses, quality give-away and unnecessary inventories.
- Maximize on-stream times of complete process units or individual pieces of equipment.
- Adjust operating plans to maximize return on investment (perhaps via computer technology).

In the second category must be considered such things as:
- Operating manpower costs, both line and staff.
- Maintenance costs, including contracted services.
- Fuel efficiency as influenced by amounts of heat exchange, furnace efficiency and turbine heater rates, etc.
- Chemical and catalyst usage.
- Use of all utilities, whether self-generated or purchased.
- Degree of sophistication in instrumentation (e.g., use of on-line or off-line computers).

How much should be spent in order to load a tanker within a period of 15 hours? Is this requirement really firm?

Government restrictions, local law, codes and practices are given as firm requirements, but are they

really firm? Is it possible to negotiate a better design with advantages to both parties?

The matter of product specifications is usually considered to be quite firm. But is it possible that some specifications requiring large capital investment are not really required after all?

The number of products to be made is usually considered a firm requirement. Yet, including certain low-volume products may actually cause a reduction in the profitability of a project.

Each scope item that is considered to be a firm requirement should be examined to make sure that it is really as firm at it has been stated.

The Problem of the Intangibles

In the list of intangible scope items, many will be matters of company policy—e.g., projecting the company image and completing the work in a manner that will create a good public impression.

There is also the matter of employee amenities. What should be provided in the way of extra facilities? Should a cafeteria be included, for example, or just a lunchroom? In either case, how fancy should it be?

Adequate facilities for fire protection, safety, and pollution control are firm items, but where do you draw the line on marginal items?

In determining how much money should be spent for "company image" and intangible items, a most difficult problem must be faced. There is no quantitative way to establish exactly what should be spent for each. It is this type of question that must be decided by management edict; but it takes careful consideration to avoid excessive spending. If the final decisions on these matters result in costs beyond those contemplated, the changes must be shown as soon as possible in the current updating of the estimate.

Economic Justification Items

Most "scope" items that can be justified on the basis of economics fall into three categories representing the company's desire for flexibility, reliability and efficiency. Economic criteria must be established at the outset of a project so that each item falling into these classifications can be independently justified. (To aid in understanding these three terms, each is defined in the box on the previous page.)

Calculations of economic criteria vary from company to company and from industry to industry. Regardless of whether you use payout time, discounted cash flow, net present value, or some other criterion, the goal is the same; develop an appropriate yardstick and use it!

Of the three categories described above, *efficiency* is the easiest to screen. The calculations are the most straightforward. Assume, as a project yardstick, a payout of two years before taxes for additional investments that will enable operating cost reductions. It is relatively easy to decide to spend an incremental $7,000 when purchasing a $90,000 steam-turbine-driven centrifugal compressor, if the turbine efficiency thereby can be increased to reduce the steam cost by $4,000 per annum.

Not so easy are the decisions concerning *reliability* factors. These calculations should include statistical data derived from historical records of similar installations. One of the more frequently encountered examples involves the determination of how far to go to ensure an uninterrupted electrical power supply to an operating plant. If power is purchased from an outside power company, should two feeders be installed? Should there be access to two separate grids? Should power be self-generated on either a primary or standby basis? In a $20-million project, as much as $5 million can be spent or saved depending on the decisions in this area alone.

Frequently, the most difficult of all decisions falls in the *flexibility* category. Investments for additional facilities required to assure flexibility and accommodate possible future growth are only easy to estimate if the facilities can be identified. Usually, the difficulties in evaluating and calculating the risk factors cause the decision-bakers to fall back on experience, judgment and other intangibles. How much money should be spent to provide alternate methods of delivery of either raw materials or finished products? In the process industries—where the materials are gases or liquids that can be handled by pipeline, tanker, tank cars, tank trucks or barrels—the impact of decisions concerning receiving, storage and shipping facilities also can change the total investment by 25%.

In establishing economic criteria to determine the extent of expenditures for these three categories of scope items, a determination of the business risk must be established. These risks involve such possibilities as technical obsolescence, changes in zoning laws and in other pertinent legislation, evaluation of penalty involved in utilizing alternate sources of supply, stability of wages and prices, and expected changes in duties and taxes. In foreign locations, there also may be the possibility of nationalization or expropriation. Only after an alert management has established economic criteria for all the above listed items, should the serious work of designing the facilities begin.

Meet the Author

Robert P. Sturgis is Manager of Engineering for Mobil International Oil Co., 150 E. 42 St., New York City. A mechanical engineering graduate of Lehigh University, he joined Mobil as a junior engineer in 1938, and has handled assignments of increasing responsibility in construction, maintenance, project engineering, project management and engineering management.

Key Concepts for This Article

Active (8)
Defining*
Designing

Passive (9)
Projects, capital*
Scope*

Means/Methods (10)
Economic evaluation*
Purpose of (4)
Controlling
Costs*

A Fresh Look at Engineering Construction Contracts

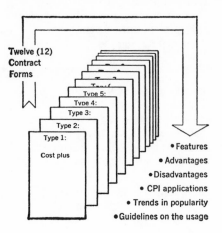

Twelve (12) Contract Forms

Type 1: Cost plus
Type 2:
Type 3:
Type 4:
Type 5:

• Features
• Advantages
• Disadvantages
• CPI applications
• Trends in popularity
• Guidelines on the usage

Here is a critical analysis—based on today's economic conditions—of twelve different types of agreements that are used between contractors and operating companies.

JOHN T. GALLAGHER, *Mobil Oil Corp.*

Although several good discussions of engineering-construction contracts have already appeared in the literature, enough has happened in the last two or three years to warrant a new and timely look at the features, advantages, disadvantages and relative popularity of various contract types. We will make a critical survey of these aspects and offer a contract selection chart that can be used as a concise, general guide for the technical personnel of contractors and clients alike.

First, a word on how the level of business activity has affected the types of contracts most prevalent in the chemical process industries today.

The number of fixed-price design and construction contracts awarded in a given period is very definitely a function of the general level of business activity. Contractors are not favorably disposed toward preparing competitive bids on fixed-price projects when their workload is heavy, because of the high bid-preparation cost and inflationary risk associated with lump-sum work.

This fact is supported by recent studies of the contract industry. In the late fifties and early sixties, the volume of fixed-price projects bid by engineering-contracting firms was extremely heavy. A quick review of the general economic indicators reveals that this was a period of slow economic growth.

Several recently published technical articles and the author's personal experience indicate that the cost for the preparation of a competitive design-construction lump-sum bid averages 0.5% of the project capital cost. Consider a project where four contractors bid on a competitive lump-sum basis. The contract value is assumed to be $10,000,000. The total bid preparation cost for all four companies is 0.005 × 4 × $10,000,000, or $200,000. The profit earned by the selected contractor on such a competitive bid is usually 1.0 to 2.0% of the project value, or $100,000 to $200,000. Thus the contractors, considered as a group entity, broke even or possibly lost money. This was a chronic problem for the construction industry in the period of 1958 to 1963, and it persuaded many contractors to stay away from competitive lump-sum bidding as much as possible.

In some cases, of course, a contractor may still want to make a competitive lump-sum bid in order to enter a new processing field or to maintain his position there if competitors are eager to enter it. The contractor may also be willing to use the lump-sum approach for standard-capacity, proprietary-process plants for which most of the cost information is at his fingertips. However, even with such standard plants, the possibility of runaway field costs in a particular region may deter the contractor from fixed-price agreements in periods of construction-labor scarcity, such as the present.

Thus, during the 1964-1967 period, there was a significant drop in the number of fixed-priced contracts awarded in the chemical process industries. Prior to this period, one large construction firm that is typical of the industry averaged 20 lump-sum bids per year. From 1964 to 1967, this number was reduced to approximately 5 per year.

From the client's viewpoint, during periods of strong economic expansion he is more anxious to begin production as quickly as possible. By awarding a cost-plus contract he can eliminate the several-month delay associated with the preparation of lump-sum proposals, thus improving the lead time from project approval to production of marketable product.

As for work outside the U.S., the reluctance of most international contractors to make lump-sum bids on foreign projects stems from bitter experience. Until 1966, foreign procurement markets were less stable than their U.S. counterparts. Foreign prices are usually quoted subject to escalation, and delivery schedules are often uncertain. Except in some Western European and other industrially developed areas, the scarcity of skilled construction labor and the possibility of political or financial instability has also been

Contract Form; Project Definition (PD)	Primary Advantages	Primary Disadvantages	Typical Applications	Comments
COST-PLUS PD: Minimal. (Scope of work does not have to be clearly defined.)	1. Eliminates detailed scope-definition and proposal-preparation time. 2. Eliminates costly extra negotiations if many changes are contemplated. 3. Allows client complete flexibility to supervise design and/or construction.	1. Client must exercise tight cost control over project expenditures. 2. Project cost is usually not optimized.	1. Major revamping of existing facilities. 2. Development projects where technology is not well defined. 3. Confidential projects where minimum industry exposure is desired. 4. Projects where minimum time schedule is critical.	Cost-plus contracts should be used only where client has sufficient engineering staff to supervise work.
COST-PLUS WITH GUARANTEED MAXIMUM PD: General specifications and preliminary layout drawings.	1. Maximum price is established without preparation of detailed design drawings. 2. Client retains option to approve all major project decisions. 3. All savings under maximum price remain with client.	1. Contractor has little incentive to reduce cost. 2. Contractor's fee and contingency is relatively higher than for other fixed-price contracts, because price is fixed on preliminary design data. 3. Client must exercise tight cost control over project expenditures.	Where client desires fast time schedule with a guaranteed limit on maximum project cost.	– – – – – –
COST-PLUS WITH GUARANTEED MAXIMUM AND INCENTIVE PD: General specifications and preliminary layout drawings.	1. Maximum price is established without preparation of detailed design drawings. 2. Client retains option to approve all major project decisions. 3. Contractor has incentive to improve performance since he shares in savings.	Contractor's fee and contingency is relatively higher than other fixed-price contracts, because price is fixed on preliminary design data.	Where client desires fast time schedule with a guaranteed limit on maximum cost, and assurance that the contractor will be motivated to try for cost savings.	Incentive may be provided to optimize features other than capital cost — e.g., operating cost.
COST-PLUS WITH GUARANTEED MAXIMUM AND PROVISION FOR ESCALATION PD: General specifications and preliminary layout drawings.	1. Maximum price is established without preparation of detailed design drawings. 2. Client retains option to approve all major project decisions. 3. Protects contractor against inflationary periods.	1. Contractor has little incentive to reduce cost. 2. Contractor's fee and contingency is relatively higher than other fixed-price contracts, because price is fixed on preliminary design data. 3. Client must exercise tight cost control over project expenditures.	1. Project involving financing in semi-industrialized countries. 2. Projects requiring long time schedules.	1. Escalation cost-reimbursement terms should be based on recognized industrial index. 2. Escalation clause should be negotiated prior to contract signing.
BONUS/PENALTY, TIME AND COMPLETION PD: Variable, depending on other aspects of contract.	1. Extreme pressure is exerted on contractor to complete project ahead of schedule. 2. Under carefully controlled conditions, will result in minimum design and construction time.	1. Defining the cause for delays during project execution may involve considerable discussion and disagreement between client and contractor. 2. Application of penalty under certain conditions may result in considerable loss to contractor. 3. Pressure for early completion may result in lower quality of work.	Usually applied to lump-sum contracts where completion of project is absolute necessity to client in order to fulfill customer commitments.	1. Project execution should be carefully documented to minimize disagreements on reasons for delay. 2. The power to apply penalties should not be used lightly; maximum penalty should not exceed total expected contractor profit.
BONUS/PENALTY, OPERATION AND PERFORMANCE PD: Variable, depending on other aspects of contract.	Directs contractor's peak performance toward area of particular importance to client.	1. Application of penalty under certain conditions may result in considerable loss to contractor. 2. Difficult to obtain exact operating conditions needed to verify performance guarantee.	Where client desires maximum production of a particular byproduct in a new process plant, to meet market requirements.	Power to apply penalties should not be used lightly.

a deterrent to the use of general firm-price contracts. In parts of Western Europe, equipment prices have now stabilized and, in some cases, have actually declined during the past year; coupled with competitive pressures, this could lead to some increase in firm-price work.

If the firm-price contract stands at one end of the spectrum, the "cost-plus" type of agreement stands at the other; in between there are many combinations and variations. The basic concepts of these contract forms permeate much of the engineer's day to day activities; not only does he participate in the selection and drawing up of contracts, but many of the decisions he must make afterwards will either directly or indirectly involve the contractual agreement.

Let us now review the most important contract types, referring to the selection chart above for a summary of features and applications.

Contract Forms: Cost-Plus

The cost-plus contract is the most flexible of all contract types. The contractor is reimbursed for all direct costs incurred plus a percentage fee to cover his overhead expenses and profit.

Under the terms of a cost-plus contract, the client can exercise the option to retain complete control over all aspects of project execution. He also reserves the right to approve all drawings, providing he exercises sound engineering judgment. (Most agreements will contain a clause protecting the contractor against being forced to comply with the client's request if, in the opinion of the contractor, said request could result in an unsafe design. This clause is a protection for both the client and the contractor.)

Cost-plus contracts are usually criticized on the grounds that the contractor has little incentive to

Contract Form; and Project Definition (PD)	Primary Advantages	Primary Disadvantages	Typical Applications	Comments
LUMP SUM, BASED ON DEFINITIVE SPECIFICATIONS PD: General specifications, design, drawings and layout — all complete.	1. Usually results in maximum construction efficiency. 2. Detailed project definition assures client of desired quality.	1. Separate design and construction contracts increase over-all project schedule. 2. Noncompetitive design may result in use of over-conservative design basis. 3. Responsibility is divided between designer and constructor.	Where client solicits construction bids on a distinctive building designed by an architectural firm, or where a federal government bureau solicits construction bids on project designed by an outside firm.	Clients are cautioned against use of this type of contract if project is not well defined.
LUMP SUM, BASED ON PRELIMINARY SPECIFICATIONS PD: Complete general specifications, preliminary layout, and well-defined design.	1. Competitive engineering design often results in cost reducing features. 2. Reduces over-all project time by overlapping design and construction. 3. Single-party responsibility leads to efficient project execution. 4. Allows contractor to increase profit by superior performance.	1. Contractor's proposal cost is high. 2. Fixed price is based on preliminary drawings. 3. Contract and proposal require careful and lengthy client review.	1. Turnkey contract to design and construct fertilizer plant. 2. Turnkey contract to design and construct foreign power-generation plant.	1. Bids should be solicited only from contractors experienced in particular field. 2. Client should review project team proposed by contractor.
UNIT-PRICE CONTRACTS, FLAT RATE PD: Scope of work well defined qualitatively, with approximate quantity known.	1. Construction work can commence without knowing exact quantities involved. 2. Reimbursement terms are clearly defined.	1. Large quantity-estimate errors may result in client's paying unnecessarily high unit costs or contract extra. 2. Extensive client field supervision is required to measure installed quantities.	1. Gas-transmission piping project. 2. Highway building. 3. Insulation work in process plants.	Contractor should define the methods of field measurement before the contract is awarded.
UNIT-PRICE CONTRACTS, SLIDING RATE PD: Scope of work well defined qualitatively.	1. Construction work can commence without knowing quantity requirements. 2. Reimbursement terms are clearly defined.	Extensive client field supervision is required to measure installed quantities.	1. Gas transmission piping project. 2. Highway building. 3. Insulation work in process plants.	Contractor should clearly define the methods of field measurement before the contract is awarded.
CONVERTIBLE CONTRACTS PD: Variable; depends on type of contract conversion.	1. Design work can commence without delay of soliciting competitive bids. 2. Construction price is fixed at time of contract conversion, when project is reasonably well-defined. 3. Over-all design and construction schedule is minimum, with reasonable cost.	1. Design may not be optimum. 2. Difficult to obtain competitive bids, since other contractors are reluctant to bid against contractor who performed design work.	1. Where client has confidential project requiring a balance of minimum project-time with reasonable cost. 2. Where client selects particular contractor based on superior past performance.	Contractors selected on this basis should be well known to client.
TIME AND MATERIALS PD: General scope of project.	1. Client may exercise close control over contractor's execution methods. 2. Contractor is assured reasonable profit. 3. Reimbursement terms are clearly defined.	1. Project cost may not be minimized. 2. Extensive client supervision is required.	Management engineering services supplied by consulting engineering firm.	Eliminates lengthy scope-definition and proposal-preparation time.

control costs, since his fee is proportional to the total cost of the project. This criticism is true only in the situation where insufficient control is maintained over the contractor. In most instances, the client is well represented, and maintains close control over project expenditures, having constant access to the contractor's cost records. Under this form of contract, the contractor acts almost as the engineering division of the client.

Cost-plus contracts are recommended where the basis for design is incomplete or where many changes are contemplated. Typical examples include development projects or major revamping of existing units involving significant construction efforts.

Cost-Plus With Fixed Fee—This contract form involves a modification to the cost-plus type of contract wherein a fixed fee, rather than a percentage fee, is established at the beginning of the work. Regardless of the final cost of the project, the fee is restricted to this amount.

In order for the contractor to agree to a fixed fee, the scope of work and preliminary basis for design must be established prior to contract signing.

Cost-Plus With Guaranteed Maximum

This type of agreement, which incorporates features of both the cost-plus and lump-sum contract, reimburses the contractor on the basis of cost plus a percentage fee, with the total price not to exceed a predetermined amount.

Guaranteed maximum contracts are advantageous to clients because they set an absolute maximum on the cost of the project. This type of contract insures the client against paying excessive fees, if the cost of the work turns out to be less than originally ex-

pected. However, the contractor has less incentive to maintain tight cost control than under a lump-sum agreement.

The scope-of-work definition, specifications and layout must be reasonably well defined before a guaranteed maximum price can be established. The project definition need not be as rigorous as that required for a lump-sum proposal.

The contractor's fee for a "guaranteed maximum" form of contract is usually equal to, or greater than, the fee included in a lump-sum price. One reason is that the contractor must accept the risk of absorbing all overruns of the guaranteed maximum price, and this price is normally based on a preliminary project specification containing less definition than a lump-sum specification.

Incentives—The major argument against the use of guaranteed maximum contracts can be eliminated by the use of an incentive-fee contract. The contractor is encouraged to reduce the project cost by being allowed a share of the savings below the guaranteed maximum price. The savings are often shared on a 50-50 or a 75-25 basis, with 75% going to the client and 25% to the contractor. There are many variations

of the incentive, such as the sliding scale where the contractor's fee increases with decreasing costs.

Incentive contracts may be written to encourage emphasis on a variable of particular importance to the client—e.g., construction time or operating cost.

Escalation Clauses—These may be added to all types of fixed-price contracts, and are quite common with "guaranteed maximum" contracts. The escalation is usually tied into variations of such cost-indexes as the Nelson (refinery-construction), Marshall and Stevens (equipment), Engineering News-Record (building and construction), and CHEMICAL ENGINEERING (plant cost).

Escalation is particularly important where project execution covers a period of one or more years. Contracts involving international financing should also have provision for escalation protection because there is a greater likelihood of delays and work stoppages.

It is not uncommon for small companies to underfinance projects, thereby resulting in work stoppages of one or more years until additional financing is obtained. A contractor involved in such a project would be forced to absorb escalation costs over the length of the project, unless protection were given by a well-phrased clause.

Construction escalation in recent years has averaged 3 to 5% per year, depending on location and economic conditions; work stoppages may stretch a project time schedule to 3 or more years.

It is not uncommon for escalation to approach several hundred thousand dollars over the life of a multi-million dollar design and construction contract.

Bonus/Penalty Incentives

In today's highly competitive markets, most operating companies will not release funds for new manufacturing facilities until they are reasonably certain that there is a profitable market for the new production. In the chemical process industries, the manufacturer usually secures long-term contracts for the sale of at least 50% of the new production before funds are allocated for new plant facilities. Once these supply contracts are signed, there is considerable pressure on the manufacturer to begin production as soon as possible. Failure to meet delivery promises to customers will force the manufacturer to temporarily fulfill orders by buying the product at market prices for resale at cost or even at a loss, until the new plant is successfully commissioned.

Consider the potential loss due to the delayed startup of a new 600 million lb./yr. ethylene plant. Assume the manufacturer had a contract to supply 75% (or 450 million lb./yr.) of this production to a manufacturer of polyethylene plastics. The contract called for delivery to begin Jan. 1, 1967. Plant startup was delayed three months until Apr. 1, 1967. The seller was forced to secure ethylene by buying at the market price of 5¢/lb. and reselling it at cost to the new customer. The ethylene production cost for new plants of this size is conservatively estimated at 3¢/lb. The loss to the seller due to delayed plant startup is com-

---

Project Financing, Here and Abroad

Generally speaking, the type of financing employed on domestic projects has little influence on the choice of contract form, because U.S. companies usually undertake financing to increase corporate cash flow for general expansion purposes, rather than for each individual project. The same reasoning can be applied to large American- and Western-European-owned international companies. The main exception occurs when financing is broken into two or more phases to accommodate a critical shortage in the cash-flow balance. In such situations, the design and construction could be separated into two contracts—or, alternatively, executed under a convertible type of contract.

Project financing does have a big influence on contract selection for small foreign companies or government-owned monopolies in semideveloped countries. Quite often, such projects are financed through one of the international monetary funds that have been established primarily to aid industry in underdeveloped countries. Before such loans are approved, the financier usually requires a detailed project execution plan that might include market surveys, scope descriptions and specifications for the project, information about the selected engineer-contractor, and a sample design-construction contract. Where such documentation is required, the lending agency has a considerable influence on contract form, the degree of influence depending primarily on the credit rating of the borrower.

In rare cases, a small foreign company has awarded contracts for design, procurement and construction services where payments have been financed with the contractor and deferred until after plant startup. Normally, however, contractors are not in a position to finance capital ventures.

puted as follows (excluding any byproduct credit):
450,000,000 lb./yr. $\times$ $\frac{1}{4}$ yr. $\times$ (0.05-0.03) $/lb.= $2,250,000.

This staggering loss of gross profit to the seller —$2,250,000 before taxes—could have been reduced or possibly eliminated by on-time completion of construction. Thus, in situations where manufacturers are fortunate enough to have a ready market for the expanded production, consideration should be given to selecting a contractor on a noncompetitive basis in order to eliminate proposal preparation time, and then using some form of bonus/penalty clause.

The term "bonus/penalty" means exactly what it implies. The contractor is paid an additional fee (bonus) if the plant is commissioned ahead of schedule. If the plant is completed behind schedule, the contractor pays a penalty by forfeiting a portion of his fee. The magnitude of the bonus/penalty is usually limited. The bonus/penalty rate is usually expressed as x thousands of dollars per day, week, or month.

The concept of "bonus/penalty" may be used to provide an incentive to the contractor in other areas of project execution. The contractor's fee may be based on the efficiency of operation designed into the new plant. The terms of reimbursement could be computed from the raw material consumption and/or utility requirements necessary for a specified amount of product. This form of bonus/penalty clause may be applied independently or in conjunction with other types of incentive. Contractual clauses applying this form of the bonus penalty are usually found in the sections titled "Product Yield or Utility Performance Guarantees."

Lump-Sum Agreements

There are two types of lump-sum contracts, and the difference reflects the degree of project definition. In the first type, the contractor bases his bid on completely defined specifications, designs, drawings and layouts; the agreement usually stipulates that he has full control over the operations pertaining to his scope of work. The most rigid of all the contract forms in regard to allowable variations in project construction, this type of lump-sum agreement is not common in the processing industries except for certain structural applications shown in the contract selection chart.

The second type is based on preliminary specifications, and is much more common (although, as pointed out earlier, contractors are seldom enthusiastic about bidding on any sort of lump-sum contracts during active, inflationary periods). The bid definition takes the form of a lengthy and complex technical-proposal volume. For the client, this type of contract has wide appeal if the project is based on known technology, and if he is buying design, procurement and construction services from one contractor.

The possible pitfall in this application of the lump-sum contract is the lack of a complete set of detailed drawings and specifications describing the work. Quality factors are sometimes hard to evaluate when comparing bids. The client must rely to a certain extent on the contractor's reputation, a detailed study of the proposal, and a site inspection of similar projects recently completed by the contractor.

Since each contractor must incur considerable expense in preparing quotations, clients should carefully preselect bidders on the basis of reputation in the particular field, and limit the number of bidders to 3 or 4. Soliciting bids from more than 4 contractors will result in excessive proposal costs to the contracting industry. In the long run, this excessive cost is paid for by the client.

For the contractor, the advantages that sometimes outweigh the high cost of proposals and the difficulty of predicting field productivity are:[1]
- Opportunity to win awards by process innovations and superior design capability.
- Control of coordination and cost optimization.
- Opportunity to increase return by superior performance.

Design changes after award of the contract will require that the lump-sum price be adjusted accordingly. The contractor submits a detailed estimate with cost substantiation, in accordance with a mutually agreeable "extra" formula that is usually negotiated with the contract.

A typical "extra" formula generally contains the following components:

1. Materials @ cost = $ ....
2. Field labor @ cost = $ ....
3. Total materials and labor = $ ....
4. Indirect field cost @ X% of line 3 = $ ....
5. Direct office cost = $ ....
6. Office overhead @ Y% of line 5 = $ ....
7. Total cost of extra = $ ....
8. Contractor's fee @ Z% of line 7 = $ ....
9. Total value of contract extra = $ ....

Depending on the value and complexity of the extra, lines 1, 2 and 5 should be substantiated with vendor quotations, commodity material takeoffs and detailed manhour estimates. In order to avoid disagreements, the client should approve all extras before releasing the work to the contractor.

The "contract extra" computation format may be varied to fit particular situations where different cost factors predominate. In other instances, certain of the lines are not included in the computation—e.g., the purchase of spare parts would require the use of lines 1, 5, 6, 7, 8 and 9 only.

When projects are well defined, and contractors are willing to prepare competitive proposals, the use of a lump-sum contract usually results in a minimum capital expenditure with a maximum of efficiency in project execution.

Unit-Price Contracts

Unit-price contracts are preferred where the design is well defined qualitatively, but where exact quantities are subject to wide variation. Examples include pipeline work, highways, pipe and equipment insulation, and industrial painting.

Unit prices are also quite often quoted as addenda to lump-sum or other fixed-price proposals. A civil contractor may quote a lump sum for all concrete work on a given project. As part of his proposal, he may state that all additional foundation work will be performed at $60/cu. yd. Usually there would be several unit prices quoted, since the cost of concrete varies considerably with the forming and reinforcing requirements. The unit-price "contract extra" concept is, of course, applicable to many other types of construction work.

To establish firm unit prices, the contractor must know the approximate quantities involved. The per-unit allocation of indirect costs—e.g., field supervision, transportation of construction equipment to the site—will vary considerably depending upon the number of units.

In order to protect themselves, contractors usually quote unit-price rates based on a minimum quantity. A typical quotation may read:

> The XYZ Co. will furnish all construction equipment, materials delivered, supervision, direct labor and testing required, in accordance with the specifications, for the installation of the black-top surfacing. The cost for the above services is 35¢/sq. ft., providing a minimum of 10,000 sq. ft. of surfacing is installed.

Clients are cautioned against the use of flat unit-price contracts, unless the approximate quantities are known within a variation of approximately 20 or 30%. If the final installed quantity greatly exceeds the estimated amounts, the client will be paying a high unit rate based on the original estimate. On the other hand, if the final quantity is considerably under the original estimate, the contractor will request a contract extra, since the agreement is based on specified minimum quantities.

Sliding-Scale Clauses—Where the order of magnitude of the quantities is unknown, the client may protect himself by soliciting unit-rate proposals based on a sliding scale. A typical sliding-rate quotation may read:

> The XYZ Co. will furnish all construction equipment, materials delivered, supervision, direct labor and testing required, in accordance with the specifications, for the installation of the black-top surfacing. The cost for the above services is based on the following slide-rate scale:

Quantity	Unit Rate
0-5,000 sq. ft.	45¢/sq. ft.
5,001-9,999 sq. ft.	40¢/sq. ft.
10,000 ft. and over	35¢/sq. ft.

The specifications, qualitative scope of work and sliding unit rates enable the client to maintain good cost control over projects where the volume of work at the onset is unknown. Competitive bidding ensures low price and high construction efficiency.

Convertible Contracts

Where clients are anxious to begin design work, without the delay of obtaining competitive fixed price bids, consideration should be given to the use of a two-step contract. Such a contract begins on a cost-plus basis, where the project lacks precise definition. During Step 1, the project is clearly defined with the development of definitive specifications and preliminary in-depth drawings. The plant layout is fixed. Following completion of Step 1, the contractor prepares a cost estimate of the project and establishes a lump-sum or guaranteed maximum price. The cost-plus contract is terminated and a fixed-price contract is established. At the end of Step 1, the client may also exercise the option of obtaining competitive bids on Step 2.

Convertible contracts offer the dual advantage of an early completion date coupled with a reasonable cost. The client is cautioned to use this contract form only where the contractor is well known from past experience, because this type of contract requires a high degree of mutual trust between client and contractor.

Time and Materials Contracts

Here materials and associated items are charged to the client at cost. Manhours expended by the contractor's employees are charged at agreed-upon rates, which usually include a percentage for overhead expense and profit. The contract form is popular where consulting engineering services are solicited from engineering-contracting firms.

So much for the various contract forms. I would like to close on a philosophical note by pointing out that the effectiveness of any contract is only as good as the intentions of the participating parties—and that the groundwork for developing mutual understanding should be laid during the proposal preparation phase.[2]

References
1. Hackney, J. W., Pros and Cons of Construction Contracts, *Chem. Eng.*, June 21, 1965, p. 160.
2. For discussion of negotiation procedures, see: Bresler, S. A., Hertz, M. J., Negotiating With Engineering Contractors, *Chem. Eng.*, Oct. 11, 1965, p. 209.

Meet the Author

John T. Gallagher is an associate engineer in Mobil Oil Corp.'s international project engineering department, 150 E. 42 St., New York City. His earlier experience—obtained with General Aniline and Film Corp., Lummus Co. and Foster Wheeler Corp.—included assignments in project and proposal management in the petroleum and chemical fields. He has M.E. and M.S. degrees in mechanical engineering from Stevens Institute of Technology, and an M.B.A. from The City University of New York.

Key Concepts for This Article

Active (8)	Passive (9)	Means/Methods (10)
Comparing	Contracts (Agreements)*	Selection guide*
Evaluating*	Construction*	
Using	Design*	
Selecting	Engineering	

(Words in bold are role indicators; numbers correspond to EJC-AIChE information retrieval system. Asterisks mark key concepts suggested for indexing. Others are added to improve reading as an abstract. Indexing is described in *Chem. Eng.*, Oct. 11, 1965, p. 187.)

Rapid Estimating of Engineering Cost

These two statistical correlations provide shortcuts that help you estimate
a project's design and engineering costs from a minimum of technical data.

JOHN T. GALLAGHER, *Mobil Oil Co.*

In the preliminary stages of a new process-plant project, the need often arises for an early, accurate appraisal of engineering and design costs; yet there is seldom enough technical information to use conventional estimating methods.

Consider a hypothetical situation in which an operating company desires to award a lump-sum engineering design contract, with the balance of the project executed on a cost reimbursable basis. Such a contract is often used where an operating company has a particularly pressing market commitment, requiring an accelerated project schedule. Engineering contractors can prepare "engineering only" bids in a matter of days, thus eliminating the lengthy time delays associated with the preparation of definitive proposals for the entire project.

Accurate engineering-design estimates can be prepared from a minimum of information. The method described here, the "Weighted Unit Man-Hour Rate," takes advantage of two statistical principles:

1. The correlation between the number of major equipment pieces and drafting-room man-hours.

2. The relationship between drafting-room cost and total direct engineering design cost.

Compiling the Drafting Man-Hour Cost

Drafting man-hours can be computed by counting and/or estimating the items of major equipment as defined below, and then multiplying this number by a statistical unit man-hour rate, which is based on the average of many process-plant projects. The unit is expressed as X drafting man-hours per item of major equipment and, according to my files, a good average value for X is 250. In other words, if a plant includes 100 items of major equipment, the approximate drafting man-hour requirement equals 100 x 250, or 25,000 hr.

A major equipment item may be generally classified as a physically identifiable object, where a specific work function or unit operation is performed on the fluid, such as a pump imparting energy to a liquid, or a tower where a mixture of chemicals is separated by distillation. There is a sharp distinction between major equipment as defined above, and "commodities" —such as piping, instruments, insulation, wiring, structural steel, etc.—which serve as fluid carriers, supporters or analytical measuring devices.

For instance, major items of direct-fired equipment would include gas- and oil-fired heaters, kilns, air heaters, inert-gas generators, steam boilers, incinerators, etc. Major heat-transfer items would include such types as exchangers, evaporators, vacuum jets, barometric condensers, and cooling towers. In the separation equipment category, major items would be various types of filters, crystallizers, separators, dryers, precipitators, vibrating screens, and so on. Then we would find all kinds of major-equipment items under such categories as vessels, rotating equipment, solids processing and handling units, and water treatment facilities.

The distinction between major equipment and commodity items as defined, will suffice for most of the components of an industrial plant. The few exceptions to the rule are normally based on unusual size or cost characteristics.

When establishing the number of major equipment items, it is preferable to use piping-and-instrumentation flowsheets. Installed equipment items such as spare pumps should be considered as separate items and added to the major equipment total. Motors and turbines are considered an integral part of the driven equipment and must not be counted separately.

The number of items assigned to a package unit should be established with care. A completely designed package can be considered as one item. However, this is seldom the case, and a judgment safety factor is usually applied to package units.

Where piping-and-instrumentation flowsheets are not available, process flowsheets may be used. Unfortunately, preliminary issues of process flowsheets do not show all major equipment items such as spared pumps or multi-shell heat exchangers. When the number of major equipment items is established from preliminary process flowsheets, I would recommend that the counted number of items be increased by 15 to 25%, depending on the confidence level attached to the completeness of the flowsheet.

Capacity Effect—Although 250 is a good figure to use for most types and sizes of fluid-processing plants, the average man-hour value will increase slightly for very large plants, and decrease for very small ones. The effect of the plant-size parameter is best illustrated by estimating the drafting man-hours needed to completely engineer 100-ton/day and 1,000-ton/day ammonia plants, each containing 180 items of major equipment. Hypothetical unit drafting-rates might be 225 and 275 man-hours per item, respectively. The

direct drafting budgets would be 40,500 hr. for the small plant and 49,500-hr. for the large one.

Because it is difficult to quantify an exact drafting man-hour relationship with plant capacity, judgment will have to be used if the plant is near either extreme of the capacity spectrum for a given process.

Utilities and Offsites—The correlation between drafting man-hours and equipment items has been developed from data obtained from the engineering of process units. The relationship is slightly different for utility and other offsite facilities. Because the density of major equipment items per unit of plot area is considerably lower for offsites, and the number of design drawings required increases with increasing plot area, the number of drafting man-hours per major equipment item is higher for offsite units. Experience has indicated that the unit rate is 300 or more drafting man-hours per item, and that the accuracy of the major-equipment piece-count technique is lower for non-process facilities because of the greater variation in equipment type and density.

Drafting Cost vs. Total Engineering Cost

Having obtained the drafting man-hour requirements by the methods of the previous section (or by some other method, such as using historical figures taken from recent similar projects), the next step is to multiply the drafting man-hours by $ Y, which includes a weighted portion for all office man-hours. In Table I, the value developed for Y is $13.89.

The simplicity of the Weighted Unit Man-Hour Method is best illustrated by considering a typical case study. What is the total direct office cost associated with the complete engineering, design and procurement of a new methanol plant? Assume that 150 major items of equipment are required.

Solution: Total Direct Drafting Man-Hours
= 250 MH/item x 150 items = 37,500 MH
Total Direct Office Cost
= 37,500 MH x 13.89/MH = $522,000

The above example considers direct cost only. In order to complete the home-office cost estimate, a contractor will include a percentage override for indirect charges such as overhead. (See: Bromberg, I., "A Look at an Engineering Contractor's Overhead Cost," p. 225.

Using Table I—The weighted cost tabulated in Table I reflects the average costs obtained from conventional type process-plant projects where the basic process technology is known at the onset of the engineering work. Other types of projects can be accommodated by making adjustments. For instance, the total unit direct cost of $13.89 per drafting man-hour would be higher where the project involves process development. The process, project and specialist engineering cost centers would increase significantly. The effect of these cost centers on the total unit cost might be as much as $1 to $2 per drafting man-hour.

As another example, if procurement is excluded from the contract, the cost centers for purchasing,

Engineering costs expressed as weighted unit rate per drafting man-hour—Table I

Department	Average Hourly Rate $	Man-Hrs. per 1,000 Drafting Man-Hrs.	Weighted Payroll Cost per DMH, $
Drafting, direct....................	6.26	1,000	6.26
Support staff....................	3.06	150	.46
Specialist engineering............	8.25	90	.74
Project management and engineering..................	10.00	135	1.35
Support staff....................	2.80	25	.07
Process engineering..............	9.07	80	.73
Cost-control engineering and estimating..................	8.08	100	.81
Support staff....................	3.50	30	.10
Inspection-expediting............	6.05	45	.27
Support staff....................	4.30	40	.17
Purchasing agents................	5.18	120	.62
Home Office Construction staff....	7.75	70	.54
Payroll Cost....................			12.12
Supplies.........................		.11	
Communications...................		.50	
Reproductions....................		.66	
Travel...........................		.50	
Non-payroll cost................			1.77
Total direct cost................			13.89

Notes: DMH=Drafting Man-Hours; hourly costs include burden and are based on 1966 figures; man-hour requirements are based on author's data for fluid processing plants.

expediting and inspection would be reduced by approximately 90%. The balance of the accounts might be decreased 10 to 15%. In general, where a particular office service is excluded from the work, the weighted unit rate should be reduced accordingly.

When using Table I, the reader is free to develop his corresponding company cost, consistent with its own internal accounting practice. Where other data are not available, the values in this table may be used as typical representative engineer-contractor costs. The figures may be periodically updated by using any recognized technical-salary escalation cost study, such as the salary surveys published by CHEMICAL ENGINEERING.

Meet the Author

John T. Gallagher is an associate engineer in Mobil Oil Co.'s international project engineering department, 150 E. 42 St., New York City. His earlier experience—obtained with General Aniline and Film Corp., Lummus Co. and Foster Wheeler Corp.—included assignments in project and proposal management in the petroleum and chemical fields. An engineering graduate of Stevens Institute, he has an M.B.A. from The City University of N.Y.

Key Concepts for This Article

Active (8)	Passive (9)	Means/Methods (10)
Estimating*	Design*	Statistical
	Engineering*	Correlation
	Costs*	

Analyzing "Cost Plus" Engineering Bids

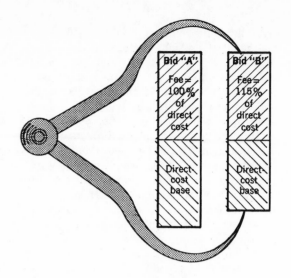

Which of these bids is really lower? It all depends on the direct-cost base. Here is a convenient new technique for quantifying this base so that a meaningful comparison of total engineering cost can be made. Also covered are the qualitative aspects that should be clearly understood when analyzing engineering-contractors' price structures.

JOHN T. GALLAGHER, *Mobil Oil Corp.*

A bid comparison involving the procurement of $1-million worth of machinery is often executed with much greater detail and care than one involving $1-million worth of engineering-design services. Yet most engineers and technical managers will agree that the impact of the engineering-design on the success of the over-all project may be greater than that of a large equipment order. Operating management should therefore analyze bids for engineering-design work at least as carefully as those for equipment.

In this area, one of the toughest problems is how to quantitatively analyze "Cost Plus Percentage-Fee" (CPPF) bids.* Each engineering company has its own bookkeeping procedure for allocation of office costs; no two companies have exactly the same definition for direct and indirect costs.

To compound the confusion, accounting procedures may change within a company from time to time—in fact, changes may occur more than once within a period of a year. Among the many reasons why a contractor may wish to change his "cost plus" check-list is the pressing need to be more competitive with other contractors and their check-lists.

(Check-list is the industry term applied to the list of reimbursable costs prepared by the contractor, which allocates his cost into direct, indirect or out-of-pocket accounts. The costs computed by using this check-list serve as the basis for invoicing clients.)

How does the allocation of charges into the various categories affect CPPF bids? In evaluating competitive engineering quotations, it is important to recognize the differences in bidder pricing formulas, which vary from contractor to contractor, depending on how the direct and indirect costs are defined. Where most of the office costs are defined as direct charges, a lower overhead percentage can be applied in the formula, thus giving the appearance of a low percentage-fee bid. Correspondingly, the reverse is true where most of the office cost is considered as overhead. Here is an illustrative example.

Contractor A bids direct cost, plus 100% to cover indirect cost and profit. Contractor B bids direct cost, plus 115%. Assuming equal operating efficiencies, at first glance it appears that Contractor A has submitted the lowest bid. As with many other aspects of cost analysis, first appearances can be deceiving.

A more precise analysis of the cost structure for our two companies may reveal differences in their definition of direct charges. Firm A considers accounting, legal and stenographic services, and payroll benefits as direct cost whereas Firm B includes these items in its overhead. All other cost allocations for the two firms are identical. Assuming that the cost for the payroll benefits and other classifications enumerated above represent 30% of all the other direct charges for Firm A, the total relative cost for the two contractors is computed as shown in Table I.

A review of the bid comparison confirms that the first conclusion was indeed premature and that Firm B had really submitted the lowest bid. The effect of defining the payroll benefits and miscellaneous services as direct cost broadened Firm A's direct cost base, leading to an accelerated increase in the total office cost.

This acceleration effect can best be observed by con-

---

* Although our discussion and examples will emphasize CPPF bids, it should be noted that the procedures described are also applicable to "Cost Plus Fixed-Fee" or any other form of "cost plus" reimbursable contract.

Two typical quotations for "Cost Plus Percentage Fee" engineering services—Table I

	Firm A	Firm B
Basic technical payroll....	1.00X*	1.00Y*
Accounting, stenographic and legal services; payroll benefits........	.30X	......†
Total relative direct cost..	1.30X	1.00Y
Overhead and profit	@ 100.0% 1.00(1.3X)=1.30X	@ 115.0% 1.15(1.0Y)=1.15Y
Total relative cost..	2.60X	2.15Y

* X and Y represent the total direct technical office cost for firms A and B, respectively, to engineer, design and purchase materials for the project. Since the operating efficiency was assumed equal for both companies, X must equal Y.

† The cost for these accounts is included in the overhead charge.

sidering two companies having the same overhead percentage of 100%, but with one company having a 25% broader direct cost base. Here, the total relative cost equals:

$$1.25 + 1.00(1.25) = 2.50 \text{ (for Company 1)}$$
$$1.00 + 1.00(1.00) = 2.00 \text{ (for Company 2)}$$

Note, that a difference of 0.25 units in the base cost accelerates to a differential of 0.5 in the total cost.

Engineering Company Check-List

All CPPF calculations are structured around some form of check-list; let us now take a look at the checklist components and some of the important definitions.

Most check-lists are subdivided into three primary categories: direct cost, overhead cost, and out-of-pocket cost. The best way to define each of the three cost sectors is to list the types of charges that are commonly found in each classification. This is done in Table II, which compares the check-lists of three hypothetical contracting companies. The trends shown—e.g., the similarities and differences among the three firms—are typical of the actual commercial environment.

Except for occasional special treatment for outside consultants, note that the definition of out-of-pocket cost (i.e., expenses for which the contractor is reimbursed at cost and that do not form part of the cost base) is identical for all three firms. We will therefore concentrate on the other two sectors.

Direct Cost

A study of how direct costs are defined reveals considerable variation among contractors. There exists, however, a common base for all three; namely line technical salaries. This is to be expected, because technical personnel are in effect the production workers in the engineering office, and production costs are normally considered direct charges.

Agreement on allocation of direct cost seems to end with technical salaries. The treatment of procurement, clerical, stenographic, accounting, legal, and office services, departmental management and payroll-benefit costs is not consistent in the contract industry. Such costs are considered as direct or indirect charges, depending upon the contractor in question. The bid reviewer must carefully weigh the effect of this categorization on the expected total engineering cost to his company.

Once the bidder supplies information on how this direct cost is compiled, the evaluation of his bid is straightforward, with the possible exception of two classifications: departmental managers and payroll benefits. Payroll benefits represent a significant segment of the total cost. For many U.S. companies, this represents as much as 25% of salary cost. For foreign companies, this percentage is considerably higher. The way this relatively large cost is allocated should be carefully examined.

The widespread use of exotic job titles, while common practice in the modern business world, tends to complicate the bid analysis. From the maze of job titles, the bid reviewer must decide which title in one contracting firm corresponds to its counterpart in another firm. This comparison becomes particularly difficult at the lower or departmental management level. Table II shows that department managers may fall into the direct cost sector, whereas managers above the departmental level may fall into the indirect-cost category. But, one might ask, what is the definition of a "department" or even what is a "manager"?

One tool that is very useful in classifying job levels is a manual of detailed job descriptions. With the aid of such manuals from all bidders, job titles may be ranked, and a uniform base can be established from which to evaluate the bids.

Indirect Cost: Understanding the Overhead

Common to all indirect-cost structures are: upper management; building utilities, furniture and depreciation; and commercial cost. The other categories overlap with direct cost, as mentioned previously.

A full appreciation of the dynamics of the indirect-cost structure cannot be obtained by simply studying the contractor's check-list. Also needed is an understanding of the various factors that influence his overhead.

Consider the overhead distribution of a typical engineering-construction company. Many such firms today are divisions of larger, diversified corporations. The overhead expenses consist of the indirect costs incurred within the division, and a weighted portion of the corporate indirect charges applicable to the division in question. The corporate overhead charge could be allocated on the basis of divisional sales as a percent of total corporate sales; or, on the basis of divisional payroll cost and depreciation as a percent of total corporate payroll and depreciation charges. (It could be argued that the latter method is the preferred, since it takes into account the varying amounts of physical

Contractor Office Service	FIRM X			FIRM Y			FIRM Z		
	DC	IC	OPC	DC	IC	OPC	DC	IC	OPC
Project management	X			X			X		
Project engineering	X			X			X		
Process engineering	X			X			X		
Specialist engineering									
Electrical	X			X			X		
Structural	X			X			X		
Architectural	X			X			X		
Mechanical	X			X			X		
HVAC	X			X			X		
Other specialty eng	X			X			X		
Home-office construction-advisers*	X			X			X		
Cost-control engineering	X			X			X		
Cost estimating	X			X			X		
Purchasing		X		X			X		
Expediting		X		X			X		
Inspection		X		X			X		
Startup operators†—while in home office	X			X			X		
Drafting and design, direct	X			X			X		
Clerical and stenographic		X		X			X		
Scheduling	X			X			X		
Department managers		X		X			X		
Payroll benefits									
Social Security taxes		X		X				X	
Insurance		X		X				X	
Workmen's Compensation		X		X				X	
Vacation, sick leave		X		X				X	
Pension allowance		X		X				X	
Profit sharing		X		X				X	
Management above departmental level		X			X			X	
Technical information service	X				X			X	
Accounting, legal, office services		X		X				X	
Building utilities		X			X			X	
Building furniture and depreciation		X			X			X	
Commercial cost									
Advertising		X			X			X	
Sales		X			X			X	
Proposal preparation		X			X			X	
Travel and living allowance			X			X			X
Reproductions			X			X			X
Communications			X			X			X
Scale models			X			X			X
Office supplies			X			X			X
Other sundries			X			X			X
Outside consultants‡			X			X			X

Notes

Columns **DC**=Direct Costs
Columns **IC**=Indirect Costs
Columns **OPC**=Out-of-Pocket Costs

* Field supervisors, while sometimes charged on a cost-plus basis, are excluded from this analysis, since office costs only are being considered.

† Startup operators are normally charged at a per diem rate while in the field. Time spent in the office is considered like any other office charge, with contractor reimbursement established from the cost-plus formula.

‡ Some contractors do not consider this as an out-of-pocket cost, their contention being that the administration of such personnel incurs indirect expense. Where overhead expense is considered it is usually much lower than the standard check-list override. As an example, if the standard overhead percentage is 75%, the indirect charge for outside consultants might be 30%.

facilities per sales dollar in the manufacturing and contract divisions of a diversified company.)

The most volatile of the overhead expenses is the commercial cost (e.g., proposal preparation, advertising, other marketing expenditures, etc.) This cost is substantial, and tends to rise in periods where low economic activity forces the contractor to increase his advertising and to bid on more lump-sum work.

In general, during economic lulls, all overhead factors increase as a percentage of direct cost, because a lesser amount of engineering-design work is performed for customers, thereby reducing the direct cost base.

Quite often, the contractor must arbitrarily reduce quoted overhead cost in order to win a competitive contract, hoping that the unabsorbed overhead can be earned on a future project. The logic behind purposely bidding at below cost is defended by contractors as a sound maneuver to allow continued employment for the skilled technical staff. Once experienced engineers are discharged, rapid rebuilding of a competent staff is difficult.

The reverse is logically true for periods of high

| | Bid (in $1,000) by Contractor: | | |
Account	X	Y	Z
Common Direct Cost Base			
Drafting (37,500 MH × $ 5.02/MH)..	188.3	188.3	188.3
Specialist eng. (37,500 MH × $0.59/MH)..................................	22.1	22.1	22.1
Proj. mgt. and eng. (37,500 MH × $1.08/MH)..........................	40.5	40.5	40.5
Process eng. (37,500 MH × $0.59/MH).................................	22.1	22.1	22.1
Cost estimating and control eng. (37,500 MH × $0.65/MH)........	24.4	24.4	24.4
Home-office const. cost (37,500 MH × $0.43/MH)......................	16.1	16.1	16.1
Total common direct cost.............	313.5	313.5	313.5
Other Direct Cost			
Purchasing, expediting and inspection (37,500 MH × $0.71/MH)..........	—	26.6	26.6
Clerical and stenographic (37,500 MH × $0.70/MH).......................	—	26.3	26.3
Department managers (5% Direct Salary Cost*)........................	—	18.4	18.4
Technical-information service (1% direct salary cost*)...................	3.1		—
Accounting-legal services (8% direct salary cost*)........................	—	30.7	—
Total salary cost......................	316.6	415.5	384.8
Payroll burden (@25% salaries)......	—	103.9	
Total direct cost......................	316.6	519.4	384.8
Indirect cost......................... @100%	316.6	@75% 390.0	@85% 327.0
Total cost............................	633.2	909.4	711.8
Fee................................. @25%	158.0	@15% 136.0	@25% 178.0
Out-of-pocket (37,500 MH × $1.77/MH).................................	66.2	66.2	66.2
Total estimated contract value........	857.4	1111.6	956.0

Detailed comparison of three contractors' CPPF bids for engineering work—Table III

Notes

MH=man-hours

Except as indicated with *, all unit cost data has been obtained from the **CE** Cost File-126 "Rapid Estimating of Engineering Cost", **Chemical Engineering**, June 19, 1967, p. 252:

* Denotes author's personal files.

activity where the indirect cost is spread over a larger direct cost base, thus allowing contractors to reduce overhead percentages. Maintaining a constant overhead factor during both economic lulls and highs will result in wider swings in net profit.

Bid Analysis: A Sample Problem

Up to now, the qualitative aspects of bid analysis have been stressed. To obtain a total-dollar cost-comparison among the bidders, let us now do a detailed quantitative analysis of three "Cost Plus Percentage Fee" engineering bids, via this example:

The ABC Fertilizer Co. had greatly increased its share of the market for nitrogen fertilizers. Pressing sales commitments required that its ammonia plant be doubled in capacity within the next 15 months. The tight schedule did not allow for the soliciting of competitive lump-sum proposals. Consequently, it was decided that "CPPF" engineering bids would be obtained

from three qualified engineering-contracting firms.

The essential part of the bids relating to terms of reimbursement for engineering design were:

• Contractor X proposed to perform all required engineering, design and procurement services for reimbursement of direct cost, plus a fixed overhead percentage of 100%. The fee would be calculated as 25% of the sum of the direct and indirect costs. Out-of-pocket expenses would be reimbursed at cost.

• Contractor Y quoted direct cost, plus 75% for indirects, plus 15% fee.

• Contractor Z quoted direct cost, plus 85% for indirects, and 25% for fee.

The definition of the terms for these contractors' direct, indirect and out-of-pocket costs are shown in Table II.

The evaluation of the estimated total costs is predicated on the assumption that the three contractors are of equal efficiency. While efficiency may vary slightly among firms, it is not unreasonable to make

this assumption, providing the companies' design offices are in the same general region of the U.S.

Before comparing bids, the total drafting man-hour requirement must be estimated. Remembering that equal efficiencies were assumed, the direct drafting man-hours will serve as the pricing base in the bid comparisons. In comparing CPPF bids, the accuracy of the total enginering cost estimate is not of primary concern. Since the reviewer is more concerned with accurate differential cost comparisons, an approximate estimate for total engineering cost is sufficient, providing the same direct cost base is used for all bidders.

Techniques are available for developing rapid engineering cost estimates.[1] The total direct drafting time required to design the ammonia plant in our example is estimated to be 37,500 man-hours; this will serve as our common base in developing the bid tabulation.

In order to evaluate the effect of each account as it relates to the direct or indirect column of the checklist, a statistical breakdown of cost by department is required. Many of the components for such a breakdown are available in the literature.[2]

The Formula for Total Cost

The end result of our calculations will be the evaluation of the bid formula for the three contractors. The formula may be generally stated as:

$$C_T = C_D[(1 + F)(1 + P)] + C_E$$

where C_T = total cost; C_D = direct cost; F = overhead factor; P = fee factor; and C_E = out-of-pocket cost. The derivation is as follows:

$$C_T = C_D + C_D F + (C_D + C_D F)P + C_E$$
$$C_T = C_D(1 + F) + C_D P(1 + F) + C_E$$
$$= C_D[(1 + F) + P(1 + F)] + C_E$$
$$= C_D[(1 + F)(1 + P)] + C_E$$

For a given contractor, both F and P are constants; i.e., $(1 + F)(1 + P) = K$. Therefore:

$$C_T = KC_D + C_E$$

The first step in compiling the total engineering cost is to calculate the value of the portion of direct cost that is common to all three check-lists. In compiling the magnitude of the common direct cost base, use is made of the "Weighted Unit Man-Hour Method."[2] In Table II, the accounts common to all three check-lists are drafting, project engineering, process engineering, cost engineering, specialist engineering, scheduling and home-office construction consulting. The approximate cost is computed from statistical units summarized on p. 252 of Ref. 1. Note that since "payroll burden" is not in the common direct cost base, the weighted man-hour costs listed in that reference must be decreased by 25% in the bid comparison, which is shown in Table III.

All other factors being equal, Contractor X should be awarded the contract, on the basis of cost (Table III). Although he has the highest percentage override for indirect cost, its percent is applied to the lowest direct-cost base.

Efficiency Variations

Up to this point, the bid analysis has been restricted to companies with equal engineering-design execution efficiencies; variations become significant when comparing bids from companies with engineering offices in different countries. Consider the following hypothetical situation:

A chemical firm in South America solicits CPPF bids from contractors in France and the U.S. Because of the significant differences in salaries between the latter countries, one might surmise that the French bid would be considerably lower than the U.S. quotation. This is not always the case. The two major factors working in favor of U.S. engineering firms are their higher productivity and lower indirect-cost percentages. The primary negative feature is, of course, the high salary structure.

Indirect costs in Europe are a much larger percentage of the total cost than in the corresponding U.S. price structure. In Europe, most of the components of the indirect cost (such as the building, utility, administrative and commercial accounts) are equal to or higher than comparable U.S. costs. To recover these costs from a low direct-cost base, European contractors must charge relatively high percentage fees. With this in mind, let us develop a typical bid comparison, using the following given data:

1. Weighted, direct, unit payroll cost equals $13.89 per drafting man-hour in the U.S. and $7.80 per drafting man-hour in France.

2. Drafting efficiency: 1.3 French man-hours equal 1 U.S. man-hour.

3. Indirect cost percentage equals 75% in the U.S. and 120% in France.

The total unit cost is computed as follows:

	U.S. Contractor	French Contractor
Weighted direct unit cost...	$13.89	$7.80
Efficiency factor..........	@ 1.0	@ 1.3
Adjusted direct unit cost...	$13.89	$10.10
Indirect cost..............	@ 75% 10.40	@ 120% 12.20
Total unit cost/drafting man-hour	$24.29	$22.30

The relative efficiencies have had the effect of almost equalizing costs between the two competitors. Thus, other factors may determine the selection of the contractor.

The most difficult feature of the analysis is establishing the efficiency factor. This figure is best obtained by comparing the time required to prepare similar design drawings in the respective countries. If an operating company does not have previous experience on which to base such man-hour data, the productivity analysis will have to be directional and approximate rather than really quantitative.

In short, while productivity analysis is not an exact science, its application should not be ignored in compiling international bid tabulations.

References

1. Gallagher, J. T., Rapid Estimating of Engineering Cost, *Chem. Eng.*, June 19, 1967, pp. 250-252.
2. Ibid., p. 252.

A Look at an Engineering Contractor's Overhead Costs

In working with contractors, it is important to understand how their overhead costs are calculated and charged to individual jobs.

IRWIN BROMBERG, *Foster Wheeler Corp.*

For the engineering contractor, overhead costs may either be indirect ones that cannot be charged exclusively to any single project, or they may be direct ones that could be charged against the job but are rather treated as overhead because of convenience. Of course, this convenience can be misused by arbitrary allocation; items such as legal fees, laboratory expenses, and interest on delayed payments should be costed directly if they apply to a particular job.

The definition of overhead varies from contractor to contractor, as it does among operating companies. In order to evaluate "cost plus" bids of different contractors for design services, and to avoid misunderstandings about billing procedures, the customer must study the way overhead will be treated. When comparing bids, he must reduce the various overhead definitions to a common denominator.

Fundamental to the overhead concept is the choice of a base to which an overhead rate can be applied. To the engineering contractor, this base is usually the manhours of technical labor directly applied to the job, because technical labor is his major salable product.

Each of the contractor's major divisions usually has a separate overhead budget so that a balance is attained between the cost of running that division and the liquidation of labor against jobs.

To illustrate Foster Wheeler's approach, I have selected some actual data from our Process Plants Div.; these data are particular to the year chosen and very much related to business volume.

The first category of divisional overhead involves labor—i.e., the technical manhours that the division expends on activities that are not directly chargeable to any one project. A comparison of these manhours with the directly productive ones is shown below:

	Percent of Total Manhours	Cents per Productive $
Direct Charges:		
Productive work on contracts	60	(100)
Overhead Charges:		
Unreimbursed proposals.....	15	25
General R & D..............	7	12
Work on standards..........	4	7
General divisional liaison.....	3	5
Holidays, vacations, time off.	11	18
Total labor overhead......	40	67

The prior example shows that for every 60 manhours of technical labor productively employed on specific jobs, 40 manhours are spent on various overhead activities that are necessary to stay in business. To be reimbursed for these labor overhead costs, a rate of 67¢ must be applied to each dollar of productive labor.

The second category of divisional overhead involves expenses other than for manhours (although manhours may be indirectly involved):

Divisional-Level Expense	Cents per Productive $
Payroll burden—statutory.............	8
(e.g., Workmen's Compensation)	
Payroll burden—other..................	12
Reproductions, communications......	8
Computer expenses...................	5
Office rental, etc.....................	10
Travel................................	3
Total divisional expense overhead..	46

Of course, only the divisional expenses that are not directly charged to contracts are shown in the above listing. For instance, the travel expense shown is the amount that cannot be charged to specific contracts because the travel may have involved unreimbursed proposals or other general departmental business.

In our sample, it can be seen that divisional overhead expenses adds 46¢ to our overhead rate, so that a full overhead application of $1.31 per dollar of productive labor is required to break even on a divisional level.

General and Administrative Overhead—This expense must be allocated among the various divisions. In our example, the Process Plants Div.'s share is as follows:

Expense	Cents per Productive $
Administrative......................	20
Legal................................	13
Accounting..........................	12
Sales................................	22
Secretarial and clerical..............	15
Total general and admin. expense..	82

Adding this 82¢ to the $1.13 of divisional overhead, we see that the Process Plants Div. requires a return of $1.95 over and above each dollar of productive salary in order to break even during the year chosen.

Of the various overhead items we have discussed, the only one that is confined to the engineering contracting business is the cost of making proposals. This cost is a substantial one, and tends to rise in periods of economic lull, when the contractor with a slim backlog may be forced to spend more money on proposals at a time that his productive salary base is small.

Key Concepts for This Article

Active (8)	Passive (9)	Means/Methods (10)
Distribution*	Burden (cost)*	Engineering
Charging (accounting)*	Expenses*	Contractors*
	Labor	

Analyzing Field Construction Costs

In examining the relationship among direct labor and the other elements of field costs, the author zeroes in on cost-estimation problems and practices—and also on some facets of construction-cost control that are particularly timely.

JOHN T. GALLAGHER, *Allied Chemical Corp.*

Field costs represent 25% to 40% of the total investment for a capital project. Concern with this important and rather volatile cost item cuts across many job lines in the chemical process industries. For instance, an understanding of the elements that affect these costs is a must, not only for operating-company engineers who are involved with inhouse construction projects, but also for operating-company personnel who must work with contractors—i.e., in evaluating bids, reviewing progress, authorizing changes, analyzing budgetary reports, etc. It is for this broad spectrum of engineers and technical managers (rather than for the full-time specialist in field-cost estimation or control) that this article is primarily intended.

Direct Field Labor

As shown in the "pie chart" on this page, field costs can be subdivided into four major cost centers. Of these, direct field labor is not only the biggest,* but also the hardest to estimate and control. In fact, under-estimating direct labor cost is the prime reason for capital budget overruns, many of which can be attributed to improper analysis of such factors as working

conditions, area labor productivity, unavoidable delays, seasonal weather conditions and area labor cost variations.

Working Conditions—The significance of this factor can be readily seen by visualizing the work conditions for two extreme, fictitious case studies:

Site A is assumed to be an oil tank farm. A team of pipefitters is installing distribution headers on sleepers. A second team at Site B is installing comparable pipe headers in confined quarters in the middle of a crowded processing area.

The difference in labor efficiency between two such sites might be as much as 100%[1] (i.e., the installation of the pipe at Site B will require twice the man-hours as the equivalent installation at Site A).

Area Labor Productivity—The impact of labor productivity on field-cost estimates is dramatically illustrated by the wide variation in labor productivity between U.S. and Indian workers. Studies based on recently constructed projects have revealed that the U.S. Gulf Coast craftsman is two to three times more productive than his Indian counterpart, depending on the trade.[2] This variation can be attributed to:

1. Superior physical stamina of the U.S. worker, on average.

2. Lower experience level of the average Indian worker.

*The relative magnitudes shown in the chart are for a $20-million chemical-plant project. For smaller projects, although direct labor would occupy a narrower slice of the total field-cost pie, it would still comprise the largest single item (as discussed later).

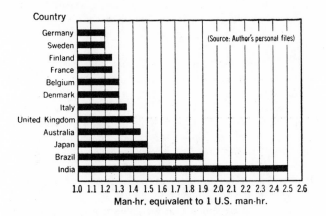

Country

1.0 1.1 1.2 1.3 1.4 1.5 1.6 1.7 1.8 1.9 2.0 2.1 2.2 2.3 2.4 2.5 2.6
Man-hr. equivalent to 1 U.S. man-hr.

TYPICAL worldwide construction-labor productivities—Fig. 1

3. Greater use of power tools and automation techniques in the U.S.

4. A higher degree of industrial standardization in the U.S., permitting more-efficient field erection techniques.

Typical labor productivities for some of the more industrialized nations are shown in Fig. 1.

Labor productivities are dynamic parameters that will vary with level of economic activity and naturally, with the level of construction mechanization. Productivity indexes can therefore be expected to vary over even a relatively short period of time. It should also be noted that Fig. 1 reflects average national productivities; this can vary geographically within a given country.

While the productivity variation within the borders of the U.S. is not as great as international comparisons, it can be significant.

The recent process-plant construction boom, particularly in the Gulf Coast area, has been accompanied by a decrease in productivity. U.S. labor productivity in the post-war period from 1946 to 1963 was normally referenced to the Gulf Coast area; this region was considered as the highest efficiency labor area during this period. But commencing with 1963, there was a general deterioration (that does not as yet show any signs of improvement), which was primarily caused by the shortage of skilled field craftsmen.

An analysis of the extreme geographic-area efficiencies reveals that labor productivity currently has a national variation of as much as 40%. Since labor efficiency is subject to relatively rapid swings, the best and perhaps only truly reliable data source is a careful site survey involving current or very recent construction projects in the area.

Unavoidable Delays—These are losses in time that are beyond the control of the worker, and include factors such as setup time, travel time, or idle time spent awaiting materials. Although these delays are often difficult to identify during the estimate phase of the project, many of them can be recognized by a careful study of the plant layout and the project execution plan. The plot plan should be studied for accessibility of personnel and equipment.

In order to highlight the potential magnitude of these delays, consider the following case study:

An industrial construction project lasting approximately 300 working days, required a labor force peaking at 1,000 men and averaging 600 men per month. The construction site was situated within the confines of a large chemical complex. Plant security required that all construction workers check in at the gate and walk to the site. The new unit was situated almost 1 mile from the gate. The average time to check in and walk to the site was 10 minutes. Craft union rules stated that this time was to be considered working time, for which the laborer must be paid. The total loss time to the construction company was computed as follows:

Loss time = 600 men/day × 300 days × 0.167-hr. delay × 2
= 60,000 man-hr.

Cost = 60,000 man-hr. × $5.00/man-hr. = $300,000

Needless to say, the contractor must try to recognize such delays in the planning stages, so that sufficient allowance can be made in compiling the budget estimate.

Seasonal Weather Conditions—Depending on the geographic area and the time of year, weather conditions can significantly affect the field labor budget. Contingency allowance can be logically developed by studying the weather history at the site over a period of several years. The average number of bad weather days during the calendar period of construction should be added to the schedule. Additional direct labor cost can be compiled by assuming that the bad weather days are worked at $x\%$ efficiency, which is considerably less than good weather productivity.

Area Labor-Cost Variations—Labor rates at the actual jobsite are normally estimated by means of a local wage survey, or from cost information generated by recent projects in the area. Excellent published data are also available. The U.S. Dept. of Labor periodically publishes prevailing labor rates for the major U.S. metropolitan areas.[3] The Aug. 1967 rates in 42 cities were reported in the Sept. 21, 1967 issue of *Engineering News-Record*. A more detailed listing that includes almost 150 U.S. and Canadian cities (also as of Aug. 1967) has been published by the Builders Assn. of Chicago.[4]

The labor rates in the U.S. vary from region to region. The geographic area variations, taking the Southeast as a base, are summarized below:[3]

Area	Relative Labor Cost
New England	1.16
Middle Atlantic	1.34
Border States	1.06
Southeast	1.00
Great Lakes	1.20
Middlewest	1.15
Southwest	1.01
Mountain	1.13
Pacific	1.27

As can be seen, the regional average labor rates vary as much as 34%. Actually, the labor rate variations are

considerably higher when comparing individual cities rather than regional areas. Electricians in the New York area currently earn approximately $5.93/hr. (exclusive of fringe benefits)[4] whereas their counterparts in Charlotte, N.C., earn $3.90/hr.

Taking another example, a sheet-metal worker in Chicago earns about $5.60/hr., whereas his counterpart in Columbus, Ohio, earns only $4.75/hr., although these two cities are considered as being in the same geographic region.

Often overlooked in the compilation of capital budgets are known labor escalations. When surveying a new site and ascertaining prevailing labor rates, one must investigate current union labor contracts. The effect of expiring labor agreements during the life of the project should be assessed, and sufficient adjustment should be made to the budget to provide for the required escalation. This is a known cost, and is in addition to normal project contingencies.

Subcontracting vs. Direct Hiring

On every construction project, the general contractor must decide whether to subcontract a portion or all of the field work, or whether to hire the labor directly. Although the economics governing such decisions often depend on individual circumstances, certain characteristic factors generally come into play, e.g.:

Size of Over-All Project—A broad general relationship applying to this factor is that the extent of subcontracting decreases as the dollar size of the project increases. An exception to this rule may be a very large project where one general contractor cannot staff the entire job or desires to share cost liability.

For very small field projects, the general contractor normally cannot absorb the overhead associated with the establishment of his own field staff and headquarters. For such projects, all field work might be subcontracted to local firms.

Size of Individual Task—The smaller the individual job or suboperation, the more likely it is to be subcontracted (unless it is nonspecialized and can be readily handled by craftsmen already at the site).

Proficiency of General Contractor—Where general contractors are not proficient in a particular specialized field, the work would logically be subcontracted to a specialized construction firm.

Proficiency of Local Contractors—The amount of subcontracting will increase with the increasing competency of the local contractors.

Local Labor Availability—Where there is a local shortage of skilled craftsmen, the general contractor will tend to increase the amount of subcontracting. Craftsmen generally prefer to work for a local firm, rather than a large national construction company; they feel that when the national contractor finishes his project, he will move on to another section of the country, whereas the subcontractor who restricts his work to local regions can provide the craftsmen with greater continuity of employment. Under these conditions, the general contractor may have difficulty in hiring skilled workmen, and will thus turn to subcontracting.

Construction Schedule—Under the pressure of a tight construction schedule, the general contractor is apt to decrease the amount of subcontracting, on the theory that job progress can be more rigidly controlled by directly supervising as much as the labor force as is practical.

Field Cost Optimization—Unless a noncompetitive situation exists, the use of subcontractors in selected areas normally results in some savings to the general contractor.

Availability of General Supervisory Personnel—During peak construction activity, the general contractor may find himself short of experienced field supervisory personnel. In such situations, the contractor will endeavor to utilize subcontractors, thus reducing the amount of direct supervision required.

Where two or more factors influence a given situation, a generalized approach cannot be developed. Each case would have to be judged on its own particular set of circumstances.

Over-All Labor Ratio

To conclude our discussion of direct labor-costs, let us take a brief look at the over-all labor ratio (R). Normally defined as the cost of direct labor divided by the cost of all plant equipment, commodity materials and material supply and erection subcontracts, this ratio is stated mathematically as:

$$R = \frac{L}{M + S_c}$$

This ratio, which is quite often referred to by top management in its final review of the over-all project cost, will vary considerably depending on the extent that subcontracting is used for labor and materials. Note that R will decrease rapidly with increasing use of subcontracts, since the numerator decreases simultaneously with an increasing denominator. The ratio will also vary for different types of processing units and plant size.

The analysis of the over-all labor ratio for a given project may be undertaken as follows: Compare the current project ratio with corresponding values obtained from similar past projects. Abnormal field-cost mixes (i.e., the proportion of direct vs. subcontract labor) will be readily identified.

What is the significance of a ratio that is not within the expected normal deviation? Consider a budget review that reveals an abnormally high over-all labor ratio. Aside from such an obvious possible cause as an unrealistically high estimate, more-subtle factors may be at work. For instance, since subcontract cost is in the denominator, a high ratio may indicate excessive use of direct labor as opposed to subcontract labor. The optimum split between direct and subcontract labor for a given type of plant may often be established from historical trends; such trends are the result of many past economic evaluations of the alternatives. The actual mix must, of course, be verified during project execution by obtaining competitive subcontract quotations and comparing them to inhouse estimates for the

equivalent material and direct labor. The evaluation should include the incremental field supervision cost associated with the use of direct labor.

A nonlinear regression analysis of the return cost from the construction of 15 process plants showed some correlation between the over-all labor ratio and the dollar magnitude of materials plus subcontracts. While the correlation is at best marginal, the resulting regression curve gives some indication of the variation of labor cost with over-all plant cost. The regression analysis showed that the ratio R equals about 0.35 when $(M + S_c) = \$1,000,000$; and 0.15 when $(M + S_c)$ equals $\$9,000,000$.

Field Office and Supervision Costs

The most significant characteristic of this cost center—which includes field supervision, office personnel, field office supplies and communications—is its relationship to the number of direct labor man-hours required. The range is usually 15 to 25% of the direct labor budget.[2]

Field Supervision—Where budgets require trimming, the tendency quite often is to arbitrarily reduce the planned field-supervision cost, but such reduction must be approached cautiously. Inadequate field-management staffing will result in increased field errors, low labor productivity, poor cost control and schedule delays. The saving in supervision cost may be more than offset by the commensurate loss in over-all field efficiency.

To ensure adequate supervisory staffing without overstaffing, some of the major factors to be considered include geographic separation or integration of new process units, complexity of construction effort, extent of subcontracting, local strengths and weaknesses in the various crafts, degree of shop fabrication, general availability of skilled personnel, and priority of time schedule.

A field supervisory task force is shown in Fig. 2. This staffing is typical for a project involving approximately 300,000 man-hours of direct labor. It should be noted that the chart considers only supervisory personnel. Additional staffing such as clerks, secretaries, guards, draftsmen and inspectors would also be required.

Supervisory costs tend to be lower when local people are used in secondary supervisory positions. This is particularly important where the construction site is outside the U.S. As an example, the U.S. personnel staffing of a foreign multimillion-dollar construction project might consist solely of a construction manager, construction supervisor, general craft foreman, chief field engineer and chief purchasing agent. Subordinate local supervisory personnel would include section supervisors, craft foremen, field engineers, office manager and security officer. Strong U.S. staffing at the top of the organization can minimize possible inefficiencies that might occur with the use of native personnel. Salary, transportation and living-cost savings commensurate with the use of local staffing will generally be significant.

Field Office Costs—This portion of the field indirect cost decreases slightly percentagewise with increasing size of the field effort. Included in this cost account are the office staff, supplies and communication.

Aside from the obvious tasks confronting the office staff (such as the preparation of weekly payroll checks, keeping of the records, maintaining field inventories and procurement of field commodities), the office performs the important task of cost control. This function, which is often underrated and not given enough recognition, serves a twofold purpose: cost control of the project, and development of unit costs and efficiencies

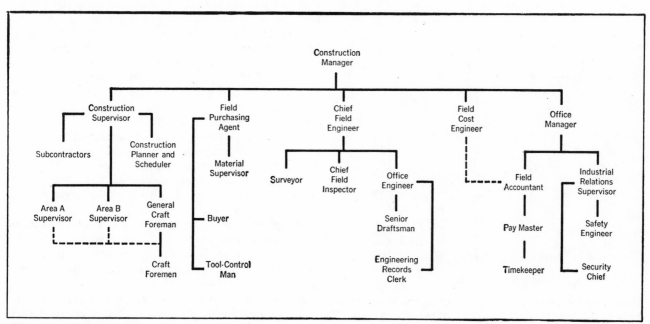

FIELD SUPERVISORY task force for a medium-size construction project may be organized along these lines—Fig. 2

for future field labor estimates. (A fuller discussion of these cost activities can be found in the literature.[5])

Field office staffing is often established from a subjective analysis of the needs; each construction manager may have his own ideas regarding the functions that the field office staff should fulfill. Before undertaking the planning of such a staff, the field manager should discuss areas of responsibility with the office project manager. Typical questions that might arise in these discussions are:

• What detail is required in the accounting records?

• What is the dollar limit and volume anticipated for field purchase orders?

• What is the extent of subcontracting? (The size of the office staff varies inversely with the extent of subcontracting.)

• What amount of field engineering is anticipated?

• What cost-control procedures will be utilized?

• What is the extent of transportation, communication and import difficulties? How much effort will be required to tailor the arrival of material and equipment so as to coincide with field needs?

• What scheduling and planning techniques are to be used?

The key to structuring the field office staff is that it should fulfill all the requirements as dictated by the needs of the construction organization, while at the same time not duplicate the functions best undertaken by the home office.

Construction Equipment and Supplies

Construction equipment, construction supplies and temporary facilities represent more than half of the indirect field cost, and equal 25 to 40% of the direct labor budget.

Typical costs for construction equipment and tools range from approximately $115,000 (1967 cost) for projects involving 100,000 direct U.S. labor man-hours, to $750,000 for projects involving 1 million U.S. labor

Summary estimate of field costs for a typical $20-million chemical-plant project—Table I

Indirect Field Costs

Field supervision and office cost:

Supervision	$87,000
Office personnel	287,000
Office supplies and communication	56,000
	$430,000

Construction equipment:

Large equipment	$447,000
Small tools	51,000
Freight	75,000
Crating and loading	35,000
Repairs and maintenance	42,000
	$650,000

Temporary facilities:

Temporary buildings	$65,000
Temporary utility lines	$125,000
Temporary construction supplies	165,000
Security	125,000
	$370,000

Total Indirect Field Cost	$1,450,000
Total Direct Labor	4,000,000
Field Testing	90,000
Total Field Costs	**$5,540,000**

man-hours. Minimum cost for this account will be assured by obtaining competitive bids on construction tool rentals, and by accurate timing of the movement of large construction equipment to and from the site.

The use of large construction equipment varies considerably with the local direct-labor rates. In general, construction equipment is used more extensively in areas of high labor cost. Consequently, most of the construction projects in the U.S. are built with a maximum of equipment and a minimum of labor, whereas the reverse is true in many nonindustrialized nations. In countries such as Pakistan, where the labor rate is $.40 to $.50 per hour,[2] construction-automation expenditures are hard to justify, and use of heavy construction equipment is minimized.

In countries where a high unemployment rate is prevalent, the government is anxious to maximize use of manual labor, even to the point where such labor cannot be economically justified. Local labor is also protected by the leveling of high tariffs on imported construction equipment.

The preparation of construction equipment budgets is relatively straightforward for domestic projects. However, for reasons cited above, extreme care should be exercised in estimating these costs for foreign projects.

Temporary Facilities—The budget for such items as field office buildings, warehouses, workshops, temporary

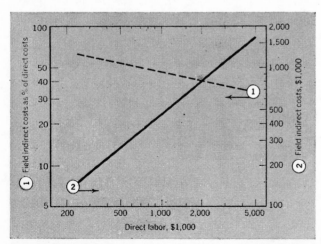

RELATIONSHIP between field indirect costs and direct labor costs for construction projects in a range of sizes—Fig. 3

utilities, roads and site security can be substantially reduced by careful planning.

Temporary facilities that are required may range in size from simple trailer offices to large prefabricated buildings. Before providing for the purchase or rental of such buildings in the budget, management should investigate the possible use of existing buildings. Quite often, an old abandoned structure is near the site and can be readily adapted for use as the construction headquarters.

Where no usable buildings exist, consideration should be given to use of the permanent project buildings. If a warehouse or similar type building is included in the project plan, it is often practical to construct the permanent warehouse on a crash-program basis. This building would serve as the construction headquarters for the rest of the project construction phase, and at the completion of construction, it would be renovated for its permanent service.

Finally, where it becomes absolutely necessary to plan for a separate new construction building, consideration should be given to the purchase of a prefabricated reusable building. The cost for the building can then be easily allocated to several construction projects.

Indirect Cost vs. Direct Labor and Project Size

The grouping of all the field office, field supervision, construction supplies and temporary facilities costs into one category is common industry practice. The summation of all these accounts represents the total indirect field cost. As shown in Fig. 3, there is a strong correlation between field indirect and direct labor cost.

Note that as a project gets larger and the direct cost goes up, the indirect cost also goes up, but at a slower rate so that it accounts for a smaller and smaller percentage of the total construction effort.

The Field-Cost Summary

Following completion of construction, all systems must be tested for mechanical strength and operability —e.g., by hydraulically pressurizing the piping systems and equipment up to the test pressure specified by code, by checking rotating equipment for dynamic balance and freedom from rotational interference, etc. Field testing statistically averages 0.5% of all materials and labor.

The end result of the proper evaluation of the various elements of field construction costs, together with the corresponding estimate, will yield a field cost summary similar to the tabulation shown in Table I.

The format for this table can serve as a checklist of all the significant costs associated with a major construction project. A study of the table points up that office personnel, large construction equipment, temporary utilities and construction supplies will constitute more than two-thirds of field indirect cost for the project analyzed, if the estimate is correct. The relative dollar amounts of each of the subaccounts can readily pinpoint the areas of significant cost that should be the focal point for detailed estimating, economic evaluation of alternatives, and cost control. Additional guidance may be obtained by comparing this estimate of field costs with similar estimates for other current projects—or with a summary of actual field costs for recently completed projects.

References

1. Perry, L. L., Sr., "How to Estimate Piping Installation Cost," *Petroleum and Chemical Engineer,* Mar. 1963, p. 206.
2. Author's personal files.
3. U.S. Dept. of Labor, Bureau of Labor Statistics, Bulletin No. 1547, July 1, 1966. A new bulletin is scheduled to be published shortly.
4. Compiled and issued by R. H. Winslow, Executive Secretary, Builders Assn. of Chicago, Inc., 228 N. La Salle St., Chicago, Ill. 60601. Copies of this comprehensive chart were recently distributed to members of the American Assn. of Cost Engineers.
5. For instance, see: Lutz, J. H., "Estimating Project-Completion Costs," *Chem. Eng.,* Jan. 30, 1967 (Cost File No. 121).

Meet the Author

John T. Gallagher is manager of cost engineering for Allied Chemical Corp.'s Wilputte Coke Oven Div. (40 Rector St., New York, N.Y. 10006). His experience includes project- and proposal-management assignments in the petroleum, chemical and petrochemical fields. Mr. Gallagher has M.E. and M.S. degrees in mechanical engineering from Stevens Institute of Technology, an M.B.A. from the City University of New York, and is a doctoral candidate at New York University. He is a registered P.E. in New York State.

Purchasing Procedures for Engineers

Unless the engineer works effectively with the purchasing agent, the details of equipment buying will take up too much of his time. Here is a survey of how equipment is often bought, plus suggestions for improving current practices.

WILLIAM J. DAVIS, *Chemical Engineer, St. Louis, Mo.*

Most companies encourage close communication between their engineers and purchasing departments when process equipment is being selected. Engineers, therefore, often find themselves deeply involved with the procedural as well as the technological aspects of purchasing equipment. If delays, breakdowns in communication, jurisdictional squabbles and confusion are to be avoided, engineers and purchasing agents must work out mutually satisfactory purchasing procedures.

To shed additional light on how well this is being done, I surveyed 150 engineers of the chemical industry who work in process design and development, production and maintenance. My questionnaire obtained 130 replies. The results are presented along with some recommendations of mine.

Steps in the Procedure

The survey showed that 85% of equipment purchases involve two or more persons besides the purchasing agent and his staff. When more than one person is engaged in a function that requires a single decision, cooperation is obviously important.

These are the steps that are generally followed when routine equipment is purchased: (1) the engineer asks the purchasing agent to get competitive bids; (2) after the bids are received, the engineer compares them to decide which is acceptable; (3) the purchasing agent writes the purchase order for the equipment; (4) to answer questions that the supplier might have, the engineer keeps in contact with him through the purchasing agent until the equipment is delivered; (5) if necessary, the equipment is inspected after it is received to make sure it meets specifications.

Obtaining Competitive Bids

The majority of the engineers surveyed (68%) stated that a supplier was usually contacted, either directly or through his sales agent, for information before a bid was requested. Here is when coordination between the purchasing agent and the engineer must begin.

When one purchasing agent serves many engineers time does not permit him to arrange this first contact with the supplier. If this is the case, the engineer must make such contact, but he should not go beyond preliminary discussions with the supplier. The purchasing agent should always be the one who makes purchases and keeps in contact with the supplier until the equipment is received. However, both the engineer and the purchasing agent have a responsibility to keep each other informed about what either is doing if it affects the other's job.

During the preliminary discussion, the engineer is sometimes tempted to ask a supplier's sales agent for quotations. If he anticipates doing this, the engineer should first contact the purchasing agent to find out about policy. This avoids situations in which a vendor has to negotiate one transaction twice with different persons.

The purchasing agent will usually let the engineer request quotations if these will be sent directly to himself. Their receipt and distribution starts the purchasing agent's file of the purchase.

This procedure has three advantages: (1) the engineer can discuss the design and operation of the equipment with the supplier to determine whether it meets the specifications (before he even asks for quotations); (2) the sales agent can obtain the information needed by the manufacturer to make a quotation without either a second call or a letter from the engineer; and (3) there is less chance for a misunderstanding when design information is transmitted if the equipment manufacturer clearly understands what is needed at the time of the purchase.

Although 95% of the engineers who were surveyed obtain bids through the purchasing agent, they also believe that going directly to the manufacturer is preferable in these specific cases: when the engineer needs technical help and a definite order will not be placed within 60 days; when emergency repairs arise and time is important; and when the engineer is preparing estimates for new installations or for other evaluation purposes.

The purchasing agent will usually understand the engineer's problems in such instances and, as long as there is coordination between them, this procedure should be acceptable.

When asked, "Do you generally specify to the purchasing agent the manufacturers from whom bids are to be obtained?" 11% of the engineers said "always,"

72% said "most of the time," 14% said "seldom," and only 3% said "never."

Comparing and Accepting Bids

In making the final choice of a supplier, 70% of the engineers believed that the engineer and the purchasing agent should make a joint decision, 26% thought that the engineer alone should make the final decision, 3% would let the purchasing agent do so, and 1% had no opinion.

Engineers and purchasing agents tend to evaluate suppliers from different perspectives. Engineers consider the design and adaptability of the equipment to the job, equipment quality, initial and probable maintenance costs, delivery, past experiences with the product, and the technical know-how of the manufacturer. Purchasing agents, on the other hand, consider initial cost, delivery date, the economic condition and financial policies of the manufacturer, cooperation in past purchases, and reciprocity agreements.

If the competitive suppliers are all acceptable to the purchasing agent, the engineer should make the choice, on the basis of engineering considerations. He is more likely to know what is needed for the design problem. This procedure satisfies the requirement for a joint decision.

Purchasing departments often attempt to set up standards for certain types of equipment so that price concessions can be obtained through mass purchasing. This is a tedious task because of the vast number of materials that are purchased. If possible, the engineer should cooperate with such a program.

The actual placing of an order should be the sole responsibility of the purchasing agent. Unless the engineer is acting under emergency conditions, or if there is no purchasing agent, he should never enter into an official agreement with a supplier.

Supplier Contact Prior to Delivery

Equipment is seldom purchased and delivered without any contact between the engineer or the purchasing agent and the supplier. Communicating with a supplier after an order has been placed is probably the most time-consuming chore for the engineer. The supplier, on the other hand, spends a minimum amount of time to get an order because he gets many inquiries and the time he spends on orders that he never realizes is wasted. Only after the order has been placed, does the supplier consider the specific details of specifications and application.

A host of problems arise between the placing of the order and the final delivery: certified drawings must be checked; minor changes in the design might be made; information not included in the drawings must be obtained; and the materials of construction in parts of the equipment may have to be changed.

The engineer must correspond at length with the supplier to resolve these matters. Because of this, administrative procedures are necessary. The engineer should route his correspondence through the purchasing agent, because it is important that the latter be aware of any changes that might affect cost or delivery. The survey revealed that 50% of the engineers followed this procedure always, while 33% did so most of the time.

There are two acceptable methods for the engineer to correspond with the supplier after the order has been placed: by memorandums to the purchasing agent, who then writes to the supplier, or by writing directly to the supplier, with copies going to the purchasing agent.

The first method is preferable. No misunderstanding will occur if the engineer approves the purchasing agent's letter before it is sent. The second method may be acceptable in smaller companies, but should not be the policy of companies that have well-organized purchasing departments. The survey showed that 51% of the engineers followed the first method, and 38% the second. The other 11% contacted the supplier directly, without the knowledge of the purchasing agent.

Equipment Inspection

Inspecting equipment during its fabrication and prior to its delivery can be an important phase of the purchasing procedure. Such inspections offer several advantages: (1) the inspector can discuss the drawings with the fabricator to assure his understanding of what is needed; (2) faulty workmanship can be corrected while the equipment is still in the shop; (3) the engineer can make last-minute changes if he notices anything in the design that has been overlooked; and (4) the supplier's workmanship may be better if he knows that shop inspections will take place.

Of the engineers surveyed, 31% stated that they had inspectors follow equipment fabrication; of the 69% who did not have inspectors, 59% thought that they should have them. They also indicated that approximately 60% of the major equipment items are inspected prior to delivery.

Need for Better Coordination

The survey suggests that better coordination between the engineering and purchasing departments is needed. In answers to three questions, 51% of the engineers thought that engineers should be in purchasing; 46% recommended that a staff engineer be designated in the engineering department who would coordinate engineering with purchasing; 41% were for having a member of the purchasing department located with the engineers.

Key Concepts for This Article

Active (8)	Passive (9)	Means/Methods (10)
Buying*	Equipment	Bids
Purchasing*		Quotations
		Specifications
		Engineers*
		Purchasing agents*

5

Estimating, Reporting and Control of Manufacturing Costs

A Guide to Clearer Cost Estimates

These guidelines are designed to avoid omissions and misdirected effort, improve the presentation, and prevent the estimate from being misinterpreted.

GERALD TEPLITZKY, *W. R. Grace & Co.*

Accurate calculations and detailed data do not necessarily assure that a cost estimate will be interpreted correctly. The guidelines that follow should help minimize this problem; they will be illustrated by a hypothetical case history that compares the economics of an experimental resin with those of a commercially available one.

Let's first consider the obligations of the cost estimator. There are times when the cost-estimate information he is asked to obtain does not represent the real needs of the person requesting the estimate. This person may be taking too narrow a view of the problem, knowing what the "symptoms" are but not necessarily the "disease." As such, the estimator should analyze, interpret and discuss the problem with him to see exactly what is needed.

To carry this one step further, the cost estimator must point out the significant factors that affect the estimate, although they may not have been requested. For example, assuming a division manager has asked him to analyze the process costs of making a given polymer that uses a critical raw material, the estimator must delve into the background of this raw material, to consider price, price trends, availability and purity. Although he may not have specifically been asked for this information, it could be as important to the understanding of the economics as the detailed calculation of the manufacturing costs. Failure could lead the division manager to make a wrong decision.

The estimator also has an obligation to anyone who might like to review and update his estimate several years hence. Therefore he must clearly present the economics so that someone unfamiliar with the estimate can review it and use it in performing the necessary updating.

An additional obligation of the cost estimator is to clearly interpret the results of the estimate for the reader. Listing a group of figures and allowing the reader to decide what they mean is not adequate, and can lead to misinterpretation. Furthermore, the estimator should qualify the accuracy and intent of the estimate to the reader. A preliminary order of magnitude estimate should not carry the same weight as a detailed engineering estimate; the "type" of estimate should be made clear to the reader.

Example: Which Resin Is Cheaper to Make?

In the hypothetical case considered, the estimator was asked to determine whether experimental Resin A was cheaper to manufacture than commercially available Resin B. In order to do this, he performed a cost estimate; the costs are summarized and shown in Table I.

Due to the early stage of the laboratory work, it

Summary of economics for two resins in which reliability of cost data varies greatly—Table I

	Experimental Resin A		Commercially Available Resin B
Capacity, million lb./yr.		20	20
Total capital, million $		6.5	6.2
A. Fixed	5.5		5.2
B. Working	1.0		1.0
Total manufacturing cost, ¢/lb.		21.5	22.0
Selling price, ¢/lb.			
A. 10% after-tax return on investment		28.0	28.2
B. 15% after-tax return on investment		31.3	31.3

Calculation of manufacturing costs—Table II

Item	Calculation	Manufacturing Cost, ¢/Lb. Resin A
Raw-materials cost	See detailed calculations	15.58
Direct labor	Assume 5 men/shift, 3 shifts/day, $6,000/man = $$\frac{90,000 \text{ \$/yr.}}{20,000,000 \text{ lb./yr.}} = 0.45 \text{ ¢/lb.}$$	0.45
Indirect labor	55% of direct labor 0.55 × 0.45¢/lb. = 0.25¢/lb.	0.25
Maintenance	5% of fixed capital per year. Based on fixed capital of $5.5 million. $$\frac{0.05 \times 5,500,000 \text{ \$/yr.}}{20,000,000 \text{ lb./yr.}} = 1.38 \text{ ¢/lb.}$$	1.38
Plant supplies	15% of maintenance 0.15 × 1.38¢/lb. = 0.21¢/lb.	0.21
Utilities	Assume 0.3¢/lb.	0.3

This format is similar for the calculation of the remaining manufacturing cost elements such as packaging, spoilage, depreciation, overheads, sales expense, etc.

was necessary to make a large number of assumptions in order to calculate the manufacturing costs for Resin A. These manufacturing costs, therefore, represent only a very preliminary estimate that is largely based on the judgment of the estimator. The assumptions used were enumerated and explained in the text of the estimate.

Since detailed manufacturing costs were available for the commercially available Resin B, a direct comparison of the two without any qualification would have been deceptive, because it would have indicated that the economics for both resins were calculated with the same accuracy.

In order to put the economics into proper perspective and present the reader with the proper conclusion, the cover letter for this estimate stated that, "The preliminary order of magnitude economics for experimental Resin A appear to be approximately the same as those for commercially available Resin B. The economics for experimental Resin A were based on preliminary laboratory data available to date; the economics for commercially available Resin B were based on detailed cost estimates calculated by engineering company XYZ."

Although the numbers in Table I show that the total manufacturing costs for Resin A are ½¢/lb. cheaper than Resin B, this conclusion would not be compatible with the accuracy and type of estimate performed on Resin A. In this case, the estimator interpreted the results, pointing out that the major conclusion was that the cost of resins A and B was about the same.

Let's look at some other aspects of manufacturing costs. The first step in doing an estimate is to work up the best approach to the problem. One way of accomplishing this is to write an outline. The estimator should have clearly in mind what he is trying to do, and how he hopes to do it, before he gets involved in the detailed calculations.

Interpreting the Cost of Raw Material

The next step—obtaining the needed information—often requires a great deal of judgment and interpretation. For example, in calculating an estimate where the raw-material costs constitute a major percentage of the total manufacturing costs, the use of raw-material prices from trade journals may not be adequate. There are many examples where the prices of chemicals purchased in large volumes are significantly less than those quoted in the journals. For these cases, it is essential that the estimator do some delving into the price history and future of the raw materials, and present the reader with a more factual picture of the raw-material costs.

Considering our example of experimental Resin A, one of the major raw materials used in producing this resin is commercially available; however, it is only produced by one U.S. company. The price shown in the trade journals was 18¢/lb. However, after investigating the availability of this raw material, contacting the vendor, and explaining the large usage contemplated, it was possible to obtain a quotation of 14¢/lb. The difference in cost had a significant effect on the over-all economics of producing Resin A, corresponding to a variance in after-tax return on investment of three percentage points. If the trade-journal price had been used without investigation, the reader would have been brought to an erroneous conclusion.

In instances where raw-material prices cannot be narrowed down, it is advisable to pick a reasonable price, and prepare curves and graphs that show the effect of price variations on total manufacturing costs or return on investment. This philosophy of investigation and interpretation also applies to other significant factors that can affect the over-all economics.

Capital Investment

Calculation of required capital investment again requires a great deal of interpretation on the estimator's part. The accuracy of this calculation should fit the type of estimate being performed. For a preliminary order-of-magnitude estimate, it does not make sense to do a detailed fixed capital estimate that entails the drawing of flowsheets, sizing of equipment, estimating of equipment costs, and so on. For these cases, the capital investment costs can be related to existing cost information, available either in cost-estimating manuals or in previous estimates. Conversely, this type of approach will not be suitable for more-detailed, engineering-type estimates.

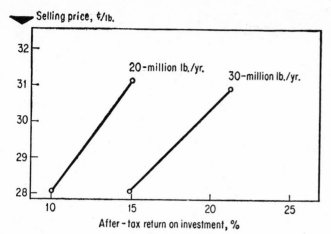

Selling price, ¢/lb.

Effect of selling price and plant size on investment return is best presented via a graph—Fig. 1

The specific format used in calculating manufacturing costs is a very individual matter, as long as the estimator follows a systematic, consistent plan of attack that clearly shows the calculations and assumptions used. This is particularly important if he has to go back and revise the estimate due to a change in assumptions. Some type of tabular format can be most helpful, since this provides a convenient mechanism for showing the calculations and assumptions; Table II illustrates a simple, but effective, format that was used in calculating the manufacturing costs for experimental Resin A.

Presentation of the Complete Estimate

Calculation of a cost estimate, regardless of how fine a job has been done, is only half of the estimator's responsibility. Proper presentation and interpretation are equally as important. The estimator should clearly put into perspective for the reader the type of estimate being done, its purpose, what the conclusions are, and what actions the reader should take as a result of the conclusions.

The actual format of the estimate should follow a certain basic order. The first page, perhaps the first paragraph or two, should summarize the whole story to the reader—he should not have to read page after page before understanding what the conclusions are. Putting the conclusions into his mind initially will help him understand the numbers and calculations that follow.

The results of the estimate can simply be summarized by a table that includes only the key figures. Too many figures tend to confuse the reader and detract from the over-all conclusions. Table I illustrates the format used in presenting the economics for Resins A and B in our hypothetical case.

In presenting a cost estimate, the estimator should not assume what the reader knows or does not know, but rather deliver a complete package that tells the reader what the process is about, why it is being looked into, what the results are, what the possible alternatives are, and what steps are recommended for future action.

Some estimates entail the study of many variables that directly affect the manufacturing cost. These should be incorporated into the estimate via charts, graphs and tables. This is illustrated by Fig. 1, which shows the effect of selling price and plant size on return-on-investment for Resin A.

Some estimates require a detailed explanation of the process in order to bring home specific points. Still other estimates require information on competitive processes if they relate to the economics being considered. The presentation letter should have appended to it the major assumptions used, and a detailed breakdown of the manufacturing costs for the process considered. This information is necessary to indicate to the reader just how the estimator has approached the problem.

Before formally presenting his conclusions, the estimator should of course analyze them to make sure they make sense. He can accomplish this analysis by comparing figures such as selling price, manufacturing costs, required capital, after-tax profit and return on investment with similar figures for known processes, to see whether the results are reasonable. The estimator himself should be convinced before he presents his conclusions.

In summary, one of the most important ingredients in any cost estimate is common sense, for which there is no formula, prescribed procedure, or price listed in a cost estimating book. The person requesting the estimate can only ask for what he thinks is required. The estimator must, however, analyze the request, decide what factors need to be considered, put the factors into proper perspective and present the estimate in such a manner that misinterpretation of the facts is minimized.

Meet the Author

Gerald Teplitzky is business manager at W. R. Grace & Co.'s Washington Research Center, Clarksville, Md. His work includes the calculation of cost estimates.

Key Concepts for This Article

Active (8)	Passive (9)	Dep. Variable (6)	Indep. Variable (7)
Preparation*	Costs*	Analysis	Clarity
Interpreting	Estimates*	Research	Reliability
	Reports*	Presentations	

(Words in bold are role indicators; numbers correspond to EJC-AIChE information retrieval system. Asterisks mark key concepts suggested for indexing. Others are added to improve reading as an abstract. Indexing is described in *Chem. Eng.*, Oct. 11, 1965, p. 187.)

Sources of Error in Operating-Cost Estimates

An apparently correct estimate of operating costs may contain hard-to-detect errors. Here are examples of such errors, and suggestions on how to avoid them.

PAUL R. WALTON, *Mobil Research and Development Corp.*

Operating-cost estimation has not received as much attention as capital-cost estimation or profitability analysis, although operating cost is a large part of the total production cost for many major chemicals, as illustrated below:

Chemical	Distribution of Costs, %		
	Net Raw Material*	Operating Costs	Capital Costs
Ammonia...................	30	40	30
Ethylene..................	0**	65	35
Ethanol...................	45	30	25
Isopropanol...............	50	20	30
Phenol....................	40	25	35
Sulfuric Acid	50	25	25
Styrene..................	60	20	20

*Includes credit for byproducts.
**For ethylene manufacture, byproduct credit equals raw material costs.

Most operating costs are estimated by referring to literature articles or company records, or by calculating operating requirements and estimating how each contributes to cost (as shown in Table I). The techniques are not complicated, and results can be precise. Nevertheless, operating-cost estimates frequently contribute to improper conclusions; the errors can usually be categorized as follows:

- Reference data are misunderstood.
- Effect of a problem environment is overlooked.
- Effect of profitability estimating method is not taken into account.

Although we will discuss each category separately, more than one category is often involved.

Reference Data

Literature data are usually correct; errors in use result mostly from the estimator not understanding the base for these data.

For example, a technical article may describe ethylene cost as follows: "Production cost has been calculated at about 3¢/lb., including all direct and indi-rect costs." The statement implies that operating costs are 3¢/lb., but details of the estimate, outlined below, show that this is a total that includes several costs in addition to operating cost:

	Cost, ¢/Lb. of Ethylene
Raw materials........................	3
Operating cost.........................	2
Allowance for capital recovery...........	1
Byproduct credit.......................	(3)
	—
Total cost............................	3

Company data often are unsuitable bases for estimating. For example, Fig. 1 shows operating cost vs. capacity for a number of similar plants. The detailed data were corrected for unusual costs, but the scatter is still too large for accurate estimating—probably because of different labor rates, utility costs, stream times, recovery efficiencies and the like.

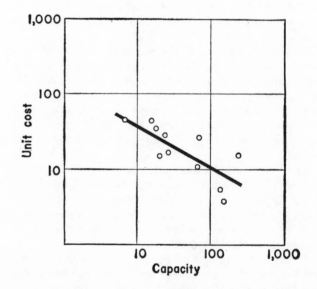

OPERATING COSTS: Correlation of company data—Fig. 1

Based on a paper presented by the author at the Twelfth Annual Meeting of the American Assn. of Cost Engineers, Houston, June 19, 1968.

Problem Environment

Environmental errors may stem from such factors as an extremely favorable or unfavorable location, a process that is unique or very complex, or the difference between an isolated plant and a plant adjacent to existing facilities. *Example*: Table II shows how the operating-cost estimate could be in error if cost elements for a plant adjacent to existing facilities were estimated from correlations better suited to an isolated "grass-roots" plant. Costs are lower for the plant adjacent to existing facilities because:

• Technical and clerical support is reduced.

• Existing facilities can provide steam and cooling water (new distribution facilities must be installed, however).

• Some maintenance for the new plant can be supplied by existing shops and labor force.

• Plant overhead, mainly salaries for support personnel (e.g., guards), will be available from the existing facilities.

Table II also shows how errors in judging environment can be reflected in capital requirements and in turn, in capital-related operating costs.

Profitability Estimating Method

Most profitability analyses compare a proposed process with an existing process, but sometimes estimated product cost from the proposed plant is compared with an established product price. With this technique, conclusions are very sensitive to the cost estimate. *Example*: Table II showed how incorrect estimating correlations increased the product cost, thus decreasing the expected margin over the established product price. The method of analysis exaggerated the error; if estimated costs for the new plant had been compared with costs estimated similarly for an alternative process, the improper correlations would have caused similar errors in both estimates, with little effect on the conclusion.

Errors can also occur because the comparisons are not on the same technical bases. *Example*: When an estimate for a new process was compared with cost data from an old plant for an established process, the new process showed a definite advantage. However, an engineering comparison of the technical features of the processes showed theoretical energy requirements, material balances and kinetics to be essentially the same for both. These results caused suspicion that the

---

Simple operating-cost estimate for a continuous process—Table I

	Amount/Day	Unit Cost	$/Day
Operating payroll and associated costs	9 man-shifts (3 men/shift; 3 shifts/day)	$50/man-shift	450
Utilities:			
Fuel	120 million Btu.	30¢/million Btu.	36
Steam	200,000 lb.	65¢/1,000 lb.	130
Power	2,400 kwh.	1¢/kwh.	24
Water	1,500 million gal.	3¢/million gal.	45
Catalysts	10 lb. (average)	$1.25/lb.	12
Chemicals	Calculated		20
Operating supplies	Estimated		10
Maintenance payroll and associated costs	3% of invest./yr.		80
Maintenance supplies	3% of invest./yr.		80
General plant overhead	5% of total payroll + 1% invest./yr.		50
Taxes and insurance	2% of invest./yr.		53
Total operating cost per calendar-day			990

Effect of plant environment on operating-cost estimate—Table II

	Costs Estimated for Isolated Plant, $1,000		Cost for Plant Adjacent to Existing Facilities, $1,000	
Capital Investment:				
Process units..	1,000		1,000	
Offsite facilities..	400		150	
Raw materials, per yr..................................		300		300
Operating costs, per yr:				
Operating payroll & assoc. costs...................	150		130	
Utilities..	85		45	
Catalyst, chemicals, oper. supplies................	10		10	
Maintenance..	85		70	
General plant overhead...............................	20		10	
Local taxes, insurance................................	30		25	
Total operating cost per yr........................		380		290
Allowance for 10% DCF rate of return (per yr.).....		280		230
Estimated annual selling expense....................		100		100
		1,060		920
Estimated minimum selling price of product, ¢/lb...	10.6		9.2	
Selling price of established product, ¢/lb...........	11.0		11.0	
Advantage for proposed plant, ¢/lb..................	0.4		1.8	

cost estimate for the new process reflected a design close to the process potential, while the cost data for the established process reflected inefficiencies in an old plant, and conservative early design. As suspected, further study identified unreasonably high costs in the old plant, and also uncovered recent major technical improvements in the established process, probably the result of normal "learning" as new plants were built. When these were incorporated into a cost estimate for the established process, the advantage for the new process disappeared.

Inconsistent-comparison errors like these are difficult to identify, and analyses containing them are sometimes used to promote new processes. A good way to detect such errors is to identify and analyze significant technological differences.

Utilities deserve special attention because their cost is often estimated on a unit basis (ϕ/kwh., etc.), and it is easy to forget that unit utility costs are subject to the same errors as the primary cost estimate, i.e., source of data, problem environment and method of analysis. *Example*: In a recent study, investment included steam generation facilities. The operating cost estimate included steam at a cost of $0.65/1,000 lb., and charges for maintenance, taxes and insurance as a percent of investment. The profitability analysis included an allowance for return on investment. Reviewing the analysis disclosed that $0.65/1,000 lb. for steam included all operating charges as well as a return on investment. Thus, steam costs related to investment were "double counted."

The table below illustrates this, and shows how the elements of a unit cost for utilities can vary, depending on the specific manufacturing situation and estimating procedure:

Situation	Cost of Steam, ϕ/1,000 Lb., Manufactured in:		
	Existing System	New System	
Costs included in unit cost estimates..............	Variable costs	Operating costs	All costs
Fuel.....................	20	20	20
Payroll...................	..	20	20
Other utilities, chemicals, etc..	5	5	5
Maintenance, local taxes, insurance, overhead, etc..	..	5	5
Return on investment, plus federal income tax......	..	..	15
Total...............	25	50	65

Avoiding Errors

It is clear that operating-cost estimates have an interdependence with problem definitions and profitability analysis, which must be recognized by the estimator. Errors that occur when this interdependence is overlooked are difficult to detect, especially when the parts of the problem are solved by different persons or groups, as they often are in large organizations.

One aid to preventing such errors is to develop operating-cost correlations for the types of situations that might occur. Tables, like the one shown above for

steam cost, or that show how maintenance relates to investment as complexity increases, are examples. Another approach is to develop checklists for operating costs to guide the estimator in selecting factors suitable for the situation under study. An example list is shown on this page. All the items apply when operating costs are being estimated for an isolated plant; but some may not apply, or may be reduced, when the new plant is adjacent to existing facilities. Such a list will usually be more detailed than the one shown, and major variations will occur between industries and companies. Further expansion of the list can be visualized when still other situations are covered—for example, when increasing throughput in an existing plant.

In any case, a careful assessment of the most likely situation, the supporting information, and the estimates' ultimate use are absolute prerequisites to avoiding these errors.

Meet the Author

Paul R. Walton is an engineering associate with Mobil Research and Development Corp., (a subsidiary of Mobil Oil Corp.), Paulsboro, N.J. 08066. A chemical engineering graduate of Bucknell University, he is a registered professional engineer in New Jersey, and has several patents to his credit. His society affiliations include AIChE and American Assn. of Cost Engineers.

Estimating Manufacturing Costs For New Processes

Here are two essential tools for an accurate estimate of manufacturing costs: a checklist of important items and a typical worksheet for calculating costs.

A critical stage in the development of any new process is reached after some laboratory work has been completed and a flowsheet is available. Usually, a preliminary estimate of manufacturing cost is calculated at this point. If the estimated sales price of the product, based on the manufacturing cost is much higher than competing or similar products already on the market, the project has little chance of progressing much further.

While accuracy is a debatable term at this stage of a project's life, it is important to develop a "reasonable" estimate of manufacturing costs even if these costs are based on rough data. In a number of postmortems on R&D projects, it was found that the major errors in estimates of manufacturing cost were not in under- or overestimates of any particular items, but in completely overlooking items that should have been included in the total.

For this reason, what follows is a list of important items (with applicable comments). This is not an exhaustive list, but overlooking any of these cost items will probably result in a misleading estimate of manufacturing costs.

Direct Costs

Raw Materials—Prices of purchased raw materials can be estimated from published sources. But captive raw materials present a more difficult problem: either a replacement value or an equivalent sales price has to be used. Requirements for raw materials are estimated by calculating the unit consumption factor, which is the stoichiometric ratio of reactant to product, divided by yield.

Labor—Labor requirements can be estimated based on a unit-operation approach.[1,2]

Another technique involves use of a correlation of labor in man-hr. per ton per processing step with plant capacity in tons/day.[3,4]

In most chemical processes, direct labor usually is less than 15% of total manufacturing cost. Where high labor requirements are expected, supervisory personnel should be consulted before working up a labor estimate.

Utilities—These items include steam, process and plant water, fuel, refrigeration, inert gas, compressed air, electricity, etc. As a general rule, utility consumption can be 0.5 to 1¢ per lb. of product when utilities are known to be a minor portion of total operating costs. Costs of utilities have been published.[5]

Repair and Maintenance—If data based on company experiences are not available, maintenance labor and materials can be estimated as 2% of plant capital investment for each. If severe operating conditions are anticipated, maintenance costs could be in the range of 3 or 4% for each.

Factory Supplies—These items include wiping clothes, test chemicals, gaskets, instrument charts, housekeeping supplies. A usual allowance is 0.5 to 1% of the plant capital investment.

Laboratory Charges—These involve costs of laboratory analyses for process control and quality control. Laboratory work is usually related to production volume and requirements for product purity. A reasonable estimate would be $1 per ton of product where analytical requirements are not excessive. Another source suggests 20% of direct labor.[6]

Packaging and Shipping Expense—Packaging and shipping includes cost of containers and rentals of tank cars. Container costs can be obtained from local manufacturers or from the literature.[7] An estimate of $1 per ton of product is reasonable for P&S when shipping distances are small and shipment is in bulk.

Royalties—Royalty expenses can be handled either in a lump sum or as a continuing payment. Lump-sum royalties are considered as part of the plant investment. Usually, information on the cost of royalties is available from the company granting the license. Although some general figures are given for estimating royalty expenses,[8] it is not recommended practice to use such data, since in most cases the actual royalty payment can be estimated without great difficulty.

Development Costs—Some companies recover the development costs of a process by including them as part of production costs. Other companies include them as part of overhead. Therefore, company policy should be consulted.

Overhead—Direct overhead usually includes plant supervision other than shift employees. It can vary between 10 and 25% of direct labor, depending on the complexity of the process.

PRODUCTION COST ESTIMATE

ESTIMATE NO. 2-18 PRODUCTION OF Vinyl Acetate Monomer
BY _____ PRODUCTION RATE 47,500,000 lb/yr. SERVICE FACTOR 90 %
DATE _____ LOCATION Chemtown, U.S.A. COSTS PER yr. (TERM), lb. (UNIT)

INVESTMENT $2,750,000

	Unit	Unit Price	Units Per Term	Unit Consumption	Item Cost Per Term	Group Cost Per Term	Item Cost Per Unit of Product	Group Cost Per Unit
RAW MATERIALS						$5,276,000		$0.1111
Acetic acid	lb.	0.10	34.6MM	0.727	3,440,000		0.0724	
Acetylene	lb.	0.11	15.1 MM	0.318	1,661,000		0.0350	
Catalyst and misc. Chem.	lb.	0.003	47.5MM	1.0	155,000		0.0032	
UTILITIES						125,000		0.00263
FUEL								
STEAM	Mlb.	0.50	120,000	0.00252	60,000		0.0026	
POWER	Kwh.	0.006	3,500,000	0.0736	21,000		0.0041	
WATER	Mgal	0.04	1,100,000	0.0232	44,000		0.0009	
AIR								
LABOR						90,000		0.00190
OPERATING	Manhr.	2.60	24,960		65,000		0.00137	
MONTHLY	Manhr.	3.00	8,320		25,000		0.00053	
OVERHEAD								
REPAIR & MAINT.						137,500		0.00290
LABOR	2% Capital investment				55,000		0.00116	
MATERIALS	2% of capital investment				55,000		0.0016	
OVERHEAD	50% of repair labor				27,500		0.00058	
SUPPLIES & MISC.	1% of Capital investment				27,500		0.00058	
LABORATORY	$1/Ton product				23,750		0.0005	
TRANSPORTATION	$1/Ton product				23,750		0.0005	
TOTAL OPERATING COST					427,500		0.00901	
DEVELOPMENT								
OVERHEAD						326,500		0.00687
DIRECT	1 superintendent				10,000		0.00021	
INDIRECT	50% of labor + repair labor + direct				77,500		0.00163	
TAXES & INSURANCE	2% cap. investment overhead				55,000		0.00116	
DEPRECIATION	15 yr. straight line				184,000		0.00387	
MANUFACTURING COST						754,000		0.01588
PRODUCTION COST						6,030,000		0.127
CREDIT FOR BYPRODUCTS								
NET PRODUCTION COST						$6,030,000		$0.127
NOTES								

Indirect Costs

Depreciation—Depreciation is an allowance, as an expense, for wear and tear on property and equipment used in a business. For preliminary estimates, straight-line depreciation can be used, although it should be realized that for accounting purposes the company may use another method. For greater accuracy, the estimator should follow company policy.

Property Taxes and Insurance—Tax rates for new locations can be developed by checking with local sources. For preliminary estimates, these costs may be regarded as 2 to 3% of the plant capital investment (installed plant).

Plant Overhead—Indirect overhead includes such items as plant office expense (plant-management salaries; plant engineering, protection, accounting, purchasing; maintenance on roads, sewers; cafeterias; dispensary). This can vary over a wide range, but may be approximated as 40 to 60% of the sum of direct labor, repair labor and direct-overhead labor costs, with 50% as a "typical" figure.

Byproduct Credits—Company policy usually has to be consulted on costing byproducts. If there is a cost involved in making the byproduct salable then the sales value of the byproduct should be decreased by this cost.

The total of items discussed above goes into what is generally called "manufacturing cost." Unfortunately, there is a considerable amount of confusion in the literature over such terms as "manufacturing costs," "operating costs," "production costs." In the worked-out example shown on the worksheet,[9] operating costs include utilities, labor, maintenance, supplies, analytical costs, packaging and shipping. Manufacturing costs include operating costs, development, overhead, taxes and depreciation. Production costs include raw materials and manufacturing costs.

Another definition[3] of manufacturing costs assumes its includes all costs for making a product and preparing it for shipment; and operating costs include manufacturing costs, with selling and distribution costs, company administration, and other general overhead expenses.

Work an Example

The example shown on the worksheet involves the synthesis of vinyl acetate: $CH_3COOH + C_2H_2 \longrightarrow CH_3COOCHCH_2$.

The process was described in FIAT and BIOS reports issued after World War II. It consists of reacting acetic acid and acetylene in a catalytic reactor and separating byproducts and unreacted feed materials by distillation. Capital investment was estimated by the method of Buchanan and Zevnik.[10] The raw-material consumption factors correspond to 96% and 95% yields for acetic acid and acetylene, respectively. It should be noted that the high raw-material costs (90% of manufacturing costs) is unusual; about 50-70% is the average range for volume chemical production.

Other items would have to be added to the net production costs shown on the sheet, to arrive at an estimated sales price. These include: sales, general and administrative expenses—or the costs of selling and research and administrative functions in the company general office—and profit. A useful rule is that manufacturing cost (or production as shown on the worksheet) should be 70-80% of sales price for a volume chemical. A lower figure would apply to a new chemical, or one sold in small quantities to a large number of customers.

References

1. Aries, R. S., Newton, R. D., "Chemical Engineering Cost Estimation," pp. 118-184, McGraw-Hill, 1955.
2. Weinberger, A. J., *Chem. Eng.*, Dec. 23, 1963, p. 85.
3. Wessel, H. E., *Chem. Eng.*, July 1952, pp. 209-210.
4. O'Connell, F. P., *Chem. Eng.*, Feb. 19, 1962, p. 150.
5. Weaver, J. B., Bauman, H. C., Heneghan, W. F., Section 26, "Chemical Engineers' Handbook," p. 26-29, 4th ed., McGraw-Hill, 1963.
6. Chilton, C. H., *Chem. Eng.*, June 1951, p. 10.
7. Dowling, T. E., *Chem. Eng.*, Oct. 3, 1960, pp. 85-96.
8. Von Noy, C. W., "Guide for Making Cost Estimates for Chemical-Type Operations," U. S. Bureau of Mines Report of Investigation 4534, Wash., D. C., 1949.
9. Caro, P., Hooker Chemical Corp., Niagara Falls, N. Y., paper "Manufacturing Cost Estimating," delivered at Feb. 25, 1964 meeting of Niagara Frontier Section of Amer. Assn. Cost. Engs.
10. Buchanan, R. L., Zevnik, F. C., *Chem. Eng. Progr.*, Feb. 1963, pp. 70-77.

Key Concepts for This Article

Active (8)	Passive (9)
Estimating	Costs, production
Evaluating	Costs, operating
	Research and development
	Projects

(Words in bold are role indicators; numbers correspond to AIChE information retrieval system. Indexing details are described in *Chem. Eng.*, Jan. 7, 1963.)

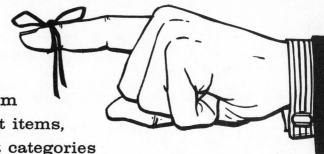

Convenient summary form
prevents the overlooking of significant items,
fits cost data into correct categories
and can be used as a basis of programming
for computer calculation.

Estimate Production Costs Quickly

JOHN W. HACKNEY, Mobil Oil Corp.

Manufacturing costs are raw-material costs plus costs of processing, packaging, loading and shipping the product. A typical form in which manufacturing costs can be organized and tabulated is illustrated on p. 181.

Costs are computed on an annual basis, since this is usually the clearest way to prepare and present the data, especially when seasonal variations are involved.

The costs tabulated are those incurred because of added production and investment, changes in raw materials, processing, packaging, loading or shipping. In other words, they are incremental costs. Computation by increments is usually the best and easiest way to determine the change a project will make in a company's over-all income. The objective of economic analysis is the determination of this income change and of the investment required to produce it.

Annual costs should be rounded off to the nearest $100. Totals are rounded to not more than three significant figures.

Degree of Investigation

Considerable judgment is required in determining the extent to which individual items of manufacturing costs should be investigated. Depending on the type of operation, almost any of them may be dominant and deserve fully detailed consideration. A preliminary order-of-magnitude check will sometimes indicate which elements of cost dominate and therefore deserve complete study. There will be other cost elements that, even if drastically misjudged, will not produce sig-

nificant-figure changes in the totals. No time should be wasted on making an elaborate computation of these.

Considerable detailed computation is required for the entries on the cost form. The necessary computation sheets, with notes as to data sources, should be carefully organized and filed.

Use of Computers

The form for estimating manufacturing costs is in effect a "program" of computation. Entries on the left half of the form of such items as fuel usage and cost constitute the input data. All of the figures on the right hand side of the form, with the exception of the rate at 100% capacity, can be automatically computed. A computing program of this type will contain built-in standard factors for calculating such items as supervision, general works expense and so forth. These factors will be used unless the machine is instructed to employ special values for the project.

Initial costs of writing such a program and setting up the factors and forms to be used are substantial. Once the program is complete, however, the computer can be used to great advantage in cases where it is desirable to determine quickly and inexpensively the effect of variations in yields, investment and labor rates on manufacturing cost. Complete cost sheets are printed out in a very short space of time for each case to be studied, or the machine need report only the key

Turn page for
sample estimating form ......

245

figures required to prepare graphs. Use of qualified consultants in setting up such programs can reduce their cost considerably.

For most projects, costs should be computed at 100%, 75% and 50% of added plant capacity. However, when added capacity is not large as compared with existing capacity, operating levels of 100%, 50% and 0% of added capacity should be used.

In determining total costs at various production levels, individual items fall into one of the following groups:

• Variable, where the cost per year varies directly with the production rate. This is usually the case with raw materials, fuel and containers. Cost of these items per unit of production tends to be constant, regardless of the number of units produced.

• Fixed, where the cost per year is not affected by the production rate—e.g., depreciation, taxes and insurance. General works expense is also usually assumed to be unaffected by the production rate. Cost per unit of production decreases with an increase in the units produced.

• Semivariable, where the annual cost decreases somewhat as production decreases, but not in direct proportion. For example, annual repair and maintenance costs tend to be lower at low production rates, but will be substantial even at zero production. In general, the cost per unit for semivariable items tends to increase as production goes down, but not in true inverse proportion to the production rate.

The "mix" of costs falling into these three categories determines the slope and shape of the production-return curve. It is an important project characteristic to be weighed when deciding whether or not the investment should be made.

Raw Material and Fuel Cost

The added manufacturing cost form has spaces in which are listed the raw materials and fuels required. The periodic makeup of losses of such items as catalyst and electrolytic-cell mercury is included as raw material. If the initial charges of these materials will have a long useful life (more than about one year), they are nondepreciable capital investment.

Opposite each item should be indicated the units in which the material is usually purchased, the unit cost and the usage in units of raw material or fuel per unit of production.

For raw materials and fuels not currently being purchased, approximate prices can be obtained.*

Costs of materials and fuels are entered on the cost sheet on a plant-delivered basis. Many price quotations will be fob. vendor's plant, or some other basing point, and freight to the plant must be added. Assistance in obtaining freight information can be obtained from the company's traffic department or the railroads that

*From *Oil, Paint, and Drug Reporter* or special costs issues of *Chemical and Engineering News.*

ADDED MANUFACTURING COST FOR ADDED PRODUCTION
ANNUAL BASIS

	Appropriation Number			Supp.
	06	41	4120	--

Title

    50 T/D Kiln for New Ores

Plant Production, principal product				(1)			
Added Tons of Roast			T / Year		17,500	13,130	8,750
Added, in % of added capacity					100 %	75 %	50 %
Total production rate						Same	
Total rate, in % of new total capacity					%	%	%

Raw Material and Fuel Cost

Type	Units	Unit Cost	Usage T/ Roast			
"B" Ore	Tons	$20.00	1.32	$ 461,000	$ 347,000	$ 231,000
Na O H, Anhyd	100 lb.	$ 5.00	3.00	262,000	197,000	131,000
Limestone	Tons	$ 5.00	0.15	13,100	9,800	6,600
Natural Gas	M.c.f.	$ 0.25	52.0	227,000	171,000	113,500
Subtotal, Raw Materials and Fuel				$ 963,100	$ 724,800	$ 482,100

Utilities

Type	Units	Unit Cost	Usage / T			
Electricity	KWH	$0.02	50	17,500	13,100	8,800
Steam	-			-	-	-
Water	M.Gal.	$0.03	10	5,300	3,900	2,600
Subtotal, Utilities				$ 22,800	$ 17,000	$ 11,400

Labor

	Units	Unit Cost	Usage / Year (2)			
Operating	M.H.	$3.00	35,000	105,000	105,000	105,000
Repair 60% of 5.2% of $370,000 (3)				11,500	11,500	11,500
Supervision 20% of operating labor				21,000	21,000	21,000
Indirect Payroll Cost @ 20 % H., 20 % S.				27,400	27,400	27,400
Subtotal, Labor				$ 164,900	$ 164,900	$ 164,900

Supplies & Miscellaneous

Operating Supplies 6% of operating labor			6,300	6,300	6,300
Repair Supplies 40% of 5.2% of $370,000 (3)			7,700	7,700	7,700
Laboratory	-----------		-	-	-
Royalties & Rentals $1.00 /T			17,500	17,500	17,500
	-----------		-	-	-
Contingencies, 3 % of non-Fixed costs			35,400	28,400	20,700
Subtotal, Supplies & Misc.			$ 66,900	$ 59,900	$ 52,200

Fixed Costs

General Works Expense, inc. Taxes & Ins. (4)	72,900	72,900	72,900
Depreciation 8 % of $352,000	28,100	28,100	28,100
Subtotal, Fixed Costs	$ 104,000	$ 104,000	$ 104,000

Loading, Packing & Shipping

Materials, inc. 3 % cont. $.80 /T	14,000	10,500	7,000
Labor 1900 hr. at $2.70	5,100	5,100	5,100
G.W.E. 45 % IP 20 % Sup. 20 % Cont. 3 % of labor	5,400	5,400	5,400
Subtotal, Loading, Packing & Shipping	$ 24,500	$ 21,000	$ 17,500

Total Added Manufacturing Cost	$1,346,000	$1,092,000	$ 826,000

Notes (1) 350 operating days at 50 T/D, 15 days kiln re-bricking (2) Four men, around the clock.
(3) Repair cost essentially constant for all production levels. (4) 3.0% of $370,000 = $11,100,
plus 45% of $137,500 = $61,800, total $72,900.

Prepared by		Date	Checked by		Date
J.W.H.		6/28/59	R.F.E.		6/29/59

will be concerned (unless this is a "hush-hush" project). Perry's "Chemical Business Handbook,"[4] Section 9, is a good source of basic information on transportation costs in general.

Transferred raw materials and fuels are those obtained by transfer from another company division, which is operated as a separate entity. Transfer price is the going market price plus freight and other transportation costs from the nearest production point, whether competitive or company-owned, to the unit receiving the materials.

Prices of raw materials, whether transferred or purchased from outsiders, will vary considerably depending upon the form in which they are received. Delivered prices are quite different, for example, for 50% caustic in tank cars as compared with anhydrous flake caustic in steel drums. The estimator should make sure that the most economical type has been selected.

Fuel prices vary not only with the type of fuel being purchased, but also with the number of Btu.'s per unit. Other fuel characteristics (and the corresponding prices) are dictated by the ability of the process unit to utilize them. Special heating equipment is required, for example, when the heavier types of fuel oil are burned. A separate analysis of fuel economy is sometimes required in order to select the proper type and quality of fuel.

Utilities

The company's standard manufacturing costs can be used to price utilities provided by a "works," and for utility increments of reasonable size at other operating plants. When large increments are required, the extra utility costs are determined from rate schedules of local public utilities or from a separate new estimated manufacturing cost sheet for the utility itself.

It is especially important to analyze utility costs for large increments, since unit costs often decrease substantially as demand goes up. When projects involve small increments of utility consumption, the cost per unit can usually be assumed constant.

Labor

Operating Labor—The best means of establishing operating-labor requirements is to prepare a complete manning table showing additions or reductions made necessary by the project. This should be reviewed with personnel familiar with operating this type of unit, preferably the ones who will later run the plant. The manning table should be established for operating at 100% capacity, with notes as to reductions that can be made if the plant is consistently operated at lower rates. Round-the-clock and week-end coverage must be included when necessary. From the manning table, the annual man-hour requirements can be computed.

When the project is a novel one, some idea of labor requirements can be obtained from publications such as T. B. Haines' "Direct Operating Labor Requirement for Chemical Processes."[1]

Having established the number of man hours of operating labor required annually for the project, the operating labor cost is determined by multiplying man hours by the average hourly rate for the geographical area in which the new unit is to be located. Operating labor is usually a semivariable expense. In some cases, however, and especially in continuous-process plants, there can be no reduction in the operating force as production is reduced.

Repair Labor—This includes the hourly wages and average premium pay (if any) of men engaged in the maintenance and repair of the added installations. It does not include general repair supervision, depreciation on repair or service equipment, maintenance shop overhead or indirect payroll costs. These are taken care of elsewhere.

Repair labor is listed separately from repair supplies, for easier computation of indirect payroll costs. Usually, about 60% of total repair cost is labor.

Standard manufacturing cost tables for the company's existing plants are the best source of repair cost information. Very few usable data have been published. A good start in this direction has been made in two articles published in *Chemical Engineering* in 1959.[2,3] In these articles, the following formulas were developed:

Pulp and paper plants:
(Source not stated)

$$M = 0.009 \sum_{0}^{n} (I_n \times t_n) + 149,000$$

Coke plants
(J. A. Williams, Wyandotte)

$$M = 0.004 \sum_{0}^{n} (I_n \times t_n) - 83,091$$

Cement plants
(J. A. Williams, Wyandotte)

$$M = 0.011 \sum_{0}^{n} (I_n \times t_n) - 300,000$$

Silicone products plants:
(R. Cutoff, Gen. Elect.)

$$M = 0.068 \sum_{0}^{n} (I_n \times t_n/L_n)$$

Electrolytic plants:
(Author's notes, not included in articles)

$$M = 0.083 \sum_{0}^{n} (I_n \times t_n/L_n)$$

In the above formulas, M is the annual maintenance expense, I_n is the investment in a plant unit, t_n is the years that the unit has been installed and L_n is the estimated total life of the unit. Σ is the summation (between limits of 0 and n) for all of the units being maintained. For computing the average annual repair cost, t_n can be assumed at half the estimated life of the unit. It is obvious that the first three formulas are valid only for large units because of the sizable constants in them.

When a unit is operating at 75% capacity, the rule-of-thumb is that repair cost may be roughly 85% of the repair cost at full capacity. When operating at half capacity, the repair cost may be about 75 percent of the repair cost at full capacity.

Supervision — Additional operating and maintenance supervisors and foremen made necessary by the project are included in this category. Managers, supervisors and foremen for general plant services, such as storerooms and accounting are not included, however. They are charged to general works expense.

For quick estimates, and in the absence of other information, supervision can be assumed to average between 10 and 25% of the operating labor. Higher figures are for complex processes or a multiplicity of small units.

Supervision is considered to be a fixed expense at the level required for 100% capacity operation, except when for some special reason plans are to operate at less than capacity for a substantial length of time.

Indirect Payroll Costs—These include costs to the company of pensions, paid vacations, group insurance, disability pay, Social Security (company's portion), unemployment taxes, and so forth.

For existing plants, percentage figures are available in the company's standard cost tables. A check should be made to be sure all indirect costs of employing labor are included, but none duplicated.

In Canada for 1959, Industries Relations Counselors Service of Toronto found the following fringe cost percentages prevailed:

```
All manufacturing...........................22.8%
Paper.......................................17.6%
Chemicals...................................23.0%
Petroleum...................................24.9%
Iron and steel..............................20.2%
```

Comparisons made by the service showed no major difference of pattern between these figures and corresponding ones for the U. S.

In the absence of other information, indirect payroll costs can be estimated for new projects as being equivalent to 20% to 25% of the labor items to which they apply, that is operating labor, repair labor and supervision. Indirect payroll costs for loading, packing and shipping are handled separately.

Supplies & Miscellaneous

Operating Supplies—Included here are such things as wiping cloths, lubricating oil, instrument charts, and any other items used in the normal operation of the plant, excepting those listed as raw materials or packaging materials.

Standard costs for this item should be used where available. For new plants, it may be assumed to be about 6% of the operating labor, although consideration should be given to any special process conditions

that will tend to increase this percent of labor cost.

Repair Supplies—This includes the various supplies such as emery cloth, nuts and bolts, gaskets, welding rod, oxygen, acetylene, and so forth, used in conjunction with repair work. It will usually be found that this is about two thirds of the repair labor, or 40% of total repair costs as previously discussed under "repair labor."

Laboratory and Other Service Costs—Depending on company practice and the type of project, manufacturing costs may include special service charges by other company units such as laboratories.

Laboratory costs in percent of operating labor for typical plants manufacturing the following materials:

```
Soda ash....................................3%
Silicates...................................3%
Chromates...................................7%
Chemical caustic...........................10%
Chlorine-caustic...........................13%
```

Typical laboratory costs for the more complex processes are usually between 10% and 20% of operating labor. Other service charges will of course depend on the type and value of service rendered.

Royalties and Rentals—Royalty and other patent-right payments required for the added production should be included in this section of the manufacturing cost tabulation if they are paid periodically or on the basis of units of production. Under these conditions, they are operating expense. Single-sum patent or know-how payments, however, are usually capital investment but not subject to depreciation. In some cases, they are amortized. The company's tax experts should be consulted in cases of doubt as to how to handle specific situations.

Rental payments are dependent on the value, number and type of items rented. The company's purchasing department can provide information on current rental rates for many items.

Contingencies—Since it is virtually impossible to be sure that all items have been included in the manufacturing cost, a contingency is added to the estimate. This should be small for installations like those the company currently has in operation, larger for novel processes in their initial development stages. Until further information is available, the following rates are suggested:

• Installations similar to those currently used by the company, for which standard costs are available. 1%
• Installations common to the industry, for which reliable data are available......................2%
• Novel installations that have been completely developed and pilot planted....................3%
• Novel installations that are in the development stage. 5%

The above percentages should be applied to the sub-total of all the items of the preceding manufacturing costs. This excluded depreciation and general works

expense, which already include a contingency since they are based on investment including reserve.

 Fixed Costs

General Works Expense—GWE is the cost of plant overhead items such as insurance, property taxes, plant management, plant engineering, general maintenance supervision, plant technical staff, personnel services, plant protection, maintenance shops, tool rooms, storerooms, accounting, purchasing, traffic, and other related items. It also includes the depreciation, operation and maintenance costs of railroads, roads, sewers, parking lots, cafeterias and other general facilities serving the process units.

General works expense percentages are difficult to establish and compare, even between plants of the same company. Comparable results can be obtained only if accounting definitions are precise and subject to uniform interpretation. Best results are obtained by an analysis of company records, segregating the items making up general works expense into two groups, one chiefly dependent on investment, the other chiefly dependent on labor.

Some rough order-or-magnitude figures follow:

	Investment Factor	Labor Factor
Soda-ash plants.............	1.5%	45%
Power plants................	1.8%	75%
Electrochemical plants........	2.5%	45%
Cement plants..............	3.0%	50%
Silicate and chromium plants...	4.0%	46%

As an example, take a case as follows:

1. Direct investment plus transferred facilities less forced retirements plus reserve.*.........$1,000,000.

2. Increase in annual labor (operating, repair, supervision, loading, packing and shipping).....$30,000.

Assuming that the project is an electrochemical plant, the general works expense would be computed as follows:

1. Investment portion, $1,000,000 at 2.5% = $25,000.
2. Labor portion, $30,000 at 45% = 13,500.
3. Total general works expense   38,500.

In some cases, it will be necessary to make a detailed study of present general-works-expense charges prior to installation of a new unit in order to make a realistic estimate of its effect. The company's accounting department can provide assistance with such studies. This check is especially advisable for modernizations and replacements, or when the formula indicates the general works expense will decrease.

Depreciation—Many process equipment installations can be assumed to have a depreciation rate of 8%, indicating an effective life expectancy of 12½ yr. Installation subject to rapid deterioration or obsolescence, however, should be depreciated at rates up to 20%, or possibly higher. Justification of these higher rates should be included in the appropriations request.

Complete process plants, including the usual proportion of long-life improvements, such as foundations, roads, railroads, sewers, buildings, and so forth, may have a depreciation rate on the order of 6⅔%, representing a life of 15 yr.

The depreciation rate is applied to direct investment plus transferred items at book value plus reserve; depreciation for supporting utility investment is included as a part of the utility costs.

Depletion—If applicable, depletion is one of the service costs of a project. This is an annual allowance for eventual exhaustion of some natural resources such as a salt deposit, clay pit, limestone quarry, gas well or oil well. From an income tax standpoint, the maximum depletion allowance is set by law. Currently, in the U. S. it is 27½% of the annual production of material, valued at market in the first usable, salable form into which it is processed.

Amortization—Another service cost, amortization resembles depreciation and depletion. It is applicable to any intangible asset that has a limited legal life, such as a patent, license or concession. A charge is made each year against the operation so that at the expiration of the legal life of the asset, the entire cost has been written off.

Loading, Packing & Shipping

Standard manufacturing costs from other similar company operations, suitably modified to present requirements, are the best source of data for estimating this expense. Fortunately, these costs are fairly easy to measure and most companies have cost-control systems that provide actual cost information which can be used in these estimates.

Published data are available in Section 8, pp. 25-31 of Perry's "Chemical Business Handbook,"[4] and in "Chemical Engineering Cost Estimation."[5]

Indirect payroll costs, general works expense, supervision, and contingencies for loading, packing and shipping are handled as indicated on the form. General works expense and indirect payroll percentages are added, then multiplied successively by the supervision and contingency percentages.

Loading, packing and shipping is segregated in this way to best show the effect on costs of various forms and containers for shipment. The total of the preceding items on the sheet is bulk cost. When loading, packing and shipping costs are added, we have our "total added manufacturing cost."

References

1. Haines, T. B., *Chem. Eng., Progr.*, **53**, 556-62 (1957).
2. Robbins, M. D., *Chem. Eng.*, Feb. 9, 1959, pp. 140-2.
3. Robbins, M. D., *Chem. Eng.*, July 13, 1959, pp. 172-8.
4. Perry, J. H., Ed., "Chemical Business Handbook," McGraw-Hill, New York, 1954.
5. Aries, R. S. and R. D. Newton, "Chemical Engineering Cost Estimation," McGraw-Hill, New York, 1955.

* The exact definitions of these terms will appear in an article by Mr. Hackney in a future issue of *Chemical Engineering*.

Chart Gives Operating Labor for Various Plant Capacities

If you know your labor requirements at one capacity, this handy chart will give an estimate of process operating labor for any other capacity.

FRANCIS P. O'CONNELL, *Asst. Prof., Dept. of Chemical Engineering*
University of Detroit, Detroit, Mich.

It is often necessary to make an economic study of a process at various capacities. Calculating a rigorous labor-requirement estimate at each capacity can be a very time-consuming job. The chart (left) can be used to overcome this difficulty, since it quickly gives process operating labor requirements.

This chart is based on earlier work by Chilton[1] and Wessel.[2] Chilton gathered economic data from the literature on man-hr. of operating labor per ton of product for 52 different processes. Wessel, using the data, obtained typical plant capacities for these processes from the literature, especially from the book "Industrial Chemicals."[3] He developed a correlation by plotting operating labor in man-hr. per ton of product per technological step, against plant capacity in tons/day. The technological steps were defined as unit operations, unit processes, and any other technical steps necessary for operation of the process.

Straight-Line Plot

A plot of these points on logarithmic coordinates produced a straight line with a negative slope of 0.75. Spread of data at a given capacity was over 100%, but most points were well within that band of data.

It may be noted that plant capacities in this correlation ranged from 2 to 1,500 tons/day and covered a tremendous variety of processes. Some of these had a high- or low-labor content. For example, processing of fluidized material with a high degree of automation typified a low-labor content per ton of product, and some batch process requiring a lot of operator handling typified a high-labor content per ton of product. Processes that gave points above the line drawn were high labor, and those processes that gave points well beneath the line were low labor. However,

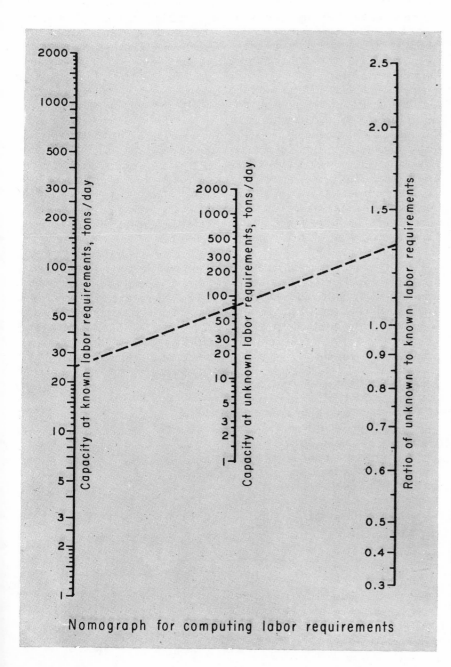

Nomograph for computing labor requirements

most processes were well within these two extremes.

Wessel also constructed extreme upper and lower lines of approximately the same slope to represent processes of high- and low-labor contents.

Scaling Up or Down

These concepts for estimating the labor requirements of a given process at a given capacity can be used for going from the labor requirements at one capacity to those at another capacity for the same process.

Frequently a chemical engineer has fairly accurate data on the labor requirements at one capacity, as a result of operating experience or a careful study. But in making an economic study covering various capacities, he finds it inconvenient, or too time-consuming, to make a complete study at each. The chart we have developed should be useful in such situations.

There is one question that does arise, however, when using this concept for jumping from one capacity to another. "Typical" plant sizes were used in Wessel's plot and the capacities an engineer wants to investigate might not be typical in the industry in question. However, it seems reasonable to assume that capacities outside the typical range will fit the Wessel correlation—because plant size is governed by economic circumstances that have little relation to the operating labor requirements of the process.

One-Fourth Rule

To demonstrate application of these concepts, let:

P_k = Plant capacity for which labor requirements are known, tons/day.

P_i = Any plant capacity for which labor requirements are unknown, tons/day.

M_k = Known labor requirements at plant capacity, P_k, man-hr./day.

M_i = Unknown labor requirements at plant capacity, P_i, man-hr./day.

From the Wessel[2] plot:

$$(\text{man-hr.}) \ (\text{tons})^{-1} \ (\text{steps})^{-1} = K \ (\text{tons/day})^{-0.75}$$

where K is a proportionality constant.

Since we are only considering one process, the number of technological steps can be assumed constant and included in the proportionality constant:

$$(\text{man-hr.}) \ (\text{tons})^{-1} = K' \ (\text{tons/day})^{-0.75}$$

and multiplying both sides by (tons/day) we get

$$(\text{man-hr./day}) = K' \ (\text{tons/day})^{0.25}$$

This extension of Wessel's relationship has been indicated by Vilbrandt and Dryden.[4]

Now we can write:

$$M_k = K' \ (P_k)^{0.25}$$
$$M_i = K' \ (P_i)^{0.25}$$
$$\text{and} \ (M_i/M_k) = (P_i/P_k)^{0.25}$$

The relationship has been summarized in the alignment chart. Capacities vary from 1 to 2,000 tons/day, a range that covers most typical processes.

To use the chart, align capacity of known labor requirement, P_k, with capacity of unknown labor requirement, P_i, using a straight edge. Then read on the third scale the ratio of unknown manpower requirements, M_i, to the known manpower requirements, M_k. The unknown manpower requirements, M_i, are readily obtained by multiplying known manpower requirements by this ratio.

This relationship could be called the "one-fourth rule" by analogy to the "six-tenths rule" governing the relationship between plant capacity and fixed investment.

Work an Example

A chemical engineer is making an economic study of a process for the manufacture of fluorine at various capacities to compare manufacturing cost and return on investment. From previous data, he knows that the operating-labor requirements for a capacity of 25 tons/day is 290 man-hr./day. He now wishes to compare this capacity with 75 tons/day and 10 tons/day. Thus, we have:

$$P_k = 25 \text{ tons/day}$$
$$M_k = 290 \text{ man-hr./day}$$
$$P_1 = 75 \text{ tons/day}$$
$$P_2 = 10 \text{ tons/day}$$

By aligning P_k and P_1 on the chart :

$$(M_1/M_k) = 1.32$$

and $M_1 = M_k \ (M_1/M_k) = 290 \times 1.32 = 383$ man-hr./day.

By aligning P_k and P_2 on the chart, we get:

$$(M_2/M_k) = 0.79$$

and $M_2 = M_k \ (M_2/M_k) = 290 \times 0.79 = 229$ man-hr./day.

Some Limitations

Before using this method, the engineer should make sure that the labor-content nature of the process does not change with capacity, since this could lead to serious error. Wessel's plot shows that labor requirements can vary over 100% because of the nature of the process.

It is suggested that to obtain best accuracy by this method the labor requirements not be extrapolated for more than a 10 to 20-fold range of capacity. In a multicapacity economic study, the range of capacities is not usually more than 10-fold.

It may be interesting to note that the two greatest sources of error in Wessel's original method cancel out in the method developed here. One of these errors lies in judging the nature of the process—whether it is an inherently low-labor or high-labor process. Any error due to judging this intangible factor is eliminated, because we are treating only one process at a given time.

The other error is in arriving at the number of technological steps in the process. This is also eliminated, because at different capacities we are dealing with the same steps, whatever the number may be.

References

1. Chilton, C. H., "Process Labor Requirements," *Chem. Eng.*, Feb. 1951, p. 151.
2. Wessel, H. E., "New Graph Correlates Operating Labor Data for Chemical Processes," *Chem. Eng.*, July 1952, pp. 209-210.
3. Faith, W. L., Keyes, D. B., Clark, R. L., "Industrial Chemicals," John Wiley, New York, 1950.
4. Vilbrandt, F. C., Dryden, C. E., "Chemical Engineering Plant Design," McGraw-Hill, New York, 1959, pp. 234-235.

Do You Know What You're Getting For Your Repair Dollar?

Because of the fundamental link between operations and maintenance, being over or under budget is not necessarily a meaningful measure of maintenance effectiveness. Here's how misleading statistics and false economies can be avoided.

R. J. STRATMEYER, *Monsanto Chemical Co.*

Assume for a moment that you're the plant manager of a moderate size process plant, and that your budget for repairs last year was $3 million; but you spent only $2 million. How can you evaluate the results? Possibly, this year's costs may soar to $4,500,000. Perhaps production losses due to insufficient maintenance have cost the company more than the $1 million saved in maintenance. How can you therefore be sure that the saving is real?

The answers to this question and the general question of "What is your plant getting for its repair dollars?" lies in the Repair Transaction. This article is intended to explain that transaction and to show how it can be used as a year-round yardstick in a variety of situations.

Reduced to its most basic terms the Repair Transaction means this: each plant spends repair dollars to buy (1) equipment on-stream time and (2) an acceptable condition for nonproducing assets such as building and roads.

Let's see what this involves in some typical plant situations.

For our first example, let's say your plant is close to obsolescence. Predictions are that within three years it will be torn down and replaced by newer facilities. Costs are critical; but for the next two years, sales are expected to be close to the plant's demonstrated capacity. What you want to buy is a level of maintenance that is adequate for dependable operation of the processing equipment. You are not trying to preserve equipment beyond the three years. Neither are you trying to preserve buildings, roads, or other fixed assets beyond that period.

In this situation, you would replace a corroded tank or heat exchanger, but you would not replace structural steel unless it would be near to failure within the three-year period. Very little painting, and little if any tuck-pointing would be done. The problem is to translate this into the terms of the Repair Transaction.

Or, your plant might be selected as a showplace unit. Possibly your products are such that plant tours by customers are part of the selling strategy. There's a difference between a very high level of maintenance

for production and preservation of assets, and showplace maintenance. For example, the structural steel in a department may be perfectly protected by a paint film, yet the paint may be seriously discolored by fumes. In a showplace plant, a repair job would be necessary.

Or take insulation. Whereas in a normal plant, it is economical to tolerate some noncritical breaks in insulation or some uninsulated portions of line until enough work builds up to warrant a general overhaul, in a showplace plant any defect must be promptly corrected and then repainted. The expense of these small jobs, with the accompanying travel and setup time, can be great. Again it is necessary to put this in terms of the Repair Transaction.

Neither Obsolescent Nor Showplace

Your plant, like the majority, may fall somewhere between the showplace and obsolescent extremes. The plant is to operate indefinitely. It is necessary to maintain equipment on stream 96% of the time, and preserve assets from deterioration. While incipient paint failures in this plant would be spot repaired, the discolored paint we have mentioned would be left untouched.

A fourth and also typical case is the combination of the obsolescent, showplace, and in-between approaches in the various departments of a single plant.

Key to the Repair Transaction

These widely different situations all fall within the Repair Transaction. In each case, the key is (1) to determine the level of maintenance that will yield the needed production rate for process equipment, and (2) to determine the needed level of maintenance for other (nonprocess) assets.

In each of the cases we have considered, these two statements define what plant management is trying to buy for its repair dollars.

In other words, dollars of repair cost buy production on-stream time and preservation of other assets. We can proceed from this statement once we define its elements in a more precise, quantitative manner:

• Dollars of repair cost is defined as total dollars (labor, materials, and overhead) spent to restore facilities to mechanical operating condition, or to functional condition in the case of assets such as buildings. Expense items to revise a department for a process modification, and revisions for safety or for any reason other than restoring facilities to operating conditions, are not included.

• Production on-stream time refers to the time the equipment is available to operate from a repair point of view. For example, if during a given month

of 720 hours, a department is down 24 hours for repairs, the repair availability is 96.67%. Obviously it may be down for other reasons in addition to repairs, but this does not affect the repair availability.

In a large plant, the downtime index provides an adequate measure of the state of production equipment. For example, in one department a fractionating tower installed in 1957 may have had essentially no downtime until it failed in 1962. In another department, a converter installed in 1952 may have had very little downtime until its failure in 1963. But if a plant includes sufficient departments and sufficient equipment, this random pattern of failure will balance out. If, in such a plant, inadequate repairs are made, the frequency of failure will increase and this will be quickly reflected in the downtime index.

In a smaller plant, where the number of equipment items is such that a single failure will throw results for the month, and possibly for the year, out of balance, it is necessary that such major items be excluded from the calculation of downtime index. The condition of these items should be made a separate category under the "preservation of assets" heading. Many of the newer techniques of nondestructive testing can be used to define the state of equipment items.

• Preservation of assets refers to all assets not directly related to producing a product. Masonry, roofs, service facilities, floors, roads, sewers (usually), some instruments, lighting, painting systems, much insulation, etc., are included here.

Measures (Indexes) of the Repair Transaction

Any measure of repair costs must include a measure of each of the above three elements, otherwise it can be misleading and dangerous to use.

For example, in the typical plant we talked about earlier, if repair cost and production on-stream time were measured, but preservation of other assets were not, many of these assets could be deteriorating. Major repairs or high expenditures might be needed in the near future. In this situation, management might feel things were going well from its indexes of cost and on-stream time. But, in reality, costs were low only because some vital work was not being done.

This situation is the inevitable result of inadequate measurement. The effort and money go into the areas that are measured. Those that are not measured tend to be neglected.

The need, then, is for reliable indexes to each of the three elements of the repair transaction.

Measures of Repair Cost and Downtime

The index to repair cost is total repair dollars, i.e., direct labor, materials, replacement equipment, overhead, and contract-repair costs.

A typical problem: Last year, Plant A spent $3 million on repairs. This year, Department 123 is expected to operate for only 6 months. Department 890 is now under construction, and is expected to go on-stream July 15. Mechanics' wage rates will go up 2¢ on April 1, and another 5¢ on August 1, and so on.

Preparing a budget from last year's repair costs—Table I

BUDGET

Repair costs—1963		$3,000,000
Less: Dismantled departments	200,000	
Reduced schedule	50,000	
	250,000	250,000
		2,750,000
Add: New departments	400,000	
Increased schedules	20,000	
	420,000	420,000
		3,170,000
Additional insulation	30,000	
Additional painting	50,000	
	80,000	80,000
		3,250,000

BUDGETED IMPROVEMENT*

Cost increases: Wages and salaries		50,000
Materials and supplies		35,000
Miscellaneous		15,000
		$100,000

* To be absorbed through increased efficiency.

There are salary increases, new fringe benefits, changes in production rates, and other factors that further complicate the picture. In addition, management has decided that to adequately preserve assets, the level of maintenance should be raised.

Sound confusing? It really isn't. To begin, take the $3 million for last year as a base. Add to this the estimated amount needed to maintain new facilities, or to maintain departments that will operate longer periods next year. Subtract dismantled departments or departments operating less than last year. Decide how much of the cost increases (wages, salaries, fringes, materials, etc.) should be absorbed through improvement. A simple goal and one that has proven to be a practical in a variety of situations, is to absorb all wage, salary, and materials increases.

A second decision must be made regarding the amount of money to be spent upgrading assets. This figure will be added to the budget mentioned above, as shown in Table I.

Once a budget has been constructed in this manner, it becomes the index, or the control for the cost portion of the repair transaction.

As for the measure of repair downtime, this has already been defined as the percent of total scheduled hours a department could not operate because of repairs.

In a multi-department plant, some means must be found to average the downtimes of each department

to obtain a plant average. The average operating labor charge is a convenient way to do this, as shown in Table II.

Measure of Other Assets

The primary measure of the condition of assets such as roofs, roads, masonry, sash and some insulation is inspection.

At first, this may seem like an expensive way of obtaining an index. However, the inspection does more than just yield an index—it is an important part of a planned maintenance program. The goal of such a program is to economically control the maintenance of assets at the proper level. The goal of an index is to measure if this is being done.

Obtaining a plant average for downtime—Table II

Department No.	A Production Labor Cost, $/Hr.	B Repair Downtime, Hr.	C Labor Lost Due to Repair Downtime (A×B), $	D Production Labor Cost, $/Mo.
847	21	20	420	15,120
423	80	40	3,200	57,600
167	15	15	225	10,800
Total			3,845	83,520

Average repair downtime = C/D = 4.6%

Preparing specifications for this year's work, based on inspection of building—Table III

SCOPE OF WORK

The proposed work consists of replacing portion of parapet wall, sealing column caps and tuck-pointing Bldg. BD.

To be included are all incidental functions such as the temporary supporting of pipes, wires and equipment, and the return of temporarily supported items to permanent supports as soon as feasible.

All work is to be done in such a manner as not to create a safety hazard to plant people or interfere with the production facilities housed in this building. Also, contractor shall make every reasonable effort to protect manufacturing facilities (electrical especially) during the performance of the work.

GENERAL SPECIFICATIONS

1. Remove and dispose of present brick parapet (Bays 1, 2 and 3, shown on photos 1, 2 and 3).
2. Sand blast and prime existing 8-in. I-beams above windows in Bays 1, 2, 3, 4 and 5.
3. Have plates and angles fabricated per sketch No. 7.
4. Weld plate and angle to I-beam with continuous weld (Bays 1, 2, 3, 4 and 5).
5. Touch prime welded area and paint two coats per attached paint spec. 502.
6. Replace brick parapet and seal roof flashing to new wall.
7. Remove, clean and reinstall copings around entire building.
8. Tuck-point entire Bldg. BD.
9. Seal caps on each column.

An example may make this clear. Exterior masonry is an important phase of building maintenance. If tuck-pointing is neglected, it can lead to deterioration of the walls of a building. It can also cause building steel to rust to the point that major repairs or rebuilding may be necessary. Of course, the deterioration is normally slow, so that if a repair budget is tight, and performance is judged by downtime or service to production, tuckpointing is one element of repairs that one is tempted to put off until deterioration has advanced to serious levels.

An inspection begins with a listing of all the assets to be covered. As the inspector makes his rounds, he lists detailed information regarding the state of repair of all masonry in the plant.

From this inspection, he then develops a program. The program lists all the jobs to be done during the current year, and can be in the form of specifications (Table III). Listed separately are all the jobs to be done for the foreseeable future—usually a 3 to 5-year period.

The least critical job to be done in 1964 compared with the most critical planned for in 1965 provides a visual definition of the minimum acceptable level of maintenance for tuck-pointing this year. The same technique applies to most of the other types of assets included in this category.

With this approach, it is possible to decide how much to upgrade or downgrade assets in a given year, and to estimate the decision's effect on repair costs.

In the tuck-pointing example, it might be decided that some of the work planned for 1965 is too critical to delay. It is a simple matter to move specific jobs forward to 1964 so as to raise the level of maintenance for this year. The estimated costs of these jobs should then be added to the budget.

Usefulness of the Repair Transaction

The repair transaction concept is a tool that can be used for most plant situations. It provides sound measures of performance from foremen to top management, avoids the pitfalls of partial measurement, and keeps continual but reasonable pressure on costs.

Returning to the case of the plant manager who only spend $2 million of his $3 million repair budget, the Repair Transaction tells him what he bought for the $3 million.

It may tell him that under the greatly reduced expenditures, downtime rose from 4% to 8%. Some sales were lost. Some material had to be purchased on the outside. The cost of the added downtime was estimated at $580,000. An inspection of assets revealed general deterioration. Layers of paint were lost, some building repairs had become both larger and more critical, insulation was in poor shape, freeze-ups had occurred and steam use had increased. The total cost of these losses was $920,000. So, in this case, the $2,000,000 bought $1,500,000 less than the $3,000,000 expenditure would have.

The concept of the repair transaction could put this plant manager well on the way to making a better maintenance buy this year.

1450

1966

How Accounting Helps the Chemical Engineer to Control Product Costs

In this first article of a three-part series, the author describes development and use of bookkeeping systems, balance sheets, income statements, process and product costs.

EDWARD B. NITCHIE, *Ernst & Ernst*

Do you hear from your production and engineering management . . . "What am I supposed to do with all these reports? Read them? Burn them? File them? Or what?" . . . "Are we producing figures for accounting or products for customers?" . . . "How can the company justify all those clerks just to keep books? Are we getting more clerks than production workers?" . . . "How could anyone in his right mind justify that big computer and turn down our proposal for the new fluidized catalyst process?" . . . "How can they hold me to a budget when my production varies from one million to two and a half million gallons a month?" . . . What do they mean 'burden'; if the Accounting Department is a burden why not get rid of it?" . . . "What is the use of all this accounting?"

These or equivalent questions have been heard many times. Yet a reasonable man such as an engineer or manager realizes that there must be many valuable uses for the rising tide of accounting data. The tide of paperwork may seem about to engulf us, yet the most prosperous and successful companies are increasing their uses of accounting.

Since we live and function in a competitive society, we know that businesses either prosper or die according to the wisdom of their management decisions. Much of our reasoning must be based on the facts of economic life, for survival in a competitive world. For certain levels of quality and of economic and sound benefits to consumers, employees and the public, there must be economic returns to justify investment in, and future growth of the business. The levels established must be comparable to, or even better than, competition to insure continued growth. Future existence of the business may even be at stake. The controls that keep the business profitable must be reasonably operable by humans. Only in this way can we be sure that

the business may grow competitively, and improve, and that we may grow with it.

Reasons for Accounting Systems

Some of the functions of accounting are well known to those in plant operations, usually on the basis of: "somebody has to keep the books!" We know that control and improvement of profits must be based on some measurement of profits or income to owners. Also, there are well-known legal requirements to report profits and property valuations for tax purposes. The means of deriving this information are more or less mysterious, depending on our degree of contact with them. The results of an accounting system, in many cases, are known only to a few top executives, which situation adds to the mystery.

Our purpose here is to eliminate some of this mystery and thereby answer the question: "What's the use of accounting?", and to enable chemical engineers to evaluate and use accounting systems. To develop these concepts, we will show how the growth of a business and its technology almost of necessity includes the growth of useful tools and techniques for accounting and management controls. To do this, we will use a fictitious chemical company as an example. [Some background material of how accounting grew from simple bookkeeping into management control systems may be helpful and is given on the next page.]

How the Exemplary Chemical Company Grew

During the last fifty years many chemical companies have grown in size and maturity along with the growth of new accounting techniques and other management controls. Rather than detail the history of many companies to tell the story of accounting, we will take one example. While it may be a typical com-

The Growth of Accounting Systems

Accounting is probably as old as man's use of numbers. The basic concept of one person's responsibility or accountability to another in exchange transactions also involves the concept of "how much." From counting on fingers (digits), we have progressed to digital computers with each step of the process enlarging the scope and value of accounting while keeping the basic concept.

Records of receipts and expenditures have been found going back to the early Greek and Roman civilizations. While these records seem primitive in comparison with present complex and modern accounting records, they represent in a simple basic manner the accountability for money transactions between people, or groups, or governing bodies.

During the second half of the thirteenth century, double entry bookkeeping came into use in several cities of Italy. This medieval double entry accounting was far superior to the statistical record keeping of earlier days. Generally, the double entry system may be compared to the balancing of a chemical equation, with entries into both "sides" of the books of account in numerical balance. The equation expressing this balancing of the books indicates that the sum of the debits (charges) equals the sum of the credits, or:

$$\Sigma \, Dr = \Sigma \, Cr$$

This principle of balancing accounts has several advantages over the simple listings used until that time, especially as a means of providing the accuracy of entries and financial results by balancing the equation. An orderly and somewhat uniform basis for recording and analying financial information was developed with double entry bookkeeping, and an integrated system for evaluating both profit and loss data and the ownership values of assets, liabilities and equities.

The development of the improved type of reporting also created a demand for refinements in accounting to recognize prepaid and accrued items, such as assets paid for in one year, but with a useful life of many years. Entries of this type date to the fifteenth century.

During this period, shipping ventures were common, wherein several individuals would pool their capital in a common cause for a single sailing. These were commonly handled as individual ventures with participants sharing profits or losses on an individual trip only, and records were generally kept on a similar basis. Thus, there was no real continuity to the partnership relationship. During the seventeenth century, stock companies (the forerunners of our present day corporations) were established. These stock companies issued exchangeable shares and provided a continuity of life of the company.

Cost accounting, and the accounting for operations of manufacturing as a function separate from the other operations of the company, was introduced in the latter part of the nineteenth century. At first such records were maintained entirely separately from the general accounting records, and served only as memorandum type records for rough analysis of factory operations. These were the first answers to: "How much does it cost to make this product?" and "How much does it cost to operate the factory?"

In the 1880's the theory of combining cost and factory records with the general accounting records was introduced. During the next several decades the art of cost accounting flourished, since it provided some reasonable historical measurement, and the means for analysis of the vital manufacturing operations and of product costs.

During the early part of the twentieth century, the concept of "Standard Costs" was introduced. This concept bases the evaluation of operations on a standard of what costs should be. Acceptance of this standard cost accounting concept was limited until around 1930, at which time violent changes in economic conditions created extreme interest in the cost and profit relationship. From then to the present time, the concept of standard cost accounting has flourished. Many varying and valuable techniques have been developed with standard costs in order to aid management in the recording, analysis and control of manufacturing costs.

---

pany, no two people are apt to agree on what or who is typical—so it will be a fictitious company. This therefore is an example of what could have happened to one chemical company.

Since our fictitious company is to be an example we have named it "The Exemplary Chemical Company", and of course it makes fictitious chemicals that we will call "exemplary" chemicals. Their fictitious, or exemplary, founder is named, of course, John Doe. He has been made Chairman of the Board, where he succeeds his father who helped him to found the company in 1915. At that time he had just graduated from college with a degree in chemistry. His father had migrated from Europe a few years before, and had brought with him both a wealth of information and a wealth of money derived from the European chemical industry. This was at a time that the supply of dyestuffs and other fine chemicals was cut off in the United States because of the war in Europe.

John Doe and his father decided to start their own chemical company, which in the beginning was to be only a research laboratory. The research laboratory phase lasted only briefly as they were quickly able to develop some products that were greatly in demand at the time. The next stage was to expand rapidly into pilot-plant equipment in which they could manufacture small batches. The output of this pilot plant was quickly sold, and at a profit that rapidly paid off in investment and equipment. The opportunities for expansion were self-evident, but the problems proved to be many.

Up to this time, the family had been able to operate the entire business without outside help. Father and son ran the laboratory and pilot plant and did what little selling was involved. Their technical records were in excellent shape, but their financial records were somewhat too simple to tell them whether they were making or losing money. The wives had helped by keeping the books, and the records consisted primarily of a checkbook. A statement at the end of the year

fortunately indicated that they had more money at the end of the year than at the start.

In a decision to expand to full-scale production, it was necessary to know somewhat more about potential profits, and the costs of capitalization for the expansion. At the same time, it was also necessary to know more of the engineering and the construction costs involved. It had been found that in the press of running both a laboratory and a pilot plant, it was necessary to hire some help to do the heavier chores, to leave the professionally oriented founders of the business for their important contribution to his technology.

While considering all of these factors it was decided to add suffcent staff, at least to cope with and evaluate the future possibilities and problems. The first to be added was a chemical engineer who was to assist with the design, construction and eventually with the operation of the first plant for Exemplary Chemical Co. The next step was to consider how much it would cost to make the projected quantities of exemplary chemicals, and from this to estimate how much profit could be expected to pay for the investment in the proposed plant. To this end an accountant was hired, and his first duties were to project costs and profits based on what little information was available. With the help of these two professional men, Exemplary Chemical Co. was able to make its first step toward being the large manufacturer of a wide variety of chemicals that it is today. A marketing specialist who knew the needs of the chemical market thoroughly was next added to take over some of the task of selling, and competent staff members were also added as they became necessary.

As all of these functions became larger, the responsibilities were spread to competent staff members as they became necessary. Eventually "functional vice presidents" were appointed, with large departments specializing in the control of each of these major functions. However, it is noted that the first two professional requirements in the business as it began to expand were engineering and accounting.

What's in A Name?

The original question that started us on this quest to find the uses of accounting was "How can we get engineers and accounts to understand each other or even to talk the same language?" Shakespeare may have answered part of the question when he says that "A rose by any other name would smell as sweet" but it leaves open a bigger question of how anyone would know what you are talking about if a rose is called something else. If you referred to a rose as a "perfume fractionating tower", would anyone else know what you were talking about? The Bible tells us that during the building of the Tower of Babel, the Lord said "Come let us go down and there confuse their language that they may not understand one another's speech." The influence of the builders of the Tower of Babel is evidently still with us today! This problem of communication is both fundamental and complex. Unless we can derive some common languages, such as

the common languages that enable computers to "talk" to each other, it will be almost impossible to communicate from one technology to another.

While there are many dictionaries and glossaries that give accounting terms, it is not expected that an engineer with an interest only in the uses of accounting would be consulting these. Nor would the definitions in such references necessarily be of any significance unless the application were immediately evident. The following definition is a good general description of the science of accounting, but it can only give the haziest notion of what accounting is all about, unless you pursue the matter a lot further:

"**Accounting—a. The art or system of making or stating accounts; the body of scientific principles underlying the keeping and explantion of business accounts. b. The application of such principles in practice; an instance of applying such principles in practice; the act of stating accounts. See COST ACCOUNTING.**
Syn. ACCOUNTING, BOOKKEEPING. BOOKKEEPING deals with the work of making proper entries and proper balances. ACCOUNTING explains the results furnished by the bookkeeper and draws the necessary inferences as to the condition and conduct of the business . . ." (Webster's New International Dictionary of the English language, Second ed., Unabridged, G. & C. Merriam Co.)

A similar definition might be helpful in telling the accountant what a chemical engineer does for a living, but it would take many definitions, and hours of wading through them, to learn much about either engineering or accounting this way.

Another problem that we find in looking up formal definitions is the fact that definitions depend on each other, because we must describe a word with other words. For each word that we use in describing other words, there are many possible meanings. Here we will quote from a mathematician who evidently had given some thought to the meanings of words:

"When I use a word," Humpty Dumpty said in a rather scornful tone, "it means just what I choose it to mean—neither more nor less."

"The question is," said Alice, "Whether you can make words mean so many different things."

"The question is," said Humpty Dumpty, "which is

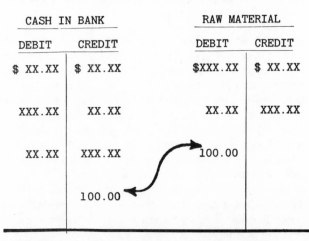

CASH IN BANK		RAW MATERIAL	
DEBIT	CREDIT	DEBIT	CREDIT
$ XX.XX	$ XX.XX	$XXX.XX	$ XX.XX
XXX.XX	XX.XX	XX.XX	XXX.XX
XX.XX	XXX.XX	100.00	
	100.00		

ACCOUNTS record flow of funds for materials—Fig. 1

to be master—that's all." ["Through the Looking-Glass," (Chapter 6), Lewis Carroll]

So, if we are to master the art of communication between accountants and engineers, we must at least find out what accountants mean when they use a word. This also has proved somewhat difficult because different accountants mean different things with the same word.

The best we can do is to tell the meaning of accounting as it was developed by the Exemplary Chemical Co. This may not exactly duplicate your experience, any more than your design of chemical processing equipment will exactly duplicate that of your competitors. It will at least give an example of how accounting was used in a fictitious company, and therefore how accounting might be used in an industry.

Fact and Fiction

Although the Exemplary Chemical Co. is fictitious, we will find in the following story of its growth many facts that are applicable to many chemical businesses.

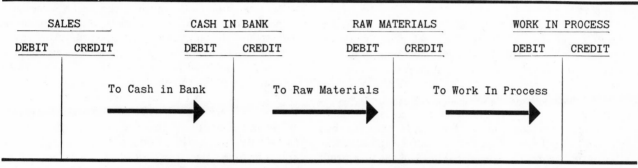

FLOW DIAGRAM traces transactions through several accounts—Fig. 2

BALANCE SHEET

THE EXEMPLARY CHEMICAL COMPANY

September 30, 1930

CURRENT ASSETS:				CURRENT LIABILITIES:			
Cash			$ 30,000	Accounts Payable			$ 90,000
Accounts Receivables, Less Reserve For Doubtful Accounts			45,000	Income Taxes			40,000
Inventories:					TOTAL CURRENT LIABILITIES		130,000
Finished Goods		$ 60,000					
Work in Process		15,000		OTHER LIABILITIES AND DEFERRED CREDITS:			
Raw Materials, Containers and Supplies		40,000	115,000	Deferred Income Taxes			25,000
				Miscellaneous			10,000
Prepaid Expenses			10,000				
TOTAL CURRENT ASSETS			$200,000	NET WORTH			
				Paid in Surplus		$110,000	
OTHER ASSETS:				Retained Earnings		55,000	165,000
Plants, Equipment and Facilities at Cost		$150,000					
Less: Accumulated Depreciation and Amortization		30,000	120,000				
Deferred Charges and Other Assets Patents, Licenses, and Miscellaneous Assets			10,000				
TOTAL ASSETS			$330,000	TOTAL LIABILITIES AND SHAREOWNERS EQUITY			$330,000

BALANCE SHEET establishes status of business at a particular time—Fig. 3

In fact, the fiction represents truths derived from many businesses in the expectation that we may find our own somewhere therein.

In the beginning Exemplary Chemical Co. had neither much of a bookkeeping system, nor an accounting system. At the present, they have a highly sophisticated and very much mechanized accounting information and management control system that has contributed greatly to their profits and their success in business. In telling the story of "EXCO" [as we shall hereafter call them], we may gain some insight into the nature of accounting. Possibly we may also help to answer our initial question "What's the use of accounting?"

What accounting system EXCO did have in the beginning consisted primarily of a check book, plus some memoranda or statistical records that John Doe kept himself. At least he did have a separate checking account for the company. At the end of the year, he found that his bank balance was higher than at the beginning of the year. This increase in bank balance was true even considering the funds that he had pumped into the business from his other investments. His natural conclusion was that he had made a profit for the year, and that he and his father had been successful in launching the business.

The next step was to consider the further expansion of the business, and to see if they would be better off to invest in enough equipment to make small production lots instead of continuing the laboratory and semi-pilot plant operation that they were conducting. A young engineer was hired to help them design a small operation, since their extensive knowledge of chemistry did not extend into the engineering information necessary to build or operate plants. Along with design work, the engineer was to estimate both costs of equipment and costs of operation, in order to determine the possible profitability of a small-scale production operation.

It was soon evident that there was not enough financial information available to predict anything about the business! About all they could tell from their simple bookkeeping records was that the then current market for dyestuffs and fine chemicals was high enough so that there was a "good margin" over the direct cost of the raw materials that went into the process. But, raw materials are only one of many costs in a chemical business. Was the "good margin" enough to insure profits with larger scale operations and a much larger investment?

These founders of the business had some limited understanding of the general principles of finance and accounting, and at this point they decided to seek expert advice. They called in a representative of a public accounting firm, who subsequently joined the company. He recently retired as their Vice President for Finance, after a successful career in helping to make their growing business profitable.

The accountant's first step was to set up a bookkeeping system. He went back to the start of the company as a partnership. From the records of the partnership, and of the individual partners, he was able to reconstruct a rough statement of their previous year's operations. In this process, he discovered many calculations that were missing from the partner's previous calculations of profit. For one thing neither of the partners had paid himself a salary, but each was content to take his share of the "profits" at the end of the year. Fortunately there was enough family income from other investments so that this was not an interim problem to the partners, but if allowances were made for a normal salary for each of the men, it was found that the "profit" was reduced considerably.

Another omission in this original calculation of profit was the lack of any differentiation between the cash expenditures for materials and supplies, and investments in equipment and inventory. During the previous year, a considerable amount of raw materials had been purchased in a market that was growing increasingly tighter. Much of this material had already been paid for, and this was considered an expense for that year—even though most of the material was still on hand at the end of the year. An allowance for the remaining value—this year-end inventory—considerably increased the profit showing.

There were similar adjustments that had to be made because purchase of the laboratory and some of the pilot-plant equipment had been shown as "expenses" during the year even though the equipment was expected to have a useful life of many years. A large gas-fired kettle had been purchased from other funds and not charged to the operation of the business at all. Much of the equipment had many years of useful life if the process or similar processes continued. It was pointed out to the partners that this durable equipment with a fairly high value should be charged

STATEMENT OF INCOME

THE EXEMPLARY CHEMICAL COMPANY

Year Ended September 30, 1930

INCOME		
Net Sales		$2,440,000
Other Income		10,000
		$2,450,000
COSTS AND EXPENSES		
Cost of Product Sold	$1,862,500	
Selling and Advertising Expense	125,000	
Administrative and General Expense	237,500	
Research Process Development Expense	162,500	2,387,500
INCOME BEFORE FEDERAL INCOME TAXES		$ 62,500
PROVISION FOR FEDERAL INCOME TAXES		7,500
NET INCOME		$ 55,000

INCOME STATEMENT traces monies and costs—Fig. 4

THE EXEMPLARY CHEMICAL COMPANY
MANUFACTURING EXPENSE

PROCESS DEPARTMENT — PLASTICS

MONTH OF September 30, 1930

	DOLLAR AMOUNT	% OF TOTAL	% OF RAW MATERIALS
RAW MATERIALS AND			
CHEMICAL SUPPLIES	$ 98,553	63.5%	100.0%
PAYROLL	26,695	17.2	27.1
SUB TOTAL	$125,248	80.7%	127.1%
SERVICES, SUPPLIES AND			
OTHERS			
Payroll Taxes and Insurance	$2,018	1.3%	2.0%
Stores and Supplies	4,811	3.1	4.9
Containers and Packing Supplies	4,501	2.9	4.6
Maintenance	4,035	2.6	4.2
Fuel Oil	1,862	1.2	1.9
Electricity Purchased	931	0.6	0.9
Depreciation	10,864	7.0	11.0
Property Taxes	931	0.6	0.9
SUB TOTAL	$ 29,953	19.3%	30.4%
TOTAL MANUFACTURING			
EXPENSE	$155,201	100.0%	157.5%
TOTAL CONVERSION COST	$ 56,648	36.5%	57.5%

MANUFACTURING expense statement—Fig. 5

THE EXEMPLARY CHEMICAL COMPANY

PRODUCT COST REPORT ON Plastic #13

MONTH OF September 30, 1930

MATERIALS:	COST	COST PER UNIT
Sodium Hydroxide	$ 1,914	$0.38
Toluol	6,430	1.29
Acetic Anhydride	6,201	1.24
Sodium Cyanide	765	0.15
TOTAL RAW MATERIALS	15,310	3.06
PACKAGING MATERIAL	700	0.14
MISCELLANEOUS SUPPLIES	748	0.15
TOTAL MATERIAL COSTS	$16,758	$3.35
PAYROLLS	4,150	0.83
FACTORY COSTS	3,203	0.64
TOTAL COSTS	$24,111	$4.82
TOTAL PRODUCTION (Units = Gallons)		5,000

COST REPORT allocates expenditures for a product—**Fig. 6**

---

off to expenses (as "depreciation") over the years of its useful life, rather than charged all at once in the year in which it is purchased. To charge all the large equipment investment expenses in the first year of its useful life, however, or to refrain from charging it to expense at all, would make it very difficult if not impossible to determine how much profit, if any, had been made in any one year.

When these various adjustments and evaluations had been completed, the profit picture for the first year of operation was considerably different than the original estimates. There was still an indication of very satisfactory profitability, which, taken together with the engineer's estimates on the equipment costs and operating costs, pointed to expansion as the next logical move.

Developing the Accounting System

At this time, it was decided that it would be logical to set up a bookkeeping and accounting system that would provide the necessary information. An accountant was put on the payroll to work with the engineer in developing cost estimates. He was also to set up the accounting system, and subsequently to provide all of the operations of the business with the numerical and dollar data that were needed for management decisions.

The accountant's first chore was to set up a double entry bookkeeping system. This system had the double advantage of providing a complete record of the flow of funds, and of also providing the means of checking accuracy by balancing out the accounts. The procedure was simple, and consisted of a sequence of accounts through which the funds might be considered to flow. Each time a transaction involving the flow of materials or inventory (as represented by funds) takes place, two entries are made. One of these will "credit" the account from which the funds are taken, and the second will "charge" or "debit" the account to which funds are flowing. Thus in a very simple set of accounts such as was started at EXCO in their early years, the "Cash in the Bank" and the "Raw Material" accounts might list the entries for a $100 purchase of a chemical as shown in Fig. 1.

In describing an accounting system, a simplified notation is used to show this double-entry bookkeeping. A flow diagram for this transaction, in a greatly simplified form, starting with the funds coming in from the sales of finished products and going as far as getting the raw materials into the kettles might look as shown in Fig. 2.

At the end of each period, whether it be month or quarter or year, these accounts are balanced out. The net difference between the two sides of each account, or set of accounts represents the amount theoretically left in that account. This may be cash in the bank, or inventory, or raw materials, or remaining value of equipment (undepreciated). It is easy (at least, theoretically) to check the cash in the bank with the bank balance to verify the accuracy of that account. Similarly, it is possible, although considerably more

difficult, to verify raw materials and other inventories by an actual physical count and an evaluation of the value of the material at the time of the inventory.

With all the verifications, the balances in each major account are summarized in a "balance sheet." This shows the net position of the company at the time of balancing out, and summarizes all of the totals remaining in the main series of accounts to give the static picture of where the business stands at a particular time.

This balance sheet gives substance to the accounting equation, "the sum of the debits (charges) = the sum of the credits." In the form of an equation (which, like a chemical equation, must balance), this would be:

$$\Sigma\ Dr = \Sigma\ Cr$$

A typical balance sheet for EXCO for one of their early years is given in Fig. 3. Note that the balance sheet shows not only what was on hand in the way of assets that could be either cash or turned into cash, but also what was owed under the term of "liabilities." By netting out these various assets and liabilities, it was possible to come up with a good evaluation of the worth of the business at that time. This is shown as "Net Worth."

By comparing the changes in the balance sheet from year to year, it is possible to get some picture of what happened in the business. EXCO soon found that this comparison of balance sheets, which represented two static pictures of the business a year apart, did not adequately represent the flow of funds.

A dynamic picture of flow was needed to give them a more complete view of what had happened in the business during the past period. The next financial tool to be used was the "Profit and Loss Statement." This operating statement (as it was also called) indicated the sources of funds for the operation of the business, as well as where the funds went. For the early stages of the business, when the operation was fairly simple, all that was required was a simple profit and loss statement. An example of an early EXCO statement is shown as Fig. 4.

The operating statement shows where the money came from and where it went during the period, and thereby indicates the flow. It shows enough detail for a very simple operation, or can be expanded for a total company operation where it is backed up with more-detailed statements.

As EXCO expanded and diversified into other types of chemical operations, it was found increasingly necessary to know both the cost to operate a process, and the cost to produce a product. With only one product being produced by each process, this was fairly simple. But as the product mix became more complex, it was frequently found that the same piece of equipment was producing a variety of products having widely different requirements for time, temperature, pressure and control. All of these variables made significant differences in the costs of the various products that were produced.

Careful batch records had always been kept for each batch or each product, primarily for quality control purposes. It was found that some additions to the batch records could provide the much needed cost information in simple form. At the end of each month, the cost information was assembled for all of the batches of each product that had been run in each department (or each process, or operation) during the month. This provided a much simpler method for management control than the scanning of hundreds of batch records.

Also the supervision of each department, process, or operation received a report showing the total cost of operations for the previous month. Together these two sets of data provided the necessary information for evaluating the cost and profitability of each process, and of each product that was made. Examples of these reports for EXCO for a typical operating period are shown in Figs. 5 and 6.

This summary information in both the process-cost and product-cost reports is analyzed each month by the department foremen, division supervisors and plant superintendents to determine whether or not the costs were as low as they should be; and if not, to determine what should be done about it. Also the product-cost reports are used by the sales and marketing departments at EXCO to determine the profitability of product lines and products, and as guides to the pricing of new products.

These fairly simple information systems were adequate for the early years of the business, and did provide controls. However, as the business grew more complex and more competitive, EXCO found need for sharper tools for control, for reduction of costs, and for financial management. During the last 40 years, they have developed these relatively simple accounting techniques into a complex information and control system, including the means to measure, analyze and improve cost and profit performance.

Meet the Author

Edward B. Nitchie is manager in charge of trade association and statistical services for Ernst & Ernst, New York. His work has included projects in cost reduction, cost accounting, statistical controls, industrial engineering and management for a wide range of chemical process industries. Mr. Nitchie has a B.S. in industrial engineering from Yale University and is a licensed engineer in Connecticut. He is a member of the American Statistical Assn., American Institute of Industrial Engineers, Armed Forces Management Assn., and Sigma Xi.

Key Concepts for This Article

Active (8)	Passive (9)	Purpose Of (4)
Interpreting	Accounting*	Analyzing
Using	Systems*	Costs*
Developing		Expenditures*
		Profits*

(Words in bold are role indicators; numbers correspond to EJC-AIChE information retrieval system. Asterisks mark key concepts suggested for indexing. Others are added to improve reading as an abstract. Indexing is described in *Chem. Eng.*, Oct. 11, 1965, p. 187.)

In this second article of a three-part series, chemical engineers can learn how to understand and use depreciation techniques, cash flow and discounted cash flow accounting, standard costs, break-even analysis, direct costing for processes and products.

Accounting Data and Methods Help Control Costs and Evaluate Profits

EDWARD B. NITCHIE, *Ernst & Ernst*

The growth of an accounting system as it developed parallel to the growth of a fictitious chemical company, Exemplary Chemical Co. (EXCO), was described in our first article (*Chem. Eng.*, Dec. 19, 1966, pp. 95–101, or p. 256 of this volume).

In its early years, the company had to establish bookkeeping systems, balance sheets, income statements, and process and product costs.

Now EXCO faces new problems and must apply new techniques in a competitive environment—techniques that emphasize depreciation methods and standard costs.

By the 1930's and increasingly into the present day, the U.S. chemical industry has become larger, more progressive, and more competitive. Chemical processes now become obsolete in a few years rather than a few decades, as was true when EXCO was founded. Frequently, while chemical engineers are designing a plant, competitors are developing a new process that will produce better-quality products, or the same products at lower cost. All of these factors have combined to make it far more difficult to predict and realize profits in the chemical industry.

As the industry has developed new processes to meet competition, it has also developed new financial and economic tools. Competition is as intense in the financial aspects of the business as it is in process design and process operations. The new techniques include many that are based on accounting information and accounting methods.

EXCO's management soon found that it needed some new tools to stay competitive, and has been applying or trying all of the practical new methods. It has also been experimenting with the mathematical techniques of "operations research," which are based on good numerical information supplied by the firm's accounting department.

In the previous article, the only asset transaction that we discussed for EXCO was inventory. It should

all reasonable charges to expense should be made as soon as possible to ensure future profits. Also, a fourth factor entered into this depreciation accounting: Since replacement cost of equipment was going up rapidly, a long-term writeoff of the investment in equipment became increasingly unrealistic. The most-realistic approach for cost-accounting purposes is normally a straight-line writeoff during the useful life of the equipment. This, however, may conflict with tax and financial considerations.

The U. S. Internal Revenue Service also recognized the advantages to companies who increase their depreciation rates and charge off the capital investments as rapidly as feasible to postpone tax payments while still retaining adequate company profits.

At the same time, it was evident that postponement of tax payments was disadvantageous to the collection of taxes. Therefore, the accountant's ability to set depreciation rates for the optimum benefit to the company has been somewhat limited by the regulations of the Internal Revenue Service. Included in these regulations have been rules and guidelines on the number of years in which a capital investment can be written off to expense.

Accelerated Depreciation

With the above limitations in mind, several methods have been developed to improve the accounting for depreciation in order to: postpone the payment of taxes; recognize all the factors in the depreciation equation; and improve future profits.

Two of these techniques permitted under liberalized laws and regulations of 1954 are accepted means for accelerating the depreciation rate during the early years of the life of equipment or other capital assets. Neither of the methods changes the total depreciation that can be charged off as an expense, nor will they alter the depreciation life of assets. However, they do recognize the generally accepted concept that the value of any piece of equipment or machinery (such as

that the straight-line depreciation method of writing off the cost of equipment purchases would give a reasonable indication of the actual cost of operations.

Modern depreciation accounting practice (one of the first of the new techniques) was influenced not only by the special needs of the chemical processing industry, with its rapid obsolescence of process equipment, but especially by the effect of increasingly higher taxes on the income of the corporation. As tax rates increased, it became imperative to charge off as much of the investment account as possible, or as legally allowable, to have the optimum expense charges for reducing taxable profit. A second factor that influenced the handling of this depreciation account was the rapid obsolescence of chemical processes. While a chemical process and its equipment may at one time have had a useful life of ten years, this might currently be five years, or less, if the process were predictably obsolete in that time.

Still a third factor in depreciation accounting was the conservative attitude of EXCO's management— all reasonable charges to expense should be made as soon as possible to ensure future profits. Also, a fourth factor entered into this depreciation accounting: Since replacement cost of equipment was going up rapidly, a long-term writeoff of the investment in equipment became increasingly unrealistic. The most-realistic approach for cost-accounting purposes is normally a straight-line writeoff during the useful life of the equipment. This, however, may conflict with tax and financial considerations.

The U. S. Internal Revenue Service also recognized the advantages to companies who increase their depreciation rates and charge off the capital investments as rapidly as feasible to postpone tax payments while still retaining adequate company profits.

At the same time, it was evident that postponement of tax payments was disadvantageous to the collection of taxes. Therefore, the accountant's ability to set depreciation rates for the optimum benefit to the company has been somewhat limited by the regulations of the Internal Revenue Service. Included in these regulations have been rules and guidelines on the number of years in which a capital investment can be written off to expense.

Accelerated Depreciation

With the above limitations in mind, several methods have been developed to improve the accounting for depreciation in order to: postpone the payment of taxes; recognize all the factors in the depreciation equation; and improve future profits.

Two of these techniques permitted under liberalized laws and regulations of 1954 are accepted means for accelerating the depreciation rate during the early years of the life of equipment or other capital assets. Neither of the methods changes the total depreciation that can be charged off as an expense, nor will they alter the depreciation life of assets. However, they do recognize the generally accepted concept that the value of any piece of equipment or machinery (such as

your own automobile) declines more rapidly in the first few years of its useful life because of such factors as maintenance cost and rapid obsolescence.

In addition, accelerated depreciation methods serve to reduce tax payments during the early life of the asset, although increasing them during the latter part of the asset life. Hence the total tax payment is virtually equivalent to that under the straight-line depreciation method. The advantage of having the funds available at an earlier date can be considerable for a growing company such as EXCO.

The first of these two methods of accelerated depreciation is the "Multiple-Declining Balance Method." Its most frequent application is as the "Double-Declining Balance" technique, in which the depreciation rate is double that which would be required by a straight-line depreciation procedure. The double rate is applied only to the undepreciated asset balance, or remaining value, each year.

Thus for a piece of equipment whose useful life is considered to be ten years, the annual rate of depreciation would be 20% of the remaining balance. If the value of the equipment were $100,000, the first year's depreciation would be $20,000, leaving a balance of $80,000 to be depreciated the second year. The second year's depreciation would be 20% of $80,000 or $16,000, leaving an undepreciated balance of $64,000, and so on up to ten years. It is obvious that if this formula is continued for ten years, or any number of years, there will always be a remaining balance to be depreciated. That is, the balance would reach zero only at infinity. It is customary practice to switch to straight-line depreciation somewhere during the life of the asset or, as an alternative, to write off the entire remaining balance during the last year of the depreciation life.

The other commonly used technique that provides

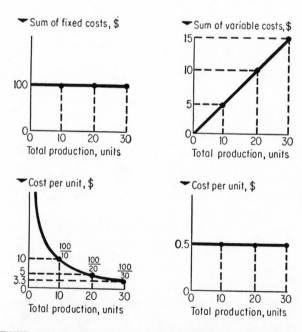

FIXED costs may mean total costs or unit costs—Fig. 1

more-rapid depreciation during the early life of the asset is known as the "sum of the years digits" method. In this method, a fractional amount of depreciation is taken each year and is based on the years of life of the asset. The numerator of this fraction is the number of years of remaining depreciation life, and the denominator is the sum of the digits of all the years in the originally established life. Thus for ten-year depreciation, the first year's depreciation will be:

$$\frac{10}{1+2+3+4+5+6+7+8+9+10} = \frac{10}{55}$$

The second year would be 9/55, the third year would be 8/55, and so onto the final year when the undepreciated value of the asset will reach zero.

Either of these formulas has a very similar effect on the depreciation rate, and either achieves the desired effects of a more-rapid writeoff and a postponement of income taxes. Each has minor advantages over the other. EXCO, after much debate, arrived at the optimum combination of rates at the time the asset accounts were put on its new computer.

Cash Flow Accounting

During the development of its accounting system, EXCO found that accounting only for cash (as it had done in the first year of its existence) was completely inadequate to represent the financial progress and position of the business. Many accrual and allocation methods had to be developed to represent current expenditures or commitments to subsequent years, and likewise to represent previous expenditures such as capital investment and plant equipment for future years. Depreciation accounting was the one biggest difference between cash accounting and accrual accounting, although there were other minor adjustments that had to be made at each accounting period.

In recent years, with the growing complexity of the business, EXCO has found it more necessary than ever to have some supplemental information on the actual cash flow in the business. Changes in depreciation accounting, for one, have altered the financial reporting requirements. Another factor in the company's consideration of cash flow accounting was the necessity for providing cash funds, not only for operating expenses but also for further capital investment.

The major difference between cash and accrual accounting is the inclusion of plant and equipment depreciation. To get a reasonable representation of cash flow, EXCO was able to adjust the appropriate accounts by the amount of depreciation charged each year. From this, and some other minor adjustments because of accruals, it set up a cash flow budget and accounting system for use by the corporate executives in their determination of financial policy.

Discounted Cash Flow

This first use of cash flow accounting proved to be an important financial control, and soon led EXCO to investigate a new series of techniques, grouped under the heading of "discounted cash flow." These techniques were designed to evaluate and compare future capital investments, especially those for new equipment. EXCO's engineers and accountants worked together to develop these techniques. In this way, new-equipment proposals are evaluated at all levels, and all phases of development, by methods that provide the best-known means of comparing alternative uses of available capital.

The general concept of discounted cash flow recognizes that money received today is worth more than money received in the future. It is assumed that money can be earning from the time of receipt by either internal or external investment, and that the difference between the value of present and future money can be estimated on a probable rate of earning.

If two projects have the same amount of yield on total investment, but one will return this in five years and the other in ten years, it is possible to compare them by discounting each year's future earnings at a theoretical compound interest for that period of time. The rate of return for the five-year project is higher than the rate of return for the ten-year project, especially when the shorter payout time permits reinvestment of these funds at a much earlier date.

EXCO found that many variations are possible for evaluating investment projects on the basis of the present value of future income, or conversely on the future value of present income. Calculation of the annual rate of return can be converted into the value of an annuity that would pay out at that rate. This annuity value gives a measure of the value of the asset that can be compared to investments at other rates of return, regardless of the size or life of the project.

Another use of the principle is the development of a "Profitability Index" or "Desirability Index" based on an expected rate of return that can be used to compare projects of different sizes and durations, for a rough measure of the relative desirability of the investment in the project.

Since all of these techniques depend on ratios of income to capital investment, it is evident that they must be based on cash flow only. In evaluating an investment, the depreciation calculation can only be used to determine its effect on income taxes, and then only to the extent that it determines net income for tax calculations. Net-income calculations by which we evaluate the investment payout cannot logically include the writeoff of the investment itself as a cost calculation.

In investigating the uses of discounted-cash-flow analysis, EXCO found some problems and pitfalls in each of the techniques. The main problem was that of forecasting the future, especially the rate of return on investment in plant and equipment, and the actual expenditure for investment. The company eventually adopted a combination of the techniques in its engineering and accounting operations that eliminated as many as possible of the unknown variables and provided more than one basis for comparing the probable values of equipment investments.

Standard Costs

By watching both the long-term trends in costs as well as the month-to-month and year-to-year deviations from previous averages or norms, much value has been derived. Exploring the reasons for these changes and trends, and putting the findings to use, should result in reduced costs and improved profits.

Especially in the manufacture of a chemical, it frequently happens that equipment failure or operation error causes unusually high costs. If the equipment

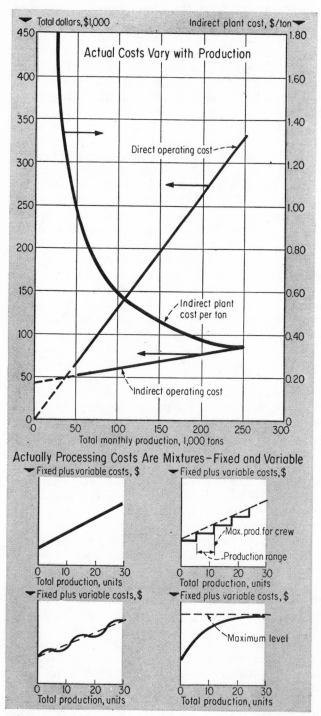

Actually Processing Costs Are Mixtures—Fixed and Variable

TOTAL costs have fixed and variable components—Fig. 2

is used for other products as well, then higher than normal costs on a specific product could be pure happenstance. Knowing the actual cost of a product, then, is of limited value to management. It is more important to know the cause of unusual costs, to correct the problems, and to improve the operations.

EXCO soon decided that the questions raised by historical cost-accounting information would be much easier to answer if the norms could be established as "standards" for each of the products and for each of the processes. Such standards should eliminate the need for estimating the effect of a change of product mix, or a change in the process, or any unusual event that might have significantly affected the cost in a previous period.

When EXCO first investigated the potentials for standard cost accounting in the chemical business, it looked at cost systems developed in other types of business, primarily in metal-goods manufacturing. While there was obviously a great deal of applicability to chemical processing, there was a wide variation in definitions of what a standard cost system is.

"Standard costs" would sometimes roughly and loosely describe a system in which a rough "bogey" cost is set by judgment of past history as some measure of control and performance. From these rough controls ("bogeys," "pars," or "norms"), some standard cost systems have developed into complete systems in which every item of labor, material, and overhead cost is fully measured, using engineered standards. All of these systems were to some extent "standard cost systems." It was up to EXCO to determine the optimum value of the precision, accuracy and completeness of the system that it was to establish.

Instead of starting with only averages or historical norms, it was decided by EXCO that par for the course could be more firmly and finely established as standards by using the available engineering talent. Industrial engineers were called in to establish labor standards for each process. These techniques were found difficult but by no means impossible for the process operations, and were fairly simple for the more-mechanical operations such as packing or filling. The chemical engineers and the industrial engineers worked with the accountants to establish standard costs for each of the operations. The application or allocation of these process costs to each product was studied. Purchasing agents also supplied information on which standard material costs could be based. With this information, a complete standard cost system could be set up.

With a standard cost system, it now became possible to report each month, or even for each batch, on what the cost of the operation was in comparison with what it should have been. From these two costs, a percentage "performance factor" was derived:

$$P.F. = \frac{\$ \text{ Standard Cost}}{\$ \text{ Actual Cost}} \times 100$$

By using this standard and comparing it with the actual cost incurred in the accounting period, EXCO management obtained a "variance" that when analyzed

supplied the means for taking action. A dollar variance for the difference between the standard and actual costs was reported as a loss or gain vs. the standard:

$$\text{\$ Variance} = \text{\$ Standard Cost} - \text{\$ Actual Cost}$$

The dollar variances were reported separately for the direct costs of both materials and labor (that are variable with volume of operations), for the semi-variable indirect costs (that contain elements of both variable and fixed costs), and for the fixed (for a time period) or overhead costs. These variances provide a measure of the size of the problem, just as the performance factor shows relative performance.

During its study of standards, EXCO decided that there were many basic types of costs to be included in its standards, depending on how the costs actually applied. Later refinements were added for using standards to obtain performance factors and variances for the volume of production or operations; for the price of materials and material usage; and for the rate of pay as well as for the hours of work.

Variation of Costs

In its development of standard costs systems, EXCO discovered the many ways in which costs may vary. Since costs are the results of many factors, their variations are exceedingly complex. To help us visualize some of these, a few simple examples will serve to illustrate basic concepts that EXCO found important.

The terms "fixed" and "variable" costs appear in almost every discussion of costs. Yet frequently there is disagreement (even at EXCO) as to which is which.

As normally used, the term fixed cost indicates a cost in which total dollar cost remains fixed for a time period but cost per unit produced varies inversely with production. One of the best examples is that of depreciation costs of a factory building, which may continue at the same rate year after year regardless of production. If production is doubled, the depreciation cost per unit will be halved. This is shown in Fig. 1.

The normal definition of a variable cost is that the total dollar amount varies with production (or with sales, or with usage, or with the particular factor).

A primitive example would be that of a man, digging a ditch with a shovel, who can dig 10 ft. in 1 hr., 20 ft. in 2 hr., and so on. Taking just the cost of direct labor, it would cost ten times as much to dig 10 ft. of ditch as to dig 1 ft. So the cost is "variable," even though cost per foot remains constant.

Another example may be of a man who fills steel drums with a liquid chemical. In 1 hr., he can load 10 drums, in 2 hr., 20 drums, in 3 hr., 30 drums. The cost is "variable" in that the total direct cost of drum loading varies even though the unit cost per drum loaded is constant.

Very few costs are so simple. Usually they are combinations of many factors. The most frequently encountered combination is a fixed cost plus a variable cost, where part of the cost remains a fixed total

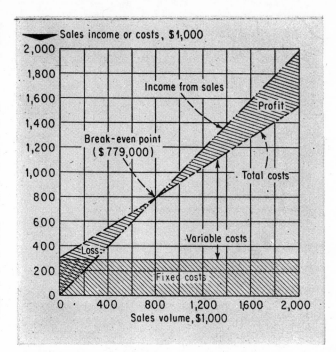

BREAK-EVEN analysis establishes profit levels—Fig. 3

amount regardless of production, and part varies directly as the units produced.

As a simple example, let us suppose that we now have a whole crew of men available to dig our ditch as required (and only as required), all at the standard rate of 10 ft./hr. The crew will be supervised by one man on a fixed salary who is on hand whether or not ditches are dug. Total daily cost for the ditch will be the total daily salary of the supervisor plus daily direct labor cost for the number of feet dug. Cost per unit will be these dollars divided by the number of feet dug. Graphically, this will be a combination (fixed and variable costs) of the two previous sets of curves, and is shown in Fig. 2.

In plant practice, there is a frequently noted tendency for costs to remain constant at certain levels of operation, and then to go up or down by steps as production varies. For example, a crew under the pressure of increasing requirements may turn out increasing amounts of work up to the point where no more load can be taken, or no more waste time can be eliminated, and an additional crew is required. This is repeated as each crew increases productivity up to its maximum, and more people are needed for an additional increase in production. Graphically, this is the step pattern. Total cost may also vary as straight steps, or as a curve, or level off slowly at a maximum. These patterns are diagrammed in Fig. 2.

These are but a few of the common examples of how costs may vary. Enough accurate statistics on actual costs will normally form patterns and trends of cost variation. For ease of use, standards are normally set on simple patterns of fixed and variable costs, aware that these patterns will never precisely fit all situations but will represent a model of most situations.

EXCO found that to analyze costs, or to form the

PRODUCT   Plastic # X 313

	Dollars	% Ratio To Sales
Sales Projection (S)	$1,000,000	100.0%
Variable Costs		
Direct Labor	$75,000	7.5%
Material	400,000	40.0
Variable Overhead	40,000	4.0
Other Variable Costs (Commissions, royalties, etc.)	100,000	10.0
TOTAL VARIABLE (DIRECT) (V)	$615,000	61.5%
MARGIN FOR FIXED COSTS (S-V)	$385,000	38.5%
Fixed Costs		
Manufacturing Overhead	150,000	15.0
Administration	100,000	10.0
Sales & Advertising	50,000	5.0
TOTAL FIXED COSTS (F)	$300,000	30.0%
PROFIT (LOSS) AT SALES VOLUME	$85,000	8.5%
BREAK-EVEN VOLUME		

$$\frac{300,000 \times 1,000,000}{385,000} = \qquad \$779,000 \qquad 77.9\%$$

Prepared   10–28–66   by

Approved   11–1–66   by

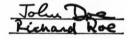

COMPUTATIONS and data for break-even analysis—Fig. 4

---

basis for more-accurate standards, statistical and graphical studies proved valuable tools for cost management in setting standards and forecasting trends.

Break-Even Analysis

This thorough analysis of the variation characteristics of costs and profits led EXCO management to use additional tools for analysis and control. The first of these was "Break-Even Point Analysis." Generally it was noted that material and direct labor costs vary directly with the volume of production or sales. Similarly, certain "departmental overhead" costs may vary with output over a longer period of time. And such costs as commissions or royalty payments obviously vary in proportion to the sales of a product.

On the other hand, there are some "fixed overhead" costs such as rental or depreciation of equipment that are fixed for a period of time. Some costs are combinations of variable and fixed costs, and may be broken down into these elements, or assumed to be predominantly either variable or fixed. Therefore if we add variable to fixed costs for each level of output, we would get Fig. 3, which graphically shows profit, loss

and a break-even point. Fig. 3 helps us to visualize that each unit sold above the break-even point contributes an increasingly large profit, and that below this point there is an increasingly large loss for each unit not sold. All necessary data and calculations are shown in Fig. 4 for determining the break-even volume for a product whose estimated annual sales are $1 million. The formula for calculating the break-even volume of sales can be expressed as:

$$B = FS/(S - V)$$

where B is the break-even volume of sales, S is sales volume, V represents variable costs, and F represents the fixed costs.

Uses of this formula are fairly evident in projecting the profitability of a new product or process. It can be used in preliminary engineering estimates or at the start of development of a new process in order to estimate economic feasibility. However, this analytical tool is not confined to engineering estimates but is also applied in planning for capital investments, controlling operating costs, and establishing sales quotas or budgets.

Direct Costing

From its use of break-even analysis, EXCO has recently developed some internal controls and analytical tools that it calls "direct" costing. The first part of the break-even analysis as shown in Fig. 4 provides data on a margin for fixed costs. Up to this point, the allocation of specific cost to a product or process can be reasonably precise and accurate. The allocation of fixed costs may be reasonably close, especially if there is a separate plant, department or process involved in only one product. However, the allocation of fixed or overhead costs in a multiproduct operation may have to be somewhat arbitrary and may be of doubtful accuracy.

From these considerations, EXCO developed a system for comparing the profit contribution of various product lines, based on the "direct" (or directly variable with volume) cost.

The calculation is the same as for deriving the "margin for fixed costs" in the break-even analysis that is shown in Fig. 4. We arrive at a margin of contribution to overhead and profit by deducting all direct (variable) costs from the sales dollars. Since we may not know the exact amount of applicable fixed costs, we have a method for comparing products and processes on their proportional contributions.

Key Concepts for This Article

Active (8)	Passive (9)	Purpose Of (4)
Interpreting	Accounting*	Analyzing
Using	Systems*	Costs*
Developing		Depreciation*
		Cash flow*
		Discounted cash flow*
		Breakeven point*
		Direct costing*
		Profits*

(Words in bold are role indicators; numbers correspond to EJC-AIChE information retrieval system. Asterisks mark key concepts suggested for indexing. Others are added to improve reading as an abstract. Indexing is described in *Chem. Eng.*, Oct. 11, 1965, p. 187.)

Modern Accounting Methods Supply Vital Cost Information For Decision Making

In this third and last article in a series, Exemplary Chemical Co. uses the latest budgeting techniques and develops an overall information system.

EDWARD B. NITCHIE, *Ernst & Ernst**

As Exemplary Chemical Co. (EXCO)† grew from a father-son operated research laboratory into a modern chemical company, it required more-sophisticated bookkeeping and accounting systems to provide necessary information for decision-making.

Now EXCO is a full-grown, modern, diversified company involved in the manufacture of a variety of chemical products. The company's management uses all the latest accounting techniques, including flexible budgeting and organized flow of data and information.

Flexible Budgets

To some old-time accountants, the term "flexible budgeting" is anathema, since budgeting was once considered only as a set of rigid limits on dollar expenditure. Yet progressive Exemplary Chemical Co. found very early in its study of fixed and variable costs that some better technique than rigid budgeting was needed for a host of cost items that are combinations of fixed and variable costs. The flexible budget recognizes the influence of variation in volume of production on fixed costs, when these costs are translated into unit costs.

To establish the flexible budget plan at EXCO required three levels of cost control. The first was a standard cost system for direct labor and direct material, which theoretically vary directly with output and are largely controllable by departmental foremen and supervisors.

At the other end of the scale of variability are those costs that are fixed for a period of time and are at least reasonably predictable for a year ahead. These costs are largely controlled by a "fixed budget."

For semivariable costs, a flexible budget system establishes dollar budgets for costs at a series of levels of operations. This recognizes the tendency for total costs—fixed plus variable—to rise as "steps" or along a sinuous curve.

The directly variable portions of a departmental budget are handled at EXCO as a percentage of standard direct labor, according to the established standard cost system. And those costs known to vary with different probable levels of production are so budgeted. Costs that are fixed for the year ahead, but still partly controllable, are budgeted for each level of operation.

Material Cost Controls

The relative magnitude of raw, and intermediate, material costs in a chemical process operation is so great that special attention was devoted to this aspect of cost control in the development of EXCO's standard cost and budget system. Effective controls were developed for the two prime factors in the cost of materials: usage and price.

The term usage obviously includes the highly important factor of yield. Material standards include both the theoretical, or formula-based, calculations of yield from input, and the expected yield for each process. It was found feasible and worthwhile in some cases to establish separate standards and performance factors to control both the possible overuse of certain materials and the yield from the process.

The price paid for materials, both raw and intermediate, is also an extremely important factor in the profitability of a process. In addition to the normal techniques, the purchasing department is now provided with material price standards and a Purchasing Performance Report showing price variances and price performance for material purchases.

Also, EXCO's purchasing department works closely with the technical director to investigate potential uses of substitute materials. Any such use of lower-priced substitute materials is shown in a separate section of the Purchasing Performance Report to highlight to management savings made by substitution.

Responsibility Budgeting

Controls are of great value at EXCO in keeping costs low and profits high. Some of the secrets of

*To meet your author, see p. 262.

†A fictitious company used to illustrate accounting methods in this three-part series on accounting for chemical engineers. The first article (pp. 256–262) discussed bookkeeping systems, balance sheets, process and product costs. The second article (pp. 263–268) discussed depreciation methods and standard costs.

PRODUCT NAME Fictitious Plastic 202										PRODUCT NO. 1115	

EFFECTIVE DATE 11-1-66	SIZE AND KIND OF MATERIAL AND CONTAINER	FORMULA FP-1369
PREPARED BY R.H.	Fine granules in 100 lb. paper bags.	UNIT Pounds
APPROVED BY G.T.	STANDARD COST RECORD	LOT SIZE 2,000 lb.

OPR. NO.	DEPT. NO.	WORK CENTER	OPERATION DESCRIPTION	ALLOWED HOURS			LABOR COST		MFGR. EXP.	MATERIAL	TOTAL COST
				SET-UP	OPERATION	PER UNIT	STD. RATE	AMOUNT			
1	A	2	Mixing	1.0	0.0239	0.0240	$2.00	$0.048	$0.054	$0.370	$0.370
2	B	5	Packing	2.0	0.0301	0.0303	1.75	0.053	0.060		0.102
											0.113
	TOTAL							$0.101	$0.114	$0.370	$0.585

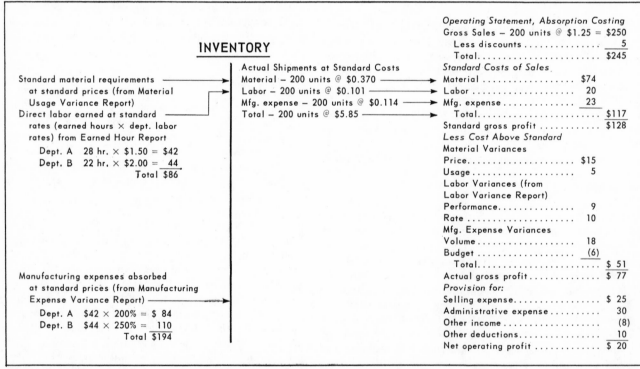

INVENTORY

Standard material requirements
at standard prices (from Material
Usage Variance Report)
Direct labor earned at standard
rates (earned hours × dept. labor
rates) from Earned Hour Report

   Dept. A   28 hr. × $1.50 = $42
   Dept. B   22 hr. × $2.00 = 44
                    Total $86

Manufacturing expenses absorbed
at standard prices (from Manufacturing
Expense Variance Report)

   Dept. A $42 × 200% = $ 84
   Dept. B $44 × 250% = 110
                 Total $194

Actual Shipments at Standard Costs
Material – 200 units @ $0.370
Labor – 200 units @ $0.101
Mfg. expense – 200 units @ $0.114
Total – 200 units @ $5.85

Operating Statement, Absorption Costing
Gross Sales – 200 units @ $1.25 = $250
   Less discounts   5
   Total $245
Standard Costs of Sales
   Material $74
   Labor 20
   Mfg. expense 23
   Total $117
Standard gross profit $128
Less Cost Above Standard
Material Variances
   Price $15
   Usage 5
Labor Variances (from
Labor Variance Report)
   Performance 9
   Rate 10
Mfg. Expense Variances
   Volume 18
   Budget . . . . (6)
   Total . . . . $ 51
Actual gross profit . . . . $ 77
Provision for:
   Selling expense . . . . $ 25
   Administrative expense 30
   Other income . . . . (8)
   Other deductions 10
Net operating profit . . . . $ 20

STANDARD-COST report (top) is part of information "flowsheet" that provides data for the preparation of up-to-date profit statement (bottom) for a typical shipment of 200 units of Fictitious Plastic 202.—Fig. 1

EXCO's success are in applying these control techniques. For example, all levels of management cooperated both in establishing and administering the budget and cost-control systems. Also, budget control is tied in with responsibility: each executive or manager is responsible for that portion of the cost or budget or profit that he can control or influence. Maximum use is made of the principal of control by exception, where out-of-line costs are highlighted for analysis and action. And controls are easy to use.

With these developments, EXCO management now has a set of practical control techniques for managing chemical processes. But also, the same tools are useful! for EXCO's design engineers in cost estimating and evaluation of equipment. Not only are standard costs available as data for estimating on similar processes, but historical and current data are fed back to show actual performance of previously designed equipment.

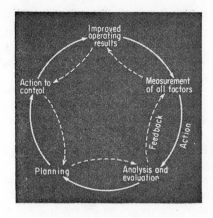

FEEDBACK for improving operations is an important element in EXCO's cost information system. —Fig. 2

Costing formulas, depreciation formulas, tax calculations, break-even and profit calculations are all available as standard accounting information and techniques. And for final comparisons of potential investments in equipment or new processes, the various methods grouped under "discounted cash flow" are proving of great value.

Communicating Via Reports

Data become information only when communicated to people. They become useful information only when communicated in usable form to the person who can take proper action.

One way to communicate data is with reports, which can be either "worth their weight in gold" or merely more waste paper. EXCO's reports are more the former, but they require constant effort to keep them useful and usable. The following principles are the basis for EXCO's management reporting system:

Information—The sole purpose of a management information report is to inform those in management responsible for taking action to improve operations or costs.

Responsibility—Reports are designed so those responsible for control will have the information they need for control (with a minimum of extraneous information).

Feedback—While the reports are designed to be a feedback system for management on the operations of its responsibility, it is also necessary for management to feed back into the system information based on its analysis and actions to improve operations. Thus, reports are designed to encourage action, and feedback from any action taken.

Teamwork and Specialization—The reporting system is designed to encourage teamwork in using the information, while acknowledging the contribution of individual specialists. This is one of the most difficult design features in any management reporting system.

Simplicity—Reports are designed to be simple and easy to use, even at the sacrifice of completeness. It was found that the mere presentation of masses of data (such as a computer can easily spew out) was very ineffective for control purposes. When backup data are needed for deeper analysis, they are provided

"as required," but reports do not contain masses of supporting data.

Exceptions—Reports are designed to highlight and point out exceptions, so that analysis will be easy and effective.

Timeliness—Reports must be available in time to permit analysis and to take action; otherwise they have little value.

Mechanization—The mechanization of reports is important not merely as a labor-saving device but to present required information as rapidly as possible.

Analysis—All EXCO reports are designed so management or engineering personnel can easily analyze them and take necessary action.

Management Information Systems

It was evident early in the development of these reports that each depended on other sources of information. Sometimes reports are a summary or excerpt of other reports, sometimes they require obtaining new data. A great part of this information is needed by accountants "just to keep the books," so that such information becomes a byproduct of bookkeeping. But more than bookkeeping is needed, and can most readily be derived simultaneously with accounting information. In this way, accounting has, over the years, turned into a comprehensive information system. Data have multiple uses in accounting, management and engineering design, and must, therefore, be part of a system for orderly flow and use of the information.

Flow of Information

Even as materials flow through a process, it is necessary that data and information flow through a management information system. If data become useful only when communicated in usable form to the person who can take the proper action, then data must flow, or be communicated from their source to that person. Data that are recorded and filed at the point of operation—whether it be pencil notations or an instrument chart—are of little use to a design engineer or manager unless there is some feasible way to communicate this information.

Therefore, an information system cannot be merely a set of reports but must also include a communication system. Since EXCO's management is heavily staffed with chemical engineers accustomed to using flow-charts, it leans heavily on flow-charting techniques for the design of its management information system (Fig. 1 is an excerpt from such a flowchart).

From Pencils to Computers

When EXCO was small—with a staff of six, for example—it was possible to review all of the operating results at their original source. Data processing was done by hand or on a slide rule, or on a simple hand-cranked calculator; and output was recorded in pen or pencil. As the company grew in size, scope, and geographic location, these simple techniques became

increasingly slow, expensive and time-consuming. Fortunately, as the company grew, more-sophisticated techniques of data processing and communications were developed.

Today, EXCO's computers have taken over the routine tasks of processing and communicating accounting data. In a new process, a computer also does routine process control at the same time that it feeds both control and management information data into the data system. Thus, computers have saved many man-hours of routine calculations while improving the accuracy and speed of information flow.

Control and Feedback

EXCO's management information system was adapted to a basic system of control feedback used to solve problems. This technique has been expressed in many different ways and forms, and the one EXCO adopted is one of the simplest and most understandable (Fig. 2).

The objective may be improved design or improved products or improved operating results. But the first step is to establish measurements, criteria, or the numerical designations of objectives and the means to get there. These measurements may be laboratory or pilot-plant data for design, or standard cost or financial budgets for control. Whatever the measurements may be, they have to be sorted out, picked apart, evaluated and analyzed.

From this analysis, plans are made for action to be taken. This is the synthetic step where the analytical information is reassembled (mentally or physically) in form for action.

The next step is action—and it may be management action to control, or the execution of an engineering design. Whatever the details of the step, this action should result in improved design, improved operating results, or improved rewards for the individual and the company.

Improvement, however, is a continuous cycle. The one best way is only "best" for a brief point in time. Tomorrow we will find a better way, or try to. And one of the ways to improve is by having feedback of information at all points. As the first results are achieved, we compare our original measurements with our previous analysis and with our plans. Then, with this feedback, we can decide on future action that, we hope, will bring even greater improvement.

With this concept in mind, EXCO aimed its entire information system at the goal of constant improvement.

What Hath EXCO Wrought?

Obviously the end is not here yet. Many other refinements were added to the EXCO cost system as the company grew, to provide even better information.

Many improvements were made possible as the combined standard cost and flexible budget system was developed. The first and most obvious was improved control by operating personnel, especially foremen, who now had a good basis for analyzing and controlling their process or departmental costs. The influence of the purchasing of raw materials on the cost of product could be readily determined by material price variance, and yields and usage of the materials were separated as an operating responsibility. The effect of volume of sales on cost of operating the plant was also clearly shown by volume variance. The marketing department no longer scanned the monthly Product Cost Reports but used instead the Standard Cost for each product as a guide to marketing policies and profitability.

The budgeting principles that were derived from the standard cost system are being extended at EXCO to every department in the company, including sales. The accounting department uses flexible budgeting principles in projecting its own year-to-year growth. Similar applications are now being considered in the engineering departments.

Meanwhile the product-development and process-engineering departments are using such tools as standard costs, accounting information, break-even formulas, and discounted-cash-flow formulas, to develop more-profitable products and processes.

With the growth of the company, the so-called overhead (or "burden" in older accounting terminology) departments such as engineering and accounting grew very rapidly. It was necessary to continuously evaluate these additions to be sure that each increment was adding its share to the profit of the company.

In the case of the accounting department, the demand for information of all sorts and kinds increased at an exponential rate as the business became more complex. Much of the information needed by management and engineers for the operation of the business was available somewhere in accounting records. The problem was to find ways to get the information to management in a form that could be used, and in time to be usable. With the dual purposes of speeding flow of information and of reducing the cost of obtaining the information, it was found profitable to mechanize the accounting system at an early date. This resulted in the present data processing and computer system that starts with a pencil at the process control desk, and with input information directly from process control instruments. All of the accounting information needed for management control purposes, and the operating control data required by all types of management, is available from the accounting system.

The secrets of EXCO's success are obvious. Rather than depending solely on inventive genius, it developed a management philosophy and a management information system that is the basis for success in all phases of the business.

Key Concepts for This Article

Active (8)	Passive (9)	Purpose Of (4)
Interpreting	Accounting*	Analyzing
Using	Systems*	Costs*
Developing	Budgets	Expenditures*
	Information	Profits*

(Words in bold are role indicators; numbers correspond to EJC-AIChE information retrieval system. Asterisks mark key concepts suggested for indexing. Others are added to improve reading as an abstract. Indexing is described in *Chem. Eng.,* Oct. 11, 1965, p. 187.)

Manufacturing Costs
For Batch - Produced Chemicals

Estimating manufacturing costs for a new chemical is a difficult problem when the product is to be made by a batch or semi-continuous process in equipment already used to produce other chemicals. In such a situation, how do you assign the costs of utilities and other services, depreciation charges and numerous overhead charges to the new product?

In a majority of cases, it would not be worth the effort required to develop a highly accurate manufacturing cost estimate for a small to medium volume, batch-produced chemical. Actually, for rough estimating, many companies making dyes, pharmaceuticals, plastic materials, etc., use a simple technique for developing an estimate of manufacturing costs for a new chemical.* The method involves four basic elements: yield of product per batch; raw-material costs; direct-labor requirements; and the ratio of overhead to operating labor.

Yield per batch can be estimated from laboratory data, after the batch size has been selected. The yield calculation also supplies information on raw-material usages (with allowances for solvent losses and plant efficiencies). If a raw material is made "in house," it is common practice to charge the new process based on the raw material's selling price minus sales expenses if an outside market exists; where no other market exists, the raw material is usually charged at "factory cost."

Estimating direct-labor requirements calls for some information on throughputs of the equipment necessary to produce the chemical. Typical labor requirements are shown in the table below.

One way to handle overhead charges is to distribute them on the basis of the direct labor required for each product. This is a reasonable assumption in a multi-product batch-process plant, where labor is a major component of production costs, since indirect costs in such a plant are tied closely to the size of the operating labor force.

A useful ratio for rough cost estimates is: (Total conversion costs)/(operating wages) = 5.6/1

Total conversion costs include cost of utilities (water, steam, electricity, effluent treatment); operating salaries and wages and supplies; maintenance wages, salaries and supplies; control laboratory; quality control; research and engineering technical assistance; all overhead charges (plant, maintenance shop, division, parent company); property taxes and insurance; and depreciation. It does not include raw-material costs, selling expenses or income taxes.

With an average gross-earnings figure for operating labor (the July 1965 figure from the U. S. Dept. of Commerce for the chemicals and allied products industry was $2.89), one can calculate manufacturing cost.

Let's assume a new product is estimated to require 10 man-hr. of direct labor per batch, and the yield is estimated as 1,000 lb. The cost calculation is:

$$\frac{10 \text{ man-hr./batch} \times \$2.89/\text{man-hr.} \times 5.6}{1,000\text{-lb./batch}} = \$0.16/\text{lb.}$$

This is the cost of labor and overhead. When raw-material costs, royalties and container costs are added, the result is total manufacturing cost.

Of course, this is a rough estimate. While the 5.6 ratio is surprisingly constant for low to medium-volume batch operations (up to about 1 million lb./yr.), it should be checked against some known data. Typical information on manufacturing costs for various production rates is shown in Perry's "Chemical Business Handbook," pp. 2-23 to 2-30. Subtracting raw-material costs from manufacturing cost, and dividing by operating wages, gives the conversion-to-labor ratio.

---

* See paper "The Estimation of Manufacturing Costs in the Chemical Industry," by R. L. Horton, given at Amer. Assn. Cost Engrs. national meeting, June 30, 1965.

Operation	Range of Man-Hr./Batch
Preparation of reactor	½-2
Charging reactants	½-2
Heating or cooling reactor	½-1
Clarification	½-1
Distillation	½-2
Extraction	½-1
Neutralization	½-1
Centrifugation (200-lb. load)	2-2½
Drumming	½-1
Drying	2-4
Grinding and packaging	½-2

Based on a charge to a 500-gal. reactor. Data based on assumption that operating labor will be assigned to other work during periods requiring no operator attention. Physical properties of material will determine actual operating labor required—a coarse, crystalline solid will require much less centrifugation time than a claylike material.

Key Concepts for This Article

Active (8)	Passive (9)	Means/Methods (10)
Estimating*	Costs, production*	Ratios
	Costs, direct*	Factors*
	Burden (cost)*	

(Words in bold are role indicators; numbers correspond to EJC-AIChE information retrieval system. Asterisks mark key concepts suggested for indexing. Others are added to improve reading as an abstract. Indexing is described in *Chem. Eng.*, Oct. 11, 1965, p. 187.)

Direct Costing Aids in Making Operating Decisions

Direct costing gives lower "cost" for producing chemical product A

Direct costing		Absorption costing	
Direct material	0.20¢/lb.	Direct material	0.20¢/lb.
Direct labor	0.40	Direct labor	0.40
Variable overhead	0.20	Overhead	0.80
Direct cost	0.80¢/lb.	Product cost	$1.40/lb.

If it cost $1.00/lb. to rework some off-grade "A" material, should you rework or scrap? Direct-costing analysis shows the material should be discarded and a new batch prepared to replace it.

F. C. BUNK, *Silicone Products Dept., General Electric Co.*

Although the ordinary method of costing chemical products (absorption costing) is adequate for an over-all summary of costs over a long period of time, it leaves much to be desired when used for day-to-day economic decisions.

Absorption costing involves allocations of numerous fixed costs in arriving at the cost of manufacturing a product. A typical example would be the $2.50 of over-head (indirect costs) charged for every hour of direct labor used to make a chemical. This overhead consists of some costs that change with product volume and other costs that are fixed and occur regardless of how much is produced. For this reason, it is difficult (but not impossible), to determine the true variable cost (sometimes called out-of-pocket cost, or incremental cost) of a product. It is primarily to overcome this problem that the direct-costing method was devel-oped. Use of direct costing in making economic deci-sions has grown substantially in recent years.[1,4]

Absorption Vs. Direct

The concept, useful in many ways, differs from ab-sorption costing in the following respects:

• In direct costing, a distinction is made between period (fixed) costs and direct (variable)* costs. Separate accounts are kept for these costs.†

• Period costs are not applied to production pounds. Thus, inventory is valued as a summation of variable costs only. Inventory is, therefore, held at a lower value.

• The terms marginal income or contribution mar-gin(sales price minus direct cost) are used to show

---

* A more appropriate name for direct costing might be "vari-able costing."
† Not all elements of costs fit neatly into a fixed or variable category, so this distinction is not as easy to apply in practice as it sounds.

the amount of funds available for fixed costs, taxes and profits.

• Since inventory is carried at variable-cost values only, profits tend to be directly related to sales. Normally, as sales rise, profits rise. In comparison, using absorption costing, profits generally relate to production rate or inventory buildup.

• It is easier to assay the effects of changing selling price or sales volume when direct costing is used.

• In direct costing, operating variances are determined by comparing actual variable costs with standard variable costs. There is no "volume variance" (a correction used in absorption costing for variance from "normal output").

It must be noted that over a period of time, if there is no change in inventory, both direct costing and absorption costing will yield the same net profit. However, with inventory change, the net profit will differ. If actual expenses are the same in both firms, the direct-costing firm will show decreased profits during inventory buildup.

Fixed and Variable Costs

In a direct-costing system, it is necessary to break all costs down into fixed and variable components. For some costs, this is straightforward. For example, real-estate taxes, fire insurance and plant management are all period or fixed costs that are constant for a period of time once the decision has been made to produce. These costs do not vary with production volume. However, other costs (direct labor, direct materials, power to run machinery, etc.) vary when production varies. Such variable costs must be placed into different accounts from the fixed costs.

In some cases, this separation is not easy to make. It must be done by one of the following methods:

• Careful inspecton of a chart of accounts for a product to determine what costs are included. The individual components of these costs can often be broken down into fixed or variable costs by inspection.

• Listing of past expenditures and plotting of a "scattergraph." This is a plot of cost incurred versus production volume. The best straight line connecting the points is then drawn by simple inspection or by a more refined statistical method. Cost at a projected volume of zero represents the fixed cost component. All costs above this are assumed to be proportional to production volume, and therefore variable.

• Statistical analysis of past expenditures and plotting of cost versus production volume by the method of least squares.[2] Once plotted, the same consideration above is used to get fixed and variable components.

• Plotting of highest and lowest volume statistics in the recent past on a production volume versus cost chart. A straight line is drawn between them, and fixed and variable costs are determined as above.

• Detailed industrial engineering studies to determine the makeup and character of the individual components of cost.

It is important to recognize that all costs involved in making a product can be broken down into fixed and variable components. This must be done to place a direct-costing system into effect.

Aid in Manufacturing Decisions

To study the advantages of direct costing for making day-to-day manufacturing decisions, let us look at a potential problem in a chemical plant and decide on the proper course of action. The table (p. 234) shows the standard costs of one pound of a chemical fluid under two different costing methods.

In addition to the 80¢ lb. direct cost shown, there are certain fixed expenses that must be calculated and paid from the sale of goods. In this example, at budgeted volume, these fixed expenses would average out to be 60¢/lb. of fluid, since at budgeted volume the total product cost must be the same with either method. It should be noted that this extra 60¢ is not specifically applicable to this product. This expense could be incurred whether or not the pounds of this particular fluid were produced.

Let us assume that 1,200 lb. of the fluid leaks from a broken valve at the bottom of a storage tank into a diked area. The plant has ample excess capacity for making this fluid. Fluid is pumped from the dike into drums and the resulting mixture weighs 1,200 lb. (1,000 lb. of fluid and 200 lb. of water).

A quick cost estimation shows that the fluid could be distilled from the water and then purified by filtration. Over-all yield is expected to be 85% of the available fluid. The cost of doing this "rework" is estimated to be $400 for cleanout materials, $160 for direct labor and $80 for other variable expenses. The probability of obtaining satisfactory quality by this rework procedure is estimated to be 75%. If the rework procedure is attempted and satisfactory quality is not attained, the resultant material must then be scrapped.

The question facing the plant operating manager is whether or not to attempt the rework. A quick look at the alternates is easily accomplished when a direct-costing system is used. The calculation to get the most probable yield of satisfactory product on rework is:

1,000 lb. available fluid × 85% yield × 75% probability of a success = 638 lb.

This is obtained at a cost of $640, yielding an average rework cost of just under $1/lb. By comparing the direct-cost figure, the decision is made to scrap the contaminated fluid and proceed to make a new batch, since virgin material can be produced for 80¢/lb. out-of-pocket expense.

There are additional advantages to the "scrap" decision. The rework procedure, being nonroutine, would require extra supervision and additional engineering over and above the supervision and engineering required for processes already reduced to a routine. And making new material will result in a better quality product.

For the cost engineer looking at the same problem in a plant using absorption costing, the decision is much more difficult. He cannot easily distinguish be-

tween fixed and variable costs. He is apt to look at the cost of $1.40/lb. as the replacement cost for the lost fluid and make the erroneous decision to rework for $1/lb., thus costing his firm money.

The point of the comparison is that a direct-costing approach brings the pertinent variable costs out in the open and makes it easier to scrap questionable material. The operating manager is less likely to send good money after bad.

Aid in Marketing Decisions

As an example of using direct costing in making marketing decisions, let us look at the effect on profit of a sudden and unplanned change in volume of the fluid cited above.

Assume that the marketing manager has an opportunity to make a one-shot sale of 10,000 lb. of this fluid, over and above the budgeted volume, for $1.40/lb. and that adequate plant capacity exists. Since fixed manufacturing overhead, R&D expenses, marketing and other overhead are completely written off as period costs at budgeted sales volume, the contribution of this extra sale is $6,000 (sales price of $14,000 minus a variable cost of $8,000). This $6,000 is available for taxes and profit and is readily obtained from a direct-costing system. An absorption-cost system, however, would show this to be a break-even sale and therefore not desirable, although a closer analysis would indicate that a "black" (or favorable) volume variance would result from the sale. Extent of this variance would not be obtainable without a careful study.

In the short run, with adequate plant capacity, dynamic marketing decisions based on the direct costing of a product will bring in extra sales dollars. But, of course, no firm can survive in the long run unless all costs, variable as well as fixed, are fully covered in the selling price. It is the product cost that must be used as a guide for establishing normal sales prices. The availability of direct-cost data just allows marketing people to take bolder pricing actions.

Some Advantages

Advantages of direct costing can be sumarized as follows:[3, 5]

1. Profits increase and decrease with sales when direct costing is used. This is true whether inventory increases, decreases or remains constant.

2. The impact of fixed costs on profits is emphasized because these costs are listed separately in the income statement. Costs are not hidden in allocations against production.

3. Products, territories, cost centers, etc., can more easily be directly compared since the listed direct costs do not include the arbitrarily allocated joint fixed costs. The costs listed are specific to the product, territory or cost center.

4. Direct costing relates to and facilitates application of standard costs and flexible budgets.

5. Inventory cost ties in closely with current out-of-pocket production expenditures.

6. Better service to customers (in the form of better delivery through higher inventory, better pricing, and higher quality) is more likely to occur in a firm using direct costing.

7. In a growing concern, with growing inventories, lower taxes will result from direct costing. This is really a delay in tax payment.

8. A firm is less likely to send good money after bad in rework of questionable material, and will thereby reduce over-all costs.

No Panacea

However, direct costing poses some problems:

1. In the breakdown of fixed and variable components of semi-variable costs, some decisions will tend to be arbitrary.

2. Long-range policy and pricing decisions must be based on the complete manufacturing costs. With direct costs available, there is a tendency to underprice.

3. Court decisions do not provide a clear answer to the question of acceptability of direct costing for determining taxable income. To convert from absorption costing to direct costing requires coordination with tax authorities. The alternative to this is use of direct costing for internal costing, and conversion of figures to absorption costing for tax purposes.

References

1. Anon., *Business Week*, Mar. 24, 1962.
2. Matz, C., et al., "Cost Accounting," p. 540, Southwestern Publishing Co., Cincinnati 37, Ohio, 1962.
3. From National Assn. of Cost Accountants Bulletin No. 23, p. 1,127, Apr. 1953, N. Y. 22, N. Y.
4. National Assn. Cost Accountants Research Report 37, Jan. 1, 1961, N. Y. 22, N. Y.
5. Wright, W., Kollaritach, F., "The Concept of Direct Costing," *The Controller*, p. 322, Financial Executive Institute, Inc., N. Y. 16, N. Y., July 1962.

Meet the Author

Frank Bunk is a production engineer on silicone resins, fluids and specialties at the General Electric plant at Waterford, N. Y. He is responsible for product-cost estimation and control. Former assignments have included education course supervisor, shift foreman, general foreman, process engineer, manager-pilot plant semi-works, and manager-finished products subunit, all with General Electric at locations in Schenectady, N. Y., Pittsfield, Mass., Decatur, Ill. and Waterford, N. Y.

He is a chemical engineering graduate of Rensselaer Polytechnic Institute, and a graduate of General Electric's Advanced Process Technology Program.

Key Concepts for This Article

Active (8)	Passive (9)	Purpose of (4)
Comparing	Direct costing	Pricing
Uses	Absorption costing	Economic evaluation
	Costs, fixed	Production
	Costs, variable	Manufacturing

(Words in bold are role indicators; numbers correspond to AIChE information retrieval system. Indexing details are described in *Chem. Eng.*, Jan. 7, 1963.)

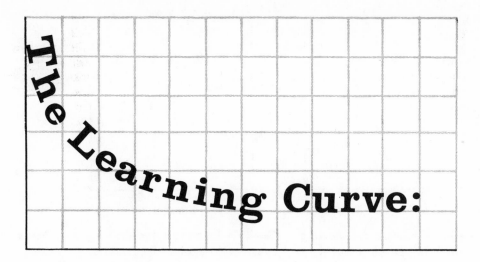

The Learning Curve:

A method of measuring past and present experience in any repetitive operation produces a quantitative relation that permits the prediction of future costs, manpower requirements and production schedules.

WINFRED B. HIRSCHMANN, *American Oil Co.*

Any operation can always be done better each succeeding time. It is apparently not common knowledge that the pattern of improvement can be sufficiently regular to be predictive. Such a pattern characterizes not only individual performance, but also the composite performance of many individuals organized to accomplish a common task.

The industrial learning curve quantifies such performance. It evolved from experience in airframe manufacture. In this industry, the number of man-hours per plane declined at a regular rate over a wide range of production. Such continuing improvement is so common in the aircraft industry that it is the normal expectation. Hence production and other types of performance are customarily scheduled on some basis of progressive betterment.

Although learning curves have been found in other industries, they have not received equivalent acceptance. Instead, the usual prediction is level performance with constant costs. Nevertheless progress constantly occurs. Since this progress reflects learning how to do things better, the paths traced by progressive improvement may be learning curves.

Learning was first observed for manufacturing operations in 1925 by the commander of Wright Field.[1] Later studies of aircraft assembly showed that the fourth plane required only 80% as much direct labor as the second; the eighth plane, only 80% as much as the fourth; the one hundredth, only 80% as much as the fiftieth, and so on.[2] Thus, the rate of learning was concluded to be 80% between doubled quantities. On linear coordinate paper (Fig. 1), the relationship is a curve with a rapid initial decline that later trails off. On log-log paper (Fig. 2), it is a straight declining line, which reflects a constant rate of reduction.

No Single Learning Curve Exists

There is no such thing as a universal curve that fits all learning. There are great variations in level at which a curve starts, i.e., the cost of the first unit.

Mathematics of the Learning Curve

Learning curves evolved from the plotting of unit assembly labor vs. cumulative number of airplanes manufactured in aircraft companies. From these data, it was discovered that man-hours per unit tend to decline with increasing experience during production. This improvement was calculated as the ratio of unit man-hours every time production doubled. In operations requiring a high proportion of direct labor for assembly, the learning curve was typically found to be 80%. Every time production doubled, the unit man-hours decreased to 80% of the previous rate.

Learning curves will have different ratios depending on: the operation involved, the proportion of direct labor for assembly to that for machine work, and the repetitive nature of the operation. When plotted on log-log paper, the learning curve is a straight line that is represented by an equation of the type:

$$c_2/c_1 = (x_2/x_1)^{-b}$$

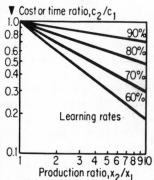

where c_2/c_1 is the ratio of unit labor or unit cost, x_2/x_1 is the ratio of production, and b represents the exponential relation between the two ratios. The accompanying graph shows learning curves based on a decrease in unit costs of 90%, 80%, 70% and 60% between doubled quantities. —EDITOR

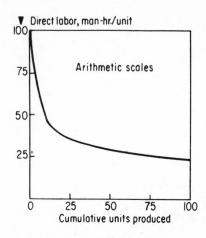

An 80% learning curve showing a rapid initial decline—Fig. 1

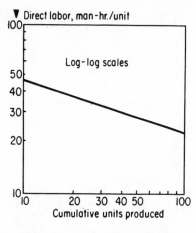

The 80% learning curve showing a constant rate of reduction—Fig. 2

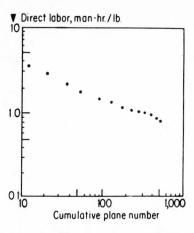

Actual performance in airframe production shows learning—Fig. 3

This is simply because of the different ranges of complexity of items. The slope of the curve, however, is common to a wide variety of experience. In fact, it was the regular finding of an 80% curve for aircraft that started speculation about a general theory of learning curves.

Operations paced by people have steeper slopes than those paced by machines. In airframe manufacture, for instance, three-fourths of the direct labor input is assembly; the balance is represented by men engaged in machine work. In such a man-paced operation, an 80% curve is commonly found. If the ratio of assembly to machine work is 50-50, the learning curve is about 85%. If the ratio is one-fourth assembly and three-fourths machine work, the operation is largely machine-paced, and the curve is around 90%.

These various percentages provide a means on which to base predictions, and such predictions have proved more valid than assuming level performance, i.e., no learning.

Successive points of a plot of unit labor vs. production units as shown in Fig. 3 do not fall on a sharp line but within a band defining a declining trend.[1] The learning curve is the line that fits the points; and the boundaries of the band are the upper and lower limits. Whenever points fall outside these limits, there is usually an assignable cause.

For example, a leveling off or even an increase can occur such as at the end of a contract when closing-out operations become inefficient. A rise in the curve can occur in the middle of a contract too because of changes in a model, or moving operations to a new building, or halting operations for a while. After the transfer, or as skill in handling the changes is acquired, the curve declines rapidly to approach the old slope.

Learning Curves Are Not in General Use

It is surprising that a technique used so long, and validated so well in one industry, has not been adopted generally. One reason for this may be a lack of aware-ness that improvement patterns can be quantified. Although everyone has had the experience of improving his skill by repetition, this experience is generally assessed only qualitatively. Improvements can seem irregular and fluctuating at the time they occur. It is not obvious that over the long term they can be tracing a trend. Studies of such empirical trends in the aircraft industry indicate that they approach a reliability comparable to that of engineering construction estimates.

There may also be a lack of awareness that the learning curve can describe group as well as individual performance, and that groups can comprise not only direct labor but also behind-the-scenes people. In airframe manufacture, this may include tool engineers contriving new jigs and fixtures. In other types of manufacture, there may be staff groups working on different combinations of operating conditions to improve yields, or devising instruments to improve control. Technological progress is a kind of learning. Assigning specialists to seek it and incorporating it in operations helps bring improvement. The industrial learning curve thus embraces more than the increasing skill of an individual by reptition of a simple operation. Instead, it describes a more complex organism—the collective efforts of many people, some in line and others in staff positions, but all aiming to accomplish a common task progressively better.

This broader concept may be why the phenomenon has many names. Among them are manufacturing progress function, cost-quantity relationship, cost curve, experience curve, efficiency curve, production acceleration curve, improvement curve, and performance curve.

Skepticism that improvement can continue may be another factor that has limited more general use. After arduous effort to achieve an increment of advance, it is natural to feel that the last ounce of betterment has been wrung out of an operation. The expectation of still further improvement then seems quite unrealistic.

These reservations are embodied in the belief that "our business is different," and consequently that such

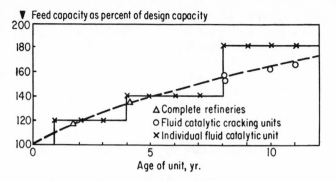

Increases above the design capacity of refineries or their units suggest a form of learning—Fig. 4

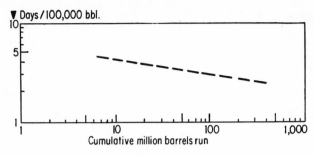

Replotting of refinery data yields a 90% learning curve that is indicative of the actual improvement in the performance of complete refineries—Fig. 5

curves do not apply to its operations. Credence is given this conclusion by the observations of some experienced practitioners who expect the learning curve to be inapplicable or have little value in such industries as basic chemicals, plastics and petroleum refining.[3]

If the premise is valid that learning curve performance can be an underlying natural characteristic, then such performance should be found not only for more types of activities already recognized as responsive, but also for operations not previously reported or believed susceptible. Petroleum refining is considered to have such operations. It is characterized by large investments in heavy equipment, and is so highly automated that learning is thought to be nonexistent or too small to be of value.

The Learning Curve in Petroleum Processing

Petroleum refining comprises such process operations as distillation, cracking and reforming. In 1951, the worldwide, installed design capacity of fluid catalytic-cracking units was 1,200,000 bbl./stream day, but the actual throughput was about one-third greater —1,600,000 bbl./stream day. This aggregate throughput is composed of the individual cracking units at a point in time. It does not show how these throughputs changed over time. Plotting the performance of individual units at a point in time against their age at that time provides a clue to this pattern, as shown in Fig. 4. The points are the ratios of the achieved capacity to the design capacity as determined from published tabulations. (Although capacity is not the basic measure of performance of a fluid-cracking plant, it is related to performance and is the statistic available.)

Some refineries have a single unit for a particular process such as crude-oil distillation, cracking, and so on, in which reported refinery capacity reflects individual unit capacity. The selected points of Fig. 4 were calculated from these data. Performance rapidly improves in the first few years, and continues at a slower rate in later years. Successive annual points for an individual unit indicate that growth occurs in a stepwise fashion.[5]

The pattern of improvement indicated by the dashed

line resembles the inverse of a learning curve on linear coordinate paper. If the parameters are changed so that the number of days to process 100,000 bbl. is plotted against cumulated throughput on logarithmic paper (Fig. 5), a declining straight line can be drawn through the points. The learning curve here is about 90%, as might be expected from a machine-paced operation that involves little direct labor.

The rationale for such performance lies in progressively removing bottlenecks. Safety margins for critical equipment are included during design stages of a project to insure getting required (design) performance. This practice suggests that expected performance should be higher than the design target. Equipment not considered critical in design and without extra safety margins, however, may limit initial performance to the design target. Removal of such a bottleneck could result in marked improvement.

Operators learn to take advantage of built-in safety margins. But as time passes, fewer and fewer bottlenecks remain to be uncovered, so progress slows. The tapering off may reflect a limit in the physical capacity for absorbing improvement.

Rearranging or enlarging some equipment to remove these limits usually requires substantial capital. A substantial immediate improvement should result, followed by the same slower growth as before modernization, because the unit is basically the same one from which most bottlenecks have been removed.

These circumstances suggest a relatively rapid early growth, gradually slowing to approach some asymptote that may be displaced upward by substantial effort and capital. The improvement curve appears to reflect technological resourcefulness. It seems reasonable to believe that this technical skill will continue to result in such enhancement patterns so long as incentives occur to prod the search for improvement.

Startup Operations

If learning is greatest where the most people are involved, it should be found in refinery operations that have a high labor content, such as startup of units after shutdown periods for repairs. Specialists are on hand to assure proper functioning of instruments;

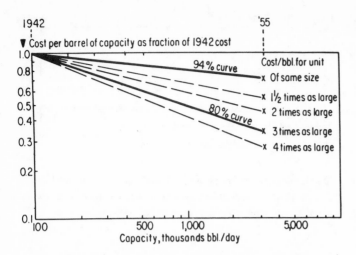

Construction of fluid catalytic-cracking units may reflect experience along a learning curve—Fig. 6

extra craft people may be assigned to handle emergencies; and more supervisors are present to give guidance. The regular crew is busy in routing flows, opening and closing valves, lining out system components, and operating manual controls before the automatic ones are cut in. Shift workers may at times "double over."

This relatively high human activity should be susceptible to a significant degree of learning. Actual experience confirms this surmise. Over a ten-year period, the time necessary to put a Whiting refinery fluid cracking unit of the American Oil Co. on stream dropped to less than half the time initially required.

Construction

Building new items of heavy equipment also appears to be characterized by learning. The 245,000-bbl. refinery at Fawley, England, which went on stream in the early 1950's, could have been duplicated five years later at 70% of the original cost. This decrease represents a rate of decline of about 7% per year. The per barrel investment costs of units for some individual processes (thermal cracking, polymerization, catalytic cracking and catalytic reforming) also decline progressively from year to year.

A learning curve measure of the rate of decline for fluid catalytic-cracking units is suggested by the observation that the steel required and the investment cost in 1955 were estimated to be one-third of those required to duplicate the capacity of the original downflow fluid plant that was built in 1942. During the intervening years, about 3 million bbl. of fluid cracking capacity were built. When these two cost points are plotted on logarithmic paper against the cumulative installed design capacity of fluid catalytic-cracking plants and connected by a straight line (Fig. 6), the slope of the declining line represents a learning curve of about 80%.

It might seem surprising to find that the construction of multimillion-dollar processing units, which are

comparatively few in number and built to order, is characterized by this same 80% curve that is commonly found in manufacturing many different items in large numbers on production lines. However, since such construction is largely assembly work, this finding is consistent with the general learning curve experience that operations with similar ratios of assembly to machine work have similar learning curves.

This decline reflects learning by construction people in how to build in order to reduce unit costs; and by research, engineering and operating people in what to build. It provides a clue to the relative contributions of capital and of technology to learning curve performance. The circumstance that a plant to duplicate the performance of an original one could be built with one-third the steel indicates that the what-to-built contributions of the progressively improved technology embodied in successive plants greatly outweighed the how-to-build contributions of capital for the better tools and equipment employed in building them.

Just how much more dominant technology can be than investment has been shown by three recent studies. They covered somewhat different time periods, but each independently concluded that about 90% of the U.S. growth in output per man-hour was due to technological change, and only about 10% to increased investment in capital equipment.[4]

Effects of Inflation and Obsolescence

The estimate on which the 80% line of Fig. 6 is based reflects costs in constant dollars. Price rises since 1942 wiped out a large part of the technological benefit. If Nelson's construction cost index for petroleum refining equipment is applicable, the actual dollar cost in 1955 of building a 1942 unit would have been 2.2 times as much as in 1942. Reflecting the technology that enabled it to be duplicated in 1955 for one-third this cost gives a unit cost in 1955 dollars of $2.2 \times 33.3\%$, or 73% of the 1942 dollar cost. This cost reduction over a 13-yr. period corresponds to the 94% curve in Fig. 6.

Technology also contributed to building larger plants with attendant economies. Doubling the capacity of a plant does not double the cost but increases it by some lesser amount, which can be represented by an exponent of the size. A plant twice as large may cost about $2^{0.7}$, or 1.6 times as much as the base size. If this 0.7 cost exponent applies to fluid cracking plants, then the per barrel investment costs of various multiples of size decline, as shown by the column of crosses on the right side of Fig. 6. Since plants have been built two or more times larger than originally, the economies of size brought about by progress in research and engineering also contributed to offsetting the effect of inflation.[5]

In this instance, technological progress decreased costs more than inflation increased them. This circumstance might suggest that depreciation allowances have been more than adequate to provide the capital necessary to replace the units. Actually, rapid obsolescence is concomitant with rapid technological progress. A

competitor with a new, low-cost unit has the advantage of a smaller capital charge for its products, and can thereby profitably sell them at lower prices.

To survive in the face of such competition, firms may be forced to replace existing units before they are fully amortized, and before depreciation reserves have become large enough to pay for the replacements. This is particularly true for units depreciated by straight line and other methods for equipment installed before 1954.

Industry Performance

We can reason that if learning in components is widespread, it should be reflected in aggregate performance. A logarithmic plot (Fig. 7) of man-hours per barrel vs. cumulated barrels of crude oil refined in the U.S. since 1860 results in the regular type of decline that such reasoning would lead us to expect.

Such advances in aggregate productivity reflect the joint effect of many interrelated influences. Among these are technological advance, increased capital investment, better methods of management, increased health and education of workers, and improved communications. The over-all aim, however, is progress, and such declines can be looked upon as the result of learning how to do things better.

Construction, maintenance, processing, and startup operations are common in industry generally. Where not previously reported or expected to be significant, learning curves can characterize such activities, and may already be contributing to the progress occurring. The credibility of this generalization is reinforced by understanding the elements of learning and the practices that promote it.

Learning Curve Doctrine

There are two main factors that affect learning: the inherent susceptibility of an operation to improvement; and the degree to which that susceptibility is exploited. Inherent susceptibility is related to the human content of an operation. It is reflected in the ratio of assembly to machine work. The greater the human content, the greater the susceptibility for improvement. The degree to which susceptibility is exploited is related to the dynamic content of the environment—the drive and resourcefulness of management, and its skill in stimulating supervisors and technical people to be creative and workers to be productive.

One of the factors affecting the dynamic content of environment is faith. If progress is believed possible, it will likely be sought; and if looked for, there is some possibility of finding it. Conversely, if improvements are considered unlikely, there is little urge to seek them. Industrial engineers have long known that once a quantitative objective is imposed on an organization, there are strong forces created to achieve the objective. There are a number of studies showing that progress ceased when the actual unit time reached that originally estimated.

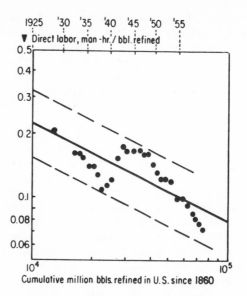

Learning may reflect aggregate performance in refining—Fig. 7

The limiting effect of "ceiling psychology" shows up in the performance of new equipment. If it is designed to operate at a given rate and doesn't perform as expected, great efforts are made to bring it up to target. It is taken for granted that rated or specified performance can be achieved. When that level is reached, attention is directed elsewhere on the presumption that rated output is the limit of capacity.

Such were the circumstances with a new machine having a given rated capacity. Initial production was less than half the rated capacity. Some had opposed the selection of this equipment on the basis that they didn't believe the quality and quantity would be acceptable. They now recommended that the machine be replaced with one they had favored. However, those who had selected the original one were still convinced it could perform adequately, and asked for more time.

The management engineering consultant supported them. He had noted some improvement along a line that suggested learning. The machine was retained for further trial. Output gradually rose to the rated capacity in about four weeks. With "normal" output achieved, no further improvement was expected and concerted efforts were to be stopped. But the consultant noted that the improvement continued to trace a learning curve. He reasoned that progress might continue, and therefore engineers were kept assigned to the machine. The result was a further increase to a level of triple rated capacity.

Thus, not putting a ceiling on expectations may permit improvements to continue. It may also result in a faster rate of improvement. Some firms report more rapid progress when the operating force is not informed of the target rate. Conversely, if betterment is not believed possible, the incentive to look is reduced and an atmosphere of maintaining the status quo encouraged.

If there is a limit to learning, it could be expected

to be reached after quite a period of time during which skill and technology would have improved performance to the ultimate level. This is what seems to occur frequently in an established plant. In a Cornell University study,[6] two separate items that were thought to have reached an achievement plateau were found to show continued learning when transferred to another firm.

Merely expecting progress, however, is not enough to bring it about. Companies with much experience in learning report occasions when it hasn't occurred as predicted. However, the general knowledge of the pervasiveness of learning and the actual experience of having had it occur leads to the conviction that it can and will occur, and even to the demand that it must occur. One aircraft firm states categorically that if a foreman doesn't get the expected amount of improvement, he is fired. The foreman, however, doesn't control all his environment. If he is provided with improper tools, or an insufficient budget, or not enough staff assistance—in other words, if the system isn't right—it is the management that some say should be replaced.

Contribution of Learning Curve Studies

It is the universal experience that it is reasonable to expect all life activity to be susceptible to improvement. The contribution of learning curve studies has been to validate this expectation by finding curves for both simple and complex organisms, and for various levels of organized activity. These studies have brought the concept into sharper focus and put it into a more meaningful and useful framework by introducing a measure of quantitative expectation where previously there was only qualitative. The results can be summarized as learning curve doctrine:

1. Where there is life (people), there can be learning.

2. The more life, the greater is the rate of learning. Man-paced operations are more susceptible to learning, or can give greater rates of progress than machine-paced operations.

3. The rate of learning can be sufficiently regular to be predictive. Operations can develop trends that are characteristic of themselves. Projecting such established trends is more valid than assuming level performance or no learning.

4. Learning is related to the dynamic context of the environment. Faith and incentive stimulate looking for progress, and provide the drive to exert the energy, resourcefulness, skill and persistence needed to bring it about. Conversely, "ceiling psychology" and the tendency to maintain the status quo inhibit learning.

If learning has been occurring and the dynamic content of the environment remains the same, it is reasonable to believe that learning can continue. It is prudent to include this potential of existing operations for continued improvement in expansion plans and other longer-term forecasts. In capital expenditure plans, it could influence the size of the next increment of capacity, or defer the increment to enable more of the evolving technology to be eventually incorporated.

Another application lies in choosing between modernizing existing plants or replacing them. It is reasonable to expect more potential for improvement in new units than in existing ones. This difference in potential is an advantage to the replacement alternative that has generally been considered intangible. It can now be evaluated.

There is also an incentive for a firm to refine its practices so as to capitalize more fully on the potential inherent in its daily operations. If historical data are available, they can provide an experience base for predicting further improvement. If the data are not available in terms of unit labor vs. cumulative output, but as costs vs. time or other parameters, historical data may still be useful if they define a trend. Once established, a trend tends to persist. But since past records reflect past practices, they should be used only as a point of departure.

The most important ingredients in learning curve performance are vision and leadership. Continued improvement is a series of influences, which starts with the conviction that progress is possible, continues with the creation of an environment and support of work that promotes it, and results in a flexibility and willingness to change established practices for better ones as they continually evolve.

References

1. Reguero, M. A., "An Economic Study of the Military Airframe Industry," p. 213, Wright Patterson Air Force Base, Ohio, Oct. 1957.
2. Wright, T. P., "Factors Affecting the Cost of Airplanes," J. Aero. Sci., Feb. 1936, pp. 122-128.
3. Andress, F. J., "The Learning Curve as a Production Tool," Harvard Business Review, Jan.-Feb. 1954, p. 96.
4. "The Growth Force That Can't Be Overlooked," Business Week, Aug. 6, 1960.
5. Hirschmann, W. B., "Profit From the Learning Curve," Harvard Business Review, Jan.-Feb. 1964, p. 125.
6. Conway, R. W. and Schultz, Jr., A., "The Manufacturing Progress Function," J. Ind. Eng., Jan.-Feb. 1959, p. 48.

Meet the Author

Winfred B. Hirschmann is an economic generalist on the research staff of the American Oil Co., Whiting, Ind. His work includes the application of business economics to operations by creating and using methods of comparing the financial merits of alternative courses of action. Prior to 1950, he was an independent chemical and business consultant for various firms, and also served on the staff of the Chief of Ordnance, U. S. Army. He has a B.S. in chemical engineering from the University of Illinois, and an M.S. in physical chemistry from the University of Chicago, and has had additional training in engineering and business management.

Key Concepts for This Article

For indexing details, see Chem. Eng., Jan. 7, 1963, p. 73 (Reprint No. 222). Words in bold are role indicators; numbers correspond to AIChE system.

Active (8)	Passive (9)	Means/Methods (10)
Forecasting	Production	Learning Curve
	Performance	
	Costs	
	Manpower	
	Schedules	

6

Techniques for Economic Evaluation and Venture Analysis

How to Appraise Capital Investments

In this "profit-squeeze" economy, having really good systems for appraising and controlling new capital investments becomes more important than ever for process companies. Here is some practical advice.

JOHN W. HACKNEY
Mobil Oil Corp.

There are many powerful forces at work today that tend to reduce the take-home pay of chemical process companies, despite rising sales.

One major opportunity for improving profits presents itself when a company is considering new capital investments. Here is where mediocre and poor projects from an investment standpoint can be eliminated promptly, and where good projects should be confidently identified and pushed to early completion.

Admittedly, this is a difficult task, particularly on large, complex jobs. But it seems clear that for the chemical process industries, sustained profits in the future will go to those companies having sound systems for realistic appraisal of their investments.

Basically, control of capital ventures for maximum profit includes these five important elements:

- Estimating the investment required.
- Forecasting profitability.
- Selecting projects of best return, considering risks involved.
- Monitoring costs and progress during construction.
- Reviewing performance for further improvement.

In this report, we will discuss reasons why formal procedures are needed for the control of appropriated funds for capital ventures; when and how requests should be made for funds; how to measure a company's true total investment in a project; how to forecast income and return; profitability and risk measurements; progress, control and performance reports.

When applied in a practical way, with a positive, creative attitude, these procedures can play a major role in improving corporate profits.

Why Justify and Control?

A new project is not started because it's a "good idea." It must be formally requested, justified and controlled.

Almost every large corporation finds it necessary to set up formal procedures for requesting, approving and controlling funds used for purposes outside the usual day-to-day operations.

These procedures for committing venture capital accomplish several things:

• A written request for funds provides a concise, complete project description. Responsible individuals can tell exactly what is proposed and how it will affect their activities.

• The request gives information that will justify the expenditure. A project should not be started just because it is a good idea. It is necessary, whenever possible, to put down on paper just how many dollars will be added to the company's income and how this compares with the required investment.

• An appropriation request provides a check list of items to be included in estimates of cost and return. Preparing these estimates is difficult at best and the company must be sure that all factors are included. There are a multitude of special cost items that, when overlooked, nibble away at corporate profits.

• The request provides information for company financial planning—required so that funds will be on hand when needed.

• Approval procedure permits authorized individuals to exercise their delegated responsibility for expenditure of company resources.

• Regular status reports allow management to monitor progress and boost projects along if they appear to lag.

• These regular reports also call attention to projects that are getting out of line financially. If a project shows signs of exceeding the authorization, steps can be taken to reverse the trend. If necessary, supplementary funds may have to be appropriated, or the project dropped, before too much damage is done.

• Procedure also spells out the steps for closing completed appropriations promptly. This permits an early start on depreciation and amortization allowances. It also cuts down on loose ends.

• And the procedure provides for comparison of actual project cost and return with the estimates in the appropriation request. These comparisons are useful for improving appraisals of future requests and checking to see that everything possible is done to achieve the forecast profitability of current projects.

When Appropriation Requests Are Needed

Most companies prepare appropriation requests for any expenditures to be capitalized. In general, therefore, they cover all acquisitions that have a life of over a year. Tangible expense items are included along with associated nontangible expenditures for such things as engineering, labor, know-how, permits. Minor items are excluded, with company accountants determining what should be expensed.

In general, projects submitted for construction approval should be well defined, to a point where completion of engineering cannot be expected to increase the accuracy of the capital cost estimate by more than 5%. This requires complete, thorough preliminary engineering for the project, but not the preparation of any detailed design drawings.

How to Prepare Requests

A typical form for submitting appropriation requests is shown in Table I. The graph on the form shows return on investment for three selected production rates. Arrows indicate expected sales for the first, third and fifth years. Point at zero return is the breakeven point.

Many entries on the form may be omitted for smaller appropriations, and attachments added for major ones. Common sense has to be applied in liberal doses when determining the amount of necessary detail. Presentation should be as simple as possible, consistent with the complexity of the project.

This single-sheet form consolidates all key information required to make intelligent decisions on approving or rejecting a project. Attachments merely supply details substantiating key data on the summary sheet. In presenting an appropriation request to groups such as an appropriation committee or board of directors, data can be put on flip-charts, with a separate card for each section.

Attachments Supply More Details

Attachments are added as required for particular projects.

Cost Estimate—Any cost estimate should have as much detail as is consistent with the status of engineering design. See *Chem. Eng.*, Mar. 7, 1960, p. 113 for estimating methods (pp. 25–58 of this volume).

When estimates are not prepared in detail because design is very preliminary, they should show distribution of costs to primary sections and, where possible, to subdivision groups, as described in the *Chem. Eng.* article referred to above. Subaccount numbers are needed for any items scheduled for early purchase.

Estimates should be either prepared or reviewed by the group responsible for construction. This uncovers omissions at an early stage and permits the construction group to do some preplanning. It also improves validity of the estimate as a project control.

Accounting Information—The upper section of the accounting information sheet (Table II) details the makeup of gross investment, cash investment and ap-

Table I	REQUEST FOR APPROPRIATION	Date to Committee Aug. 1, 1959	Appropriation Number 06 | 41 | 4120

Division, Department or Subsidiary: Chemical   **Location:** Chicago   **Plant:** Ore Roasting

Descriptive Title: 50 Tons/Day Kiln for New Ores   **Appropriation Amount:** $440,000

Purpose: Provide 50 tons/day of roasting capacity to convert new-type ores uncovered at Northwest Mines to saleable product.

Summary Description: Remove existing No.1 kiln, accessories and foundations. Install new kiln to roast high-yield ores using knowhow from Ware Institute of Research. Product to be sold to extraction companies in the Chicago area.

Project Type	Std. to Co.	New to Co.	Novel	Capital Estimate Summary*		Cumulative Cash Flow**	
Raw Material			✓			1st Yr.	$148,000
Process			✓	Gross Investment	$722,000	2nd Yr.	348,000
Product		✓		Cash Requirement	$497,000	3rd Yr.	595,000
Packaging	✓	✓		Reserve included	20 %	5th Yr.	1,178,000
Sales		✓				10th Yr.	2,291,000

Justification: New ores being obtained from our mines cannot be handled by existing kilns. When properly roasted by a newly developed process, can be converted to a highly saleable product. Return favorable.

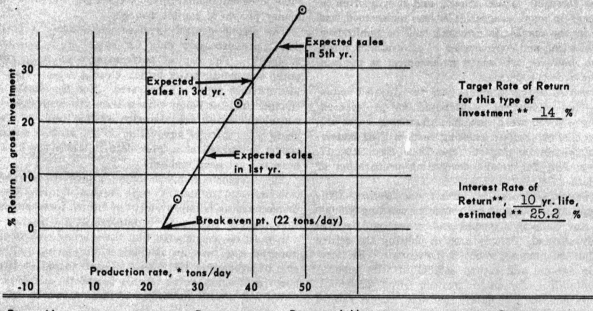

Target Rate of Return for this type of investment ** 14 %

Interest Rate of Return**, 10 yr. life, estimated ** 25.2 %

Prepared by:	Date	Recommended by:	Date
R. J. Collins	6/30/59	B. L. Byers, Div. Mgr.	7/12/59
Checked by: J. L. Johnson	7/4/59	**Recommended by:** App'n Comm. per J.B.L.	8/1/59
Sales concurrance: B. D. Ross	7/6/59	**Approved by:** R. C. Ruth	8/6/59

*Indicates no entry required for projects under $25,000. ** Indicates no entry for projects under $250,000. Attach where appropriate: Accounting information sheet, cost estimate, flowsheets, extended description, market forecast, manufacturing cost, return estimate, risk rating.

propriation amount for the project. The definition of these items and methods of computing them are discussed later.

Facilities to Be Transferred or Retired—This section of the accounting information sheet should be completed when appropriate. It includes all existing company assets used exclusively by the project and all assets forceably retired because of the project.

Information must have enough detail so facilities can be identified and a basis established for proper accounting treatment. Transferred and retired items should be listed and totaled separately.

Market Information—Appropriation requests for producing new products or expanding existing production when an investment of over $250,000 is involved should be accompanied by market information covering all products and byproducts.

Reference should be made to detailed reports and other sources, such as divisional sales forecasts. Where applicable, and in detail proportionate to the importance of the project, the following information should be provided in summary form:

- Estimated volume and rate of growth of sales.
- Estimated prices and anticipated price changes.
- Competitive producers, existing and potential.
- Competitive products, existing and potential.
- End uses, breadth and location of market.
- Sales, distribution and service methods.
- Seasonal and weather effects.
- Penetration of market, existing and potential.
- Other pertinent information.

Such information should be brief and simple for moderate plant expansions of well-established products, increasing in detail and scope for the larger and more novel projects. Difficulty of making price and volume forecasts is well known, and it will often be necessary to quote ranges. It is also understood that much of the market information will be qualitative. The collection and condensation of all available information, however, will assist management in appraising the project's potential.

Estimated Manufacturing Cost for Added Production—This attachment is prepared for projects exceeding $25,000 when new manufacturing costs will be established or when existing costs will be materially affected by the project. See *Chem. Eng.*, Apr. 17, 1961, p. 181, for details concerning preparation of this data (p. 245 of this volume).

Estimated Increase in Income and Return—This information is prepared for projects exceeding $25,000 when the company's income and return will be materially affected. Data is used in plotting the return chart on the summary sheet of the request. (See forecasting income and return section of this report.)

Profitability—For major projects (over $250,000), computations for cumulative cash flow, target rate of return and interest rate of return are attached. See profitability and risk section for details and examples.

Other Attachments—If the project is a major one, requiring a project management group to supervise design, procurement, construction and startup, a preliminary table of organization and a construction schedule are attached. Projects of this size also justify preparation and attachment of operating tables of organization and a startup schedule. Other material in support of the request is attached as required.

Charts for Control of Capital

Two types of reports are required for control of appropriations.

The first is the monthly analysis of appropriations. The purpose of this sheet, and the supplementary cost and progress sheet, is to provide an accurate picture of project status promptly at the end of each month so that difficulties may be anticipated and early action taken. When properly prepared and promptly distributed, the monthly analysis of appropriations provides excellent management control for the time and money phases of capital projects.

The second control involves performance reports. These are necessary only for larger projects (over $250,000). The capital performance report is prepared two months after project closing. The operating performance report is prepared after the elapse of enough time for dollar return from the project to be considered stabilized. Usually a time limit of two years after project completion is set, with a second report submitted at a later date if stable conditions have not yet been reached.

These reports compare actual project cost, execution time and profitability with forecast figures. Their prime purpose is improvement of future forecasts and they must be kept entirely impersonal and factual.

It must be remembered that the original request for appropriation was an attempt at the fortune-teller's role of predicting the future. There should be little surprise if actual results fail to be exactly as forecast.

We will now discuss some of these forms in detail.

Determining Investment

Questions that come up: What is gross investment? Appropriation amount? Cash requirements? Operating expense?

Computing actual investment in a large project is not a simple matter. Many factors must be considered, each in a consistent way, if forecast profitabilities for various proposed projects are to be comparable.

Investment in a major project should be viewed in three separate ways in appraising its impact on a company's finances and financial records.

The first, the long-term view appropriate for profitability analysis, involves determining the project's gross investment. This is the value of all company resources devoted to the project. It includes not only funds directly expended for construction, but also:

- Value of items already owned and transferred to the project's use.
- Operating costs directly related to construction.
- Funds to be tied up in working capital.
- Allocation of general facilities and utilities serving this and other production units.

- Value of installations destroyed because of the project.
- A reserve, which may or may not be used.

A second viewpoint considers only immediate dollar commitments for the project. This is the project's cash requirement and it is usually less than gross investment. For example, some necessary equipment may be already on hand that can be used without cash outlay. It may not be necessary to provide new cash to build roads or other general facilities, even though the project cannot function without them. (From a long-term viewpoint, roads must be charged by a general facilities allocation.) However, this particular project may trigger an unavoidable cash requirement for a new boiler, office building or set of pipe-racks that could not, in fairness, be charged entirely to the current project's gross investment. Cash requirement may, therefore, be substantially greater or less than gross investment. Primary use of this figure is in planning for capital funds.

The third method views investment with respect to the portion of the total that can be effectively recorded and controlled by those responsible for carrying out the project. This appropriation amount includes direct investment, operating expense that will be charged to the appropriation, book value of items transferred for the project's use, cash to be spent for utilities and general facilities, and other similar items. It does not include tax credits or working capital or allocations, since these are usually outside of the control of the project group. Appropriation amount, therefore, is the target investment that can be controlled during the project's progress. For this reason, it also does not include the reserve, which is a part of gross investment and cash requirement.

Accounting Data on Separate Form

An accounting-information form with sample entries showing how project investment is made up is illustrated in Table II. This is the important second sheet attached to almost all appropriation requests. Each entry will be discussed in detail.

In tabulating these investment items, all figures should be rounded to the nearest $100 ($1,000 for large projects). They should be adjusted so gross investment, cash requirement and appropriation amounts are rounded to three significant figures.

Table II ACCOUNTING INFORMATION	Date to Committee Aug. 1, 1959	Appropriation Number 06 | 41 | 4120
Division, Department or Subsidiary Chemical	Location Chicago	Plant Ore Roasting
Descriptive Title 50 Tons/Day Kiln for New Ores		App'n. Am't. $440,000

Investment Required	Gross*	Cash*	App'n. Am't.
Direct Capital Investment	$ 320,000	$ 320,000	$ 320,000
Supporting Utilities & General Facilities	(alloc.)66,000	(cash) 18,000	(cash)18,000
Transfer of Existing Capital Items	(value) 50,000	–	(book)32,000
Operating Expense	20,000	20,000	20,000
Forced Retirements	(value) 95,000	(salv.) 5,000	–
Less: Tax Credit @ 52 % of $20,000 + $45,000	(34,000)	(34,000)	–
Working Capital	35,000	35,000	–
Other Items: Know-how purchase	50,000	50,000	50,000
Subtotal	$ 602,000	$ 414,000	$ 440,000
Reserve 20 % (minimum 15%)	120,000	83,000	–
Total	$ 722,000	$ 497,000	$ 440,000

Facility Number	Facilities to be Transferred or Retired	Year Inst.	Estimated Value	Depreciated Investment	Estimated Salvage	Disposition*
461-20 461-01	Firing hood from No. 1 kiln. No. 1 kiln, installed complete with foundations, piping, wiring and instruments, less firing hood. (Value destroyed $100,000 - 5,000 = $95,000) (Tax writeoff 50,000 - 5,000 = $45,000)	'56 '46	$50,000 $100,000	$32,000 $50,000	– $5,000	T R

Budget Data: To be completed 52 weeks after authorization: Included in five-year plan in the amount of $500,000 cash in 1960. Forecast cash expenditure by quarters after authorization: $100,000, $100,000, $130,000, $130,000, $17,000, plus $20,000 operating expense.

Prepared B. L. D.	Checked J. L. V.	Approved R. B. K.

*Indicates no entry required for projects under $25,000. **T is abbreviation for transfers; R for retirements. This form is not necessary for projects under $25,000 with no facilities or transfer requirements.

Considerable judgment is required in deciding just how much detailed investigation and effort should be applied in estimating various investment components. Individual items, such as inventories, when they bulk large in the totals will deserve much more study and investigation than when they are minor factors. No time should be wasted on elaborate computation and detailed checking of items that cannot appreciably affect project investment.

What Are Battery Limits?

"Battery limit" is a term used in segregating direct from supporting investment. It is that continuous line circumscribed about any project, plant, department or division that encloses and lies at least ten feet outside of all the buildings, structures, equipment, roadways and other facilities dedicated to its sole use. In general, this line will parallel main roads, railroads, pipe racks, plant boundary fences and natural terrain features. The proposed battery limit should be indicated on general layout drawings when they accompany appropriation requests (Fig. 1).

Many times, the product of one unit is the raw material for another. Here, the producing division is considered to include within its battery limit pipelines or other special facilities required to move the material to the battery limit of the using company division.

An appropriation committee should act as an advisory body in making recommendations concerning equitable battery-limit locations.

Capital Investment in Battery Limit

Direct capital investment includes all capitalized expenditures for installations within the project's battery limit. In addition to the more obvious items, it includes:

• Value of store items that must be replaced for stock. Issues not to be replaced, however, are considered as transfer items and handled somewhat differently.

• Cost of detailed design and field supervision by associate engineers.

• Cost of detailed design and field supervision by company engineers for major new facilities. Major facilities usually involve more than $100,000 in direct capital investment, exclusive of engineering.

• Cost of utilities and facilities required for the project, within the project's battery limit, are included in direct capital investment. When outside the battery limits, they are supporting utilities and general facilities, and are handled as will be described later in this report.

A cost estimate substantiating estimated direct capital investment is attached to the request. For projects of large size, this estimate may be in summary form, with an item for each major project division. See *Chem. Eng.*, Mar. 7, 1960, p. 113 and Apr. 4, 1960, p. 119, for information and procedures on preparing capital cost estimates (p. 25 of this book).

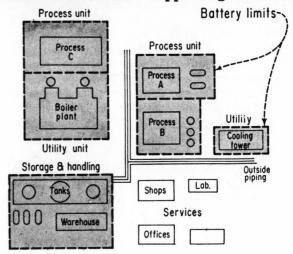

Figure 1 Battery limits around unit exclude "supporting" facilities

For the example in Table I, direct capital investment is $320,000.

Handling Supporting Utilities

Supporting utilities are defined as serving a project but being outside of the project's battery limits. They serve, or are planned to eventually serve, more than one project.

For multiproduct plants, it is desirable for utilities serving the various units to be set up as an independent operating department. This department charges each production unit for utilities used, in proportion to usage, and is charged with all of its own operating expenses, including those related to investment. It is expected to show a modest profit, similar to that of a small private utility company.

Supporting-utilities investment, therefore, is not normally included in gross investment for a production unit. If actual supporting-utility investment is required, its cost is included in cash requirement and in the appropriation amount. This provides a basis for cash-requirement planning and for control of progress and costs on this portion of the program.

Supporting utilities included in the program may be substantially more or less than will actually be required for the operating unit, depending on excess capacity available or the new capacity required in the system. When substantial expenditures are involved, an estimate of the effect on the profitability of utility operation should be attached to the request for appropriation.

When utility investments unrelated to a specific production project are required, they should be on a separate appropriation request, drawn up to show the effects on profitability of the utility's operation. In such appropriations, the utility system is treated like any other producing unit, with allocation of general facilities and so forth.

Utilities include installations for the production,

collection, processing and distribution for general use of the following: Water, steam and condensate, compressed air, electricity, pulverized coal, refrigeration media, natural or manufactured gases, fuel oil.

From Cafeterias to Lighting

General facilities lie outside of project battery limits, and support or are planned to eventually support more than one production unit.

They include the following, when outside limits:

Cafeterias, change houses, docks, fencing, first aid buildings and equipment, garages, guard houses and patrol vehicles, laboratories, land, maintenance buildings and equipment, offices, parking lots, personnel buildings and equipment, pipe racks, planting, railroads, recreation facilities, roads, safety equipment, sewers, sewage treatment facilities, sidewalks, shops, storerooms, track scales, trucks, truck scales, waste ponds, yard lighting.

When a project is served by existing or proposed general facilities, a capital allocation is included in gross investment for the new project, to cover use of these items. Cash requirement and appropriation amount, however, include only actual dollars expended in connection with the current program. This may be either more or less than the allocation. In some cases, adequate facilities are already available to satisfy part or all of the needs of the new process unit. In others, facilities must be built that will serve not only the present unit but planned future units as well.

In the "justification" for such projects, previous and probable future appropriations that include allocations for general facilities in excess of cash requirements should be mentioned.

Transfer of Existing Capital Items

Transfer of existing capital items includes the value of any existing equipment, buildings, piping and wiring that will be transferred from other service or surplus to the new project. Items are normally transferred at depreciated investment value (book value). This value is used in the appropriation amount column of the request. In the gross investment column, however, book value may be increased or decreased in proportion to the actual effective worth of the transferred item when such variation is substantial and acceptably documented.

Surplus and storeroom items not to be reordered for stock are treated as transferred facilities on the forms.

All transferred items are listed on the form (Table II) titled Accounting Information. The facility number on this form comes from property records, and aids in identifying the unit and its records. "Year installed" and "depreciated investment" are also from property records. Depreciated investment includes the depreciated original installation cost if the item is to be left in place. Original installation cost is listed separately as a "retired facility" if the transferred

item must be reinstalled for the current project. Cost of dismantling and removing the transferred unit is part of project operating expense. Cost of reinstalling it is part of the direct capital investment in the project.

In the sample accounting information sheet, the firing hood could be sold for $50,000, so this is the value inserted in the gross investment column. The book value of $32,000 is entered under appropriation amount. Funds to dismantle the hood are included in operating expense. Funds to reinstall it are included in direct capital investment.

Project's Operating Expense

Operating expense, as used in investment computation, includes any extra operating dollars spent on construction and starting up of the project. Operating expense is detailed in the cost estimate attached to the request and is set out as a separate item or group of items in the cost-estimate summary. The company's accounting department should be consulted in cases of doubt as to whether specific items are expense or capital.

Some typical items of operating expense for a project are:

• Site clearing. Removal of existing buildings, equipment, foundations, piping, conduit, timber, brush, rubbish, stored material and other items that would obstruct the new installation.

• Temporary process connections and enclosures. This is temporary work required to keep existing equipment in safe, reliable production during construction of new facilities.

• Repairing, relocating and reconditioning existing structures. This includes any cleaning, painting, dismantling, repairing and reassembling required to put existing facilities into good operable condition. Modifications that will result in substantial increases in life, capacity or efficiency must be capitalized and are not included in operating expense.

• Startup costs. For very large projects (over $1,000,000) and for novel projects, startup costs are usually large enough to warrant estimation and inclusion in gross investment. They include operator training, operating manual preparation, startup overtime premiums, special meals and transportation during startup.

Not included in startup costs are any modifications made to increase the life, capacity or efficiency of the unit beyond that intended in the design. Startup costs are included in gross investment and cash requirement, but not in appropriation amount, since necessary funds are expended through the operating budget and normally are not controlled by the project group.

• Extraordinary production and sales costs. These are included in operating expense of the project where they cause actual out-of-pocket losses. Profit on lost sales and any additional cost of filling orders from remote or outside sources is included if the construction program has prevented normal operations. Other

unusual operating expenses, such as operating department overtime premiums resulting from construction conditions, are also included.

• Preliminary project engineering. Work to analyze new ventures, to determine whether or not the company can profitably engage in them, is part of the normal operating expense. It is carried in engineering and other departmental budgets, which go to make up administration expense, and does not appear as an item in the operating expense of a specific project. Preliminary engineering includes all work required for the preparation of a preliminary design report.

• Detailed design work by company engineers for minor modifications. This is an item of operating expense. To qualify, modifications must involve existing facilities—and direct investment for the project, exclusive of engineering cost, must be less than some arbitrary amount, usually $100,000. Otherwise, detailed design is capitalized as previously discussed.

How Forced Retirements Affect Investment

Forced retirement of facilities from company service because of an upcoming project has an influence on over-all investment.

Any real value destroyed in this way is charged to the project's gross investment. The normal amount is the depreciated investment or book value destroyed, including residual installation value. A credit is allowed for conservatively estimated resale or scrap value, which may occasionally exceed book value.

Book value is normally assumed to be representative of the actual worth of items. It is reasonably easy to determine, and accurate enough for most situations. In special cases where obsolescence or deterioration has abnormally reduced the value of retired items, or where circumstances have made their actual dollar worth substantially greater than book value, destroyed value should be appraised for economic study at estimated actual value. One measure of actual value is the present, or reasonably probable, net income after taxes from the existing unit, capitalized at, say, 6%. Another measure is sale value to an outsider.

Value of retired facilities destroyed is not included in cash requirement for a project, since no expenditure is involved. Cost of dismantling and removing is part of gross investment, cash requirement and appropriation amount for the project and is included in operating expense. Salvage expected from the sale of retired facilities is handled as a credit to cash requirement. It also serves to reduce the value destroyed in the gross-investment column.

In the sample accounting information sheet, estimated installed value of No. 1 kiln, less the transferred firing hood, is $100,000. In other words, the kiln in its present condition and service is considered to be worth more than the $50,000 book value. The $100,000 value less $5,000 estimated salvage, i.e., $95,000, goes into the gross-investment column. The $5,000 salvage is a credit item in the cash column.

Cost of tearing out the kiln is part of a $20,000 operating expense.

Don't Forget Tax Credit

Tax credit is the amount that income tax is reduced because of any retirement losses or special operating expenses incurred in connection with the project. Because incremental effect is considered, credit for most corporations is assumed to be 52% of operating-expense items, plus tax credit for the loss on items physically retired because of the project.

Estimating the tax effects of retirements is sometimes a complicated matter. The general rules, however, are currently as follows:

• When sale value is less than book value: (1) Item sold as scrap—tax credit is 52% of book-less-scrap value. (2) Item sold as an operable unit—if the company has offsetting capital gains in the same calendar year, tax credit is 25% of book-less-sale value. Otherwise, tax credit is 52%.

• When sale value is more than book value: (1) Item sold as scrap—tax to be paid is 52% of scrap-less-book value. Gross investment and cash requirement are both increased by this tax amount. (2) Item sold as an operable unit—tax to be paid is 25% of sale-less-book value. As before, this item increases gross investment and cash requirement.

Because of the many and changing tax rulings, any important, difficult or doubtful tax situations should be referred to tax experts.

In the example, the kiln retired and sold as scrap has a book value of $50,000. Deducting the $5,000 salvage, there is a $45,000 operating loss. This, added to the $20,000 operating expense for the project, makes a total of $65,000 subject to the 52% tax credit.

Working Capital Easily Overlooked

Working capital is one of the more subtle elements of capital investment and probably the one most easily overlooked. Failure to provide for it is a classic cause of bankruptcy in small and growing businesses.

Working capital is made up of money and goods that must exist in an organization at any given time to enable it to function. Working capital chargeable to a given project is the net increase (or decrease) in the working capital of the corporation as a whole due to addition of the project.

For appropriation request purposes, working capital is the average annual value of all working capital items listed below, assuming operation at 100% capacity. This assumption is only slightly conservative, since working capital usually does not drop nearly so fast as operating rate. In a few special cases, it may be necessary to establish several working capital estimates corresponding to operating levels that will be maintained for a substantial length of time.

Working capital includes:
• Process inventories.
• Supplies inventory.
• Accounts receivable.

- Current liabilities (a credit).
- Other current assets.

Process inventories include inventories at estimated cost—or transfer price if from another company division—of: raw material; in-process material; finished product in storage; product enroute to customers, but not billed; product in customers' or retailers' hands

> *"Working capital is one of the more subtle elements of capital investments. . . . Failure to provide for it is a classic cause of bankruptcy in small and growing businesses."*

on consignment; stored fuel and stored packaging material.

Any one of these items may either be small enough to ignore or so large as to be a major investment item for the project. Inventories deserve special attention when seasonal, high-value or bulk-shipment materials are involved. In such cases, it is a good idea to make a complete analysis of the inventory situation, considering expected production, shipping and sales fluctuations, and the inventory required to meet them. Successful solutions to these problems often require the joint efforts of sales, operating, purchasing, traffic and accounting departments.

The table below gives some indication of inventories in order-of-magnitude figures:

Type of Plant	Raw Material and Fuel, in % of Annual Mfg. Cost at 100% Capacity	In-Process and Finished Goods, in % of Annual Mfg. Cost at 100% Capacity
Heavy chemicals (caustic, soda ash, chlorine).....	14 to 19	2 to 4
Sodium silicate.........	3 to 5	4 to 6
Metallic salts..........	13 to 22	4 to 6
Cement...............	variable	10 to 12
Chlorinated products....	1 to 5	3 to 19

For the entire chemicals and allied products industry, total inventories have averaged as follows at year end:

	In % of Annual Sales	In Months of Sales
1956........................	16.7	2.00
1957........................	17.1	2.05
1958........................	17.1	2.05
1959........................	16.3	1.95

Individual plants and projects may require inventories considerably greater or smaller than these averages.

The usual supplies-inventory criterion is total direct capital investment plus replacement value of transferred items. Company experience should be collected and analyzed to determine the ratio between average inventory of supplies and this investment. Usual values in the process industries are on the order of 3%.

The accounts receivable amount varies with type of

business and with credit policies. It is usually expressed in terms of months of production and valued at gross sales less profit. The value of product transfers within the company is excluded from gross sales, since accounts receivable do not exist with respect to this portion of production.

Actual average values for the chemical and allied products industry have been:

	Receivables in Months of Sales
1st quarter 1959	1.43
1st quarter 1960	1.45

Current liabilities are included as a credit against working capital. This is money that the company need not possess at a given time, payment not being due, even though goods or services have been received. The amount of current liabilities credit depends on practice in a particular trade, company financial policy and rules for tax payment. It can be estimated as a percentage of total receivables and inventory.

Past experience in similar endeavors is the best guide to the percentage credit to be allowed. For the chemical and allied products industry, percentage of current liabilities to total receivables and inventory in recent years has been as follows:

	% Current Liabilities to Rec. and Inv.
1st quarter 1958.....................	48
1st quarter 1959.....................	44
1st quarter 1960.....................	49

Other current assets are primarily cash on hand or in banks, government bonds or other readily convertible securities. Such funds are necessary to take care of surges in the company's flow of capital. Amount tends to be proportionate to the current liabilities of the company and is affected by company financial policies. For the chemical and allied industries this percentage of cash and government securities to current liabilities has been:

	Cash and Gov'ts. to Cur. Liab.
1st quarter 1958.....................	68%
1st quarter 1959.....................	80%
1st quarter 1960.....................	71%

As usual, a study should be made of company practice and records before estimating this item.

Other Important Items

Other items of investment, particular to a specific project, may include single-sum process or know-how rights, initial catalyst or solvent charges (with a life of over one year, such as mercury charges for electrolytic cells), major uninstalled spare equipment or parts (over $1,000 per item) and any other items of gross investment required for the success of the project. Items of this type often require special treatment with respect to depreciation, as will be discussed in a later section, and therefore must be segregated from the normally depreciable items.

Reserve is an amount by which the project may overrun if a great many things go wrong. It is computed

on a percentage basis, proportionate to the degree of novelty of the project and the extent to which project investigation and engineering have been completed. The procedure for getting this percentage is described on p. 47.

In no case should the reserve be less than 15% of the subtotal of the other items of gross investment or cash requirement. Projects so sensitive to capital investment that they will be unattractive if they have an overrun of 15% are usually not justifiable.

Forecasting Income and Return

To get income and return, you have to estimate sales, distribution and manufacturing costs.

Income and return are two of the most important items used in appraising a new project. Both are computed on an annual basis as shown in Table III on p. 295.

In setting up a standard arrangement of data, existing recording and reporting practices should be considered. And once established, the "standard" should be used throughout the company so that presentations can be easily understood and compared by everyone concerned.

All data are incremental and represent economic changes produced by the project as realistically as possible. Temporary incremental advantages must be handled as such. The study's purpose is to forecast economic effects over the full life of the particular project.

These incremental return economics will not always coincide exactly with returns reported in regular monthly reports for operating control. Such reports show status after project completion, not changes produced by the project.

Income and return results are plotted on the graph in the lower section of the appropriation request summary sheet. Additional points are computed if necessary, to define the curve.

How Sales Affect Costs

Sales-increase effects are most important when evaluating a project. They are, also, most difficult to estimate.

Spaces are provided in Table III for tabulating products that will be made and their expected sales prices. All products, byproducts and major wastes should be noted. Special attention is required for byproducts. Their price and sales volume are often very hard to predict and may have a controlling effect on the profitability of the project.

The accounting department can provide sales prices for products the company is currently selling. These may or may not be list prices. If the product is one the company is now buying but will be producing in the future, a good purchasing department can provide information on going prices. Sales taxes should be deleted from these figures. For items the company neither sells nor buys, preliminary information can be obtained from the *Oil, Paint, and Drug Reporter* and

from the quarterly price issues of *Chem. and Eng. News.* These sources give general information and should never be used for construction authorization.

A check should be made for possible uses of any major wastes resulting from a proposed project. Potential development and suggested research in this direction should be noted at the bottom of the sheet if such possible income is not reliable enough to be included in the tabulation.

An excellent source of information on market-research methods, together with a very complete listing of sources for prices and other market data: sections 5 and 6 of "Chemical Business Handbook," edited by John H. Perry, McGraw-Hill, 1954.

In many cases, your company's sales or commercial development department will have information on prices and volume at which new products may be sold. A resume of information, prepared by the originating parties, should be included in the appropriation request.

For all projects over $25,000 that involve return computation, increase in sales is estimated for the first, third and fifth years. Production levels for each of these years are shown by arrows on the return curve of the summary sheet. For projects over $250,000, sales price and volume are forecast for each year of project life, before calculating interest rate of return. (To be discussed in the next part of this report.)

Calculating Distribution Costs

Distribution costs include equalized freight, freight taxes, allowance for returns, absorbed freight, warehousing costs, import duties (less drawback) and other expenses related to physically moving the product to the purchaser.

Equalized freight is the difference between (a) freight from the actual manufacturing point to the customer and (b) freight from the customer's basing point to the customer. Basing point is a producing plant nearest the customer, whether or not owned by the company. In many cases, therefore, the company's own plant may be the basing point and no equalization is required.

Some process products, such as plastics, are sold on the basis of free delivery to the customer, rather than what is essentially free delivery to the customer's basing point. In these cases, the entire freight from producing point to customer is paid by the company. This is absorbed freight. Assistance in determining

Table III	ESTIMATED INCREASE IN ANNUAL INCOME & RETURN	Appropriation Number			Supp.
		06	41	4120	

Title

50 Tons/Day Kiln for New Ores

Sales Rate, principal product

Added Tons of Roast / Yr.			17,500		13,130	8,750
Added in % of added capacity			100 %		75 %	50 %
Total sales rate Same						
Total rate in % of new total capacity			___ %		___ %	___ %

Increase In Sales

	Price	Units			
Hi - roast	$1.20	T	$ 2,100,000	$ 1,575,000	$ 1,050,000
Slag (0.6 T/T Hi-R.)	$3.50	T	36,800	27,600	18,400
Kiln dust (0.05 T/T Hi-R.) [1]	–		–	–	–
Total increase in sales			$ 2,136,800	$ 1,602,600	$ 1,068,400

Increase in Distribution Costs

Freight equalization Only producer				
Warehousing None				
Total increase in distribution costs	$ 0	$ 0	$ 0	

Increase in Net Sales	$ 2,136,800	$ 1,602,600	$ 1,068,400
Increase in Manufacturing Cost (from Added Manufacturing Cost Sheet)	$ 1,346,000	$ 1,092,000	$ 826,000
Increase in Gross Profit	$ 790,800	$ 510,600	$ 242,400

Adm., Sales & General Expense
(based on net sales less trans. raw mat. @ capacity)

Hq. 2%, Div. 6.5 % of $1,675,800 [2]	$ 142,800	$ 142,800	$ 142,800
Increase in Net Profit before I.T.	$ 648,000	$ 367,800	$ 99,600

Income Tax

State -- %	–	–	–
Federal 52 %	337,000	191,000	51,700
Total Income Tax	$ 337,000	$ 191,000	$ 51,700
Increase in Net Profit after I.T.	$ 311,000	$ 176,800	$ 47,900
Return, in % Net Profit after I.T., on Gross Inv. of $722,000	43.1 %	24.5 %	6.7 %

Notes:

[1] Very fine, almost pure silica. No present use known.
[2] $2,136,800 less $461,000, "B" ore from Loco Mines Div.

Prepared by:	Date	Checked by:	Date
P. J. P.	6/29/60	R. P. E.	6/29/59

equalized and absorbed freight can usually be obtained from traffic and accounting departments. Railroad assistance is also available. Cost of warehousing can be determined from standard cost tabulations or by separate computation when required. Reference is again made to Perry's "Chemical Business Handbook" for general information on distribution costs.

Net sales is total income from sales less distribution costs, as described above.

Need Manufacturing Costs

For a discussion of manufacturing costs see p. 245 of this volume.

Manufacturing cost is not the same as cost of sales. Cost of sales is manufacturing cost for a given period, plus or minus manufacturing cost of any products drawn from or added to inventory. Since return estimates for appropriation requests cover a full year, it can usually be assumed that inventory effects cancel out and manufacturing cost equals cost of sales for the typical year.

Gross profit is net sales less manufacturing cost.

Administrative, Sales and General Expense

A major company overhead item comes under the heading of administrative, sales and general expense (A.S.&G.). This includes: cost of general company management, accounting, purchasing, personnel service, traffic, sales advertising, public relations, engineering (portion not capitalized), research, development, corporate taxes (other than income taxes), office expense, interest on corporate debts* and all other items of general corporate expense. State income tax is sometimes included, although if significant it should be a separate item in the return tabulation.

A. S. & G. expense is, in general, proportionate to net sales, the percentage tending to be less for large corporations than for small ones. For a process company of moderate size, whose principal sales are to other manufacturing companies, incremental A. S. & G. may be about 8% of net sales. The figure will be substantially larger when direct selling to retailers, consumer advertising or heavy research expenditure is involved.

For new products that will have continuing promotion costs not consistent with established products, a separate calculation of advertising and selling expense should be made.

In a divisionalized company, separate A. S. & G. determinations are necessary for headquarters and for each division. The headquarters portion may be 2-3%, depending on how individual items are assigned by accounting, the nature of the business, and management policy. Variations between individual divisions of a company will come from the same sources that produce variations between individual companies. Percentages can range from 3% for a large well-established bulk-product division to about 25% for a small, newly established division serving many customers.

In analyzing and computing A. S. & G. by percentage of net sales in a divisionalized company, the value of any raw materials from other divisions is deducted from the divisional net sales figures. This eliminates pyramiding of overheads. It also, in an approximate way, gives credit to the using division for eliminating outside expenses connected with sales and special services.

In deriving percentages to be used for A. S. & G. estimating, the net sales of the company or division operating at full capacity should be used as a base. For the new unit, such percentages are applied to net sales at 100% capacity. These A. S. & G. data are assumed constant over the range of operating levels presented. Such an assumption is reasonably correct, since low operating levels often demand added sales, advertising, research and other headquarters activities, which balance the items that can be cut as production goes down.

Correct Levels for A.S.&G.

A. S. & G. percentages show interesting patterns when plotted over a period of time. Major capacity additions normally will decrease the percentages when units first come into production, but percentages drift back upward as time goes on, stabilizing at a new level somewhat below the one preceding the addition. Management changes have their effect. A new manager for an established, profitable division usually can decrease A. S. & G. by trimming "soft" practices. A new manager in a faltering division, or company, is likely to increase A. S. & G., as he pushes sales, advertising or research effort.

Correct A. S. & G. levels are a matter for decision by the best management opinion available. In many companies, these expenses could be cut, so that immediate profits per share would appear to double. Future profits are jeopardized by such action, however, since it involves cutting research, advertising, engineering and so forth. These are the tactics of the company raider. They contrast with more conservative practices that build up hidden assets of good-will, elite personnel, research potential and public acceptance. Management can therefore be a substantial factor in forecasting A. S. & G.

Net profit is gross profit less administrative, sales and general expense.

Net profit less state and federal income taxes must also be calculated. Incremental federal income tax is 52% at present for all but the smallest process companies. State income taxes are variable and specific state regulations must be consulted. If the state tax is minor, it can sometimes be included in A. S. & G., even though it is more correctly a function of net profit.

---

* The handling of interest on indebtedness, used here, assumes that the corporation obtains its funds from a relatively fixed mix of common stock, retained earnings, loans and other forms of indebtedness. Each project pays its proportionate part of the interest cost of obtaining money, with no regard to the actual cost of obtaining funds for the particular project. This assumption is reasonably correct if the funds are obtained in normal fashion and the mix of ownership and indebtedness is not seriously disturbed. Special handling may be required for joint ventures and other cases involving stock or bond issues.

What Is Return?

The return of a project at a given level of operation is the percentage of annual net profit after income taxes, using gross investment as a base for calculations.

Return for each of the three selected production levels is plotted against production rate on the graphical insert of the summary page of the request for appropriation. Shape of this curve is an indication of project sensitivity to variations in sales. Arrows are placed on the curve to show expected sales for the first, third and fifth years. The general elevation of the curve is an indication of return at varying sales levels. Point of zero return is the breakeven point for the project.

Profitability and Risk

Acceptance or rejection of a project hinges on interest rate of return at a given risk.

Return is a measure of the profit potential of a project at a given level of sales. But for important projects, it is desirable to have a broader measure for profitability; one that indicates net effective profitability of the project over its life span. This must take into account not only the return at a given level of sales, but what sales volume and price levels are likely to be, and over what period of time they will be maintained.

Obviously, this is a more difficult determination to make, and there is still much discussion about the various techniques used. The extra work involved can only be justified for projects of considerable importance to the company.

Interest Rate of Return Easily Understood

In the author's opinion, interest rate of return is the most easily understood and technically correct method of supplying a single figure indicating project profitability.

In using this method, the project is examined just as a banker considers a mortgage he is taking on a house. The banker sets up a mortgage so that at all times during its life it will return to him a suitable fixed interest on the unpaid balance of the loan. The owner has the privilege of paying on the capital amount of the loan from time to time, but always must pay interest on the unreturned balance.

Similarly, a company can value a project in terms of a constant annual rate of interest that will be produced on the unreturned balance of investment during a project's life. Any money netted by the project in excess of this constant annual "interest rate of return" is credited against project investment.

Interest rate of return is set so investment is reduced to zero at the end of the project's life. This interest rate on unreturned balance is an excellent measure of net effective project profitability.

Need Trial-and-Error

Interest rate of return must be determined by trial-and-error. For each year, estimates are made of any capital investment required and of net cash flow that the project will produce. Net cash flow is total sales income less total expenses paid out. The latter includes A. S. & G. and income taxes but not depreciation, since depreciation does not produce an outflow of cash.

A reasonable interest rate of return is assumed. Starting with the first year, the project is charged with interest at this rate on the unreturned balance of company funds invested in the project at the middle of the year. Any excess of funds received by the project and not required for this interest, or for normal expenditures, is in effect paid by the project to the company to reduce the unreturned balance.

Continuing this procedure, as indicated in Table IV, a year is reached when the entire investment, plus interest at the assumed rate less the salvage value of plant and working capital, has been returned to the company. If time to accomplish this is less than the estimated life of the project, then the assumed interest rate is too low. A higher rate is assumed and the procedure repeated. Three trials, plotting percent inter-

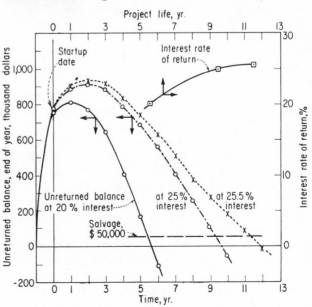

Figure 2 Interest-rate-of-return curve gives true rate for 10-yr. life

Table IV — Trial-and-error computer calculations

Yr.	COL. 0 Output Tons/ Day	COLUMN S Net Sales, (Gross Sales less Dist. Cost)[1]	COLUMN M Mfg. Cost less Ave. Dep. plus A.S. & G.[3]	COLUMN D Depreciation Sum-of-digits %	COLUMN D $	COLUMN P Net Book Profit Before Fed. In. Tax (S-M-D)	COLUMN T Income Tax, 52% of P	COLUMN F Net Cash Flow (S-M-T) or (P-T+D)	COLUMN I Investment by Company
	0	0	0	0	0	0	0	0[9]	\$687,000[4]
1	30	1,280,000	1,040,000	18.2	64,000	176,000	92,000	148,000	35,000[5]
								148,000[9]	
2	35	1,500,000	1,146,000	16.4	57,700	296,000	154,000	200,000	0
								348,000	
3	40	1,710,000	1,250,000	14.5	51,000	409,000	213,000	247,000	0
								595,000	
4	45	1,920,000	1,358,000	12.7	44,800	517,000	269,000	293,000	0
								888,000	
5	45	1,920,000	1,358,000	10.9	38,400	524,000	272,000	290,000	0
								1,178,000	
6	45	1,920,000	1,350,000	9.1	32,000	530,000	276,000	286,000	0
								1,464,000	
7	45	1,920,000	1,358,000	7.3	25,700	536,000	279,000	283,000	0
								1,747,000	
8	45	1,820,000[2]	1,358,000	5.5	19,400	443,000	230,000	232,000	0
								1,979,000	
9	45	1,720,000[2]	1,358,000	3.6	12,700	349,000	181,000	181,000	0
								2,160,000	
10	45	1,625,000[2]	1,358,000	1.8	6,300	261,000	136,000	131,000	0[6]
		17,335,000	12,942,000	100	352,000	4,041,000	2,102,000	2,291,000	-50,000
11								131,000	
12								131,000	

[1] From curve plotted from data on annual income & return sheet.
[2] Assuming price competition in latter years.
[3] From curve plotted from data on manufacturing cost sheet (Chem. Eng., April 17, 1961, p. 181).
[4] Gross investment less working capital, assumed spent over one year prior to startup.
[5] Working capital.
[6] Salvage, at end of 10th year, of working capital and scrapped plant.

est rate vs. time to return investment, are usually sufficient to establish a true rate for the expected project life (Fig. 2).

Interest rate of return for the project is noted on the initial, summary page of the request for appropriation.

The relationship between return on gross investment and interest rate of return depends on the sales level selected for quoting return on gross investment. It might be said that the sales level producing a return equal to the interest rate of return is the effective sales rate for the project. This is not strictly true, because the percentage bases are quite different.

For return-on-gross-investment computations, the base is a fixed investment over the life of the project. For interest-rate-of-return calculations, the base is an unreturned balance that usually rises in the early stages of the project, drops to salvage value at the end.

An excellent cost-engineering research project is possible in this area. Interest rate of return should be computed for a series of cases with varying relationships between unit gross investment and annual sales, and with varying ratios of unit manufacturing cost to sales. The resulting curves should be illuminating.

Cumulative Cash Flow

Cumulative cash flow is developed as indicated in Table IV and tabulated for key years on the summary page of the request. Negative values mean that financial planning must include a cash provision to cover operating losses in early years.

Also from a financing viewpoint, cumulative cash flow indicates when the project may be expected to complete the return of cash required for its execution.

Computers Permit Detailed Analysis

Although the theory of interest rate of return is simple enough, its practical application involves much laborious computation. This is especially true because:

• Best results are obtained if individual sales and expenditure figures are established for each year of project life, considering the expected pattern of growth of sales and deterioration of price with competition.

• Most firms use a policy of depreciation that writes off properties rapidly in early years. This

yield interest rate of return for 50 tons/day kiln

20% INTEREST			25% INTEREST			25.5% INTEREST		
R	C	B	R	C	B	R	C	B
Interest Returned $\frac{B_1+B_2}{2} \times 20\%$	Capital Returned (F - I - R)	Balance Unreturned[7]	Interest Returned $\frac{B_1+B_2}{2} \times 25\%$	Capital Returned (F - I - R)	Balance Unreturned[7]	Interest Returned	Capital Returned (F - I - R)	Balance Unreturned[7]
76,300	-763,300		98,300	-785,300		100,500	-787,500	
		763,300[9]			785,300[9]			787,500[9]
157,100	- 44,100		208,000	- 95,000		214,000	-101,000	
		807,400			880,300			888,500
157,300	42,700		223,000	- 23,000		230,000	- 30,000	
		764,700			903,000			918,500
142,500	104,500		224,000	23,000		232,500	14,600	
		660,200			880,000			903,900
106,800	186,200		210,000	83,000		221,500	71,500	
		408,000			797,000			832,400
58,500	231,500		186,500	103,500		201,000	89,000	
		176,500			693,500			743,400
7,500	278,500		157,200	128,800		175,300	110,700	
		-102,000			564,700			631,700
			120,000	162,200		143,400	139,600	
					402,500			492,100
			81,800	150,200		110,100	122,000	
					252,300			370,100
			46,200	134,800		81,700	99,300	
					117,500			270,800
			14,700	116,300		60,000	71,000	
					- 49,200			199,800
						39,200	91,800	
								107,600
						12,300	118,700	
								- 11,100

- - - - - 5.5 yr.[8] - - - - -   - - - - - 9.4 yr.[8] - - - - -   - - - - 11.5 yr.[8] - - - - -

For 10 yr. life, interest rate of return = 25.2%.

> [7] $B_2 = \dfrac{B_1(1+r/2)+I-F}{1-r/2}$ where r is the selected interest rate. Solve for B_2, compute R and correct B_2 to fourth significant figures. B_2 also = B_1 - C.
>
> [8] Plot "balance unreturned" to interpolate to fractional years (Fig. 2).
>
> [9] Totals at end of each year.

means that the allowance for depreciation, and therefore income tax, will be different for each year, even if all other items are constant.

- The dollar interest return to the company for an individual year is the interest rate of return times the average unreturned investment for the year. Investment at the first of the year is known, but year-end investment depends upon payment the project is able to make on the principal amount during the year. This complicates the arithmetic.

- Computation of payout time for at least three interest rates is usually required to determine interest rate of return for the assumed project life.

Although a correct solution to interest rate of return can be turned out by any trained engineer or accountant doing hand calculations, the procedure is quite time consuming so the job is best turned over to a computer.

For major projects, it is a good idea to put all the project economics, not just the interest rate of return, on punch cards. In this way, the entire manufacturing cost, net profit and interest rate of return computation can be carried out rapidly by a preprogrammed computer, once it has the basic project information.

Improved data can be substituted for earlier assumptions with little difficulty. Special studies can readily be made on the effect of variations in particular elements of the project, such as process yield, labor rates, investment or project life.

Computer Techniques

Computer programming for manufacturing-cost and net-profit computations has been developed by several chemical process companies (Diamond Alkali, Dow, Monsanto, Columbia Southern). An engineer makes appropriate entries on a form, corresponding cards are punched, and the program run. Runs can be made with input data for each successive year of project life, to arrive at net cash flow. This can be further programmed to determine interest rate of return. Complete print-outs of key data can be made for inclusion in reports. Curves can also be produced showing the effects of changing various elements of the project. The programming of these computations can be a lifesaver when last-minute significant changes must be made in presentations.

Consultants are available and should be used to cut

Figure 3 Typical target rates of return for various risk projects

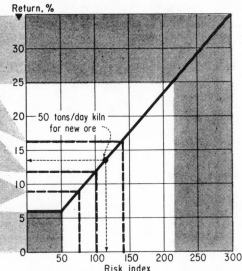

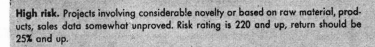

High risk. Projects involving considerable novelty or based on raw material, products, sales data somewhat unproved. Risk rating is 220 and up, return should be 25% and up.

Fair risk. Projects somewhat outside the present field of activity, or novel projects and processes that have been thoroughly investigated. Risk rating is about 140, with a return of 16%.

Average risk. Usually projects in present field of operations, but involving some novel element or lack of definite market information. Risk rating is about 100, with a 12% return.

Good risk. These are often expansions of existing operations where there's a known market. Risk rating is about 75, with a 9% return.

Excellent risk. These projects usually are designed to improve yield or reduce labor costs on a known process. Usual risk rating is under 50, with a target rate of return of 6% or less.

the substantial cost of adjusting existing computer programs to a particular company situation. If computer time is not available on company machines, it can be rented from computer service companies or from other industrial companies with idle computer time.

Once a program is established, the solution of individual problems is a straightforward matter, best handled by a company's own cost-engineering section. By direct handling of its own computation, any company will be able to wring the ultimate in project analysis from the computer program.

Despite its computation complexities, interest rate of return is a simple, precise and inclusive criterion of project profitability. It consolidates, in one figure, a broad array of information and relationships affecting profitability. The complexity of the procedures required to reach this single figure is due to the conditions under which modern industry operates and not to the method itself. Failure to consider some of the complex facts of industrial life is common. The result has often been glowing profit predictions, followed in many cases by hard-won, commonplace returns.

Risk Measurements Are Difficult

In selecting projects, consideration of apparent profitability is not enough. The risk of not obtaining this profitability must be considered.

High-risk projects should have correspondingly high returns if they are to be accepted in lieu of low-return, low-risk projects. Some criterion is required, however, for "high" and "low" in project-risk appraisals.

Measuring risk is difficult. The rating method suggested here is of the check-list type. It assumes that the possibility of failing to reach the apparent profitability of a project is a function of:

• Inherent reliability, for the particular project, of each of the items that can have a major effect on profitability.

• Degree to which each of these items has been investigated.

A project-risk-rating form is illustrated in Table V. The list of items includes these that have a major effect on profitability for process projects, excluding those that affect investment and have already been considered in establishing the investment reserve.

Space is provided for summary notations of project conditions. Inherent risk and investigation status are rated "excellent" through "poor," or with corresponding numerical ratings of one through five. Multiplying the inherent risk and investigation ratings gives a risk index for each item. Total risk index for items in the market group is doubled, reflecting its importance. The grand total is a measure of over-all project risk.

Target Rate of Return

The relationship of a given degree of risk to acceptable interest rate of return is a matter of management policy decision. The curve (Fig. 3) illustrates a typical pattern for target rate of return vs. risk index.

For minor projects, target rate of return can be approximated by a simple description. For major ones, a check list helps to determine risk index; target rate is then read from the curve. This is entered on the summary sheet of the appropriation request, for comparison with the interest rate of return forecast for the project.

Needless to say, this detailed review of risk status is justifiable only for important projects. The technique by its nature cannot be precise, but the very act of detailed review is most valuable. Areas requiring more investigation are uncovered. Avoidable risks are eliminated. Unavoidable risks are faced and openly accepted or rejected by management.

Table V — PROJECT RISK RATING

Project Titled 50 Tons/Day Kiln for New Ores No. 06-41-4120

Date 6/29/59

Rated By R. J. Collins Checked By J. L. Johnson

Item	Risk Characteristics (4)	Risk Rating (1)	Investigation Rating (2)	Risk Index (3)
Raw Materials & Fuel				
Source & Transportation....	Our mine, Lake Shore R.R.	E	E	1
Price.................	Mining costs well developed.	E	E	1
Quality................	May vary. Drilling in progress.	F	F	16
Hazards & Regulations	None expected; similar to present.	E	E	1
Subtotal				19
Process				
Yield................	Prel. pilot-plant data; may vary.	F	F	16
Utility, Labor & Supply Usage	Similar to present kilns.	E	G	2
Patents & Know-how	Search complete. Buying from Ware.	G	E	2
Obsolescence	Ware working on electrolytic method; negative results to date.	S	S	9
Subtotal				29
Products & Byproducts				
Quality...............	Pound samples tested by customer.	S	S	9
Packaging & Shipping	Similar to present.	E	E	1
Hazards & Regulations	None expected, except silicosis with dust.	G	G	4
Subtotal				14
Market				
Competitive Products				
Existing	None known.	E	G	2
Potential............	Other high-ratio halides.	G	G	4
Competitive Producers				
Existing	None.	E	G	2
Potential	Others mining similar ores.	F	G	8
Breadth & End Use	Space projects only.	F	E	4
Sales & Service Methods....	Existing.	E	E	1
Seasonal & Weather Effects..	None expected.	E	E	1
Penetration Percentage ...	100% of market.	P	E	5
Subtotal, doubled (5)				27x2=54
RISK INDEX, Grand Total				116

Notes:

(1) E, excellent conditions, little risk; G, good conditions, moderate risk; S, satisfactory; F, fair; P, poor; O, not involved.

(2) Use E thru P, as above, to indicate degree of investigation of the item to date.

(3) Product of columns (1) and (2), using numbers 1 thru 5 to replace letters E thru P.

(4) Refer to separate page for comments requiring more than available space.

(5) Doubled to give proper weight to market factors.

Reporting Performance

Keeping tabs on approved projects provides feedback for future improvements in forecasting and performance.

When an appropriation has been approved, it becomes the responsibility of project management to monitor execution of the program for speed and economy.

The principal tool for this is the monthly analysis of appropriations, with its associated cost and progress sheet. The other major tool used to monitor projects is the construction progress schedule, discussed in some detail in *Chem. Eng.*, Apr. 3, 1961, p. 155.

A form used for the monthly analysis report, with sample entries and notes on its preparation, is illustrated by Table VI.

How to Use Progress Sheet

For major projects (over $100,000) analysis of proj-

ect status is economically justifiable, involving a cost and progress sheet (Table VII). Work sheets, covering every subaccount of the appropriation, are prepared in this same form, to obtain data entered on the cost and progress sheet. For very large appropriations, status of each major section of the project is also reported.

Reporting Project Performance

Within two months after project closure, the accounting department should prepare a complete performance report for the capital phases of major appropriations (Table VIII).

Early completion of the report is desirable, at a time when all records are in usable condition and the report can be prepared by personnel completely familiar with the project. Some items, such as unsettled claims, may have to be estimated.

To the report are attached the final subaccount tabu-lations of material cost, labor cost and construction man-hours. Engineering man-hours are also reported, by project section and class of work.

This is a final, official record that can be confidently used in preparing future plans and estimates for similar projects.

Feedback From Operations

When a major revenue-producing project has been in operation long enough so dollar return is somewhat stabilized—but not later than the second anniversary of project completion—the accounting department should prepare an operating performance report (Table IX). This compares actual project return with return expected when the request for appropriation was approved.

Actual and estimated values are compared for the production level reached during the project's best operating month up to the time of the report.

Table VI — MONTHLY ANALYSIS OF APPROPRIATIONS

Division _____ Chemical
Plant/Location _____ Chicago
Month Ending _____ 4/30/60

Approp. No. (1)	Approp. Title (2)	Date Apprv'd. (3)	Expected 98% Completion Date			Paid for to Date (7)	Expected Total Cost			Expected Variance		Comments (13)
			Per Approp. (4)	Last Month (5)	This Month (6)		Reestimate from Cost & Progress Sheet (8)	Per Approp. (9)	Approp. with Supplements (10)	Over-run Dollars and Percent (11)	Under-run Dollars and Percent (12)	
06-41-4120	50 T/D Kiln for new ores	8/6/59	8/6/60	9/1/60	10/1/60	$459,000	$559,000	$440,000	—	$119,000, 27%	—	Over-run made necessary by new process information. Supplement being prepared.

(1) From appropriation. (2) From appropriation, but abbreviated to a few words. List all active appropriations for this plant. Projects under $25,000 may be eliminated at the option of the Manager, except at the end of the month in which the project is physically completed. (3) From appropriation. (4) From estimate of time of completion after approval, as stated in appropriation request. (5) From previous monthly analysis. (6) Supplied by individual in charge of construction. (7) Actual expenditures to date for the project, from accounting records. (8) Prepared from Cost & Progress sheets containing all cost information

available to date. Use for major projects (over $100,000) and for other projects when necessary. (9) From original appropriation request. (10) Appropriation amount plus all supplements approved to date. (11) The difference between "Expected Total Cost Re-estimate" from Cost & Progress sheet and the "Appropriation with Supplements". For minor projects, show over-runs only when actual expenditures exceed appropriation. Percentages are to be based on appropriated amount plus approved supplements. (12) The difference between "Expected Total Cost Re-estimate" from Cost & Progress sheet and the "Ap-

propriation with Supplements". For minor projects, show under-runs only at project completion. Percentages are to be based on appropriated amount plus approved supplements. (13) Indicate dates when projects and major divisions of projects are completed. Show status of supplement preparation, if required. Indicate projects to be dropped, curtailed or enlarged. Indicate principal progress and reasons for major changes since preceding report. Comments are prepared by accounting representative, with the help of the individuals in charge of design and construction.

Table VII — COST & PROGRESS SHEET FOR MAJOR APPROPRIATIONS

Division _____ Chemical
Location _____ Chicago
Month Ending _____ 4, 30, 60

Approp. No. (1)	Approp. Title (2)	% Completion				Material					Labor					Total	
		Detail Design (3)	Req. (4)	Purch. (5)	Field Labor (6)	Charged to Project (7)	Committed (8)	Req'd. to Complete (9)	Expected Total Cost (10)	Approp. w/Supp. (11)	Charged to Project (12)	Accrued (13)	Req'd. to Complete (14)	Expected Total Cost (15)	Approp. w/Supp. (16)	Expected Total Cost (17)	Approp. w/Supp. (18)
06-41-4120	50 T/D Kiln for new ores	98	98	94	50	380,000	4,000	25,000	409,000	300,000	64,000	11,000	75,000	150,000	140,000	559,000	440,000

(1) From appropriation. (2) From appropriation, but abbreviated to a few words. For very large projects, list major sections also. (3) Percentage that developed detailed-design man-hours are of expected total (by design engineer). (4) Percentage that dollar value of requisitions sent to purchasing is of expected total (by design engineer). (5) Percentage that value of filled requisitions is of the expected total requisition value. (Unfilled requisition value by purchasing, Balance by design engineer). (6) Must check with labor required-to-complete at the option of individual in charge of construction. (7) Material and fees invoiced and paid for, plus stores issued and freight paid

(by accounting and stores). (8) Unpaid-for material on order. Stores material billed to field storeroom but not issued. Estimated freight and escalation, if of significant size. Earned fees not paid (by accounting and stores). (9) Includes material requisitioned but not bought (by purchasing), requisitions not yet sent to purchasing (by design engineer), stores bills not yet issued to the storeroom (by design engineer) and fees expected but not yet earned (by accounting). (10) Total of three preceding columns. In early project stages, may be based in part on appropriation estimate. (11) Includes only approved supplements. (12) Company or

contractor field labor performed and paid for (by accounting). (13) Company or contractor field labor performed but not paid for. Estimated by accounting, aided by individual in charge of construction. (14) Labor required to complete the project but not yet performed. Includes labor contracted for but not yet performed (by individual in charge of construction). (15) Total three preceding columns. In early project stages, may have to be based in part on appropriation estimate. (16) Includes only approved supplements. (17) Total of expected totals for material and labor. (18) Includes only approved supplements.

Table VIII — CAPITAL PERFORMANCE REPORT

Date Dec. 15, 1960
Appropriation Number 06 | 41 | 4120

Division, Department or Subsidiary Chemical
Location Chicago
Plant Ore Roasting

Descriptive Title 50 Tons/Day Kiln for New Ores

Gross Investment	Appropriated + "Purpose" Supp.	Actual	Over-run	Under-run
Direct Capital Investment	$320,000	$418,000	$98,000	–
Supporting Utilities & Gen. Facilities	66,000	66,000	–	–
Transfer of Existing Capital Items	50,000	50,000	–	–
Operating Expense	20,000	45,000	25,000	–
Forced Retirements	95,000	95,000	–	–
Less: Tax Credit @ 52 % of $65,000	(34,000)	(47,000)	–	13,000
Working Capital	35,000	* 44,000	9,000	–
Other Items Know-how purchase	50,000	50,000	–	–
Sub-total	$ 602,000	$ 721,000	$ 132,000	$ 13,000
Reserve	120,000	- 0 -	–	–
Total	$ 722,000	$ 721,000	$ 132,000	$ 13,000
Net Change	–	–	$ –	$ 1,000
% Change	–	–	–	Negligible

Capital Performance Data

Appropriation App'v'd.	8/6/59	Supplemented	6/15/60	Closed	11/3/60
Detail Eng., Man-hr.	11,800	Started	8/10/59	Complete (98%)	6/1/60
Construction Man-hr.	25,206	Started	4/1/60	Complete (98%)	10/5/60
Production, Sched.	8/6/60	Initial	10/10/60	Commercial (75%)	11/1/60

Summary Comments (expand on attached sheets if necessary):
Process changes based on pilot-kiln tests used up all of the allowed reserve and were principally responsible for extending project completion by two months. Operating start up expense was substantially larger than expected.

Prepared B. L. C.	Checked R. J. B.	Approved B. B. C.

*Working capital should be re-estimated, based on data available to date. Attach complete report, by sub-account, of actual material and labor costs, engineering man-hours and field man-hours. This report not required for projects under $250,000.

Appropriation estimate figures include changes due to approved "purpose" revisions only. Otherwise, they are the figures on which original approval of the project was based. Interpolation is necessary to get values at actual production levels, based on the three levels computed in the estimate.

Actual figures are from operating records, but are computed on the same principles used in preparing the estimate. All costs, for example, are incremental ones.

The "Using Estimated Value" column is a recomputation of the return, using actual figures for all items except the single item in question. The estimated value is used for this. The return computed in this way indicates influence of the variance of this single item on project return.

Preparing actual figures for comparison with forecast figures is usually difficult. Data must be based on project economics considered in exactly the same way as when preparing the estimates. This seldom conforms precisely to the viewpoint used in preparing normal operating reports for project control. And arithmetic for comparisons is quite tedious. In fact, it may not be attempted unless project economics are programmed for an electronic computer.

In the "comment" space, notations should be made with the objective of pointing up areas where improvements seem possible in both project operations and in future estimating of profitability.

John W. Hackney is Manager of Cost Engineering for Mobil Research and Development Corp. His experience has included cost engineering and project management assignments with M. W. Kellogg Co., Beaco Ltd. of Montreal, Diamond Shamrock Co., and the Aluminum Co. of America. Mr. Hackney is a Fellow of ASCE and a past-president of AACE. He has received B.S., M.S., and C.E. degrees from Carnegie-Mellon University. In addition to the *Chemical Engineering* articles republished here, he has authored many other papers on cost engineering and project management. His book "Control and Management of Capital Projects" was published by John Wiley and Sons, Inc. in 1965.

ACKNOWLEDGMENTS

Since this presentation has been built on an experience of many years in various phases of appropriation work, it is impossible to acknowledge all of the individuals who have directly or indirectly contributed

to its preparation. Special acknowledgment is due, however, to Roy Glauz, Development Engineer; E. J. Isackila, Cost Engineer; S. Perkins, Research Engineer; and Ray Armor, Controller, all of Diamond Alkali Co. and to E. A. Vaughn, Manager of Fixed Asset Accounting for the Aluminum Co. of America.

Table IX — OPERATING PERFORMANCE REPORT	Date June 1, 1962	Appropriation Number 06 | 41 | 4120
Division, Dept. or Subs. Chemical	Location Chicago	Plant Ore Roasting

Title   50 Tons/Day Kiln for New Ores

Values for Best Month to Date	Appropriation Estimate*	Actual	Using Estimated Value**	
			% Return	Changes from Act.
Production, month of April, 1962	-		-	-
Raw materials & fuel	$ 67,780	$ 91,100	45.5	-18.8
Utilities	1,600	1,700	26.7	0
Labor	13,560	27,300	37.8	-11.1
Supplies & Misc.	4,960	5,100	26.7	0
Fixed Costs	8,410	16,100	33.0	- 6.3
Loading, Packing & Shipping	1,860	2,100	26.7	0
Manufacturing Cost	98,170	143,400	62.9	-36.2
Sales Value of Production	150,600	201,000	-13.4	+40.1
Distribution Cost of Production	0	12,000	36.5	- 9.8
Gross Profit	52,430	45,600	-	-
Administration, Sales & General	11,900	12,000	26.7	0
Net Profit	40,530	33,600	-	-
Income Tax	21,100	17,500	-	-
Net Profit after I.T.	19,430	16,100	-	-
Annualized Profit	233,000	193,000	-	-
Gross Investment	$ 722,000	$ 721,000	26.7	0
Return	32.3 %	26.7 %	-	-

Interest Rate of Return	Appropriation	Current Re-estimate	Cumulative Cash Flow	
			App'n.	Re-estimate
Target Rate	14.0 %	-	1) $148,000	$ 5,112
Forecast Rate	25.2 %	19.1 %	2) 348,000	200,000
Forecast Life	10 yr.	10 yr.	3) 595,000	450,000
			4) 1,178,000	810,000
			5) 2,291,000	1,750,000

Summary Comments:

Inflation substantially increased sales prices, but the gain was wiped out by increases in manufacturing cost and the necessity for incurring distribution costs to obtain sales.

Profit still well above target estimate, although not as attractive as anticipated.

Prepared R. L. Beam	Checked B. B.	Approved R. B. L.

*Appropriation Estimate is from Appropriation plus Purpose changes, interpolated to actual production.
**All items taken at actual value except the one being tested, for which estimated value is used. This report not required for projects under $250,000.

Cost Engineering

The job of today's cost engineer is not just to estimate capital or operating expenses but rather to tell management how much profit can be made from a venture, with what likelihood, and at what risk. This broader mission calls for new skills, and offers new opportunities.

G. ENYEDY, *Diamond Alkali Co.*

Cost engineering represents the merger of the technical with the business aspects of the company's operations. The man who masters this relationship will find that his cost engineering training can readily lead him into management-science areas that encompass the improvement of corporate operations, as well as the improvement of any specific plant unit.

By cutting across all engineering fields, cost engineering is really an adjunct rather than a replacement for an engineering specialty. There are no schools that give degrees in cost engineering today, and there probably never will be. And yet, the cost engineer's talents are currently very much in demand, and the scope of his mission is increasing dramatically.

A decade ago, the cost engineer was defined as one who applied scientific principles and techniques to problems of cost estimation, cost control, and profitability. To express these areas of interest mathematically:

$$\text{Profitability} = \frac{\text{Receipts} - \text{Expenditures}}{\text{Investment}}$$

Thus, in the chemical process industries, the cost engineer was the one who estimated the cost of equipment and plants (*investment*), the cost of manufacturing the output (*expenditures*) and, after receiving sales volume and selling-price data (*receipts*) from the market research group, the profitability of a particular venture.

But the cost engineer of today has found that if he is to make a meaningful profitability evaluation, he must carefully consider all factors, including the "receipts." This income information can no longer just be plugged in on blind faith, because greater depth is needed and is obtainable than was the case a few years ago. Although not actually doing market research, the cost engineer is concerned with the forecasting of sales volume and prices under various different assumptions, and with evaluating the effect of various possible sales volumes on manufacturing costs, overhead, product life and so on.

There are other reasons why the cost engineering field is becoming more challenging: the magnitude of company operations is expanding; financial commitments are greater; and any error in judgment in certain areas has great effect on the outcome of a decision. Also: The complexity of enterprises is increasing; more facilities mean more interaction between plants; competition is getting stiffer; profits are harder to earn, and the spread between costs and the earning rate is becoming less.

Career Paths of Economics-Oriented Engineers

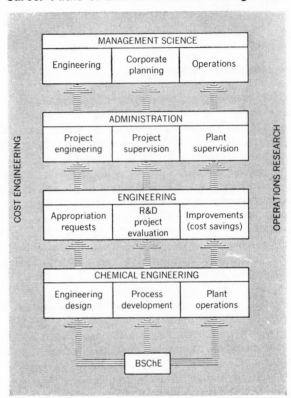

From the strictly technical level at the bottom of the chart, the paths lead to the highly economics-oriented management science level. On the left side, the job focus is on engineering (i.e., cost engineering); on the right, the focus is on the business-production problems of the company (i.e., operations research). Overlap of cost engineering and operations research activities occurs at each level, as well as along each path.

The cost engineer realizes that consideration must be given to uncontrolled factors, such as marketing idiosyncrasies and uncertainties arising in technological developments—as well as the traditional, more readily controlled factors, such as cost of plant facilities and plant operations. All these things have to be considered in a time environment in which the only thing that is certain is change.

Tools of the Cost Engineer

Today, the cost engineer can be looked upon as the engineering counterpart of the management scientist.

The chief difference is merely a matter of focus; the cost engineer works in the more technical area, but he uses the same basic tools—for example:

• The transportation-problem approach to locate a plant.

• Critical-path methods to control the plant's construction costs.

• Forecasting techniques to estimate manufacturing-cost improvements, selling-price declines, and sales-volume curves.

• The principles of economics to quantify the effect of time on the value of money.

• Statistics to assess the chances that the estimates will be reasonably correct.

• The computer to simulate and optimize proposed management strategies.

Obtaining the Background

The chemical engineering graduate of today can leave the college campus ill-prepared to do cost engineering. He has been exposed to an engineering curriculum crammed with fundamental scientific courses that have crowded out practical economics. Thus his first job assignments will be concerned with applying his technical training in such areas as process development, engineering design and plant operations. These initial assignments (which, incidentally, are recommended even for engineers with business school training) are necessary to establish a practical approach to the visualization and solution of problems—an approach that the chemical engineer must have if he is to become an effective cost engineer.

If he wants to expand his career opportunities, the would-be cost engineer has two choices. He can take courses in engineering economics, operations research and marketing at a nearby university; often this leads to a masters degree in business administration. Or he can take advantage of on-the-job training that may be offered by his own company. This sort of training assignment is invaluable because it is just not available in the classroom with the depth of coverage possible on the job; it can be combined with a study of the literature.

To supplement such training, the best sources of outside help is active participation in the American Assn. of Cost Engineers' technical programs. Here the engineer will be able to meet and exchange ideas professionally with people whose problems are similar to his own. If he is working in an engineering department, he can get help on capital-cost estimating, preparation of appropriation requests, and control of capital expenditures. In a research and development environment, he will learn to evaluate the economic as well as the technical feasibility of a new product or a new process. In plant operations, he will be concerned with cost savings arising from improved raw-materials use, process improvements, and better maintenance and operating procedures.

There are pitfalls during the cost engineering stage. Over-specialization has cut short many a promising career by its restrictive tendencies. This danger is the greatest in an engineering department where a cost engineer can over-concentrate on capital-cost estimating details at the expense of slighting or even ignoring the more sensitive factors such as marketing, pricing, and the interrelationships that exist in a profitability estimation procedure. He can forget that the optimum design is not based upon minimum capital cost alone. Said another way, he must remember always that what management really wants to know is how much profit can be made from a venture at what risk, not just how much a plant will cost.

It is this trap that has probably contributed more than anything else to the frequently encountered idea that a cost engineer is nothing more than a materials takeoff estimator. The detailed "nuts and bolts" type of estimation in the future will be handled by computers, so the material estimator's job will be virtually eliminated. He will have to get down to the true engineering aspect of his work. His judgment as to whether an estimate is proper for the particular job that he is trying to do will become paramount. He no longer will spend his time as a clerk adding up columns of figures.

Today and Tomorrow

As a group, cost engineers are primarily employed by private industry and concentrated heavily in the chemical and petroleum sectors. A 1964 salary survey has shown the range of cost engineers' salaries to be quite similar to that found in the EJC national survey for all engineers.

The cost engineer of the future has his work cut out for him. In his search for better estimating procedures, he will have to borrow highly developed techniques from many specialties. As he assimilates these various disciplines, his contributions to his employer and to society are bound to become that much greater.

Author of This Section . . .

Gustav Enyedy, Jr., is supervisor of computer services at Diamond Alkali Co.'s T. R. Evans Research Center in Painesville, Ohio. He is responsible for all scientific and engineering computer applications, including project evaluation economics, forecasting models, process simulation and control, and applied statistics. A graduate of the Case Institute of Technology, he has B.S. and M.S. degrees in chemical engineering. Last year, Mr. Enyedy was re-elected as Technical Vice-President of the American Assn. of Cost Engineers.

Business Potential of R&D Projects

Industrial R&D projects are not very meaningful unless their results lead to production profits. This is a way to analyze whether they will be commercially profitable.

GERALD TEPLITZKY, *Research Div., W. R. Grace & Co.*

The research manager has increasingly been given the responsibility of picking research projects with the greatest commercial potential. Consequently, there is a need for a more-analytical approach in selecting and evaluating research projects, so that a more-quantitative evaluation can be made early in the project.

For this, the research feasibility analysis can prove a useful tool. Such an analysis is a specific type of commercial-economic evaluation having objectives and intent somewhat broader than those normally thought of in connection with cost estimates.

Managing and directing research projects is, at best, a difficult task. A research manager must have the ability to understand the technical research objectives (and the researcher's implementations of them), as well as the means of quantitatively evaluating the commercial worth of the project. Although many techniques for evaluation of research projects have been developed, it is the judgment of the research manager that plays the major role in making the final decision.

The research feasibility analysis, like many other types of quantitative evaluation, is at best a tool and is only as good as the assumptions and judgment used in carrying out the study. This technique does not replace judgment and common sense.

Cost Estimating

There are many different ways to estimate costs, each having specific objectives and being intended to meet specific needs such as:

• Determining detailed costs for specific pieces of equipment to be used in existing facilities.

• Assessing required capital investments for new manufacturing facilities.

• Determining the cost of manufacture of new, modified and existing products.

The research feasibility analysis has objectives broader than any of the above single objectives. It is intended to evaluate the commercial-economic worth of an entire project, and takes into account such items as:
  • Market penetration.
  • Rate of market buildup to maturity.
  • Required capital.
  • Cost of manufacture.
  • Competitive and patent uniqueness.
  • Estimated likelihood of success.

Most important, the research feasibility study includes an analysis and interpretation of the results, not merely a series of uninterpreted figures that are subject to misunderstanding. This interpretation is perhaps the most important element in making this type of study a useful management tool.

Emphasis in feasibility analyses can differ, depending on which items are most important.

For some projects, the required capital may be critical to the economic success of the project, and consequently a major portion of the analysis should be devoted to obtaining an accurate assessment of capital. More often, however, required capital is not a major factor, and only contributes a small part of the total picture. Other items such as raw-material cost, rate of market buildup, operating and overhead costs, and pre-operating development and marketing expenses play much larger and more significant roles.

Still other cases require emphasis on level of production and the economics of competitive materials. It it largely up to the author of the feasibility study to determine the key factors, and to decide the emphasis to be placed on each phase of the analysis. There is no routine set of procedures that can be successfully applied to analysis of all research projects. The use of a "forced procedure," in place of judgment, can only lead to incorrect evaluation and misinterpretation of the project in question.

Types of Feasibility Analysis

In general, there are three major variations of the research feasibility analysis, which vary in objective, accuracy and implementation:
  1. Research guidance analysis.
  2. "Guesstimate" analysis.
  3. Project analysis.

Research Guidance Analysis

The most elemental type of feasibility analysis is that used for research guidance. This analysis is intended for research projects in the conceptual or early stages of development. Its specific objectives are:
  • To provide economic guidance on the effect of key variables. These variables might include raw-material costs, yield, and such specific things as effect of monomer ratio. This phase of the analysis is intended to provide direction to the research chemist.
  • To indicate the probable profitability that could

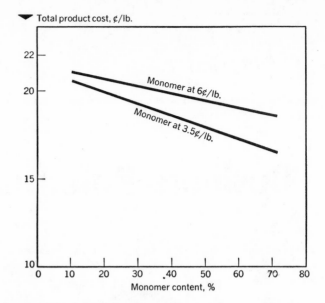

RELATION of product cost to monomer content—Fig. 1

be obtained from manufacture and sale of the chemical in question. This information is intended to assist the management of the research department in assigning a level of priority and effort to the project.

The research guidance analysis is not intended to judge the economics of a product or process that has not yet been developed, but rather to state economic guidelines that will provide direction during initial and conceptual stages.

This type of research feasibility analysis is illustrated in the following example, in which the research objective was to develop a new adhesive product with properties similar to a product already on the market and satisfactorily being used. The new adhesive would be made via a low-pressure batch process based on a minimal monomer concentration. This process would require a minimum capital investment, in contrast to the high-pressure continuous process used for the existing product, which is based on a high monomer concentration.

At the time the analysis was performed, the chemist working on the project had done some exploratory research and was interested in determining the likelihood of coming up with a product that could economically compete with the existing product, assuming technical success. The specific objectives of this research guidance analysis were to:
  • Establish the economics for the existing product and thereby establish an economic base for comparison.
  • Analyze the effect of key variables such as monomer concentration and raw-material prices on the economics of the research adhesive product.
  • Determine the potential profitability of the research adhesive as compared to existing material.

The key figures for the manufacture and sale of the proposed research adhesive are compared in Table I with those for the commercial product. This table

Research guidance analysis summary for new adhesive—Table I

	Competitive Product	Research Product
1. Capacity, million lb./yr.	50	20
2. Total required capital, million $	11.2	4.4
Fixed	7.5	2.0
Working	3.7	2.4
3. Monomer content of the product, %	70	20
4. Raw-material cost, ¢/lb.	7.8	12.0
5. Direct manufacturing cost, ¢/lb.	9.9	15.3
6. Total product cost, ¢/lb.	14.7	20.6
7. 15% ROIAT* selling price, ¢/lb.	19	23.5

* Return on investment after taxes.

shows that the major difference in manufacturing cost between the two products was in raw-material costs, which constituted a differential of about 4¢/lb. This table also indicates that, based on the assumed conditions of the research process, the existing material can be produced and sold at a price lower than the total product cost of the experimental adhesive. One additional step carried out in this research guidance analysis showed the effect of monomer concentration on the research adhesive economics (Fig. 1). This curve revealed that in order to be competitive with the existing product, a monomer concentration in the range of 70% had to be obtained.

In summary, this analysis clearly pointed out that it was not worthwhile to develop new adhesives using

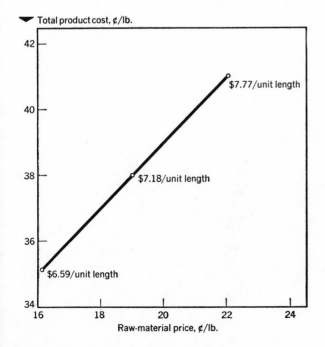

EFFECT of raw-material on economics—Fig. 2

the low-pressure process. At monomer concentrations much less than 70%, the economics for these materials would not compare favorably with those of the existing material. Hence, a 70% monomer incorporation was set as the research goal. As a result of this research guidance analysis, research department management became aware that the potentially lower capital inherent in this new low-pressure process was not enough to offset the higher raw-material cost unless the monomer concentration were raised to ~70%. Because the technical chance of success was very low, research management decided to drop the project.

Guesstimate Analysis

The research guesstimate feasibility analysis is intended to assess specific economics and commercial feasibility of a research project in the early stages of development where it is not yet possible to have specific detailed information such as flowsheets, yields, material balances, and detailed equipment schemes. This type of analysis is particularly useful in assessing the economics of potential competitive products.

The use of the research guesstimate analysis is illustrated by the following example: A research team had been working on a filament product for several years. During the latter phases of research on this project, a competitive product was introduced that might have presented a threat. Management asked whether the newly introduced product threatened the research product from the standpoints of both performance and economics. This had to be resolved before additional major research expenditures could be authorized.

Hence, a research guesstimate was carried out to determine the economics for the competitive material. The guesstimated economics were based on information obtained from journal articles and from specific product performance as measured by the research team. It became obvious that the company producing this product might have a captive advantage on the basic raw material used. Consequently, the effect of raw-material price on the competitor's economics was studied, as illustrated in fig. 2 and Table II. These data indicate that the competitive material does not have any economic advantage over the research product, and in fact has economic disadvantages due to high weight of material required to make a filament product with physical properties equivalent to the research product. These economics were predicated on the lowest assumed value for raw material illustrated in Fig. 2.

In summary, this guesstimate analysis established that the newly introduced material was not an economic threat to the proposed research product.

Management authorized additional funds to complete the research project.

Project Analysis

The most detailed type of feasibility analysis is the project analysis, which is used to evaluate projects

that are in the latter stages of development and are being considered for commercialization. For these, information is normally available on material balances, raw-material requirements, equipment requirements, and yields. The objectives of this analysis are to point out to research management the potential profitability in commercializing a process or product. Since this type of analysis will be used to authorize large amounts of spending, we must consider not only the research product in question but also competitive products, market penetrations, rate of buildup, or any other factors that could ultimately affect the potential commercial-economic success of the research product.

The economic summary from a project feasibility study, shown in Table III, was used by management in deciding whether or not to commercialize a new type of thickening product. These economics (including battery-limit capital estimates) were calculated for the production of the thickener at several different levels of capacity, and are compared with economic guesstimates of the three prime competitive products now on the market. In addition to economics, this table also shows relative thickening efficiencies of the various products, since it is necessary to consider a "use cost" (product price times quantity needed) in deciding the economic utility of the product. The range of plant capacities shown for the research product corresponds to pessimistic, realistic and optimistic estimates of market penetration. Although the economics for the competitive materials are less accurate than those for the research product, they provide order-of-magnitude numbers that can be compared.

As a result of this study, it was possible to conclude that the total product cost and selling price for the experimental thickening products were extremely sensitive to production volume. At the 2-million-lb/yr. level, based on available efficiency data, the research product would be competitive with the thickening products now on the market. Fig. 3 illustrates this conclusion. In this graph, the selling price of the research material has been plotted versus the existing competitive material at various selected efficiency parameters.

This project analysis concluded that the economics of this research thickening product appeared to be competitive with existing materials. On this basis, management decided that it was worth initiating the test-marketing commercialization stages to determine customer acceptance of the product.

Analysis and Presentation of Results

It should be emphasized that the engineer's prime function in carying out the research feasibility analysis is not the calculation of numbers. Although this is necessary, his main function is, first, deciding which factors to consider, then analyzing these factors and presenting the results in a coherent fashion.

Presenting management with a series of uninterpreted figures can cause trouble, the figures may not convey the intended meaning nor point out the assumptions inherent in them.

Comparison of research filament-product economics with competitive material—Table II

		"Guesstimate" Competitive Filament Product	Base Case Research Filament Product
1.	Pounds of material/unit length	19	11.2
2.	Current selling price		
	$/unit length	9.0	7.5 (assumed)
	¢/lb.	47.4	67.0
3.	Est. total product cost,* ¢/lb.	37.8	44.3
4.	Breakeven selling price		
	$/unit length	7.18	4.96
5.	Reqd. total capital, million $*	4.3	6.2
6.	After-tax return on investment @ current selling price, %	28	46

* Based on 25-million-lb./yr. level.

The following paragraphs explore some specific approaches in carrying out these studies. They include a discussion of the analysis of the results, and some suggestions on format together with some ideas on presentation.

There is no routine approach or procedure that will work equally well for all research analysis. The analyst must first establish the type of analysis to be performed and its related objectives. These should set the general pattern of approach.

For example, in the research guidance analysis, the prime objective is to establish the effect of key variables and set research goals, as opposed to the project analysis, where the prime objective is to establish potential profitability on which management can base a decision. Inherent in all research analyses are several common elements including:
- Process.
- Markets.
- Manufacturing cost.
- Competitive situation.
- Product performance.
- Patent situation.

Process

The amount of information available on the process used to manufacture the product can vary in importance, depending on the type of analysis. Basically, the process information provides the input for establishing the required capital investment, which ultimately affects the total product cost. For the research-guidance type of analysis, it is important to understand the process from the standpoint of process parameters (such as attainable yield) that ultimately affects the raw-material use, and other manufacturing costs that may be important. This input provides the basis for establishing the research targets necessary for suitable economics.

In the case of a project analysis, it is important to

Summary—Types of Research Feasibility Studies

Type of Analysis	Type of Research Projects To Be Used on:	Objectives
Research guidance	Conceptual-early stage	Measure effect of key variables Help set research goals Estimate probable profitability
"Guesstimate"	Early stage, competitive products	Estimate order of magnitude—economics, potential profitability
Project	Latter stage, candidates for commercialization	Establish potential profitability and commercial worth of the project

have this basic understanding of the process, along with the basis for designing flowsheets so that an accurate assessment of the required capital investment can be derived.

Market Estimates

The estimate of market size and penetration ultimately determines the quantity of product that will be sold, and consequently the profitability of the project. Market estimates for research analyses can be obtained via several routes. In the early-stage estimate, one can pick market figures based on company experience with similar products, or even guess at market penetrations, based on capacities of similar competitive products. For more detailed estimates, it is desirable to work with the company product manager and have him estimate both over-all markets and likely penetrations. Due to the uncertainty of such estimates, it is reasonable to consider a range of figures that correspond to optimistic, pessimistic and realistic estimates. For example, if the product performance is excellent it might be possible to attain the optimistic penetration; if patent protection is not obtained, perhaps the pessimistic penetration would be more realistic. These thoughts also hold true for estimate of selling price. Here again, it is well to consider a range of figures corresponding to optimistic, pessimistic, and realistic estimates.

One of the most common mistakes made in research analysis is to assume that any one company can obtain 100% of the available market without any consideration for competition, nonideal product performance, and so on. In addition, when estimating markets and market penetration, it is important to consider the rate of buildup to the mature year and the attainable selling price for each year, since these will affect both profitability and payout time. It is not usually realistic to assume that a new product introduced on the market can obtain mature-year penetration in the first few years. These market buildup figures and selling prices should be derived in conjunction with the product manager or marketing arm of the company.

Required Capital

In carrying out research studies, the analyst must decide whether to base the required capital on the assumption of either building a new plant, or using existing company facilities. This may be obvious, depending on the size of the plant being considered and also on the nature of the equipment required relative to that available in existing company facilities.

Although it is difficult to generalize, new research projects should be justified on completely allocated, new, battery-limit capital investments since it is difficult to project the availability of equipment several years hence. From the standpoint of the research manager, if he can justify his project on this basis, he can be assured that the project is on firm ground. (If, at the time of commercialization, the operating division can find a way of using existing equipment, the project will look even better.) Many research managements have been deceived by the use of "tail ended" economics that assume availability or existence of equipment. This then makes the research project dependent upon other factors not under the research management control.

For special situations of availability of equipment, the estimator should work closely with the operating division in determining exactly what is available, and at what price this equipment can be transferred to the new process.

Manufacturing Costs

The points concerning required fixed capital also apply to estimates of manufacturing costs. It is best to estimate the manufacturing costs of a new research project on a completely allocated basis rather than on a "tail ended" incremental basis. If "tail ending" can be carried out by the operating division, the project can only look better.

Raw Materials—The estimator must use judgment in deciding on the price to assume for major raw materials. It is not realistic to use quoted journal prices for large multimillion-pound quantities of bulk chemicals, since these are usually sold on a contract basis. When it becomes evident that raw-material cost is an important element in the over-all economics, and there are one or several raw materials used in large quantities, the estimator should contact suppliers and obtain indications of the likely contract price, as opposed to using the "textbook" prices, which may be misleading.

Direct Manufacturing Cost—The calculation of direct manufacturing cost should be based on a realistic assessment of required labor and associated operating expenditures. For purposes of research analyses, it is best to assume that a new process will have to be justified on a completely allocated labor and manufacturing basis and cannot take advantage of

existing associated facilities that may be available in the plant. These matters are usually beyond the estimator and would have to be worked out in minute detail by the plant manager. This cannot usually be done at the time that the research analysis is performed.

Overhead Costs—This is the area where it is most difficult to estimate required expenditures. Wherever possible, the best approach to follow is to work directly with the operating division that will commercialize the product, and use figures that are commensurate with the division's similar businesses. For cases where the new product is an extension of an existing product line, the operating division can usually estimate the incremental sales and overhead expenditures that would be required. For products that represent departures it is best to consider these overheads on a completely allocated basis.

For early-stage research guidance estimates where information is vague, it may be necessary to use correlations to establish overhead figures, as discussed earlier. If this is done, it is important to convert these dollar figures to absolute manpower, and to be sure that these manpower estimates are reasonable.

Product Performance

Many new products are sold on a performance-use-level basis. For these products, calculation of a ¢/lb. selling price figure is not entirely meaningful, since it is important to also consider the use level relative to competitive products. For such cases, the estimator should work closely with the research people in establishing typical product performance characteristics, and then show these as a variable, just as he would any other process cost parameter. A product that is 30% cheaper than a competitive product but requires a 50% greater use level is actually more expensive than the competitive product. Presentation of only the cost figure without consideration of the use level is incomplete, and can mislead the manager's ultimate decision.

Competition and Patents

Although the research analyst may have been asked to specifically look at the commercial potential of a new research product, he must also consider the question of competitive products and decide how this competition will affect the new product's introduction. This was illustrated earlier in this article in an example showing the evaluation of a filament product. You will recall that it had become apparent toward the end of the research project that a new competitive product had been introduced on the market. In order for management to make an intelligent assessment of the economic worth of the research project it also needed a relevant reading on the competitive product.

Although patents do not usually affect the economics of a research project directly, the analyst should point out the existing patent situation, since this may affect the over-all likelihood of commercial success. Since interpretation of the patent situation may be complex,

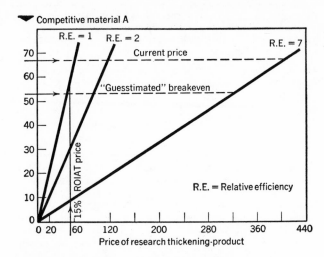

EFFICIENCY affects competitive price—Fig. 3

it is best to get this opinion from the company's patent department or the research people and patent department jointly.

Analysis

After carrying out the economic calculations inherent in the research feasibility analysis, the analyst will have an understanding of the assumptions used and also a feeling for any problem areas. He is therefore in the best possible position to qualify, interpret, and make recommendations. Without this analysis and interpretation, the study becomes a presentation of numbers that can mean different things to different people, and can lead to gross misinterpretation and possibly a wrong decision. The following examples illustrate this facet of the feasibility analysis for each of the three variations previously discussed.

The example describing a research guidance analysis on a new adhesive product had the objective of assessing the new research process, pointing out which variables were important and thereby setting research goals, and also determining the comparative economics of this product versus a competitive product on the market. Table I was presented in the final Research Guidance Report as a summary of the economic data of these two processes. A review of this Table would indicate that the competitive product is several cents per pound cheaper than the research product. In the Research Guidance Report, the following conclusion was stated:

"The major difference between the competitive and research product is in raw-material costs, which constitute a differential of approximately 4¢/lb. In order to minimize this difference, it is necessary to maximize the amount of monomer in the research product. It appears from this analysis that research goals on the new adhesive route should be directed toward developing a product with high monomer incorporation (greater than 50%) using the low-pressure process. It does not appear worthwhile to develop adhesives

of this type with monomer concentrations of less than 50%, since the economics of this material will not compare favorably with the competitive material. If a suggested goal of 70% monomer incorporation can be attained, the only remaining difference between the economics of the two processes will be based on level of production and monomer price. Under these conditions, the economics can be further improved by locating the plant near a low-cost monomer stream. The effects of monomer price and concentration are illustrated in the following figures."

This discussion interpreted the economic figures shown in Table I, pointing out why the economics of the two products are different and what research goals are necessary to bring them into comparable range. This explanation, along with the subsequent curves that were shown on the effect of the monomer concentration, was able to give research management a good enough understanding of the problem on which it was able to base a decision. As a result of this analysis, research management decided that the postulated goal of 70% was not reasonable, and the project was terminated.

The second example discussed a research guesstimate for a filament product that had the potential of being competitive with the new research filament product. The objective of this analysis was to determine the economics of the competitive product and to assess whether it would compete with the newly proposed research product. The economics for these two processes were compared in Table I.

If this table were presented by itself, it would be subject to misinterpretation because the ¢/lb. figure for the competitive product appears to be lower than that for the research product. What the table does not illustrate is that the product is not sold on a ¢/lb. basis but rather on a ¢/unit length basis. Since the research product requires fewer pounds per unit length, it is actually cheaper than the existing competitive products. The table also does not indicate that the competitive manufacturer has a captive advantage in raw materials.

Also, the cost of the competitor's product was calculated at an estimated raw-material price that was several cents below both the present market level and the price used for calculating the research product economics. The conclusion as stated in the final guesstimate report was as follows:

"The competitive filament product economics shown in the above Table is a guesstimate of Filament Chemical Company's manufacturing process cost. Since this company is captive in the major raw material, and we cannot accurately predict the price at which it is charging the process for this raw material, the effect of this price on their economics has been considered. The economics shown in the Summary Table were based on a raw-material price of 19¢/lb. This is optimistic, based on the current market price of 22¢/lb., which was used in calculating the economics of the research product.

The research filament material is economically attractive compared with the competitive product. The major contributing factor to the favorable economics is the low weight of the material required per unit length to obtain physical properties equivalent to the competitive material. It is this inherent product advantage that enables the research product to compete with other materials. Based on our guesstimate of its process, the Filament Chemical Co. cannot compete against our material at current or projected selling prices."

On the basis of this research guesstimate, management made the decision to proceed with the research and development work leading toward commercialization.

The third example discussed earlier described the project analysis of a new specialty thickening product. Table III shows the economics of the research material versus several other materials that were potential threats. By excluding a review of this table, it would be difficult to conclude the exact meaning of the economics. Several of the competitive materials have total product costs in the range of, or lower than, the research product, and this table indicates that these competitive products are equal or lower in cost. What this table does not show is that the comparative economics must be based on a "use cost" consideration (¢/lb. times amount of product used). The following major conclusions were presented in this Project Analysis Report, which amplified the point.

1. Based on preliminary economics presented in this study and available efficiency data, it appears that the research product is competitive at the 2-million-lb./yr. level with most of the specialty products now on the market under development, with the possible exception of product Z. (See Table III.)

2. The total product cost and selling price for manufacture and sale of the specialty thickening product are extremely sensitive to production volume. From the preliminary economics it appears that in order to be competitive with other materials on ¢/lb. basis, the production level of approximately 2 million lb. would be required in a new battery-limits facility. This conclusion does not take into account the relative efficiencies of the various products or the potential use of existing facilities. At the 2-million-lb./yr. level, the total product cost for the research product is approximately 43¢/lb., which is close to the guesstimated product cost of competitive materials.

3. The research-product economics are relatively insensitive to raw-material cost. At the 2-million-lb./yr. level, raw-material cost constitutes only 22% of the total product cost. This suggests that future research to develop modifications of the research material based on lower-cost raw materials should be carried out only if these new products have significantly greater efficiency than the present product. This also suggests that a captive-raw-material producer would not have any significant economic advantage, since raw materials do not constitute a major portion of the total product cost.

Two recommendations were made: that management proceed with test marketing of the research product to establish its acceptance; and that the re-

search department develop more-definitive efficiency data on the research product and each of the major competitive materials.

The above conclusions and recommendations went far beyond the original numbers shown in Table III. However, they were necessary to give management the proper understanding of the results of this analysis.

In summary, this discussion on analysis pointed out that it is the duty of the research analyst not only to provide figures and economics but also to properly qualify the economics (such as showing the degree of accuracy of the figures) to interpret the results (including any factors that are necessary for complete understanding) and, finally, to draw appropriate conclusions.

Presentation

The last segment of the research feasibility analysis that will be discussed is the presentation of results. This is equally as important to the ultimate understanding of the study as is the calculation and analysis of figures. The reader understands only what is in the report and what he is told. He does not understand what is in the mind of the research analyst. All of the important elements of the research analysis must therefore be presented so they give a clear, meaningful picture of the basis of the analysis, the elements included, the results, conclusions, and recommendations. Perhaps the best way to judge the quality of the research analyst's presentation is by the number of questions the reader must ask after its review. A good presentation will anticipate any likely questions and include these in the report.

Although there are no universal acceptable practices on the presentation of research feasibility studies, the following guidelines may be useful:

1. The writer should be honest with the reader, pointing out the basis for the analysis, indicating where guesses were made, and where more-accurate data were obtained.

2. The results of the study should be summarized within the first two or three pages of the report. This summary should tell the reader the type of study performed, why the study was performed, the specific objectives. It should include a minimum background necessary to refresh the reader's memory on the process or product, a brief summary of the economic results, pertinent conclusions, and recommendations. Providing the reader with this information within the first two or three pages will establish the proper framework for his understanding the remainder of the report.

3. The various major segments of the study—markets, patent situation, competitive products, etc.—should follow the summary and be discussed in a logical sequence. Each of these sections should include the major assumptions that were used.

4. The amount of numerical data included in the report should be commensurate with the type of analysis being performed. For example, in a very early stage analysis it is undesirable to put in too many detail figures since this may imply greater accuracy than is actually intended and could mislead the reader. Also, too much data in the wrong place could distract the reader from the main issue and result in irrelevant questions.

5. If the research report is lengthy it should be accompanied by a brief table of contents showing the reader exactly what is included and where it can quickly be located.

Economics and relative efficiencies of new thickening product—Table III

	Preliminary Estimates New Thickening-Product			"Guesstimates"		
				Competitive Material A	Competitive Material B	Competitive Material C
	Base Case					
Plant capacity, million lb./yr.	0.7	2	6	3	2	2
Manufacturing cost						
Raw-matl cost, ¢/lb.	9.3	9.3	9.2	23	9	6
Direct mfg. cost, ¢/lb.	33	21	17	33	19	17
Total prod. cost, ¢/lb.	68	43	29	54	41	38
Profitability, 15% ROIAT*						
Selling Price, ¢/lb.	89	56	36	70	53	50
Annual net profit, million $	80	130	230	250	130	130
Total capital investment, million $ (battery limits)	520	910	1,500	1,600	850	830
Payout time, yr.	—	7	—	—	—	—
Current selling price, ¢/lb.	—	—	—	67.5	100	69
Relative thickening efficiency				1.8	0.75	—

* Return on investment after taxes.

The above points are illustrated by the introduction-summary section from the project analysis on the specialty thickening product that was previously discussed.

"The new research product is a combination of raw materials X and Y, in the ratio of approximately 1:3. This product is one of a family of new materials developed by the research division primarily for specialty applications. The objectives of this study are to analyze the economic merit of the project and to advise management of the potential markets, economics and profitability. The estimates of available markets and potential penetration were developed by the marketing research department. These estimates consider several mature-year cases corresponding to pessimistic, realistic and optimistic penetrations that fall in the range of 0.7 to 6 million lb./yr. The marketing department chose a base-case level of 2-million lb./yr., which its estimates indicate could be available to this new research product. The breakdown of the total available markets by end-use, estimated penetration of the research product into each, and the total penetration for each of the three cases considered are discussed in the marketing section of this report.

"The economics presented in this study include several battery limit cases for manufacturing the material, as well as several additional cases for interim low-volume production in existing facilities. These economics are preliminary and are based on process information obtained from the laboratory and pilot-plant runs. The overheads and selling expenses used were based on actual company experience for production and sale of similar products. This information was supplied by the accounting department. Raw-material costs were based on large-volume quotes received from suppliers.

"In order to obtain a basis for comparison with other materials in the market, the economics for several of the most likely competitors were guesstimated and included in this study. Although these economics are less accurate than those for the research product, they provide order-of-magnitude numbers that can be compared. There is one new competitive product that has just been introduced, but due to the timing, economics for this material have not been assessed or included in this study. The economics of the above mentioned materials are summarized in the following Table, and are discussed in detail in the text of this report.

"For purposes of this study, efficiency is defined as pounds of material required to thicken a resin to a given viscosity. The relative efficiency of the product is defined as the ratio of the thickening efficiency of the competitive product to that of the research product. The Summary Table also includes data on relative efficiencies for the thickeners considered in this study; these are based on laboratory work and are confirmed by initial field tests. They do not, however, illustrate the three prime qualitative factors that are also considered in evaluation of thickeners: namely, drainage, settling, and stability. These qualitative factors may be overriding for some applications, but more field

evaluation is required to definitely establish this.

"The patent department opinion is that there are no patents that would be infringed by making or selling the research product. The economics shown in the Summary Table indicate that for the 2-million-lb./yr. base-case level, the research product is competitive with most existing products now on the market on a use/cost basis. As a result of this conclusion, it is recommended that management authorize proceeding with the test marketing phase of this project to determine the field acceptance of the material."

The above description gave the reader an understanding of the brief description of the product, the type of analysis made, why the study was being performed, and a synopsis of the conclusions. The subsequent sections of the report contained conclusions, recommendations and a detailed discussion of each of the various facets of the study.

One last point on the subject of presentation: It is generally desirable in presenting a study to bring out all of the controversial questions that the reader might ask, even though there are no answers to these questions. The fact that the author has considered these in itself is an indication that he had done a thorough job and has not neglected these major points. As will often be the case in research analysis there are many questions that will not be answered within the time limitations during which the study was carried out.

Research feasibility analysis requires a basic understanding and knowledge of the process or product being analyzed, and an understanding of the various techniques required to develop the internal elements of the study, but most of all, a great need for judgment by the report writer in deciding what elements should be considered, and what they mean, and in organizing them in a meaningful fashion that will present to management the total picture.

Meet the Author

Gerald Teplitzky is a supervisor at the W. R. Grace & Co. Research Div. in Clarksville, Md., where his responsibilities include supervising a group of engineers in the economic evaluation of research and development projects. Formerly he was business manager at the Grace Research Div., and was in process development and economic evaluation at the Dewey & Almy Div. of Grace. He is a chemical engineer and graduate of the Polytechnic Institute of Brooklyn, and is a member of AIChE and American Assn. of Cost Engineers.

Key Concepts for This Article

Active (8)	Passive (9)	Means/Methods (10)
Analyzing*	Potential*	Evaluating*
Estimating	Business	Comparisons*
Forecasting*	Commercial	Examining
	Profit*	Interpreting
	Return on investment*	

(Words in bold are role indicators; numbers correspond to EJC-AIChE information retrieval system. Asterisks mark key concepts suggested for indexing. Others are added to improve reading as an abstract. Indexing is described in *Chem. Eng.*, Oct. 11, 1965, p. 187.)

Using the Profit-and-Loss Statement

When deciding whether a research project will pay off commercially, the profit-and-loss statement is the basic tool for summarizing all the cost elements of the project.

GERALD TEPLITZKY, *Research Div., W. R. Grace & Co.*

The first article in this series described the research feasibility analysis, and its use in evaluating research projects. This second article deals with the components of the research feasibility analysis; their definition, and a discussion of how these various cost components are derived.

The profit-and-loss statement (P&L) is the accounting man's tool for summarizing all of the various cost elements of a given project. It provides a convenient way of grouping these cost items in one place, so that they are easily recognizable and can be used for comparison with figures from other projects. The P&L takes into account the sales volume of the product (which in effect is a measure of market penetration) and manufacturing costs that reflect many of the research process variables. It also reflects the capital investment, which shows both the cost of the physical plant and the associated working capital.

In addition to being a useful accounting tool, the P&L statement may be considered the heart of the research feasibility analysis in that it groups all of the components considered in this type of study.

The P&L statement, like many other accounting tools, can take many different formats. These are largely dependent on the specific accounting philosophies and practices of a company.

Table I illustrates a typical P&L statement, and is the basis of discussion of this article. It is general enough in format to be readily adaptable to specific company procedures.

This P&L statement summarizes the picture of a new latex product nearing commercialization. For example, the first portion gives information on the production level (which is an estimate of market penetration) and the total annual dollars that will be accrued by sale of the product. The next segment shows the direct manufacturing cost for the product, being in effect a tabulation of the various cost components that will be accrued during the manufacture of this product, including raw materials, labor, related maintenance and packaging. The next segment of the P&L deals with the overheads associated with manufacture and sale.

This information shows management the commitments that its marketing, sales and manufacturing operations must make in order to obtain the predicted gross sales. This is followed by information on the total product cost and profitability of the product.

By looking at these figures, it is possible to obtain an assessment of the mature-year profit dollars that can be expected. These figures do not take into account the associated pre-operating expenses that will be incurred during the research and development phases of the project. However, it gives management a preview analysis of the profitability of the product.

Lastly, the P&L shows the capital required in order to obtain these profits, including the dollars that will have to be invested to build the plant, and the associated working capital that will be necessary to support the plant. The after-tax return on investment shown is a measure of mature-year profitability and is the ratio of mature-year net profit to total capital investment.

The following discussion defines the terminology of the P&L statement, and suggests various techniques for deriving each of these items.

Nomenclature

Production Capacity—The mature-year production capacity of the plant. It does not necessarily reflect the actual production level at any given time, which may be less than the ultimate capacity of the plant, particularly during an interim buildup, or prior to maturity.

Gross Sales—The total dollars accrued by sale of the product (amount of sales times selling price). In order

to obtain this figure, it is necessary to estimate the assumed market penetration in any given year, as well as the corresponding selling price.

Sales Deductions—The freight charges incurred in shipping the product from the plant to the customer, as well as other customer-related costs such as returned unsatisfactory goods, special sales discounts and other miscellaneous sales-price adjustments that might be encountered. The major portion of this cost component is freight.

This sales deduction figure is generally composed by individually estimating the freight cost, returned-goods cost, and customer-deductions cost, and summing these figures. Returned goods and customer deductions can normally be obtained by means of an experience factor that most companies have developed for each particular business.

In cases where many of these figures are not readily available, it is also possible to estimate a total sales-deduction figure by taking an over-all percentage of gross sales, which is dependent on the nature of the product, type of business, and level of selling price. The percentage of gross sales generally is on the order of 5%, but can vary considerably, depending on individual circumstances.

Net sales—The difference between gross sales and sales deductions.

Raw-Material Cost—The cost of raw materials used in the process, which must take into account the yield factor of the process and any spoiled goods. In some companies, it is also customary to include the cost of packaging materials, although these are usually shown in a later category.

Direct Labor—The cost of direct manufacturing labor. It may be calculated either by estimating the number of men required per shift and multiplying this by the dollars per man-year, or by relating the estimate to a known cents-per-pound or dollars-per-year figure for similar processes. (The dollars-per-year figure to be used may or may not take into account overheads and fringe benefits, depending on the particular company's accounting practice.)

Indirect Labor—The line supervision of the direct labor shown above. In many companies, this figure also takes into account the overheads and fringe benefits for direct labor. This is often calculated as a percentage of direct labor, generally varying from 20 to 50%, depending on whether or not the overheads are considered in the original direct-labor figure.

Maintenance—An accounting of the labor and material costs necessary to maintain the plant operation. Common figures are in the range of 5 to 8% of fixed capital per year, depending on the nature and complexity of the process. An alternative to deriving this estimate may be the use of a cents-per-pound or dollars-per-year figure based on experience for similar processes.

Supplies—The supplies used during the operation of the process; e.g., charts for temperature recorders, office supplies, pails. This number is often obtained by either picking a dollars-per-year figure or, when unavailable, estimating a percent of the maintenance cost, usually in the range of 15 to 20%.

Utilities—The cost of utilities employed during operation of the process. This figure can either be estimated by experience, or derived by estimating the quantities of utilities used and their cost in the particular location.

Packaging—The cost of materials used in packaging of the product. This normally does not take into account the labor cost for packaging (considered in the direct-labor category).

Direct Manufacturing Cost

Direct Manufacturing Cost—The summation of the preceding cost components, which is a measure of the actual manufacturing cost incurred, without regard for the company and plant overheads that will be involved. This is a particularly useful number when comparing different processes that might be employed in the same existing plant equipment.

However, it is not a good comparative measure of cost when considering the desirability of investing in new facilities to produce a product. It does not take into account the required overheads, depreciation, selling expenses, and other related items that will affect the ultimate profitability of the project.

Gross Margin—The difference between net sales and direct manufacturing cost.

Sales Expenses—The cost of selling the product. In many accounting systems, the sales expense category may also include technical service requirements. This cost is best derived by actually estimating the number of men required to sell the product, and multiplying this by an average dollars/man-year figure. Many companies also use an arbitrary percent of gross sales, based on prior experience, to determine selling expenses. This can be quite misleading, depending on the magnitude of the sales price. If this technique is used, it should be cross-checked and translated to number of men per year, to make sure the figure is meaningful.

Merchandising Margin—The difference between gross margin and sales expenses.

Overhead

Overheads—Those overheads normally encountered during the manufacture and sale of chemical products. The total overhead figure is a summation of the following five items:
- Depreciation.
- General and administrative expenses.
- Factory administration.
- Taxes and insurance.
- Research and development.

Depreciation—The amortization of the equipment—normally figured on a 10-yr. life for equipment and 15 to 20-yr. life for buildings. The depreciation life actually used in the estimate may vary, depending on the nature of the equipment, the likely life-span of the project, and the incentive to purchase or invest in the equipment.

General and Administrative Expenses—The divisional overheads associated with manufacturing in-

clude such items as divisional administration, plant management, and allocated portions of home-office expenses. This figure is largely dependent on the relative size of the company; however, it is often estimated by taking a fraction of gross sales, perhaps 3 to 5%. It may be more accurately determined by following existing company practice.

Factory Administration—The factory costs that have not previously been considered in the indirect-labor categories—for example, the costs of running the shipping and receiving facility, and the quality-control lab. This figure may be derived by estimating the cost of each of these components required by the new process and summing them. Many companies have established figures based on experience that may be more meaningful for early-stage estimates.

Taxes and Insurance—The cost of real-estate taxes and insurance, which is normally based on a percentage of fixed capital per year, commonly 2 to 3%.

R&D—The sustaining research that will be required to maintain the product once production has been initiated. It does not take into account the research funds required during the initial research phase prior to production. These are considered under pre-operating expenses.

New latex product raw-material balance—Table II

Raw Material	Lb. Solids	Lb. Solution	Lb. Solution/Lb. Polymer Solids	¢/Lb. Soln.	¢/Lb. Polymer Solids
Monomer A	50	50	0.505	20	10.1
Monomer B	48	51	0.515	13	6.7
Catalyst	0.05	0.05	0.051	100	0.5
Emulsifier	3	3	0.303	23	0.7
Total (Based on 98% Yield)	99.0				18.0

The sustaining R&D figure is best determined by estimating the number of man-years that will be required to maintain the product, multiplying this by an average $/man-yr. rate, and dividing by the number of pounds produced per year. Many companies have established typical R&D figures based on a percentage of gross sales; however, in order to be sure that these are meaningful, they should be converted to the equivalent of man-years and compared to actual man-year estimates. In some company practices, technical-service expenditures are also considered under the R&D category rather than under the sales-expense category previously mentioned (this should be specified by the author of the P&L statement to make sure the point is clear).

Total Product Cost

The summation of all the direct and indirect manufacturing and overhead costs incurred during the manufacture and sale of the product. This term is often misused and can lead to gross misinterpretation. Many research managers use it to mean the sum of the direct manufacturing costs exclusive of overheads. Naturally, the costs associated with this figure will be lower than for the total product cost. When using the term "total product cost," it is important to clearly define the items included to eliminate any misinterpretation.

Profits

Before-Tax Profit or Gross Profit—Gross profit is obtained by taking the difference between gross sales and total product cost. This term may also be obtained by taking the difference between merchandising margin and overheads.

Taxes—The portion of gross profit paid to taxes. This figure generally ranges from 46 to 50%, depending on the corporate tax structure.

Net Profit—The difference between gross profit and taxes. The net profit figure is the $/yr. profit that is ultimately used to derive the return-on-investment or payout-time calculations.

Profit-and-loss statement, mature year—Table I

Process – new experimental latex
Production – 5 million lb./yr.

	Thousand Dollars	¢/lb.	% Gross Sales
Gross sales	2,000.0	40.00	100.0
Sales deductions	60.0	1.20	3.0
Net sales	1,940.0	38.80	97.0
Direct mfg. cost	1,183.8	23.68	59.2
Raw matl. cost	900.0	18.00	45.0
Direct labor	44.0	0.88	2.2
Indirect labor	36.1	0.72	1.8
Maintenance	91.0	1.82	4.5
Supplies	2.7	0.05	0.1
Utilities	90.0	1.80	4.5
Packaging	20.0	0.40	1.0
Gross margin	756.2	15.12	37.8
Sales expenses	120.0	2.40	6.0
Merchandising margin	636.2	12.72	31.8
Overheads	228.6	4.57	11.4
Depreciation	63.6	1.27	3.2
General and administrative	66.0	1.32	3.3
Factory administration	25.0	0.50	1.2
Taxes & insurance	14.0	0.28	0.7
R. & D.	60.0	1.20	3.0
Total product cost	1,592.4	31.85	79.6
Before-tax profit	407.6	8.15	20.4
Taxes @ 48.0%	195.6	3.91	9.8
Net profit	211.9	4.24	10.6
Total capital employed, thousand $	1,099.9		
Fixed capital, thousand $	700.0		
Working capital, thousand $	399.9		
After-Tax return on investment	19.3%		

New latex product—calculation of total product cost—Table III

Item	Calculation	Cost, ¢/Lb. Latex Solids
Raw materials	See detailed raw-material balance	18.0
Direct labor	2 men × 3 shifts = 6 men/yr. Assume 10% overtime = 0.6 men/yr. Total = 6.6 men/yr. @ 6,600/man yr. = \$44,000/yr. $\dfrac{44,000/\text{yr.}}{5 \text{ million lb./yr. solids}} = 0.88$ ¢/lb.	0.88

Indirect labor	Assume 50% direct labor. This includes fringe benefits and vacations
	    $0.5\,(0.88¢/\text{lb.}) = 0.44$ ¢/lb.
	Due to high supervision required on this process, a portion of the lead-shift supervisor will be required per shift, est. @ 50%.
	50% × 3 shifts = 1.5 man-yr.
	This lead-shift supervisor and associated overhead = \$9,300/yr.
	Therefore, allocation to this process
	$= \dfrac{1.5\,(\$9,300/\text{yr.})}{5 \text{ million lb./yr.}} = 0.28¢/\text{lb.}$
	Total indirect labor = 0.44 + 0.28
	    = 0.72 ¢/lb.
		0.72

Maintenance	Total fixed capital = \$700,000 Assume maintenance = 8% fixed capital/yr. due to complex mechanical nature of this process. 8% (700,000) = \$56,000/yr. In addition to this charge, special replaceable stainless steel parts will be required, which we estimated at \$35,000/yr. Total maintenance = \$56,000 + \$35,000 = \$91,000/yr. $= \dfrac{\$91,000/\text{yr.}}{5 \text{ million lb./yr. solids}} = 1.82$ ¢/lb.	1.82
	The remaining components of the P & L are calculated in a similar manner.	
Total product cost	Summation of direct mfg. cost, sales deductions, sales expenses, and overheads = 31.9¢/lb.	31.9

Capital

Total Capital—The summation of required fixed capital necessary to build a plant and the working capital needed to operate it.

Fixed Capital—The investment required to construct and build the physical facilities to produce the product. This figure may be derived via several different techniques, depending on the type of estimate and accuracy required. Specific techniques for calculating and estimating the capital investment are available throughout the literature.

The basic approach to capital estimation need not be the same for each of the variations of research feasibility analysis.

Research Guidance Analysis—In this case, fixed capital is likely not to be an important criterion and, if possible, may be estimated by relating the process in question to known processes where fixed capital estimates are available.

Research Guesstimate—For this type of analysis, it would also be reasonable to relate the fixed capital to that of similar processes available in the literature or in company records. For example: In estimating the cost of a competitive filament process, which was discussed earlier, it was possible by reviewing the literature to obtain order-of-magnitude capital investment figures for comparable processes.

Where it becomes evident that a variation in capital has a significant effect on the ultimate economics, this approach for obtaining fixed capital may not be suitable, and it will be necessary to actually derive a capital estimate via equipment components, as discussed in the next paragraph.

Project Analysis—For this type of analysis, it may be necessary to actually derive a detailed fixed capital. This can be done by first establishing a flowsheet for the visualized process, sizing and estimating the cost of each of the pieces of equipment, multiplying these by appropriate factors to obtain an over-all installed cost, and summing these to obtain the fixed capital.

It is largely up to the judgment of the author of the feasibility analysis to decide the importance of fixed capital, and to choose the approach.

Working Capital—Working capital is made up of several items, including accounts receivable, raw-materials inventory, work in process, and finished-goods inventory. The definition and method of calculation of each of these terms follows:

Accounts Receivable—The period of time that the customer takes before paying for the product. This working-capital component is calculated by the following formula:

$$\frac{\text{No. days of accounts receivable}}{\text{Operating days/yr.}}\ (\text{Prodn. capy.})\ (\text{Selling price})$$

Raw-Materials Inventory—The number of days of raw-material supplies kept on hand. This figure is derived in a similar fashion to accounts receivable and is illustrated by the following formula:

$$\frac{\text{No. of days raw material}}{\text{Operating days/yr.}}\ (\text{Prodn. capy.})\ (\text{Raw material cost})$$

Work In Process—A measure of the amount of material which is in process at any given time. This figure is calculated by the following formula:

$$\frac{\text{No. of days work in process}}{\text{Operating days/yr.}}\ (\text{Prodn. capy.})\ (\text{Direct mfg. cost})$$

Finished-Goods Inventory—A measure of the amount of finished goods held on hand ready for shipment to customers. This figure is calculated by the following formula:

$$\frac{\text{No. of days of inventory}}{\text{Operating days/yr.}}\ (\text{Prodn. capy.})\left(\begin{array}{c}\text{Total product cost}\\ -\text{sales expense}\\ \text{and deductions}\end{array}\right)$$

The accounts-receivable category constitutes the largest segment of the working capital, since the cost

figure is derived by using the selling price of the product as opposed to any of the segments of the total product cost. For many typical processes, it is common to take 30 to 45 days for accounts receivable, 2 weeks to a month for raw-materials inventory, 2 to 5 days for work in process, and approximately 2 weeks to 1 month for finished-goods inventory. The exact figures to be used are largely determined by existing company practices.

Return on Investment

After-Tax Return on Investment—A measure of the mature-year profitability of the project, which is the ratio of mature-year profitability to total capital employed. Return on investment is one measure of the profitability of a project; however, this does not take into account two important factors.

Pre-operating expenditures are required to get the project to commercialization. These include R&D expenses, marketing expenses, plant startup and engineering expenses.

Most chemical operations do not start out at maturity but rather build up over a period of years, during which time the annual net profits are lower than for the mature year.

Since the mature after-tax return on investment calculation does not take into account either of these two factors, it can be misleading as a decision-making gage. If the mature-year profitability is poor, then the over-all project profitability that takes into account the above two factors can only be worse. It is in this connection that the payout-time calculation is used. This is discussed later in this article.

Calculation Format

Previous portions of this article have dealt with the definition and nomenclature of the P&L statement as related to the research feasibility analysis. This next discussion deals with techniques for calculation of the P&L.

It is important to use a method that will clearly relate to the reader the assumptions and techniques used, since it is quite likely that questions will arise concerning them.

The P&L statement shown earlier in this article is for a latex product that would be produced in a new battery-limits facility. Because of the similarity to existing products, it was possible to relate the required fixed investment to an existing process that Company XYZ was presently using. The following two tables illustrate the format used in deriving some of the components of this P&L.

The format shown in the *Total Production Cost Table* (Table III) can be used for all types of P&L cost calculations, and provides a convenient mechanism for tabulating each of the assumptions and calculations. Similar format tables can also be used for the calculation of capital investment, where it is necessary to go through a vigorous derivation of the cost of each piece of equipment, and the associated installation costs, to obtain a total capital figure.

Mature-year profit calculations—Table IV

Assumed selling price

Assume selling price = 40¢/lb.
Total product cost = 31.9¢/lb.
Before-tax gross profit = 8.1¢/lb.
Annual after-tax profit = 8.1¢/lb. (5 million lb./yr.) (50% tax rate)
= $203,000/yr.

After-tax return on investment

$$= \frac{\$203,000/\text{yr.}}{\text{Total required capital} = \$1.1 \text{ million}} = 19\%$$

Assumed after-tax return on investment

To obtain a predetermined 15% after-tax return on investment

$$\text{pretax return} = \frac{15\%}{50\% \text{ tax rate}} = 30\%$$

Total required capital = $1.1 million
Required pretax profit – 30% ($1.1 million) = $330,000/yr.

$$= \frac{\$330,000/\text{yr.}}{5 \text{ million lb./yr.}} = 6.6¢/\text{lb.}$$

Required selling price = Total product cost + profit
= 31.9¢/lb. + 6.6¢/lb. = 38.5¢/lb.

Profitability Calculations

Mature-Year Profit Calculations—Derivation of the total product cost has already been described. From this figure, the mature-year net profit may easily be calculated by either assuming various selling prices and calculating the respective return on investments or by arbitrarily setting the required after-tax return on investment and solving for the mature-year selling price. An example of both of these calculations for the previously mentioned latex example is shown in Table IV.

Payout-Time Calculations—Prior discussions on profitability have pertained to mature-year return on investment. This calculation neglects the required pre-operating research and development expenditures, and the lower plant capacity in the early periods. In order to take account of both of these phenomena when considering project profitability, a payout-time calculation is normally performed. The payout time measures the number of years required to regenerate, via profits and depreciation, the total investment of the fixed assets and pre-operating expenditures that were required to get the product into commercialization. Definitions of the key terminology used in the cash-flow calculation are illustrated in Table V. Table VI illustrates a typical cash flow for the latex product described earlier in this chapter. It should be noted that the pre-operating R&D expenditures shown in this cash-flow table are entered after tax; i.e., at approximately 50¢ on the dollar actually spent. Also, the working capital shown for each year of the payout-time calculation is incremental capital; i.e., differential capital between the present and prior year. In the case of the first year (where there is no prior year), the incremental working capital and working capital are equal.

The payout time of 5.8 yr. illustrates that it will take this number of years to recoup the required fixed capital investment of $700 thousand plus the asso-

Cash-flow definitions—Table V

Total source of funds	= Net profit after taxes + Straight-line depreciation + Tax savings × (1 – tax rate)
Tax savings	= Straight-line depreciation – sum-of-the-years-digits depreciation
Application of funds	= Preoperating expenses (after taxes) + Fixed assets + Incremental working capital
Net cash flow	= Total source of funds – Application of funds
Cumulative cash flow	= Sum of all previous annual net cash flows + the current net cash flow

Projected cash-flow statement—Table VI

Process – New experimental latex

	Thousands of Dollars			
	Yr. 1	Yr. 2	Yr. 3	Yr. 4
Percent of capacity	25.0	50.0	75.0	100.0
(million lb./yr.)	(1.2)	(2.5)	(3.7)	(5.0)
Net profit after taxes	12.6	110.1	149.5	211.9
Depreciation	63.6	63.6	63.6	63.6
Tax savings on sum-of-digits-depreciation	27.6	22.1	16.5	11.0
Total source of funds	103.8	195.8	229.7	286.6
Pre-operating expenses (After taxes)				
Total	78.0			
Fixed assets	700.0			
Working capital/yr.	127.5	101.3	90.8	80.3
Total application of funds...	905.5	101.3	90.8	80.3
Net cash flow	–801.7	94.5	138.9	206.3
Cumulative cash flow.......	–801.7	–707.2	–568.3	–362.0
Payout time equals	5.8 yr.			

ciated pre-operating expenditures of $158 thousand (pretax).

While it is difficult to generalize on the desired level of payout time for a new research project, figures in the range of 3 to 5 years are generally considered reasonable, whereas those approaching 10 years are considered high and unreasonable.

This discussion of payout time and cash flow was based on a nondiscounted basis; i.e., the face value of the cash flows for each year without any regard for the time value of money. In many company practices, it is deemed desirable to consider the effect of time value; the fact that $1 today is worth less several years hence. For this practice, each of the yearly cash flows are discounted by appropriate factors (depending on the interest rate chosen) to derive a cumulative discounted cash flow, and consequently a discounted payout time.

Computer Programs

In order to satisfactorily carry out feasibility analyses, it is often necessary to consider many variations of a basic theme to determine the effect of key variables, and to show ranges of high-low estimates. If mature year P&L statements and pay-out time calculations are required for these variations, a considerable amount of burden in the form of "arithmetic" is required of the author.

The most time-consuming and difficult portion of the payout time calculation is the determination of net profits for each of the interim years prior to maturity. In order to facilitate the carrying out of this type of calculation and to alleviate this burden, the use of computer programs for cost estimating is becoming more prevalent. The purpose of these programs is not to replace the judgment of the estimator but rather to allow the running of many duplicate cases; the "what-if" type approach with a minimum of work. The computer lends itself to this type of operation and has been satisfactorily employed for this use.

The general theme of Grace's program is to pick assumptions for a mature case which will allow the calculation of a mature year P&L. Subsequent assumptions are chosen for each of the interim years prior to maturity in order to develop interim year P&L's. These interim year assumptions are based on predetermined relationships between interim and mature year which are inherent in the computer program.

The output of this program is: a list of both mature and interim year assumptions, a mature year P&L statement, a P&L statement for each of the interim years, and finally a cash flow statement and pay-out time calculation. The P&L statements and cash flow table shown earlier in this section are direct outputs of this computer program. These outputs are printed in a directly usable form, which also facilitates their use in reports and minimizes the amount of required retyping.

The purpose of discussing this computer program is to illustrate the fact that the burden of calculation can be minimized so that the author may spend his time analyzing as opposed to "cranking out" numbers.

As a closing statement to this series of articles, it is the author's feeling that research feasibility analysis requires a basic understanding and knowledge of the process or product being analyzed, an understanding of the various techniques required to develop the internal elements of the study, but most of all, a great need for judgment by the author in deciding what elements should be considered, what they mean, and organizing them in a meaningful fashion that will present to management the total picture.

Key Concepts for This Article

Active (8)	**Passive (9)**	**Purpose of (4)**
Using*	Profits*	Forecasting*
Employing	Losses*	Profitability*
	Summaries*	

(Words in bold are role indicators; numbers correspond to EJC-AIChE information retrieval system. Asterisks mark key concepts suggested for indexing. Others are added to improve reading as an abstract. Indexing is described in *Chem. Eng.*, Oct. 11, 1965, p. 187.)

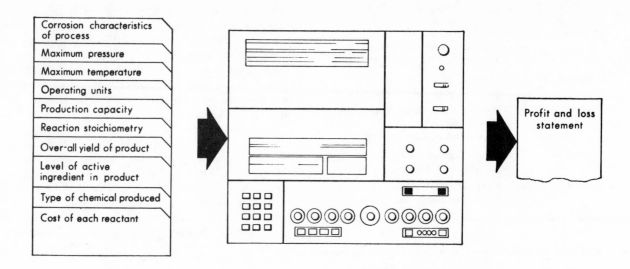

Corrosion characteristics of process
Maximum pressure
Maximum temperature
Operating units
Production capacity
Reaction stoichiometry
Over-all yield of product
Level of active ingredient in product
Type of chemical produced
Cost of each reactant

Profit and loss statement

These ten easy-to-obtain data inputs form the crux of a simple, versatile project-screening program. Here are examples of how to apply such a program in the . . .

Economic Evaluation of Research Projects—By Computer

ROBERT W. DeCICCO, *W. R. Grace & Co.*

The methods of economic evaluation in research are well developed, and their application to improve profitability has been discussed by a number of authors. But how do you make the best use of only a limited technical staff that must evaluate a host of projects?

Here, the use of computers to remove the burden of routine calculations is becoming more prevalent. Versatile computer programs can generate the necessary analytical information in a fraction of the time normally required, and thereby allow the evaluator to concentrate upon his primary functions of analysis and interpretation.

This article describes one such computer program, effectively employed for the evaluation of R&D projects at the Research Div. of W. R. Grace & Co. It is called the Project Screening Program because it is used to assess the potential economic worth of those projects in the early stages of development, when a minimum of information is available. It applies empirical functions of known reliability to translate the available data into a complete profit and loss statement encompassing all of the cost and capital factors associated with manufacture of the product. The pro-

gram is particularly useful for the research-guidance type of evaluation, since it can readily identify and show the effect of changes in the factors controlling process economics.

Program Description

The Project Screening Program was designed to be as versatile as possible. It does not require specific knowledge in regard to operating and overhead costs, capital and selling price. It consists of two subprograms (which calculate fixed capital and raw-material costs) coupled to a master display program.

The ten simple inputs shown on this page are the only data required. The program uses the first four of these inputs to determine the fixed capital investment. The reaction stoichiometry, molecular weights, over-all yield and reactant costs are necessary to calculate product raw-material cost. The type of chemical produced is a signal input by which the master display program establishes the sales deductions, packaging expense, sustaining R&D cost, sales expense and profitability criteria. The percent of active ingredient adjusts the basic sales deductions and packaging expense for variations in product concentration.

Fixed-Capital-Investment Subprogram

The computer calculates fixed capital investment by the empirical method devised by Zevnik and Buchanan.[1] This technique has proved itself reliable for approximating direct plant cost when only a conceptual flow diagram is available. It determines fixed capital as a function of production capacity, process complexity and operating units.

Process complexity is measured by a factor F_c that represents the degree of equipment sophistication required by the process:

$$F_c = 2 \times 10^k$$

where k is the sum of the following factors:

$F_t \equiv$ Process temperature factor
  $= 1.8 \times 10^{-4} (t - 27)$ for $t \geq 27$ C.
  $= 2.0 \times 10^{-3} (27 - t)$ for $t < 27$ C.
$F_p \equiv$ Process pressure factor
  $= 0.1 \log p$ for $p \geq 1$ atm.
  $= 0.1 \log 1/p$ for $p < 1$ atm.
$F_a \equiv$ Process alloy factor
  0   for cast iron, carbon steel, wood
  0.1 for aluminum, copper, brass, Type 400 stainless steel
  0.2 for Type 300 stainless steel, nickel, Monel, and Inconel alloys
  0.3 for Hastelloy alloys
  0.4 for precious metals

The fixed capital investment, in millions of $, is then calculated from the following equations, the first of which applies to production capacities (c_p) of less than 10 million lb./yr., and the second to capacities equal to or above this number.

$$I_{FC} = 0.01 \, (F_c) \, (c_p)^{0.5} \, (N) \, (I_{ENR}/300) \qquad (1)$$

$$I_{FC} = 0.008 \, (F_c) \, (c_p)^{0.6} \, (N) \, (I_{ENR}/300) \qquad (2)$$

In the above, I_{ENR} is a program constant of 455, corresponding to the 1967 average value of the *Engineering News Record* construction cost index. It is updated periodically to account for the rising cost of plant construction. As noted in the nomenclature, N is the number of operating units, one such unit being defined as a piece of equipment that, with its associated service units, performs an operation or carries out a process step on the product stream.

This method of estimating fixed capital has both drawbacks and advantages. The principal advantages are that not very much data are required, and that various process routes to a product can be compared simultaneously. However, the definition of an operating unit is necessarily subjective, because, for example, a heat exchanger could be either a service unit to a distillation tower or an operating unit itself, depending upon its purpose in the process.

As mentioned previously, the Project Screening Program was designed to be as versatile as possible. In this interest, the subprogram for determining fixed capital can be bypassed at the user's option, and fixed capital entered directly at a data input to the master display program.

Raw-Material-Cost Subprogram

The product raw-material cost C_{RM}, in ¢/lb., is calculated from the stoichiometric moles of each reactant per mole of product, over-all yield, molecular weights, and the cost of each reactant. (In the equa-

Nomenclature

c_p	Production capacity, lb./yr.
C_{RM}	Raw-material cost, ¢/lb.
$(C_{RM})_a$	Raw-material cost for reactant a, etc.
C_T	Total product cost, ¢/lb.
F_a	Process alloy factor
F_c	Complexity factor
F_p	Process pressure factor
F_t	Process temperature factor
I_A	Percent active ingredient in product
I_{ENR}	Current value of the Engineering News-Record construction cost index (Base Year 1939 $= 100$)
I_{FC}	Fixed capital investment, million $
I_T	Total capital investment, equal to fixed capital plus working capital, million $
k	Equal to $F_a + F_p + F_t$
M	Molecular weight
N	Number of operating units
p	Maximum process pressure, atm.
P_S	Selling price, ¢/lb.
R_{AT}	Return on investment, after taxes, %
S	Stoichiometric moles of each reactant per mole of product
t	Maximum process temperature, C.
T_R	Income-tax rate, corporate
Y_O	Over-all yield of product, mole %

tion below, the subscripts $a, b, \ldots n$, refer to the reactants; the subscript p refers to the product.)

$$C_{RM} = \frac{(S_a)(M_a)(C_{RM,a}) + (S_b)(M_b)(C_{RM,b}) \cdots + (S_n)(M_n)(C_{RM,n})}{(Y_O)(M_p)}$$

The subprogram has provision for six reactant inputs, including catalyst consumption. It is recognized that the over-all yield may be difficult to estimate for a product of many reactants of varying molar conversions. However, for the project-screening type of analysis, the assumption that one reactant is controlling is not unreasonable. Again, to provide the cost estimator with the ability to treat product raw-material cost as a direct data input, this subprogram can also be bypassed at will.

Master Display Program

The master display program uses the outputs of the subprograms, together with a built-in procedure for determining working capital and total product cost, to solve for the selling price corresponding to a specific return on investment. It then generates a complete profit and loss statement. The only data inputs required are the type of chemical produced and the percent of active ingredient in the product.

The type of chemical (bulk or specialty) refers to the performance characteristics and marketing requirements of the product. A bulk chemical is a standard commodity item sold in a highly competitive market. This type of chemical usually has many applications and is produced in large volumes. A specialty chemical is normally produced in small quantities and sold on a relatively noncompetitive basis since performance cannot be easily duplicated. The percent

Profit and Loss Statement Format—Table I

Particulars	¢/Lb. Basis	Particulars	¢/Lb. Basis
Gross sales	In terms of ¢/lb., equivalent to selling price	Sustaining R&D	2% of gross sales for bulk products and 5% of gross sales for specialty products
Sales deductions	The expense of freight out, return goods and customer allowances is a program constant at 0.5¢/lb. for bulk products, and 1¢/lb. for specialty products. Correction for product potency is made by dividing these basic rates by the percent active ingredient in product	Overheads	Depreciation + G&A + factory administration + taxes and insurance + R&D
		Total product cost	Sales deductions + direct mfg. cost + sales expense + overheads
		Gross profit	Gross sales minus total product cost
Net sales	Gross sales minus sales deductions	Income taxes	50% of gross profit
Raw-material cost	Determined by Subprogram 2	Net profit	Gross profit minus income taxes
Direct labor	(Operators/shift) x (3 Shifts/day) x ($7,000/Man-yr.)/$(c_p)$. The latter term represents production capacity; we divide by this term in order to get direct labor cost in ¢/lb. The number of operators per shift is programmed as a function of the number of operating units; specifically:	Accounts receivable	(No. of days) × (Production capacity) × (Selling price)/(Operating Days/Yr.) The number of days of accounts receivable is a program constant of 35 for bulk products and 30 for specialty products. The number of operating days per year is a program constant of 330 for bulk products and 250 for specialty products.

Operating Units	Operators/Shift
1– 5	2
6– 8	3
9–10	4
11–12	5
13–16	6

Particulars	¢/Lb. Basis	Particulars	¢/Lb. Basis
Indirect labor	40% of direct labor	Raw-materials inventory	(No. of days) × (Production capacity) × (Raw-material cost)/(Operating days/yr.) The number of days of raw-material inventory is a program constant of 15 for bulk products and 30 for specialty products.
Maintenance	(5% of fixed capital)/(c_p)		
Supplies	15% of maintenance		
Utilities	3.5% of direct labor		
Packaging	Program constant at 0.5¢/lb. for bulk products, and 1.0¢/lb. for specialty products. Correction for product potency is made by dividing these basic rates by the percent active ingredient in product	Work-in-process inventory	(No. of days) × (Production capacity) × (Direct mfg. cost)/(Operating days/Yr.) The number of days of work-in-process inventory is a program constant of 10 for both bulk and specialty products.
Direct manufacturing cost	Raw-material cost + direct labor + indirect labor + maintenance + supplies + utilities + packaging	Finished-goods inventory	(No. of days) × (Production capacity) × (Mfg. cost + overheads)/(Operating days/Yr.) The number of days of finished goods inventory is a program constant of 30 for both bulk and specialty products.
Gross margin	Net sales minus direct manufacturing cost		
Sales expense	5% of gross sales for bulk products, and 15% of gross sales for specialty products	Working capital	Accounts Receivable + raw material inventory + work-in-process inventory + finished goods inventory. Note: current liabilities are considered negligible and, therefore, working capital is equivalent to current assets.
Merchandising margin	Gross margin minus sales expense		
Depreciation	It is assumed that 80% of the fixed capital investment is allocated for equipment, and 20% for buildings. The equipment portion is amortized over 10 yr. and the buildings over 30, in accordance with the straight-line method	Total capital investment (I_T)	Fixed capital + working capital. Fixed capital is determined by subprogram 1.
General and administrative	3.5% of gross sales	Return on investment after taxes (R_{AT})	(Net profit)/(Total capital) Return on investment is a program constant of 10% for bulk products and 20% for specialty products.
Factory administration	40% of direct labor		
Property taxes and insurance	(2% of Fixed capital)/(c_p)		

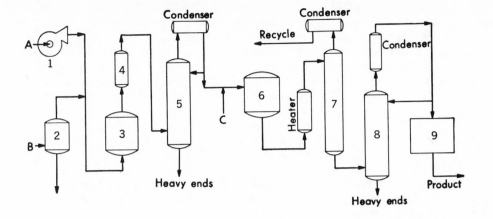

Operating Units
1. Compressor
2. Vaporizer
3. Vapor-phase reactor
4. Condenser
5. Refining column
6. Liquid-phase reactor
7. Stripper
8. Rectifier
9. Product storage

Reactant	Molecular Weight	Stoichiometric Moles/Mole of Prod. (inc. lab excess)
A	55	1.1
B	17	2.2
C	50	1.1
Catalyst	22	0.1

Molecular weight of product — 122
Over-all yield — 92 mole %
Active ingredients in product — 98%
Maximum process pressure — 5 atm.
Maximum process temperature — 300 C.
Corrosion characteristics — predominately Type 316 stainless steel construction. Alloy factor 0.2.

CONCEPTUAL FLOWSHEET, and basic data, for the economic evaluation of a specialty polymer additive—Fig. 1

of active ingredient in the product corrects the basic sales deductions (freight, returned goods and allowances) and packaging expense for variations in concentration. For example, a latex containing 30% solids will cost more to ship and package per pound of active ingredient than will one containing 50%.

Table I summarizes the functional relationships and empirical constants that are built into the master display program from which the various components of the P&L statement are established. These components are not defined in detail because this subject has been adequately covered in the literature by Weinberger[2] and Teplitzky.[3]

Selling-Price Determination

The selling price, P_S, is calculated by equating net profit as dictated by return on investment after taxes (R_{AT}) with that attained from operations:

$$(R_{AT})(I_T) = (P_S - C_T)(1 - T_R)(c_p)$$

or:

$$P_S = \frac{(R_{AT})(I_T)}{(c_p)(1 - T_R)} + C_T$$

The components of the above equations are defined in the nomenclature table, and they are all either data inputs, program constants or a function of selling price. Therefore, the computer is confronted with solving one equation, albeit a rather arduous one, in one unknown. Since it can be readily derived from the information presented, the equation is not reproduced here. Suffice to say that it is the type of expression that the computer can solve in a matter of seconds.

Options Provide Flexibility

When designing any computer program, it is wise to recognize the need for options in regard to input data. To ignore this need is to deny the reality that chemical products have widely different cost characteristics and profitability criteria. As such, the master display program has a dropout feature that permits the user to bypass the built-in inputs. This feature effectively transforms the program into a mathematical tool for generating a complete P&L statement that reflects specific knowledge on the part of the user. Moreover, it permits rapid generation of sensitivity analyses data. The only restriction upon the user is that he must maintain the functions the display program employs as a format for calculation of selling price. For example, indirect labor must still be specified as a percentage of direct labor; maintenance as a percentage of fixed capital.

Data Sheets and Printout

Because even the best of computer programs will gather dust if the data sheet cannot be readily understood, the program designer must give considerable thought to the input format. Also, the printout should not only provide the economic information sought but should also be complete in regard to all input assumptions so that a permanent record other than the data sheet is always available.

The Project Screening Program uses either one or two data sheets, depending upon the extent of the user's knowledge of the product under evaluation. To

provide a better understanding of the use and versatility of this program, two sample cases are presented: the first illustrates how it is employed to screen the potential profitability of a product when only the physical and chemical properties of the system are available; the second is an example of a case in which the manufacturing and overhead costs, selling expense and capital requirements are known.

Evaluating a New Specialty Chemical

Fig. 1 shows a conceptual flowsheet of a process for a specialty polymer additive. The potential market for this product was estimated to be between 1 and 5 million lb./yr. However, if the selling price exceeded $1/lb., it was feared that the customer's use cost (selling price × addition level) would destroy his incentive to purchase, and this potential market would not materialize.

To ascertain if there was any economic incentive to produce the product by this process, certain physical and chemical data were compiled; these are shown in Fig. 1, below the flowsheet.

The completed Standard Data Sheet is given in Fig. 2. Note that four cases can be evaluated with this single sheet. Sensitivity analyses that disclose the effect of variations in process temperature, pressure, or operating units can be prepared should these data be open to conjecture. On the other hand, if the parameters are known, as in this example, they may be held constant to show how total product cost and selling price vary with production capacity.

The raw-material economics printout (Fig. 3) documents the raw-material input data and presents the pounds and cost of each reactant per pound of product. This information enables the evaluator to rapidly distinguish the reactants that control product raw-material cost. The assumptions upon which the computer has determined total product cost and working capital are printed out after the raw-material economics. The final printout is the P&L statement for each case submitted. The income statement for the 1-million-lb./yr. plant is shown in Fig. 4 on the next page.

Selling price and total product cost are plotted in Fig. 5 as a function of production capacity. With the knowledge that selling price should not exceed $1/lb., this curve indicated the importance of obtaining an accurate appraisal of the market. As a sales volume greater than 1 million lb./yr. was considered feasible, additional development work on this additive was authorized.

An Experimental Latex

This example illustrates the utility of the Project Screening Program to evaluate a product for which there is a minimum of information readily available. The next example shows how the program is used to evaluate a product with which an operating division has had considerable experience regarding manufacturing and overhead costs, sales expense and capital.

In this case, we were asked to determine whether an experimental latex could be sold for less than a commercially available latex selling for 28¢/lb. To avoid unnecessary speculation in regard to manufacturing and overhead costs, our group consulted with an operating division that was currently selling latex products.

The raw-material cost of the experimental latex was known to be 18¢/lb. Examination of company records documenting capital expenditures indicated that a 20-million-lb./yr. plant would require a fixed capital investment of $1 million. However, it was highly probable that existing production facilities could be employed, reducing the fixed capital expendi-

DATA SHEET provides input on polymer additive—Fig. 2

CONCISE PRINTOUT shows raw-material economics—Fig. 3

PROFIT & LOSS STATEMENT
MATURE YEAR

PROCESS**   POLYMER ADDITIVE
PRODUCTION**   1.0 MILLION LB. PER YR.

	DOLLARS—M	CTS./LB.	% GROSS SALES
GROSS SALES	1056.3	105.63	100.0
SALES DEDUCTIONS	10.2	1.02	1.0
NET SALES	1046.1	104.61	99.0
DIRECT MFG. COST	379.8	37.98	36.0
RAW MATL. COST	216.1	21.61	20.5
DIRECT LABOR	84.0	8.40	8.0
INDIRECT LABOR	33.6	3.36	3.2
MAINTENANCE	28.6	2.86	2.7
SUPPLIES	4.3	0.43	0.4
UTILITIES	2.9	0.29	0.3
PACKAGING	10.2	1.02	1.0
GROSS MARGIN	666.3	66.63	63.1
SALES EXPENSES	158.4	15.84	15.0
MERCHANDISING MARGIN	507.9	50.79	48.1
OVERHEADS	184.4	18.44	17.5
DEPRECIATION	49.6	4.96	4.7
G & A	37.0	3.70	3.5
FACTORY ADMINISTRATION	33.6	3.36	3.2
TAXES & INSURANCE	11.5	1.15	1.1
R & D	52.8	5.28	5.0
TOTAL PRODUCT COST	732.9	73.29	69.4
BEFORE TAX PROFIT	323.5	32.35	30.6
TAXES @ 50.0 %	161.7	16.17	15.3
NET PROFIT	161.7	16.17	15.3

TOTAL CAPITAL EMPLOYED, M$ 809.5
FIXED CAPITAL, M$ 572.7
WORKING CAPITAL, M$ 236.8

AFTER TAX RETURN ON INVESTMENT   20.0%

FINAL PRINTOUT is P&L statement for additive—Fig. 4

Project Screening Program

Project No. _11m_

STANDARD DATA SHEET

Process Description:
Name of Product: _Experimental Latex_
   Product
Date: _1-13-48_

	Description	Input
GENERAL	% Active Ingredient in Product	45
	Molecular Weight of Product	1
	Overall Yield of Product - Mole %	100
	Type of Chemical Produced* — WRITE For: 1 Bulk / 2 Specialty / 3 Special Case	3
	Corrosion Characteristics of the Process — WRITE For: 0.0 Non / 0.1 Slightly / 0.2 Moderate / 0.3 Very / 0.4 Extreme	

REACTANTS** (include Catalyst)	Name of Reactant	
	Stoichiometric Moles/Mole Product	
	Molecular Weight	1
	Cost (¢/lb)	

| | | Product RMC (¢/lb) | 18 |
| | | Number of Cases to Run | 4 |

FIXED CAPITAL***		Case No.			
		1	2	3	4
	Production Capacity (MM lb/yr) (between 0→100)	20	20	20	20
	Max. Process Temperature (°C)	0	0	0	0
	Max. Absolute Process Pressure (atm)	0	0	0	0
	Operating Units				
	Fixed Capital ($MM)	0.5	0.7	0.9	1.0

*If (3) is used, complete and submit the Special Data Sheet also.
**If product raw material cost is known, insert in space provided. Enter under INPUT the numeral 1 for molecular weight of product and reactants; enter 100 for % yield.
***If fixed capital is known, insert in space provided. Enter the numeral 0 for maximum process pressure and temperature. The number of Operating Units must be specified if Special Case is not checked as this number determines the operators per shift.

SECOND EXAMPLE: Standard Data Sheet for latex—Fig. 6

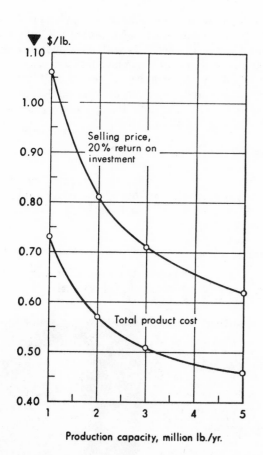

CAPACITY vs. product cost and selling price—Fig. 5

Project Screening Program
SPECIAL DATA SHEET

Process Description: _Experimental Latex_

Cost Area	Description	Units	Special Case
General	Return on Investment After Taxes	%	15
	Construction Cost Index		
Sales Deductions	Freight, Returned Goods, and Other	¢/lb	1
Conversion Costs — Labor	No. of Men/Shift	Men/Shift	2
	No. of Shifts/Day	Shifts/Day	3
	Annual Salary	$/Man Year	7800
	Indirect Labor	% Direct Labor	55
Conversion Costs — Other	Maintenance	% of Fixed Capital	4
	Supplies	% of Maintenance	17
	Utilities	% Direct Labor	15
	Packaging	¢/lb	0.4
	Selling Expenses	% Gross Sales	6
Overheads — Depreciation	Rate A	%	10
	Rate B	%	3.3
	Fraction B of Fixed Capital	%	35
Overheads	General & Administration	% Gross Sales	3.3
	Factory Administration	% Direct Labor	52
	Taxes and Insurance	% of Fixed Capital	2
	Research & Development	% of Gross Sales	3
Working Capital	Accounts Receivable		40
Working Capital — Inventory	Raw Materials	Operating Days/Yr for Each Category	15
	Work in Process		5
	Finished Goods		30
	Operating Days/Year	Days/Year	333
Income Taxes	Income Tax Rate	%	52

SECOND EXAMPLE: Special Data Sheet for latex—Fig. 7

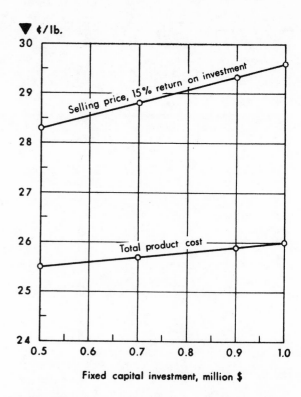

PROFIT & LOSS STATEMENT MATURE YEAR			
PROCESS** EXPERIMENTAL LATEX PRODUCTION** 20.0 MILLION LB. PER YR.			
	DOLLARS—M	CTS./LB.	% GROSS SALES
GROSS SALES	5928.6	29.64	100.0
SALES DEDUCTIONS	444.4	2.22	7.5
NET SALES	5484.2	27.42	92.5
DIRECT MFG. COST	3904.1	19.52	65.9
RAW MATL. COST	3600.0	18.00	60.7
DIRECT LABOR	46.8	0.23	0.8
INDIRECT LABOR	25.7	0.13	0.4
MAINTENANCE	40.0	0.20	0.7
SUPPLIES	6.8	0.03	0.1
UTILITIES	7.0	0.04	0.1
PACKAGING	177.8	0.89	3.0
GROSS MARGIN	1580.1	7.90	26.7
SALES EXPENSES	355.7	1.78	6.0
MERCHANDISING MARGIN	1224.3	6.12	20.7
OVERHEADS	494.4	2.47	8.3
DEPRECIATION	76.5	0.38	1.3
G & A	195.6	0.98	3.3
FACTORY ADMINISTRATION	24.3	0.12	0.4
TAXES & INSURANCE	20.0	0.10	0.3
R & D	177.9	0.89	3.0
TOTAL PRODUCT COST	5198.7	25.99	87.7
BEFORE TAX PROFIT	730.0	3.65	12.3
TAXES @ 52.0 %	379.6	1.90	6.4
NET PROFIT	350.4	1.75	5.9

TOTAL CAPITAL EMPLOYED, M$ 2337.2
FIXED CAPITAL, M$ 1000.0
WORKING CAPITAL, M$ 1337.2

AFTER TAX RETURN ON INVESTMENT   15.0%

PROFIT AND LOSS printout for proposed latex—Fig. 8

SELLING PRICE and product cost vs. investment—Fig. 9

ture to no more than $600,000. The required return on invested capital for a project of this type was set at 15% after taxes.

Because manufacturing and overhead costs, working capital, and profitability criteria differ from the values built into the master display program, both the Standard Data Sheet and the Special Data Sheet (Fig. 6 and 7) are used. Note that the options to bypass both the raw-material cost and fixed-capital subprograms have been exercised, because the raw-material cost is known, and a sensitivity analysis to show the effect of variations in specific fixed-capital requirements is desired.

The P&L statement for the case in which fully allocated facilities would be required is shown in Fig. 8, while selling price and total product cost are plotted as a function of fixed capital in Fig. 9. The graph indicates that even if existing facilities were made available, the latex could not be sold at a price below that of the competition. Since the product did not offer any significant advantage in performance characteristics, it was not considered a candidate for commercialization.

No Panacea

The Project Screening Program should not be thought of as a replacement for the cost engineer, or a panacea for those concerned with the problems of economic evaluation. Rather, it is a labor-saving device that, when combined with the ingredients of personal experience and good judgment, can provide rapid order-of-magnitude evaluations sufficient for assessing the economic worth of research projects when little data are available. It is with this understanding that such a program is recommended. The functional methods and factors can be altered to suit the particular process and economic characteristics of your industry to provide a reliable tool for the rapid screening of research projects.

Acknowledgment

Development of the computer program described herein has been made possible by the contributions of many fellow employees, particularly R. Vaughn, B. Paul and G. West.

References

1. Zevnik, F. C., Buchanan, R. L., *Chem. Eng. Prog.*, Feb. 1963, p. 70.
2. Weinberger, A. J., *Chem. Eng.*, Dec. 23, 1963, p. 81.
3. Teplitzky, G., *Chem. Eng.*, June 19, 1967, p. 215.

Meet the Author

Robert W. DeCicco is a research engineer in the Central Research Div. of W. R. Grace & Co. (Clarksville, Md. 21029), where he specializes in the technical and economic evaluation of R&D projects. He gained experience in the development, design and startup of process engineering equipment during employment with the Podbielniak Div. of Dresser Industries, and with Babcock & Wilcox Co. Mr. DeCicco received a B.S. in chemical engineering from Illinois Institute of Technology, and an M.B.A. from Roosevelt University.

Should Your Pet Project Be Built?
What Should the Profit Test Be?

You have a pet project that shows a 10% estimated rate of return. Let's say your company either has plenty of working capital, or can borrow money at a low 6%. Does this mean your project should get the green light? Not necessarily, says a leading financial authority; his reasons may give you new insights into the true cost of capital and the justification for go and no-go investment decisions.

JOHN F. CHILDS, *Irving Trust Co.*

Our economic system has helped produce the world's highest standard of living. Of course it has faults, but it is an incentive system that appeals to our basic human self-interest. It permits the maximum opportunity for individual initiative and freedom. One of its characteristics is the free market place, where everyone can produce what he chooses and buy what he wants. Prices are set by the interplay of supply and demand. Competition, which is maintained through our antitrust laws, acts as the regulator of prices and protects the consumers.

What determines whether there should be an added supply? You engineers, working with management, must decide whether a product can be turned out at a profit on the basis of the price expected to be obtained with the estimated demand. In other words, new supplies come on the market when there is sufficient demand to provide a price that will more than cover all costs. When there is an oversupply, prices are driven down below costs; then the supply should be gradually withdrawn so that prices will rise again. In this way our free enterprise system determines what is produced: The consumers "vote" for what they want by their demand, which, through price action, induces management to either increase or decrease the production of goods.

Costs, Capital, and Cost-of-Capital

We generally think of profit as earnings after all costs: wages, materials, taxes, depreciation, etc. Actually, capital also has a cost, and a true economic profit is not earned until more than the cost associated with capital is recovered. There is much mystery about cost-of-capital; major mistakes are made in this area when deciding whether or not to authorize a project.

First, what do we mean by capital? In order to answer this question, consider how capital flows into a corporation and how it is put to use. A look at a balance sheet will give us the best picture. Assume that you start a new industrial company with the simplified balance sheet below:

Assets	$	Liabilities		$
Current assets	100	Current liabilities		50
Plant	50	Long-term capital:		
		Debt 20		
		Common Equity .. 80	100
Total	150	Total		150

When determining cost-of-capital, we will be looking at the right-hand side of the balance sheet, and dealing with the total long-term capital of $100. This capital is provided by investors through their purchases of long-term securities, consisting principally of debt and common equity. Surplus arising from retained earnings is equivalent to raising new common equity. In fact, as earnings are plowed back, a company regularly raises new common equity. All common equity, whether it be the stated or par value, capital surplus or retained earnings, should be treated alike. There is no distinction so far as the common stockholders are concerned. It all represents their ownership.

There is a competitive market for capital just as there is a competitive market for goods and services. All capital has a price or cost. To be successful, a company should earn something above all costs.

When applying the idea of cost-of-capital, we will be working with assets on the left-hand side. It is necessary to visualize the relationship between the two sides of the balance sheet so that the rate obtained on the right-hand side will be correctly associated with the assets on the left. On the basis of our simplified balance sheet, the $100 of long-term securities provides the money for $50 of plant, and $50 of working capital (the latter represents current assets of $100, less current liabilities of $50). For example, if we decided that the return on long-term capital should be 10%, then the equivalent return on plant would have

to be 20%, on total assets 6⅜% and on working capital plus plant, 10%. Of course, the relative amount of working capital and plant will vary depending on the type of business.

Cost-of-capital would be easy to understand and calculate if all capital consisted of debt, because its cost would then equal the interest rate. The problem becomes difficult because of the common equity part of our capital structure. Does it have a cost, and how can it be calculated?

The Case of the All-Debt Capital Structure

Before we try to answer the question of the common cost rate, let's take another view of the long-term capital in the simplified balance sheet—that is, the $20 of debt and $80 of common equity. For the moment, let's reason in terms of debt securities, since we can probably agree that debt has a cost.

If we assume an interest rate of 5% on the $20 of debt, the picture would be as shown below:

	Amount	Rate	Cost
Debt	$20	5%	$1.00
Common	80		

At this point, forget about the tax savings that interest produces; we are now talking only about the rates that investors require, and they do not pay the corporate tax. We will cover the effect of interest on taxes subsequently.

Now let's change the composition of our capital structure and substitute a layer of $30 of junior debt for $30 of common equity. Since the added debt would be junior and in a more risky position, it would require a higher rate. For illustrative purposes let's use 6%.

For an industrial company, we already have added much more debt than would be wise from the viewpoint of sound financial policy. However, purely for the purpose of illustrating the idea of cost-of-capital, let's go further and use a third layer of debt; let us substitute $49 of junior-junior debt for $49 of common so that we only have $1 of common stock remaining. Certainly, no ordinary investor would purchase such debt with practically no common equity for protection, and any interest rate we assign to it would be a pure guess. It is conceivable that the rate on this third layer of debt might be 15%.

With the two additional layers of debt, the picture would be as shown below:

	Amount, $	Rate, %	Cost, $
Debt	20	5	1.00
Jr. Debt	30	6	1.80
Jr., Jr. Debt..........	49	15	7.35
Common	1		
	100		10.15

Thus, leaving out the cost for the remaining $1 of common and just using the debt costs as above, the over-all cost for 99% of our capital is about 10%.

We used this approach in explaining cost-of-capital to show that:

1. All capital has a cost, whether it be debt or common equity. Merely substituting debt for common equity or vice versa does not change that fact.

2. Capital is not cheap. As the debt becomes more junior, and is protected by less in the way of common equity or other junior securities, the interest rate must rise because of the added risk.

Cost of Common Equity

With this background, we can discuss the difficult task—determining the cost of the common equity. Common stock does not have a cost in the sense that debt has a cost, but common stock has an economic cost. Some people may prefer such a term as "required earnings rate."

In essence, when a person buys a common stock, he looks forward to receiving two benefits: dividends and capital appreciation. It is future expectations of these two elements that induce stockholders to pay a given price for common stock. But both dividends and market appreciation come from future earnings per share. Therefore, the common-cost rate is based on the relationship of market price of the common stock to the earnings per share expected by investors.*

The word "expected" is the all-important one in the definition. People inexperienced in the field may believe that the current price-earnings ratio represents the common cost, but current earnings are over the dam once a stock is purchased; all gains to the stockholders come from future earnings. Past earnings may well be a guide to future earnings, but they are not what benefit the common stockholders. This is particularly important to keep in mind for growth stocks that sell at high price-earnings ratios.

Our explanation has admittedly compounded the problem of actually figuring the common cost rate. Investors' expectations may be very difficult to measure. At various times they may be above, below or the same as is actually realized by the company in the future. The space of this article does not permit detailed examples of the common cost, so we will take another approach from the broad point of view of our economy as a whole. If our free enterprise system works correctly, and competition drives prices down to the level at which all costs are covered, then we might ask what has industry earned? Here we are talking about all industry and not one particularly depressed or very profitable industry. According to Table I, the average return on year-end common equity was 10.5% for the most recent ten-year period. On average beginning and year-end capital, the average figure would be approximately 11%.

It is interesting to note the average returns on common equity for the 500 largest companies included in the *Fortune* magazine survey, as shown in Table II.

*To obtain a refined calculation of the common cost, the market price should be adjusted to allow for financing costs in connection with the sale of new stocks. Also, the calculation should be made over a period of years, to include both favorable and unfavorable market conditions, so that the resulting figure represents average conditions.

Profit on year-end common equity, all manufacturing companies*—Table I

Year	%
1966	13.1
1965	12.6
1964	11.5
1963	10.2
1962	9.7
1961	8.8
1960	9.2
1959	10.4
1958	8.6
1957	10.9
Average % profit, 1957–1966	10.5

*Includes all manufacturing corporations except newspapers. Source: Quarterly Financial Report for Manufacturing Corporations, Federal Trade Commission, Securities and Exchange Commission. Figures for preferred stock are included with common equity, but the amount of preferred is relatively small.

Return on year-end equity, 500 largest companies—Table II

Year	Median Return, %
1966	12.8
1965	11.8
1964	10.5
1963	9.1
Average, 1963–1966	11.1

Reprinted from Fortune Directory by permission; © 1966 Time Inc. Year-end equity is referred to as invested capital by Fortune magazine; it includes capital stock, surplus and retained earnings.

Thus, if industry as a whole has earned its common cost, both of these tabulations suggest a figure of about 11% for common equity. As a matter of fact, calculations for common cost based on our competitive capital market for a large, well-capitalized industrial company do show an average figure close to 11% over a ten-year period.

Composite Cost-of-Capital

As a group, the manufacturing companies referred to in Table I had long-term capital consisting approximately of 19% debt and 81% common equity. If we arbitrarily use a 5% rate for the debt, and the 11% figure suggested above for the common equity, the over-all cost-of-capital, after taxes, for a typical large industrial company would be about 10%, as shown below:

	Amount	Rate	Total
Debt	19%	5%	0.95%
Common	81%	11%	8.91%
	100%		9.86%

Some Composite Rates

Cost-of-capital depends on the risks of the business as assessed by investors. For an individual project, the magnitude should be commensurate with the risk of that project. To give you an idea of the figures for certain types of business with different risks, a few approximate figures are presented in Table III in summary form without proof. We include figures on regulated industries because much work has been done in this field, and they give us a good reference point for comparison. Because of the relatively stable nature of their businesses, their capital costs are among the lowest we can expect to find. The rate for any industrial company should be substantially higher.

Table III shows two rates in addition to cost-of-capital. The "return-on-total-capital target" represents what a company might wish to earn on its long-term capital. The "profit-goal for expansion" represents what a company might use as a goal on an individual project.

The rates shown in Table III are based on average rates for debt securities. There has been a substantial increase in the general level of interest rates in recent years. If this increase continues, it will boost the composite rates required. This will have the greatest effect on the utilities, because a higher proportion of their long-term capital consists of debt. Also, it is possible that a prolonged high level of interest rates may have some effect on increasing the rate that common stockholders require because of the alternative investment considerations.

The spread between the rates shown in the three columns of Table III depends on many factors. In regulated industries, the spread is small. On the other hand, managements of industrial companies may set the expansion profit goal well in excess of cost-of-capital for many reasons. How much the spread should be is a matter of judgment; one must take into account that when a company goes into a new venture, there is added risk because of the lack of familiarity with the business. If a company used the cost-of-capital as its goal in each new venture, and some of the

Composite rates, after taxes, for various types of businesses—Table III

	Cost-of-Capital	Return-on Total-Capital Target	Profit Goal for Expansion
Industrial companies, average	10%	10% plus 1%, 2%, 3% or more	Up to 20%
Telephone		8%	
Electric utility		6¾% to 7%	

ventures failed, the average return would be below the cost-of-capital; some allowance must be made for this. Furthermore, the company should seek to make some profit above all economic costs.

It has been reported* that some major industrial companies have used 20% after taxes as an expansion profit goal.

Understanding the Cost-of-Capital

The subject of cost-of-capital is a most difficult one. Much more discussion than has been presented in this article is required to show how to calculate the cost of common and to cover other related subjects. However, I hope I have left some useful thoughts with you. Here are additional points:

1. The rate for cost-of-capital is the over-all composite cost of all the capital. It is the composite rate, representing the risk of the particular project, which must be used as a basis for the goal; i.e., in deciding how much should be earned on the capital committed to that project. The method of financing the project doesn't matter.

2. It is not always possible to calculate the common cost for a publicly owned company because it may be impossible to determine investors' expectations. In such a situation, we must arrive at a figure based on a judgment appraisal of the risk of the particular business as compared to the risk of some other business for which we are able to calculate the cost-of-capital. Because a privately owned company has no public market for its stock, it must of necessity use such an approach.

3. The composite cost-of-capital calculations were based on rates, after taxes, which have to be paid or earned to attract investors. Corporate taxes reduce the burden of interest charges. Therefore, when considering the rate from the point of view of the company, appropriate adjustments can be made to reflect the net cost of interest, after allowing for the tax reduction due to interest.

4. Cost-of-capital should represent the risk of the business. The appropriate use of debt can reduce the cost-of-capital, but too much leverage will increase the over-all cost-of-capital in the long run. Too much debt requires not only higher interest charges but also a higher earnings rate on the common because of the financial risk that has been added, which will be felt particularly when adverse conditions are encountered.

5. The business risk increases substantially for small companies and foreign investments.

6. A company that has divisions with different risks will require different rates for each.

7. The concept of cost-of-capital is basic to proper acquisition pricing. Today, incorrect emphasis on the effect of acquisitions on current earnings per share is leading some managements into acquisitions that may prove to be millstones around the acquiring company's neck.

8. Some companies show good profits and, through efficiency, earn well above the cost-of-capital for the marginal producer. They are the companies that benefit all elements in our economy the most—the employees, stockholders, tax collector and especially the consumers.

Applying the Concept

In evaluating a proposed capital project, the standard procedure is to make a forecast of sales, expenses and capital outlays, and then to make a profitability calculation based on the forecast. (Allowance must be made for the time value of money. This can only be done by some form of discounted-cash-flow technique. Such methods as pay-back, return on original investment, and return on average investment are inadequate.)

Then, to apply the ideas discussed in this article, it is necessary to establish a profit goal as a cutoff rate of return on capital. This is the rate with which the project's profitability can be compared in order to determine whether there is sufficient profit to justify the project. The project goal must be based on cost-of-capital and be somewhat above that cost.

The concept of cost-of-capital is basic to capital expenditure decisions. Understanding and applying this concept will result in wide benefit:

• Management will not be inhibited in capital expenditure decisions by the mistaken belief that the existing return on capital must necessarily be increased. Regardless of the effect of earnings generated by a new project on the existing return, the stockholders will benefit, providing the return realized on the new project is above the cost-of-capital for that particular project.

• A company's stockholders will avoid being hurt by the decline in the market value of their stock that is ultimately sure to follow if management makes capital expenditures with returns below the cost-of-capital.

• Companies may be prevented from causing overexpansion and excessive competition that depresses prices and profits unreasonably.

• Capital will tend to be used where it can be most beneficial, and goods will not be produced that are not really needed. This will reduce the possibility that capital may be wasted, and will assure the highest possible standard of living for our nation.

Meet the Author

John F. Childs is author of "Profit Goals and Capital Management," a book just published by Prentice-Hall that elaborates on the ideas presented in this article. He is Senior Vice-President of Irving Trust Co. (1 Wall St., New York City), and in charge of the Corporate Services Div. Mr. Childs is a graduate of Harvard Business School and Fordham Law School, and is a member of the New York Bar. His first book, "Long-Term Financing," was published in 1961 by Prentice-Hall.

---

*"Variability of Private Investment in Plant and Equipment." Materials submitted to the Joint Economic Committee, Congress of the United States (87th Congress, 2nd Session) ; Part II, Some Elements Shaping Investment Decisions, p. 14.

Probability Technique Improves Investment Analysis

A profitability anaylsis for a new capital investment can be greatly improved by careful application of this statistical technique that gives probabilities for various profit levels.

J. T. THORNGREN, *Ethyl Corp.*

Investment analyses have become quite sophisticated in the past several decades. No longer must the decision maker determine the attractiveness of a new venture by how many years the investment will be "paid out." The time value of money can be used to determine a more realistic measure of profitability.

However, until recently, even the best methods of investment analysis employed single-valued numbers. Since uncertainties are always present, several methods are used to give the decision maker some indication of the risks involved in a new investment. Two of the more popular methods consist of changing the major economic variables one at a time (sensitivity analyses) or changing all the variables from maximums through averages to minimums and computing the measures of profitability. Hertz[4] gives an excellent analysis on the limitations of these approaches. He points out that, although they show the ranges on possible results, they do not show what the chances are of obtaining any of these results.

Applying Statistical Methods

The concept of chance or odds on investment risks brought about the application of statistical methods. Statistical methods involve specifying that each economic variable (sales price, labor costs, etc.) in an investment analysis can assume any value throughout the projected period of the analysis. Certain values would, of course, be more probable than others. With this concept, that certain values of the economic variables have a greater chance of occurring, one can then determine what the chances will be for obtaining certain ranges on a measure of profit.

To mathematically set up a statistical investment analysis, one must predict the type of frequency distribution curve for each of the economic variables. For the sales price of a product, a frequency distribution curve would be a plot of sales price vs. the percent of time that this sales price occurred.

Two quantities that can be determined from this curve are the mean value of the sales price and its

variance. Though these two terms do not specify the shape of the curve, they are, however, all that are required to hand calculate a statistical investment analysis. The mean value of the sales price is simply the sum of all the values observed divided by the total number of observations. To obtain the variance, each observed value is subtracted from the mean value, each quantity is squared and then the total obtained. This total is divided by the number of observations minus one to give the variance. Another useful term is the standard deviation. The standard deviation is simply the square root of the variance and has the same dimensional units as the mean value.

The probability distribution curve is proportional to the frequency distribution curve. When one talks about a "distribution" curve, one is generally referring to a probability distribution curve. A cumulative probability distribution curve for the sales price would be a plot of the sales price vs. the percent probability that the sales price occurring would be less than or equal to that read from the curve.

The preceding paragraphs give a simplified explanation of the mean, variance, and various types of probability curves. For a more detailed explanation, refer to Hald.[2]

Using Distribution Curves

The normal distribution curve is easy to manipulate and is used frequently in investment analyses.[1, 5, 11] But when the distribution curve is specified as normal (the familiar bell-shaped curve), certain restrictions occur. The first is that there will be a 50% chance of obtaining a value below the mean and a 50% chance of obtaining a value above the mean. The second restriction is that there would be a 68.26% chance of obtaining a value between the mean plus the standard deviation and the mean minus the standard deviation.

To remove these restrictions, computer programs using Monte Carlo techniques have been developed to calculate the probability of obtaining ranges on the measure of profit. This scheme allows the economist to actually specify the distribution curves for each of the variables in the analysis. He is no longer restricted to using normal distribution curves.

The first and most logical question is whether an individual can truly specify the distribution curve on an economic variable with little or no previous statistics on the variable. The answer is definitely no.

Certainly no mathematical measure is any better than the estimates on which it is based.[9] However, the more background information available on the variable in question, particularly about how it is affected by factors from outside and within the company, the closer one can come to predicting the "true" distribution. Development of such probability distributions must be classified as an art that requires experience and good judgment.

This article presents a method for specifying cumulative curves that lend themselves to relatively easy hand calculation. The method produces the probability of obtaining ranges on a measure of profit. The technique can also be programmed for computer solution and does not require use of random numbers.

Making Polymetholine

To explain the procedure, let us set up a fictitious example. Batchem Chemical, a small specialty chemical company, has been approached by Mammoth Industries concerning a chemical called polymetholine. Mammoth would be willing to sign a ten-year contract in which they agree to purchase from Batchem a minimum of 10 million lb./yr. of polymetholine. The sales price would be fixed at 9.2¢/lb., but would be adjusted yearly to reflect any market price change in the major raw material, metholine-A.

Mammoth has indicated that there is some small chance that they would take as much as 15 million lb./yr. However, at the end of the ten-year period, Mammoth would not renew the contract. By then, they expect to be either out of the business that requires polymetholine or will have developed their own polymetholine production facilities. Batchem must decide whether or not to accept the contract within three months.

Twenty years ago, Batchem had a polymetholine production unit. The process chosen for the speculative design is identical except that more modern equipment will be used.

The various departments of Batchem have been working on the Mammoth project for 2½ months and

Single-valued variables for producing polymetholine—Table I

Sales volume, 10.0 million lb./yr.

Product price, $0.92/lb.
Variable costs, $1,000/million lb.

Metholine-A	30.0
Metholine-B	10.4
Catalyst	3.2
Utilities	5.1
Total	48.7

Fixed costs, $1,000/yr.

Operating labor with direct overhead	105.0
Repair labor with direct overhead	41.0
Repair material	42.5
Plant supervision	40.0
Indirect overhead	80.0
Insurance	3.5
Laboratory	10.0
Local tax	7.0
Operating supplies	3.0
Technical service	50.0
Total	382.0

Other costs, $1,000

Investment	350.0 (occuring at end of year 1)
Start-up	60.0 (during year 2)
Investment credit	0.0
Working capital in	60.0 (middle of year 2)
Working capital recovered	60.0 (end of year 11)
Salvage value	20.0 (middle of year 12)
Depreciation	Sum of the years
Fed. income tax (FIT)	48%

a directive has been issued to perform an economic evaluation. The economic evaluation is assigned to Joe Sharps. This is Joe's big chance to impress management with his diligence and ability to handle an assignment. He carefully collects the necessary information from the various departments. At the end of the week, he performs calculations to determine the percent interest rate of return using annual discount factors. His results are summarized in Tables I and II. The values of his economic variables at this point are referred to as "single-valued" since they are incapable of changing (cannot take on random values).

Monday morning the management review group convenes. Also present is Joe's supervisor, who has just returned from a trip and has not looked over Joe's results. Joe anxiously hands out copies of his analysis and settles back in his chair. Glancing up from his copy at the furrowed brows of his reviewers, he thinks, "Yes, it is unfortunate that the Mammoth project doesn't appear attractive, but that's the way it goes sometimes."

Unanswered Questions

Joe's supervisor is the first to break the silence "This is real good, Joe. I can tell you've done a lot of work. But what would the rate of return be if the sales price of metholine-A should increase later on: what would happen if our actual investment on the 15-million lb./yr. plant turned out to be less than

$350,000; what if Mammoth should purchase more than the minimum 10 million lb./yr., what if . . .? How about trying to have that for us by the end of the week?"

There is only one furrowed brow in the room now, and Joe is trying hard not to show it.

Back in his office, Joe begrudgingly starts to vary the major costs in his analysis, up and down and all around. After several hours of labor, he wonders just what his results will tell management. Another hour is spent with his boss discussing these results.

"I must agree, Joe, your results show what the rate of return would be if these conditions occurred, but they would in no way indicate to management what the chances of obtaining the rates would be. Go ahead and try your statistical approach."

Random-Valued Factor

Joe must now transform all of his single-valued economic variables into random-valued variables. As mentioned earlier, all that is required for the hand calculation is the mean and variance of each variable. To obtain these, Joe will introduce a new term called a random-valued factor. This random-valued factor will transform a single-valued variable into a random-valued variable such that the mean and variance can readily be determined.

Let the variable Z (such as operating labor, maintenance, etc.) be given by

Percent return on singled-valued variables—Table II

	1	2	3	4	5	6	7	8	9	10	11	12
Sales volume, MM lb.	0	10.0	10.0	10.0	10.0	10.0	10.0	10.0	10.0	10.0	10.0	0
Product price, M$/MM lb.	—	92.0	92.0	92.0	92.0	92.0	92.0	92.0	92.0	92.0	92.0	—
Variable costs, M$/MM lb.	0	−48.7	−48.7	−48.7	−48.7	−48.7	−48.7	−48.7	−48.7	−48.7	−48.7	0
Variable revenue, M$/MM lb.	0	43.3	43.3	43.3	43.3	43.3	43.3	43.3	43.3	43.3	43.3	0
Production revenue, M$	0	433.0	433.0	433.0	433.0	433.0	433.0	433.0	433.0	433.0	433.0	0
Fixed costs, M$	0	−382.0	−382.0	−382.0	−382.0	−382.0	−382.0	−382.0	−382.0	−382.0	−382.0	0
Start-up costs, M$	—	−60.0	—	—	—	—	—	—	—	—	—	—
Profit before FIT, M$	0	−9.0	51.0	51.0	51.0	51.0	51.0	51.0	51.0	51.0	51.0	0
Investment, M$	−350.0	—	—	—	—	—	—	—	—	—	—	—
Depreciation factors	—	0.1818	0.1636	0.1454	0.1273	0.1091	0.0909	0.0727	0.0545	0.0364	0.0182	—
Profit after FIT, M$	0	25.86	54.01	50.96	47.90	44.85	41.79	38.74	35.68	32.63	29.57	0
Working cap., M$	—	−60.0	—	—	—	—	—	—	—	—	60.0	—
Salvage, M$	—	—	—	—	—	—	—	—	—	—	—	20.0
Continuous cash flow, M$	0	25.86	54.01	50.96	47.9	44.85	41.79	38.74	35.68	32.63	29.57	0
Instantaneous cash flow, M$	−350	−60.0	0	0	0	0	0	0	0	0	60.0	20.0
Total cash flow, M$	−350	−34.13	54.01	50.96	47.9	44.85	41.79	38.74	35.68	32.63	89.57	20.0

Percent interest rate of return:

$$\sum_{n=1}^{q} \frac{\text{Total cash flow in the year } n}{(1.0 + r_a)^n} = 0.0$$

$$[-350.0/(1.0 + r_a)] + [(-34.13)/(1.0 + r_a)^2] + \cdots [20.0/(1.0 + r_a)^{12}] = 0.0$$
$$r_a = 0.02844; \quad \% \text{ return} = 2.8\%$$

$$Z = S\{Z\} X_f \qquad (1)$$

where $S\{Z\}$ is the single value of Z, and X_f is a random-valued factor. It is easier to work with factors rather than the whole variable.

The mean value of Z will be given by

$$m\{Z\} = S\{Z\}\, m\{X_f\} \qquad (2)$$

and the variance on Z, by

$$\gamma\{Z\} = (S\{Z\})^2\, \gamma\{X_f\} = \sigma^2\{Z\} \qquad (3)$$

where the variance is always positive and equal to the square of the standard deviation.

The procedure will be to determine a cumulative probability distribution curve for the random-valued factor X_f so that the mean and variance can be calculated. Eqs. (2) and (3) will then give the mean and variance of the economic variable.

It is always possible to transform variable X_f that is continuously distributed, but not necessarily normally distributed, into a function $g(X_f)$ that is normally distributed. Generally, the variable can be written in a form

$$u = \frac{g(X_f) - g(\epsilon)}{\sigma\{g(X_f)\}} \qquad (4)$$

where $g(X_f)$ is normally distributed about a constant $g(\epsilon)$ with standard deviation σ, and u is a standardized variable. The variable u is normally distributed with mean zero and variance one. A simple form for $g(X_f)$ can be

$$g(X_f) = \log_{10}(X_f + \alpha) \qquad (5)$$

Thus, Joe's cumulative probability curve for X_f will follow the laws of a normal cumulative probability curve when he makes a transformation to $\log_{10}(X_f + a)$. This should become clearer when Joe determines the mean and variance for the polymetholine plant investment.

To determine a cumulative probability distribution for X_f to be used on the $350,000 single value for the plant investment, Joe must first consider two pertinent facts:

1. The $350,000 investment was obtained from a scaled-up number for the same kind of plant built 20 years ago, which no longer exists. The number does include escalation resulting from inflation.

2. Last year, money was appropriated for another new plant based on the Chief Estimator's investment figures. The actual cost of that new plant significantly overran the appropriation.

Joe might consider $350,000 to be quite high because:

1. Design technology and innovations would probably result in a less costly plant.

2. The Chief Estimator might be "making sure" this estimate would not be overrun.

Would the economist feel confident in using the normal distribution approach?

Cumulative Probabilities

Joe now estimates several cumulative percent probabilities as shown in Table III. Here and elsewhere

Distributions on plant investment—Table III ·

X_f	$P[X_f]$
1.5	99.0%
1.0	90.0
0.5	5.0

$P[X_f]$ will be defined as the cumulative percent probability that X_f will have a value equal to or less than X_f. The cumulative percent probability that X_f will have a value greater than or equal to X_f will then be $100 - P[X_f]$. The data in Table III show that he expects there is a 99% chance that the random factor will be less than or equal to 1.5 x $350,000 = $525,000, a 90% chance it will be less than or equal to $350,000, and a 5% chance it will be less than or equal to $175,000.

Nomenclature

A_n	Before-federal-tax profit = (Sales price-variable costs) × (Sales volume) − Fixed costs − Startup costs = (Variable revenue) × (Sales volume) − Fixed costs − Startup costs = (Production revenue) − Fixed costs − Startup costs.
C_n^T	Total cash flow in the year n.
C_n	After-federal-tax profit and uniform cash flow through the year n.
C'_{in}	A miscellaneous cash flow in the year n occurring instantly at fraction X_i of the year n.
C'_n	Sum of all miscellaneous cash flows in the year n.
f_D	Yearly depreciation factor on the investment.
$g(X_f)$	Transform function of random-valued factor X_f.
I	Investment.
j	Any specific year.
M	$\log_{10} e = 0.4343$.
$m\{\,\}$	Mean value of a random variable.
$P[\ ]$	Percent probability of having a value less than or equal to the given value.
$p\{\,\}$	Probability distribution function for a random variable.
q	Final year of the investment analysis.
r	Fractional interest rate of return.
r_a	Fractional interest rate of return determined by annual discount factors.
r_m	A specified minimum fractional interest rate of return.
$S\{\,\}$	A single value of a variable.
t	Fractional federal income tax rate.
u	Standardized random variable, normally distributed.
X_f	Random-valued factor.
X_i	Fraction of year n at which C'_{in} occurs instantly.
Z_i	Variable in an investment analysis.
α	Constant in transform function for X_f.
$\sigma\{\,\}$	Standard deviation of a random variable.
$\gamma\{\,\}$	Variance of a random variable.
ϵ	Constant for function of the mean of $g(X_f)$.

Simplified exponential functions—Table IV

Range on β	Values of $\exp[1/2\,(\ln\beta)^2]$
$1.00 \leqq \beta < 1.2\ldots\ldots$	1.0
$1.2 \leqq \beta < 1.8\ldots\ldots$	$0.3\,(\beta - 0.942)^2 + 0.996$
$1.8 \leqq \beta < 7.0\ldots\ldots$	$0.13\,(\beta - 0.308)^2 + 0.899$
	Values of $\exp[(\ln\beta)^2 - 1.0]$
$1.00 \leqq \beta < 1.08\ldots\ldots$	$(\beta - 1.0)^2$
$1.08 \leqq \beta < 1.8\ldots\ldots$	$0.52\,(\beta - 0.912)^2 - 0.00889$
$1.8 \leqq \beta < 7.0\ldots\ldots$	$[e^{1/2(\ln\beta)^2}]^2 - \left[1.0 + 1.7\left(\dfrac{\beta - 1.8}{\beta}\right)^2\right]$

If he plots the values in Table III on normal probability paper (Fig. 1), he would not obtain a straight line. This excludes a normal distribution.

Joe plots these values on logarithmic probability paper (Fig. 2) as implied by Eq. (5) with α equal to zero. A straight line does not occur and hence a value of α must be determined. To determine α, he simply moves to the right or left of his original points by an estimated value for α. In general, for straightening concave downward curves as in Fig. 2, α will be negative. For concave upward, α will be positive. A negative value of α must, however, be restricted; when X_f is approaching the absolute value of a negative α, $P[X_f]$ must be approaching zero. One should not expend the effort to straighten the curve completely by estimating α to more than one digit strictly because of the nature of subjective distributions.

Let ϵ be the value of $X_f + \alpha$ at $P[X_f] = 50\%$ and let λ be the value of $X_f + \alpha$ at $P[X_f] = 15.87\%$. Then

since the group $\log_{10}(X_f + \alpha)$ is normally distributed,

$$m\{\log_{10}(X_f + \alpha)\} = \log_{10}\epsilon$$

and

$$\sigma\{\log_{10}(X_f + \alpha)\} = \log_{10}\epsilon - \log_{10}\lambda$$

The standard deviation of $\log_{10}(X_f + \alpha)$ is determined from the symmetry of the normal distribution curve. The mean occurs at a probability of 50.0%. The difference in percent probability between the mean and the mean minus the standard deviation would be $68.26/2.0 = 34.13\%$. Hence λ is evaluated at $P[X_f] = 50.0 - 34.13 = 15.87\%$. Letting $\beta = \epsilon/\lambda$, the mean of X_f is given by

$$m\{X_f\} = \epsilon \exp[\tfrac{1}{2}(\ln\beta)^2] - \alpha \qquad (6)$$

and the variance of X_f by

$$\gamma\{X_f\} = (m\{X_f\} + \alpha)^2 (\exp[(\ln\beta)^2] - 1.0) \qquad (7)$$

The exponential functions in Eqs. (6) and (7) are somewhat awkward to evaluate. Also, for values of β approaching 1.0, the right parenthesis in Eq. (7) will yield a serious round-off error if not properly evaluated. The simplified equations given in Table IV may be used if desired. The maximum inaccuracy on the $m\{X_f\}$ and $\gamma\{X_f\}$ determined via the equations in Table IV will be $2\tfrac{1}{2}$ and 5% respectively.

For Fig. 2,

$$\alpha = -0.3$$
$$\epsilon = 0.41$$
$$\beta = 0.41/0.26 = 1.5769$$
$$\exp[\tfrac{1}{2}(\ln\beta)^2] = 0.3\,(0.6349)^2 - 0.996 = 1.117$$
$$\exp[\ln(\beta)^2] - 1.0 = 0.52(0.6649)^2 - 0.00889 = 0.2210$$
$$m\{X_f\} = 0.41\,(1.117) - (-0.3) = 0.7579$$
$$\gamma\{X_f\} = [0.758 + (-0.3)]^2\,(0.221) = 0.04635$$

and from Eqs. (2) and (3).

$$m\{\text{investment}\} = 0.7579\,(\$350,000) = \$265,300$$
$$\gamma\{\text{investment}\} = (\$350,000)^2\,(0.04635) = 5.677 \times 10^9\;(\$)^2$$

It should be emphasized that the mean and variance

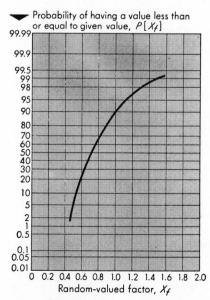

Curve indicates data do not fit normal distribution pattern—Fig. 1

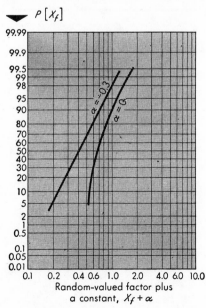

Applying a constant to Fig. 1 data yields a straight line—Fig. 2

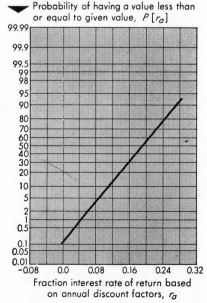

Interest-rate-of-return calculations produce straight line—Fig. 3

Summary of random-valued factors—Table V

	Years	(Random-Valued Factor)/Cumulative Probability)			α	ϵ	β	Mean Value of Variable	Variance of Variable	Notes and Comments
Sales volume...........	2–4	1.0/5	1.05/80	1.1/99	0	1.04	1.020	10.40	0.04155	Discussion with Mammoth representative.
	5–6	1.0/5	1.2/70	1.3/95	0	1.13	1.077	11.30	0.7414	
	7–9	1.0/5	1.3/75	1.5/99	0.1	1.29	1.093	11.90	1.362	
	10–11	1.0/3	1.3/65	1.5/99	1.0	2.20	1.048	12.00	1.097	
Product price..........	2–3	1.0/10	1.05/70	1.1/97	0	1.03	1.020	94.76	3.520	Random-value factors identical to those for Metholine-A cost.
	4–9	1.0/4	1.1/30	1.2/95	1.0	2.10	1.029	101.2	32.26	
	10–11	1.0/10	1.05/70	1.1/97	0	1.03	1.020	94.76	3.520	
Metholine-A...........	2–3	1.0/10	1.05/70	1.1/97	0	1.03	1.020	30.90	0.3743	Review with process designers; utilizations are firm. Investigated firm supplying Metholine-A.
	4–9	1.0/4	1.1/30	1.2/95	1.0	2.10	1.029	33.00	3.431	
	10–11	1.0/10	1.05/70	1.1/97	0	1.03	1.020	30.90	0.3743	
Metholine-B...........	2–8	0.95/2	1.0/40	1.05/99	0	1.00	1.020	10.40	0.04501	Investigated firm supplying Metholine-B.
	9–11	1.0/10	1.05/50	1.1/95	0	1.05	1.040	10.92	0.1870	
Catalyst..............	2–11	0.9/1	1.3/30	1.5/90	5.0	6.35	1.032	4.320	0.4361	Discussion with process design group; new catalyst will be used that may be 50% above present cost.
Utilities..............	2–11	0.9/5	0.95/45	1.0/80	0	0.960	1.032	4.896	0.02501	Expanding utility company expected to lower charges.
Oper. labor...........	2–4	1.0/2	1.05/40	1.1/90	0	1.06	1.039	111.3	19.04	Expected wage increases.
	5–11	1.0/1	1.1/30	1.2/98	1.0	2.1	1.024	115.5	28.95	
Repair labor...........	2–6	0.9/10	0.95/60	1.0/80	−0.2	0.75	1.064	38.95	3.849	Supervisor of former plant expects increasing costs with increasing age of plant.
	7–11	1.0/3	1.1/35	1.2/90	0	1.11	1.057	45.51	6.753	
Repair material.........	2–6	0.9/10	0.95/60	1.0/80	−0.2	0.75	1.064	40.38	4.136	Similar to repair labor.
	7–11	1.0/3	1.1/35	1.2/90	0	1.11	1.057	47.18	7.256	
Plant supervision.......	2–5	1.0/4	1.05/35	1.15/98	0	1.08	1.049	43.20	4.390	Production people expect less supervision after some experience with new plant.
	6–11	0.8/2	0.9/40	1.1/96	−0.5	0.43	1.194	37.20	9.638	
Indirect overhead......	2–6	1.0/1	1.05/30	1.1/90	0	1.08	1.039	86.40	11.06	Increasing costs at first, but expect decrease later with new plants sharing the burden.
	7–11	0.9/15	1.0/35	1.05/95	4.0	4.95	1.010	76.00	16.31	
Insurance.............	2–11	0.5/5	1.0/90	1.5/99	−0.3	0.41	1.577	2.653	0.5677	Based on investment.
Lab..................	2–3	1.05/5	1.1/20	1.2/90	0	1.12	1.037	11.20	0.1717	Based on discussion with lab supervisor.
	4–11	0.9/10	1.0/50	1.05/99	3.0	3.95	1.013	9.500	0.2556	
Local taxes............	2–11	0.5/5	1.0/90	1.5/99	−0.3	0.41	1.577	5.306	2.271	Based on investment.
Oper. supplies.........	2–11	1.0/8	1.05/20	1.2/95	−0.2	0.88	1.067	3.24	0.03101	From production and purchasing depts.
Tech. service..........	2–3	1.0/1	1.2/30	1.3/90	3.0	4.20	1.024	60.00	26.26	Discussion with tech. service supervisor.
	4–11	0.95/5	1.0/40	1.2/99	−0.5	0.505	1.074	50.25	3.539	
Investment............	1	0.5/5	1.0/90	1.5/99	−0.3	0.41	1.577	265.3	5,677	
Startup cost...........	2	1.1/10	1.2/40	1.3/95	1.0	2.20	1.048	72.00	39.48	Discussion with production and tech. service depts.
Invest. cred...........	2							0	0·	
Working cap. in	2	0.9/10	1.0/30	1.1/90	2.0	3.0	1.017	60.00	9.254	Discussion with sales and production depts.
Recovered	11	0.9/10	1.0/30	1.1/90	2.0	3.0	1.017	60.00	9.254	
Salvage...............	12	1.0/5	2.0/40	3.0/90	5.0	7.0	1.308	45.08	1,533	Discussion with tech. service and design groups.
Fed. income tax........	2–5	1.0/3	1.1/80	1.15/95	0	1.07	1.039	0.5136	0.0003992	Projection by Joe Sharps.
	6–11	1.0/3	1.05/60	1.1/99	0	1.04	1.03	0.4992	0.0002198	
Depreciation factors....										Sums of years. Identical to those for single-valued variables.

for the now random-valued investment do not have the connotations of a normal distribution, i.e., 50% chance of a value above or below the mean, etc.

Our economist estimated only three cumulative probabilities for X_f. Certainly no more than four points, preferably three, should be estimated. It would be practically impossible for an individual to approximate the "true" distribution by more than four points.[10] These three or four points should include essentially the whole expected range on X_f.

Obtaining Sales Picture

Joe talked with Batchem's sales representatives and found there was absolutely no chance of selling poly-metholine to other companies. He then talked with a representative from Mammoth concerning the odds on how much more than the minimum amount Mammoth might take through the ten-year period. He concluded that during the first three years, the chances of selling more than 11 million lb./yr. were essentially nil, and the chances were very slight that it would exceed the minimum. In the next two years the chances would be very, very small that sales would exceed 13 million lb./yr., but it could easily be over 10 million. In the remaining years, the chances of exceeding 13 million would increase.

As to the sales price of metholine-A, Joe investigated the recent annual reports of the company that would supply Batchem. During the first two years of production, the chance of an increase appeared small and if it did occur, the size of the increase would be small. However, reviewing the economic picture of this company, he could see that an increase was imminent during the fourth year. During the tenth year, the price would probably fall to its original value.

Table V summarizes Joe's cumulative probabilities and calculated X_fs on each term in the investment analysis. The simplified exponential formulae (Table IV) were used in his calculations. Note that the distributions on the random-valued factors applied to the sales price of metholine-A are the same as those

used on the sales price of the product. Also, since insurance and local taxes are based primarily on investment, the same distribution on the random-valued factor for the investment is used on these terms. In each year for the sales volume, a small percent is allowed for the chance that sales could fall below 10 million lb. because of unforeseen circumstances. The units that Joe used on the variables are important. To avoid unit problems, he chose units such that

$$\frac{\text{(variable cost units) (sales volume units)}}{\text{(money units for fixed costs, startup costs, etc.)}}$$

is dimensionless. For this example, money units are in thousands of dollars, sales volume in millions of pounds, and variable costs in thousands of dollars per million pounds.

Cash Flow Calculations

Now that each single-valued variable in the investment analysis has been changed to a random variable, Joe is ready to perform the cash flow calculations. In the mathematical operation to obtain the cash flows where

$$R = f(Z_1, \quad Z_2, \quad Z_3 \ldots) \qquad (8)$$

the mean and variance of R are given by

$$m\{R\} = f(m\{Z_1\}, \quad m\{Z_2\}, \quad m\{Z_3\} \ldots) \qquad (9)$$

$$\gamma\{R\} = \left(\frac{\partial R}{\partial Z_1}\right)^2 \gamma\{Z_1\} +$$

$$\left(\frac{\partial R}{\partial Z_2}\right)^2 \gamma\{Z_2\} + \left(\frac{\partial R}{\partial Z_3}\right)^2 \gamma\{Z_3\} \ldots \quad (10)$$

where the partial derivatives are evaluated using the mean values of Z_i. Eqs. (9) and (10) are completely valid provided R is linear over the range of variation of (Z_1, Z_2, Z_3) and every Z_i is stochastically independent. Certainly neither of these criteria is completely fulfilled in an investment analysis. However, the error introduced by the linear assumption should be small. The dependence between variables is usually minor. When dependence is significant, a compensating effect can be achieved by using identically or similarly distributed X_f on those variables, as was done for insurance and local taxes in the example.

For all mathematical operations, mean values of the variables are manipulated exactly as though they were single-valued variables. For addition and subtraction operations, variances are directly added and never subtracted. Manipulation of variances in multiplication and division should be determined by Eq. (10).

The cash flow in a year n can be separated into an after-tax flow

$$C_n = A_n(1-t) + tf_D I \qquad (11)$$

and a miscellaneous flow

$$C'_n = \sum_i C'_{in} \quad \text{where } i \text{ refers to investment credit,} \atop \text{incremental working capital etc.} \qquad (12)$$

The variance for C_n via Eq. (10) will be given by

$$\gamma\{C_n\} = (1 - m\{t\})^2 \gamma\{A_n\} + (m\{A_n\} - \atop f_D m\{I\})^2 \gamma\{t\} + (f_D m\{t\})^2 \gamma\{I\} \quad (13)$$

Variance for total cash flow $(C_n{}^T = C_n + C'_n)$ is, of course, the sum of the variance for C_n and C'_n.

Some of the cash flow calculations are shown in Table VI. It should be emphasized that mean values are handled exactly the same as the single values were in Table II. Let us examine the determination of the variances for year number two. The variance of the product price (3.52) is added directly to the sum of the variances (0.88) for the variable costs to give the variance of the variable revenue (4.4). For single-valued terms, the variable revenue is multiplied by the sales volume to obtain a production revenue. For the variance of the production revenue, this multiplication operation calls for the use of Eq. (10).

Production revenue =
(variable revenue) (sales volume)

$$\gamma\{\text{prod. rev.}\} = \left(\frac{\partial (\text{prod. rev.})}{\partial (\text{var. rev.})}\right)^2 \gamma\{\text{var. rev.}\} +$$

$$\left(\frac{\partial (\text{prod. rev.})}{\partial (\text{sales vol.})}\right)^2 \gamma\{\text{sales vol.}\}$$

where the partial derivatives are evaluated at the mean values.

$$\gamma\{\text{prod. rev.}\} = (m\{\text{sales vol.}\})^2 \gamma\{\text{var. rev.}\} + \atop (m\{\text{var. rev.}\})^2 \gamma\{\text{sales vol.}\}$$
$$= (10.4 \text{ MM lb.})^2 4.4 \text{ (M\$/MM lb.)}^2 + \atop (44.24 \text{ M\$/MM lb.})^2 (0.04155) \text{ (MM lb.)}^2$$
$$= 557.3 \text{ (M\$)}^2$$

The sum of the variances for the fixed costs (71.77) along with the variance for the startup costs (39.48) are added to the variance of the profit before federal taxes (668.6). The after-tax flow variance is obtained via Eq. 13.

$$\gamma\{C_2\} = (1 - 0.5136)^2 (668.6) + [-14.49 - \atop 0.1818 (265.3)]^2 (0.0003992) + \atop [0.1818 (0.5136)]^2 (5,677) = 209.3$$

Calculating Return on Investment

To obtain the variance of the total cash flow, the variances of the miscellaneous cash flow are added directly to the variance of the after-tax flow. Once the mean and variance of the cash flows have been computed, Joe is ready to determine the mean and variance for his measure of profit.

The mean fractional interest rate using annual discount factors will be given by

$$\sum_{n=1}^{q} \frac{m\{C_n{}^T\}}{1 + m\{r_a\}^n} = 0 \qquad (14)$$

which as noted is exactly the same as for a single-value interest rate. The variance on r_a will be given by

$$\gamma\{r_a\} = \frac{1}{\delta^2} \sum_{n=1}^{q} \frac{\gamma\{C_n{}^T\}}{(1 + m\{r_a\})^{2n}} \qquad (15)$$

where

$$\delta = \sum_{n=1}^{q} \frac{nm\{C_n{}^T\}}{(1 + m\{r_a\})^{n+1}} \qquad (16)$$

The mean value of the fractional interest rate becomes 0.1904.

To obtain the mean interest rate, Joe used the mean cash flows in exactly the same manner as he did for the single-valued cash flows. For the variance,

$$\delta = \frac{1.0\,(-265.3)}{(1.0+0.1904)^2} + \frac{2.0\,(-42.27)}{(1.0+0.1904)^3} \cdots \frac{12\,(45.08)}{(1.0+0.1904)^{13}}$$
$$= 1{,}135$$

$$\gamma\,\{r_a\} = \frac{1}{1{,}135^2}\left[\frac{5{,}677}{(1.0+0.1904)^2} + \frac{218.5}{(1.0+0.1904)^4} \cdots \frac{15.33}{(1.0+0.1904)^{24}}\right]$$
$$= 0.004171$$

The standard deviation is the square root of the variance and is equal to 0.06459.

Now that the mean and variance of the measure of profit r are available, what type of distribution does r follow? It is felt that the central limit theorem can be readily applied to the distribution of r. The central limit theorem in its simplest form states that the distribution of a variable that is the sum of an infinite set of similarly or dissimilarly distributed random variables, all having finite means and variances, will be normally distributed. Since r has been determined from the sum of cash flows over a period of years, and each cash flow is made up of the sum of many variables (raw material costs, labor, sales, etc.), then it is expected that it would tend to be normally distributed.

To determine the cumulative probability distribution on the interest rate, Joe uses the following two facts concerning the normal distribution:

1. The mean value (0.1904) occurs at $P[r_a] = 50.0\%$.

2. The mean value (0.1904) minus the standard deviation (0.06459) occurs at $P[r_a] = 15.87\%$ or the mean value plus the standard deviation occurs at $P[r_a] = 100\text{-}15.87 = 84.13\%$.

Thus he can construct a straight line on normal probability paper as shown in Fig. 3.

With the cumulative probability distribution for r_a drawn on normal probability paper (and letting $r \geqq 0$ be defined as making a profit), the following risks can now be determined:

1. Percent probability of having an interest rate greater than or equal to the mean interest rate.

2. Percent probability of making a profit.

3. Percent probability of having an interest rate greater than or equal to that determined from single-valued variables.

4. Percent probability of having an interest rate greater than or equal to a specified minimum (r_m).

5. Assuming a profit will be made, the percent probability the interest rate will be greater than or equal to (a) the mean rate, (b) single-valued rate, (c) the specified minimum rate.

The first risk, as previously noted, is 50%. The second through fourth risks can be determined from the plot on normal probability paper. The three risks under number 5 can be calculated as conditional probabilities by respectively dividing risk number 1 by risk number 2 and multiplying by 100, risk number 3 by risk number 2, and risk number 4 by risk number 2.

The risk terms and calculations as defined are summarized in Table VII.

Joe's results are presented to the second management review on the Mammoth project. Everyone is in complete agreement that the odds of obtaining a profit are exceptionally good.

Five years later, after the completion of the polymetholine plant, Joe, now resident manager, is again calculating a random-valued factor. He has chosen a single value of two years and is estimating the chances when Batchem will be called Sharps Chemical Co.

Continuous Discount Factors

Many economists feel that continuous discount factors more realistically represent the actual flow of cash.[6] The mathematical operations required to obtain

Typical cash flow calculations—Table VI

	\multicolumn{2}{c}{1}	\multicolumn{2}{c}{2}	\multicolumn{2}{c}{3}	\multicolumn{2}{c}{4}	\multicolumn{2}{c}{8}	\multicolumn{2}{c}{9}	\multicolumn{2}{c}{12}
	Mean	Variance	Mean	Variance	Mean	Variance	Mean	Variance	Mean	Variance	Mean	Variance	Mean	Variance
Sales volume			10.40	0.04155	10.40	0.04155	10.40	0.04155	11.90	1.362	11.90	1.362		
Product price			94.76	3.520	94.76	3.520	101.2	32.26	101.2	32.26	101.2	32.26		
Total variable costs			50.52	0.8805	50.52	0.8805	52.62	3.937	52.62	3.937	53.14	4.079		
Variable revenue			44.24	4.401	44.24	4.401	48.58	36.20	48.58	36.20	48.06	36.34		
Production revenue			460.1	557.3	460.1	557.3	505.3	4,013	578.1	8,341	572.0	8,292		
Total fixed costs			402.6	71.77	402.6	71.77	391.2	49.14	392.3	75.57	392.3	75.57		
Startup costs			72.0	39.48										
Profit before FIT			−14.49	668.6	57.51	629.1	114.1	4,062	185.8	8,417	179.6	8,368		
Investment	−265.3	5,677												
Profit after FIT			17.73	209.3	50.27	189.0	75.32	995.1	102.7	2,124	97.18	2,109		
Working capital			−60.00	9.254										
Salvage value													45.08	1,533
Continuous cash flow			17.73	209.3	50.27	189.0	75.32	995.1	102.7	2,124	97.18	2,109		
Instantaneous cash flow	−265.3	5,677	−60.0	9.254									45.08	1,533
Total cash flow	−265.3	5,677	−42.27	218.5	50.27	189.0	75.32	995.1	102.7	2,124	97.18	2,109	45.08	1,533

Probabilities of making a profit—Table VII

1. Probability that interest rate will be greater than or equal to the mean interest rate

$$= P[0.1904] = 100 - P[0.1904]$$
$$= 50.0\%$$

2. Probability of making a profit

$$= 100 - P[0.0]$$
$$= 99.87\%$$

3. Probability that interest rate will be greater than or equal to the single valued rate

$$= 100 - P[0.02844]$$
$$= 99.50\%$$

4. Probability that interest rate will be greater than or equal to a specified minimum rate $(r_m = 0.10)$.

$$= 100 - P[0.10]$$
$$= 92.5\%$$

5. Assuming a profit is made, the probability that
   (a) rate will be greater than or equal to the mean rate

$$= \frac{100 - P[0.1904]}{100 - P[0.0]} \times 100 = \left(\frac{50.0}{99.87}\right) \times 100$$
$$= 50.06\%$$

   (b) rate will be greater than or equal to the single-valued rate

$$= \frac{100 - P[0.02844]}{100 - P[0.0]} \times 100 = \left(\frac{99.50}{99.87}\right) \times 100$$
$$= 99.63\%$$

   (c) rate will be greater than or equal to the specified minimum

$$= \frac{100 - P[0.10]}{100 - P[0.0]} \times 100 = \left(\frac{92.5}{99.87}\right) \times 100$$
$$= 92.62\%$$

---

the variance of the interest rate are somewhat more involved than those for annual discount factors. These mathematics are not prohibitive for hand calculation and can be simplified if one makes the following assumptions:

1. All C_n are uniform throughout the whole year n.
2. Every C'_{in} is instantaneous at fraction X_i of the year n or is uniform throughout the whole year n, in which case C'_{in} is treated as a C_n flow.

General Comments

It is noted that in the equations to determine the interest rate of return (using either annual or continuous discount factors), the first year in the summations can correspond to any year prior to or beginning with that year in which the first amount of cash flow occurs.

An investment analysis always presents a measure of profit that is relative to an alternate investment, whether the alternate is explicitly stated or not. To place these "alternates" on a more common basis, all analyses should be discounted to the year in which the analysis is made.

The example given was calculated on an IBM 1130 computer. Running time was approximately one minute. The data shown in Tables V and VI represent computer print-out values rounded to four significant figures. The results for the interest rate using simplified exponential functions (Table IV) and continuous discount factors were 0.03003, 0.1897, and 0.003641 for the single value, mean, and variance, respectively. Results for the interest rate using actual exponential functions and continuous discount factors

were 0.03003, 0.1913, and 0.003711 for the single value, mean, and variance, respectively.

For a real problem as compared to the example, one usually does not have a captive market and also could not specify such narrow ranges on the cumulative probabilities for the X_i's that were used. Wider ranges would, of course, result in a larger variance for the interest rate. Also, one could expect for some problems to actually obtain a negative single-valued or mean interest rate. Although a negative interest rate might appear somewhat meaningless, it nevertheless is correct mathematically and the risk terms can still be obtained. No normal distributions were specified for any of the variables in the example. This was not intended to suggest a total exclusion of normal distributions for the variables, and certainly one should use them wherever they are believed to be applicable in a particular situation.

Acknowledgment

The author gratefully acknowledges the invaluable assistance received from W. L. Schuette, Engineer Economist at Ethyl Corp.

References

1. Gore, W. L., *Petrol. Refiner,* **37**, p. 142 (June, 1958).
2. Hald, A., "Statistical Theory with Engineering Applications," Wiley, New York, (1952).
3. Happel, John, "Chemical Process Economics," Wiley, New York, (1958).
4. Hertz, D. B., *Harvard Business Review,* **42** (1), p. 95 (1964).
5. Hicks, J. S., Steffens, L. R., *Chem. Eng. Prog.,* **52**, p. 191 (May, 1956).
6. Hirschmann, W. R., Brauweiler, J. R., *Chem. Eng.,* July 19, 1965, p. 210.
7. Parzen, E., "Modern Probability Theory and Its Applications," Wiley, New York, (1960).
8. Perry, J. H., "Chemical Engineers Handbook," 4th Ed. McGraw-Hill, New York, (1963).
9. Quinn, J. B., *Harvard Business Review,* **38** (2) p. 69 (1960).
10. Raiffa, H., Schlaifer, R., Applied Statistical Decision Theory, Graduate School of Business Administration, Harvard University, Boston, (1961).
11. Reynard, E. L., *Ind. Eng. Chem.,* **58** (7) p. 61 (1966).
12. Schenck, H., "Theories of Engineering Experimentation," McGraw-Hill, New York, (1961).

Meet the Author

John T. Thorngren is a process engineer in the Process Design Department of Ethyl Corp. in Baton Rouge, La. He has also served in the Departments of Production, Technical Services, and Economic Evaluation. He joined this firm in 1963 after receiving his B.A. in math and B.S. in chemical engineering from the University of Texas. Mr. Thorngren is a member of Omega Chi Epsilon and AIChE.

Key Concepts for This Article

Active (8)	Passive (9)	Means/Methods (10)
Evaluation	Project*	Return on investment*
Comparison*	Alternatives*	Cash flow
	Investment	Probability*
	Profits*	Statistics
	Economics	

(Words in bold are role indicators; numbers correspond to EJC-AIChE information retrieval system. Asterisks mark key concepts suggested for indexing. Others are added to improve reading as an abstract. Indexing is described in *Chem. Eng.,* Oct. 11, 1965, p. 187.)

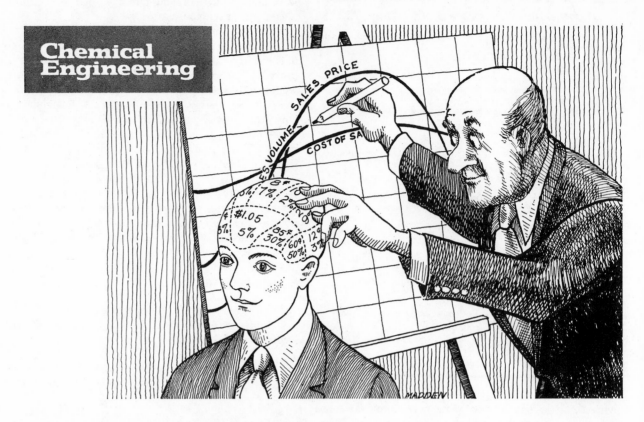

Chemical Engineering

Decision-Making Through Opinion Analysis

When making investment decisions, executives must rely on data that they know are uncertain, but they generally do not know the degree of uncertainty. Here is a method of evaluating and presenting the uncertainties that are always present.

PARK M. REILLY and HARI P. JOHRI,
University of Waterloo, Ontario, Canada

The most difficult job in management is in making investment decisions. According to one estimate,[1] in 1968 U.S. business spent $65 billion in new capital investments.

The uncertainty associated with these decisions is very high. But it is rarely appreciated that project-evaluation criteria such as simple rate of return, present worth, discounted cash flow, etc., as presently used, do not properly reflect this uncertainty. As a result the decision-making process in the face of uncertainty is little understood.

R. Brech[2] once noted that managers act empirically and often rather illogically, since information given to them is misleading: "In his [the manager's] mind, he assesses the wrongness of the information, acts in what seems to be an illogical manner, and comes to a right decision."

How can we improve the quality of information, so that it is not misleading? How can we reduce the empiricism in investment decision-making? How can we improve the probability of a right decision? We will be concerned with these questions in this article.

In the first place, we feel that although a great deal of work has been done in developing techniques[4, 5, 6] of project evaluation, very little has been said about the information required for such analysis. For example, calculation of even a simple rate of return on investment requires precise information on the cost and benefits of the project. In calculating either cost or benefits, several elements such as sales volume, sales price, cost of production, etc., have to be calculated. In principle, data of this

nature contain uncertainties. Recognizing the uncertainty in estimates is a step in the right direction; however, it is not sufficient. The criterion used for evaluation should incorporate a method of incorporating such uncertainties in the estimates. The methods developed in this article expose to management the full extent of the uncertainty inherent in information. It also proposes to show the consequences of this uncertainty on criteria used to evaluate projects.

The Conventional Approach

These concepts may be illustrated by considering an analysis of the conventional type. We will consider an investment decision facing the executives of XYZ Chemicals Inc. It is considering a research and development type proposal. To analyze profitability, certain estimates are developed, which are given in Table I. The logic for the decision process is given in Fig. 1, where we have also explained the steps involved in the decision-making process.

We feel that this conventional approach is not satisfactory because it gives only one "best" estimate that is certain not to be absolutely correct—yet there is no way of telling how far wrong this estimate might be. If a second estimate is provided, as is done here, to show the pessimistic side of the story, just how pessimistic should it be? If it is assumed that everything will turn out badly as is illustrated here, it is most certainly too pessimistic and represents an unrealistic picture. With no prior knowledge of future events, one does not know what combination of events is most likely to happen. The mapping of the entire uncertainty region is therefore very helpful in decision-making.

Opinion Analysis

The main problem in evaluating the business potential of any project is that of gathering data for analysis.

In all such cases, an attempt is made to assess the future behavior of some system. Since it is impossible to measure the future, the only way in which this assessment can be made is to use the opinions of people who are in a position to have opinions on the subject. Of course, these opinions involve more or less uncertainty; but there is necessarily some uncertainty present.

Because the data entering analyses of this type must be the opinions of people, we call our method of dealing with the accompanying uncertainty "opinion analysis." It entails the use of subjective probability to show the range of uncertainty in the opinions.

Subjective probability differs from the familiar type of probability that measures relative frequency. The latter is called objective probability and is measured by the number of favorable events divided by the possible events. When we say that a coin has 50%

XYZ Chemicals Inc., R & D proposal—Table I

	Best Estimate	Pessimistic Estimate
Capital investment	$100,000	$100,000
Cost of sales	15% of total sales	20% of total sales
Sales price	$1.40/lb.	$1.20/lb.
Yearly sales volume	50,000 lb.	40,000 lb.
Cost of production	$0.70/lb.	$0.80/lb.
Gross profit	$24,500	$6,400
Net profit (50% of gross), assumed	$12,250	$3,200
Return on investment	12.25%	3.2%

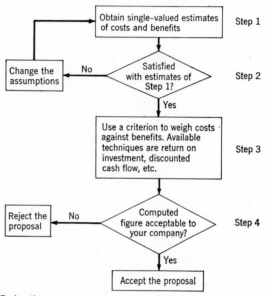

Explanation:

Step 1: Costs are net cash outflow. Benefits are net cash inflow. To calculate these, some estimates as in Table I are needed.

Step 2: The estimates obtained in Step 1 are reviewed and changed if necessary.

Step 3: Several criteria for evaluation of projects are available. Generally each company has its own standard procedure for evaluation.

Step 4: The computed figures are then presented to higher management which in turn takes the decision to either accept the proposal or to reject it.

DECISION-PROCESS logic for conventional analysis—Fig. 1

probability of showing heads when flipped, we mean that about one half of a very large number of flips will show heads.

Subjective probability is closely related. It is the concept involved when a physician says that a person has a 50% chance of living. What he means is that he feels as much certainty about the person living as he does about a flip of a coin coming up heads. That is, the 50% figure measures his opinion in

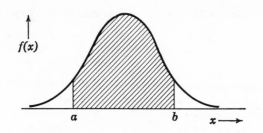

PROBABILITY is measured by area under curve as shown—Fig. 2

terms of his degree of belief in the patient living. It cannot be an objective probability because we only have one patient and cannot (even in theory) repeat his illness to see in what fraction of illnesses he will live.

The mathematical laws followed by objective probability are also followed by subjective probability. For instance, we may define the probability density function as we do for objective probability. A probability density function for a continuously varying quantity (called a random variable by the statistician) is a non-negative function $f(x)$ whose integral over the entire axis is unity. For any pair of values a and b such that $a < b$ (see Fig. 2):

$$p(a \leqq x \leqq b) = \int_a^b f(x) \, dx$$

In other words, the probability is measured by the area under the curve between the points a and b. The logic of the technique of opinion analysis is simple and illustrated in Fig. 3. We will now explain the individual steps of the logic diagram.

Obtaining Subjective Probability Estimates

Step 1—The first step of the analysis is to prepare an opinion distribution function of the input variates. By selecting an "expert" whose opinion best describes the distribution of the particular input variate, an opinion distribution function is developed. The use of the word expert requires some clarification. Its usage here implies a person within a firm who is responsible for the knowledge in a particular field. Other problems of selecting an expert are discussed later in this article.

The opinion distribution diagram is prepared by deciding on a maximum value, a minimum value and a most favored value. The maximum and minimum are the highest and lowest values that are expected, within reason, for the quantity. The most favored value is the one of all the possible values that is believed most likely to occur, whether or not it is sharply defined. The most favored value is often called the mode of distribution. At this stage, it is helpful to lay out a scale covering the desired range and to mark a point at an arbitrary height above the point where the mode occurs on the scale as shown in Fig. 4.

The remainder of the diagram is now filled in by estimating the degree of belief in each point along the scale relative to belief in the mode (the most-favored value) and positioning the curve such that this relative degree of belief is shown. The way in which relative degree of belief is estimated is to think of the odds at which a small bet on the values concerned would be reasonable. This cannot be assessed exactly, of course, but a sufficiently good estimate is usually fairly easy to obtain. The ratio of the height of the curve at the mode to the height of any other point therefore represents the

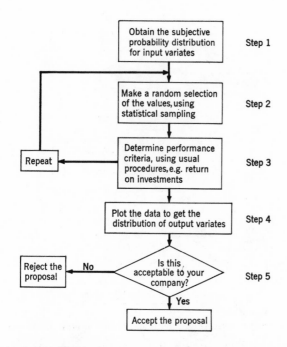

LOGIC DIAGRAM for approach using opinion analysis—Fig. 3

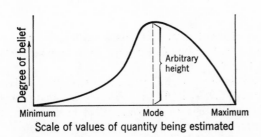

OPINION ANALYSIS diagram is prepared this way—Fig. 4

fair odds in favor of the modal value over the value near the point in question. It may be noted that equal odds imply no ability to distinguish between the alternatives while long odds (preference towards one value) imply considerable knowledge of the outcome. It is important that the full degree of uncertainty which is felt, be shown; consequently if there is any doubt about the position of the curve above a particular point, it should be placed at the highest position (relative to the mode) of those positions being considered. A completed distribution diagram is shown in Fig. 4.

Strictly speaking, degree of belief is measured by area under the curve rather than by the height of the curve. Since areas of narrow strips are proportional to their heights, however, the ratio of two heights gives the odds in favor of values near those where the curve has greater height, over those near where it has the smaller height.

Calculating Probability Distribution for Output Variates

Step 2, 3 and *4*—The output variates are the performance criteria that we want to calculate, and that we produce in the form of subjective probability distributions. In comparison to the input variates, which are the expression of opinion that enter the calculation, the output variates are the answers.

The output variates are generated from the input variates, using statistical sampling (commonly called Monte Carlo Simulation). The ordinary methods of distribution theory become hopelessly complicated (for other than the simplest calculations) if one uses quantities described by probability distribution. The use of experimental sampling is a simple general approach to problems of this type (in some simple cases, it may not be necessary to use the general approach—elementary methods can be employed).

To use the method, a number is drawn at random from the distribution of each of the quantities that are described by probability distribution. With these numbers, the desired calculations are carried out exactly as if they were the definite values of the quantities involved.

When the answer is obtained, it is recorded as one count in the appropriate cell of a frequency distribution. The procedure is then followed repetitively to produce many individual values for each performance criterion and a frequency distribution of these values becomes a closer approximation to the subjective probability distribution of the performance criterion, as the calculations are repeated more and more times.

Drawing Inferences from the Output Distribution Curves

Step 5—The distribution curve for output variates depicts the whole range of uncertainty associated with the final outcome. It represents the collective opinion

■■■

Opinion distribution for various variables— Table II

1. **Investments**	We assume that the most likely value for investment is $100,000. It can vary within a range of $45,000, as shown in Fig. 6.
2. **Cost of sales**	The opinion distribution diagram for cost of sales is shown in Fig. 7.
3. **Sales volume**	Let us assume that the opinion distribution for volume can be expressed by the following relationship. Sales volume $=k+p$ where k is dependent on the selling price. We will assume that the relationship between sales volume and sales price is linear as shown in Fig. 5. Variable p is the independent component and explains the component of sales volume not dependent on selling price. The opinion distribution diagram for p is shown in Fig. 9.
4. **Cost of production**	We assume $0.70/lb. as the most likely value for cost of production. Its range and distribution are shown in Fig. 10.
5. **Sales price**	The opinion distribution diagram for sales price is shown in Fig. 8.

■■■

of experts in their different fields as applied to the output quantities. To draw inferences from the output distribution curve, various alternative approaches are possible. From our experience, we prefer to present the complete output distribution to management for decision-making. In this way, we ensure that the decision-maker has all the information that is relevant to a particular output variate.

Some argue that the complete output distribution presents so much information to a decision-maker that it cannot be properly evaluated. Hertz[3] has employed the mean of this distribution as a point estimate of the output variate for use in decision-making. However, we favor the use of the mode if a single value of the output variate must be produced, since an expert generally has this value in mind while giving his opinion as to single preferred value of an input variate. It is also possible to calculate a range of values to denote an interval of plausible values, for instance the upper and lower limiting values for the central 95% of probability.

Opinion Analysis Applied to XYZ Chemicals

Let us now recalculate the return on investment for the example of XYZ Chemicals Inc. While we retain essentially the same figures used in the conventional approach, we introduce the notion of dependency of sales volume with the selling price in

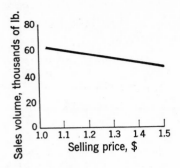

SELLING PRICE vs. sales volume relationship—Fig. 5

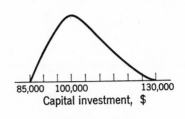

CAPITAL INVESTMENT opinion distribution diagram—Fig. 6

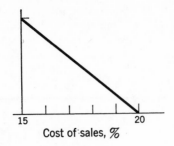

COST OF SALES opinion distribution diagram—Fig. 7

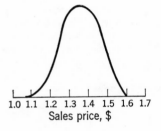

SALES PRICE opinion distribution diagram—Fig. 8

VARIABLE "p" opinion distribution diagram—Fig. 9

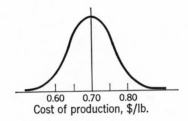

COST OF PRODUCTION opinion distribution diagram—Fig. 10

order to show that fairly complicated situations may be handled simply. We will assume Sales Volume = $k + p$, where k is dependent on selling price, while p is independent of sales price.

To solve this example, using opinion analysis, we require additional information in the form of opinion distribution curves for the various variables shown in Table II.

Opinion analysis approach, XYZ Chemicals— Table III

Input Variates	Mode	Range	Figure Number
Capital investment	$100,000	$45,000	6
Cost of sales	15%	5%	7
Sales price	$1.33	$0.54	8
Sales volume			
k	......	......	5
p	$0.00	$15,000	9
Cost of production	$0.70	$0.30	10

Table III summarizes the additional data required for the analysis. It may be noted that these data are on the same quantities as those used for the analysis that employs conventional approach. The major difference lies in depicting the uncertainty in the form of distributions.

For the example of XYZ Chemicals Inc., the output distribution curve shown in Fig. 11 was derived (by computer) by taking 10,000 random samples from the distribution curves of input variates (Fig. 5, 6, 7, 8, 9 and 10).

The output-variate distribution curve depicts the full range of uncertainty associated with this decision situation. For example, it indicates that the return on investment is within the range of 0.1629 (min.) and 21.5479 (max.). The output distribution curve is relatively flat (except the hump at the modal value of 8.7%) within the range of 6.00%-11.26% return on investment. This implies nearly equal degrees of belief for all values.

The importance of this analysis lies in the information contained in the output-variate distribution curve, whereas the conventional approach gave only two values of the return on investment, with no indication of how likely these events were. The

opinion analysis approach covers the whole range of possibilities, and indicates the degree of belief in the occurrence of each event. Through computer simulation, we have in effect created artificial experience to guide us in decision-making situations of this type.

It may be useful to discuss some of the problems relating to the use of opinion analysis approach:

Scope of Opinion Analysis

The technique of opinion analysis is not a replacement for scientific investigation based on objective data collected through research, where such data are available or can be gathered within the limitations of time and financial resources. In a physical science such as chemical engineering, it is quite common to reduce uncertainty in engineering design by means of direct experimentation on a pilot plant. Even after experimentation, however, there is still some residual uncertainty.

In investment-making situations, uncertainty is always present. When the decision-maker has direct experience with similar projects (this should be carefully assessed for relevance of data), the uncertainty is low. If the novelty of the project increases, the uncertainty associated with it is high. When there is true novelty, high uncertainty is present, and no amount of research can reduce the uncertainty since it lies in expectation of future events. Evidence from direct observation or past data is not fully applicable in these situations. Shackle[7] has called these situations unique and nonseriable, nondivisible experiments. It is here that some means of mapping the uncertainty range is most useful. Note that by analyzing uncertainty, one does not reduce it. The analysis only gives fuller appreciation of the uncertainty, and thereby helps the decision-maker in decision-making.

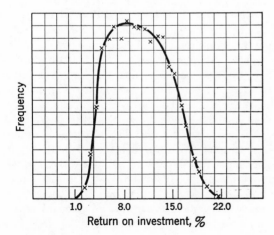

OUTPUT VARIATE distribution curve—Fig. 11

Selection of Expert

The word "expert" is used here in a very special sense. It does not imply any necessarily high level of knowledge on the part of the person concerned. It only means that he is the one from whom the information for the study is obtained.

Of course, it is essential in a useful study that the expert be a good one, that is that he possess real knowledge. However this is equally important in the conventional approach. The only difference in opinion analysis is that we attempt to measure how sure the expert is of his knowledge.

Knowledge alone on the part of the expert is not sufficient. He must be able to bring it to bear effectively on the problem in hand, which is often difficult. Helmer,[8] for example, suggests that an expert be a person whose predictions with regard to hypotheses in a certain field show a record of comparative successes in the long run. Hence, some contend that methodology for selecting an expert should include an assessment of reliability, where reliability is measured as the relative frequency of cases in which, when confronted with several alternative hypotheses, the expert ascribes the highest subjective probability to the correct one.

In present practice, the knowledge of a person is the only criterion used in selecting him as an expert. Sometimes it is useful to have a team of experts to give their opinion on one variable. If there is consensus in their opinions, their joint opinion distribution curve could be used in the analysis. When no consensus exists, the output variates can be calculated based on the opinion of each individual expert. It may be mentioned here that the DELPHI technique[10] developed by Rand Corp. could be used to bring about the consensus. The technique replaces direct debate (to avoid influences of certain psychological factors such as unwillingness to abandon publicly expressed opinion) by a carefully designed program of sequential individual interrogation, in which a loop for information and opinion feedback after each cycle of interrogation of experts has been provided. Winkler[11] has also recently discussed several techniques, including ranking for combining the opinion distribution diagrams while using several experts.

Obtaining Opinion Distribution Curves

In obtaining accurate opinion distribution curves, the ability of an interviewer is most challenged. While there is no substitute for skill and imagination, some practical suggestions given earlier (Step 1) may be helpful. It is only recently that the researchers in this field have addressed themselves to this problem. Winkler[9] has compared the different techniques like CDF (Cumulative distribution function), HFS (Hypothetical future sampling) etc., for obtaining opinion distribution diagrams. Helmer[10] is another useful reference.

The discussion which follows is by no means com-

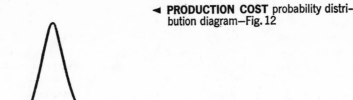

◄ **PRODUCTION COST** probability distribution diagram—Fig. 12

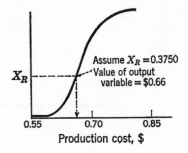

Assume $X_R = 0.3750$
Value of output variable = $0.66

X_R

Production cost, $

PRODUCTION COST cumulative distribution diagram—Fig. 13 ►

Production cost, $

prehensive, but will answer some of the questions which may arise in the reader's mind regarding the skill of the interviewer and the structuring of a typical interview with the expert.

First of all, in our opinion the interviewer besides being skilled in the techniques of Opinion Analysis should be reasonably informed on the structure of problems on which he is seeking the opinion of the expert. He should be able to communicate freely with the expert and should have enough perseverence to understand the uncertainty felt by the expert. The interviewer and expert should feel jointly responsible that all the relevant information is available.

The Interview

The interview consists basically of two parts. The first is devoted to familiarising the expert with the concepts of subjective probability. In this period the problem at hand may also be introduced and an attempt should be made to solicit the support of the expert in analysing the problem using opinion analysis. The interviewer may encounter several psychological barriers that must be removed to gain the support of the expert. Typically the objective of the study is often misunderstood; there is also hesitancy on the part of an expert to indicate his uncertainty.

In the second part of the interview the opinion distribution curves are obtained (Step 1). One problem of using this technique is the misleading ease with which these distribution diagrams can be obtained from many experts. The expert sometimes does not think carefully while giving the values of subjective probability for use in construction of opinion distribution diagrams. We use a hypothetical betting situation as a means to confirm the data provided by the expert. Whenever inconsistencies are noticed these are resolved with the help of the expert. In all cases it is essential that the full range of uncertainty felt by the expert be shown.

Use of Standard Distributions

In actual use of opinion analysis, standard shapes such as beta, triangular, normal or uniform distribution may be used for particular opinion distributions. The four types of distributions mentioned above are all of the type where any numerical value over a considerable range is possible. They occur commonly and usually describe such quantities as selling price, production costs, sales volume etc.

It is convenient to use standard distributions rather than purely arbitrary ones, though of course a standard distribution should never be used for convenience alone if it is not realistic. Even in those cases where standard distributions are used, it is often helpful to sketch out the form of distribution as described earlier before choosing which particular form will be used as the final distribution of opinion.

The Computer Program

Because calculation of the output variates is based on repetition of a single calculation many times, it is necessary to use a computer. It should be pointed out that some calculations in Opinion Analysis can be done by methods that do not require computers, but these will not be discussed here. The computer method is completely general and can be used in all cases. A general program in FORTRAN IV is available on application from the authors. It requires that a program be written by the user only for the actual calculation of the values of the output variates.

In this section, we will explain the logic of the computer program. Earlier we explained the Opinion Analysis approach (Fig. 3, and discussion of various steps of the analysis). In Step 1, we have indicated how to obtain the subjective probability distributions for input variates. Some of these distributions may be of the type we have called standard distributions—for example, cost of production for XYZ Chemicals Inc. (Fig. 10) is a normal distribution with a mean of $0.70 and a standard deviation of 0.05. Others may be completely arbitrary distributions—for example, sales price for XYZ Chemicals Inc. (Fig. 8). The computer input that describes the input variates to which have been assigned standard distributions is just the parameters of the distributions. For example, for a normal distribution, the mean and standard deviation are given; for the beta distribution, the mode, range, a and b, etc.

The input for the arbitrary distributions consists of

the ordinates of the probability density function at a number (usually 25) of variate values along with the mode and range. The computer converts these to the cumulative distribution function (see Fig. 12, 13) and when random choices of variate values are required it uses the Inverse Transformation Technique, which is fully described in textbooks on simulation. In this technique, a uniformly distributed random number in the range 0 to 1 is applied as an ordinate to the cumulative distributions described by the cumulative distribution function that has been used. As shown on Fig. 12, if a random choice of production cost is required, a uniformly distributed random number in the range 0 to 1 is obtained. This value, let us say, 0.3750, is applied at X_R in Fig. 13 and the variate value for production cost $0.66 is read off.

Similarly, samples are taken from other distributions.

Output-Variate Distribution

The frequency distribution for the output variates becomes a better and better approximation to the corresponding subjective probability distribution as the repetitive calculations (cases) increase. The number of cases required depends on the desired precision and smoothness of the output curves, their shape, and number of cells used for frequency counts. For a program with 25 cells, the number of cases required for good resolution is between 10,000 and 50,000. The latter figure seems ample for all practical situations but may be considered somewhat wasteful. A simple problem, as for example, that of XYZ Chemicals Inc.,

takes less than 2 min. of computer time on 360/75 I.B.M. Computer when calculations are run for 10,000 cases.

Conclusion

We have used opinion analysis for analyzing several problems of venture analysis and corporate planning. It has also been used effectively as a tool for short-term profit forecasting where uncertainty exists in sales volume, sales mix, sales price, cost of sales, etc. Opinion analysis is applicable where human judgment must be used to supplement facts in engineering and business. In such situations, it provides a tool for quantifying the opinions of experts, and depicting uncertainty in the opinion of experts. By using statistical sampling, it provides a means for combining the uncertainty in basic estimates.

There is another advantage of using opinion analysis. It helps to put things in right perspective. In evaluation of projects, often the analyst unwittingly conceals the source of information from the searching eyes of a decision-maker. (See, for example, Ref. 5). It is this hide-and-seek game that results in empiricism in decision-making or what Brech has described as a decision-maker's attempt to assess the 'wrongness of the information.' We believe that it is misleading to hide the source of information.

In this article, we have explained the art and science of using opinion analysis to aid decision-making. We believe that a proper mapping of uncertainty goes a long way to help a decision-maker in properly choosing among the alternatives. ∎

References

1. Hertz, D. B., "Investment Policies That Pay Off," *Harvard Bus. Rev.*, Jan., Feb. 1968.
2. Brech, R., Simulation Studies of Industrial Operations, *J. Royal Statistical Soc.*, 122, p. 507.
3. Hertz, D. B., Risk Analysis in Capital Investment, *Harvard Bus. Rev.*, Jan., Feb. 1964.
4. Teplitzky, Gerald, Business Potential of R&D Projects, *Chem. Eng.*, June 5, 1967, p. 136.
5. Teplitzky, Gerald, Using the Profit and Loss Statement, *Chem. Eng.*, June 19, 1967, p. 136.
6. Allen, D. H., Two New Tools for Product Evaluation, *Chem. Eng.*, July 3, 1967, p. 75.
7. Shackle, G. L. S., "Expectation in Economics," in "Uncertainty in Business Decision," Ed. by Carter, others, Liverpool University Press, 1962.
8. Helmer, Olaf, Rescher, Nicholas, On the Epistemology of the Inexact Sciences, *Management Sci.*, 6, No. 1 (1959).
9. Winkler, Robert L., The Assessment of Prior Distribution in Bayesian Analysis, *J. Amer. Statistical Assoc.*, 62, No. 319 (1967).
10. Helmer, Olaf, "Social Technology," Basic Books, New York, 1966.
11. Winkler, Robert L., The Consensus of Subjective Probability Distributions, *Management Sci.*, 15, No. 2, Oct. 1968.

For More General Information

Mood, A. M., Graybill, F. A., "Introduction to the Theory of Statistics," 2nd ed. McGraw-Hill, 1963. *An excellent book for statistical concepts.*
Schleifer, R., "Probability and Statistics for Business Decisions," McGraw-Hill, New York, 1969. *Deals with concepts of subjective probability applied to business decisions.*
Savage, L., "The Foundation of Statistics," Wiley, New York, 1954.
Bowman, E. H., Fetter, R. B., "Analysis for Production Management," 1st ed., R. D. Irwin, Inc. *Chapter 11 discusses the Monte Carlo Method.*
Hammersley, J. M., Handscomb, D. C., "Monte Carlo Methods," Wiley, New York, 1964. *An advanced treatise on the subject.*

Meet the Authors

Park M. Reilly is a professor in the Dept. of Chemical Engineering, University of Waterloo, Waterloo, Ontario, Canada. Prior to entering academic life, he was Principal Chemical Engineer with Polymer Corp., Sarnia, Ontario, Canada. He has a B.A.Sc. in chemical engineering from the University of Toronto (Canada), a D.I.C. in statistics from Imperial College, and a Ph.D. in statistics from Imperial College, University of London. He is a registered professional engineer in the Province of Ontario, and is president of the management consulting firm of Plandex, Inc., 200 King St. East, Kitchener, Ont.

Hari P. Johri is a doctoral candidate in the Dept. of Chemical Engineering, University of Waterloo. He has had industrial experience with Hindustan Insecticides (India), and with Du Pont of Canada. He has a B.Ch.E. and a D.I.M (post-graduate diploma in industrial management) from the University of Delhi (India), and an M.A.Sc. in management and systems engineering from the University of Waterloo. He is a registered professional engineer in Ontario and is associated with the firm of Plandex, Inc.

Guidelines for Evaluating New Fields and Products

This candid look at how Du Pont uses economic evaluations in deciding whether or not to enter a new area of business, offers an insight into the risks as well as rewards of diversification, and provides useful advice.

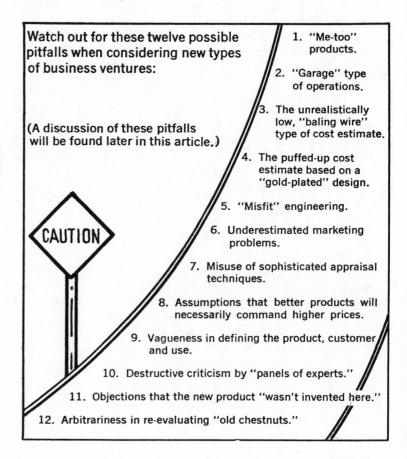

Watch out for these twelve possible pitfalls when considering new types of business ventures:

(A discussion of these pitfalls will be found later in this article.)

CAUTION

1. "Me-too" products.
2. "Garage" type of operations.
3. The unrealistically low, "baling wire" type of cost estimate.
4. The puffed-up cost estimate based on a "gold-plated" design.
5. "Misfit" engineering.
6. Underestimated marketing problems.
7. Misuse of sophisticated appraisal techniques.
8. Assumptions that better products will necessarily command higher prices.
9. Vagueness in defining the product, customer and use.
10. Destructive criticism by "panels of experts."
11. Objections that the new product "wasn't invented here."
12. Arbitrariness in re-evaluating "old chestnuts."

E. A. RANSOM, *E. I. du Pont de Nemours & Co.*

In discussing the use of economic evaluations in research and development for diversification of business, much of what I will say is based on work in Du Pont's Development Dept.; let me therefore start with a little bit of background.

About four years ago, an R&D division was organized in the Development Dept. to study and undertake diversification work, including not only paper studies, but also research, development, and even manufacture and sales. The ultimate intention is to carry new ventures to the point where industrial departments (existing or new) can take over.

Du Pont is divided into a number of departments, twelve of which are industrial ones that manufacture and sell products and engage in research in support of their businesses. These departments are: Textile Fibers, Photo Products, Pigments, Plastics, Organic Chemicals, Industrial and Biochemicals, Film, Explosives, Elastomer Chemicals, Electrochemicals, Fabrics and Finishes, and International. Most of these departments do research aimed at new products, but such efforts are usually extensions of present areas of businesses.

In addition to the industrial departments, there are thirteen staff departments such as Legal, Advertising, Employee Relations, and so on. Four of these do research, and their efforts are distinct from those of the industrial departments: the Central Research Dept. does chemical research in new areas—for example, fluorocarbon chemistry; the Engineering Dept. does research to aid in development, design, and construction, and offers consulting and general engineering services; the Employee Relations Dept. runs a toxicology laboratory for company-wide use; the Development Dept. conducts research aimed at diversification and includes in this program assistance to the staff departments when they have research findings that should be considered for commercialization. The Development Dept.'s activities come into play if company interest is broader than that of a single industrial group.

This department's R&D Division has already transferred to other groups a number of projects that were

Based, in part, on a paper presented on July 12, 1967 at the National Meeting of the American Assn. of Cost Engineers, held in Cleveland.

Exploratory Research is a 15 to 20-man effort with a manager, up to a dozen laboratory research men (not counting technicians), four patent specialists, two specialists responsible for market evaluations, and one responsible for engineering and economic evaluations (the latter represents my job). The laboratory team fills the need of the group to have a research facility for finding and developing vital information for new venture evaluations. Patent studies are important, as will be emphasized later when discussing venture selection bases. Other help may be contracted for outside the company, or it may be provided internally by other departments.

The ideas to be screened come from the staff or industrial departments as already mentioned, from the group's market studies or product ideas, from outside the company via unsolicited suggestions, or from encouraged relationships involving other organizations or individuals, or from studies contracted with various organizations.

The studies, as is probably obvious, try to determine what we can sell and what profit we might make. Market needs, existing products, prices, practices, etc., are investigated. After defining process and product possibilities, estimates are made of manufacturing costs and other business expenses or requirements. In some cases, prices may not be critical because a market need is great and value-in-use is high while probable costs are low. But more often, prices and costs are critical and the profitability becomes a key question in appraising venture candidates. There are, of course, other questions such as proprietary position, nature of the business versus company experience, effects on customers, effects on other company products, etc.

One is always on the lookout for guiding principles in this kind of work; particularly in diversification where the technology and business considerations are often unfamiliar. Some of these will be described here for what they may be worth to you, and some specific examples will be given. It should be emphasized that there are no pat formulas or procedures; there is no detailed pattern or mold into which a new venture must fit.

Basic Justifications

One should consider first either the market need or the product definition. If one has a product discovery or at least a specific product idea, then a substantial and important use for it must be found, and vice versa. In these days of advancing technology there are lots of ideas and lots of products; hence, the problem may well be one of finding or creating a matching need. It is also very worthwhile to undertake market-oriented research, that is, to study the market and identify opportunities based on large needs, and then to look for product ideas or discoveries to fill the needs. Both of the above approaches are followed.

The next thing to look for is attractive economics. This, of course, requires estimates of the market opportunity, sales, manufacturing costs, and investment, both permanent and working capital. An estimate of

in various stages of development. Some ventures that are still in the department, but that are well along with actual sales of products, are plastic tube heat exchangers and building products. A small group in R&D, called Exploratory Research, is concerned with screening various new venture candidates. If appropriate, these may become separately managed efforts in R&D or elsewhere. Three of these programs have recently moved on to other company laboratories and a sizable number have been considered and dropped.

the economics is important at an early stage, even if only a very rough one is possible. Further criteria often depend upon it. Assuming economic studies point up a potentially big money-maker, one might be willing to take chances and enter a field that would not otherwise be considered.

With the first cut at economics one can ask: "Will the venture significantly increase company sales and earnings, and is it big enough to warrant further study?" If the venture is in a field new to the company and offers real diversification, it ought to be large enough to make a noticeable impact on company performance. If the venture is not this large, it should at least lead, potentially, to large opportunities.

For a company as large as Du Pont, raising earnings a significant fraction is difficult. It requires filling large needs with products that will really sell; and in the early evaluation stages, this takes optimism.

If a new product fits into or directly complements a going business, it does not have to be so large. In fact, the product could be highly desirable even though it might afford no significant increase in earnings. If you are in the paint business, a new color might be just what your salesmen need to justify new calls on the same old stores. If competition is stiff, or if your customers need prodding, the new color might bring in more business for the rest of the product line. If you're not in a business even associated with the product, maybe the last thing you need is a new colored paint.

Now if a large, attractive opportunity has been found, the next consideration is the proprietary position. Because patents may offer the protection needed in starting a new business, they are usually considered essential in diversification. Protection can also be afforded by unique know-how not easily duplicated, or by a unique position in a key raw material, or by other advantages that competition can't readily match. Economics can help sort out the relative protection afforded by such potential positions.

An important consideration is what the competition can do. Of course, if one worried too much about what others may invent in the future, one might never undertake anything; but, if a way for the competitor to catch up or pull ahead is pretty obvious, then the situation warrants careful economic studies—of value-in-use, of other products' costs and profits, of obvious new or alternative products, as well as analysis of trends in the products' uses and prices.

Other Reasons for Entering a New Field

If the hoped-for new venture looks large, attractive and secure, there ought still to be other reasons for entering the new field.

One such reason involves allied technology. For example, if the company is in the business of manufacturing films such as cellophane (which involves coating operations), and if it is in other fine chemicals businesses, and has extensive know-how in web handling, testing, packaging, etc., it might combine all and extend operations into making photographic films.

Another reason often advanced involves existing

The Business Plan: Available Alternatives

One thing that can drastically affect the profitability of a venture is the business plan—that is, how the business is to fit in with suppliers, contractors, distributors, customers and competitors. One has to decide how far backward to integrate and what suppliers will provide; how far forward to integrate toward the ultimate user; what intermediate operations will be contracted to others, and how competition might react with modified products and lower prices. So, one can use venture economics to help determine what to make, what to buy, and what business partners to consider.

As an example, suppose one had a new candidate material for replacing silver salts in photographic film.

The first cut at the economics might indicate that one could sell the new material as a powder to film manufacturers at a high price; if manufacture of the powder is not too complex, a high profit should result. If present film-manufacturing costs, profit, and distributing markups are high, the potential profit for the new powder might really look wonderful.

However, if one tried to sell this new chemical at a high price, maybe the film manufacturer would not buy a large volume. He would be tempted to use the chemical only for a low-volume, premium film that could be sold at a high price. The basic material supplier couldn't force him to use it broadly. In fact, a situation might arise where the leaders in the film business feel little incentive to adopt the new product, and the small companies in the business hesitate if the product were to be pushed and sold broadly to all manufacturers.

If the basic material is really good, a better way to obtain a high profit in this situation might be to make the photographic film oneself. Then one could capitalize on the new value-in-use, sell broadly, charge appropriate prices, and at the same time, participate in film manufacturing's supposedly high profits.

Of course, film making and selling capabilities would be required. This could call for a joint effort with someone who has these capabilities, or some other way to procure them might be found. Studies of these operations can uncover all sorts of problems and important considerations requiring evaluations.

For instance, questions involving yield and quality might drastically affect costs. Economics could guide studies of reproducibility, coating-machine control, etc., directed at optimizing yield. Product testing might also involve high costs; special instruments for large-scale work may have to be developed.

Another consideration might be broadening the scope from just photographic film-making into camera manufacture and selling. Suppose the new film had unusual exposure latitude and speed, permitting a very inexpensive camera with a small lens and a fast shutter to produce good snapshots. Now, one might use the film to sell the cameras at high prices and maybe make less money on the film itself. Of course, one might not want to do this if the camera business were harder to protect from competition than the film business. Obviously, these are marketing problems, but the economics can be controlling and should be brought into planning as early as possible.

capabilities that are not fully utilized. For example, a company that does a lot of its own design and construction work in regard to plant equipment, and that is strong in the development of new materials and processes, might have an engineering capability for making mechanical products.

I was talking with a sales engineer from a large equipment manufacturer at a technical society meeting about diversification, and how everyone seems to think that the other grass is greener. He laughed and said his firm would like to diversify out of its unprofitable heavy equipment business, and supposed that maybe I would like to see Du Pont diversify into it.

A diversification cannot only allow present knowledge to be used more fully, but can lead to new capabilities of broad value, even beyond the scope of the new product field. For instance, a diversification into an engine-driven product might lead to new opportunities in petroleum products, alloys, bearings, materials, etc. These might combine existing with new capabilities in many ways.

Economic evaluations can help to appraise these "other reasons" for diversification. If what the company has to offer in a new field is not worth much economically, then there is probably no good reason to enter the new field.

Sample Evaluation

You might be interested in how results of studies of economic effects of variables were presented for a particular venture candidate.

Sales forecasts and manufacturing cost estimates were made and presented in considerable detail. Then some key variables in the cost assumptions were plotted as parameters on separate graphs of return on investment vs. selling price. (Many say there are better economic criteria than return on investment, but it turned out to be the easiest, and adequate in this particular case.)

The effect of yield was presented, as shown in Fig. 1. The three significant figures in the "current best estimate" of 64.4% yield weren't really justified by estimate accuracy, but retaining them helped to identify which calculation they came from, and facilitated checking. (As long as the approximate accuracy is kept in mind, it helps not to round off numbers.)

Another item of cost uncertainty was that for manufacturing a key ingredient. A graph of this one is shown in Fig. 2, with several estimates of mill cost per pound and the investment involved in each case. The graph really helped to focus attention on the process development of this material; it enables us to show more cases than in a tabulation, including some that might otherwise be rejected as unduly optimistic or pessimistic.

Operating labor—like yield, a variable that everyone always asks about—turned out to have only a small effect on this particular product. We plotted this variable in the same way as the preceding ones.

We set up a catchall category for expenses that might have to be added to the cost of sales—e.g., possible extra outlays for overhead, research or selling. A graph (Fig. 3) helped to show how far we could be off in these areas before the venture would become impractical.

The key variable other than manufacturing costs was permanent investment, as shown in Fig. 4. Actually, this graph demonstrated that doubling the permanent investment would not cut the return on investment in half, because working capital requirements and some manufacturing costs are not directly associated with investment.

In addition to these rather simple presentations, a number of others were tried so as to display the results of many runs of a complex computer program. This program was designed to study a number of variables in marketing, as well as in manufacturing cost and investment.

These results were expressed as plots of important variables—with 10, 50, and 90% certainty levels indicated—against the estimated financial value of the venture. This value counted the estimated investment costs, earnings, and funds to be recovered by an assumed liquidation after ten years, all converted to beginning-time dollars discounted at a normal rate of interest. This has sometimes been called "venture worth". Papers have already been published on this value estimating procedure, and there will probably be more in the future on this subject.

Possible Pitfalls

Here are some things to avoid, or at least watch out for in economic evaluation work of this type; they are gleaned from our experience (and were summarized in the little "box" on the first page).

The first caveat involves entering a new field where you have no advantage over competition. Diversification is inherently risky to begin with, but it can be made somewhat safer if the price can be protected. "Me-too" products may be all right where capabilities exist or can be readily established, but one doesn't enter a new field just because others seem to make money at it. The entering company usually needs real advantages to make up for some disadvantages. However, there are exceptions to this rule. Sometimes things sell just because they're different; sometimes people want another source of supply; and sometimes prices hold even when competition moves in. It is important for cost estimates to take into account the necessary differences in products, and the requirement to establish an advantage.

Another thing to watch out for is the "garage" operation. We get many suggestions to start a small business, off to the side, and to put one man in charge, with authority to make and sell such items as seat cushions for use at football games. The economics can look attractive until you add company overhead, freight from a central factory to widely scattered customers, working capital, trouble and expense to sell a highly seasonal product with perhaps a low volume per salesman, etc. Many ventures that are really too small or too "unscientific" come into this category.

When this diversification venture was presented to management, we prepared graphs to show how the estimated return on investment would be affected by...

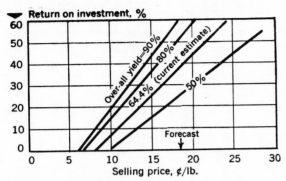

Manufacturing yield — Fig. 1

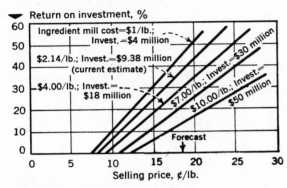

Cost of key ingredients — Fig. 2

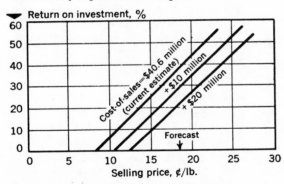

Extra cost-of-sales expense — Fig. 3

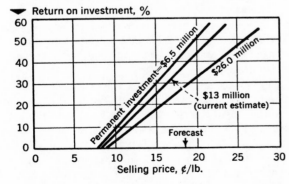

Permanent investment in manufacturing — Fig. 4

Baling-Wire vs. Gold-Plate

One must also be on the alert for "baling-wire" proposals—i.e., those that are based on unrealistically low expenses. Many new ventures can be made to seem fabulously profitable if one uses a $1.50 wage rate, a $3/sq. ft. building, no fence, no change house, no gate house, no fire protection, no research to keep the product up to date, no extra depreciation allowance to take care of impending obsolescence, and so on. Because such estimates look so good at first glance, one can waste a lot of time trying to make the package salable when it should have been discarded early.

A hazard of an opposite nature involves unnecessary expenses in such things as design features. "Gold plate" can kill a good prospect. For example, a process where shutdowns are expensive can lead one to go overboard on exotic and expensive materials of construction, premium-quality equipment, installed spares, and elaborate controls, even to the point of putting tile on the walls to keep the dirt out.

Businesses that don't need and cannot afford this kind of insurance may still find it creeping into plant designs—and into estimating. If motors, starters and wiring are customarily too elaborate, it's easy to get into the habit of using $1,000 for the cost of small pumps. This generosity is compounded when one applies a $5\times$ or $6\times$ factor to go from purchased equipment to plant cost. Should the plant indeed turn out to cost that much, everybody may think such estimating is sound. But woe to the venture that can't afford it.

More Caveats

The next item on the list is misfit engineering. For example, an engineer may be skilled in the design of industrial mixing equipment, but he may fall victim to poor judgment when trying to design a household mixer. Something that is "sound engineering practice" in the industrial field may not be sound practice at all in the consumer field.

Marketing problems present hazards, too. These are generally outside the scope of this paper except for the point that every cost-of-sales estimate includes an item for selling expense. This can vary widely, from less than 5% to perhaps 30%, or in some cases much more. For example, household items pass through several hands to a small retailer. If you are going to manufacture and sell directly, your expenses could be practically all selling expenses.

Misuse of sophisticated appraisal techniques is another danger. Many appraisal techniques have been developed to calculate the profitability or worth of a venture; for instance, they may involve time value of money and probability predictions for market penetration vs. many variables. These techniques are valuable where they are needed and where the inputs of essential information are available. They should not be belittled as they often are by "seat-of-the-pants" estimators and by some others, too. But sophisticated appraisal techniques can present a hazard; they can lead many users astray.

Complex math and computer programming, for ex-

ample, can lead an unsuspecting or inexperienced analyst into an area of difficult communications and easy mistakes. The originator of a so-called best guess that is fed into the computer sometimes fails to realize what he is doing. Combinations of these guesses can result in computed answers that are far from reality, particularly with the less likely cases in a Monte Carlo program. The computer estimates may thus gyrate wildly and the resulting print-out may possibly be viewed with a serious face by one who should really be seeing only his own reflection.

Of course, where the necessary data and the relationships can be supplied, these new tools can provide valuable insights. For comparison of projects from many different departments, a sophisticated and standardized technique is probably essential. The trick is to use them without misusing them.

Another possible pitfall involves the assumption that a better product will necessarily command a higher price. This assumption does not always hold true. For instance, a longer-lived phonograph record will not command a higher price. Although the extra cost would be but a very small fraction of the retail price, it represents real money that the manufacturer apparently doesn't feel he would recoup in the form of more record sales. In fact, the change might even be viewed as probably reducing sale of records. Costs for unwarranted product quality have a way of sneaking into venture evaluations. If the cost comes out high, one is tempted to decide the product is better, when the decision should really be that the cost has to be lowered.

A situation that is particularly dangerous at early evaluation stages is one in which product, customer and use are not specifically defined. Often, it is not known exactly what one will sell, or to whom the sale will be made, or for what the product will be used. Accurate estimates of economics cannot be made without these specifics. For example, suppose research finds a stronger fiber. Should the fiber or the ingredient polymer be sold? Would the customer stop buying a highly profitable product to use the new one? Or has he no need for a stronger fiber? Could he make his product stronger with present-day fibers if he wanted to? Or is it possible that the only good future market for a stronger fiber precludes using organic polymers and requires inorganic fibers?

The Objective vs. the Subjective

Another pitfall may be relying on panels of experts. It is remarkably easy to misuse specialists who are available for consultation. Take a *problem* to an assembled group of them, and they will try to solve it. But take them an *idea*, and they may try to kill it; they may unconsciously seek individual recognition by being first to find a flaw. Sometimes they may try to compensate for not having thought of such ideas themselves by belittling the invention and the inventor. To bring out constructive rather than destructive contributions, the person consulting the panel must really want to find a way to use the idea, and inspire others to look at it that way.

One should also be alert to proprietary attitudes and the "not invented here" factor. At early stages, evaluations are often subjective. The definitions aren't good enough to eliminate people's attitudes. A person who thinks a new plastic bottle is a great idea estimates its cost at 3¢, while someone who is convinced that the proposed process is no good or that the product won't sell, estimates it at 11¢. It takes detailed study to tell who is right.

"Old chestnuts" pose another hazard. If a company has many people who screen ideas and developments, and has many economic evaluation groups, some old ideas can be brought up many times for review and evaluation. If one tries to be optimistic about a concept and then finds out it was studied and rejected by someone else, there is sometimes a temptation to adopt the earlier conclusion. Then the promoter has to find a new champion. Many innovators are very persistent; their research ideas may be promoted seemingly endlessly. The problem becomes one of utilizing the older work so it will not be repeated unnecessarily, but avoiding giving it undue weight, because conditions may by now have changed or a mistake may have been made.

In summary, the economic evaluation is an important tool in R&D planning, particularly in diversification studies. As may be the case with any tool, its use can be dangerous as well as valuable. It can save a lot of time and effort or it can waste them. It can be interesting and satisfying, and it may sometimes be burdensome and difficult, or even impractical.

The use of economics studies should be strengthened; and the people who make them need to broaden their knowledge, skills, interests, and their communication with others. Everyone needs to apply some basic economics in his work—researchers and other engineers, as well as technical managers and accountants. There is a need for more evaluations specialists, too, and it appears likely that the career opportunities for these specialists will continue to grow.

Meet the Author

Edwin A. Ransom is a specialist in the Research and Development Div. of E. I. du Pont de Nemours & Co.'s Development Dept. (Wilmington, Del. 19898). He joined Du Pont in 1946 as an engineer, and held various supervisory and consultant positions in the company's Engineering Dept. until 1963, when he took up his present assignment. Mr. Ransom attended the University of Wichita, Kan., and Oklahoma State University, where he received a B.S. in chemical engineering in 1942.

Key Concepts for This Article

Active (8)	Passive (9)	Means/Methods (10)
Analyzing*	Potential*	Economic evaluation*
Forecasting	Products*	Venture analysis*
Estimating	Ventures*	**Purpose of (4)**
	Profit*	Commercial
	Return on investment*	Business

(Words in bold are role indicators; numbers correspond to EJC-AIChE information retrieval system. Asterisks mark key concepts suggested for indexing. Others are added to improve reading as an abstract. Indexing is described in *Chem. Eng.*, Oct. 11, 1965, p. 187.)

****

Guidelines for Estimating Profitability

All profitability formulas are worthless if you are using incorrect data. Here is some frank, practical advice on how to compile and apply various costs to produce a meaningful estimate of profitability.

USEFUL ADVICE ON COMPILING AND APPLYING:

Incremental costs

Revenue

Fixed capital investment

Working capital

Manufacturing and distribution costs

Overhead

Income tax and depreciation

Value and cost of money

Risk factors

****

J. ROSS, *Thompson Ramo Wooldridge Inc.*

F. V. MARSIK, *Celanese Chemical Co., Celanese Corp. of America**

F. R. DOUGLAS, *Texaco Inc.**

R. L. WAGNER, *Diamond National Corp.**

Much has been written on the use of rigorous financial and economic analysis methods for solving profitability problems. However, most of the literature deals primarily with better ways of conducting the analysis and has little to say about the problems involved in compilation and application of the required data. Consequently it is in this area that many profitability estimates go astray.

The purpose of this paper is to (1) indicate possible sources of data and their uses in profitability estimates, and (2) present some fundamental guidelines for the profitability estimator.

What Is Profitability?

To determine kinds and forms of data needed to make a profitability estimate, it is first necessary to clarify the meaning of the term "profitability."

Profitability can be defined as: a measure of the desirability of risking either additional capital or expense (or both) for new proposals or projects; the basic considerations being capital, expense, income and time. It is the economic result stemming from the expenditure of capital and/or expense.

In general, profitability estimates take the form of economic justifications for the outlay of funds. These outlays may be for productive growth or for improving existing operations.

The profitability measurement can be expressed in many ways. Among these are total dollars, percent of sales, ratio of annual sales to investment, or return on investment. This latter measurement is favored by most profitability estimators. The methods for calculation of return on investment are many and varied. Some of the more common are (1) payout period, (2) return on original investment, (3) return on average investment, (4) present worth, and (5) interest rate of return. Each of these methods has been discussed widely in the literature.

Although the choice of method is too broad in scope for adequate discussion here, it should be recognized that the method used will have important effects in profitability estimating, especially where the timing of expenditures and revenues is significant.

Profitability Estimating Goals

In selecting a method of analysis, the profitability estimator must be sure that the results of the analysis are consistent with management's goals of profitability.

Return on invested funds, regardless of the method used to measure it, is the most important standard of profitability. Therefore the primary goal of any profitability estimate should be to predict the net change in the company's over-all cash position, or the basic profit and loss statement that will occur as a result of the project under study.

To reach this goal, the estimator must be able to make studies of clear alternatives, stripped as much as possible of arbitrary and conventional cost allocations. It is the *change* in cost and revenue that must be considered, rather than the burden that the project may, in some cases, ultimately bear as a result of pricing and accounting policies.

Types of Data Required

To satisfy the purpose and goals of a profitability study, the estimator will be confronted with the

* This paper was prepared as an effort of the Profitability Committee of the Metropolitan N. Y. section, American Assn. of Cost Engineers. Paper was presented as a panel discussion at an annual meeting of the N. Y. section.

proper use of accounting, profitability, and engineering cost data.

Accounting data are assembled from records to indicate profits, losses, operating costs, net worth, etc., of a business or a business operation for a past period or periods of time. They are usually assembled in standardized formats and usually do not emphasize or indicate the timing of the cash flows involved. Accounting data may also include certain distributed costs that are required for cost control purposes, compliance with federal, state, and local laws, and company policies.

Engineering cost data are assembled from material balances, heat balances, kinetics, flowchart information, and time and motion studies, so the analysis of costs and profits can be related to the physical variables and operation of the project. Capital costs (including construction and erection) also are considered engineering cost data.

Profitability data are assembled from records, projection of records, and forecasts, to indicate or predict the future income, profits, losses, operating costs, etc., of a proposed or existing operation. They are presented in a format designed to emphasize the effect the proposal will have on cash flow and profits or losses of the operation.

Thus profitability data are concerned with future cash effects that result from a decision either to spend or not to spend money. While based in many instances on accounting data and records, profitability data do not have the same validity as accounting data, since future projections are involved.

A profitability estimate should recognize the nine items listed at the beginning of this article.

Incremental Cost Concept

Incremental cost is the most controversial concept involved in a profitability study. Yet this concept is the basic difference between a true profitability estimate and an average cost estimate.

The following simplified example will demonstrate the difference between an incremental cost and an average cost. Assume a proposed increase in productive capacity will increase the purchase of raw-material requirements so that volume discounts can be realized. The raw-material costs for the present and total proposed operation are as follows:

	Present	Additional Proposed (Incremental)	Total
Consumption, thousand tons/mo.	100	100	200
Cost of raw matls., $/ton	1.00	(0.92)	0.96
Total monthly cost, $..	100,000	(92,000)	192,000

The problem is to determine total monthly cost for the additional capacity. The most obvious path is to simply state: consumption of 100,000 additional tons/mo. at the price of $0.96/ton results in a cost of $96,000/mo. for the new operation. In fact, if separate

* *

''Incremental cost is the most controversial concept involved in a profitability study. Yet this concept is the basic difference between a true profitability estimate and one based on average costs.''

* *

accounts are kept for the two operations, this is exactly what accounting cost reports will show.

If this $96,000 was entered on line 3 in the table, total monthly costs would equal $196,000. However, since 200,000 tons of material could be purchased at $0.96/ton, the total cost should be only $192,000. If the difference between this total cost ($192,000/mo.) and the cost before the capacity increase ($100,000/mo.) is considered, then the total additional cost amounts to $92,000/mo. On this basis, the incremental unit cost will be only $0.92/ton and not $0.96/ton as the average purchase price would indicate.

At this point, there appears to be a conflict between profitability data and accounting data. Actually this conflict is only apparent and not real. Total cost for raw materials to the company will be $192,000, and this will be the total amount that shows on the company's records. The difference arises because average unit cost does not take into account the over-all effect of the new operation.

Of course, the estimator is costing and not pricing the additional production. When sales prices are determined, average costs could be used.

In using an incremental cost analysis, the temptation of understating costs must be avoided. Only direct purchase price of raw material was shown in our example. In a real analysis, the possibility of additional purchasing, storage, and handling costs must be investigated. These costs must also be determined on an incremental basis.

An ideal way to determine an incremental cost is to subtract total costs before the capacity increase from the total costs after the increase. Unfortunately a practical cost estimate usually requires the incremental unit costs first, in order to predict effect on total costs.

The incremental cost is usually extracted from cost records of the company's existing operation. The estimator attempts to determine those existing costs that will not be affected by the new proposal. This requires a thorough knowledge of the company's operation, policies and future plans.

Estimating Revenues

Revenue can be generated by various means. The most common source of manufacturing revenue is the

sale of the company's products, or services. In general, profitability estimates will be concerned with:
- Additional sales of existing products.
- Additional sales of new products.
- Maintaining present sales volume.

Revenue from the first two items is generally estimated from sales-volume forecasts. These forecasts are usually furnished by sales or market-research groups. All products and byproducts must be priced before an estimate of total revenue can be made.

Market prices of products currently on the market can be obtained from the accounting department (for products the company is currently selling), the purchasing department (for products the company is currently buying), and literature sources (for common items).

In using prices obtained from these sources, it is important to determine whether the planned increase in production will affect current market prices. Market research groups can be helpful in developing possible price-volume relationships.

For products not currently on the market, possible price range and sales-volume forecasts can be developed by commercial development or market research groups.

In cases where the price-volume relationship is quite diverse, estimates should be prepared based on the highest, lowest and median revenues. From this information, optimum price can be developed for the new product and byproducts. The estimator can determine the price at which the resulting sales volume will give a total revenue yielding the best rate of return on the over-all investment. He can also determine the break-even price for the new product.

Where a project is required to protect existing revenues, the probable loss in revenue if the project is not adopted must be estimated. Saved revenue is used in the same manner as additional revenue to determine the profitability of the project under consideration.

Fixed Capital Investment

Consider the problem of new facilities in a grassroots plant. Where only battery-limits investment* is being considered, required capital is included in the calculation of return on investment and in calculating manufacturing costs. For new off-sites† one of at least two philosophies applies: either capital is included in the calculation of the return on investment or it is not. The philosophy behind the latter is that off-site facilities need not have as high a return on the investment as battery-limits facilities since a smaller risk of obsolescence is involved. In the calculation of manufacturing costs related to new off-site capital, this capital is usually included.

The problem becomes more difficult when an addition to an existing facility is being considered and it is planned to use part of some idle facility. The incremental part of such an investment presents no

---

* Another term is direct manufacturing facilities.
† Another term is service facilities.

Fundamental Guidelines

- Profitability is the measure of the desirability of risking funds for new ideas—the basic considerations being capital, expense, income, and time. The profitability estimator must recognize the differences between accounting, engineering and profitability data when using them in his estimates.

- End result of every profitability estimate should be to predict net change in the company's cash position as a result of the project under study.

- The profitability estimator should recognize the concept of incremental cost.

- Capital value assigned to idle facilities should reflect their actual value to the company.

- Change in scope of a company's operations will not necessarily result in a proportional increase in the amount of overhead expenses incurred.

- The method used to calculate depreciation should be identical to that used for tax purposes.

- Some recognition of risk should be made in all profitability studies, and so stated.

- Changes in working capital required due to the project should be estimated and included in the profitability estimate.

- Minimum rate of return that is used in an economic evaluation should be greater than the cost of money, to justify the risk.

- Since management policy will have an effect on profitability, the estimator should be aware of his management's long and short term goals and policies.

---

problem: it is included both in the return on investment calculations and the manufacturing costs related to this incremental investment.

However, the existing battery-limits investment either is or is not included in the calculation of the return on investment or manufacturing costs. The philosophy behind including this investment is that full account of the project's requirement is thus taken. In this way, if there are many projects competing for this idle equipment (known or unknown), each will stand on its own. The philosophy behind not including this investment is that this idle equipment can be considered paid out by the project for which it was originally built. Which of these philosophies to use depends largely on whether or not the company is under expansion and many projects are competing for idle equipment.

Another approach to assigning capital investment, particularly for existing idle off-site facilities such

as power production, may be to consider that the new project will force expansion of the power facilities sooner than would otherwise be necessary. Here, time-value of money can be used by showing the timing of the expenditure for the new facilities in cash flow.

Value to be assigned to idle existing facilities must reflect actual value to the company. This value can be determined on the basis of book or depreciated value, salvage value, replacement cost or remaining useful lives.

Working Capital

Accountants and economists think of net working capital as the difference between current assets and current liabilities. Cost engineers sometimes estimate working capital as the allowance in a capital estimate for necessary operating inventory (raw, in-process, and finished materials), cash, and net receivables. The primary reason for considering these nondepreciable and/or nonreturnable items in profitability estimates is to highlight their effect on cash flow. For example, inventory increases require immediate cash outlays that delay cash flows from generated sales revenue.

If a more detailed analysis of working capital is justified, the estimate of receivables should be figured in consultation with responsible credit experts. Real-istic inventory allowances must be set with the knowl-edge of firm sales forecasts and production levels.

Working capital is required to operate new plant facilities or to market resulting additional production. The following should be tabulated and totaled to obtain total working capital: change in current payables; change in accounts receivable; change in inventory of raw materials and supplies; change in cash-on-hand balances; change in current liabilities and nonde-preciable assets.

Increase in working cash would be the amount neces-sary to support the operations resulting only from the investment. Ordinarily, a higher requirement for cash on hand occurs during periods when operations are increasing; a lower requirement exists when op-erations are decreasing; when operations are steady there theoretically would be no need for a change in cash on hand. An exception would be investment that decreases working capital requirements by increasing efficiency, such as data-processing equipment for billing and inventory storage improvements.

Working capital is as important to consider as other capital in a profitability study; it should not be ig-nored simply because it may be recovered at the close of the project life. A credit for working capital at its estimated value on termination date of a project's life should be considered.

Manufacturing and Distribution Costs

Direct manufacturing costs include raw materials, operating and maintenance labor, operating and main-tenance materials, utilities, royalties and depreciation. Distribution costs include freight, terminal charges, warehousing costs, and other expenses related to the actual task of physically moving the materials.

Raw-materials costs and byproducts credits are very often the most critical portion of the estimate.

For profitability estimating purposes, raw materials fall into two classifications:

• Materials from outside sources. These can be charged at present market price delivered to the con-suming point.

• Materials from within the organization from existing facilities.

In the case of raw materials with clear-cut manu-facturing costs from within a company, two ap-proaches can be used. If there is excess capacity for manufacturing these raw materials and a ready mar-ket does not exist in the foreseeable future, incre-mental or out-of-pocket costs may be used.

If there is no excess capacity, or if a likelihood exists that future markets could use the excess ca-pacity, then the net value based on market price or replacement cost should be used. This approach re-flects the actual cost of the raw material to the com-pany in terms of reduced sales revenue.

Consideration may also be given to expanding the existing facilities to take care of raw-material re-quirements for the proposed venture. In this case, cost of expansion should be included.

If the raw material is one of several manufactured concurrently, the cost problem is more complex:

• If a ready market exists for the byproduct, then the byproduct may be charged at its net value based on market price.

• If no market exists for the byproduct but an alternate use is possible, for example as fuel, then an alternate use value may be assigned.

• If the raw material is an existing waste ma-terial, then its cost of disposal should be credited to the new use.

In many cases, a profitability analysis will encom-pass a range of values where unknown market and other situations exist.

Overhead Costs

In addition to direct costs, cost of administrative, selling and general expenses can also have an effect on a profitability estimate.

When determining incremental cost of these activi-ties, it must be remembered that a certain minimum level is required to conduct the company's operation. A change in scope of a company operation, therefore, will not necessarily result in a proportional increase in the amount of these expenses.

To determine the exact change in overhead costs, it is necessary to determine which overhead items may be affected. Since accounting systems vary widely in internal alignment of accounts, it is impossible to make a clear-cut listing of those accounts subject to change. However, there are certain areas of any or-ganization that are relatively unaffected. Some of these are: executive management; corporate level sales, engineering, and operations administration; comptroller's and treasurer's administration; cor-

porate legal and tax administration; public relations; donations and endowments.

In general, these areas will be more affected by changes in management policy and philosophy than by limited changes in the scope of a company's operations. It is usually unnecessary to consider these costs in an incremental analysis.

Some of the overhead functions more affected by operational changes are: employee benefits; plant administration; employee training programs; medical department expense; district sales expense; engineering and development expense; insurance; property taxes. Some of these overhead costs are also very susceptible to management policy. Personnel and labor relations, research and engineering budgets, emphasis on long- or short-range planning, and maximum-short-term vs. optimum-long-term profit policy all have an effect on overhead costs. Once management has fixed its policy on these matters, however, the costs will vary with operational changes.

To accurately determine the relationship between variable overhead costs and direct costs, it is usually necessary to make a thorough analysis of the company's operation. If possible, the analysis should determine the rate of change in overhead costs as compared to direct costs. To illustrate, assume that a hypothetical company has the following past and projected sales volume:

	1955	1961	1964 (Projected)
Sales, thousand $/yr....	1,000	2,000	2,500
Sales force, men........	10	22	29
District sales expense, thousand $	100	230	350
Ratios:			
Expense, % of sales....	10	11.5	14
Sales/man, thousand $..	100	91	86
Cost/man, thousand $...	10	10.5	12

Sales costs in the past have risen faster than sales income and will continue to rise at an accelerated rate. Dollar sales per man are dropping and cost per man is increasing.

With these trends in mind, a projected ratio is selected that will best reflect future costs. By using the best economic barometers available, the estimator must determine if present trends will continue, level off or reverse. The best sources of information on this subject are the company's economists or business journals.

If the project under analysis will result in a product line to be marketed under similar competitive conditions to the existing line, and it is expected that sales costs will level off, the 14% ratio may be used to determine sales expense for 1964 and later. If, however, the new product will be in a different competitive position, sales expense can be estimated by estimating the increase in sales manpower required. New expenses can then be computed.

The same approach is used to determine other overhead cost ratios. Extent of the analysis will depend upon sensitivity of the profitability estimate. If a preliminary analysis shows a certain variable overhead cost is small compared with the over-all running costs, then a very rough approximation can be used. If overhead costs appear to be large enough to have a significant effect on the estimate, a more thorough analysis must be performed.

Depreciation and Income Taxes

The American Assn. of Cost Engineers has defined depreciation as: "The allocation of the cost of capital assets less salvage (if any) over the estimated useful life of the unit in a systematic and rational manner; a non-cash expense."

Depreciation is important in profitability studies since it affects income tax liability. Since tax liability

Meet the Authors

Jack Ross, is district sales manager for the TRW Computer Div. of Thompson Ramo Wooldridge, Inc., in New York City. He is involved in marketing control computers to the process industries. Mr. Ross previously worked for Arabian American Oil Co. in process engineering, process evaluation, and economic and production planning. He has a B.S. in chemical engineering from Purdue University.

Fred V. Marsik is an administrative assistant in the Planning Dept. of the Celanese Chemical Co., Celanese Corp. of America, N. Y. His previous experience included technical and economic evaluation work, pilot-plant and production operations. Mr. Marsik has A.B., B.S. and Ch.E. degrees from Columbia University. He is a registered professional engineer in New Jersey.

Fred R. Douglas is a research chemical engineer with Texaco Inc., Beacon, N. Y. His work at Texaco has centered around economic evaluation of research projects. Mr. Douglas has a B.S. in chemical engineering from Newark College of Engineering and an M.S. in chemical engineering from Polytechnic Institute of Brooklyn.

Richard L. Wagner is a valuation engineer at Diamond National Corp.'s staff headquarters in New York City. He previously worked at Diamond's research and engineering center in Stamford, Conn. He has a civil engineering degree from Cornell University and is a licensed professional engineer in Ohio, California and New York.

is calculated after allowances for recovery of invested capital, the rate of this recovery and its timing will have an effect on after-tax profits in any given period of time.

The government recognizes several methods of calculating depreciation. Profitability is particularly affected by the choice of method if the time value of money is considered. Although the ultimate recovery of the original investment is the same for all methods, the timing of this recovery is different in each case. Choice of method depends upon project life, cost of capital, and management policies.

Management policies can affect depreciation by the method of accounting treatment of depreciation, or by liberal or conservative capitalization policies (selection of length of useful lives).

Adjustment for corporate income taxes can affect the relative profitability of capital projects. Depending on the timing of cash flow of projects, relative after-tax profitabilities could be significantly different from before-tax. After-tax earnings are the only tangible basis to judge the relative attractiveness of proposed investments.

Value and Cost of Money

There are always alternate uses for a company's funds or physical assets. Money and assets are a limited commodity since they can be used up in many ways. Funds can be distributed to the owners as cash dividends or returned to the business as new investment. Assets can be liquidated or put to a different use.

In some cases, it may be to the company's best interest to liquidate assets and distribute funds. This situation will occur when funds and assets cannot be used to yield an acceptable return.

There is a difference between value and cost of money. The existing value of money to a company is equal to the interest rate of earnings on the equity capital of the company. The goal of most managements is to increase this return. This is different from the cost of money, which is usually less than the value of money.

Value of money can be approximated by using the company's annual report. Total equity capital of the company can be found on the balance sheet. Net earnings after taxes are shown on the profit-and-loss statement. The percentage ratio of net earnings to total equity capital is the interest rate on the company's equity capital. This is the bare minimum that the new project must earn if it is to break even with current investments.

Cost of money is made up of the charges incurred when money is borrowed. It includes interest charges and administrative expense of carrying out the loan contract.

Cost of borrowed money should be considered as a cost in the project being evaluated.

Minimum rate of return in an economic evaluation should be greater than the cost of money to justify the risk involved. If the minimum rate of return is the same as the cost of borrowed money, the owner will gain no advantage by investing in the project as opposed to investing as a lender.

Risk Elements

Every capital expenditure entails some amount of risk. This risk element can vary from a negligible to a considerable amount. Roland Soule in his "Design of Capital Structures" * classifies projects into three types, rather than attempting to assign probability factors to each:

• Safest type—outcome can be predicted. It is characteristically a product-improvement or cost-reduction project.

• Less safe type—a plant expansion project, involving increased capacity to make an existing product. Public acceptance is not a problem, manufacturing costs are known. The doubt is whether to expand by 10, 50 or 100%.

• Riskiest type—a new process or product.

When management considers an investment, it will consider the amount of risk involved. It is reasonable to assume that management would be satisfied with the lowest rate of return on the safest projects and the highest rate of return on the projects involving the greatest risk.

Another approach, as noted above, would be to assign each project with a risk factor based on probability. This factor would include such items as:

• Chance of the technical success of the process.

• Reliability of the capital requirement estimate.

• Reliability of sales forecasts and estimates of net revenues.

• Process obsolescence time.

• Present and future availability and costs of raw materials.

Management policies can affect the reliability of a profitability study if investment decisions are necessarily made on limited information. Under these circumstances, the more-refined methods of profitability estimating may not reduce the risk involved. More-refined methods tend to place greater importance on reliable sales forecasts, firm prices, and accurate cost estimates. Less-refined methods such as percent of sales measurements or break-even analysis may indicate to management the relative attractiveness of a proposal.

* "Design of Capital Structures," presented before the Chemical and Economics Group, N. Y. section, American Chemical Society, Oct. 13, 1955.

Key Concepts for This Article

For indexing details, see *Chem. Eng.*, Jan. 7, 1963, p. 73 (Reprint No. 222). Words in bold are role indicators; numbers correspond to AIChE system.

Active (8)	Passive (9)
Compiling	Data
Applying	Profitability
Using	Return on investment
Estimating	Expenditures
	Costs
	Costs, production
	Costs, incremental
	Depreciation
	Taxes
	Burden (cost)
	Risks (cost)

Economic Justification for Equipment

Need to make quick evaluations and justify economically new pieces of equipment? Try these simple graphical and differential methods.

DAVID STUHLBARG, *The Fluor Corp., Ltd.*

Engineers must often justify installations that may be as small as an item costing a few dollars or as great as a complete new process or plant costing many millions. Sophisticated methods of evaluation, such as the continuous discounting method of Hirschmann and Brauweiler[1] may be used; usually, however, most engineers cannot spend too much time on any one of the many decisions to be made, particularly those involving choice between alternates or optimization of size. To reach such decisions, he must resort to procedures that are rapid and easy, such as the differential and graphical methods that follow.

Because of the risk involved, the acceptable rate of return on an investment in production facilities is almost invariably much higher than the usual interest rates on secured investment. How much higher depends somewhat on the method of evaluation. If a comprehensive method is used, the rate is usually lower than when shortcut ways are adopted. It should be possible, however, to judge and approve a project based upon a careful study and then, by working backwards, establish an equivalent rate of return or payout period.

Two methods are used for justification of an expenditure. One is called rate of return on investment, the other is termed payout. The two methods should be reciprocally equivalent, but because of differences in definition this is not always so.

Return on investment is defined by Osburn and Kammermeyer[2] as net profit after taxes plus depreciation, and rate of return as the annual return divided by the investment. Payout is defined as the true reciprocal:

Payout = investment/return or $n = P/(R + D)$ (1)

Return on investment is recommended by Schweyer[3] and Perry[4] for justification studies, but a great many companies prefer to use the payout concept because it has a realistic significance for engineers, and can be directly related to project life. Some times return on average investment (average yearly net profit divided by average fixed investment plus working capital) is used with payout to evaluate a potential investment.

Schweyer differentiates between the payout defined by Eq. (1) and the economic payout time, which he defines by the equation:

$$n' = \frac{-\log\left[1 - iP/(R+D)\right]}{\log\left(1 + i\right)}$$ (2)

where i is chosen as the earning rate, the going rate for money, or the going rate corrected for income tax.

For the usual engineering decisions, the writer prefers a formula based on cash flow without regard to financing considerations, which are usually not available anyway to the individual engineer:

$$n = P/[(R_G - E)(1 - r) + rD]$$ (3)

where R_G is the gross return or savings, E is the actual operating expense—not including depreciation or interest on borrowed capital, r is the applicable income tax rate, and D is the average annual depreciation allowable as an income tax deduction.

The expense term E includes maintenance, insurance, property tax, etc. Eqs. (1) and (3) are equivalent because the term R in Eq. (1) includes depreciation.

Substitution of n for $P/(R + D)$ in Eq. (2), and rearrangement, gives:

$$n = (1/i)(1 + i)^{n'} - 1/i$$ (2a)

Companies using economic payout time as the criterion for investment will need to find n by this relationship for the equations that follow.

Eq. (3) is completely general and may be applied to any investment. If R_G, E, or D are expected to vary from year to year, then n may be replaced by $1/R_n$, where R_n (the net return for any one year) is calculated on a year by year basis until:

$$\sum_1^{(n-1+f)} R_n = P$$

In this relationship, f is the fractional part of the nth year necessary to obtain equality.

If depreciation is figured on a straight-line basis, and if E is assumed to be constant from year to year, then E and D may be replaced by $E'P$ and $D'P$, where E' and D' are constants. Substituting in Eq. (3), and rearranging, gives:

$$P = \left[\frac{(1 - r)}{1/n + (1 - r)E' - rD'}\right]R_G = jR_G$$ (4)

where j represents the function within the brackets.

If n is preset by policy to some exact number $\bar{n}$, then

$$\bar{P} = \bar{j}R_G$$ (5)

$\bar{P}$ in this case is the maximum investment permitted by company policy to effect R_G as the return.

Subtracting Eq. (5) from the identity $P = P$ gives

$$P - \bar{P} = P - \bar{j}R_G$$ (6)

If $P - \bar{P}$ is negative, the project is justified.

If x is the variable upon which the investment and the return both depend, then both P and R_G are functions of x. To illustrate, if the saving of heat by the addition of insulation were the reason for the investment, then x would be the amount of heat saved, R_G the annual savings that would result, and P the installed cost of the insulation. Variation of heat recovery x could be accomplished by varying the insulation thickness (therefore cost P) and result in a change in the annual savings R_G.

Determination of Optimum

Graphical Method—If the function $P - \bar{j}R_G$ is graphed vs. x, a concave upward curve results (see figure). The lowest point on this curve is the optimum. Plotting this curve is generally the best and simplest method of obtaining the optimum.

Analytical Method—The optimum, or minimum point on the curve $P - \bar{j}R_G$ vs. x, can also be found by setting the first derivative equal to zero:

$$\frac{d}{dx}(P - \bar{j}R_G) = 0$$ (7)

This method depends upon P and R_G both being con-

Nomenclature

A	Heat-transfer area, sq.ft.
AMTD	Arithmetic-mean temperature difference, deg. F.
C	Cost of equipment, \$/sq.ft.
D	Depreciation, \$/yr.
D'	Depreciation rate, 1/years
E	Operating expenses, \$/yr.
E'	Operating expenses, rate %
H	Operating time, hr./yr.
i	Interest or earning rate, %
j	A function of n, r, E' and D', as defined by Eq. (4)
$\bar{j}$	A specific value of j
LMTD	Log-mean temperature difference
n	Payout, years
n'	Economic payout time, defined by Eq. (2)
P	Principal or capital investment, \$
$\bar{P}$	Allowable capital investment for a payout of $\bar{n}$ yr.
q	Amount of heat, Btu./hr.
R	Net return on investment, \$/yr.
R_G	Gross return on investment, \$/yr.
r	Income tax rate, %
T_2	Outlet temperature of fluid being cooled in interchanger
t_2	Outlet temperature of fluid being heated in interchanger
x	The variable to be optimized

Subscripts:

c	cooler
h	heater
i	interchanger
p	piping and/or other capital items independent of x
u	utility

tinuous functions of x. In practice, these functions must also be simple enough so that the derivative is easily determined and easily solvable for x.

Discontinuous Functions—Very often, the changes in the investment and the return occur in some stepwise fashion. One example would be the distillation problem cited by Quigley and Weaver[5] wherein the yield of a product and the capital investment vary with the number of plates in a fractionation column. The graphical method noted above can be used for this problem, with x representing the number of plates, but the optimum obtained must be adjusted to a whole number value for x. This being the case, it is unnecessary to draw the curve; it is sufficient to set up a table of values of $P - \bar{j}R_G$ vs. corresponding whole number values of x.

Application to Heat Recovery

The preceding methods are completely general and applicable to any type of equipment or even to complete processes. It is convenient, however, to modify Eq. (6) for specific applications.

Consider the case where a process stream is heated, processed in some way, then cooled. The problem: Is the installation of a heat recovery interchanger justified, and if so what is its optimum size?

The installation of such an interchanger may have secondary effects. The heater and cooler involved are usually reduced in size, and if the amount of heat involved is appreciable, utilities such as a boiler or cooling tower may also get smaller. The equation for the investment P, then, is as follows:

$$P = P_p + P_i - \Delta P_h - \Delta P_c - \Delta P_u$$

where P_p = cost of extra piping required by an interchanger plan; P_i = cost of the interchanger; and ΔP_h, ΔP_c and ΔP_u are the differences in costs of the heater, cooler and utilities, respectively.

The equation for the annual savings R_G is

$$R_G = H\,(C_h\,x + \Delta E)$$

where C_h = cost of heat, $/Btu.; x = heat saved, Btu./hr; ΔE = reduction in other operating expenses (such as cooling-water pumping costs), and H = hours per year.

Substituting in Eq. (6),

$$P - \bar{P} = P_p + P_i - \Delta P_h - \Delta P_c - \Delta P_u - \bar{j}\,R_G \qquad (8)$$
$$\text{or } P - \bar{P} = P_p + P_i - \Delta P_u - \Delta P_c - \Delta P_i - \bar{j}\,H\,(C_n\,x + \Delta E)$$

Using these equations, the graphical method is usually the only practical way to find the minimum for $P - \bar{P}$.

Graphical-Method Example

A process stream containing 1% moisture is to be heated from 120 F., dried in a vacuum dryer to less than 0.005% moisture, then cooled from 300 to 120 F. For a 25,000-lb./hr. bone-dry fluid flow-rate and 7,800 hr./yr. of operation, determine if a heat interchanger is justified, and if so its optimum size.

The specific heat of the dry fluid is 0.5. Steam is available at 150 psig. at the boiler house at a cost of 50¢/1,000 lb., and tower water-temperature is 85 F. The depreciation rate is 5% per year; maintenance, insurance and property taxes average 4% per year; and income tax is 50%. Changes in minor operating expenses, such as cooling water costs may be ignored. Boiler and cooling tower already exist but policy allows a six-year payout (longer than normal) for installations that would preserve existing boiler and tower capacity.

Preliminary studies on this problem gave a coefficient of 45 Btu./(hr.)(sq.ft.)(°F.) for the interchanger, 120 for the heater, and 85 for the cooler. A 200-sq.ft. fixed tubesheet all-carbon-steel interchanger would cost $1,250, a 187-sq.ft. all-steel heater would cost $1,400, and a 215-sq.ft. carbon-steel cooler with 304 S.S. tubes would cost $1,970. All exchangers should be sized with a 10% safety factor. Heat-exchanger costs may be assumed to vary with the six-tenths power of the size, installation costs may be taken at 20% of

Heat interchanger data for various values of x—Table I

x Btu./Hr.	T_2 °F.	t_2 °F.	AMTD °F.	A_i Sq. Ft.	P_i $	$-\bar{j}R_G$ $
1,250,000	200	218.0	81	375	2,960	−17,420
1,500,000	180	237.6	61.2	600	3,910	−20,900
1,750,000	160	257.3	41.3	1,038	5,440	−24,390
1,875,000	150	267.0	31.5	1,452	6,640	−26,130
2,000,000	140	276.9	21.5	2,274	8,680	−27,880
2,062,500	135	281.8	16.6	3,050	10,400	−28,740

Heater data for various values of x—Table II

x Btu./Hr.	q_h Btu./Hr.	Inlet Temp. °F.	LMTD °F.	A_h Sq. Ft.	P_h $	$-\Delta P_h = P_h - \$2,480$ (when $x=0$) $
0	2,520,000	120	107	216	2,480	0
1,250,000	1,270,000	218	76.5	152	1,990	− 490
1,500,000	1,020,000	237.6	69.6	134	1,850	− 630
1,750,000	770,000	257.3	62.4	113	1,670	− 810
1,875,000	645,000	267.0	58.9	100	1,570	− 910
2,000,000	520,000	276.9	55.0	87	1,430	−1,050
2,062,500	457,500	281.8	53.2	79	1,350	−1,130

the purchase price, and engineering and construction overhead at 35% of the installed cost. Interchanger piping will cost \$4,500, including engineering work and overhead charges.

Interchanger Calculations:

For an interchanger sized to cool the hot stream from 300 to 200 F.:

$$x = q_i = (25,000) (0.5) (300 - 200) = 1,250,000 \text{ Btu./hr.}$$

Heat capacity of cold stream =

$$(25,000) (0.5) + (250) (1.0) = 12,750 \text{ Btu./(hr.) (°F.)}$$

Outlet temperature of cold stream,

$$t_2 = 120 + (1,250,000/12,750) = 218 \text{ F.}$$
$$\text{AMTD (counterflow)} = [(300 - 218) + (200 - 120)]/2 = 81 \text{ F.}$$
$$A_i = (1,250,000) (1.10)/(81) (45) = 375 \text{ sq. ft.}$$
$$P_i = (375/200)^{0.6} (1,250 \times 1.20 \times 1.35) = \$2,960$$
$$\bar{j} = (1 - 0.5)/[1/6 + (1 - 0.50) (0.04) - (0.50) (0.05)] = 3.093$$
$$\bar{j}R_G = (3.093) (1,250,000) (7,800) (0.50)/ (866.1 \text{ Btu./lb.}) (1,000) = \$17,420$$

(See Table I for other calculated values of $\bar{j}R_G$)

Heater Calculations:

Assume that steam is 355 F. (approximately 145 psia.) at the point of use. To allow for flash cooling in the dryer, based on 100% water removal:

Heater stream outlet temperature =

$$300 + [250 (1,194.8 - 269.5)]/[(25,000) (0.5) + (250) (1)] = 318.1 \text{ F.}$$

Total heat requirement = (12,750) (318.1 − 120) = 2,520,000 Btu./hr.

For $x = 1,250,000$, $q_h = 2,520,000 - 1,250,000 = 1,270,000$ Btu./hr.

Inlet temperature of stream (from interchanger calculation) = 218 F.

$$\text{LMTD} = (318.1 - 218)/\ln [(355 - 218)/(355 - 318.1)] = 76.5 \text{ F.}$$
$$A_h = (1,270,000) (1.10)/(76.5) (120) = 152 \text{ sq. ft.}$$
$$P_h = (152/187)^{0.6} (1,400 \times 1.20 \times 1.35) = \$1,990$$

(See Table II for other calculated values of P_h)

Cooler Calculations:

Total cooling requirement = (12,500) (300 − 120) = 2,250,000 Btu./hr.

For $x = 1,250,000$, $q_c = 2,250,000 - 1,250,000 = 1,000,000$ Btu./hr.

Stream inlet temperature (as it leaves the interchanger) = 200 F.

For a 20 F. rise in cooling-water temperature:

LMTD (counterflow) =
$$[(200 - 105) - (120 - 85)]/\ln [(200 - 105)/(120 - 85)] = 60.2 \text{ F.}$$
$$A_c = (1,000,000) (1.10)/(60.2 \times 85) = 215 \text{ sq. ft.}$$
$$P_c = (215/215)^{0.6} (1,970 \times 1.20 \times 1.35) = \$3,190$$

(See Table III for other calculated values of P_c)

Value of $P - \bar{P}$ for various values of x (see plotted figure)—Table IV

x	$P - \bar{P}$
0	0
1,250,000	−11,260
1,500,000	−14,220
1,750,000	−16,710
1,875,000	−17,640
2,000,000	−17,820
2,062,500	−17,270

Cooler data for various values of x—Table III

x Btu./Hr.	q_c Btu./Hr.	Inlet °F.	LMTD °F.	A_c Sq. Ft.	P_c \$	$-\Delta P_c = P_c - \$4,000$ (when $x = 0$) \$
0	2,250,000	300	93.2	313	4,000	0
1,250,000	1,000,000	200	60.2	215	3,190	− 810
1,500,000	750,000	180	52.6	184	2,900	−1,100
1,750,000	500,000	160	44.3	146	2,550	−1,450
1,875,000	375,000	150	39.9	122	2,260	−1,740
2,000,000	250,000	140	35.0	93	1,930	−2,070
2,062,500	187,500	135	32.5	75	1,700	−2,300

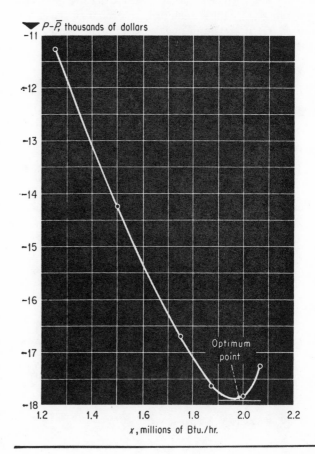

$P - \bar{P}$, thousands of dollars

x, millions of Btu./hr.

Calculation of $P - \overline{P}$:

From Eq. (8), $P - \overline{P} = P_p + P_i - \Delta P_h - \Delta P_c - \overline{j}R_G$

At $x = 1{,}250{,}000$,

$P - \overline{P} = 4{,}500 + 2{,}960 - 490 - 810 - 17{,}420 = -\$11{,}260$

As $x \to 0$, $P - \overline{P} \to \$4{,}500$, therefore $P - \overline{P} = 0$ is a discontinuous point. To determine the optimum, plot $P - \overline{P}$ vs. x as shown in the accompanying graph (see Table IV for other calculated values of $P - \overline{P}$).

Conclusions:

1. At the optimum, $P - \overline{P}$ is negative, therefore heat recovery is justified.

2. From the graph, it is seen that the optimum occurs at about 1,960,000 Btu./hr. heat recovery.

Differential-Method Example

Assume for the conditions given in the foregoing example that the process is an existing one where a heater and cooler are already installed and in operation. Using an allowable payout period of three years —and the analytical method—determine the optimum interchanger size and the actual payout.

Calculations:

For this problem, Eq. (8) is reduced to $P - \overline{P} = P_p + P_i - \overline{j}R_G$, and according to Eq. (4)

$\overline{j} = (1 - 0.50)/[1/3 + (1 - 0.50)(0.04) - (0.50)(0.05)] = 1.524$

Here it is convenient to change the variable from x (the heat recovered) to T (the outlet temperature of the fluid being cooled). Also, to keep the mathematics from being too cumbersome, the interchanger cost will be used on a constant dollar per square foot basis. Thus:

$$x = (25{,}000 \text{ lb./hr.})(0.5 \text{ specific heat})(300 - T)$$
$$= 3{,}750{,}000 - 12{,}500\,T$$

For the outlet temperature (t_2) of the fluid being cooled in the interchanger,

$t_2 = 120 + [(3{,}750{,}000 - 12{,}500\,T)/12{,}750]$
$\quad = 414.12 - 0.98\,T$
AMTD $= [(300 - 414.12 + 0.98\,T) + (T - 120)]/2$
$\quad = 0.99\,T - 117.06$
$A_i = (3{,}750{,}000 - 12{,}500\,T)(1.10)/(0.99\,T - 117.06)$ (45)
$\quad = (92{,}590 - 308.64\,T)/(T - 118.24)$
$P_i = C_i A_i = C_i(92{,}590 - 308.64\,T)/(T - 118.24)$

For an interchanger cost of C_i dollars per square foot:

$\overline{j}R_G = (1.524)(3{,}750{,}000 - 12{,}500\,T)(7{,}800)(0.50)/$
$\quad\quad (866.1 \text{ Btu./lb.})(1{,}000) = 25{,}734 - 85.78\,T$
$P - \overline{P} = 4{,}500 + [C_i(92{,}590 - 308.64\,T)/(T - 118.24)] -$
$\quad\quad (25{,}734 - 85.78\,T)$

$\dfrac{d}{dT}(P - \overline{P}) =$

$0 + C_i\left[\dfrac{(T - 118.24)(-308.64) - (92{,}590 - 308.64\,T)}{(T - 118.24)^2}\right] + 85.78$

$\quad = C_i[(-55{,}978)/(T - 118.24)^2] + 85.78$

For $\dfrac{d}{dT}(P - \overline{P}) = 0$:

$(T - 118.24)^2 - 652.6\,C_i = 0$
$\quad\quad T = 118.24 + (652.6\,C_i)^{1/2}$

For a first cut, assume the interchanger to have 2,000 sq. ft., then:

$C_i = (1/2{,}000)(2{,}000/200)^{0.6}(1{,}250 \times 1.20 \times 1.35)$
$\quad = \$4.04/\text{sq. ft.}$
$T = 118.24 + (652.6 \times 4.04)^{1/2} = 169.59$
$A_i = [92{,}590 - (308.64 \times 169.59)]/(169.59 - 118.24)$
$\quad = 976 \text{ sq. ft.}$

For a second cut, assume the interchanger to have 976 sq. ft., then:

$C_i = (1/976)(976/200)^{0.6}(1{,}250 \times 1.20 \times 1.35)$
$\quad = \$5.36/\text{sq. ft.}$
$T = 118.24 + (652.6 \times 5.36)^{1/2} = 178 \text{ F.}$
$A_i = [92{,}590 - (308.64 \times 178)]/(178 - 118.24)$
$\quad = 805 \text{ sq. ft.}$
$P_i = (805)(5.36) = \$4{,}320$

Gross savings:

$R_G = (12{,}500)(300 - 178)(0.50)(7{,}800)/$
$\quad\quad (886.1 \text{ Btu./lb.})(1{,}000) = \$6{,}850/\text{yr.}$

Expense term:

$E = (0.04)(4{,}320 + 4{,}500) = \$353$

Depreciation:

$D = (0.05)(4{,}320 + 4{,}500) = \$441$

Applying Eq. (3):

$n = (4{,}320 + 4{,}500)/[(6{,}850 - 353)(0.50) + (0.50)(441)]$
$\quad = 2.54 \text{ yr.}$

Conclusions:

The interchanger is justified; the payout is 2.54 yr. Note: For more accurate results, an exchanger can be designed and priced for $T = 178$ F., and the above procedure repeated for the new coefficient and cost.

References

1. Hirschmann, W. R., Brauweiler, J. R., Continuous Discounting for Realistic Investment Analysis, *Chem. Eng.*, pp. 210-214, July 19, 1965, and pp. 132-137, July 26, 1965.
2. Osburn, J. O., Kammermeyer, K., "Money and the Chemical Engineer," Prentice-Hall, Englewood Cliffs, N. J., 1958.
3. Schweyer, H. E., "Process Engineering Economics," McGraw-Hill, N. Y., 1955.
4. Perry, R. H., others, eds., "Chemical Engineers' Handbook," McGraw-Hill, New York, 1963, p. 26-5.
5. Quigley, H. A., Weaver, J. B., Optimum Equipment Sizing, *Ind. Eng. Chem.*, **53**, p. 55A, Sept. 1961.
6. Williams, R., Jr., "Six-Tenths Factor" Aids in Approximating Costs, *Chem. Eng.*, Dec. 1947, p 124.

Meet the Author

David Stuhlbarg is senior project engineer with The Fluor Corp., Ltd. of Los Angeles. He was formerly senior process engineer for The Ralph M. Parsons Co. and, prior to this, project engineer and project manager for Processes Research, Inc., of Cincinnati. He holds a Ch.E. degree from the University of Cincinnati and has done graduate work in both chemical engineering and mathematics. He is a member of AIChE and has several published articles to his credit.

Key Concepts for This Article

Active (8)	Passive (9)	Purpose of (4)
Justification*	Equipment*	Return on
Optimization*	Processes*	investment*

(Words in bold are role indicators: numbers correspond to EJC-AIChE information retrieval system. Asterisks mark key concepts suggested for indexing. Others are added to improve reading as an abstract. Indexing is described in *Chem. Eng.*, Oct. 11, 1965, p. 187.)

Use Capital Ratio

Invested dollars
Sales dollars

- To keep investment/sales in line
- Avoid over engineering
- Catch errors in estimates

LAWRENCE LYNN and R. F. HOWLAND, Engineering Research and Commercial Development Laboratory, General Foods Corp., Tarrytown, N. Y.

How can we best distribute our capital among our various activities? As companies grow bigger to remain competitive, as the minimum economical size of company operations increases, this question in management minds becomes more and more pressing. There is a need for tools that will help answer the question if answers are not to be based on hunches or guesstimates.

Capital ratio is one such tool for rapidly sizing up or screening investment alternatives. Though it can be helpful in assisting management to make order-of-magnitude appraisals, it is not used as widely or as often as it might be. This neglect stems in part from a lack of data and a lack of awareness concerning its usefulness as a screening tool, and in part from inherent limitations in its reliability.

Capital ratio can be defined most simply as the ratio of fixed investment capital to annual sales volume for a given company. The inverse of capital ratio is the ratio of sales value to unit investment. This ratio, usually called turnover ratio, is sometimes used as a measure of the rate of sales generation by capital.

With accurately known capital ratios for various industrial activities, you have a basis for making quick order-of-magnitude capital cost estimates. Ideally, you can approximate the capital investment required for a venture by projecting sales prices and volume—the sales rate—and then multiply the annual sales rate by the capital ratio applicable to the industry involved. You can also use known capital ratios for quick checks on more careful estimates of either capital or volume. We will discuss these and other applications for capital ratio.

The thesis that capital requirements for process industry ventures can be related to the annual value of products produced is not new. Over a generation ago, Tyler[1] mentioned that the average investment in chemical plant facilities was about $1 per dollar of sales, with a spread from about $0.75 to $1.25 per dollar of sales. More recently, Nichols[2] and Schweyer[3] suggested that, if the desired annual capacity and market value were known, one could approximate the necessary capital value. Kiddoo[4] pointed out that the nature of the product, the maturity and stability of the market and other factors affect the turnover ratio for various chemical investments.

In our study[5] for the period 1939 to 1952, we showed that the capital ratios for different process industries could be statistically correlated within reasonable confidence limits. However, the average capital ratio for this period—1.27 —included such diverse values as 2.51 ± 0.30 for petroleum and 0.48 ± 0.07 for soaps and detergents.

We have undertaken the present study to appraise capital ratio for essentially the same group of industries as we covered in the 1939-1952 study, but over a more recent time period. We wanted to compare recent capital ratios with previous ones to determine if and where changes might have occurred. We also wanted to learn whether or not any trends in capital or turnover ratio exist over recent years and, if so, whether the trends are uniform or non-uniform among the various process industries.

Definitions

We will define capital ratio, R, as the ratio of current or required real value, C, of total plant to the volume, S, of business which we might expect from the investment fixed in this plant.

Obviously, both the investment and the sales volume must be expressed in monetary units of the same year. Here, we consider capital as the estimated replacement value for a given year of the facilities normally included in fixed assets. In general, it is this investment which would have to be allocated and made in up-to-date dollars when considering a major expansion, entry in a new field, or acquisition versus its alternatives.

Thus capital, to a large extent, is made up of money invested in land, production facilities, warehousing and other manufacturing

Successive Studies Show Gradual Rise in Capital Ratio, Variance by Industry—Table I

	1939–1952 Period			1948–1958 Period		
	Average[2] capital ratio	S.D.[3]	Number[4] of Cases	Average capital ratio	S.D.	Number of Cases
Chemicals, general[1]	1.25	0.36	310	1.67	0.69	297
Carbon black	3.37	0.41	21	3.73	0.04	22
Explosives	1.20	0.17	20	1.33	0.06	22
Glass	0.97	0.29	40	1.09	0.36	55
Fibers, synthetic	1.79	0.30	51	2.40	0.68	44
Foodstuffs processed	0.63	0.39	72	0.61	0.34	88
Inorganics, heavy	2.06	0.79	129	2.04	0.84	143
Nonferrous metals	1.87	0.90	84	2.37	1.26	66
Petroleum	2.51	0.30	75	2.53	0.70	88
Pharmaceuticals	0.53	0.15	138	0.82	0.45	110
Pigments, paints, inks	0.70	0.31	55	0.81	0.33	55
Pulp and paper	1.74	0.34	69	1.77	0.39	77
Resins and plastics	0.77	0.29	61	1.55	0.92	55
Rubber	0.75	0.15	38	0.77	0.25	55
Soap and detergents	0.48	0.07	18	0.66	0.01	22
Steel	1.60	0.38	70	1.96	0.51	55
Sulfur	1.12	0.34	20	1.17	0.58	22
Average	1.37		1271	1.61		1276

components. However, we must also include in "plant" the research and development facilities needed to support the operation (even at maturity), the administrative facilities and elements of the distribution system including, in some cases, captive transportation facilities. Investments in these and other related nonmanufacturing components are all necessities for supporting sales positions in business today.

It is important to include these factors when considering a realistic picture of the total fixed assets in a given venture. This is true even though some of these items are not always included in many detailed engineering estimates, but persist rather malevolently in making their appearance via the overhead route.

By sales in capital ratio, we mean net sales volume per year. Net sales volume means the total annual sales less cash discounts, the value of returned goods and similar items, taken over the calendar or fiscal year depending on the practice of the particular company studied. Where possible, we obtained this information from annual reports or other published financial data.

Scope and Bases of Study

This study covers the period from 1939 to 1958. We have determined capital ratios for 17 process industries or industry groupings based upon the data published by 114 major companies during this period. The fact that large and relatively integrated companies were included in the survey may introduce a limitation on the validity of the results—especially if they are taken as an indicator for small projects or companies. However, we did not feel that determination of a size bias effect on the results was within the scope of this study.

We selected companies on several bases. These were:

• Accessibility of product and financial data.

• Size in an absolute sense and relative to its industry. In general, preference went to the largest and most representative companies.

In some process industries, such as rubber products, we found that a sampling of only five companies was enough for industry coverage and for a consistent capital ratio with reasonable standard deviations. In the case of general chemical manufacturing, on the other hand, we expanded the industry sample to 30 companies because of both the size and complexity of the industry and the wide diversity of product lines and technologies represented.

The listing on page 135 shows the companies included in the survey and the industry groupings. As you might expect, grouping of companies by process industry type often is not simple because a single company may participate in many widely diverse areas. For example,

1953–1958 Period			1958 [5]		
Average Capital ratio	S.D.	Number of Cases	Average capital ratio	S.D.	Number of Cases
1.76	0.71	162	2.02	0.68	27
3.76	0.30	12	3.98	0.45	2
1.42	0.01	12	1.64	0.02	2
1.17	0.40	30	1.46	0.57	5
2.84	0.90	24	3.44	1.08	4
0.63	0.32	48	0.66	0.26	8
2.04	0.90	78	2.24	0.75	13
2.57	1.38	36	3.31	1.68	6
2.71	0.76	48	3.08	0.91	8
0.91	0.58	60	0.92	0.58	10
0.88	0.32	30	1.04	0.41	5
1.81	0.38	42	2.01	0.56	7
1.74	1.14	30	1.90	1.01	5
0.82	0.27	30	1.04	0.41	5
0.72	0.01	12	0.69	0.09	2
2.08	0.56	30	2.78	0.73	5
1.35	0.59	12	1.97	0.65	2
1.72		696	2.01		116

References to Table I

1. See list for companies included in each grouping.
2. Ratio of estimated replacement value at the given year for all facilities normally included in fixed assets to net sales for the same year.
3. Standard deviation, during period stated, for capital ratios.
4. Company-year cases used to compute average capital ratio for period stated.
5. Calendar or fiscal depending on company usage.

ference between reported fixed assets for the given year and the prior year. New funds so invested were not corrected by index since the value of the dollar in the year these funds were invested should be equal to the dollar expressed in net sales.

Thus the calculation followed the general equation below:

$$R_n = (C/S)_n = \frac{(C_{n-1})\, i + \Delta C}{S_n}$$

Where C = fixed assets, \$.
S = net sales, \$.
R = capital ratio
n = calendar (or fiscal) year for R.
$n-1$ = previous calendar (or fiscal) year.
i = construction cost index correction from year $n-1$ to year n.
ΔC = change in C from year $n-1$ to year n.

Replacement value of facilities in operation in 1939, built or acquired prior to 1939 but expressed in 1939 dollars, was estimated from general information on the type and age of these assets in 1939 and from the *ENR* indexes prior to 1939. Since precise information from years before 1939 was often not available, this approach introduces an indeterminate error. Where a company had a large percentage of its present assets in operation before 1939, the error will be greatest, but if the company expanded greatly during the 1940's and 1950's, this initial error in the 1939 replacement value is not significant. The error is particularly small in R_n values for the later years.

We considered the value of R_n for any company during any year over the period chosen as one company-year case. For each year, we averaged R_n for all cases in each industry category to obtain R_n for the industry in that year. Since in our previous study[5] we statistically checked compilations of both total assets to sales and fixed assets to sales, we included all years and all cases in this study. Table I gives

Abbot Laboratories is easily classified in "pharmaceuticals." Similarly, despite some diversification, General Foods is as readily classified under "foods, processed." But with Celanese, which produces synthetic fibers (⅔ of sales), a wide line of organic chemicals (⅙ of sales) and plastics (the remaining ⅙ of sales), classification is difficult. We considered each company individually and classified it into one or more categories. These classifications were based upon either an analysis of assets and sales, when the composition of these factors was available, or on the character of its major product lines when insufficient data for analyses were available.

Primary data sources included standard financial references,[6–9] annual industry summaries,[10–12] and the *Engineering News Record*[13]. We supplemented these sources with information from the annual reports of the individual companies.

With the index corrections of the *Engineering News-Record* construction cost index, we corrected the total fixed assets for a given company to the undepreciated replacement values for each year during the period considered. Money invested in new fixed assets during any given year, where not reported per se, was determined by the dif-

Relative Trends In Capital Ratios, 1948–1958   **Table II**

	Relative Slope, Capital Ratio vs. Time
Foodstuffs, processed	1.0
Rubber	1.4
Pulp and paper	1.5
Pharmaceuticals	1.8
Soap and detergents	2.1
Carbon Black	2.2
Glass	2.4
Pigments, paints and inks	2.5
General chemicals	2.6
Explosives	3.6
Steel	4.3
Inorganics, heavy	4.5
Nonferrous metals	4.6
Petroleum	5.0
Sulfur	5.0
Resins and plastics	5.4
Fibers, synthetic	10.0

the resultant values of R by industry for the time periods chosen.

R Varies by Industry

As you can see from Table I, the rule of a dollar of capital per dollar of sales no longer holds—at least, not for the chemical process industries. Table I also shows that there is no good average value of R for the process industries. This is espe-

cially true for recent periods. For example, for the 1953-58 period, there are values as low as 0.6 for processed food products and up to 2.8 in synthetic fibers and 3.8 in carbon black.

Though we developed no correlation, and doubt that any really rigorous correlation can be developed, it is evident that a number of factors indicate whether R will be high or low for an industry. Gen-

erally, R will be high, more than 2.0 for 1958, in the following cases:

• Very large volume operations, producing low unit sales value items, are typical to the industry —there are few small operators. These are usually the industries that are the most difficult for the nonparticipant to enter. Examples might include aluminum, steel, petroleum and newsprint pulps.

• Operations are technologically complex. This complexity extends to the distribution pattern for sales as well as in raw materials handling and production. There is heavy dependence upon aggressive, broad research and development programs aiming at new processes and products, but successful investments of effort lead to handsome payoffs. Novelty and rapid obsolescence create an urgent need for flexibility in facilities. Examples include plastics, heavy organic chemicals, synthetic fibers and petroleum.

• Products produced by high R value operations are workhorse items furthest removed from the consumer. Many undergo much further processing, such as heavy inorganic chemicals, nonferrous metals, and carbon black.

• There tends to be a high degree of vertical integration among successful participants which extends to most of the significant raw materials. Again we might cite the synthetic fiber, petroleum, copper and aluminum industries.

Aside from the inverse of the above characteristics, there are several other factors that tend to exist where R values are low—below 1.5 for 1958. These factors would include:

• Total size of the industry may be large, but there are many companies engaged in it, some of them small regional businesses. These industries have the most diverse product lines and contain many successful small companies as well as large enterprises. Profit margin on the sales dollar is frequently small. Industry domination is rare or impossible because of the scatter of participants in size, geography, and product line. Examples might include food products, pharmaceuticals, pigments, paints and inks.

• Relatively speaking, the operations are not technologically complex. Even where scientific complexity exists, processing requires comparatively simple equip-

Increase in R and Industry Variations Easily Seen in Graphs

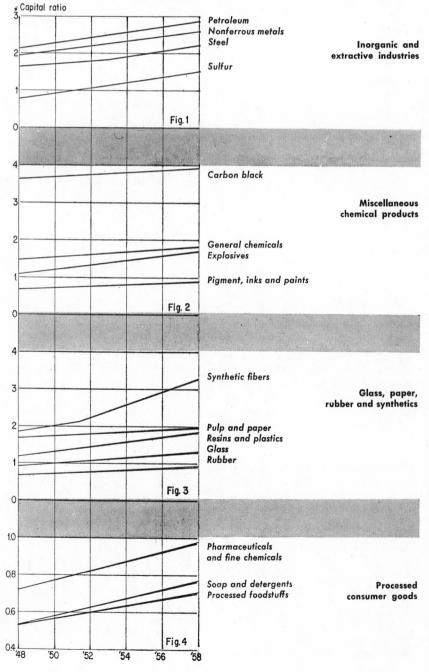

ment, such as kettles, or transformations are by physical rather than chemical means. Here we might include glass products, pharmaceuticals and rubber products.

• Operations produce relatively low volume specialty products, frequently with high unit sales value, such as inks, pharmaceuticals and fine chemicals.

• Products produced in these operations are relatively close to the consumer. These would include soaps, detergents and food products.

What Are the Trends?

Besides identifying recent capital ratios for various industries, we wanted to find out if there was a trend in capital ratio over a period of time. Assuming such a trend to exist, we were interested in learning whether this trend was relatively uniform or divergent from industry to industry.

We charted average R_n values by year for each of the industries or groupings shown in Table I. Figs. 1 through 4 show the curves for the period 1948 through 1958. The R_n vs. time curves in this period are almost straight line extrapolations of the earlier 1939-1948 curves with only rather small divergence during the World War II years.

You can see from these charts that there is a trend for R values with time and that it is a persistently upward trend. The trend is up, we believe, because of inflation in the cost of facilities rather than in the products they produce. We base our belief on the nature of the increase and its variation from industry to industry. This phenomenon is closely related to the profit squeeze.

Another factor that increases R values is the tendency toward higher quality, including aesthetic considerations, in new facilities. Landscaping, cafeterias, complex waste disposal systems and more costly materials of construction all contribute to this trend. Architecture is an important first cost.

These considerations, however, do not explain the widely variant rates at which R has been increasing in different industries. In some instances, it has been a sharp and steady increase, even an accelerating increase in a few industries, while in others the increase has been small to almost indiscernable.

Survey Covered 132 Companies in 17 Different Industry Groupings

Chemicals, general
Allied Chemical and Dye
American Cyanamid
American Potash
Celanese Corp. of America
Commercial Solvents
Davison Chemical*
Diamond Alkali
Dow Chemical
E. I. du Pont de Nemours
Eastman Kodak
Food Machinery and Chemical
General Aniline and Film
Hercules Powder
Heyden-Newport
International Minerals and Chemicals
Koppers
Olin-Mathieson Chemical
Merck
Minnesota Mining and Manufacturing
Monsanto Chemical
Pennsalt Chemical
Pittsburgh Plate Glass
Pfizer, Chas. and Co.
Pittsburgh Coke and Chemical
Publicker Industries*
Rohm and Haas
Tennessee Products and Chemical*
Union Carbide and Carbon
Victor Chemical
Virginia-Carolina Chemical
Carbon black
Columbian Carbon
United Carbon
Explosives
Atlas Powder
Hercules Powder
Fibers, synthetic
American Viscose
Celanese Corp. of America
E. I. du Pont de Nemours
Industrial Rayon
North American Rayon*
Foodstuffs, processed
Borden
Corn Products Refining
General Foods
General Mills
Penick and Ford
Pillsbury Mills
Quaker Oats
Standard Brands
Glass
Anchor-Hocking
Corning Glass Works
Libby-Owen-Ford Glass
Owens-Illinois Glass
Pittsburgh Plate Glass
Inorganics, heavy
Air Reduction
Allied Chemical and Dye
American Agricultural Chemical
American Potash and Chemical
Davison Chemical*
Dow Chemical
Diamond Alkali
E. I. du Pont de Nemours
Hooker Electrochemical
International Salt
Olin Mathieson Chemical
Lindsay Chemical*
National Lead
Pennsalt Chemical
United States Potash*
Union Carbide and Carbon

* Included only through 1952

Nonferrous metals
Aluminum Co. of America
Aluminum Ltd.
American Smelting anl Refining
Anaconda
Kaiser Aluminum and Chemical
Kennecott Copper
Reynolds Metal
Petroleum
Atlantic Refining
Cities Service
Phillips Petroleum
Shell Oil
Sinclair Oil
Socony Mobil Oil
Standard Oil (Indiana)
Standard Oil (New Jersey)
Texaco
Pharmaceuticals and related products
Abbott Laboratories
American Cyanamid
American Home Products
Bristol-Myers
Heyden-Newport Chemical
Mead Johnson*
Parke, Davis
Pfizer, Chas. and Co.
Sharpe and Dohme*
Squibb, ER and Sons*
Sterling Drug
Vick Chemical
Pigments, inks and paints
Eagle-Picher
Glidden
Interchemical
National Lead
Sun Chemical
St. Joseph Lead*
Sherwin-Williams*
Pulp and paper
Champion Paper and Fibre
Crown-Zellerbach
Kimberly-Clark
International Paper
Marathon*
Rayonier
St. Regis Paper
West Virginia Pulp and Paper
Resins and plastics, synthetic
Catalin Corp. of America
Celanese Corp. of America
Dow Chemical
E. I. du Pont de Nemours
Durez Plastics and Chemicals*
Rohm and Haas
Rubber
Firestone Tire and Rubber
Goodrich, B. F.
Goodyear Tire and Rubber
General Tire and Rubber
United States Rubber
Soap and detergents
Colgate-Palmolive
Procter and Gamble
Steel
Bethlehem Steel
Crucible Steel
Inland Steel
Republic Steel
United States Steel
Sulfur
Duval Sulfur*
Freeport Sulphur
Texas Gulf Sulphur

We assumed straight line behavior for R_n vs. time and tabulated the slopes of these industry curves for easy comparison. Table II, on page 133, shows these slopes for the 1948-1958 period on an arbitrary 1 to 10 scale where 10 represents the sharpest rate of increase in R_n.

Table II shows that, with some exceptions, those industries with the highest initial R values have the highest slopes. It appears that the more complex the industry, the more rapid the pace with which this industry is becoming even more

complex because of higher capitalization per worker, more technologically complex processing, higher volume requirements, and so forth. Difficulty of entry probably parallels the R trend.

Careful scrutiny of the variant R trends reveals a number of factors that separate those industries or operations with rapidly increasing R from those in which the rate of increase is much smaller. Generally, R increases rapidly where:

• Technology is developing rapidly with larger volume units,

lower unit costs and a major trend toward automation.

- Labor is either more strongly organized or labor union pressures are greatest.

- In special cases, recessionary tendencies during a large part of the period and/or import trade from low cost areas have restrained and may continue to restrain the price of goods.

As with the absolute value of R, there are recognizable factors common to industries where the rate of increase in R has been lowest:

- The industries, and the main products and operations in them, are relatively mature. The impact of technology either has not been major, cannot be dominant, or has not started to strongly assert itself for most of the industry's participants.

- The industry is very large and composed of many small as well as large companies. The product line is comparatively stable. Ease of entry, at least regionally, with small capital permits sharp price competition.

- The product line is close to

R. F. Howland   **L. Lynn**

ROBERT F. HOWLAND, B.S. in Chemical Engineering, Brooklyn Polytechnic, 1954; M. Ch. E., N. Y. U., 1958. He has been with General Foods Research Center since 1954, spending most of his time on research economics. Mr. Howland is now project leader in a group in the Engineering and Commercial Development Laboratory concerned with economic analysis.

LAWRENCE LYNN, (B.S.) ('48) and M.S. ('49) in Chemical Engineering, Columbia. After various assignments with Celanese Chemicals, he was briefly Group Leader, Exploratory Economics Group at Larkwood, Tex. Later, he headed Celanese Plastic's Plant Engineering Section and was Manager of the Economic Planning Dept. for Celanese Plastics Division. In 1955, Mr. Lynn joined General Foods Engineering and Commercial Development Laboratory as Ass't. Director and is now Laboratory Director.

the consumer. This factor exerts strong pressures on product price.

How to Use Capital Ratio

Assuming its limitations are not exceeded, capital ratio presents management with an effective tool for first or rough-cut screening of proposed ventures. Whether the problem at hand is a possible internal expansion, a venture into a related or new field or an expansion by acquisition, questions of capital needs and factory door cost arise early. Known R values can assist in establishing order-of-magnitude estimates for each situation.

Some Examples

Assume you want to expand your operations, integrating vertically backwards toward several large volume inorganic chemical raw materials, A, B and C. Knowing the volumes required, V_A, V_B and V_C, and the price of these raw materials, P_A, P_B and P_C, we may derive the sales rate. With the capital ratio, R, for inorganic heavy chemicals, we can approximate the fixed capital investment needed thus:
$C_{ABC} = R_{ABC} (P_A V_A + P_B V_B + P_C V_C)$.

If we know how these raw materials are derived, the price of the basic materials needed and the yields involved, we know the materials cost of A, B and C. But to know the factory door cost of A, B and C, we must know the cost of labor, maintenance, depreciation, utilities and services, taxes and insurance and plant general overhead.

If we know that the raw material cost is high, we can quickly estimate the unit costs for all fixed charges. We can estimate a rough factory door price for A, B, and C from our approximation of C_{ABC} above and from a general acquaintance with depreciation rates, maintenance costs, labor costs and so forth in the industry as a percentage of plant cost.

Suppose a detailed process engineering estimate or perhaps a final engineering capital request has been prepared for a proposal undergoing consideration. We can rapidly check this detailed estimate for reasonableness by comparison with competition as measured by the R value for this type of industry. We may now ask selected questions based on this check. Is the estimate reasonable or has the installation

been over-engineered? Does the estimate appear too low because of the ommission of some important element not within the scope of the engineering assignment? Is the sales volume target too conservative for a reasonable production cost, considering the high capital per unit sales required?

Watch the Limitations

In using capital ratio, you should observe a few precautionary points.

1. Capital ratio is not valid for estimates of new plant investment in modifications to existing facilities, or in improvements or small additions to existing plants.

2. If old facilities, salvage equipment or rented facilities form a major part of the intended venture, capital ratio should not be used to approximate new capital needed.

3. You must consider not only the type of company or industry but also the type of operation which may or may not be typical for the given industry. A food processor building a new gelatin operation would be constructing a general chemical type plant rather than a processed foods facility, hence using the R for processed foods would be misleading.

Finally, estimates of capital requirements or factory door costs derived from capital ratio are inherently only approximations or checks for validity. They do not take the place of the more detailed and carefully prepared estimates that are the necessary bases for final decisions by management.

REFERENCES

1. Tyler, C., "Chemical Engineering Economics," 2nd Ed., p. 51, McGraw Hill, N. Y., 1938.
2. Nichols, W. T., *Chem. Eng.*, June 1951, pp. 248-50.
3. Schweyer, H. E., *Chem. Eng.*, Jan. 1952, p. 164.
4. Kiddoo, G., *Chem. Eng.*, Oct. 1951, p. 145.
5. Lynn, L., *Chem. Eng.*, Apr. 1954, pp. 175-6.
6. Moody's Industrial Manual. American and Foreign. 1954-1959. Moody's Investors Service, New York.
7. Moody's Manual of Investment. American and Foreign. Industrial Securities. 1946-1953, Moody's Investors Service, New York.
8. Standard and Poor's Corporation Records, 1939-1952, Standard and Poor's Corp., N. Y.
9. Monthly Stock Digests (various in 1959), Data Digests, Inc., N. Y.
10. Anon., *Ind. Eng. Chem.*, 42, 948-73 (1950).
11. Anon., *Ind. Eng. Chem.*, 44, 1209-22 (1952).
12. Anon. *Ind. Eng. Chem.*, 46, 1094-1108 (1954).
13. *Engineering News-Record*, "Construction Cost Trends," Reprint from 20th annual report on construction costs (10/17/57).

For economic selection based on project rate of return . . .

Use Discounted Cash-Flow Method

$$V^{n} = \frac{1}{(1+i)^{n}}$$

Where alternative uses of funds are available, financial yardsticks provide a concise way to summarize project income to investment cost.

HERBERT E. KROEGER, Socony Mobil Oil Co., Inc., New York, N. Y.

Many companies have adopted payout period and the discounted cash flow rate of return as financial tools in the economic evaluation of proposed investment projects. It is the purpose of this article to explain these yardsticks, to show why they are needed, what they mean and why they are being adopted widely by many industries.

First, we shall discuss why financial yardsticks are needed and what they are expected to do. To remain in business and to meet the ever-increasing demand for petroleum products, it is necessary for large petroleum companies to spend huge sums of money for capital expenditures and exploration expenses. These sums are frequently in excess of the net income earned during similar periods.

Future earning power of large petroleum companies will depend largely upon channeling such sums to projects which will maintain and improve their competitive ability and will generate the highest earnings or largest savings commensurate with prudent management.

The effects of selecting a project which is not consistent with these aims cannot be overcome easily once the funds have been committed. Consequently, such projects often have an adverse effect on profits for many years.

An important method for measuring the relative profitability of alternative investment opportunities is the discounted cash flow rate of return. By using this method, available funds can be apportioned to the most profitable projects in a better manner than is possible with the more conventional methods.

Of course, we recognize that the actual calculation of the payout and rate of return is only one step in the preparation and evaluation of investment proposals. At this point let us review the steps usually involved in the over-all evaluation of an investment.

1. Basic factors related to the specific investment are developed. This involves determination of factors such as estimated sales volumes, prices, material costs, operating expenses, capital investment, strength and nature of competition, rates of depreciation, obsolescence or depletion and other related economic factors.

2. Arrange in convenient form for evaluation the basic estimates of required investment, project life and annual income. This step enables us to perform the mechanics involved in calculating payout and rate of return.

3. Managerial judgment must be exercised to determine whether or not: estimated expected return is adequate to justify the risks; a particular investment opportunity is attractive compared to alternative investment opportunities; and timing of the investment is right relative to anticipated developments in the near future.

In this article, we are primarily concerned with the second step. We wish to emphasize that there is nothing in these mechanics which will make it simpler to prepare the basic estimates or which will improve the accuracy of these estimates. Furthermore, there is nothing in these mechanics which will relieve management at any level of the responsibility for exercising judgment with respect to the matters outlined in the third step.

A financial yardstick is merely a standardized, concise way of summarizing the information relating the expected project income to investment. Regardless of what financial yardsticks are used, the same basic studies such as engineering, design, market research and analysis of competition are required. However, these studies must be developed in terms of the expected cash flows in and out of the company. Once these cash flows are developed, it is a simple calculation to develop payout and discounted cash flow rate of return.

To compute the payout and discounted cash flow rate of return

Formula for present value v^{n} of one when interest i is compounded annually.

Payout Measures Liquidity of Proposed Investments—Table I

	Project 1	Project 2	Project 3
Cash investment	$40,000	$40,000	$40,000
Project life	10 yr.	10 yr.	10 yr.
Cash intakes over 4 yr.	$40,000	$40,000	$40,000
Total cash intake	$70,000	$100,000	$120,000
Average annual profit	$3,000	$6,000	$8,000
Financial yardsticks:			
Payout	4 yr.	4 yr.	4 yr.
Avg. earnings rate of return	7.5%	15%	20%

Payout Is Not Enough to Evaluate Projects—Table II

	Project 4	Project 5	Project 6
Life	5 yr.	15 yr.	30 yr.
Cash investment	$40,000	$40,000	$40,000
Total cash intakes	$60,800	$139,200	$256,800
Total profit	$20,800	$99,200	$216,800
Financial yardsticks:			
Payout: $\frac{\text{cash investment}}{\text{annual cash intake}}$	3.3 yr.	4.3 yr.	4.7 yr.
Avg. earnings rate of return	10.4%	16.5%	18.1%

Consider Time Factor in Proposed Investments—Table III

Year	Project A Cash Flow	Project B Cash Flow	Project C Cash Flow	Project D Cash Flow
0	($40,000)*	($40,000)	($40,000)	($40,000)
1	$16,927	$12,000	$5,927	....
2	14,535	9,000	6,535	....
3	13,142	10,000	7,142	$12,000
4	11,749	9,000	7,949	9,000
5	10,356	8,000	8,756	10,000
6	8,964	12,000	9,664	9,000
7	7,571	9,000	10,571	8,000
8	6,578	11,000	12,178	12,000
9	5,685	10,000	14,285	9,000
10	4,493	10,000	16,993	11,000
11	....	....	....	10,000
12	....	....	....	10,000
Total cash intake	100,000	100,000	100,000	100,000
Total profit	60,000	60,000	60,000	60,000
Avg. annual profit	6,000	6,000	6,000	6,000
Financial yardsticks:				
Payout	2.6 yr.	4 yr.	5.4 yr.	4 yr.
Avg. earnings rate of return	15%	15%	15%	15%
DCF rate of return	29.1%	21.7%	17.2%	15.5%

Project A: Decreasing cash flow
Project B: Variable cash flow
Project C: Increasing cash flow
Project D: Two-year construction period

* Cash investment shown as ($40,000)

yardsticks, we need estimates of:

1. Cash outlays for the project and the time periods when these outlays will be made.
2. Economic life of the project.
3. Cash intakes anticipated from the project and the time periods when these cash intakes will be received.

It cannot be overemphasized that the best available estimates must be used. We have pointed out that each individual annual amount affects both payout and rate of return. Consequently, the profitability indicated by the rate of return depends entirely upon the basic assumptions which make up each proposal.

Define Standards

We have mentioned that the rate of return yardstick provides a basis for arranging prospective investments in order of rate of profitability. If the company is engaged in a number of different operations and the prospective investments in these diverse operations involve varying degrees of risk, then different payout and rate of return standards should be used for each operation.

Payout should be the maximum length of time which management considers acceptable for the recovery of the initial investment for a particular type of investment.

Rate of return should be the minimum earnings rate considered acceptable by management for a particular type of investment.

How to Compute Payout

In addition to discounted cash flow rate of return, payout is also computed by many companies for prospective investment proposals. Since payout is a familiar yardstick, let's discuss its advantages and disadvantages first.

Payout measures the time required to recover the money estimated to be expended for the proposed investment. The payout period begins when the proposed project starts operating, and ends when the last dollar invested in the project has been recovered. In Table I, we find payout amounts for three projects.

Payout is computed by accumulating the annual cash intakes until the accumulation equals the amount

How to Compute Discounted Cash Flow Rate of Return for Project A—Table IV

Year	Cash Earnings Before Income Taxes	Depreciation Charges [2]	Taxable Income	Income Tax (54%)	Net Cash Flow [3]	Discount Factor at 29%	Present Value [4]	Discount Factor at 30%	Present Value [4]
0	($40,000) [1]	. . . . . .	. . . .	. . . .	($40,000)	1.0000	($40,000)	1.0000	($40,000)
1	$28,261	$7,272	$20,989	$11,334	$16,927	0.7752	$13,122	0.7692	$13,020
2	23,913	6,546	17,367	9,378	14,535	0.6009	8,734	0.5917	8,600
3	21,740	5,818	15,922	8,598	13,142	0.4658	6,122	0.4552	5,982
4	19,565	5,091	14,474	7,816	11,749	0.3611	4,243	0.3501	4,113
5	17,390	4,364	13,026	7,034	10,356	0.2799	2,899	0.2693	2,789
6	15,219	3,636	11,583	6,255	8,964	0.2170	1,945	0.2072	1,857
7	13,044	2,909	10,135	5,473	7,571	0.1682	1,273	0.1594	1,207
8	11,739	2,182	9,557	5,161	6,578	0.1304	858	0.1226	806
9	10,652	1,454	9,198	4,967	5,685	0.1011	575	0.0943	536
10	8,913	728	8,185	4,420	4,493	0.0784	352	0.0725	326
Totals	170,436	40,000	130,436	70,436	100,000	. . . . . .	40,123	. . . . . .	39,236

1. Investment ($40,000).
2. Sum of the years-digits depreciation method.
3. Net cash flow equals earnings before taxes less income tax.
4. Present value is product of net cash flow and appropriate discount factor.

spent for the project. Each project shows accumulated cash intakes at the end of four yr. of $40,000 which is equivalent to the amount invested. Therefore, the payout period for each of these projects is four yr.

From these examples, we note that payout measures the length of time required to recover the money spent for the project and does not evaluate the profitability of the project.

If payout alone is used as the measurement device in project evaluations, we may make poor investment decisions because the three projects have an identical payout period. However, since the total profits earned by Projects 1, 2 and 3 amount to $30,000, $60,000 and $80,000 respectively, our decision based upon profitability would call for investing in Project 3 before Project 2, and Project 2 before Project 1.

An even stronger reason why payout should never be used alone in project evaluations is shown in Table II. In this table, total profits rank Projects 4, 5 and 6 in exactly the reverse order from the ranking indicated when payout alone is used. The three projects have different economic lives and different average annual cash intakes but have identical investment costs.

The assumption in Table II that cash intakes are received in equal annual amounts is made only for the purpose of simplicity in presenting the data. As a practical matter the annual cash intakes never would be equal. Use of accelerated depreciation for tax purposes would cause these annual amounts to vary.

Note the average annual cash intakes of $12,160; $9,280 and $8,560 result in payouts of 3.3, 4.3 and 4.7 yr. On the basis of payout, these projects would be ranked in the order of Project 4, 5 and 6. When we examine the total profit figures, we find that Projects 4, 5 and 6 earn $20,800, $99,200 and $216,800 which result in average earnings rates of 10.4%, 16.5% and 18.1% respectively. Therefore, ranking of these projects based on rate of profitability is in exactly the reverse order of the ranking based on payout.

Consider Rate of Return

Projects shown in Tables I and II indicate that in addition to payout which measures the liquidity of proposed investments, another yardstick is needed which will measure the profitability of proposed investments. This yardstick is rate of return.

Rate of return is the percentage relationship between annual profits earned by a project and the capital invested in that project. The con-

ventional method of computing rate of return is to divide the average annual profit after depreciation and income taxes by the initial investment cost for the project. This method as illustrated in Table I shows that a series of investment proposals can be ranked according to profitability.

Average earnings rate of return has two significant shortcomings. First, it fails to account for the timing of the cash outlays and the cash intakes applicable to proposed investments. This shortcoming frequently results in ranking alternative proposals in an incorrect order of profitability and may lead to poor investment decisions.

Secondly, this rate of return fails to account for the capital amounts recovered over the project's economic life. Since average earnings rate is based on initial investment cost, this rate is unrealistically low.

Since money has earning power, a dollar received or spent today is worth more than a dollar received or spent at some time in the future. This is another way of saying that profitability is a matter not only of how many dollars are spent for a project as opposed to how many dollars are returned, but it is also a matter of when these dollars are spent and when they are returned. The rate of return that evaluates this time factor applicable to cash outlays and intakes in prospective

investments is the discounted cash flow rate of return.

Discounted Cash Flow

In Table III, we shall compare the average earnings rate with the discounted cash flow rate of return for each of four projects. This will give us some idea of the inadequacy of the average earnings rate for measuring the profitability of proposed investments.

In Table III, we illustrate four projects in which the over-all investment factors are identical—the $40,000 investment cost, the $100,000 total cash intakes and the 10-yr. economic life—but the only difference is in the timing of the cash intakes.

In Project A, money is returned each year in decreasing amounts; in Project B, money is returned in increasing and decreasing amounts; in Project C, money is returned in increasing amounts; and in Project D money does not start coming back until two years after the $40,000 has been invested and then in increasing and decreasing amounts.

Note that the average earnings rate of return, which does not recognize the different timing of the cash flows, is 15% for each project. However, the discounted cash flow rate of return, which takes into account the difference in value between near and distant dollars, is different for each project. It is highest for Project A in which money is returned fastest and lowest for Project D in which money is returned slowest.

Based on the rates calculated by the average earnings method, the four projects appear to be equally attractive. However, the analysis in terms of the discounted cash flow rate shows that the projects should be ranked in the following order: Projects A, B, C and D.

Now let us see what this discounted cash flow rate of return means. We will use the 29.1% rate computed for Project A as an illustration. Assumptions used in calculating this rate are:

• Estimated investment outlay for the project is $40,000.

• Estimated life of the project is 10 yr.

• Total estimated cash intakes of $100,000 will flow into the company in the annual amounts shown as net cash flow in Table IV.

To help us get a better understanding of what the discounted cash flow rate of return really means, let us think of the $40,000 investment as money that will be loaned to this project by the company.

At the beginning of the first year, the project will owe the company $40,000. Interest computed at 29.1% on this amount is $11,640. Since the project will return $16,927 to the company in the first year, this will leave $5,287 for amortization and reduce the investment balance outstanding at the beginning of the second year to $34,713. The interest computed at $29.1% on this amount is $10,101; and since the project will return $14,535 to the company in the second year, this will leave $4,434 for amortization in the second year and will reduce investment balance outstanding at the beginning of the third year to $30,279.

If these computations are followed through to the end of the tenth year, we find that the estimated annual cash intakes consist of interest computed at the constant interest rate of 29.1% on the unrecovered balance of the investment outstanding at the beginning of each year plus amortization amounts which will recover fully the estimated investment outlay over the estimated 10-yr. life of the project.

Having illustrated the significance of the discounted cash flow rate of return, let us look at Table IV and see how the 29.1% rate for Project A is calculated. Our first step is to determine the annual net cash flow as follows:

1. Calculate the annual depreciation charges based on the method used for income tax purposes. In this example, the sum of the years-digits method is used.

2. Deduct the annual depreciation charges from the cash earnings to get the taxable income.

3. Multiply the taxable income by the current income tax rate to find the annual income tax amounts.

4. Subtract the annual income tax amounts from the cash earnings to get the annual net cash flow.

Once we have determined the annual net cash flow, we are ready to compute the discounted cash flow rate of return. This is done by finding the rate whose discount factors will discount the annual net cash flow down to a total present value

that is equal to the $40,000 investment. The rate is found by trial and error.

However, a short-cut enables us to find the starting rate quickly. In this example, the starting rate is found by averaging the $66,709 net cash flows for the first 5 yr. of the project and then by dividing the average of $13,341 by the $40,000 cash outlay for the project. This gives us a 33% starting rate. When the $100,000 net cash flows anticipated over the 10-yr. life of the project are discounted at 33%, the total present value gives a figure which is smaller than the total outlay. Hence, the actual rate is lower than 33%.

When the annual cash intakes are discounted at 29% as shown in Table IV, the total present value is $40,123. On the other hand, when these annual cash intakes are discounted at 30%, their total present value is $39,236. By interpolating between these two rates, we determine that the rate of return for this project is 29.1%.

We have just seen that the discounted cash flow rate of return for a proposed investment is computed by discounting the stream of future cash intakes down to a total present value which is equal to the investment outlay for the project. We have noted also that the rate which will discount the future cash intakes down to this amount is the discounted cash flow rate of return for the proposed investment.

HERBERT E. KROEGER is financial analyst for the budgeting department of Socony Mobil Oil Co., Inc. His work requires the preparation and analysis of financial data for the management reports and studies group. Prior to his current assignment, Kroeger was methods analyst for the company. He has the B.S. degree in accounting from N.Y.U.

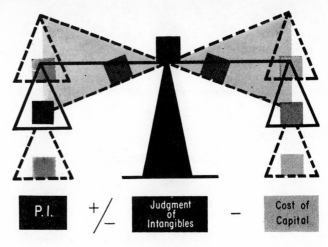

The Profitability Index is one of the most logical and useful tools to help you evaluate an investment. You may find there's more to the P. I. than you realize.

WILLIAM D. McEACHRON, *Standard Oil Co. (Indiana).*

Let's say you have three cost-reducing proposals in front of you. Each one involves considerable investment in new machinery—it is up to you to choose the most promising. If the costs and savings estimates have been made carefully, if you know not only how to calculate the Profitability Index but also its full significance, and if furthermore your judgment in assessing intangibles is good, you will probably make the right decision.

The Profitability Index (P. I.) goes under various names—other authors have called it Discounted Cash Flow Rate of Return; the Investor's, or the Interest, Rate of Return; and in one case, the True Rate of Return.

The P. I., you may recall, involves setting up a cash flow schedule on which the various outlays have minus signs and the cash returns (including salvage value) have plus signs. The discounting rate that will reduce this schedule to a present value of zero is the P. I.

Most other common investment yardsticks are easier to compute but have less meaning.

For instance, take payout times—very easily calculated by accumulating the annual cash returns until the accumulation matches the amount spent. However, payout times only tell you how soon the investment will be recovered, and this is of less importance than the extent of the anticipated profits.

Even relating the total savings to the investment to obtain the Average Earnings Rate of Return may be misleading, as the precise timing of costs and returns may be just as important as their magnitude.

For example, let's say that Project A brings in $10,000 a year for ten years while Project B, involving the same investment and risks, brings in $120,000 in the tenth year as its sole return. Although B would show a higher average rate of return, Project A may really be more attractive because the money starts coming in sooner.

The Profitability Index takes this time factor into account to arrive at a meaningful rate of return.

The one other general investment evaluation procedure that is based on the concept of the time value of money is the Present Value technique. Here, the discounting is done at a prescribed rate, usually the cost of capital, and the answer represents the present value in dollars of the total project cash flow. While

This article is based on a paper "The Profitability Index Technique" given by the author at the Chicago Chapter meeting of the American Assn. of Cost Engineers.

The Role of P.I. in Investment Evaluation

the calculations are thus simpler—no trial-and-error or interpolation is involved—the Present Value lacks the full power of the P. I. as an investment tool.

This power and usefulness derive from two distinct but related characteristics of the P. I. estimate. The first of these is that the P. I. is expressed in terms that have meaning in their own right and are consistent not only from project to project, but from project evaluation to corporate accounting as well. The second favorable characteristic is that the P. I. lends itself particularly well to the weighing of risks at the decision-making level.

The P.I. and Corporate Accounting

Although both measure business economics, P. I. and accounting represent two quite different cross sections of the total corporate enterprise.

This difference is illustrated in Fig. 1. There, the vertical dimension represents the total investment that goes to make up the company, separated into individual projects or investment decisions. The horizontal dimension represents a portion of the time spectrum, which in real life extends indefinitely into the future, as well as going back to the inception of the company.

Corporate accounting, as summarized in a company's annual report, represents an annual cross section of the total firm, and measures the current health of the enterprise. It is illustrated by the shaded vertical rectangle in Fig. 1. Since the time dimension is very short—only one year or less—the problem of time weighting is not directly involved.

The relationship between P.I. and corporate accounting—Fig. 1

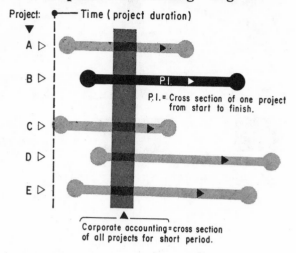

Project: ●—Time (project duration)

P.I. = Cross section of one project from start to finish.

Corporate accounting = cross section of all projects for short period.

In contrast, Profitability Index—or any investment evaluation for that matter—represents quite a different cross section. We have shown this by the black horizontal rectangle. Here, we are concerned with a single investment venture throughout its life, and the time value of money becomes of crucial importance.

These two different cross sections or viewpoints can be compared by a simple illustration, as shown in Table I. We will assume a single depreciable investment of $100,000, which after a one-year startup delay earns an after-tax operating profit (or cash income before depreciation) of $40,000 for four more years. After this, revenue ceases and the investment has no further value. Total net profit is $60,000, equal to total cash income less the investment.

To determine the actual return on investment, we would have to divide profit by investment on an annual basis. Although we know the total net profit over the five-year period, we don't really know the net annual profit, as we have not yet decided on a depreciation schedule. The investment to be used is even more obscure. Initially it was $100,000. At project termination, it was zero. Hence, we obviously can't use the same value all the way through.

We can solve both these problems by drawing up a year-by-year schedule for the life span of the project. To do this, we must decide how to handle depreciation —how to charge off revenue to recover the investment.

Under conventional accounting, the year-by-year depreciation charge is computed solely on the basis of time, without reference to income. Straight-line depreciation would yield the schedule shown in Table II. Note that after the loss in the first year, the annual rate of return rises with time, because a constant profit has been related to a declining investment.

Such a varying rate of return is of little value to the decision-maker who must match the value of the project *in its entirety* against the cost of capital. The average value of 24% is also inadequate since the

higher rates of return occur late in point of time and carry less weight than the earlier, lower, values.

Constant Rate of Return Needed

For evaluation purposes, we can set up a schedule for the return of capital any way we want to without changing the intrinsic value of the project itself. Hence, we can develop a depreciation schedule that year by year provides a constant return on investment. Table III shows this schedule for our example. Note particularly that we have a negative value for depreciation during the first year. This situation is analogous to plowing back earnings into the project.

For the usual type of investment situation, there is only one depreciation schedule that will yield a constant rate of return through time. Likewise, there is only one rate of return that can be so generated. It represents the average rate of return for the project in its entirety and, since it is constant through time, there is no problem with time-weighting. This rate of return is the Profitability Index. Thus, the P. I. for our example is 16.3%.

The indicated closing investment shown in Table 3 represents what the investor has at stake in this particular project at the end of each year. The $100,000 is only the initial outlay. For a year, the investment increases in value as the period of earnings draws closer. Thereafter, the investment is gradually recovered as the project is liquidated.

What is done with the capital recovered has no bearing on the attractiveness of this particular project. The income from investing it in some other venture concerns only that other venture—not this one.

Some companies actually do develop the P.I. estimate by setting up a schedule by trial-and-error as we did. However, most have found it more convenient to use a discount factor approach.

Margin for adversities increases with time—Fig. 2

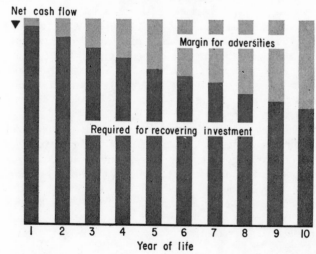

Net cash flow

Margin for adversities

Required for recovering investment

Year of life

As shown in Table IV, calculations using discount factor tables yield the 16.3% answer with little work:

An investment proposal—Table I

Year	Investment, $	Cash Income, $	Net Profit, $
0	100,000		
1		0	
2		40,000	
3		40,000	
4		40,000	
5		40,000	
Total	100,000	160,000	60,000

Application of conventional straight-line-depreciation—Table II

Year	Cash Flow, $	Income, $	Deprecia- tion, $	Closing Investment, $	Rate of Return, %
0				100,000	
1	0	−20,000	20,000	80,000	LOSS
2	40,000	20,000	20,000	60,000	28.6
3	40,000	20,000	20,000	40,000	40.0
4	40,000	20,000	20,000	20,000	66.7
5	40,000	20,000	20,000	0	200.0
Total	160,000	60,000	100,000		
Average		12,000		50,000	24.0

Note: Rate of return shown is the ratio of income to the average of opening and closing investments for each year.

P.I. can also be considered as the time-weighted average rate of return over the life of the project, regardless of what depreciation schedule is used.

To illustrate, let us reconsider the rate of return schedule developed by conventional straight-line-depreciation accounting, as shown in Table V.

The first three columns of this table repeat the earlier calculation of annual rate of return using accounting data. The last two columns show book profit and investment in terms of their present value, using a 16.3% continuous discount factor.

We can see that the ratio of average profit to average investment in these terms is also 16.3%, representing the P.I. In this example, 16.3% is the only value that when used to derive the average present value of profit and investment will also yield itself as the ratio between these two derived values.

In summary, therefore, Profitability Index has a really basic value for the decision-maker—it represents the effective average ratio of profit, after provision for return of capital, to outstanding investment. In effect, it represents an extension of the concepts of corporate accounting to cover the life span of the project in such a way as to recognize the time value of money.

The Role of P.I. in Investment Decision

So much for what the P.I. means. Now, let's look more sharply at how it affects investment decisions.

These decisions must be made in the light of the corporate profit goal. Just what is this profit goal, specifically? It is not maximizing income, nor even maximizing return on investment. If it were the former, we should try to achieve every last penny of net income regardless of how much investment it took to do it, and regardless of how low a return was earned on that investment. In contrast, if maximizing return on investment were the objective, we should select only that single investment opportunity that afforded the highest return and forego all others.

Instead, the true profit goal is maximizing income above the cost of capital invested to generate that income. Our objective should therefore be to undertake only those investments that, after consideration of all risks, promise to earn more than the costs of capital, and to reject all those investments that are expected to earn less than the cost of capital.

From this, it follows that the problem of making an investment decision is a matter of testing the profits anticipated against the cost of capital. This is essentially a balancing operation in the hands of the decision-maker.

Suppose we are confronted with a project that carries a P.I. estimate of 16%. Suppose further that our

Application of time-adjusted depreciation—Table III

Year	Cash Flow, $	Income, $	Deprecia- tion, $	Closing Investment, $	Rate of Return, %*
0				100,000	
1	0	17,740	−17,740	117,740	16.3
2	40,000	17,360	22,640	95,100	16.3
3	40,000	13,300	26,700	68,400	16.3
4	40,000	8,590	31,410	36,990	16.3
5	40,000	3,010	36,990	0	16.3
Total	160,000	60,000	100,000		
Average		12,000		73,650	16.3

* Based on average of opening and closing investment for year.

P.I. solution using discount factors— Table IV

		16% P.I.		17% P.I.	
Year	Cash Flow, $	Discount Factor	Present Value, $	Discount Factor	Present Value, $
0	−100,000	1.000	−100,000	1.000	−100,000
0–1	0	.924	0	.920	0
1–2	40,000	.788	31,500	.776	31,000
2–3	40,000	.671	26,800	.655	26,200
3–4	40,000	.572	22,900	.552	22,100
4–5	40,000	.487	19,500	.466	18,600
Total			+700		−2,100

By interpolation, P.I. = 16.3%

Note: Year designations shown as 0-1, etc., denote uniform cash flow during the year starting at time 0 and ending at time 1, etc. This convention helps to insure use of the proper discount factors. The latter are obtained from discount tables based on continuous compounding. (For instance, see *Engineering Economist*, Vol. 5, Nos. 2, 3 and 4.)

Time-weighting of accounting rate of return—Table V

Year	Income, $	Average Invest-ment, $	Rate of Return, %*	Present Value at 16.3 Income, $	Investment, $
1	−20,000	90,000	LOSS	−18,500	83,000
2	20,000	70,000	28.6	15,700	54,900
3	20,000	50,000	40.0	13,300	33,300
4	20,000	30,000	66.7	11,300	17,000
5	20,000	10,000	200.0	9,600	4,800
Average	12,000	50,000	24.0		38,600
Time-weighted average			16.3	6,280	

* Based on average of opening and closing investments for each year.

capital cost is 10%. Direct comparison of the two would, of course, favor undertaking the project.

In actual practice, we know that the problem is not this simple. The decision-maker does much more than merely compare two numbers that are handed to him, however meaningful those numbers may be. He must take into account the uncertainties in the P.I. estimate, the possibility of operating failure, the prospects of premature obsolescence, the sensitivity of the project to environmental factors such as product demand and pricing, and a host of even more intangible factors.

All of these combined go to make up the element we call judgment. If we think of the evaluation process in terms of a balance, judgment acts as a counterpoise, which can either subtract from or add to the P.I. estimate of the project. This is symbolized in our opening sketch. Instead of two components in the economic balance, we now have three: the P.I. estimate, the cost of capital, and the element of judgment. For the marginal project, judgment precisely offsets the difference between the P.I. estimate and the cost of capital.

Hence, the key question facing the decision-maker is whether unforeseen adversity not provided for in the P.I. estimate will be enough to swing the balance against the project. The evaluation is really a question of testing the economic balance of the project in the light of this judgment.

Significance of the Driving Force

Returning to our hypothetical project with the P.I. of 16%, we know that the P.I. estimate indicates this project can earn the cost of capital plus something in addition. It is this something extra—expressed as a differential of six P.I. percentage points—that represents the driving force favoring investment. What is the nature of this driving force?

Mathematically, the P.I. for any project will be reduced by six percentage points by reducing the cash flow schedule by 6% per year. Thus, at the end of the first year, cash flow would be reduced to 94% of its original value, at the second year to 94% squared at the third year to 94% cubed, and so on.

Fig. 2 shows this 6% attenuation applied to the cash

flow for our 16% project. The total height of the bars depict, of course, the cash flow of the original estimate. The lower solid black portion of the bars represents a cash flow schedule that would yield a project P.I. of exactly 10%, matching our assumed cost of capital.

The remaining gray areas of the bars can be considered to be the margin for adversities embodied in the original estimate. This much of the predicted cash flow could be lost before the project would fall below the cost of capital. Or, if the predicted cash flow were actually realized, this would be the profit above and beyond the cost of capital.

Thus, in terms of the P.I. estimate itself, we can see that the apparent profit is equivalent to the margin available to absorb unforeseen adversities in the future, assuming that the cost of capital must and will be met.

This is not to imply that when adversity strikes, it does so in an idealized geometric pattern. For the individual project, many possibilities exist. For the aggregate of many projects, the geometric pattern of decay in real earning power seems rational. As risk is a function of investment magnitude and time, expectations in the far-distant future are more uncertain than those that are short term.

In summary, the decision-maker who has to inject his own considerations of judgment into the economic balance is likely to find the P.I. an extremely useful concept. If the problem is purely analytical, the Present Value approach discussed earlier may be less cumbersome and more straightforward. But until the day comes when we can quantify intangibles and forecast the future with much more precision than we can today, Profitability Index should remain a powerful investment evaluation tool.

Meet
the
Author

WILLIAM D. McEACHRON *joined Standard Oil Co. (Indiana) shortly after obtaining his B.S. in chemical engineering at Purdue University in 1939.*

He has held a number of increasingly responsible positions in the field of economic evaluation and planning with that company, and is currently manager of the Long-Range Planning Department, in Chicago.

He is a member of AIChE, American Petroleum Institute, Society for Advancement of Management, and the Institute of Management Sciences.

Continuous Discounting for Realistic Investment Analysis

Continuous discounted cash-flow analysis provides a realistic basis for appraising capital projects. It can take into account changing conditions such as eroding profit margins, increasing costs, and improved performance over the life of the plant.

W. R. HIRSCHMANN, *American Oil Co.*

J. R. BRAUWEILER, *National Dairy Products Corp.*

Everybody is aware that our economy is far from stationary, yet it is common practice to assume stable conditions when analyzing capital investments. Why does this practice persist? Primarily because everyone knows that the future is uncertain. This uncertainty applies both to technical and economic projections. Forecasts have a record of being wide of the mark; and the further into the future they peer, the more unreliable they become. Even engineering estimates of current construction, commonly expected to be within plus or minus 10% of actual costs, are sometimes much further off target because of the infirmities of estimating practices and of unstable environment.* Consequently, even though future changes are expected, estimates of their magnitude and direction are considered sufficiently unreliable, so that continuing performance (measured as profits) at immediate levels is used as the "best" projection.

This is both unfortunate and unnecessary. Unfortunate because a firm that regularly embarks on new projects while failing to allow for significant changes in our economy may be set on a disaster course; unnecessary because there is a better way, both simple and realistic, to come to grips with change.

Better Forecasting Method

A realistic method involves three main approaches:

• Forecast trends of cash flows instead of annual incomes and expenses. Trends may be an extension of the recent past, or they may represent patterns of performance that can be expected to repeat. In either case, forecasting trends is simpler than making a series of annual forecasts. It is also likely to be more reliable because, as considerable experience indicates, trends once established tend to persist.

• Explore a range of projections. Extending a recent trend into the future can give a good first projection but, as with investment estimates, it should be used only as a point of departure. It is prudent to explore a range on both sides of the first forecast, with emphasis on deviations that actual studies or experience indicate are probable. Such a range is not only more likely to embrace the truth, it can also indicate

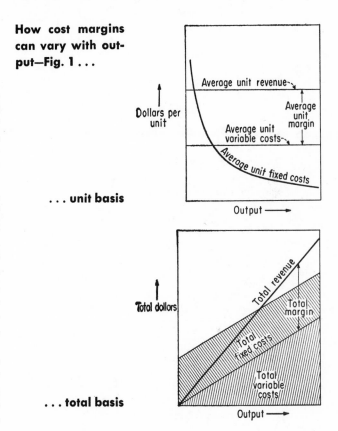

How cost margins can vary with output—Fig. 1 . . .

. . . unit basis

. . . total basis

the sensitivity of the total evolution to variations in the particular item being forecast.

• Combine cash flows having similar trend patterns. A particular investment proposal may require quantifying dozens of different items. Since most changes in cash flow are responses to the same forces, combining similar ones and projecting them as a group is simpler and likely to be just as accurate as totaling the projections of the separate items. It also highlights the circumstance that only a small number of different trends need to be projected. The three most generally important ones in our present business climate are:

• Rising costs, which reflect increasing wage rates and other pressures.

• Declining profit margins, which reflect the fact that unit selling prices are not increasing as fast as costs, usually because of competition.

• Improving performance, which reflects increased skill of workers, better equipment or processes, and advances in technology.

* See A. A. Alchian, "Reliability of Cost Estimates: Some Evidence." RM-481, RAND Corp., Oct. 30, 1950.

Cash flows change with changing conditions—Fig. 2

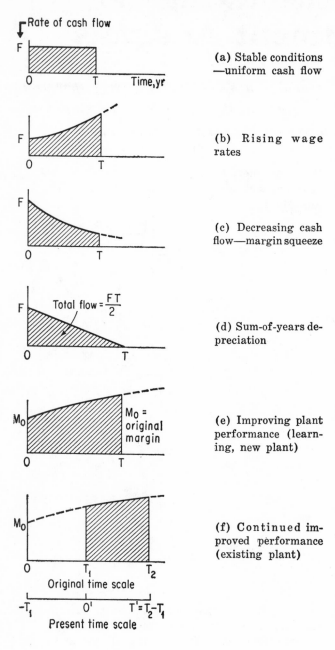

(a) Stable conditions —uniform cash flow

(b) Rising wage rates

(c) Decreasing cash flow—margin squeeze

(d) Sum-of-years depreciation

(e) Improving plant performance (learning, new plant)

(f) Continued improved performance (existing plant)

We believe that this forecasting framework suitably handles the cash-flow patterns commonly encountered with investments. When properly quantified and projected, these patterns can result in dynamic cash flows considerably different from those predicted by assuming static conditions. And more important, they can lead to different, more-accurate investment decisions.

Cost Trends

Various published indexes provide a reasonable basis for determining cost trends. The U.S. Bureau of Labor Statistics, for instance, indicates that hourly wages have risen steadily since the early 1930's. These data must be modified upward, however, to reflect the addi-tional costs of fringe benefits not included in these basic data.

Various indexes of material costs have similar increasing trends. Because variations within each industry give somewhat different average annual rates of change, projecting a firm's own historical data should be more reliable than using broad averages. A comparatively new firm, however, may well find industry trends quite useful.

While projecting each cost component separately is most desirable, it will likely be reasonable and reliable to combine and project several cost elements at an average rate. For instance, fixed costs (those that do not change with the level of operations) comprise mainly labor and materials, so projecting them as a group is simpler than projecting the separate elements, and probably as realistic.

Profit Margin Changes

Profit margins decline when competition keeps sales prices from rising, as fast as costs. Meaningful data for estimating margin squeezes are generally not available. Published reports on over-all operations are too gross for a particular problem; and detailed data on specific processes are generally proprietary, or are not available in a form amenable to analysis.

But for dynamic cash-flow analysis, costs can be segregated into those that are relatively fixed, and those variable costs that change with the level of operations. These variable costs are directly related to production and thus are usually constant on a unit basis. Subtracting this constant unit cost from unit realization gives a margin that is also constant on a unit basis at all levels of production, but that varies linearly with production for the operation as a whole, as shown in Fig. 1.

The rate of decline of this unit margin is generally not large. For one thing, it is affected only by changes in the variable-cost portion of total costs. Consequently a cost-price squeeze results in smaller changes in this unit margin than in unit profit. Furthermore, a decline of only 4% per year continuously compounded for 20 years reduces a unit margin to 44% of its initial value. A rate of decline of this magnitude will surely prompt remedial measures to arrest it.

One fact is important: the absolute and not the percentage margin is what counts. Consider, for instance, an item that sold for $10 in 1946 and cost $9 to make and sell. The $1 profit is conventionally spoken of as a margin of 10%. Suppose that inflation and other factors raised the price of the item to $20 now, and the same factors increased costs to $18.80. The corresponding margin of $1.20 is 6% and it can be considered to have been squeezed from 10%. But the profit in dollars has risen from $1 to $1.20. Thus, while the profit margin has decreased in percent, it has increased in dollars; and it is the dollar margin that is used in economic analyses.

Even if the unit margin increases, the annual rate of increase will likely be small. A growth of 4% per year results in a total increase of 50% after ten years.

Increases of this magnitude are not likely. Either existing competition prevents such a situation, or new firms are attracted to the field, and the added capacity to supply the market holds down prices.

In the absence of definitive data, the prudent course is to explore the effect of assumed changes of realistic magnitude. If margin change has little influence on the analysis, such rough approximations are satisfactory. If it has large weight, more-definitive projections are indicated.

Typical Factors for Continuous Discounting
$R \times T$ = Annual Interest Rate $\times$ Number of Years in Time Period Involved

Instantaneous

Factors for Cash Effects That Occur at
A Point in Time After the Reference Point.

$R \times T$	0	1	2	3	4	5	6	7	8	9
0	1.0000	0.9901	0.9802	0.9704	0.9608	0.9512	0.9418	0.9324	0.9231	0.9139
10	0.9048	0.8958	0.8869	0.8781	0.8694	0.8607	0.8521	0.8437	0.8353	0.8270
20	0.8187	0.8106	0.8025	0.7945	0.7866	0.7788	0.7711	0.7634	0.7558	0.7483
30	0.7408	0.7334	0.7261	0.7189	0.7118	0.7047	0.6977	0.6907	0.6899	0.6771
40	0.6703	0.6637	0.6570	0.6505	0.6440	0.6376	0.6313	0.6250	0.6188	0.6126

Uniform

Factors for Cash Effects That Occur Uniformly
Over a Period of Years Starting With the Reference Point.

$R \times T$	0	1	2	3	4	5	6	7	8	9
0	1.0000	0.9950	0.9901	0.9851	0.9803	0.9754	0.9706	0.9658	0.9610	0.9563
10	0.9516	0.9470	0.9423	0.9377	0.9332	0.9286	0.9241	0.9196	0.9152	0.9107
20	0.9063	0.9020	0.8976	0.8933	0.8891	0.8848	0.8806	0.8764	0.8722	0.8681
30	0.8639	0.8598	0.8558	0.8517	0.8477	0.8438	0.8398	0.8359	0.8319	0.8281
40	0.8242	0.8204	0.8166	0.8128	0.8090	0.8053	0.8016	0.7979	0.7942	0.7906

Years-Digits

Factors for Cash Effects Declining to Zero at a Constant Rate
Over a Period of Years Starting With the Reference Point.

$R \times T$	0	1	2	3	4	5	6	7	8	9
0	1.0000	0.9967	0.9934	0.9901	0.9868	0.9835	0.9803	0.9771	0.9739	0.9707
10	0.9675	0.9643	0.9612	0.9580	0.9549	0.9518	0.9487	0.9457	0.9426	0.9396
20	0.9365	0.9335	0.9305	0.9275	0.9246	0.9216	0.9187	0.9158	0.9129	0.9100
30	0.9071	0.9042	0.9013	0.8965	0.8957	0.8929	0.8901	0.8873	0.8845	0.8818
40	0.8790	0.8763	0.8736	0.8708	0.8682	0.8655	0.8628	0.8602	0.8575	0.8549

Performance Trends

A decline in unit costs with time is reflected either by a decrease in manpower for the same output, or by an increase in output with the same work force. The improvement is rapid initially, and later tapers off; it resembles the performance of individuals learning a given task. A plot of the improvement is considered a learning curve. Many factors other than worker skills are involved—technological advance, rate of operation, state of labor relations, need or incentive for improvement, and resourcefulness of management in integrating operations, stimulating creativity, encouraging innovation, and learning how to use its resources better. Continual striving for improvement can be expected to result in progressively improved operations. Studies show that this expectation can be quantified by a learning curve.*

Fig. 2 shows learning curve cash-flow patterns and

* For additional information on improvement potential, see: Winfred B. Hirschmann, "The Learning Curve," Chem. Eng., Mar. 30, 1964, pp. 95-100.

others commonly encountered in assessing investment merit.

Dynamic Cash Flows

The essence of a discounted cash-flow analysis is to determine the earnings rate that discounts all future net cash income to a present value equal to the project's cost. Although discounting can be done on either an annual or continuous basis, we find that continuous has several major advantages over annual.

It is usually more representative of the actual flow of funds. Annual discounting assumes that cash flows occur in a lump sum at the beginning or at the end of each year. In most businesses, however, transactions occur throughout the year. Dividends are paid quarterly; bills are due in 30 days or sooner; and salaries and wages are paid semimonthly or weekly. In fact, every business day requires a sizable exchange of cash. Although transactions may be seasonal or come in surges, these patterns indicate that continuous discounting more realistically reflects the actual flow of business cash than annual discounting.

Continuous discounting is computationally simpler. The present worth functions for instantaneous cash flow are as follows for both the annual and continuous discounting methods, where R is the annual interest rate as a decimal:

	Present Worth Functions	
	Annual Discounting	Continuous Discounting
Flow P, at an instant in time T years hence,	$= P(1+R)^{-T}$	Pe^{-RT}

Note that although both R and T occur in the functions for both types of discounting, they occur as the product $R \times T$ with continuous discounting. Because of this, a continuous discount function has the same value for each combination of interest rate and time period that has the same product. Consequently, continuous discounting requires only one table of factors (based on the product $R \times T$), while annual discounting requires many tables, one for each interest rate. The abbreviated table (left) illustrates this simplicity. A single sheet is equivalent to dozens of pages of annual factors.

Continuous discounting also handles trends readily. This characteristic is not generally recognized, but each of the cash-flow patterns in Fig. 2 can be discounted with a single factor, or at most with a combination of two factors.

Stable Conditions: Uniform Flow

If the rate of cash flow at any instant t is F, the flow over any increment of time Δt is $F \Delta t$, and its present worth at interest rate R is $F \Delta t (e^{-RT})$.

The present worth of the total cash flow over $T =$

$$\int_0^T Fe^{-RT} dt = F\left[\frac{e^{-RT}}{-R}\right]_0^T = \frac{F(1-e^{-RT})}{R}$$

If the total flow FT is used instead of the rate of flow F, the discount function can be expressed in terms

of one variable RT instead of two variables, R and RT. The present worth then is: $FT(1 - e^{-RT})/RT$. The table of discount factors for uniform cash flows on p. 213 is the evaluation of $(1 - e^{-x}/x)$, where $x = RT$. In terms of that table, the present worth of the total cash flow is FTU_{RT}, where U_{RT} is the factor for discounting uniform cash flows at the rate R for T years.

Rising Wage Rates

The rate of flow at any instant t is Fe^{gt}, where g is the annual percent increase, and F the initial rate of flow (at $t = 0$). Present worth of the flow over the increment of time dt at instant t is $Fe^{gt}e^{-RT}dt = Fe^{-(R-g)t}dt$. The present worth of the total flow over T years is:

$$\int_0^T Fe^{-(R-g)t}\, dt = FT\frac{[1 - e^{-(R-g)T}]_0^T}{(R-g)\,T} = FTU_{(R-g)T}$$

where $U_{(R-g)T}$ is the factor for discounting uniform cash flow over T years at interest rate $(R - g)$.

Margin Squeeze: Flow Decreasing

For exponentially decreasing cash flow, the rate of flow at any instant t is Fe^{-xt}, where x is the annual percent decrease and F is the initial rate of flow ($t = 0$). The present worth of the flow at instant t is $Fe^{-xt}e^{-RT}dt = Fe^{-(R+x)t}dt$. The present worth of the total cash flow over T years is

$$\int_0^T Fe^{-(R+x)t}\, dt = FT\frac{[1 - e^{-(R+x)T}]_0^T}{(R+x)\,T} = FTU_{(R+x)T}$$

where $U_{(R+x)T}$ = factor for discounting uniform flow over T years at interest rate $(R + x)$.

Sum-of-Years Digits Depreciation: Straight Line

If an initial flow F declines to zero over T years by a constant amount each year, the flow f at any instant t is given by the proportional relationship of right triangles:

$$\frac{f}{T-t} = \frac{F}{T}; \quad f = F\left(1 - \frac{t}{T}\right)$$

The present worth of the flow over an increment of time is $F(1 - t/T)e^{-RT}dt$, and over T years is

$$\int_0^T F\left(1 - \frac{t}{T}\right)e^{-RT}\, dt = \frac{F}{R}\left(1 - \frac{1 - e^{-RT}}{RT}\right)$$

The total flow is $a = FT/2$, from which $F = 2a/T$. In terms of the total flow a over T years, the present worth is $(a)(2/RT)[1 - (1 - e^{-RT}/RT)]$. A table of discount factors for such a flow is that evaluation of $(2/x)[1 - (1 - e^{-x}/x)]$, where $x = RT$. The present worth at interest rate R of the total flow a which declines to zero over T years can be represented as aY_{RT}, where Y_{RT} is the factor for discounting such decreasing flows at the rate R for T years. This type of cash flow approximates sum-of-years-digits depreciation. The table gives factors for discounting it.

Performance Improvement at New Plant

Assume that performance will trace an achievement curve such as sketched, which can be represented by $P_t = P_0(2 - e^{-kt})$, where

P_t = performance at time t in terms of output.
P_0 = initial performance ($t = 0$).
t = age of plant.
k = constant determined empirically.

If the difference, m, between realization and variable costs per unit produced remains constant (i.e., unit margin remains constant), the total difference M or margin for the plant at time t is

$$P_t m = M_t = M_0 (2 - e^{-kt})$$

The present worth at R interest rate on such flow over T years is

$$\int_0^T M_0 (2 - e^{-kt})\, e^{-Rt}\, dt = M_0 T\, [2U_{RT} - U_{(R+k)T}]$$

If the unit margin decreases by a constant fraction x each year, the unit margin at any time is $m_t = m_0 e^{-xt}$. Total margin of the operation at any time is $P_t m_t = M_0(2 - e^{-kt})e^{-xt}$, and its present worth at interest rate R over T years is:

$$\int_0^T M_0 (2-e^{-kt})\, e^{-xt}\, e^{-RT}\, dt = M_0 T\, [2U_{(R+x)T} - U_{(R+x+k)T}]$$

Learning at Existing Plant

The chart represents any portion of such a flow. The present worth of the flow from T_1 to T_2 can be found in either of two ways:

1. Subtract the present worth of the flow from 0 to T_1 from the present worth of the flow from 0 to T_2.

2. Consider T_1 as a new zero (i.e., 0') and find the present worth in terms of the flow M_0' at 0'. The present worth at time 0' of the flow from 0' to T' (i.e., from T_1 to T_2) will be

$$M_0'\, T'\left[\frac{2U_{(R+x)T'} - I_z\, U_{(R+x+k)T'}}{2 - I_z}\right]$$

where $Z = kT_1$

These examples illustrate how cash-flow patterns can be discounted with continuous discounting. The next article, to appear Aug. 16, will show how a dynamic investment analysis is used on a typical problem.

Meet the Authors

W. B. Hirschmann is business economics supervisor in the research and development department of American Oil Co., Whiting, Ind. He holds a B.S.(Ch.E) from the University of Illinois and an M.S. from the University of Chicago.

J. R. Brauweiler is manager of systems, National Dairy Products Corp., Chicago, Ill. Previous to his present position, he worked for American Oil Co. Mr. Brauweiler holds an M.E. from Notre Dame and an M(B.A.) from the University of Chicago.

Key Concepts for This Article

Active (8)	Passive (9)	Ind. Var. (6)	Dep. Var. (7)
Evaluating	Return on	Costs	Investments
Analyzing	investment	Wages	Return on investment
		Depreciation	**Means/Methods (10)**
		Productivity	Discounted cash flow

(Words in bold are role indicators; numbers correspond to AIChE information retrieval system. Indexing details are described in *Chem. Eng.*, Jan. 7, 1963.)

Realistic Investment Analysis--II

Application of continuous discounting methods illustrates how estimated changes in profits, work performance and costs can be incorporated into a profitability analysis.

W. B. HIRSCHMANN, *American Oil Co.*
J. R. BRAUWEILER, *National Dairy Products Co.*

Business operates in a world of change: competitive practices erode profit margins; quality improves; old markets disappear and new ones arise; increasing wage rates continually press prices upward; and improved technology makes existing equipment, practices and processes obsolete.

The previous article* presented a framework for coming to grips with such changes:

• Capitalize on the experience that changes usually move in trends that establish characteristic patterns.

• Forecast a range of trends.

• Combine cash flows having similar trend patterns into major categories such as costs, profit margins, and investment performance.

• Use continuous instead of annual discounting because it realistically describes the flow of business cash and simplifies the handling of evaluation computations.

This second article illustrates the application of these concepts.

Assume Typical Problem

A company completed a plant ten years ago. It was expected to be serviceable for twenty-five years, but technical advances and accumulated know-how suggest that obsolescence may have progressed faster than expected, so it may be profitable to replace the plant now. Assume for simplicity that a new plant would have the same capacity that the old one has at present, and would produce the same array of products—initial revenue for both would be the same. The advantage of the new is then reflected in its lower operating costs. A study indicates that the savings in operating costs are $180,000/yr. A conventional evaluation assumes these savings to be the same each year. Straight-line depreciation on the old plant is $35,000/yr.

If the investment required for the new plant is $1 million and if depreciation of the new plant would be on a sum-of-the-years digits basis, the discounted-cash-flow (DCF) rate of return promised by the replacement is 7.7% (see Table II). If the firm requires new proposals to promise at least a 10% rate of return, the replacement proposal is not attractive.

History, however, shows that the performance of the existing plant has traced a 90% learning curve. (This is a fairly good rate of learning. A 95% curve is more conservative, while a 100% curve indicates

no learning at all.)† Similar plants built at other times have, on the average, also followed such a curve. It, therefore, seems reasonable to assume that the performance of the existing plant will continue to improve along the established curve from year ten on, but that the new plant will start a similar new curve (Fig. 1).

Since margin varies linearly with production, a breakdown of the costs of the existing and proposed

---

† If the fourth unit requires only 90% of the direct labor of the second unit, the eighth unit only 90% of the fourth, the hundredth unit 90% of the fiftieth, the process is following a 90% learning curve—rate of learning is 90% between doubled quantities. See *Chem. Eng.*, Mar. 30, 1964, p. 95.

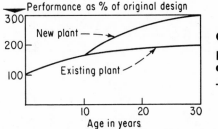

Old and new plants improve at a 10% rate —Fig. 1

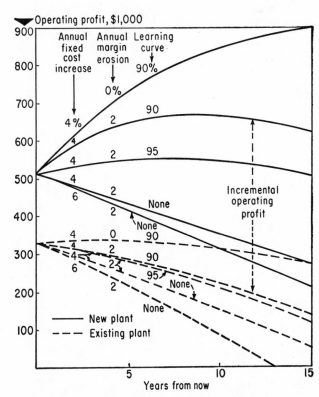

Earning patterns for various trends—Fig. 2

Factors for Continuous Discounting—Table I

Instantaneous
Factors for Cash Effects That Occur at A Point in Time After the Reference Point.

Uniform
Factors for Cash Effects That Occur Uniformly Over a Period of Years Starting With the Reference Point.

Years-Digits
Factors For Cash Effects Declining to Zero at a Constant Rate Over a Period of Years Starting With the Reference Point.

R × T = Annual Interest Rate × Number of Years in Time Period Involved

Dynamic cash-flow analysis gives brighter investment picture—Table II

The proposed replacement investment of $1-million for a new plant has the following cash effects during the first year:

	Existing Plant	New Plant	Savings With New Plant
Realization (revenue)	$1,500,000	$1,500,000	
Variable costs	990,000	890,000	$100,000
Margin	510,000	610,000	
Fixed costs	180,000	100,000	80,000
Operating profit	330,000	510,000	180,000

Conventional analysis assuming uniform annual income and expenses:

Time, Yrs.	Action	Cash Consequences		Discount Factor*	Trial Rates			
					5%		8%	
		Before Tax	After 50% Tax		Factor	Present Worth	Factor	Present Worth
0	Investment	-$1,000,000	-$1,000,000	I_{R0}	1.0	-$1,000,000	1.0	-$1,000,000
0-15	Total straight-line dep. effects lost on old plant	-525,000	-262,500	U_{R15}	0.7035	-184,700	0.5823	-152,900
0-15	Total sum-of-the-years-digits dep. effects on new plant	+1,000,000	+500,000	Y_{R15}	0.7906	+395,300	0.6961	+348,000
0-15	Annual fixed cost savings	+80,000	+40,000	$15\,U_{R15}$	10.5525	+422,100	8.7345	+349,400
0-15	Annual margin on new plant	+610,000	+305,000	$15\,U_{R15}$	10.5525	+3,218,500	8.7345	+2,664,000
0-15	Annual margin on old plant	-510,000	-255,000	$15\,U_{R15}$	10.5525	-2,690,900	8.7345	-2,227,300
					Total	+$160,300		-$18,800
					\multicolumn{4}{l}{$R = 5 + [160{,}300/(160{,}300 + 18{,}800)] \times 3 = 7.7\%$}			

Dynamic cash flow analysis. Fixed costs increase 4%/yr.; margin eroded 2%/yr.; performance improvement follows curve $2 - e^{-0.10T}$:

Time, Yrs.	Action	Cash Consequences		Discount Factor*	Trial Rates			
					15%		20%	
		Before Tax	After 50% Tax		Factor	Present Worth	Factor	Present Worth
0	Investment	-$1,000,000	-$1,000,000	I_{R0}	1.0	-$1,000,000	1.0	-$1,000,000
0-15	Total straight-line dep. effects lost on old plant	-525,000	-262,500	U_{R15}	0.3976	-104,400	0.3167	-83,100
0-15	Total sum-of-the-years-digits dep. effects on new plant	+1,000,000	+500,000	Y_{R15}	0.5355	+267,700	0.4555	+227,700
0-15	Fixed cost savings, first year	+80,000	+40,000	$15\,U_{(R-4)15}$	7.3455	+293,800	5.6835	+227,300
0-15	Margin, new plant, first year	+610,000	+305,000	$15[2U_{(R+2)15} - U_{(R+2+10)15}]$	7.206	+2,197,800	5.658	+1,725,600
0-15	Margin, old plant, first year	-510,000	-255,000	$15\dfrac{[2U_{(R+2)15} - I_{(10)(10)}U_{(R+2+10)15}]}{2 - I_{(10)(10)}}$	5.8241	-1,485,100	4.667	-1,190,100
					Total	+$169,800		-$92,600
					\multicolumn{4}{l}{$R = 15 + [169{,}800/(169{,}800 + 92{,}600)] \times 5 = 18.2\%$}			

*See Chem. Eng., July 19, 1965, pp. 210-214 and Table I for definitions of these discount functions.

plants indicates their respective initial margins to be as follows:

	Existing Plant	Proposed Plant	Savings
Realization (revenue)	$1,500,000/yr.	$1,500,000/yr.	
Variable costs	990,000	890,000	$100,000
Margin	510,000	610,000	
Fixed costs	180,000	100,000	$80,000
Operating profit	$330,000	$510,000	$180,000

The record also indicates that the unit margin has been eroded over the past ten years at a rate of about 2%/yr. It seems reasonable to assume as a point of departure that this rate of erosion will continue. However, while fixed costs have risen at the rate of about 5½% to 6%/yr., there is some feeling that cost pressures are subsiding so fixed costs will rise at a smaller rate. An increase of 4% per year seems a reasonable initial assumption. When the analysis reflects these assumptions, the rate of return promised by the advantages of the new plant rises from 7.7% to about 18%, well over the minimum level of 10% desired for new investments (Table II).

Management might feel that average improvement in performance along a 90% learning curve is too optimistic an expectation; or that even if the plant is capable of such improvement, the market picture is such that all the potential for improvement will not be needed as it evolves. If conservative improvement is assumed at a rate about one-half as great—a 95%

learning curve—the rate of return is 13%, which is still over the minimum level required.

Analysis of Key Factors

Of the three factors responsible for change, the major one is improvement in performance. It predominates because it rises rapidly in the early years, which are most heavily weighted in DCF analysis, while changes in fixed costs and margin are gradual.

The effect of changing unit margin is influenced by output. Since the performance of the new plant improves faster than the existing one, changes in unit margin benefit or penalize the new plant, depending upon whether the margin increases or decreases. The effect is comparatively small, so that in this case further definition is not needed.

Rising costs for labor and materials penalize the existing plant (because its initial fixed charges are larger) and favor the new plant. This disadvantage can be expected to become progressively greater. Its effect is to benefit the new plant. Further, the benefit is relatively certain; a halt in the long-term upward march of unit labor and material costs seems most unlikely. Even if the new plant has no differential performance advantage, the benefit of increasing savings in labor and material alone is nearly sufficient to raise the promised rate of return above the 10% considered necessary to make replacement economically attractive.

Consequences that various trends will have on estimated earnings—Table III

The Situation—A new plant is proposed to replace an existing one. It would manufacture the same array of products and have the same realization of $1,500,000/yr. The new plant would cost $1,000,000 and be depreciated over 15 yrs. by sum-of-the-digits method. The existing plant would continue to be depreciated by the straight-line method at $35,000/yr. Initial margins (realization minus variable costs) and fixed costs are are follows:

	Existing Plant	Proposed Plant
Annual margin	$510,000	$610,000
Annual fixed costs	180,000	100,000

The DCF earning rates for various assumed trends for fixed costs, margins and performance are as follows (see sample calculation, Table II):

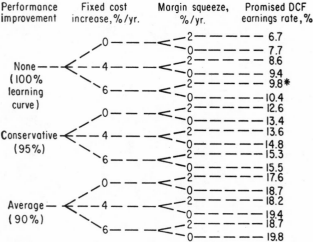

Performance improvement	Fixed cost increase, %/yr.	Margin squeeze, %/yr.	Promised DCF earnings rate, %

None (100% learning curve) — 0 — 2 — 6.7 / 0 — 7.7; 4 — 2 — 8.6 / 0 — 9.4; 6 — 2 — 9.8* / 0 — 10.4

Conservative (95%) — 0 — 2 — 12.6 / 0 — 13.4; 4 — 2 — 13.6 / 0 — 14.8; 6 — 2 — 15.3 / 0 — 15.5

Average (90%) — 0 — 2 — 17.6 / 0 — 18.7; 4 — 2 — 18.2 / 0 — 19.4; 6 — 2 — 18.7 / 0 — 19.8

* This rate is based on 13 year life for the existing plant, shown in the bottom curve of Fig. 2.

Table III shows the anticipated rates of return for a new plant under various assumptions and conditions. For example, if the expected performance is conservative (95% learning curve), the return will be 13.6% if fixed costs rise 4% annually and profit margins shrink 2%; but if expected improvement is average (90%), then the rate of return is 18.2%.

Changes in operating profit for each plant that lead to these larger rates of return are charted in Fig. 2 for several cases. Such charts help to visualize the net effect of trends on operating profit.

Note that the assumptions of no improvement in performance, a 2% annual erosion of margin, and a 6% annual increase in fixed costs, lead to a loss for the existing plant after 13 more years of life. Should these prove to be the trends, the plant will not reach its projected 25-year economic life. It is for this reason that the 9.8% rate of return was calculated on the basis of only 13 years of life. Although performance improvement offsets this loss in other cases, it remains important, as these charts illustrate, to investigate the cumulative effects of trends and of terminal cash flows when making expected life projections, especially for cases combining trend extremes.

The progressive obsolescence of existing plants points out why it is desirable to monitor the economic benefits of technological advance that can be incorporated in a new plant. A rigorous analysis will, of course, include differences in working capital requirements, investment credits, range of investment estimates, and other pertinent factors omitted from this example for simplicity.

Merits of Tableau Presentations

Various assumptions for the trends of costs, performance and profit margin yield different rates of return for a proposal. An array of these assumptions and their corresponding rates of return can give management a feeling for the relative weights of the various factors and for the consequences of various patterns of change. It may also indicate a different choice among alternatives than does the assumption of uniform annual income and expense over the life of a proposed investment.

There is another benefit of such an array. An economic analysis of a major investment proposal includes many estimates of expected costs and revenues. They are built into the analysis as it progresses, and reflect the best judgments of the various echelons through which a proposal passes on the way to top-level consideration. A single reported rate of return is a simplifying abstraction of these judgments and of the promise of a venture. However, it does not indicate the range of uncertainties nor the consequences of other judgments. Rather, it tends to obscure the circumstance that the reported rate of return is only a promise, and that this promise can be threatened by the many changes possible in a dynamic economy.

A tableau of the assumptions for the trends of costs and incomes enables a manager to apply his own judgment to some of the important uncertainties. If a computer is available, the array can be refined by developing probability profiles for the trends and investment cash flows, and from these profiles, a continuous profile of the promised rate of return can be found.*

The tableau of Table III requires about one minute of IBM 704 machine time—a small cost for the amount of extra insight obtained regarding the effect of various trends on project profitability. Whether the array is continuous or discrete, it relates the uncertainties and corresponding consequences in terms of the rate of return on the committed funds. It, thereby, can give a feeling for the risks of a venture in the context of one of the important concerns of management—the efficiency of its invested capital.

* For a description of this refinement, see David B. Hertz, Risk Analysis in Capital Investment, *Harvard Business Review*, Jan.-Feb. 1964, pp. 95-106.

Key Concepts for This Article

Active (8)	Passive (9)	Ind. Var. (6)	Dep. Var. (7)
Evaluating	Return on	Costs	Investments
Analyzing	investment	Wages	Return on investment
		Depreciation	**Means/Methods (10)**
		Productivity	Discounted cash flow

(Words in bold are role indicators; numbers correspond to AIChE information retrieval system. Indexing details are described in *Chem. Eng.*, Jan. 7, 1963.)

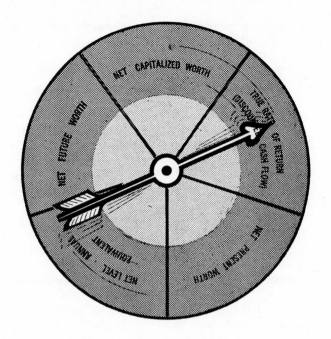

Which Investment Appraisal Technique Should You Use?

Do you believe that the selection of an investment appraisal technique can be a matter of personal preference? Can all procedures that take into consideration the time value of money provide equally reliable evaluations, if properly used? You may change your mind as you read this concise, quantitative analysis of the answers provided by various widely advocated methods.

RAYMOND I. REUL, *FMC Corp.*

The basic objective of a business enterprise can be very simply stated as "to make a profit, preferably as much as possible." In retrospect, this goal can be viewed as a simple maximizing or optimizing process—one that should lend itself to straightforward mathematical analysis. But at the time that one is trying to select prospective investments, the implementation of a specific goal is not the same simple problem; a strictly mathematical approach does not seem feasible because:

1. The selection involves prediction of future events, recognition of uncertainties as to the extent of success, and consideration of the risk of failure. With no way to make specific predictions of the odds applicable, there is no real basis for firm optimization.

2. Current commitments, if large or numerous enough, may limit the availability of funds for future commitments. Hence, even if optimization of results from currently known investment opportunities were possible, this action must be recognized as a sub-optimization, which would not necessarily lead to overall optimization.

3. Even if optimization of total earnings were feasible, this approach would not assure the exclusion of prospective projects of only marginal profitability.

4. Furthermore, the optimization of earning rates could easily result in the rejection of prospective projects with very large earnings at lower, but still acceptable, earning rates.

Since there is no reasonable basis for purely mathe-

matical optimization, the techniques designed for appraisal of prospective investments should be evaluation procedures, not decision tools. They should provide factual, quantitative data to give management the best possible measure of the interactions of prospective actions and future events. They should help managers improve the reliability of their judgment, and not attempt to provide a substitute for management's responsibility to make decisions.

The following conceptual goal is suggested as a fairly accurate formulation of good management practice in the selection of prospective investments: *"Commitments should be sought, selected and implemented so that the enterprise will maintain a reasonable rate of growth, an acceptable level of profits, and not be exposed to excessive uncertainty or unnecessary risk."*

The search for, and identification of, prospective profitable investments is a creative rather than an analytical process. This requires a great deal of experience and skill, plus a thorough knowledge of both technical aspects and business conditions. For in addition to prospective profitability, many other factors must be taken into consideration, including process know-how, status of resources, market situations, competitive pressures and the availability and cost of obtaining the required funds.

Once a proposal has been identified as a feasible investment prospect, the next step is to predict all anticipated cash flows, out and in, over the entire expected life of the project and then use these data to make an appraisal of prospective economic productivity. This is an analytical process that can and should be subject to accurate, quantitative mathematical analyses.

The selection of the means for measuring this economic performance is of critical importance. The procedure must be applicable to every possible type of prospective investment, and must yield answers that will permit valid comparisons among all types of competing opportunities. The answers must provide a consistently valid measure of economic productivity and must be directly comparable with the cost of obtaining funds.[1] Furthermore, the answers must be suitable for subjective comparison involving both the uncertainty due to possible variations of predictions and the probability and consequences of partial or complete failure.

In order to do this job, the following three factors must be built into the computations:

• All proposed outflows of cash or cash equivalent.
• All inflows of cash or cash equivalent anticipated as a result of proposed outflows.
• Time relationship between all cash flows.

The traditional ratios of return on investment, payout and average return on average investment do not comply with these requirements. The inaccuracies and unreliability of these techniques, which are based on accounting-derived data, have been extensively documented. The task of accounting is to measure the past performance or history of a complete "going business" during a specific, relatively short span of time called an accounting period. The many perfectly logical, but necessarily arbitrary, assumptions employed by accounting procedures to achieve this objective are completely incompatible with the problem of measuring the profitability of a single proposed investment over its entire life. This conclusion has been adequately presented in the literature, and need not be detailed here.

Interest-Based Appraisal Techniques

The compound-interest-based investment computations that take into consideration the time value of money are now widely accepted and used by a large segment of management. But it is not correct to assume that all methods employing interest computations are necessarily valid and useful. Some have very limited applicability. There are, however, five distinct variations that can successfully withstand the test of reliable validity—i.e., when the methods are correctly used and the answers properly interpreted, they will never give false indications. These are:

1. Net present worth (or net present value).
2. Net future worth (or net terminal value).
3. Net level annual equivalent (or equivalent annual cost).
4. Net capitalized worth (or capitalized cost).
5. True rate of return (or internal rate of return).

Every one of these methods is dependably valid under all circumstances when applied to a true investment situation, but each measures a different aspect or attribute of the proposed commitment, and thus can appear to give quite different evaluations. The methods for computing each of these yardsticks have been carefully delineated, but the differences in the meanings of the answers obtained have never been adequately explained. Because a thorough understanding of these meanings is a firm prerequisite to the objective selection of the most appropriate appraisal method, let us now attempt to explicitly define the five answers, demonstrate the application of each method to the same problem, and discuss the differences and relationship among the results obtained.

The following simple prospective commitment has been selected as the demonstration proposal:

Disbursements: Year 0 . . . . $995
Receipts: Year 1 . . . . $400
Year 2 . . . . $400
Year 3 . . . . $400

Additional sets of problems will be introduced later to illustrate the impact of specific differences in the nature of the answers. Annual compounding and end-of-the-year receipt convention have been used throughout to simplify the presentation.

The Net Present-Worth Method

This method assumes continuing availability of unlimited opportunities to invest all assets at a constant stipulated interest rate. Each prospective investment is compared with this standard alternative. The answer obtained is "The *added amount* that would have to be invested *at the start of the proposed project* at the standard stipulated interest rate in order to produce

receipts equal to, and at the same time as, the prospective investment."

With a stipulated interest rate of 5%, the net present worth of the demonstration proposal is computed by discounting all receipts to year zero at this rate and subtracting the proposed disbursement, i.e.:

		Present Worth Factor @ 5%		$
Year 1 to zero:	$400 ×	.952	=	381
Year 2 to zero:	400 ×	.907	=	363
Year 3 to zero:	400 ×	.863	=	345
		Total		1,089
Less proposed disbursement			=	995
Net Present Worth				$94

This $94 is the amount that must be added to the $995 to establish the amount that would have to be invested at 5% to achieve receipts equal to and at the same time as those predicted for the proposed commitment as is demonstrated below:

$$(\$995 + \$94) \times 1.05 = \$1,144$$
Less payment 400

$744

$$\$744 \times 1.05 = \$781$$
Less payment 400

$381

$$\$381 \times 1.05 = \$400$$
Less payment 400

0

Net Future-Worth Method

This method is based on the assumption that all *receipts* as well as all assets can be invested at a constant stipulated interest rate in continuously available investment opportunities. The prospective investment plus reinvestment of proceeds at the stipulated interest rate is then compared against the standard alternative of investing just the original commitment at the stipulated interest rate. The answer obtained is "The *added amount* that would be achieved *at the end of the project* if the project's anticipated receipts instead of the proposed disbursement were invested at the standard stipulated interest rate."

With a stipulated interest rate of 5%, the net future worth of the demonstration proposal is computed by compounding the anticipated receipts of $400 per year at this 5% interest rate to the terminal year; then subtracting from this the amount that would have resulted if the standard alternative (investing the proposed commitment at the stipulated interest rate to the terminal year) had been employed.

Receipts (compounded to terminal year at 5%)
Year 1 to 3:	$400 × (1.05)²	=	$441
Year 2 to 3:	400 × (1.05)	=	420
Year 3:	400	=	400
	Total		$1,261

Less:
Disbursement (Compounded to terminal year at 5%)
Year 0 to 3: $995 × (1.05)³ = $1,152

Net Future Worth $109

The method of computation itself demonstrates that this $109 future worth is the additional amount that would be obtained if the project is implemented and the proceeds reinvested at the stipulated 5% interest rate, instead of just investing the proposed disbursement for the same period at the same stipulated 5% rate.

The Net Annual-Equivalent Worth

This method is based on the assumption of continuing availability of unlimited opportunities for investment of both assets and receipts at a constant stipulated interest rate, plus *unlimited availability of capital* at the same stipulated interest rate. The answer is "The amount by which a *single year's level equivalent of all receipts* exceeds the *level amounts that would be generated* if the proposed commitment were invested at the standard stipulated interest rate."

With a stipulated interest rate of 5%, the net annual equivalent worth of the demonstration problem is computed by subtracting the level annual equivalent of the proposed investment from one of the already level annual receipts,* i.e.:

$$\text{Net annual-equivalent worth} = \$400^* - \frac{\$995}{.952 + .907 + .863}$$
$$= \$34.50$$

This $34.50 is the amount by which the proposed investment's anticipated receipts, leveled at 5%, exceed the annual equivalent of the proposed investment (leveled at the same 5%). The validity of this definition is demonstrated below:

Anticipated equal annual receipts $400.00
Less equal annual equivalent worth @ 5% 34.50

Level annual receipts to be generated by investing
$995 at 5% $365.50

$995 × 1.05	= $1045.00	
Less payment	365.50	
	$679.50	
$679.50 × 1.05 =	713.50	
Less payment	365.50	
	$348.00	
$348 × 1.05	$365.50	
Less payment	365.50	
	0	

The Net Capitalized Worth

Although also based on a comparison with the standard alternative of investing at a constant stipulated interest rate, this method expresses the answer as: "The amount that would have to be invested at the stipulated interest rate in order to generate a perpetual series of receipts equal to the net equal-annual-equivalent worth."

* Note: When the receipts are not uniform, they must first be converted to total level-annual-equivalent by converting to total present worth and then to annual equivalents. If factors are available, multiplication by the capital-recovery factor may be substituted for the division by the sum of the present-worth factors.

The net capitalized worth is computed by dividing the net annual equivalent worth by the stipulated interest rate of 5%.

$$\frac{\$34.50}{0.05} = \$690$$

This $690 is the amount that must be invested at the stipulated interest rate of 5% in order to generate perpetual annual receipts of $34.50 per year.

$$\$690 \times 1.05 = \$725.50$$
$$\text{Less} \qquad\quad 35.50$$
$$\overline{\hspace{1em}\$690.00, \text{ etc.}}$$

The True Rate of Return

The adjective "true" is used in this expression in the same sense as it is used in Grant and Ireson's "Principles of Engineering Economy," to distinguish this method from other, nonvalid methods that have been mislabeled rate of return. Other names applied to the true rate of return include internal rate of return, interest rate of return, discounted cash flow and profitability index.

Here there is no assumption of any alternative investment opportunities and no stipulated interest rate. The answer is the equivalent, level, average annual earning rate of funds in use, and may be specifically defined as: "The interest rate at which a sum of money, equal to that invested in the proposed project, would have to be invested in an annuity fund in order for that fund to be able to make payments equal to, and at the same time as, the receipts from the proposed investment."

Rate of return is computed by finding the interest rate at which the sum of the present worth of receipts exactly equals the sum of the present worth of disbursements. In using this calculation, there is one precaution that must be observed: since the method of computation is an algorithm, it can only be applied to problems in the format for which it is designed; i.e., true investment proposals in which all net disbursements occur prior to all net receipts. Cash flow patterns not in this format must be adjusted to meet the required pattern prior to application of the algorithm.

Solution is by repeated trials until the correct interest rate is found. (Graphical interpolation using pretabulated worksheets reduces this computation to a simple, quickly performed clerical procedure as described in the references.[2]

The answer to our sample problem with the $995 disbursement at year zero is 10%, as can be seen from the following successful trial:

Receipts	Present Worth Factor at 10%	
Year 1: $400 ×	.909	= $364
Year 2: 400 ×	.826	= 330.5
Year 3: 400 ×	.751	= 300.5
		$995

The answer, 10%, is the interest rate at which the $995 could be invested to generate sufficient funds to make payments equal to, and at the same time as, the prospective investment's anticipated receipts; i.e.:

$$\$995 \times 1.1 = \$1,094.5$$
$$\text{Less payment} \qquad 400$$
$$\overline{\hspace{3em}\$694.5}$$
$$\$694.50 \times 1.1 = \$760$$
$$\text{Less payment} \qquad 400$$
$$\overline{\hspace{3em}\$364}$$
$$\$364 \times 1.1 = \$400$$
$$\text{Less payment} \qquad 400$$
$$\overline{\hspace{3em}0}$$

As the above calculation clearly demonstrates, this answer is the actual earning rate of the funds in use, and is thus directly comparable with the cost of capital. It also demonstrates that this method does not, as is so often claimed, include implicit reinvestment of proceeds at the solution interest rate.

Comparing the Methods

The five different answers derived by these methods for the same demonstration proposal are tabulated below:

1. Net Present Worth at 5%......................... $94
2. Net Future Worth at 5%.......................... $109
3. Net Annual-Equivalent Worth at 5%............ $34.5
4. Net Capitalized Worth at 5%.................... $690
5. True Rate of Return............................ 10%

It is relatively easy to demonstrate that all of these numerically different answers are mathematically exactly equivalent. To illustrate:

• The *net present worth* of $94 can be compounded forward at the stipulated interest rate of 5% to the net future worth at that interest rate, or $109.

$$\$94 \times (1.05)^3 = \$109$$

• The *net annual-equivalent worth* of $34.5 can be obtained by multiplying the *net present worth* of $94 by the three-year capital recovery figure at the stipulated 5% interest rate.

$$\$94 \times .367 = \$34.5$$

• The net capitalized worth of $690 is found by dividing the *net annual equivalent worth* of $34.5 by the stipulated 5% interest rate.

$$\frac{\$34.5}{.05} = \$690$$

• The *true rate of return* cannot be calculated from any of the other answers. Instead, it must be computed from the original data by determining the *interest rate at which all the other answers are equal to zero.*

From the above, it is clear that the equivalency between the rate-of-return answer and the answers found by the four other methods is entirely different from the equivalency among the four. Equivalent answers for all of the first four yardsticks can be calculated at any stipulated interest rate, but the rate-of-return answer can be equivalent to all of the others at only one stipulated interest rate—the rate of return of the proposal. Thus, the answers yielded by the first four yard-

Alternatives	Project A	Project B	Project C	Project D	Project E	Project F
Investment at year 0, $	15	995	10,798	995	995	995
Receipts: year 1, $	40	400	4,000	600	1,144	0
year 2, $	40	400	4,000	571	0	0
year 3, $	40	400	4,000	0	0	1,261
Interest @5%						
Net Present Worth, $	94	94	94	94	94	94
Net Future Worth, $	109	109	109	104	99	109
Net Annual-Equivalent Worth, $	34.5	34.5	34.5	50.6	99	34.5
Net Capitalized Worth, $	690	690	690	1,012	1,980	690
Interest @4%						
Net Present Worth, $	96	115	302	110	106	126
Net Future Worth, $	108	130	340	119	110	142
Net Annual-Equivalent Worth, $	34.6	41.4	108.7	58.3	110	45.4
Net Capitalized Worth, $	865	1,035	2,718	1,457	2,750	1,135
Interest @6%						
Net Present Worth, $	92	74	−106	79	84	64
Net Future Worth, $	109.4	89	−124	89	89	76
Net Annual-Equivalent Worth, $	34.4	27.7	−39.6	43	89	23.9
Net Capitalized Worth, $	576	462	−660	716	1,483	398
True Rate of Return	261%	10%	5.5%	11%	15%	8.2%

sticks can be recognized as merely differences among the numerical equivalents of the disbursements and receipts at the stipulated interest rate, because the rate stipulated is not the true rate of return. Actually, the main difference between the answers to the first four yardsticks is the time to which the equivalency is related. The net capitalized worth has one additional built-in difference—the assumption of perpetual repetition of receipts.

What the first four methods actually measure is not the performance of the proposed project itself, but rather that of the total money committed during the entire anticipated life of the project—with the answers being various unspecified, weighted averages of the project's own performance, plus the arbitrarily valued but unstated other uses for which portions of the funds are presumed to be assigned. The true-rate-of-return answer, on the other hand, measures only the earning rate of the funds in use by the project. The only earnings included in the rate of return evaluation are those produced by the project itself; hence this answer is a true measure of the project's own profitability or economic productivity.

Since the answer is expressed as a single interest rate—the level, annual, equivalent earning rate of the funds in use—it is a number that is directly comparable with the cost of capital, which can be expressed in exactly the same fashion.[1] The usefulness of the four other answers is not only compromised by the inclusion of unrelated factors, but is also hindered by the requirement that they be stated in two figures—an interest rate, plus a remainder for which it is difficult to establish real significance.

The difficulty of interpreting the significance of

the first four answers is not as well recognized as it should be. This is because when the results from the application of these methods to a single problem are compared, the relationships between the answers obtained are relatively easy to establish and understand. It is only when one tries to compare the results obtained by application of these methods to a series of prospective investment opportunities that involve commitments of unequal magnitude and widely varying receipt patterns, that the possible confusions and misinterpretation become clearly apparent. And when the effects of even a small change in the stipulated interest rates are explored, opportunities for even greater confusion can be observed.

Table I has been prepared to show the kaleidoscope of answers that can be expected with these four methods. Now, compare these with the straightforward, single answer: the true rate of return on the bottom line of the table.

Additional Techniques

There are, in addition to the five methods discussed, two more basically different investment-appraisal approaches that appear in the literature and are occasionally used. These are:

1. The sinking fund, or Hoskold method.
2. The total return method.

In the sinking fund method (which has been reintroduced at frequent intervals, sometimes under a different name[3]), the receipts are first reduced to a level annual equivalent at the stipulated sinking-fund interest rate. The next step is to calculate the level annual sinking-fund deposit required to achieve a total

balance at the end of the project equal to the original investment. Then the annual sinking-fund payment is subtracted from the level annual equivalent of receipts. This difference, divided by the present worth of net disbursements and multiplied by 100, becomes the sinking-fund return. The application of this method to the previously used demonstration proposal is shown below. (Receipts are already uniform, so the reduction to level annual equivalent is unnecessary and the single disbursement at the start of the project is already the present worth.) With a stipulated interest rate of 5%, the answer is calculated as follows:

Level annual receipt.................... $400
Less sinking-fund deposit: 995 × .317 = 315
—————
Difference......................... $85

Sinking-fund answer: $\frac{85}{995} \times 100 = 8.5\%$

This 8.5% answer is the weighted average of 10% on the project and 5% on the sinking-fund deposits.

If an alternative such as the one shown below—i.e., one with exactly the same receipts over the three years, plus additional receipts in later years—is evaluated by this method, with the same 5% stipulated interest rate, some very peculiar results are achieved.

$ Investment—Year 0: 995
$ Receipts—   Year 1: 400 × .952 =   381
              2: 400 × .907 =   363
              3: 400 × .863 =   345
              4:   1 × .823 =   .8
              5:   1 × .784 =   .8
              6:   1 × .746 =   .7

Total present worth. $1,091.3
Level annual equivalent = $1,091 × .197 = $217
Less sinking-fund deposit   995 × .151 = 150
                                         ————
                                         $67

Sinking-fund answer: $\frac{\$67}{\$995} \times 100 = 6.7\%$

This would seem to indicate that additional receipts could lower the earning rate. It is because this method produces such distortions that I believe this approach should be classified as nonvalid and unacceptable.

Total Return Can Lead to Distortion

The total return method is generally referred to as the Baldwin Method.[4] The answer calculated is the interest rate at which the proposed disbursement would have to be invested in order to produce, at the termination of the project, an amount equal to the future worth of all proceeds at the end of the project at the stipulated interest rate.

Applied to the same two problems that served to demonstrate the sinking fund method, and using the same stipulated interest rate of 5%, the following results are achieved:

First Proposal:
$400 × (1.05)^2 =   \$441$
$400 × (1.05) =   420$
$400 × 1 =   400$
—————
Future Worth = $1,261

Since $995 \times (1.08)^3 = \$1,260$ (i.e., the future

worth), the Baldwin answer for this proposal is 8%. Now let us see what happens when there are additional receipts.

Second Proposal:
$400 × (1.05)^5 =   \$510$
$400 × (1.05)^4 =   486$
$400 × (1.05)^3 =   463$
$1 × (1.05)^2 =   1$
$1 × (1.05) =   1$
$1 × 1 =   1$
—————
Future Worth = $1,462

Since $995 \times (1.065)^6 = \$1,452$ (i.e., the future worth), the Baldwin answer is approximately 6.5%. This is an even greater distortion than that produced by the sinking fund method.

Conclusions

Returning to the five approaches discussed earlier, the following conclusions seem inescapable:

1. The true rate of return is the only one of the five valid yardsticks that can be depended upon to provide a consistent, reliable and quantitative measure of the extent of economic productivity of prospective investments. The answer can be directly compared with the cost of capital.

2. The four other valid yardsticks are really little more than decision tools with very little capability for measuring the extent of prospective project performance. While they may occasionally have marginal value in connection with certain types of analyses, the ease with which the meanings of the answers can be misconstrued makes the use of them by other than thoroughly experienced engineering economists a hazardous choice.

3. The sinking fund or total-return methods are not valid. The answers they yield do not measure the true economic productivity of proposed investments and will, at times, yield answers that would lead to ridiculous conclusions.

References

1. Childs, J. F., "Should Your Pet Project Be Built?," *Chem. Eng.*, Feb. 26, 1968.
2. Reul, R. I., "A New Way to Calculate Pay-Out," *Factory*, Oct. 1955; "The Profitability Index," *Harvard Bus. Rev.*, July-Aug. 1957; "Capital Investment and the Profitability Index," *The Manager*, Nov. 1962; "Profitability Index for Machine Justification," *Automation*, Mar. 1965.
3. Herron, D. P., "Comparing Investment Evaluation Methods," *Chem. Eng.*, Jan. 30, 1967.
4. Baldwin, R. H., "How to Assess Investment Proposals," *Harvard Bus. Rev.*, May-June 1959.

Meet the Author

Raymond I. Reul has spent the last 26 years with FMC Corp. (633 Third Ave., New York City), where he is coordinator of the firm's industrial engineering and systems engineering programs. He is well known as a lecturer and seminar leader—activities that have taken him to Canada, England, Ireland, Sweden, Norway, Holland, Germany, Spain, France and Switzerland. Mr. Reul has taught at Stanford, Rutgers and New York University, and has done considerable writing. He is a chemical engineering graduate of Lehigh University.

Two New Tools for Project Evaluation

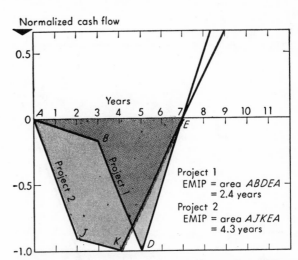

Project 1
EMIP = area *ABDEA*
= 2.4 years
Project 2
EMIP = area *AJKEA*
= 4.3 years

EMIP criterion offers an advantageous way of comparing the attractiveness of these two projects—Fig. 1

Based directly on the shape of the project cash-flow diagram, these two criteria are particularly useful for assessing projects in which prolonged research and development is needed, or in which the investment is made over a period of time.

D. H. ALLEN, *University of Nottingham, England*

The two criteria we will discuss can be used for measuring the attractiveness of a single project, for comparing several different projects, or for controlling a project in progress. Before examining their particular advantages, let us take a quick look at existing methods.

These can be divided into two categories: the short-term and oversimplified ones such as return on investment and payout time, and the more-sophisticated longer-term ones such as present worth and interest rate of return. Both types have certain drawbacks. The oversimplified ones take no account of the pattern of cash flow during a project, whereas those that involve discounting require the total life of the project to be known, together with the cash-flow pattern throughout. Although the effect of discounting reduces the importance of cash flows the further they occur in the future, the flows nevertheless have to be predicted.

It should be noted that all project-evaluation criteria are arbitrary to some extent. There is no such thing as an "absolute" criterion. Adelson[1] has pointed out that if the interest rate of return were to be universally

used, there would be a larger number of smaller plants than if present worth were always used. The project life usually has to be chosen arbitrarily and, to quote Adelson, "The cost of capital is one of those quasi-mythological creatures like the Loch Ness Monster—everybody has heard of it; a few people claim to have seen it, but nobody has ever run it to earth."

Since there are already so many economic criteria to choose from, why introduce more? The two criteria to be described here, EMIP and IRP, span the "simple" and "sophisticated" categories and, to some extent, avoid the disadvantages of each. They are simple to evaluate and do not require assumptions about project life or interest rates, but at the same time they take into account the pattern of cash flow during the earlier years—the ones most important and predictable from the viewpoint of assessing a project. They are particularly useful for projects where the investment in facilities and R&D is made over an extended time interval.

In developing EMIP and IRP, Nottingham University's chemical engineering department studied histories of several projects, including both successful and unsuccessful ones. We found EMIP and IRP to be more efficient than more-conventional criteria in distinguishing between the successes and failures. Some of this work has been reported in greater detail in Ref. 2 and 3.

Equivalent Maximum Investment Period (EMIP)

EMIP can be calculated or measured directly from the predicted cumulative cash-flow diagram, as shown in Fig. 2. The early stage *AB* represents expenditure on research and development, and this is followed by the larger expenditure *BC* on investment in full-scale plant. *CD* is a period of commissioning, and then the curve turns upwards as net income starts to come in

from sales. This income should be the net income after deduction of tax. The curve reaches the breakeven point E, where the initial expenditure is balanced by

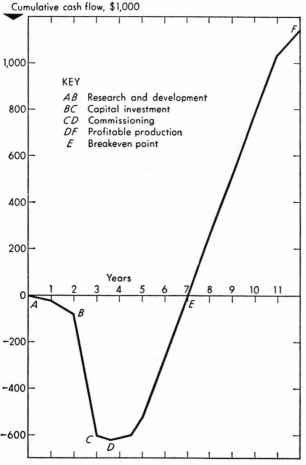

CASH-FLOW graph shows stages in a typical project—Fig. 2

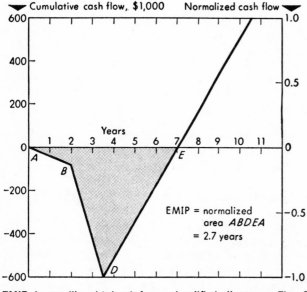

EMIP is readily obtained from simplified diagram—Fig. 3

the accumulated income, and it continues to rise for the rest of the project's profitable life.

EMIP is defined in terms of the cash-flow diagram. It is the area under the cash flow curve from the start of the project to the breakeven point, $ABDEA$ (in dollar years), divided by the maximum cumulative expenditure given by Point D (in dollars). It has the dimensions of time and is measured in years. In the simplified cash flow diagram in Fig. 3, the maximum cumulative expenditure is $600,000; the breakeven point is reached seven years after the start of the project; and the area under the curve, $ABDEA$, is 1,620,000 $-years. In this example, therefore:

$$\text{EMIP} = \frac{1,620,000}{600,000} = 2.7 \text{ yr.}$$

Alternatively the cumulative cash-flow scale can be "normalized" by dividing throughout by the maximum cumulative expenditure, as shown by the right-hand scale in Fig. 3. In this case, area $ABDEA$ gives the EMIP directly.

The EMIP value is equivalent to the length of a side of a rectangle whose area is the same as $ABDEA$ and whose other side corresponds to the maximum cumulative expenditure. As stated, EMIP stands for Equivalent Maximum Investment Period and it represents the equivalent period during which the total project debt would be outstanding if it were all incurred at one instant and all repaid at one instant.

EMIP is thus a short-term criterion analogous in some ways to payback time. But it is an improvement on payback time because it takes into account the pattern of expenditure and income from the start of a project up to the breakeven point. The shorter the EMIP for a project, the more attractive it is. For instance, consider Fig. 1 (previous page), which shows the normalized cash-flow curves for two projects that have the same total-project-payback time. Project 1 is more attractive than Project 2 because its EMIP (normalized area to the breakeven point) is less. This means that the equivalent maximum investment for Project 1 is recovered in a shorter time than for Project 2. Since the cash flows have each been normalized independently, the EMIP's are not affected by the actual size of each project. EMIP thus provides a ready means of comparing the attractiveness of projects of different magnitudes and timing.

Sensitivity analysis, to find the effect of various changes in a project, can be carried out easily using EMIP. The relationships between the different stages of a project in respect to both the relative expenditure and income, and also their timing, are clearly revealed. The project shown by Curve 1, Fig. 4, has an EMIP of 2.5 yr.; let us say this is regarded as the longest EMIP that will make such a project attractive. Consider what would be the effect if the R&D stage AB were to cost twice as much and take twice as long (AB_1 of Curve 2). If the capital expenditure is unchanged, i.e. $B_1D_1 = BD$, the income would have to increase from the original rate indicated by the slope of DE to the new rate given by the slope of D_1E_1 in order to keep the EMIP at 2.5 yr.

The position of E_1 can be found as follows. The new maximum cumulative expenditure, D_1, is $700,000, and the required EMIP is 2.5 yr. Therefore:

$$\text{Area } AB_1D_1E_1A = 700,000 \times 2.5$$
$$= 1,750,000 \text{ \$-yr.}$$

The position of the breakeven-point E_1 that satisfies this equation can now be found, and the slope of D_1E_1, i.e., the new rate of income needed to compensate for the change in R&D, can be computed.

As a further example, consider the effect on the original project of an increase in the capital investment BD from $500,000 to $700,000. This is shown as BD_2 in Curve 3 and gives a new maximum cumulative expenditure of $800,000. To compensate for this so that the EMIP remains the same, the rate of income would have to increase from that given by DE in Curve 1 to D_2E_2 in Curve 3. The position of E_2 is such that:

$$\text{Area } ABD_2E_2A = 800,000 \times 2.5$$
$$= 2,000,000 \text{ \$-yr.}$$

The project case histories that we have studied confirm that the duration of R&D in relation to the rest of a project has an important effect on its eventual success, due to such factors as competitive activity and the general rate of technological advance and technological obsolescence in an industry. Extending the duration of R&D tends to reduce the project's chance of success, unless there are corresponding additional benefits to be gained at a later stage, such as reduced capital or operating costs, easier startup, or an improved product or market. An increase in the cost of R&D will have a similar effect, although it will be less pronounced when the R&D expenditure is only a small part of the total cost of a project.

As shown in Fig. 4, an EMIP analysis brings out the relationship that should be maintained between R&D, capital investment, commissioning, production and sales if continuing a project beyond the R&D stage is to be worthwhile. Thus EMIP analysis is useful for reviewing a project during R&D, to see how the balance between the different stages is being maintained. If it cannot be maintained, the EMIP may become too long, and the analysis will show whether the project should be stopped.

EMIP analysis can also be applied in the initial selection of projects, as the relative attractiveness of projects having different investment requirements, different returns and different time scales are easily compared in terms of their EMIP's. Spier[4] regards EMIP as a much less arbitrary criterion for assessing projects than the more conventional ones. Like all criteria, however, the numerical value of EMIP that distinguishes between attractive and unattractive projects is arbitrary insofar as it must be based on experience. It depends on the industry and type of project and on the degree of risk involved.

One way to establish the critical value is to analyze data from earlier projects, the results of which are known. The analyses carried out at Nottingham on a

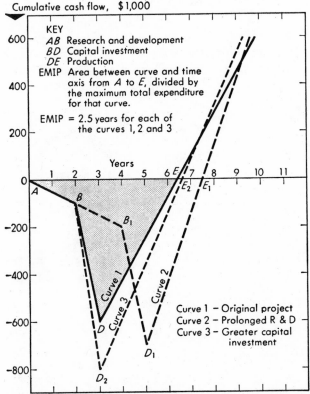

SLOPE of D_1E_1 and D_2E_2 reflects change in income needed to keep EMIP constant when project is modified—Fig. 4

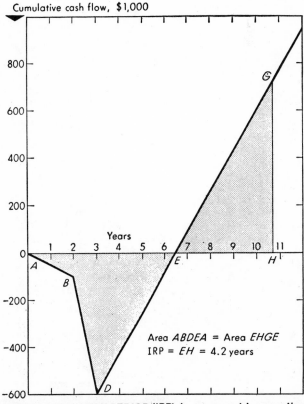

INTEREST RECOVERY PERIOD (IRP) is measured by equating the areas above and below the graph's time axis—Fig. 5

selection of British projects[2] indicate that in the chemical and allied industries, the EMIP should not be more than about six years for a low-risk project, the outcome of which can be predicted with a fair degree of confidence, and not more than three to three and one-half years for a moderately risky one. Although these figures may be taken as a general guide, it is advisable for a company to check the EMIP's of a few of its own projects in retrospect before using EMIP for project assessment. This will enable it to relate numerical EMIP values to its measurement of risk and degree of success and the effect of any special circumstances.

The Interest Recovery Period (IRP)

Although similar in many respects to EMIP, the Interest Recovery Period (IRP) is a longer-term criterion because it takes into account the pattern of project cash-flow for a period beyond the breakeven point.

Referring to Fig. 5, IRP is defined as the time from the breakeven point to the point on the curve where the area below the time axis equals the area above it—i.e., it is the time in which simple interest charges on the total investment could be recovered by investing the balance of profits at the same interest rate. IRP is a useful criterion when the pattern of income for the project can be predicted beyond the breakeven point, and it is used in conjunction with EMIP (or alternatively, payout time) in such situations.

In Fig. 6, the project shown by Curve 1 has an EMIP of 2.5 yr. and IRP of 4.2 yr. Suppose it becomes known

that the rate of income will fall four years after startup because, let us say, a competitive plant will then come on stream. Curve 2 shows the new income schedule. The EMIP is still the same, but the IRP has increased to 5.3 yr., indicating that the revised project is less attractive.

Since the basis for the IRP measurement is the area up to the breakeven point, i.e. the EMIP, a change in EMIP will tend to produce a corresponding change in IRP. The IRP provides additional useful information about the attractiveness of a project when it is known that the slope of the cash-flow curve changes significantly soon after the breakeven point.

EMIP and IRP do not give a direct measure of profitability in money terms. This does not detract from their usefulness as a means of assessing projects, especially in the early stages when uncertainty is greatest and long-term profitability estimates are in any case speculative. The relationship between EMIP and IRP on one hand and ultimate profitability on the other is a close one.

Both EMIP and IRP are very simple to measure or calculate directly from a cash-flow diagram. They can be measured by counting squares or with the aid of a planimeter, or they may be easily calculated if the cash-flow diagram is approximated to a series of straight lines.

It is hoped that this article has shown why the criteria provide a useful basis for estimating the attractiveness of a project, for comparing projects (including those with different levels of investment or with different time scales), for studying sensitivity and also for making reevaluation and project-control decisions in the light of changing developments.

References

1. Adelson, R. M., *Operational Research Quarterly*, **16**, p. 19 (1965).
2. Allen, D. H., Edgeworth-Johnstone, R., *AIChE-IChemE Symposium Series No. 7*, p. 67 (1965).
3. Allen, D. H., *Science Journal*, **1**, 10, p. 79 (1965).
4. Spier, G. L. E., *Director*, Oct. 1966, p. 25.

Meet the Author

Derek H. Allen, lecturer in chemical engineering at the University of Nottingham, England, is currently engaged in investigating the strategy of process development. He holds a B.Sc. and Ph.D. in chemical engineering from Birmingham University, England, and is a Chartered Engineer and Fuel Technologist. He gained development, operational and planning experience in the gas and petroleum industries before taking up his present appointment.

Key Concepts for This Article

Active (8)	Passive (9)	Means/Methods (10)
Evaluating*	Investment*	Cash flow*
Selecting	Projects*	Diagrams
Comparing		Curves
Controlling		Relationships

(Words in bold are role indicators; numbers correspond to EJC-AIChE information retrieval system. Asterisks mark key concepts suggested for indexing. Others are added to improve reading as an abstract. Indexing is described in *Chem. Eng.*, Oct. 11, 1965, p. 187.)

Cumulative cash flow, $1,000

Area *ABDEA* = Area *EHGE* = Area *EH₁G₁E*

Years

Curve 1 – Original project
Curve 2 – Changed income curve
IRP for Curve 1 = *EH* = 4.2 yr.
IRP for Curve 2 = *EH₁* = 5.3 yr.
Both curves have the same EMIP of 2.5 yr.

CHANGE in income curve after breakeven point is reached does not affect EMIP but does have an effect on IRP—Fig. 6

Working Capital Affects Project Profitability This Way

Project A production: 4 million lb./yr.		
Income, less marketing expenses		$905,000
Material cost	$112,000	
Manufacturing cost	$493,000	
Total product cost		$605,000
Profit		$300,000
Investment		$1,000,000

Rate of return, excluding working capital.........................32.5%

Rate of return, including working capital by

- Estimating average month's needs......27.0%
- Computing 30% of annual sales.......25.8%
- Using 15% of fixed investment........28.5%

For economic evaluations, three ways to ...

Estimate Working Capital Needs

L. R. BECHTEL, Atlantic Refining Co., Philadelphia, Pa.

UNCERTAINTY is always an element in the review of any economic evaluation. To narrow the area where judgment must be applied, it is important that the evaluator consider all economic factors.

Working capital is one such factor. Working capital is that money tied up in raw material, semi-finished and finished inventories, accounts receivable and operating cash.

LOUIS R. BECHTEL is group leader, technical service dept., of Atlantic Refining Co. He joined the company in 1954, after having worked with Firestone Tire & Rubber Co. Graduated at Case Institute of Technology in 1950, he studied chemical engineering as an undergraduate. He is a member of AIChE, active in its committee work.

Although working capital funds are continuously liquidated and regenerated, this money is unavailable for any other purpose—it should be regarded as real investment.

Where new product justifications are involved, working capital is of particular importance. Too, expansion projects should include working capital. On the other hand, working assets are less important when the capital expenditure involves minor changes in current plant operations.

The literature contains many methods to determine working capital. Of those reviewed and successfully applied, the following three methods are analyzed here:

- Inventory—one month's supply each of raw material, semi-fin-

ished material and finished material. Accounts receivable—one month's sales. Operating cash—one month's total cost.
- Use 30% of annual sales.
- Use 10-20% of fixed capital investment.

Changing Profitability

To determine the effect of working capital on rate of return of a project, two hypothetical cases have been selected—one at double the investment of the other.

Inclusion of working capital in evaluation of Project A, which returns 32% without it, lowers the return 4-7% depending on which of the three methods is used to calculate working capital. In Project B, the effect of working capital is

Cost Summary for Two 4-Million Lb./Yr. Plants—Table 1

	Project A*	Project B
Material cost, $/yr.		
Raw material debit	$155,000	$155,000
Byproduct credit	43,000	43,000
Net material cost	112,000	112,000
Manufacturing cost, $/yr.		
Labor	107,000	107,000
Maintenance	100,000	100,000
Utilities	36,000	36,000
Overhead	150,000	150,000
Depreciation	100,000	200,000
Total manufacturing cost	493,000	593,000
Total product cost, $/yr.	605,000	705,000
Annual income, less marketing cost	905,000	905,000
Profit	300,000	200,000
Investment	1,000,000	2,000,000
Expenditure period	1 year	1 year
Plant life	10 years	10 years
Profit and depreciation	$400,000	$400,000
Salvage	None	None
Rate of return†	32.5%	14.0%

* Ref. 7   † Investor's method

not so pronounced; returns were lowered 2½-3%.

Several influencing factors dictate the method to use for calculating working capital. They include

1. Seasonal variations in project operation.

2. Accuracy of investment estimate.

3. Firmness of price structure.

In projects involving low yields or conversions, large volumes of raw materials are required; average inventory days, therefore, deserve close scrutiny.

In projects with high product value, capital necessary for product inventory warrants a detailed look—it can have significant effect on return. Remember, too, that terms offered to customers—accounts receivable—affect the company's cash position.

How to Compute

To see how the three methods are used, consider the following calculations based on Project A.

Method 1—Compute inventory costs. For one month's supply of raw material: $155,000/12 = $12,900. For one month's supply of semi-finished material (based on raw material plus one-half normal conversion cost); labor — 2.7¢/lb., maintenance — 2.5¢/lb., utilities — 0.9¢/lb., conversion cost total — 6.1¢/lb. One-half of this is $3.05¢/lb. or $10,200/mo., which added to

one month's raw material cost totals $23,100. For one month's supply of finished material: $605,000/12 = $50,500.

For accounts receivable, one month's sales is $905,000/12 = $75,500. For operating cash, one month's total cost is $605,000/12 = $50,500. The total of the three items is $86,500 + $75,500 + $50,500 = $212,500.

Method 2—30% of annual sales is (0.30)($905,000) = $271,000.

Method 3—15% of fixed capital investment is (0.15)($1 million) = $150,000.

Of the methods described, the first best details the working or quick assets, *i.e.*, assets which can be readily and quickly converted into cash. The time elements involved for inventory were arbitrarily chosen as representing a typical operation.

Accounts receivable is simply capital retained to cover credit extended to customers, terms of sale. Operating cash is that fund set aside for running the business. Note that where seasonal markets exist, Method 1 is subject to error.

For Project A, the working capital of $212,500 calculated by Method 1 is about equally divided between inventory, accounts receivable and operating cash. Incorporating this sum in the rate of return calculation lowered profitability rate from 32.5% to 27%. Project B return, with Method 1 working

capital included, dropped from 14% to 11.6%.

Method 2 accuracy is limited by accuracy of income or selling price estimates. When working capital calculated by this method is included in Project A, rate of return drops from 32.5% to 25.8%. Project B return drops to 11.5% from 14%.

Using a percentage of fixed capital investment for working capital estimates is open to question where investment estimates are preliminary. Thus, in Method 3, 15% is an average and should be applied with caution. Cases reviewed indicate a variation in this figure ranging from 5-50%. For this reason, Method 3 is probably the least desirable formula and should be used only after considerable experience. Including working capital in Project A, by Method 3, lowers return from 32.5% to 28.5%; in Project B, from 14% to 11.2%.

REFERENCES

1. Bates, A. G., Weaver, J. B., "Six Profitability Methods," *Chem. Week*, June 15, 1957, p. 115-27.
2. "Chemical Inventory Growth" *Chem. Eng. News*, 35, 82, 84 (Aug. 19, 1957).
3. Chilton, C. H., "Cost Estimating Simplified," *Chem. Eng.*, June 1951, pp. 108-111.
4. Clarke, L., "Cost Estimates Answer Questions," *Chem. Eng.*, Dec. 1951, pp. 144-5.
5. Dybdal, E. C., "Engineering and Economic Evaluation of Projects," *Chem. Eng. Progr.*, 46, 57-66 (Feb. 1950).
6. Jones, H. E., Kjellmark, E. W., "How Computers Aid Economic Studies," *Petr. Ref.*, 37, 161 (June 1958).
7. Martin, J. C., "Economic Analysis," Chemical Process Economics in Practice, 83, Reinhold, 1956.
8. Perry, J. H., "Chemical Business Handbook," McGraw-Hill, New York, 1954.
9. Rothschild, R. M., Kircher, P., "Projecting Capital Needs," *Journal of Accountancy*, 100, 51-6 (Sept. 1955).
10. Sandel, M., "Allocating Existing Facilities in Evaluating Investment," *Chem. Eng.*, Nov. 1957, pp. 231-4.
11. Symposium on Estimation and Profitability, *Chem. Eng. Progr.*, 52, 399-401 (1956).
12. "That's Why Companies Go for Cash Forecasts," *Business Week*, 46-54 (Apr. 28, 1956).
13. Tielrooy, J., "Fixed Capital Cost Estimates," *Chem. Eng. Progr.*, 52, 187-90 (1956).
14. Tyler, C., "Where Cost Estimates Go Sour," *Chem. Eng.*, Jan. 1953, pp. 198-200.
15. Tyler, C., "Economics, Chemical." Encyclopedia of Chemical Technology, Vol. 5, p. 446-465, Interscience, New York, 1950.
16. Tyler, C., "Chemical Engineering Economics," 3rd ed., McGraw-Hill, New York, 1948.
17. Van Noy, C. W., Dunsville, T. C., Dressler, R. G., Chaffee, C. C., Guide for Making Cost Estimates for Chemical-Type Operations, U. S. Bureau of Mines, R.I. 4534 (Nov. 1949).
18. Vilbrandt, F. C., "Chemical Engineering Plant Design," 3rd ed., Chap. XIII, p. 410-544, McGraw-Hill, New York, 1949.
19. Wessel, H. E., "How to Estimate Costs in a Hurry," *Chem, Eng.*, Jan. 1953, pp. 168-171, 200.
20. "Working Capital Case." *Chem. Eng. News*, 35, 96 (Apr. 22, 1957).

We all take risks for several reasons: for profit, for personal satisfaction in the results, for career advancement, for thrills, and so on. But money is not the only thing we engineers risk. Personnel safety, time, and reputation are also subject to risk—and we need a method for . . .

Calculating the Calculated Risk

DAVID STUHLBARG, *The Fluor Corp., Ltd.*

An engineer's task is not merely to design but to design so as to achieve optimum economy. This is an objective usually reached by developing and evaluating various alternate plans and then choosing the "best" of these. But we all know that even the most economic plan may present some risks—and we obviously need some tools we can use when a decision must be made as to whether or not to take these risks.

This article will discuss the various factors affecting such decision, and will present a number of equations that enable us to evaluate quantitatively the "% Risk," "Weighted % Risks," "Capital Equivalent of Operating Profit," and "Capital Equivalent of Operating Loss."

A Commonsense Perspective

Not all of the risks we take are calculated risks. Some we take because we believe we must. These are the ones that are inherent in our way of life, our way of making a living, or inherent in the nature of our business.

Many of the risks we take can and should be minimized. We minimize traffic risks by driving carefully and at reasonable speeds, by using seat belts, and by keeping our cars in good mechanical condition. Most chemical companies today minimize their risks by testing new products and new processes in a pilot plant before launching full-scale production.

Important as it may be to avoid and minimize risks, however, we should not make this our prime concern because there are times when common sense tells us which risks are justified and which are not—as in the following example.

Some years ago, one of the larger manufacturers in this country studied a plan to manufacture alkyl benzene, one of the basic chemicals used in the manu-facture of detergents. This was a company whose two leading products were detergents and edible fat, or shortening. The risk was minimal since the payout was satisfactory, the process was an established one that the firm had further tested in its own laboratories, and it had a captive market available in its own well-established line of detergents. Nonetheless, the decision was against this almost fool-proof plan and in favor of the apparently riskier course of developing and marketing new products that would broaden the scope of the business.

The ultimate outcome of this decision is interesting and instructive. Alkyl benzene is now obsolete as a raw material. The public outcry against stream pollution led to the development of new, more readily biodegradable detergents, and alkyl benzene was the first material to fall.

On the other hand, the company developed and marketed a cake mix under the policy that led to the above decision, and this, in turn, took up the slack in the sales of shortening, which occurred as housewives switched, almost overnight, from home-made cakes that required shortening, to packaged mixes that already contained it.

In a business sense, what we should do, as one author puts it,[1] is to "maximize opportunity," and that is what this company did.

Presumably calculated risks are taken because there is a choice to be made. This can be a yes-or-no type, or it can be a choice between alternate plans. In the case of monetary risks, decisions are either in favor of the lowest cost plan or else based upon calculations for payout, rate of return, new worth, or some similar criterion.

To some managers, financial considerations would be sufficient. Some years ago, the writer's family, who are in the hardware business, had a customer whose

only requirement for builders' hardware was "I don't care how much it costs,—just so long as it is the cheapest you have." Fortunately for him, and for the people who bought his houses, the cheapest line carried was reasonably durable and not unattractive.

Potential Gain vs. Possible Loss

Most of us, in considering an investment, will use more than initial cost or payout or rate of return. We would consider first of all if we could afford the risk. If so, we would want to know the effect on product quality, safety, and schedule. If we were choosing between brands of equipment, we would consider reliability of guarantees, availability of spare parts, reputation of the manufacturer. Each problem would have its own set of abstract factors that should be considered in making the decision.

A calculated risk, in the final analysis, must therefore be a matter of judgment. However, though we must not overlook or ignore these abstract factors, the main thing we consider is the potential gain, and how it compares with the possible loss. For this comparison, we propose the equation: Risk = (possible loss)/(potential savings), or Eq. (1) in the box above.

The "way out" in Eq. (1) is the course of action we take if the risk should fail. Consideration of this is a step that is often overlooked; yet if we fail to consider it, we are not really taking a calculated risk (though we may call it that) but gambling on success.

The % Risk obtained by Eq. (1) is strictly a function of costs, profits and losses. By itself, it does not say whether the risk is a good one, which should be accepted, or a bad one that should be avoided. We must compare this figure with a suitable criterion, and the one we use is "chance for success." Thus, we can revise Eq. (1) by including the "chance for success," as shown in Eq. (2).

If the resultant "Weighted % Risk" from Eq. (2) is less than 100%, the risk is apparently a favorable one. The trouble is that the chance for success is usually little more than an educated guess. Therefore, to be reasonably conservative, we would expect the weighted % risk to be well under 100%. Looking at this in terms of "odds," the "weighted odds" in favor of the plan (dropping the % term) is the reciprocal of the weighted risk.

Eq. (1), however, can usually be used directly. If the chance for success (provided it is above 50%) is well above the % Risk, the risk is a favorable one. Eq. (2) may be used also in marginal cases.

To illustrate the foregoing, here is an example from the writer's experience.

In 1950, a waste disposal basin similar to one the company had recently installed at another plant was proposed for the treatment of chemical processing and electroplating wastes at a large jet-engine plant then under construction in southern Ohio. This basin was estimated at about $300,000.

Detailed analysis of the planned operations showed that the wastes would consist of: (1) continuous flows of slightly contaminated water from various rinse tanks, and (2) a very infrequent dump of one of the strong process solutions. The rinses were expected to be essentially self-neutralizing.

The alternate plan that the writer developed and proposed called for: (a) no basin, (b) polishing pH control directly in the sewer line for the continuous rinse overflows, and (c) the installation of batch treatment tanks for the strong wastes, with a subsequent slow bleed to the sewer. This plan was expected to cost about $40,000.

There was some opposition to the plan initially, particularly by the company's specialist in waste disposal, who was distrustful of theoretical calculations of wastes. It was also felt that the plan had no better than a 50% chance for success.

The way out consisted of installing the basin. To facilitate the way out, and to counter the objections, the alternate plan was amended to include routing of sewer lines alongside the area allotted to the basin, at an increase in cost to about $45,000, and several abstract factors in favor of the plan were presented: (1) stream pollution laws had been formulated and published, but had not yet been enacted into law—as written, industry would be given a reasonable period of time to comply; (2) waiting until after startup would permit the development of a more precise design basis for the basin if it should be required; (3) about half of the $45,000 cost would have to be spent, anyway.

In the face of all this, opposition was withdrawn and the plan was accepted on a calculated risk basis. Years later, it was still in good working order.

Eq. (1) had not been formulated at the time of the above study. Were we to have this problem now, we would find that:

$$\% \text{ Risk} = \frac{\$45,000 + (\$300,000 - \$22,500) - \$300,000}{\$300,000 - \$45,000} \times 100\% = 8.8\%$$

If we compare this with the estimated 50% chance for success, we see that the latter is far greater than the % risk; therefore, the same recommendation would be made now that was made then—in favor of the plan.

How to Account for Operating Costs

Operating costs were not a factor in the previous example. Most of the time, however, they are of importance in the choice of plans. Moreover, if the risked plan fails, there could be a loss in profit due to a delay in startup or a subsequent shutdown for changes. We should therefore include the potential difference in operating costs in determining the "potential savings"— and any loss of profit in the cost of the way out.

To get these all on the same footing, we use:

$$P_e = N [(L_G + E) (1 - r) - rD] \qquad (3)$$

where P_e is the capital equivalent of operating costs or losses, and N is the period over which they apply, and

$$P_e = N [(R_G - E) (1 - r) + rD] \qquad (4)$$

for the capital equivalent of specific profits.

Eq. (3) and (4) were derived from the well-known formula for payout based on cash flow,[2] as follows:

Payout = Investment/[(Net return after expenses and taxes) + (depreciation × income tax rate)]

Or:

$$n = \frac{}{(R_G - E) (1 - r) + rD} \qquad (5)$$

Practical Example of Risk Evaluation

Let us illustrate all of the foregoing by working out a problem that is based on an actual case—though all of the figures and some of the facts are fictitious.

For a high-pressure process, the decision had to be made whether to have two reactors in parallel or a single larger one. The former were of a proven design; the latter would be larger than any previously built for this type of service or pressure, and would involve some recent metallurgical advances. Relative costs were \$2,200,000 vs. \$1,500,000, respectively.

The chance for success was rated quite high—better than 90%—because similar (though smaller) reactors of the new type had been built by the vendor, and the metallurgical principles were quite sound. The risk was

that the vendor would be unable to extend his manufacturing techniques to the larger size, and that the reactor would consequently have a lower pressure rating. This, in turn, necessitated a lower operating pressure for a resultant reduction in throughput rate of about 8%. The two smaller reactors, however, would have had a 1% smaller capacity in the first place.

The way out, in case of failure, would be to operate at the reduced rate until a new smaller reactor could be purchased and installed in parallel. This would cost \$720,000 and take 15 mo. to accomplish.

Value of the product, after allowing for raw-material costs and operating expense, was estimated at \$20,000,000/yr. before income tax. Other pertinent figures were maintenance, insurance, and property taxes—or 4% of the investment per year; depreciation—say, 5%; and federal income tax, 52%. The over-all plant investment was based on a 4-yr. payout. Using these figures, we obtain:

$$E = 4\% (\$1,500,000 - \$2,200,000) = -\$28,000$$
$$D = 5\% (\$1,500,000 - \$2,200,000) = -\$35,000$$

Applying Eq. (3) and (4), we have:

P_e (profit) = 4 yr. [{(1%) (\$20,000,000) + (−\$28,000)} (1 − 0.52) + 0.52 (−\$35,000)] = \$364,960

P_e (loss) = 1.25 yr. [{(8%) (\$20,000,000) − (−\$28,000)} (1 − 0.52) + 0.52 (−\$35,000)] = \$965,950

The minus sign for E and D indicates that the risked plan resulted in a capital savings. Next:

Weighted original cost = \$1,500,000 − \$364,960 = \$1,135,040

Weighted cost of way out = \$720,000 + \$965,950 = \$1,685,950

We can now calculate the % Risk by using Eq. (1):

$$\% \text{ Risk} = \frac{\$1,500,000 + \$1,685,950 - \$2,200,000}{\$2,200,000 - \$1,135,040} \times 100\%$$
$$= 86.9 \%$$

The chance for success, 90%, is greater than the risk; therefore, the risk is a favorable one. But since the figures are close, it is wise to apply Eq. (2):

$$\text{Weighted } \% \text{ Risk} = \frac{86.9\% \times 10\%}{90\%} = 9.7\%$$

The weighted odds in favor of the larger single reactor are then 1/0.097, or 10.3 to 1.

References

1. Drucker, Peter F., "Management for Results," Harper & Row, New York, 1964.

2. Schweyer, H. E., "Process Engineering Economics," McGraw-Hill, New York, 1955.

3. Stuhlbarg, D., "Economic Justification for Equipment," *Chem. Eng.*, Jan. 17, 1966.

Meet the Author

David Stuhlbarg is a Senior Project Engineer for The Fluor Corp., Ltd. (2500 South Atlantic Boulevard, Los Angeles, Calif. 90022). He has also been a project manager, process engineer, heat-transfer specialist, chemical supervisor, and teacher. He is a member of AIChE, holds a chemical engineering degree from the University of Cincinnati and has done graduate work in both mathematics and chemical engineering. This is his fourth article for CE. He has also been a contributor to Perry's "Chemical Engineer's Handbook."

7

Economic Evaluation of Business and Technical Alternatives

What four chemical companies return on investments. Company[*]:	A	B	C	D
Gross profit as % of ownership equity	16.3	18.4	4.9	13.9
Net profit as % of ownership equity	8.3	9.3	2.6	7.2
Gross profit as % of book value	25.7	20.0	6.2	20.4
Net profit as % of book value	13.1	10.1	3.3	10.6

Evaluating the Proposed Plant

Here is a guide for determining whether that proposed new plant or plant addition will make the kind of profits your company expects.

ROBERT FRUMERMAN, *Consulting Engineer, Pittsburgh, Pa.*

Making even the toughest business decision should be a simple matter, provided that the problem is clearly stated, the objectives are well-defined, all the necessary and pertinent information is at hand, and a logical and unemotional approach is used to analyze the situation.

Since such an ideal situation rarely if ever exists, it is worthwhile to examine the tools of decision-making in order to make the best of what is generally a bad situation.

Applied to the chemical process industries, questions like these arise:

• Should we expand and modernize our present plant or build a new one?

• Should we expand our manufacturing operation to produce a new product in a new plant?

We shall examine the techniques for evaluation of problems such as those above.

Obviously, this type of decision-making requires forecasting the future, which means that all the facts will never be known until too late. Therefore, some manager in a position of responsibility must add a judgment factor to the very best possible assembly of facts and thus resolve a course of action.

The kinds of information required to make these decisions—market surveys, business forecasts, and engineering studies—are interpretations of a body of

facts that are meaningless individually. Hence, managers differ widely in their reliance on such information.

An executive might well ask (and some do) "Since I am going to be forced to make my final decision based on my intuition, why waste time and money on exhaustive studies?" One answer is that intuition is an unconscious distillation of experience, and every fact and outside opinion adds to that experience. Another answer is that a thorough investigation, using the available facts and the most able interpreter, is an inexpensive form of insurance that the final action will have a strong probability of success.

We must understand what goes into an economic evaluation in order to know how to make one, how to interpret it, and how far to trust it. Table I shows some of the requirements for an engineering study and a market analysis. However, the most important ingredient of an evaluation is people, so let us now introduce the chemical process engineer and watch his mind at work as he performs the engineering study. This will prove enlightening when interpreting the results.

The process engineer is a specialist in reducing a chemical process to a workable and economic plant. He is expert in the selection and design of all the hardware needed to transform the raw material into product. He is aware of alternate techniques to achieve the same end-result and knows how to select the best of these. He is keenly conscious of costs as well as

* From annual reports (1960 and 1961) of four diverse chemical companies.

Market and Engineering Study Checklists— Table I

A market study should show:

1. Existing and probable future demand.
2. Current and probable future prices.
3. An estimate of effects of competition.
4. An estimate of the following costs of marketing, related to the number of units sold and the producing location:
   a. Warehousing.
   b. Distribution.
   c. Selling.
   d. Advertising.

An engineering study includes a comprehensive plant design and estimates of the following:

1. Capital cost.
2. Labor cost.
3. Raw-material cost.
4. Utilities cost.
5. Any other items that may significantly affects costs, e.g.:
   a. Waste disposal.
   b. Storage.
   c. Control laboratory.
6. Effect of location.
7. Effect of plant size.
8. Effect of process alternates.
9. Effect of operating variables.
10. Effect of items that may change in the future.

Economic Study Checklist—Table II

1. Revenues
   a. Sale of products.
   b. Sale of byproducts.
2. Costs
   a. Direct labor and supervision.
   b. Administration and overhead.
   c. Raw materials.
   d. Utilities.
   e. Supplies.
   f. Maintenance and repairs.
   g. Depreciation.
   h. Rent.
   i. Licenses and royalties.
   j. Insurance.
   k. Packaging.
   l. Transportation.
   m. Distribution.
   n. Sales expense.
      1. Salaries and expenses.
      2. Advertising.
   o. State and local taxes.
3. Gross Profit
   Minus estimated federal income taxes.
4. Net Profit.

technology, and his whole job consists of forever balancing one against the other to obtain the optimum compromise for any given objective—lowest plant cost, cheapest unit cost, or any other stated goal.

He has a grasp of practical problems, such as how the operator will function in the newly designed plant, as well as the theoretical concepts necessary for good design. It sounds like a rather unspecialized sort of specialty; yet specialty it is, for a great deal of training, experience and practice underlie the solution of the more-complex problems.

The raw materials needed by the chemical process designer to produce a knowledgeable study are data. He must know:

• The physical properties of materials in process.
• The principal reactions that occur, as well as the side reactions.
• Rate and equilibrium data.
• Special information, such as corrosion rates of materials in contact with the chemicals in process.
• The process involved and the conditions required for operation, such as temperature, pressure and concentration.

Preferably, the process engineer should have enough data to evaluate the effect, on the yield and quality of product, of varying the process operating conditions. If he understands what latitude he has in designing the process, he can make the plant more economical and easier to operate.

If the study is to help the selection of plant location or size, those facts that will influence the choice must be known. For example, the choice between two plant sites is affected by the size and condition of the properties, the availability of utilities and transportation facilities, and the labor situation apt to be encountered at each site.

First Steps

Data in hand, the engineer now begins the application of technique. First, a crude flowsheet is sketched, showing the principal pieces of apparatus, the process operating conditions (temperature, pressure, concentration) and any other conditions required for proper operation.

Not all the process equipment is shown. For example, the need for pumps at an unexpected place may result from later layout studies that demonstrate the impracticability of gravity flow. In another situation, the requirement for heat addition in the process may result in the use of a fired heater or a Dowtherm heat exchanger. The choice can be made later. The purpose of the first flowsheet is to make a general picture of the process good enough for the preparation of a heat and material balance, which is the next step.

Once the heat and material balance is made and tabulated in ready reference form, the engineering flowsheet can be started. This flowsheet is the foundation of all the process engineering. As he constructs it, the engineer performs a preliminary design or selection of all process items including vessels, mechanical equipment, heat transfer apparatus, instruments, controls, and piping. The designer asks himself, "What do I want to happen? How do I make it happen? How do I control it? What can go wrong, and what do I do about it?" The answers to these questions must then be incorporated in the flowsheet.

The flowsheet[1] is a schematic representation of the plant. Every equipment piece must show. This includes spare items such as pumps—because the piping will certainly be affected by the addition of spare pumps, and because the flowsheet must indicate what steps are taken to ensure dependable continuous plant operation. All necessary lines and equipment are shown, not only for principal flow but for startup, bypass, recycle, emergency, utilities, and safety systems. The well-made flowsheet leaves nothing to the imagination.

Each item of equipment is then assigned a number (using some consistent and convenient numbering system) and an item list is prepared. A specification is then written for each item on the list. The specification must show size, service, duty, materials of construction, and design features. It must be complete enough to permit preparation of a cost estimate of each item. The cost estimate for the complete plant

is then prepared by techniques described at length in the literature.[2]

Watch Non-Process Costs

A revealing test can be applied to the process engineer's work. Has he considered the nonprocess costs as seriously as the process costs? The making of bleach is certainly a chemical process, but the investment and operating cost may be found largely in the packaging operation. In another case, a large off-battery expense may have to be incurred for installing an electric substation. Have such items been treated with the same meticulous attention as the chemical process? This test is an excellent measure of the engineer's appreciation of the full scope of his work. By the full scope of work, we mean that the cost of erecting a manufacturing plant, prepared to operate, includes everything necessary to that operation.

A checklist such as the one in CHEMICAL ENGINEERING, April 16, 1962, p. 134,[3] is helpful in preventing oversights. If freight must be moved by barge, dock facilities have to be considered. If the new plant requires modifications to the existing boiler plant and the addition of a long steam header and condensate-return header, the neglect of such a situation in an economic evaluation can be serious.

A complete engineering study yields a unit production cost made up of the items shown in Table II. To this must be added the items of marketing cost shown in the same table. Now we can see what the study has produced. The first result is a critical examination that determines if a workable technology exists. If any doubt remains, and the economics appear favorable, more development work is needed.

The study has also produced a basis for budget appropriation and financial planning, as well as yielding the basis for final engineering work. The usefulness of all this depends on whether it shows what everyone hoped for—a potentially profitable investment.

Again we have oversimplified the problem. We have already recognized that the study should show the effect of plant size and location on economics. To generalize, it must show not only if a process is economical, but what are the conditions required to make it so. Since profit is the small difference between two large numbers—revenue and cost—it is quite sensitive to minor changes in these numbers. Part of the result of a shrewd study is the recognition of which elements exert substantial leverage on the results of the study, as well as the identification of future events that could reverse the conclusions of the study.

If, for example, the plant cost is so high that depreciation figures heavily in cost of goods sold, it is plain that a poor plant-cost estimate may render the entire study useless or misleading. The leverage of plant cost on unit production cost may be reduced by considering a larger plant (provided there is adequate market for the product) since the cost of plant per unit of capacity decreases with size.

Labor is often the controlling element of cost. Perhaps for this reason there is a tendency toward self-delusion when defining labor requirements. A favorite way of doing this is to state that 4½ operators are required for each shift. This means that if, by good fortune, a man employed in the area under study can spend half his time working productively on efforts that are chargeable to a different operation, half of his wage can be justly charged to each. Close scrutiny generally shows this assumption to be unjustified and, as a result, operating labor cost has been understated by 11%.

If raw-material cost significantly affects the cost of the finished goods, a different pitfall lurks. What if the cost of raw material goes up? Can the product command a proportionally higher selling price?

Compare Alternatives

Implicit in much of the foregoing is the suggestion that an engineering study is not a single analysis made for a single set of conditions. It is apparent that a selection of the better alternative requires the comparison of at least two choices. Whether it is worthwhile going through a complete study of more than one solution (in the same amount of detail) is a question generally left to the engineer. Suppose he wishes to determine the effect of plant size. He may select one size and do a complete design, capital cost estimate, and unit cost estimate.

Now he may elect to use the "six-tenths rule" for scaling the plant cost up and down for different plant capacities. This approach requires caution. If the process in question is a hydrogenation reaction, dissociated ammonia is more economical for medium-sized plants, while reformed natural gas is usually a cheaper hydrogen source for larger ones. The engineer must recognize that even his material-balance calculations do not hold if he switches from one source of hydrogen to the other, because of the introduction of 25% nitrogen with the hydrogen from the dissociator.

There are other errors inherent in the use of the six-tenths factor. Berk and Haselbarth[4] have determined the actual factor for a variety of plants, as well as the extent of variation in the factor between a grass-roots plant, a new plant built within an existing plant site, and expansion of an existing plant. Their factors are shown in Table III. Graph shows the error incurred by using the six-tenths factor (rather than the true factor) in scaling up an estimated plant cost to determine the cost of a larger plant.

Because the complete design and costing technique is so time-consuming, the engineer could not, even if he wished, apply this procedure to every conceivable alternative. He may eliminate some possibilities by short-cut evaluations. Then, using his experience, common sense, and imagination, he selects those routes most worthy of concentrated effort. By now it is apparent that his good judgment is a prerequisite for a sound analysis.

A company that produced certain plastics considered the possibility of making its own basic raw material, the resin. The company did not wish to produce more than required for its captive needs, since such a move

would have placed it in competition with one of its best customers.

Al, the chief engineer, was asked to study the problem and make recommendations. Very shortly he discovered that building a plant just large enough to supply his own company's requirements would be a marginal investment. Not satisfied with this conclusion, he called in the supplier who furnished the resin, and

Six-tenths factor can throw you off

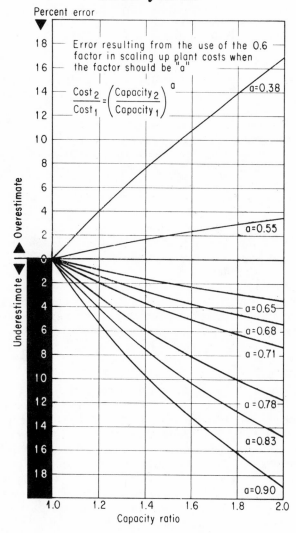

Percent error

Error resulting from the use of the 0.6 factor in scaling up plant costs when the factor should be "a"

$$\frac{Cost_2}{Cost_1} = \left(\frac{Capacity_2}{Capacity_1}\right)^a$$

a=0.38

a=0.55

a=0.65

a=0.68

a=0.71

a=0.78

a=0.83

a=0.90

Overestimate

Underestimate

Capacity ratio

Scale-Up Factors—Table III

Type of Plant	Factor		
Ethylene oxide	0.78	Chlorine	0.38
Ethanol (synthetic)	0.60	Sulfuric acid	0.90
Styrene	0.68	Hydrogen cyanide	0.82
Oxygen	0.70	Ammonia	0.88
Acetylene	0.75	Butadiene	0.65
Methanol	0.83	Ethylene	0.71
Butyl alcohol	0.55	Polyethylene (hi pressure)	0.67
Isopropyl alcohol	0.60	Polyethylene (low pressure)	0.90
Caustic	0.38		

some of the supplier's competitors too. Al explained to each the situation he was evaluating. One thing that could surely make up his mind, he pointed out, was a change in price structures for the resin.

He suggested several alternatives to the suppliers. One was a lower price. Another was that the suppliers consider locating a large new plant on an adjacent property. This, he noted, would permit the supplier to tap new markets in that section of the country and would result in Al's firm paying much lower freight costs on the resin. Al invited proposals.

His current supplier replied first. The market in the resin showed signs of softening, and no new construction would be considered at that time. However, Al's account was a valuable one, and the supplier offered a new price schedule. Al next submitted his study to management with recommendations for accepting the lower prices offered. A contract was signed, and everyone was happy. By injecting a new degree of freedom into his study, Al had achieved the best possible solution for his company.

This gets us back to one of the provisos in our very first sentence—the problem must be defined. On the surface the problem appeared to be, "Is a plant that will supply our own captive resin needs justified?" Al solved the problem by redefining it more exactly, "What is the most economical source of resin for us?"

Emotions May Mislead

Sometimes a simple problem becomes clouded by human attitudes. A small family-owned manufacturing business faced a crisis. This company produced ten products in an overcrowded facility. The manufacturing operation contained many inefficiencies, all due to lack of space. An opportunity unexpectedly occurred for acquiring a large and choice property nearby. A decision had to be made promptly, or the property would be lost. So sudden was this turn of events that the president was caught unprepared. An interview with him revealed the following:

• His only concrete and immediate plan was to move his present operation to a more spacious layout, thereby obtaining some economies.

• He had some future plans for expanding his operation, but they were vague and ill-defined.

• The move would require some disadvantageous financing and would increase his annual operating expense by about 4% of his gross, disregarding the effect of economies.

• His cost records afforded no real basis for judging the profitability of his various products.

The immediate problem had a rather simple answer. He could not justify the move, because the efficiencies anticipated would not offset the increased cost of the operation. (In a more personal analogy, we could afford to burn a lot of oil in the old jalopy before justifying the price of an engine overhaul.) The future benefits were too nebulous to be considered seriously.

More to the point, did the operation justify producing so many products, or would a cut to perhaps three products, accounting for 75% of the sales volume, pro-

duce more profit? The answer was to install a simple cost-accounting procedure that would soon yield a rational basis for such decision-making.

Again we return to a fundamental proviso for sound business judgment—a detached approach. The president felt there was a growth potential by moving into new products and was tempted to jump at the offer of more space. Had he made the move, as originally conceived, he would have tied up much of his credit without realizing genuine expansion, and would have penalized his existing profits. He, of course, was more emotionally involved than is the engineer performing an evaluation study inside a large corporation.

Nonetheless, people do form attachments for ideas, particularly their own ideas. Sometimes these are traceable to the hope for personal gain, but more often they are entirely innocent and unconscious. The motivations we shall leave to the psychologists. Our problem is to obtain an unbiased interpretation of facts.

The best way we know for reaching this objective is to have one or more competent persons, with no visible reason for prejudice, review the study and comment on the conclusions. Sometimes, if the project is sufficiently large and important, a completely independent and parallel study is warranted.

What if the results differ enough to result in opposite conclusions? Either an error has been made, or there is an area of judgment broad enough to make the study inconclusive. If this is the case, the manager may as well recognize the gamble before funds are committed. Where such a latitude for opinion exists, it is the manager's responsibility to decide.

What It All Means

At this point, we can crystallize some ideas about the economic evaluation, what it is, how we draw conclusions from it, how we judge its reliability.

The economic study has three separate and distinct parts. First, there is an engineering study to determine technical feasibility, capital investment requirements, unit production costs, and the effect of major variables on the capital and production costs. Next, there is a market study that forecasts future price, demand, competitive trends, and the requirements and costs of marketing the product. Finally, these two studies are integrated to furnish an over-all forecast of the impact of the proposed new move on the company's financial and profit picture.

It is nearly as important to say what an economic study is not, as what it is. It is not an exercise performed in a vacuum. It considers every important alternative course of action. Neither is it an inflexibly long and complex study of all factors. Certainly it does not concern itself with factors of minor significance where the influence of major variables is clearly predominant. It jumps to the heart of the problem as rapidly as possible by determining which areas of study the final decision is most sensitive to, and whether the decision can and should be made from a magnified view of these areas without recourse to a more exhaustive study.

Drawing conclusions from the study requires that an estimate be made of future events. Any conclusion to an economic study can be rephrased to say, "In view of the facts we have examined, and our estimate of future events, which is as follows . . ., we believe that the effect of the proposed investment on the company's finances will be as described herein. The following course of action is therefore recommended."

As for the manager who receives the report and will formulate the course of action, he must:

- Weigh the accuracy and completeness of the facts.
- Decide whether those facts have been expertly handled.
- Conclude whether he is in agreement with the appraisal of future events.
- Deduce what impact other possible future events such as change in process technology, in market conditions, or in raw-material supply and cost might have on the action recommended.

He may require additional help to judge the proposal, but now the nature of the risk has been defined and evaluated in the most complete terms available.

The final outcome of the action is still uncertain and must be tested by time, which proves what we knew all along—economic forecasting is an art as well as a science. But a business decision based on such evaluation becomes a calculated risk rather than an outright gamble. That's the best we can hope for.

References

1. O'Donnell, J. P., *Chem. Eng.*, Sept. 1957, p. 249.
2. Chilton, C. H., ed., "Cost Engineering in the Process Industries," McGraw-Hill, New York, 1960.
3. Frumerman, R., *Chem. Eng.*, April 16, 1962, p. 133.
4. Berk, J. M., Haselbarth, J. E., CE Cost File (Series printed during 1960 and 1961 in *Chem. Eng.*). Available as CE Reprints 172 and 199.

Meet the Author

ROBERT FRUMERMAN *is a consulting engineer specializing in process design, improvement, economics and evaluation.*

A registered engineer (Pennsylvania), he holds a B.S. from the University of Pittsburgh and an M.S. from Carnegie Institute of Technology.

His experience in industry is varied, including jobs with Blaw-Knox, Koppers Co., Elliott Co., and Nuclear Materials and Equipment Corp., Apollo, Pa.

He is a member of AIChE and Sigma Tau.

"Design Maintenance" Optimizes On-Stream Time

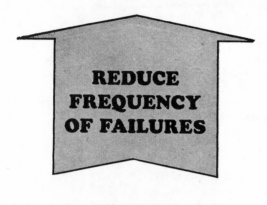

This three-pronged approach should interest design, operations and cost engineers, as well as those directly concerned with maintenance.

PETER B. ALFORD, *Nopco Chemical Canada Ltd.*

Design maintenance represents a design approach to the reduction of downtime and maintenance costs. It can be applied to the design of new plants or to the improvement of existing plants or departments.

While there is nothing revolutionary about the approach, it does represent a systematic, quantitative way of thinking that can be a great asset. At my plant, we believe it played a key part in reducing our maintenance costs about 30% after the first year, and significantly increasing our on-stream time. (It is difficult to isolate the exact savings, since other improvement programs were going on concurrently.)

The three prongs of design maintenance are:

1. Designing or modifying a facility so that failure of one unit or piece of equipment has a minimum effect on other units. This involves the use of reliability and probability concepts to evaluate the need for such things as emergency storage facilities, installed spares, and alternate paths.

2. Minimizing the number of planned and unplanned shutdowns of the components.

3. Designing the facilities so as to minimize the length of time a shutdown takes—in other words, designing for *ease* of maintenance.

In all these cases, design maintenance recognizes that where a capital investment is needed to reduce downtime, some sort of quantitative evaluation must be made to discover whether the investment is justified. That is why the title of this article refers to the "optimizing" rather than the "maximizing" of on-stream time. Design maintenance concepts are very useful in this optimizing process.

Increasing the Operational Efficiency

Operational efficiency, whether for a complete plant or a control valve, depends on two characteristics: performance and availability.

Performance is the more uncontrollable of the two characteristics—uncontrollable in the sense that it is usually fixed by the nature of the application and the range of specifications within which the operation must be carried out. The performance of a pump is usually fixed by such things as the properties of the fluid being handled, the volumetric capacity, and the dynamic head required; that of a reaction vessel by such things as the reaction environment, the space velocity, and the corrosive nature of the reactants.

The performance limits should be established early in the life of a project since all other factors are inherently affected by them. Of course, there is no substitute for past experience in this important area of design; however, this experience can usually be augmented by intelligent application of technical infor-

mation that is available in design manuals and suppliers' literature. Once the performance limits have been defined, they should be kept readily available for constant reference throughout the remainder of the design program.

Most of this article will deal with availability. The availability of a plant or piece of equipment is the percent of on-stream time that can be reasonably anticipated, assuming good but not extreme maintenance practices.* Within the scope of this definition, if the availability is expressed as a decimal, it represents the probability that the unit or plant is able to function at any given time.

The basic aim of the design study then is to determine the optimum balance between the performance and the availability for a certain system.

Designing for Ease of Maintenance

Availability is dependent on two secondary characteristics; reliability and maintainability. Ideally, a unit with a low reliability (i.e., one that is susceptible to frequent failure) should possess a high degree of maintainability (i.e., should be quickly and easily repairable).

There are many ways of increasing reliability and, particularly, maintainability. For instance:

1. The over-all process should be studied with primary emphasis on the accessibility of parts. The design should allow adequate working areas around all critical subsystems or units. If, for instance, a fork truck is the best way of moving heavy equipment to the repair shop, is there room for it to maneuver? At turnaround time, will there be space to move in and store materials and tools, or to set down disassembled equipment? Is there enough space to do an efficient job when pulling impellers, tube bundles, or other frequently removed parts? Are built-in handling facilities indicated? Are components that require frequent inspection accessible enough to make this practical? (Grouping similar equipment together where possible is sometimes helpful in this.)

2. Dismantling and disconnecting should be made as easy as possible. Quick-disconnect connectors, detachable mountings and similar devices can be major timesavers. In instrumentation, the use of modular, plug-in devices that operators can remove and take to a bench can mean faster maintenance and less downtime.

3. Where possible, equipment should minimize the necessity for specialized maintenance personnel. This is usually best achieved by simplicity in design, plus use of wide-tolerance components. Self-compensating or self-aligning features are useful.

4. Standardization of equipment and parts should be kept in mind. Standardization not only permits ready interchangeability and simplifies the stocking of spare parts but usually promotes faster servicing, since the mechanic can develop standard procedures.

5. Strategic service stations with utility outlets are

*See Gibbons, E. J., Probability, Downtime and Doomed Designs, *Chem. Eng.*, Nov. 12, 1962, p. 250.

important. They are not difficult to incorporate into the design, but are expensive to add after the plant is complete.

6. Draining of vessels and tanks can be frustratingly time-consuming, particularly if plugging occurs. Paying attention to tank-bottom slopes, drainage valves and dimensions of drainage piping can pay big dividends not only in emergencies but even in such things as chemical cleaning.

7. Nondestructive testing and monitoring can cut down on shutdown requirements, yet give advance warning of equipment deterioration and potential damage. Monitoring devices should be considered as part of the design, although they can be added afterwards.

8. Environmental control designed into a process can greatly prolong the time the equipment can go without dismantling. This can take many forms—corrosion inhibition, anti-fouling agents, injection of clear liquid, special water treatment, etc.

9. A critical evaluation of new equipment from the maintenance viewpoint can suggest modifications that should be made prior to installation. Is the equipment easy to open up, to clean, to lubricate? How easy is it to remove parts? Would the addition of a lifting lug be helpful? Can the piping connections be broken loose without handling long lengths of pipe and heavy valves? What are the likely weak points that can shut down this piece of equipment, and are there any ways of either beefing these up for the particular service, or making them noncritical so that they can be bypassed in an emergency (e.g., by going to manual control)? And so on.

10. As little equipment as possible should be located in poorly ventilated, toxic or potentially explosive atmospheres. If the wearing of gas masks, obtaining of special welding permits or application of gas blanketing can be safely avoided, maintenance is obviously speeded up. Incidentally the valving and flanging that will permit lines and equipment to be inerted and blanked off not only in the safest but also in the fastest and most convenient manner should be given careful thought. As in most of the other categories mentioned, convenience bears a price tag but investigation may reveal that the investment is more than warranted.

Probability and Availability

The ten points just discussed are based on a generalized processing plant; it is suggested that every plant (or company division) develop a more detailed checklist that is tailored to its own type of operation and that incorporates its own maintenance experience.

Having thus reviewed the design with an eye to reducing the number of shutdowns and their duration, we have carried out the second and third concepts of design maintenance. Let us now go back to the first, which involves ways of estimating and minimizing the effect of equipment breakdowns on the on-stream time of the plant as a whole.

A prerequisite for this is to estimate the availabilities for each critical piece of equipment.

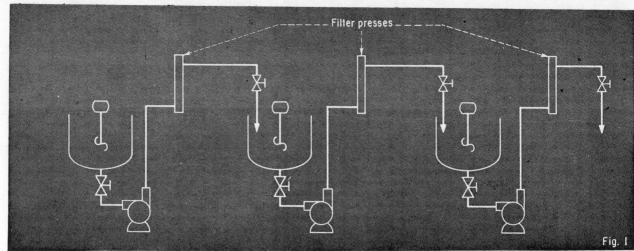

Failure of one unit can shut down this three-stage filtration process—Fig. 1

The availability for a single component in an existing installation can be readily estimated from equipment records. Let's say that for the past few years, a certain pump has been out of service an average of six days each year, that there are no obvious factors or trends that would tend to make the current year any different.

The pump's availability would be 294/300, or 0.980, assuming there are approximately 300 days in a working year at the plant. Putting it in terms of probability, the probability is 0.980 that the pump will be working at any time.

The availability for new equipment can be estimated from records of similar equipment, where such records exist. Allowances can be made for newness of equipment, design improvements by the manufacturer, etc. Where no records exist—as in a new plant being designed—there are usually maintenance records on similar equipment somewhere within the company. If so, these can be extrapolated, making allowances for changes in environment, differences in degree of skill of the maintenance crews, etc.

If there are no such records, a workable estimate can usually be made by consulting operating and maintenance people experienced on that type of equipment, as well as the equipment manufacturer. Operating men possess valuable practical experience coupled with a natural skepticism, and should be consulted in choosing components for a plant operation or in estimating their availability.

The purpose of this estimating is to form an opinion about the availability of a whole section or plant, to point up the best ways to increase this availability, and to permit investment evaluations needed to achieve this increase.

Let us say A_1, A_2, . . . A_n represent the separate availabilities of a group of components, all critical and arranged in series. By definition, these availabilities represent the probability that each unit will be functioning at any one time; the system can only function if all the components are functioning.

According to the laws of probability, this availability relationship is expressed as follows (assuming there is no failure interrelationship between the individual units):

$$A_T = (A_1)(A_2)(A_3) \ . \ . \ . \ (A_n) \qquad (1)$$

where A_T is the over-all availability of the group, and A_1, A_2, etc., are the individual availabilities of the various critical components that make up the group.

Working With Availabilities

As an example of how availability figures can be put to use, let us consider the three-step filtration process shown in Fig. 1. All units are critical to each other. To calculate the availability for the whole system, it is sometimes convenient to group similar components:

Critical Component	Individual Availability	Number of Components	Group Availability
Mixers.........	0.985	3	$(0.985)^3 = 0.955$
Valves.........	0.990	6	$(0.990)^6 = 0.941$
Pumps.........	0.985	3	$(0.985)^3 = 0.955$
Filter presses...	0.945	3	$(0.945)^3 = 0.843$

According to Eq. (1), the over-all availability of this system is the product of the availabilities of the individual critical components, which is the same as the product of the group availabilities above and is equal to 0.723.

This is a rather low figure and it would indicate (assuming our availability estimates are correct) that the system will only be operative about 217 days out of a 300-day year.

What can we do to increase this figure? First we can try to reduce the number of critical components that are arranged in series. Perhaps if we add two surge tanks to the system just discussed, the three units would no longer have to be critical to each other. Any one of the three could be isolated without shutting down the other two. A system that would accomplish this is shown in Fig. 2.

Each of the three units still consists of a mixer, a pump, a filter press and two valves. (The additional valves in the system, and the surge tanks, can be considered noncritical.) Assuming the individual availability of each component to be the same as under the old system, the availability of each of the three units would be:

$$(0.985)(0.990)^2(0.985)(0.945) = 0.898$$

Since each unit is now no longer critical to the others and can be considered self-sufficient, the 0.898 will approximate the availability of the whole process. Sometimes the nature of the processing operation

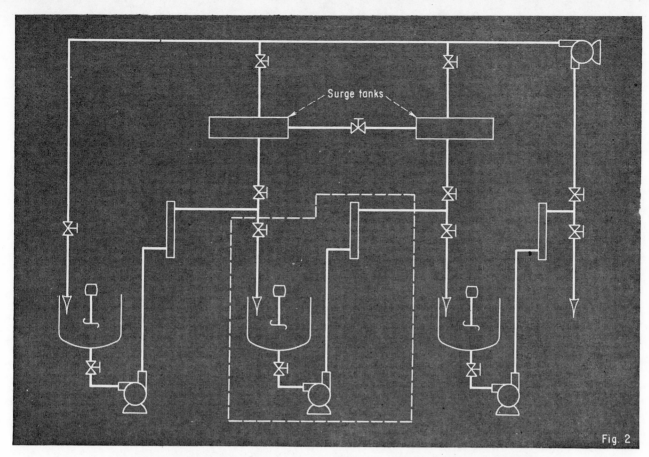

Revised filtration arrangement is less prone to forced shutdowns—Fig. 2

is such that a modification of this kind makes the units only partially self-sufficient, in which case a correction can be applied at the end to represent the degree of self-sufficiency. In our case, if we assume that the modified units are essentially self-sufficient, their estimated availability of 0.898 indicates that the system can be expected to be operative 269 days out of a 300-day year—a gain of 52 days over the old system.

Of course, to complete this evaluation, it is necessary to establish the cost of the additional equipment and then to determine whether or not the investment is justified. This can easily be done by standard investment-evaluation procedures once a realistic value is placed on the added production and on possible gains in efficiency through smoother operation.

The value placed on gaining a day of on-stream time will depend on the plant's goals. In the case we have just discussed, if the mixing-filtration system can meet its production goals by being on stream only 80% of the time, the storage tanks would obviously be harder to justify than if 90% on-stream time were called for.

Installing Standby Equipment

As a supplement or alternative to the surge tanks just discussed, let us see what happens to on-stream time when we install standby equipment in parallel with existing process equipment.

In our theoretical study, the piece of equipment with the lowest availability is the filter press. We will take one of the units shown in Fig. 2 and now assume that a standby filter press is to be used in conjunction with it. Fig. 3 shows the new arrangement; under it, both filter presses would have to be

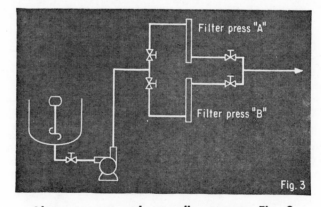

Alternate approach: standby presses—Fig. 3

down for repair simultaneously in order to force the unit to shut down.

Let us say the additional filter press will have an availability of 0.945, the same as the original (a conservative assumption). The probability of both presses being down for repair at the same time is then $(1.000 - 0.945)^2$, or 0.003. We can think of this as the *un*availability of the system of two filter presses. The availability will be $1.000 - 0.003$, or 0.997.

Multiplying this by the availabilities of the other components shown in Fig. 3, we can obtain a new availability for the unit as a whole. This will be 5.2% higher than the earlier one, and represents a gain of about 15 days of on-stream time per 300-day year.

An evaluation can now be carried out to determine whether this 15-day saving makes the cost of the additional filter press a good investment. Or, going back to Fig. 1, an evaluation will now reveal whether

surge tanks or additional filter presses are likely to offer the greatest amount of additional on-stream time for the least cost. This comparison can be put on the basis of potential return on investment for each of the two alternatives.

Design Maintenance as a Way of Thinking

At this point, the reader is reminded that in addition to emergency storage facilities, alternate paths and installed spares, on-stream time can be increased by improving the reliability and maintainability of the equipment. This takes us back to checking the design of the system or of the equipment, as discussed earlier. It also suggests how design maintenance can supply a systematic approach in deciding:

• Whether to repair or replace an old piece of equipment.

• Whether to use a standard piece of equipment or a nonstandard, perhaps cheaper, piece of equipment that might take longer to repair.

• Whether a piece of equipment is critical enough to warrant extensive preventive maintenance, monitoring, or investment in special repair facilities.

In all these cases, design maintenance does not stipulate the use of one profitability indicator over another, but allows you to use whatever measure is preferred by management. Management is basically profit-minded and it will be much more likely to appropriate funds on the basis of attractive profitability data than on intangibles. The success of the whole design maintenance program is definitely dependent on whether or not we present our results in the type of language that management people understand—in other words, whether we can convert technical data into economic recommendations.

The maintenance engineer cannot be held wholly responsible for process downtime. However capable he and his staff may be, if the process is poorly designed with little consideration for maintenance, he will likely experience difficulties in keeping it on stream. At the design stage, management is susceptible to stressing the immediate costs of the plant rather than the future costs of maintaining it because the former is tangible, the latter is not! Design maintenance and profitability data enable us to present management with more-tangible facts concerning this nebulous area of future savings.

While much of what we have talked about is applicable to plants already in existence, as engineers we fully appreciate that features conducive to low maintenance costs can be installed in a process at less expense when the process is still in the design stage. We should use every tool at our disposal to sell management on this fact and to make specific recommendations.

It may take time to document these proposals by citing availabilities, considering alternatives, and calculating profitabilities. However, the potential future savings in manpower, money and irritation should repay this initial effort many times over.

When starting out on a design maintenance pro-

gram, the availability estimates may not prove to be very accurate but they get better with increased use. Even at the start, design maintenance concepts prove to be a very useful evaluation and improvement tool.

At Nopco, we initially pinpointed those units of equipment that were considered to be critical to the over-all operation. We then estimated the component availability of each unit as a separate entity. Past equipment records were our primary source of information during this stage of our program. In cases where these records were considered inferior or questionable, we drew heavily on the practical experience of our maintenance supervisors.

The final unit availability, approximate as it may have been, was then recorded on an equipment catalog card that contained the past history of the specific unit. Since all future maintenance activity is recorded on this card, we were able to examine and revise this availability factor continuously in the light of actual performance. In this way, the accuracy of our original estimate is being continuously improved.

Once the unit availabilities had been established, we were able to look at the resultant availability of a group process containing more than one of these critical components, each interdependent on the others. Where extremely low factors were encountered on a process, we spent extra time examining possible ways of improving the resultant availability, e.g., by the reduction of system complexity, the improvement of maintainability, etc.

When this initial list of components and processes had been completely examined, we repeated the whole process with a second list of components that were considered to be slightly less critical than the original list. Subsequent cycles were completed in this way, each cycle containing components less critical than the preceding one, until virtually all of our major equipment was covered.

Meet the Author

Peter B. Alford is chief engineer for Nopco Chemical Canada Ltd., P.O. Box 68, London, Canada. As such, he is responsible for both the design engineering and plant engineering functions for Nopco Chemical Co.'s Canadian operation. His earlier experience includes process design engineering for Allerton Chemical Co., Rochester, N. Y. A registered professional engineer in Ontario, he holds a B.S. degree in chemical engineering from the University of Rochester and is a member of AIChE and the Chemical Institute of Canada.

Key Concepts for This Article

Active (8)	Passive (9)	Means/Methods (10)
Designing	Facilities	Availability
Modifying	Equipment	Reliability
Reviewing	Plants	Probability
Evaluating		Profitability

Purpose of (4)

Operations
Maintenance

(Words in bold are role indicators; numbers correspond to AIChE information retrieval system. Indexing details are described in *Chem. Eng.*, Jan. 7, 1963.)

Design For Expansion

Here are some techniques and suggestions for reconciling opposing objectives as to the use of capital, and for dealing with market forecasts and various economic factors that affect a plant's ultimate profitability.

JAMES M. ROBERTSON, *Celanese Chemical Co.*

How do you find the optimum initial installation in a plant that must be planned for future expansion?

A problem of this sort is perhaps one of the greatest challenges a process design engineer faces, and his success in meeting this challenge may decisively change his career for the better because his decisions will have long-range effects on the earnings of the company or client.

The problem of defining initial plant size is always a very real one, and becomes even more complex when a company introduces a new product whose ultimate sales potential has not yet been determined because its markets must still be developed.

Initial plant size is a critical determinant of ultimate economic success: sales potential is influenced by production cost, and production cost is affected by how one uses capital in building the plant. A too-small installation is plagued by high unit operating costs and limited capability for expansion, and one that is too large is penalized by an excessive initial capital demand and high fixed cost of maintaining unused facilities.*

The optimum strategy for investment and expansion will very likely be one that entails a minimum initial expenditure for small-volume production, without jeop-

ardizing the position of the ultimate high-volume low cost plant.

Thus, as soon as one begins to strike the best economic balance among many variables so as to minimize production costs, one immediately comes up against opposing objectives: initial capital must be conserved but expensive duplication of facilities during subsequent expansion must be avoided.

There are a number of economic considerations and techniques involved in achieving an optimum balance between these opposing objectives, and I will explore them in this article. In a future article I will tackle the optimization problem from the engineering aspect, and show the important role of plant layout and equipment selection and sizing.

*For a discussion on the cost of capital, see John F. Childs' article "Should Your Pet Project Be Built?", p. 329.

Plant Location Factors—Table 1

Tangible	Intangible
Land and construction costs	Local community attitude
Skilled labor availability	Living conditions
Wage rates and union attitude	Travel facilities
Raw material costs (at plant site)	Educational facilities
Product freight costs	Intra-company liaison
Water availability and cost	History of area labor relations
Fuel and power costs	Management preference
Pollution control requirements	
Taxes, climate	

CONGESTION due to expansion, as in this plant, can be avoided. A future article will suggest how.—Fig. 1

Manufacture of 500 Million Lb./Yr. Ethylene— Table 2

Capital, $ Million	Single Unit	Parallel Units
Battery limit cost..............	14.9	18.7
Land and offsites..............	5.2	6.5
Total fixed capital............	20.1	25.2
Costs, ¢/lb. Ethylene		
Net raw material, utilities and chemicals...................	1.279	1.279
Labor and supervision.........	0.054	0.075
Maintenance...................	0.094	0.119
Labor overhead................	0.011	0.015
Plant general, insurance, and taxes.........................	0.129	0.167
Depreciation, 12% B/L........	0.358	0.463
Total Production Cost.......	1,925	2.118
30% earnings on capital.......	1.209	1.513
Transfer Price..............	3.134	3.631

Plant Location

One of the first choices to be made in designing a new plant is its location. This is important to the ultimate economic success of the installation. While a detailed discussion of plant-site selection techniques is beyond the scope of this article, some of the factors to be considered are summarized briefly in Table 1.

Preliminary planning will have defined, at least generally, the utility and raw-material requirement growth rates, as well as the ultimate plant capacity. When making the immediate decision of plant location, the long-range impact of the envisioned expansion must be considered. For example: Is adequate space available for the ultimate plant? Will the initial plant absorb most of the labor available in the area, making future recruitment of personnel difficult? Can sufficient water resources be developed to accommodate an expanding plant and satellite plants that may evolve?

Do Not Overlook Freight Costs

Product freight rates should be studied in detail, because chemicals usually are sold on a "freight equalized" basis. Again, the long-range view must be considered. Typically, a small initial plant may have only one product—often sold to only two or three customers, or transferred internally. When the plant is expanded, however, the number of customers for the basic product is likely to increase. In addition, several new derivatives of this material will probably be produced at the location.

A plant initially producing only acetaldehyde, for example, may well evolve into a large complex producing several acetyl chemicals. Potential acetaldehyde derivatives are acetic acid, acetic anhydride, butyl alcohol, crotonaldehyde, butyraldehyde, 2-ethylhexanol, butylene glycol, and butyl acetate. Freight rates are not constant but depend upon the class of material, destination, and transportation facilities available. This may be illustrated briefly by the following, where freight rates are given per 100 lb. of product from a shipping point in upstate New York:

Destination	Acetaldehyde	n-Butanol
Houston.................	$1.20	$1.00
Los Angeles.............	$2.80	$2.05
Chicago.................	$1.05	$0.87
Charlotte, N. C..........	$1.15	$0.86

Thus long-range market planning, including potential customer location, is vital to the selection of a site for a plant that is likely to be expanded.

Why Plan for Expansion?

A plant entering a growing market situation must remain economically competitive over the long haul. Its ultimate high-volume production costs must not exceed materially the costs of efficient, high-capacity plants that might be built in the future. And, usually, minimum costs are promoted by single-line plants rather than multiple, parallel units.

Consider, for example, the manufacture of 500 million lb./yr. of ethylene by steam-cracking of ethane. Cost factors for a single unit of this capacity, and two parallel units of 250-million-lb./yr. capacity each, are compared in Table 2. The parallel units are assumed to be adjacent and to involve optimum integration. Isolated units would cost more. These cost factors were developed from recent ethylene literature (see References).

The single-train plant enjoys a 14% transfer-price advantage over its dual-unit competitor. At the same transfer value of 3.134¢/lb., the single-train plant earns 30% on its investment but the dual-unit plant earns only 20% (i.e., 1.016¢/lb., obtained by subtracting the parallel units' production cost, 2.118¢/lb., from the transfer value). Future economy, then, can be advanced by planning for expansion of a single production line rather than by duplication of facilities.

Saving Money Costs Money

In a forthcoming article, I will indicate how judicious plant layout and equipment selection can conserve initial capital without jeopardizing the economics of ultimate expansion. For the moment, what should be stressed is that in almost every case the initial installation will cost more than would a plant tightly designed for the low capacity. Similarly, the ultimate plant will be more expensive than a large plant installed initially as one complete system.

Table 3 gives an idea of the extra costs involved in planning for future expansion under two different ethylene manufacturing situations.[3, 5, 6, 8] Although the relationship between these costs will be different for each situation, Table 3 does serve the purpose of illustrating that planning for expansion: (a) adds appreciably to the initial cost; (b) adds somewhat to the ultimate cost compared with a large single installation; and (c) costs less than the total cost of parallel units that achieve the same capacity.

Key Economic Factors

Are these increased capital costs justified? Should extra money be spent initially to prepare for the future in view of the relative unpredictability of the marketplace?

The wise planner will dispense with the luxury of "hunches" and let objective economic facts determine just how far to go in providing for things to come. Some of the factors that influence his decision are availability of capital, time value of money, obsolescence of technology, reliability of market forecasts, and effects of cash flow. For convenience, I will treat each of these factors separately, but it should be understood that there is considerable interaction between the elements in an actual situation.

Availability of Capital

If capital is tight and must be conserved rigidly, there is only one course of action open: cut initial cost

Ethylene Unit Capital Costs—Table 3

(Battery Limit Only)

Case A: Inital capacity, 250 million lb./yr. ethylene.
Ultimate capacity, 500 million lb./yr. ethylene.
Total feed: ethane-propane mix.

	$, millions
Tight design 250-million-lb./yr. unit	$10.5
Same unit designed for expansion	$13.8 (added cost, $3.3)
Added capital for expansion......	$ 3.4
Plant expanded to 500 million lb./yr.............................	$17.2
Tight design 500 million lb./yr.......	$16.5 (saving $0.7)

Ultimate saving: (2 × $10.5) − $17.2 = **$3.8 million**

Case B: Initial capacity, 500 million lb./yr. ethylene from ethane. Added capacity, 700 million lb./yr. ethylene from naphtha.

	$, millions
Tight design ethane feed unit.......	$14.9
Same unit designed for naphtha expansion......................	$21.9 (added cost $7.0)
Added capital for naphtha expansion...........................	$11.2
Ultimate 1,200 million-lb./yr. unit	$33.1
Tight design 1,200-million-lb./yr. unit	$30.3 (saving $2.8)
Total for separate ethane and naphtha units (ethane $14.9; naphtha $22.2).................	$37.1

Ultimate saving: $37.1 − $33.1 = **$4 million**

Ultimate Cost of Expanded Plant—Table 4

(Expansion in Five Years)

	(a)		(b)	
Time value of money.............	7%		25%	
Cash discount factor.............	0.713		0.328	
Initial expenditure, $ million.....	10.0	12.0	10.0	12.0
Initial value in five years, $ million	14.0	16.8	30.5	36.6
Future expenditure, $ million....	8.0	4.0	8.0	4.0
Ultimate cost in future dollars, $ million......................	22.0	20.8	38.5	40.6

to the bone. In an expanding market situation, this usually means building the largest plant possible with the available money and letting the future take care of itself. It is false economy, however, to cut costs so far that the plant is highly inefficient or difficult to operate.

More-flexible capital circumstances offer greater freedom in expansion planning. In this case, the long-range gain to be realized must compete with alterna-

- Case A: Initial capital cut to bare minimum. Spare pumps omitted, resulting in low stream efficiency. Heat exchangers undersized.
- Case B: Initial capital adequate for good operation even before expansion and full-rate production are required.

In both cases, the same ultimate plant requires the same total investment. Cash flow calculations are compared in Table 6. Differences in "book" and "tax basis" depreciation schedules, investment credits, and inflation are omitted for simplicity. Future dollars are discounted at 10% compound rate. Sale price of the product declines predictably during the 10-yr. period studied. At the end of this time the plant is obsolete or worn out and has no salvage value. Investment alternatives are compared by the familiar "net present worth" method, which—by summing the annual discounted cash-flows—indicates the ultimate cash surplus generated by the project in terms of present-day dollars.

Notice that in this situation the cash-flow pattern for Case B is superior to that of Case A, even though the former requires a greater initial investment. Loss of earnings due to inability to meet sales demand more than offsets the savings achieved by cutting initial capital below the safety point. (In a real situation, some money would be spent for "de-bottlenecking" during the first two or three years, so the actual cash-flow pattern would be somewhere between the two cases shown.)

Again, these data are for a specific case and should not be taken as generalizations. Different ratios between investment schedule and cash-flow components could easily lead to the opposite conclusion that minimum initial expenditure is preferred. For example, it can be shown by calculations similar to those of Table 6 that minimum-investment Case A is superior to Case B is sales potential is 50 million lb./yr. in Year 1 and rises in equal increments to 400 million lb./yr. in Year 5.

Decision-Making Guides

Each of the above economic factors was presented as an isolated case to show the ideas involved. In an actual situation, these and many other effects are, of course, interwoven in a network of mutually influencing factors that combine to yield a total picture of the project.

What is the best course of action? Answering this question requires at least two tools: (a) a standard method by which alternate strategies can be measured and compared; and (b) a decision-making process that can allow for the uncertainties of the future. It is far beyond the scope of this article to discuss and appraise the many methods available for investment evaluation —e.g., payout time, direct earnings (before tax), return on investment (after tax), profitability index, net present worth, net annual worth, and equivalent rate of return. These, and others, are described in detail by various authors in many books and magazine articles.

When the investment evaluation technique has been selected, the results of the various strategies possible are compared on this basis to determine the ultimate course of action. This comparison may range anywhere from an intuitive glance at the data developed above to a complex "decision tree" calculation involving considerable computer time.

Optimize

If each potential variable is explored in detail, a tremendous amount of time can be required when tracing the planning of a plant from the initial location to the ultimate program of action. Sometimes the stakes involved are worth it and sometimes they are not. Computers do help, of course, but their tireless precision may deceive us by manipulating numbers with a much higher degree of accuracy than our basic data may warrant. The engineer who has made the plans and compared the alternatives, and thus understands the interrelationships between the many facets of the problem, is in the best possible position to advise management on the optimum course of action to reach the objective.

For optimization is the engineer's job. His task is to strike the best economic balance between all these factors, so that the market can grow and the ultimate product can be produced at minimum cost.

How production costs can be minimized at a plant destined for expansion is the subject of a forthcoming article in which I will explore some of the techniques for optimizing the plant layout and the choice of equipment.

References

(1) Anon., "Aspects Economiques de la Production d'Olefines"; *Revue de l'Institut Francais du Petrole*, Feb. 1967.
(2) Braber, P., "Technical and Economic Changes in Ethylene Manufacture"; United Nations Conference, Tehran, Iran, Oct. 1964.
(3) Burke, D. P. and Miller, R., "Ethylene"; *Chem. Week*, Oct. 23, 1965 and Nov. 13, 1965.
(4) Davenport, C. H., "Ethylene—What You Should Know"; *Petroleum Refiner*, March, 1960.
(5) Frank, S. M. and Lambrix, J. R., "How Important Is 'Bigness' in Modern Ethylene Plants?"; *The Oil and Gas J.*, Nov. 22, 1965.
(6) Guccione, E., "Ethylene: Where the Profits Are"; *Chem. Eng.*, Apr. 25, 1966.
(7) Herron, D. P., "Comparing Investment Evaluation Methods"; *Chem. Eng.*, Jan. 30, 1967.
(8) Knaus, J. A. and Patton, J. L., "Effect of Feed Composition on the Economics of Ethylene Production"; *Chem. Eng. Progr.*, Aug. 1961.
(9). Peters, E. H., "Ethylene, Organic Chemical Building Block"; *Chem. Eng. Progr.*, June 1966.
(10) Stobaugh, R. B., "Ethylene: How, Where, Who—Future"; *Hydrocarbon Processing*, Oct. 1966.
(11) Stuhlbarg, D., "Calculating the Calculated Risk"; *Chem. Eng.*, Jan. 15, 1968.
(12) Twaddle, W. W. and Malloy, J. B., "Evaluating and Sizing New Chemical Plants"; *Chem. Eng. Progr.*, July 1966.
(13) Will, R. A., "Finding the Best Plant Location"; *Chem. Eng.*, Mar. 1, 1965.
(14) Zdonik, S. B., Green, E. J., and Hallee, L. P., "How Operators Handle Problems of By-Products and Plant Sizing"; *The Oil and Gas J.*, Jan. 2, 1967.

Meet the Author

James M. Robertson, currently research associate at Celanese Chemical Co.'s Technical Center (Box 2768, Corpus Christi, Tex. 78403), has been involved in all phases of plant design and expansion. He is the holder of four patents and the author of various articles. Most recently, he was on the New York planning staff of Celanese for two years and, before that, was chief process engineer at his company's complex in Bishop, Tex. A chemical engineer from Texas A & M University, he began his career in 1946 at Texaco Inc.

Economics of Multiple Units

Multiple units make operations in process plants more flexible—usually, but not always, at the cost of higher investments, which depends on the "x" factor, how many units are installed, and on whether or not spare capacity is provided.

H. A. QUIGLEY, *Ingenieria Panamericana, S. A. de C. V.*

A single function is frequently performed by multiple units; for example, a plant's steam requirements may be generated in two or three small boilers rather than in a large one.

This may be done for one or more of the following reasons:

Supplying all the demand from a single unit would require one so large that it would be impractical to build it.

The operating range of certain equipment in the process may be so narrow that fluctuations in demand could be handled more efficiently by taking whole units on and off stream rather than by trying to adjust the output from a single unit.

Production would not have to be curtailed when equipment failed, or had to be shut down for maintenance, if a spare unit were available.

If, when the plant is being designed, the decision is made in favor of multiple units, how the number of units being proposed will affect investment can be determined by means of the "six-tenths" factor, which relates cost to capacity according to Eq. 1:[2, 4]

$$\text{Cost} = \text{Constant (Capacity)}^x \qquad (1)$$

When $x = 0.6$, Eq. 1 becomes the "six-tenths" rule. Each type of equipment has its characteristic exponent, which usually varies from 0.5 to 0.9. Cases when x is outside this range, however, will also be discussed. A list of good exponential values for the installed cost of several different types of equipment is presented by Bauman.[1]

Two cases need to be analyzed: when the purpose of

installing multiple units is to furnish spare capacity, and when this is not the intent. If multiple units are being installed either because of size limitations or because they can handle operating fluctuations better, it is expected—at least for the purposes of this analysis—that all of them will normally be in operation.

Multiple Units Without Spares

If all of the units will be operating, the total investment as a function of the number of units can be found by means of Eq. 1. If the total capacity is C and there are N identical units, the capacity of each unit, c, will be C/N, and from Eq. 1:

$$i = Kc^x$$
$$I_N = KN(C/N)^x = KC^x(N/N^x)$$

The cost of a single unit is, of course, $I = KC^x$, and the ratio of the cost of N units to that of one is formulated in Eq. 2:

$$I_N/I_1 = N/N^x = N^{(1-x)} \qquad (2)$$

Because x is almost always—and N is always—less than 1, the right side of Eq. 2 will usually be larger than 1 and will get still larger as N increases. Obviously, multiple units will usually be more expensive than single units—how much more expensive can be calculated from Eq. 2.

What additional investment will be required—for different values of x—when multiple units are being installed can be found from Fig. 1. Note that as x becomes larger, the cost for multiple units becomes smaller. While multiple units make operations more flexible, how much this flexibility is worth must be evaluated by the design engineer in light of the particular circumstances.

While, as has been stated, x will range between 0.5 and 0.9 for most equipment and processes, Eq. 2 is, of course, valid for any value of x. Chilton gives two instances for complete plants when x is approximately 1.0 (butadiene from butylenes, and TNT plants) and another when x is about 1.4 (high-purity oxygen plants).[2]

When x almost equals 1, the right side of Eq. 2 will also be about equal to 1, and I_N and I_1 will be about the same, which means of course that there will be no economic penalty in going to multiple units. This can be demonstrated with Chilton's data by calculating the costs for one, two and three-unit plants that each produce 300 tons/day of butadiene from butylene.[2] These costs (Table I) are essentially equal, within the accuracy that the data can be read.

When x is significantly larger than 1, the right side of Eq. 2 will be less than 1, which means, of course, that there will be a cost advantage in going to several smaller units rather than to a single large one. This can again be illustrated with Chilton's data by finding the costs for one, two and five-unit plants, each of which produces 10 tons/day of high-purity oxygen ($x = 1.39*$). The results are shown in Table II.

---

* This exponent is based on plants having capacities between 1.5 and 20 tons/day. Investment data for plants producing 99.5% oxygen at rates between 150 and 1,000 tons/day have been published by Katell and Faber.[3] At these higher capacities, and using their data, I found the exponent to be about 0.57. An exponent of 0.56 for 20 to 1,000 tons/day oxygen plants is given by Bauman (p. 146).[1]

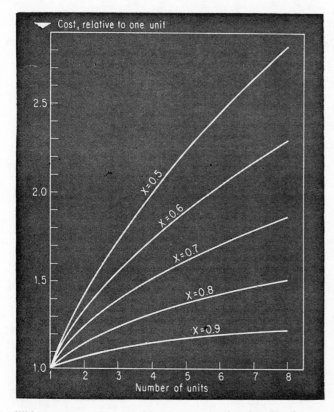

With **no spare** unit, cost changes more uniformly—Fig. 1

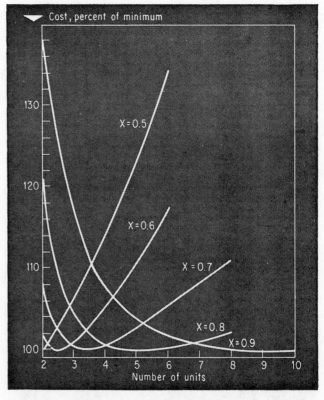

With spare, optimum number of units can be found—Fig. 2

None of this means, of course, that high-purity oxygen plants (or any plants having x values larger than 1) should be built with the largest number of units of the smallest practical size. Investment costs amount to only 15 to 20% of manufacturing costs. Besides, such a dispersed plant would make operations unwieldy and personnel costs prohibitive.

However, since investment is an important part of manufacturing costs and since it enters into the denominator of most equations for calculating profit, this potential investment savings should not be overlooked. For instance, steel manufacturers, who consume large quantities of high-purity oxygen, might well consider locating multiple units adjacent to their facilities.

Multiple Units With Spares

When multiple units provide spare capacity, the same basic approach will give slightly different results. Each unit, in this instance, will have a capacity of $C/(N-1)$ because one unit will be on standby. The derivation for Eq. 2 now gives:

$$I_N = NK [C/(N-1)]^x = KC^x N/(N-1)^x \qquad (3)$$

This time, however, an optimum N can be found that will require the least investment, I_N. If the derivative with respect to N is taken, Eq. 3 becomes:

$$\frac{dI_N}{dN} = KC^x \left[\frac{(N-1)^x - Nx(N-1)^{(x-1)}}{(N-1)^{2x}} \right] = 0 \qquad (3a)$$

Simplifying and solving for N reduces Eq. 3a to an equation for finding the optimum N:

$$N_{opt} = 1/(1-x) \qquad (4)$$

Of course, Eq. 4 only works when x is less than 1. In Table III, practical x values are related to optimum number of units when one of the units is kept back for a spare.

Five units would be best for equipment whose cost varied with the 0.8 power of capacity (when one were held back for a spare), Table III shows. But, because this many units would make for a fairly complicated arrangement, it would be useful to know how clear-cut is the optimum. If, for example, the cost penalty in going to three units would be slight, the gains in simplicity of installation and operation would probably justify paying the higher investment that going to three units would require.

The total cost, I_N, for any N can be obtained from Eq. 3. By taking the ratio of I_N to I_{opt}, the sensitivity of I to changes in N can be determined.

$$\frac{I_N}{I_{opt}} = \frac{NK[C/(N-1)]^x}{N_{opt}K[C/(N_{opt}-1)]^x} = \frac{N}{N_{opt}} \left(\frac{N_{opt}-1}{N-1} \right)^x \qquad (5)$$

Eq. 5 is convenient for computation. Eqs. 4 and 5 can be combined to eliminate N_{opt}:

$$\frac{I_N}{I_{opt}} = \frac{N}{(N-1)^x} \left[(1-x) \left(\frac{x}{1-x} \right)^x \right] \qquad (5a)$$

How cost varies with the number of units for different values of x is shown in Table IV. Note that the units on the abscissa of Fig. 2 start with 2. Having a

Investment when x approaches 1.0—Table I

No. of Units, N	Tons/Day Each Unit, c	Investment/Unit, $	Total Investment, $
1	300	68,000,000	68,000,000
2	150	33,000,000	66,000,000
3	100	22,000,000	66,000,000

Investment when x is larger than 1.0—Table II

No. of Units, N	Tons/day, Each Unit, c	Investment/Unit, $	Total Investment, $
1	100	410,000	410,000
2	50	160,000	320,000
5	20	44,000	220,000

spare naturally requires that there be at least two units.

Fewer units than the optimum could, as was mentioned, simplify operations. If x were equal to 0.8, and three units were substituted for five, the additional cost would only be about 5% (Table IV), which is probably a small enough difference to warrant going to three units. On the other hand, two units would be difficult to justify because the cost would be increased by about 20%.

Minimum Cost Output

The previous analysis depended on lowering total investment to the least amount possible; sometimes these analyses are based on getting the cost per unit of output to a minimum. While this latter method may be valid in some cases—as when spare capacity is deliberately provided—minimizing the cost per unit of output can lead to errors.

Errors arise from the fact that—because of the spare unit—installed capacity is greater than the capacity required. How much in excess is the installed capacity depends on how many units have been installed, and becomes less as the number of installed units increases (Table V). The ratio of this installed-to-required capacity is called the spare factor, S:

$$S = cN/c(N-1) = N/(N-1) \qquad (6)$$

If N equals 3, S will equal 1.5. When three units are installed, one of which is a spare, the spare capacity is, of course, 50%. For several N values, related S values are shown in Table V.

If the calculation of cost per unit of output includes investment charges that are based on the installed capacity, the large spare factor that goes with fewer units (Table V) can lead to mistakes. How such errors can occur is demonstrated in Table VI, taking data from Table IV, $x = 0.8$.

Table IV shows N_{opt} to be five units. Yet, according

Effect of Market Variation on Selling Price—Table 5

	Market Develops as Anticipated	Future Market Over- Estimated	Future Market Under- Estimated
Total fixed capital, $ million	$20	$16	$22
Sales volume, lb. million	500	300	500
Unit capital, ¢/annual lb.	4.00	5.33	4.40
Production costs, ¢/lb.			
Variable	2.00	2.00	2.00
Capital related fixed	0.80	1.07	0.88
Capital unrelated fixed	0.75	0.90	1.50
20% direct earnings on capital	0.80	1.07	0.88
Required selling price, ¢/lb.	4.35	5.04	5.26

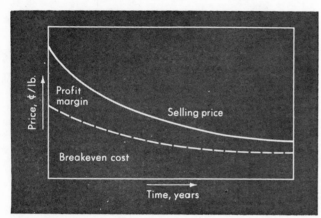

SELLING price decay curve—Fig. 2

tive investments for the incremental capital on a present worth (or other) basis.

Time Value of Money

Evaluation of alternative investments must consider (a) the time value, discount rate, or interest rate of money, and (b) the length of time before expansion, and the additional expenditure that will be required.

Let us examine a case in which an initial *minimum* plant costs $10 million. Expansion will be required in five years and will cost $8 million. However, if $2 million is spent now to prepare for the future, expansion in five years will cost only another $4 million. For this case, we will compare on the basis of future (five years from now) dollars and neglect the effect of inflation. The plant is to be financed on borrowed money that costs 7% annual compound interest, resulting in a five-year cash discount factor of 0.713. Calculations of the expanded plant's ultimate cost with minimum initial investment or with the added initial capital for expansion preparation are shown in Table 4 (column a).

Early expenditure of $2 million is seen to decrease the ultimate cost of the plant by $1.2 million. Hence the decision would be to plan ahead for expansion, provided that ultimate capital value was the only consideration involved.

Contrast this with the situation in which capital is limited and an alternate investment will yield an equivalent interest rate of 25%. Our initial investment must now compete with this alternate use for the money and therefore must be discounted at the significantly lower rate of 0.328. Calculations are outlined in Table 4 (column b). In this case, the ultimate plant would cost $2.1 million *more* if $2 million were spent now for expansion, and the proper choice would be to limit the initial expenditure to a bare minimum.

"Future" value dollars are used in these examples because the answers are a little more obvious. Discounting future expenditures back to "present worth" dollars will produce the same comparisons and lead to the same conclusions.

Obsolescence of Technology

Some segments of industry are characterized by rapid development of technology and frequent improvement in processes. The first plant to use a new process usually is modified considerably during the beginning years of operation, as improvements are made and errors in translation of pilot-scale to full-scale operation are overcome. In fact, the ability to gain information on a relatively small scale as technology evolves is in itself one of the arguments for starting a new process in a small way rather than building the ultimate plant initially. At any rate, validity of planning for expansion in a new plant is particularly sensitive to the maturity of the process.

Several questions should be resolved when evaluating the risk of pre-spending expansion capital in a plant subject to technological change. A few of them might be:

• What is the likelihood that significant change will occur before expansion is required?

• Will this change affect the part of the plant where expansion capital might be pre-spent?

• What would be the incrementally added cost of designing the expansion facilities to accommodate almost any eventuality?

• Could the added equipment be used in an alternate service if the expansion took a route different from that initially foreseen?

• What would be the salvage value of the added equipment if it could not be used for the expansion?

Economic effects of the above can be quantified to a certain extent by assigning relative probability factors. These are then included in the over-all decision-making process. The judgment of people in the technical, engineering and marketing fields will be required.

Reliability of Market Forecast

The decision to even consider expansion planning at all is based on some kind of future market forecast.

Like most prophetic activities, market forecasting is not really an exact science. There are at least three distinct possibilities: the future market will be less than forecast; or it will equal that forecast; or it will be greater than forecast.

Relative probabilities can be assigned based on maturity of the market, assured future contracts, economic growth patterns, captive requirements, and other factors.

The penalty for "guessing" wrong may not be the same in each case. For example, the added operating costs of parallel facilities if future requirements are underestimated may be more detrimental to over-all project economics than the added capital burden if requirements are overestimated.

Let us assume that the ultimate required selling price to achieve 20% pretax earning rate is the single criterion by which performance is to be judged. Inflation and time value of money will be neglected for simplicity. Table 5 summarizes the comparison of each marketing possibility.

In this case, the extra fixed costs of operating two lines of equipment (primarily labor and related costs) in the plant where expansion was not anticipated add more to the ultimate required selling price than do the capital-related costs and earnings in the plant where expansion money was pre-spent unnecessarily. Bear in mind that this is not a generalization; different relationships between the various cost elements could produce the opposite result.

An important element of market forecasting is estimation of future selling price. Here the planning engineer can make a significant contribution by assessing the probable maximum size of future plants that produce the product in a mature market, and calculating a required selling price for these conditions. Generally, selling price declines at a characteristic rate as the market matures. This is caused by larger plants producing at lower cost, and because investors are willing to accept a lower rate of return on their money in an established market situation where sales are less speculative than might be the case for a new product. Fig. 2 shows a typical price-decay curve.[12]

Cash Flow

Each of the factors mentioned so far directly influences the key economic consideration of cash flow. This is a measure of the rate at which money invested will be returned as profit. The time-discount relationship emphasizes the value of income during the early years of the project life. For this reason, it is important to commit enough capital in the initial investment so that reasonably efficient operation can be maintained.

The importance of good initial operation can be illustrated by considering two alternate cases:

Effect of Initial Operation on Cash Flow—Table 6

Cash Discount Rate 10%

Year	1	2	3	4	5	6	7	8	9	10
Sales potential, million lb	250	300	350	400	450	500	550	600	650	700
Sales price, ¢/lb	2.8	2.8	2.7	2.7	2.6	2.6	2.5	2.5	2.5	2.5
Case A—Minimum Investment										
Total investment, $ million	8.0				6.0					
Actual sales, lb. million	200	280	320	400	400	500	550	600	650	700
Net sales, $ million	5.6	7.8	8.6	10.8	10.4	13.0	13.8	15.0	16.2	17.5
Production cost	4.5	5.5	6.0	7.0	7.0	9.3	9.9	10.5	11.1	11.8
Pretax profit	1.1	2.3	2.6	3.8	3.4	3.7	3.9	4.5	5.1	5.7
Posttax profit	0.6	1.1	1.3	1.9	1.7	1.9	2.0	2.3	2.6	2.9
Depreciation reserve	0.8	0.8	0.8	0.8	0.8	1.4	1.4	1.4	1.4	1.4
Cash flow	(6.6)	1.9	2.1	2.7	(3.5)	3.3	3.4	3.7	4.0	4.3
Discounted cash flow	(6.0)	1.6	1.6	1.8	(2.2)	1.9	1.8	1.7	1.7	1.7

Net Present Worth = $5.6 million

Year	1	2	3	4	5	6	7	8	9	10
Case B—Increased Investment										
Total investment, $ million	9.0				5.0					
Actual sales, lb. million	250	300	350	400	450	500	550	600	650	700
Net sales, $ million	7.0	8.4	9.5	10.8	11.7	13.0	13.8	15.0	16.2	17.5
Production cost	5.2	5.9	6.5	7.1	7.8	9.3	9.9	10.5	11.1	11.8
Pretax profit	1.8	2.5	3.0	3.7	3.9	3.7	3.9	4.5	5.1	5.7
Posttax profit	0.9	1.2	1.5	1.8	1.9	1.9	2.0	2.3	2.6	2.9
Depreciation reserve	0.9	0.9	0.9	0.9	0.9	1.4	1.4	1.4	1.4	1.4
Cash flow	(7.2)	2.1	2.4	2.7	(2.2)	3.3	3.4	3.7	4.0	4.3
Discounted cash flow	(6.6)	1.7	1.8	1.8	(1.4)	1.9	1.8	1.7	1.7	1.7

Net Present Worth = $6.1 million

Nomenclature

c	Capacity of one unit
C	Total required capacity
i	Installed cost per unit
I	Total investment
K	Proportionality factor in cost-capacity relationship
N	Number of units installed
S	Spare factor, ratio of installed to required capacity
x	Exponent in cost-capacity relationship

Subscripts

N	Number of units installed.
opt	Optimum

to Table VI, the investment cost per unit of output will be about 30% greater than for two units if the cost per unit of output dependent on the total installed capacity is the basis for the analysis.

When multiple units are installed to provide spares—as when they are installed for flexibility—x values almost equal to or greater than 1 are of interest. The investment, when x almost equals 1, is essentially independent of the number of units, according to Table I.

In a situation such as this, the designer has almost total freedom—from an investment cost standpoint—in deciding how many units will be best for operations. For example, supplying 100 tons/day could be done with five units each having a capacity of 25 tons/day —with one of these held as a spare; or with six units each having a capacity of 20 tons/day—again with one serving as a spare. The investment in either case would be about the same.

When x is greater than 1, the right side of Eq. 4 becomes negative, which, of course, is meaningless. Eq. 3a also is only valid for values of x that are less than 1; for values of x equal to or greater than 1, there is neither a minimum value for I_N nor an N_{opt}. As N increases, I decreases, which is shown in Table II.

Base Exponent Value on Installed Cost

When using the techniques described in this article, x values must always be based on the total installed cost. Serious errors will result if x is determined solely from the purchase price. Values for x should reflect all costs, such as those for foundations, equipment settings and motor hookups. Charges that are fundamentally constant, such as small-motor hookup costs, tend to lower x, which in turn decreases N_{opt}.

This analysis has been limited to determining the optimum number of units only after the decision has been made to furnish spare capacity. This decision requires that the probability and the cost of a failure be evaluated, and that these be balanced against the cost of providing spare capacity, all of which is outside the scope of this article but which can be found in textbooks on operations research.

Optimum number of units—Table III

x	0.5	0.6	0.7	0.8	0.9
N_{opt}	2.0	2.5	3.3	5.0	10.0

How the number of units, N, affects total cost—Table IV

Exponent, x	Number of Units Installed, N							
	2	3	4	5	6	8	10	12
0.5	100.0	106.1	115.4	125.0	134.2	......	......	......
0.6	102.2	101.0	105.6	111.0	116.6	121.9	......	......
0.7	108.6	100.3	100.6	102.8	105.6	111.2	......	......
0.8	121.3	104.5	100.7	100.0	100.4	102.3	......	......
0.9	......	116.2	107.5	103.8	101.8	100.3	100.0	100.2

Spare factors related to number of units—Table V

No. of Units, N	2	3	4	5	6	7	8
Spare factor, S	2.00	1.50	1.33	1.25	1.20	1.17	1.14

Relative cost per unit of output—Table VI

No. of units, N	2	3	4	5	6	8
I_N/I_{opt}	121.3	104.5	100.7	100.0	100.4	102.3
S	2.00	1.50	1.33	1.25	1.20	1.14
$(I_N/I_{opt})/S$	60.65	69.67	75.32	80.00	83.39	89.82
Relative to N = 2	100.0	114.9	124.2	131.9	137.5	148.1

References

1. Bauman, H. C., "Fundamentals of Cost Engineering in the Chemical Industry," Reinhold, New York, 1964.
2. Chilton, C. H., Six-tenths Factor Applies to Complete Plant Costs, *Chem. Eng.*, **57**, April 1950.
3. Katell, S. and Faber, J. H., What Does Tonnage Oxygen Cost?, *Chem. Eng.*, **66**, June 29, 1959.
4. Williams, R., Six-tenths Aid in Approximating Costs, *Chem. Eng.*, **54**, Dec. 1947.

Meet the Author

Harry A. Quigley is currently managing modernization projects for both an automobile assembly and a cement plant in Mexico. For these assignments, he has been temporarily transferred to Ingenieria Panamericana by Day & Zimmermann, a Philadelphia construction company that owns 50% of the firm. Before this, he was a process and project engineer for Catalytic Construction Co., and a research laboratory administrator for Stauffer Chemical Co. He earlier spent several years with Atlas Chemical Industries, doing economic evaluations, operations research and long-range forecasting—and with General Foods and Celanese, also in economic and engineering functions. He received a bachelor's degree in chemical engineering from Villanova University in 1953, and has done graduate work in engineering at Newark College of Engineering and in business at Temple University.

Key Concepts for This Article

Active (8)	Passive (9)	Means/Methods (10)
Estimating*	Costs*	Charts
Selecting*	Spares*	Data
	Units*	Equations
	Multiple*	Formulas
	Single	

Ind. Variable (6)	Dep. Variable (7)
Capacity*	Investment*
Equipment	
Units*	

(Words in bold are role indicators; numbers correspond to EJC-AIChE information retrieval system. Asterisks mark key concepts suggested for indexing. Others are added to improve reading as an abstract. Indexing is described in *Chem. Eng.*, Oct. 11, 1965, p. 187.)

Determining Optimum Plant Size

Over-conservatism or over-optimism is usually responsible when plants for making a new chemical encounter financial difficulties. These undersized or oversized facilities are built on the basis of inadequate forecasting of sales and prices.

THOMAS E. CORRIGAN and **MICHAEL J. DEAN**
Mobil Chemical Co.

A plant's size is usually determined after market research has been done. Marketing men analyze the potential sales of the product at a reasonable price, and then decide how much of the market the company can hope to capture. They pass the word to the planners, who size the new plant accordingly. A price is guessed by referring to price of similar chemicals.

One of the problems that may be inadvertently ignored when justifying a new plant is establishing how low a price for the product the market can stand as it expands. Each competitor will attempt to increase sales by cutting price. The over-all market can probably only expand by replacement of other materials that themselves have a declining price. This may result in the market price's dropping to the point where only the optimum size plant can be profitable.

The declining price of new chemicals (or old chemicals made by new methods) is well illustrated by reference to the ABS (acrylonitrile-butadiene-styrene) plastic market, the acrylonitrile market, or the cyclamate market. In each of these markets, plants have been built that, although technically successful, have been financial "flops." Why? This happened because the plants were too small and, thus, the fixed costs were too large a proportion of the manufacturing costs.

In other instances, companies have jumped into production only to find that their product sales were insufficient for them to operate their plant at even 60% of capacity. Most plants operating below 60% of capacity probably do not do so at a profit.

This article demonstrates how a plant can become unprofitable or less profitable than the original economic studies indicated because the size was either too large or too small for the particular market conditions. This situation will illustrate the need for meticulous market research to obtain the most accurate prediction possible of future market size. It also will illustrate the folly of yielding to the temptation of being over-optimistic despite moderate market predictions and going "too big." On the other hand, there is no such thing as "playing it safe" and "being sure by going small" if market research predicts an expanding market. Playing it conservatively and building too small a plant could turn out to be even more dangerous a move than building too large a plant.

This article is written with the objective of demonstrating that the decision "how big" is one of the most important factors in the success of a new plant.

There are three characteristics of the market that corporate planners should take into consideration in trying to decide the best size plant to build. These are: (1) the static price-volume relationship, (2) the rate of growth of the class of compounds as a whole, and (3) the rate at which the new product can penetrate the market.

Static Price-Volume Relationship

A well-known characteristic of chemical markets is the price-volume curve for a class of compounds. This curve defines the size of market for a specific compound by its price. An example of this is shown in Fig. 1, which is a plot of the price vs. the market volume. This curve defines the size of market for a specific compound by its price. From Fig. 1, we can see that the engineering plastics, such as polyphenylene oxide and polycarbonates, are restricted in their sales because their high prices preclude their use in any but specialized markets. As the price of the plastic decreases, it can be used for more general purposes and, therefore, its market expands. The most versatile plastics are the least expensive and have the largest sales. Naturally, the least-expensive plastics are the ones that are made from the less-expensive raw materials and entail the simplest processing steps.

If we take a new plastic for which the manufacturing cost and the properties are known, we can forecast the market in which it can be sold. For instance, if we make a plastic costing 50¢/lb., we know that the total market for the material is probably about 150 million lb. unless it has some very unusual properties that allow it to be used for some very specialized purpose. Even in a case like this, however, it can only expand by displacing more-expensive products that have smaller markets. Thus, even here, the amount of extra sales is limited.

The chart (Fig. 1) for plastics includes an area where there are no plastics. This is the exclusion area. It is unlikely that plastics having the particular properties and cost of manufacture to put them in this volume-price range will ever be developed. Most chemical products show this type of volume-price curve. Information necessary to construct such a curve can be obtained from the following sources:

1. *Oil, Paint and Drug Reporter* (chemical prices).

427

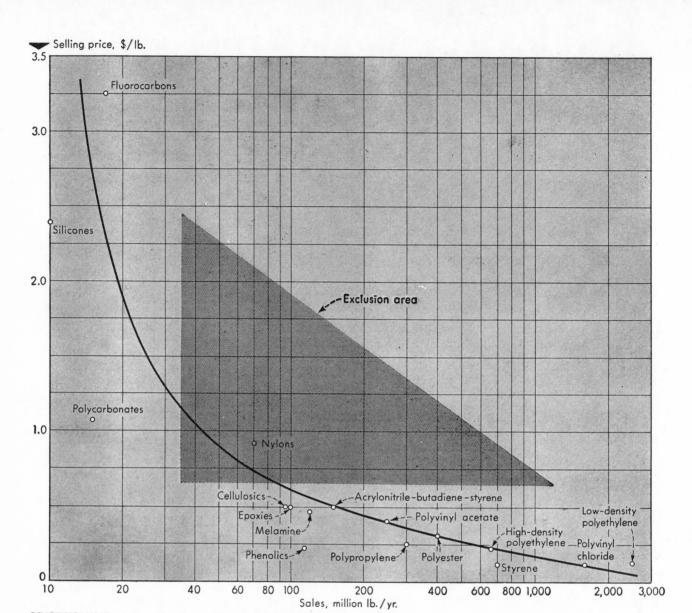

PRICE-VOLUME relationships for molding, extrusion and film plastics—**Fig. 1**

2. Dept. of Commerce Statistics.
3. Dept. of Interior Statistics.
4. The U.S. Tariff Commission Statistics.

Rate of Growth of the Market as a Whole

The way to determine the market growth rate is to generate data for various sections of the market and sum them. Again, using plastics as an example, growth has occurred because of the ability of plastics to replace metals in the automobile industry. Numerous well-documented studies of this penetration have been made. The main property that helps plastics penetrate this market is their lower density. By reference to a forecast of the number of automobiles to be made in any given period and to the rate of replacement of metal by plastics, the rate of growth of the automobile section of the plastic market can be determined. If this is done for all sections of the market, a general annual growth rate can be established. This can then be used for the new plastic that is being considered.

An alternate approach is to make a more detailed analysis of some specific section of the plastic market. This may be required to get a better idea of the potential market. For instance if we wanted to produce a new engineering plastic, we should evaluate how far nylon gear wheels have replaced metal gear wheels in light engineering industry. If a company does not have the time or manpower to evaluate markets, evaluations can be obtained from marketing companies who specialize in chemical industry forecasts.

Market Penetration Rate

The final characteristic of the chemical market must be determined by the sales force itself through customer contact. The sales department must determine how much of the product it can sell, how much support it will need from technical service, which competitors may be contemplating manufacture of the product, the geographical location of the market, and the size of package (i.e., tankcar, drum, 5-lb. bag) for product.

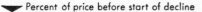

Percent of price before start of decline

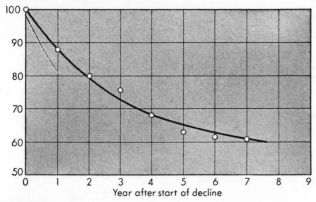

DECLINE in price for sixteen chemicals—Fig. 2

When the static price-volume relationship, growth rate and market penetration characteristics have been determined, there will be enough information to make a reasonable estimate of the best size and geogra cal location of the plant.

The curve in Fig. 2 shows the decline in price with time for a series of medium-priced chemicals. The data for Fig. 2 were prepared in the following manner: The lowest prices for nonintermediates for which there had been price declines were taken from the *Oil, Paint and Drug Reporter*. Either the year in which the commodity first appeared in this journal, or the year in which the price decline began, was taken as year zero. The ratio of the price in any given year to the price at year zero was calculated for the following seven years after which time the price had leveled out. The ratios for all of the commodities considered were averaged for each year.

The percentage of the initial price is plotted in Fig. 2. Similar curves would result whether the plot were made for intermediates as shown, or for finished products such as pesticides or herbicides.

Price declines are extremely common when patent coverage expires and the right to manufacture the product is no longer restricted to patent holders and licensees. When a decision on plant size is needed, the planners should keep in mind that this could happen in any market.

Basis for Plant Sizes

Companies A, B and C all made surveys to determine the market for their new chemical. They evaluated the total market by asking their customers whether or not they would purchase the product. They also compared their product with the growth rates for similar chemicals. Each company estimated the cost of manufacture, as shown in Table I. Table I shows the estimated capital investment and the predicted manufacturing costs and profitability index for plants of 10 million, 20 million, 60 million and 100 million lb./yr. The prediction for the 10-million-lb./yr. plant is a safe, conservative 15½% profitability index. The 100-million-lb./yr. plant shows 53%, and the 60-million-lb./yr. plant shows an intermediate 30%.

Company A did not believe that the market would expand rapidly, or that others would enter the field. Since the planners knew the company could definitely sell 10 million lb. of the product with a guaranteed profitability index of 15%, they went to their management with a proposal for a plant with a capacity of 10 million lb./yr. The proposal was accepted and the plant was built.

Company B surveyed its own customers and did an industry survey as well. It could see sure sales of 60 million lb./yr. including exports, and thought it could eventually sell another 40 million lb./yr. Since the

Manufacturing costs for all three companies—Table I

Plant Size, lb./yr. product	10,000,000	20,000,000	60,000,000	100,000,000
Fixed capital, $.	6,600,000	10,000,000	19,350,000	24,600,000
Working capital, $.	830,500	1,552,300	4,400,800	7,211,700
Total investments, $.	7,430,500	11,552,300	23,750,800	31,811,700
Manufacturing Costs, $/yr.				
Raw materials & chemicals.	1,000,000	2,000,000	6,000,000	10,000,000
Utilities.	140,000	240,000	640,000	1,040,000
Labor.	600,000	600,000	600,000	600,000
Maintenance.	330,000	500,000	968,000	1,230,000
Taxes and insurance.	165,000	250,000	483,800	615,000
Total manufacturing cost, $/yr.	2,235,000	3,590,000	8,691,800	13,485,000
Amortization, $/yr.	660,000	1,000,000	1,935,000	2,460,000
Interest on working capital, $/yr.	41,500	77,600	220,000	360,600
Total annual cost, $.	2,936,500	4,667,600	10,846,800	16,305,600
Cost of product, ¢/lb.	29.4	23.4	18.1	16.3
Expected discounted return on investment at capacity*.	15.5%	17%	30%	53%

* 10 year life. No investment credit.

Summaries of the net cash flow for each company—Table II

	Company A	Company B	Company C
Plant size, lb./yr.	10,000,000	100,000,000	60,000,000
Fixed capital, $.	6,600,000	24,600,000	19,350,000
Working capital, $.	830,500	7,211,700	4,400,800
Total capital, $.	7,430,500	31,811,700	23,750,800

COMPANY A

Year	1	2	3	4	5 to 10
Sales, lb./yr.	10,000,000	10,000,000	10,000,000	10,000,000	10,000,000
Sales price, ¢/lb.	50	45	40	40	40
Manufacturing Costs, $/yr.					
Raw materials & chemicals	1,000,000	1,000,000	1,000,000	1,000,000	1,000,000
Utilities	140,000	140,000	140,000	140,000	140,000
Labor	600,000	600,000	600,000	600,000	600,000
Maintenance	330,000	330,000	330,000	330,000	330,000
Taxes & insurance	165,000	165,000	165,000	165,000	165,000
Total manufacturing cost, $/yr.	2,235,000	2,235,000	2,235,000	2,235,000	2,235,000
Total annual cost, $.	2,936,000	2,936,000	2,936,000	2,936,000	2,936,000
Gross sales income, $/yr.	5,000,000	4,500,000	4,000,000	4,000,000	4,000,000
Net sales income, $/yr.	4,500,000	4,050,000	3,600,000	3,600,000	3,600,000
Net cash flow after tax, $/yr.	1,462,500	1,237,500	1,012,500	1,012,500	1,012,500

Discounted return on investment for 10 year life is 9%.

Expected discounted return on investment at capacity is 15.5%.

COMPANY B

Year	1	2	3	4 to 7	8	9	10
Sales, lb./yr.	10,000,000	20,000,000	40,000,000	60,000,000	70,000,000	80,000,000	100,000,000
Sales price, ¢/lb.	50	45	40	40	40	40	40
Manufacturing costs, $/yr.							
Raw materials & chemicals.	1,000,000	2,000,000	4,000,000	6,000,000	7,000,000	8,000,000	10,000,000
Utilities	140,000	240,000	440,000	640,000	740,000	840,000	840,000
Labor	600,000	600,000	600,000	600,000	600,000	600,000	600,000
Maintenance	1,230,000	1,230,000	1,230,000	1,230,000	1,230,000	1,230,000	1,230,000
Taxes & insurance	615,000	615,000	615,000	615,000	615,000	615,000	615,000
Total manufacturing cost, $/yr.	3,585,000	4,685,000	6,885,000	9,085,000	10,185,000	11,285,000	13,485,000
Total annual cost, $.	6,405,000	7,505,000	9,705,000	11,905,000	13,005,000	14,105,000	16,305,000
Gross sales income, $/yr.	5,000,000	9,000,000	16,000,000	24,000,000	28,000,000	32,000,000	40,000,000
Net sales income, $/yr.	4,500,000	8,100,000	14,400,000	21,600,000	25,200,000	28,800,000	36,000,000
Net cash flow after tax, $/yr.	1,687,500	2,937,500	4,987,500	7,487,500	8,737,500	9,987,500	12,487,000

Discounted return on investment for 10 year life is 14.5%.

Expected discounted return on investment at capacity is 53%.

COMPANY C

Year	1	2	3	4	5 to 10
Sales, lb./yr.	10,000,000	20,000,000	40,000,000	60,000,000	60,000,000
Sales price, ¢/lb.	50	45	40	40	40
Manufacturing Costs, $/yr.					
Raw materials & chemicals	1,000,000	2,000,000	4,000,000	6,000,000	6,000,000
Utilities	140,000	240,000	440,000	640,000	640,000
Labor	600,000	600,000	600,000	600,000	600,000
Maintenance	968,000	968,000	968,000	968,000	968,000
Taxes & insurance	483,800	483,800	483,800	483,800	483,800
Total manufacturing cost, $/yr.	3,191,800	4,291,800	6,491,800	8,691,800	8,691,800
Total annual cost, $.	5,346,800	6,446,800	8,646,800	10,846,000	10,846,000
Gross sales income, $/yr.	5,000,000	9,000,000	16,000,000	24,000,000	24,000,000
Net sales income, $/yr.	4,500,000	8,100,000	14,400,000	21,600,000	21,600,000
Net cash flow after tax, $/yr.	1,621,000	2,871,000	4,921,600	7,421,600	7,421,600

Discounted return on investment for 10 year life is 18%.

Expected discounted return on investment at capacity is 30%.

Total market, sales price and market share of new product—Table III

Year	Total Market Million Lb./Yr.	Selling Price ¢/Lb.	Market Share		
			Company A %	Company B %	Company C %
1	30	50	33	33	33
2	50	45	20	40	40
3	90	45	12	44	44
4	130	40	8	46	46
5	130	40	8	46	46
6	130	40	8	46	46
7	130	40	8	46	46
8	140	40	7	50	43
9	150	40	7	53	40
10	170	40	6	59	35

profitability index of a 100-million-lb./yr. plant was 53% and since even at 60 million lb./yr. it exceeded the company's minimum for new projects, a forward-looking management agreed to build a plant with a capacity of 100 million lb./yr.

Company C surveyed its own customers and the market. Since it also knew that Company A and Company B were going into the market, it decided to build a plant large enough to take advantage of large-scale manufacture, but not so large as to take a chance on having to operate at less than full capacity. Company C then built a 60-million-lb./yr. plant.

The figures for the total market are shown in Table III. The market developed as was forecast by Companies B and C, but Company B could not expand its sales. Therefore, it gave discounts. Its sales increased, but Company C matched the discounts. The net effect was a squeeze on Company A, which also had to drop its price. The over-all effect was less profit for all.

Company A did not get the return on its money it had expected. Company B did not sell what it expected even though it lowered the price. Company C proceeded unscathed and was in a position to add more capacity after ten years had passed. It was able to take advantage of the technical experienced gained in the smaller plants.

The summaries of cash flow for each of the three plants are shown in Table II. The discounted-cash-flow table for Company A is based on full-capacity operation, but with the price starting out at 50¢/lb. the first year, dropping to 45¢/lb. the second year, and to 40¢/lb. the third year, and stabilizing there. The actual profitability index for the operation is only 9% instead of the predicted 15.5%.

Table II contains the same type analysis for Company B. The price schedule is the same as for Company A, but Company B does not reach capacity production until its tenth year. Actual profitability index is only 14.5% instead of the 53% predicted.

Also in Table II is the analysis for Company C. The actual profitability index was 18%. This was higher than for either of the other two companies. It

was not as high as the predicted 30% because of the lower price, but the decrease from the predicted return was less than in the other two cases.

Conclusion

Thorough market research and a full assessment of what competition can do is an absolute necessity in arriving at the correct decision regarding potential plant size for the production of a new product. Plant size is critical, especially when fixed costs make a large contribution to total costs. The sizes used in these hypothetical examples were quite arbitrary and were for comparison only. For a heavy chemical such as ethylene, we may well have been speaking of an order of magnitude higher in production capacity, such as 100 million, 600 million or 1,000 million lb./yr. For an insecticide, we could have equally well been comparing sizes of 1, 6, and 10 million lb./yr. The order of magnitude is dependent upon the type of product and industry.

The market research and assessment of competition will decide what size plant is too large, what size too small, and what size is advisable for each chemical.

Meet the Authors

Thomas E. Corrigan   **Michael J. Dean**

Thomas E. Corrigan is with the Research and Technical Div. of Mobil Chemical Co., Metuchen, N.J. Dr. Corrigan is responsible for advising management on the technical and economic feasibility of new processes and engineering technology. He is also an internal consultant in chemical reaction engineering and design. He has taught graduate courses in engineering economics that have emphasized project analysis and management science. Dr. Corrigan has a B.S. in chemical engineering from Fenn College, an M.S. in chemistry from the University of Michigan, and a Ph.D. in chemical engineering from the University of Wisconsin.

Michael J. Dean evaluates new chemical processes for Mobil Chemical Co., Metuchen, N.J. He also advises management regarding the economic feasibility for these processes. Prior to joining Mobil, he spent six years with a major oil company in England, and two years with Polymer Corp., Ltd., in Canada. Mr. Dean graduated in chemistry from the Imperial College of Science and Technology, London.

Key Concepts for This Article

Active (8)	Passive (9)	Means/Methods (10)
Selection*	Optimization*	Profitability*
Determination*	Plants*	Return on investment*
	Size*	Discounted cash flow*
	Chemical*	Forecast*
		Sales*
		Markets*
		Costs*

(Words in bold are role indicators; numbers correspond to EJC-AIChE information retrieval system. Asterisks mark key concepts suggested for indexing. Others are added to improve reading as an abstract. Indexing is described in *Chem. Eng.*, Oct. 11, 1965, p. 187.)

Evaluating Proposed Ventures That Tie In With Existing Facilities

Here is how "sunk" costs, "opportunity" costs, and other incremental cost principles can be used to evaluate new projects that are going to draw on, feed, augment or otherwise involve existing resources.

W. P. HEGARTY, *Techni-Chem. Co.*

Although many articles on economic evaluation and profitability analysis appear in the chemical engineering literature every year, most of them consider ventures that are isolated; not much attention is given to the complications introduced by tie-ins with existing facilities. Actually, most chemical companies today are highly integrated, and few ventures stand apart from the mass of a company's existing operations. Very often, a project cannot be properly evaluated without quantifying the various links between it and facilities that already exist.

The question of how to do this causes a great deal of confusion and difficulty in economic evaluation today. This is surprising, because the economic fundamentals governing such situations are well established. The purpose of this article is to outline these fundamentals, show how they are applied in some typical situations and consider some possible pitfalls. A followup article, scheduled for the Sept. 9 Cost File, will take up a detailed illustrative example: the addition of a vinyl chloride operation to an existing ethylene-chemicals complex.

Incremental Costs in Practice

The incremental cost concept is an old one. Essentially, it states that incremental additions to a company's operations must rest on an incremental justification. To apply the incremental cost concept, the cash flows that will result if a proposed project is adopted are estimated. The cash flows that will result if the project is not adopted are also estimated. The incremental difference, then, is the true measure of the effect of the proposed venture. Determination of the discounted-cash-flow (DCF) rate of return, or the net-present-value at a preselected interest rate,[†] will then provide a basis for deciding if the proposed project is attractive.

Of course, it is fine to argue that incremental addi-

tions to a company's operations must rest on an incremental justification. In practice, determination of the incremental profits, costs and investments may be difficult. Proper definition of the scope of the proposed project is the key, and here the concept of a thermodynamic system offers a helpful analogy. In thermodynamic problems, a system is defined as a portion of the universe delineated for study. By respecting a set of rules covering all transfers across the system boundaries, the range of consideration is limited and the problem can be defined and solved in a convenient framework. Similarly, in economic evaluation, careful definition of the scope will simplify the problem, so that only net cash flows in and out need be considered. All areas in which changes will result if the project is adopted would normally be included in the scope. To the extent practical, areas where no change will result should be excluded.

With the scope defined, evaluation on an incremental basis proceeds as described earlier, and the pertinent cash flows that will result if the project is adopted, and is not adopted, are estimated. The incremental difference is then determined. Here, it should be noted that the investments and fixed costs associated with existing facilities are the same whether the project is or is not adopted, and are therefore irrelevant to the project evaluation.

The cash flows that are projected if the venture is not adopted should be based on optimum use of the existing facilities. Otherwise, the incremental advantage attributed to the proposed project will be inflated unrealistically. The incremental difference should be a measure of the advantage of the proposed addition, not of the misuse of existing facilities.

The above procedure can lead to a rigorous incremental evaluation. However, in many practical situations, raw materials for a proposed project may, for example, come from all over the company. If the source facilities for all these inputs are included in the project scope, it may balloon to include practically the whole company. It is here that the concepts of sunk costs and opportunity costs make their entrance. Basically, the utility of these concepts is that they permit a contraction of the project scope while still

---

† Unusual cash-flow patterns that can be generated in incremental evaluations can result in multiple DCF rates or none. When this occurs, net present value at a preselected interest rate is the evaluation criteria ; see Bierman, et al, "The Capital Budgeting Decision," 2nd ed., Macmillan, pp. 44-49.

permitting an incremental evaluation of the alternate investment situations.

Sunk Costs and Opportunity Costs

Sunk costs are defined as expenditures or commitments made in the past. Typical are the original or present book value of an existing facility, the depreciation or taxes on an existing facility, contracted rental payments, and the like.

The principle of sunk costs is simply that they are irrelevant to future decisions. Sunk costs are a matter of history; they are not affected by adoption or rejection of a given project, and they do not affect incremental costs. While this principle may appear obvious, in different guises its violation is a frequent source of difficulties, as we shall see.

When an existing facility is used in a project, it may mean that the opportunity for another mutually exclusive use is foregone. This would mean that the future cash flows from the alternate use would be lost; these losses are defined as *opportunity costs*, and are used in an incremental evaluation.

In this article, the opportunity cost concept will be extended to cover future outlays that would result because the current project preempted use of existing facilities. Rather than forego an alternate use for existing facilities, it may be economical to pursue the alternate use as well, by expanding existing facilities to accommodate both. In such a situation, to the extent that the expansion is necessitated by preemption by the proposed venture, the investment for expanding the existing facilities would be an opportunity cost chargeable to the proposed venture at the time that the expansion is actually required.

Applying the Principles

In evaluating a proposed project, existing resources can be excluded from the project scope if inputs from these resources are charged in at cost. This cost should exclude sunk costs, but must include opportunity costs.

In light of this principle, consider supply of a raw material X to a proposed venture. At what price should X be charged to the venture for purposes of economic evaluation? The price, of course, should be the incremental cost, and would depend on the particular circumstances—i.e.: X can be purchased from outside, in which case the price obviously is the delivered purchase price, or X can be supplied from within the company, in which case three situations are possible: X could be supplied from available excess production capacity, it could be diverted from sales, or production could be expanded.

Consider first the case where X could be supplied from available excess capacity. Here, if production of X will continue independent of adoption of the project and if there is no alternate use for the excess capacity, X should be charged to the proposed project at production plus shipping cost, with sunk cost items deleted from production cost. Thus depreci-

ation, taxes and any other sunk costs should be deleted. Note that existing labor costs are sunk costs here; there are no opportunity costs.

If X will be produced only to supply raw materials to the proposed project, the situation is changed. Labor and other operating costs are not sunk costs. Also, options of shutting down and abandoning or selling the X production facilities, with any resultant revenues or immediate tax savings, are foregone. Either net disposal value, or the net present value of the immediate tax saving foregone, is an opportunity cost to be added to the venture's over-all cost.

Note, though, that the tax advantage of an immediate writeoff is only an advantage in timing when the company is in the top corporate income tax bracket. Any tax savings produced by depreciating an investment this year would otherwise have been produced in some subsequent year. In discounted cash flow analyses, where the time value of money is considered, this is handled as follows: The depreciation writeoff tax savings that is foregone is listed as an immediate negative cash flow, with the alternate normal year-by-year depreciation tax saving as a credit. After discounting, this procedure gives a realistic evaluation to the lost opportunity for claiming an immediate tax loss.

Consider next the case where X would be supplied to the proposed venture by diverting it from sales or other uses. Here again, for evaluation purposes, X would be supplied to the proposed venture at production plus shipping cost, with sunk cost items deleted. The loss of the sale or alternate use,

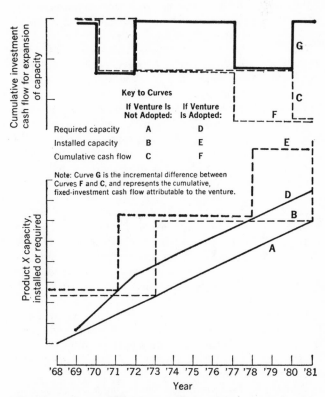

EXAMPLE OF OPPORTUNITY COSTS: Preemption of existing facilities may affect timing of future expansions—Fig. 1

however, will typically involve an opportunity cost to be added, and this cost is the net cash flow that would have been produced by the best of the mutually exclusive alternate uses that was forfeited.

Finally, consider the situation when production capacity for X is expanded. The capacity and timing of the expansion would, of course, be governed by the requirements of the future ventures to be supplied. Typically, the size of the X capacity expansion would be increased by an amount equal to the capacity preempted by the proposed venture. In this circumstance, the opportunity cost to the present venture is a capacity pro rata share of the future expansion investment chargeable at the time that the X capacity expansion is actually required.

However, another situation is also possible. For some chemicals experiencing continuous growth requirements, preempting of existing capacity may not affect the size of the expansion, but may instead advance the time when the expansion and subsequent expansions are made. This situation is developed hypothetically in Fig. 1. Curves A, B and C, respectively represent projected X capacity requirements, installed capacity, and cumulative, fixed-investment cash flow for expanding X capacity when the proposed venture is *not* adopted. Note that expansions are shown coming onstream in 1973 and 1981; the investments are made during construction the preceding years.

Curves D, E and F, respectively, represent projected X capacity requirements, installed capacity and cumulative, fixed-investment cash flows when the proposed venture *is* adopted. Note that addition of the proposed venture requirements advances the time when economic-sized X production capacity expansions are added. Subsequent additions would presumably also be affected, ad infinitum, but they would have little present value at acceptable interest rates and can be ignored.

Curve G is the incremental difference between Curves F and C, and is the incremental, cumulative fixed-investment cash flow attributable to adopting the proposed venture. It can be seen that the opportunity cost chargeable to the proposed venture is given by the Curve G cash flows. Thus, in the situation where adoption of the proposed venture advances the time when X capacity expansions are made, the entire expansion-investment cash flow over the periods that the expansions are advanced is the associated opportunity cost.

As shown in Fig. 1, the proposed venture would be charged with the full expansion investment for the years 1971, 1972, and 1977 through 1980. Note also that during these years, the production costs chargeable to the proposed venture would include all the fixed costs of the expansion, such as operating labor, maintenance, supplies, taxes, etc.

Whether the previously described capacity pro-rata investment share, or the investment timing advance, is used in evaluating a given venture would depend on the specifics of the projected X requirement and capacity expansion.

While the foregoing comments have been specific to the use of existing facilities in supplying raw materials, the principles are the same for supplying utilities, overhead items, and manpower resources.

Pitfalls in Incremental Evaluations

Some common errors now become recognizable:

Defensive Investments—Any practitioner in economic evaluation soon runs into the misapplied "defensive investment." In a typical instance, it would be argued that a new investment must be made to protect an existing investment. This could be throwing good money after bad. The point, of course, is that an existing investment is a sunk cost. No additional investment can be justified by a sunk cost. Investment could, however, be justified to protect future income from an existing facility, but the attention is on the future income being defended, not on the investment.

Allocated Investments—A long-established but often erroneous practice is addition of allocated investment to the investment base of a proposed project. As usually seen, it involves addition of a capacity pro rata investment for existing raw-material production units and other support facilities. Either original book value or present book value is used as a basis. Again, the point is that original book and present book value of existing facilities are not pertinent—they are sunk costs. Instead, as noted earlier, the actual future costs of expansions necessitated by use of existing facilities should be included as opportunity costs.

While the incremental evaluation technique is thoroughly correct in economic comparison of alternatives, it presents potential pitfalls. The evaluation is only correct within the framework of the alternatives examined, and is not a substitute for management foresight. Management must fix an evaluation framework that will properly recognize long-range requirements. The problem could be likened to that of determining the optimum path through a maze or network. Looking only one step ahead usually will not suffice. There are, for example, sad histories of corporations that examined a succession of problems on an incremental basis over a period of years, meeting new requirements by revamping existing plants, only to awake and find themselves with outmoded plants, and outclassed by the competition. For each of these problems standing alone, revamping always emerged as the best alternative. Taken in the aggregate, major modernization was the real answer.

Another pitfall may involve expansion into a new field with extensive "free rides" on existing facilities. Looking to the future, normal expansion may not be economic when such facilities are no longer available.

Incidentally, the utility of the sunk cost and opportunity costs concepts will probably decline, with increased computer use for economic evaluation. At some future time, it will be practical to develop an economic model to simulate the entire breadth of operations of the largest of companies. Using such a model, it would be relatively easy to simulate operation with, and then without, a proposed venture, and develop and discount the cash flows to yield the incremental discounted cash flow directly.

Evaluating the Incremental Project: An Illustrative Example

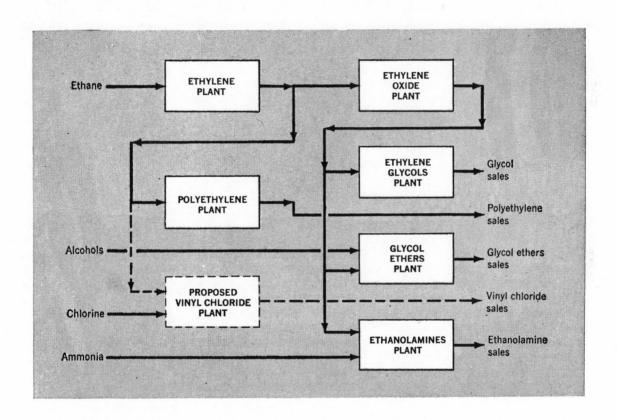

Carrying last Cost File's discussion of incremental cost techniques one step further, here is how these techniques can be applied in evaluating the addition of a vinyl chloride operation to an existing ethylene-chemicals complex.

W. P. HEGARTY, Techni-Chem Co.

In Cost File No. 142 (which appeared on pp. 432–434), we discussed how "sunk" costs, opportunity costs, and other incremental cost principles can be used to evaluate new projects that are going to draw on, feed, augment or otherwise involve existing facilities. To put this discussion in focus, and to provide some specifics, let us consider the following illustrative example:

A chemical company presently operates a Gulf Coast ethylene complex. Ethylene is produced via ethane pyrolysis and fed to polyethylene and ethylene oxide operations (see diagrams above). The poly-

ethylene is marketed directly, whereas the ethylene oxide is fed to glycol, glycol ethers and ethanolamines operations, and these products are then marketed.

The problem is to evaluate the economics of adding a 200-million-lb./yr. vinyl chloride operation to this existing complex. Specifics are as follows:

Process—A modern, balanced oxychlorination process would be used to produce vinyl chloride from chlorine, ethylene and air. No byproduct hydrogen chloride is produced in the process.

Market Forecast—Projected vinyl chloride sales volume and netback prices are shown in Table I. With initial vinyl chloride production set for Jan. 1971, capacity operation would be projected for 1975. Project life is estimated at 12 years.

Chlorine—Purchased chlorine would be used at a projected delivered price of $50/ton.

Ethylene—The existing ethylene plant has a capacity of 400-million lb./yr. This capacity will be adequate to supply the projected polyethylene and ethylene oxide requirements, in addition to the vinyl chloride needs, through 1974. Increased ethylene capacity would be required in 1975. The pro forma production cost (as it would appear on the plant accounting records) for the existing ethylene plant at capacity operation is shown in Fig. 1. The utility costs listed in Fig. 1 are based on averaged total utility unit costs; these include depreciation and other fixed costs, such as operating labor, maintenance, overhead, taxes and insurance costs. As for the incremental costs of utilities from existing facilities, these are listed below under steam, electricity, etc.

Incremental cost of ethylene production from the existing ethylene plant is the incremental unit–raw-materials plus utilities cost, minus byproduct fuel credits that total 1.47¢/lb. of ethylene.

The operating and overhead costs are sunk costs where the vinyl chloride operation is concerned. These costs will continue independent of acceptance or rejection of the vinyl chloride project, because the ethylene plant operation will be continued in order to supply the ethylene oxide and polyethylene plant requirements.

Steam—Existing steam-generation and distribution facilities will be adequate for projected requirements, including the vinyl chloride operation, through 1972. Expanded capacity will be required by 1973. Total (pro forma) production cost, including depreciation from existing facilities, is 45¢/1,000 lb. steam. Incremental production cost is 32¢/1,000 lb. steam. The incremental production cost of steam from the new facilities in 1973 will be 45¢/1,000 lb.; this cost will include operating labor, maintenance, taxes, insurance, and incremental overhead for the new steam-generation facilities.

Electricity—Power is purchased from a public utility company. Average unit cost is 0.8¢/kwh. Incremental power cost, however, is 0.6¢/kwh. Existing power distribution facilties will be adequate for future requirements, including the vinyl chloride operation, until the ethylene capacity is expanded.

Fuel—Natural gas is purchased at a cost of 20¢/million Btu.

Cooling Water—Existing cooling towers, pumping and distribution facilities will be adequate for requirements, including the proposed vinyl chloride operation, until 1974, when expanded capacity would be required. While pro forma production cost of cooling water from existing facilities is 2¢/1,000 gal., incremental production cost from existing facilities is 1¢/1,000 gal. Incremental production cost from the new facilities will be 2¢/1,000 gal.

PRODUCT: ETHYLENE LOCATION: U. S. GULF COAST

ANNUAL CAPACITY: 400,000,000 LB/YR PROCESS: ETHANE PYROLYSIS

BATTERY LIMITS CAPITAL COST (BLCC): $12,000,000

RAW MATERIALS	Unit	Annual Quantity	Unit Price	Unit Cons.	Annual Cost	Unit Cost
ETHANE	Lb.	500,000,000	0.0085	1.25	4,250,000	0.0106
CHEMICALS & CATALYSTS					520,000	0.0013
TOTAL RAW MATERIALS COST					4,770,000	0.0119
UTILITIES						
Steam @ psig		NONE				
Electricity	Kwh.	164,000,000	0.008	0.41	1,310,000	0.0033
Fuel	Million Btu.	3,240,000	0.20	0.0081	648,000	0.0016
Cooling Water	Thousand gal.	12,100,000	0.02	0.030	240,000	0.0006
Process Water	Thousand gal.	40,000	0.20	0.0001	8,000	
TOTAL UTILITIES COST					2,206,000	0.0055
OPERATING COSTS						
Labor		7 MEN/SHIFT			210,000	
Supervision					42,000	
Supplies					21,000	
Maintenance: Labor					150,000	
Materials					120,000	
Overhead					30,000	
Total					300,000	
TOTAL OPERATING COSTS					573,000	0.0014
OVERHEAD EXPENSES						
Plant Overhead					360,000	
Royalty					—	
Insurance & Property Taxes					120,000	
Depreciation @ 10 %BLCC					1,200,000	
TOTAL OVERHEAD EXPENSES					1,680,000	0.0042
SUBTOTAL = MANUFACTURING COST					9,229,000	0.0230
BY-PRODUCT CREDITS	Unit	Quantity	Price	Prod.		
FUEL	MMBTU	3,200,000	0.20	0.008	640,000	0.0016
TOTAL PRODUCTION COST					8,589,000	0.0214

Notes:

PRO FORMA cost estimate for ethylene—Fig. 1

Sales forecast for sample project—Table I

Year	Sales Volume Million Lb./Yr.	Sales Netback Price, ¢/Lb.
1971	80	6.0
1972	110	6.0
1973	140	5.5
1974	170	5.5
1975	200	5.5
1976	230	5.3
1977	270	5.3
1978	310	5.3
1979	350	5.0
1980	400	5.0
1981	450	5.0
1982	500	5.0

RAW MATERIALS	Unit	Annual Quantity	Unit Price	Unit Cons.	Annual Cost	Unit Cost
Ethylene	Lb.	97,000,000	0.204	0.484	1,980,000	0.0099
Chlorine	Lb.	135,000,000	0.025	0.674	3,380,000	0.0169
Catalyst & Chemicals					126,000	0.00063
TOTAL RAW MATERIALS COST					5,486,000	0.02743
UTILITIES						
Steam @ psig _Thousand Lb._		300,000	0.45	0.0015	135,000	0.00068
Electricity	Kwh.	21,400,000	0.006	0.107	128,000	0.00064
Fuel	Million Btu.	360,000	0.20	0.0018	72,000	0.00036
Cooling Water	Thousand gal.	6,200,000	0.02	0.0310	124,000	0.00062
Process Water						
TOTAL UTILITIES COST					459,000	0.00230
OPERATING COSTS						
Labor _5 Men/Shift_					150,000	
Supervision					30,000	
Supplies					15,000	
Maintenance: Labor					125,000	
Materials					100,000	
Overhead					25,000	
Total					250,000	
TOTAL OPERATING COSTS					445,000	0.00222
OVERHEAD EXPENSES						
Plant Overhead					100,000	
Royalty						
Insurance & Property Taxes					50,000	
Depreciation @ %BLCC					500,000	
TOTAL OVERHEAD EXPENSES					650,000	0.00330
SUBTOTAL = MANUFACTURING COST					7,040,000	0.03525
BY-PRODUCT CREDITS	Unit	Quantity	Price	Prod.		
TOTAL PRODUCTION COST					7,040,000	0.03525

Notes:

INCREMENTAL cost estimate for vinyl chloride—Fig. 2

the vinyl chloride operation when that project is accepted.

Vinyl Chloride Production Costs—The incremental vinyl chloride production cost at capacity operation for 1975 and later years is shown in Fig. 2. Note that the steam and cooling-water production costs are incremental costs from the new facilities necessitated by the vinyl chloride operation. The listed plant overhead is an estimated incremental overhead. Also note that the listed input ethylene production cost is the estimated incremental ethylene production cost from the new ethylene plant.

Depreciation Schedule—The investments involved in the vinyl chloride operation, and the depreciation schedule, are shown in Table II. Depreciation is based on sum-of-the-digits depreciation over an 11-yr. life, as now allowed by Internal Revenue rules for chemical plants and most affiliated facilities. Because choice of the sum-of-the-digit technique accelerates depreciation, it reduces taxable income in the project's years and tends to shift the income tax burden to later years.

Cash-Flow Projection—Based on the preceding figures and tables, the vinyl chloride project cash-flow projections have been developed and are summarized in Table III, which is based on a 12-yr. project life. The footnotes of Table III explain in detail the basis for entries that might require further explanation.

The net present worth of the cash flows has been determined at interest rates of 15, 20 and 25% as listed in Table III. By interpolation, it is seen that the project discounted-cash-flow rate of return (the interest rate at which the project cash flows' net present worth is zero) is 18%.

To complete the evaluation and decide whether to accept this project, the company would now compare this venture with others available, and also judge

Fixed Capital Investment—The capital requirements associated with the vinyl chloride project, and their timing, are listed in Table II. Note that the utility and ethylene-plant costs are opportunity costs, and are capacity-pro-rata portions of the new facility investments. Inherently, this assumes that the timing of these additions is determined by other requirements and is not affected by acceptance or rejection of the vinyl chloride project. It assumes, however, that the utility and ethylene plant capacity added is increased by an amount equal to that preempted by

Fixed investment requirements and depreciation schedule for vinyl chloride project—Table II

Investment	Amount, $	Comments on Timing of Investment
Vinyl chloride battery-limits plant, plus storage handling	5,500,000	40% uniformly over 1969; remaining 60% uniformly over first 8 mo. in 1970.
Steam generation and distribution expansion	300,000	Uniformly over 1972. Capacity of 37,500 lb./hr. (300 million lb./yr.); pro rata investment share of a 70,000 lb./hr. steam capacity expansion.
Cooling water tower and distribution expansion	320,000	Uniformly over 1973. Capacity of 12,900 gpm. (6.2 billion gal./yr.); pro rata investment share of a 30,000 gpm. cooling tower system expansion.
Ethylene capacity expansion	3,600,000	Uniformly over 1974. Capacity of 97 million lb./yr.; pro rata investment share of a new 400 million lb./yr. ethylene plant. Includes offsites and supporting utility requirements.

Item	Investment, $1,000	Yearly Depreciation											
		1971	1972	1973	1974	1975	1976	1977	1978	1979	1980	1981	1982
Vinyl chloride plant	5,500	917	834	750	667	581	501	416	333	250	167	84	0
Steam generation and distribution expansion	300			50	45	41	36	32	27	23	18	14	9
Cooling water tower and distribution expansion	320				53	48	44	39	34	29	24	19	15
Ethylene capacity expansion	3,600					600	545	491	436	382	327	273	218
Total		917	834	800	765	1,272	1,126	978	830	684	536	390	242

#	Year:	1969	1970	1971	1972	1973	1974	1975	1976	1977	1978	1979	1980	1981	1982	Net present worth, Jan. 1971
1	Operating capacity, 1,000 lb./yr.			80,000	110,000	140,000	170,000	200,000	200,000	200,000	200,000	200,000	200,000	200,000	200,000	
2	Gross sales, $1,000/yr.			4,800	6,600	7,700	9,350	11,000	10,600	10,600	10,600	10,000	10,000	10,000	10,000	
	Production costs, $1,000/yr. (Note b):															
3	Raw materials: Ethylene (Notes b,c)			570	780	1,000	1,210	1,690	1,690	1,690	1,690	1,690	1,690	1,690	1,690	
4	Chlorine			1,350	1,860	2,360	2,870	3,380	3,380	3,380	3,380	3,380	3,380	3,380	3,380	
5	Catalysts and chemicals			50	70	90	110	130	130	130	130	130	130	130	130	
6	Total raw materials			1,970	2,710	3,450	4,190	5,200	5,200	5,200	5,200	5,200	5,200	5,200	5,200	
7	Utilities: Steam (Notes b,d)			40	50	80	100	110	110	110	110	110	110	110	110	
8	Electricity (Note e)			50	70	90	110	130	130	130	130	130	130	130	130	
9	Fuel			30	40	50	60	70	70	70	70	70	70	70	70	
10	Cooling water (Notes b,f)			20	30	40	90	100	100	100	100	100	100	100	100	
11	Total utilities			140	190	260	360	410	410	410	410	410	410	410	410	
12	Operating costs: Oper. labor, supervn.			180	180	180	180	180	180	180	180	180	180	180	180	
13	Supplies			20	20	20	20	20	20	20	20	20	20	20	20	
14	Maintenance			250	250	250	250	250	250	250	250	250	250	250	250	
15	Total operating costs			450	450	450	450	450	450	450	450	450	450	450	450	
16	Overheads: Plant overhead			100	100	100	100	100	100	100	100	100	100	100	100	
17	Insurance, property taxes			50	50	50	50	50	50	50	50	50	50	50	50	
18	Total overheads			150	150	150	150	150	150	150	150	150	150	150	150	
19	Total production cost (excl. depr.) (6 + 11 + 15 + 18)			2,710	3,500	4,310	5,140	6,210	6,210	6,210	6,210	6,210	6,210	6,210	6,210	
20	Depreciation, $1,000/yr. (Note b)			920	830	800	770	1,270	1,130	980	830	680	540	390	240	
21	Gen'l. and admin. expenses, $1,000/yr.			390	500	560	660	760	730	730	730	700	700	700	700	
22	Startup expenses, $1,000/yr.		400													
23	Pre-tax profit, $1,000/yr. (2 − 19 − 20 − 21 − 22)		(400)	780	1,770	2,030	2,780	2,760	2,530	2,680	2,830	2,410	2,550	2,700	2,850	
24	Income taxes, $1,000/yr. (48% of 23)		(190)	370	850	970	1,330	1,320	1,210	1,290	1,360	1,160	1,220	1,300	1,370	
25	After-tax profit, $1,000/yr. (23 − 24)		(210)	410	920	1,060	1,450	1,440	1,320	1,390	1,470	1,250	1,330	1,400	1,480	
	Miscellaneous cash flows, $1,000/yr.:															
26	Fixed capital investment	2,200	3,300													
27	Opportunity costs				300	320	3,600									
28	Working-capital increases		30	450	170	130	140	180	0	0	0	0	0	0	(1,070)	
29	Net cash flow, $1,000/yr. (20 + 25 − 26 − 27 − 28)	(2,200)	(3,540)	880	1,280	1,410	(1,520)	2,530	2,450	2,370	2,300	1,930	1,870	1,790	2,790	
30	Discounted cash flow, $1,000/yr., @ 25% int.	(3,210)	(4,020)	780	880	760	(630)	830	620	470	350	230	170	130	160	(2,480)
	@ 20% int.	(2,960)	(3,920)	800	950	860	(750)	1,030	810	650	520	350	280	220	280	(880)
	@ 15% int.	(2,760)	(3,820)	820	1,020	970	(900)	1,290	1,070	900	750	540	450	370	500	1,200
31	Net present worth, interpolated @ 18% int.															0

(a) Quantities are rounded to the nearest $10,000. Negative entries are in parentheses. For simplicity, escalation effects are ignored.

(b) Depreciation is excluded from production cost, but is included in the listed depreciation on line 20 (based on the depreciation schedule in Table II).

(c) Ethylene is charged in at incremental production cost of 1.47¢/lb. from the existing ethylene plant for the years 1971 through 1974. This incremental cost is the incremental raw-materials plus utilities cost, minus byproduct fuel credit. The other production costs of Fig. 1 are sunk costs. From 1975 onward, ethylene is charged in at 1.74¢/lb., the estimated incremental production cost (excluding depreciation of 0.3¢/lb. of ethylene) from the new ethylene plant projected for initial production in 1975. The individual unit production-cost components will be essentially the same as those of Fig. 1, except that incremental electricity cost and overhead will be less.

(d) Steam is charged in at incremental production cost of 32¢/1,000 lb. from the existing steam facilities for the year 1971 and 1972. This represents incremental fuel and utilities cost; the other production cost components of the existing facilities (such as operating labor, maintenance, overheads, taxes, insurance and depreciation) are sunk costs. From 1973 onward, steam is charged in at 38¢/1,000 lb. — the estimated incremental production cost minus depreciation from the new steam facilities projected for initial production in 1973.

(e) Electricity is charged in at the incremental cost of 0.6¢/kwh.

(f) Cooling water is charged in at the incremental production cost of 1¢/1,000 gal. from the existing facilities for the years 1971, 1972 and 1973. This is incremental makeup water and electricity cost; the other production cost components such as operating labor, maintenance, overheads, taxes, insurance and depreciation are sunk costs. From 1974 onward, cooling water is charged in at 1.66¢/1,000 gal. — the estimated incremental production cost minus depreciation from the new cooling water facilities projected for initial production in 1974.

(g) This is incremental overhead for the vinyl chloride plant addition.

whether the 18% return is commensurate with the risks involved. Additional cases would also be developed to test the sensitivity of the rate of return to changes in the market forecast, project life, and other parameters involved. Different plant sizes would also be studied to assure that the optimum-size plant is built. ■

References*

1. Dean, J., "Measuring the Productivity of Capital," *Harvard Bus. Rev.*, Jan.-Feb. 1954, p. 127.
2. Grant, E. L., Ireson, W. G., "Principles of Engineering Economy," 4th Ed., Ronald, p. 288, 295, 296, 365.
3. Peters, M. S., "Plant Design and Economics for Chemical Engineers," McGraw-Hill, p. 84-87.
4. Perry, R. H., et al. (Ed.), "Perry's Chemical Engineers' Handbook," 4th Ed., McGraw-Hill, p. 26-29.
5. Robnett, R. H., et al., "Accounting, a Management Approach," Irwin, p. 240-245.
6. Peterson, S., "Economics," Holt, p. 509, 545-546.
7. Bierman, H. Jr., Smidt, S., "The Capital Budgeting Decision," 2nd Ed., Macmillan, p. 44-49.
8. Burke, D. P., Miller, R., *Chem. Week*, Oct. 23, 1965, p. 63.
9. Edwards, E. F., Weaver, T., *Chem. Eng. Prog.*, 61, No. 1, p. 21 (Jan. 1965).

*The first seven references deal with incremental cost principles, and pertain to the previous Cost File as well as to this one. The last two references are specific to the illustrative example used for the present Cost File.

Meet the Author

William P. Hegarty is engaged in process design, evaluation and development work for Techni-Chem Co., an independent organic-chemical research and development firm located in Wallingford, Conn. His earlier experience was obtained with Olin, Hercules, Standard Oil Co. (Ind.), and Esso Research. A chemical engineering graduate of the University of Michigan, he belongs to Alpha Chi Sigma, American Chemical Society, and AIChE.

Finding the Best Plant Location

*Carefully defining the requirements for a
new plant site enables you to set up a series
of screens through which you can sift all of
the possible locations.*

ROBERT A. WILL, *The Austin Co.*

You have just been handed the assignment of finding a 20-acre site for your company's new facility
somewhere on the U. S. mainland. The 48 contiguous
states have a total of nearly two billion acres, which
means your company needs 0.000001% of this area.

You may not think of your assignment in quite this
light, and stating the problem in this way may be a
bit far-fetched. But it does show that a lot of real
estate must be eliminated from consideration before
you reach that final 20 acres.

In looking for a site, you want the maximum
economic benefits that a location can contribute to the
facility. It is probably not oversimplifying the plant
location study to say that it is completed in just two
basic steps: (1) establishing as accurately as possible
those requirements of the facility that will be influenced by location, including a relative weighting of
these requirements; (2) applying these criteria to
the largest geographical area that can be considered
logically, then continuing to eliminate unqualified
locations until only the best site remains.

Both steps are essential. Without correctly estab-

lishing the "ground rules" (Step 1), the results of the screening process (Step 2) are, at best meaningless, at worst incorrect.

The importance of a systematic approach to eliminating locations cannot be overemphasized. This has been demonstrated to us time and again by clients who set out to make their own study on a hit-or-miss basis, but who end up turning the project over to us, together with reams of information collected—most of it irrelevant.

Step One: Setting Up the Screen

Just as the number of possible locations for a facility are virtually unlimited, so are the criteria that can be used to evaluate these locations. The trick is to keep the important criteria from being eclipsed by minor considerations. This becomes progressively more difficult as the study moves along.

The criteria that govern the approach to the screening process in the initial stages are primarily tangible economics; the intangibles are applied later. The tangibles, as we consider them, are the measurable costs.

Measureable costs are basically of two types: (1) the continuing costs affecting operation and (2) the one-time costs of setting up shop. The continuing costs, in most cases, consist mainly of inbound and outbound freight, labor, utilities and taxes. One-time costs are largely those of site acquisition and preparation, construction, and business organization taxes.

It is usually possible early in the study to pinpoint the lowest-cost area for shipments to customers (outbound freight), since in most situations there is but one lowest-cost area for this item. At the other extreme, the costs attributable to real estate taxes and the site are so localized that it is usually impossible to consider them until the final stages of the study. The accompanying table categorizes the more-frequent cost criteria by geoghaphical pattern of occurrence. The descending order of classification also illustrates a logical order of screening steps appropriate to most site searches.

The Intangibles

Not all important location criteria can be assigned a value as readily measurable as most of the cost considerations listed in the table. Yet, the intangibles can be of great and even overriding importance.

In our experience, we have seen intangibles range from the valid to the ridiculous. For example, a valid reason for rejecting a location, even though the cost picture looks good, is that key personnel necessary to the operation's success may refuse to move to the area because of unattractive living conditions. An example of a questionable intangible, which we see with increasing frequency, is the restriction of possible locations to those within a few miles of a commercial airport, to lessen executive travel discomfort. We would have to place in the ridiculous category a requirement, in one site search we know of, that all communities below a certain elevation be rejected because of management's belief that those people were not as likely to be as industrious as their highland cousins.

We cannot enumerate the many intangibles that should be applied to a plant location search, since they vary considerably with each situation and even with each company's philosophy. It suffices to mark the importance of intangibles, and to caution against going astray in their application.

Step Two: Beginning the Screening Process

With the selection criteria tailored and weighted for the specific operation proposed, the screening of potential locations can begin. It is axiomatic that the more stringent the requirements, the easier the second step and the more positive the results.

By its very nature, applied screening requires considering the largest geographic area within reason, whether international, national, regional, state or local. It becomes essentially a matching process: match and reject; match and consider further.

A classic example from our files illustrates how two billion acres was boiled down to 100,000. (Admittedly, we are not often called upon to find a 100,000-acre site, nor is anyone else.) In this particular case, we were looking for a remote test site, somewhere in the U. S., for the future use of a major space-age company.

Costs associated with a site can be classified by area.

AREA OF OCCURRENCE	TYPE OF FACTOR	
	Continuing	One-Time
National	Outbound Freight	—
National or regional	Inbound Freight	—
National, with many regional and local variations	Labor Power Fuel Climate (heating and air conditioning)	Construction
State variations	Business taxes	Business organization taxes; sales tax on equipment and materials
	{ Air and water Pollution control Financing programs }	
Local variations	Water Real estate taxes	Building site

NOTE: Some consideration frequently must be given to other influences that may be neither one-time nor long-continuing. Such factors are usually related to governmental procedures. Examples are tax forgiveness to new industry for a specified period, right-to-work laws and transportation regulations.

The size of the needed property established two immediate screening requirements for the survey, even before our client furnished us the specific criteria: (1) the site had to be purchasable at a very low cost per acre; and (2) since our client was a private organization without the government power of condemnation, the property had to have relatively few ownerships, to make it feasible to assemble one parcel of 150 square miles.

Our client also had a number of other requirements. For one thing, the proposed installation could not afford to be shut down or harassed by long periods of inclement weather. Also, it had to be within a reasonable distance of a fair-sized city, to provide supporting services as well as the amenities necessary to attract and retain scientists and engineers. And, finally, the client wanted a site on a navigable waterway to permit barge transportation of large space hardware.

With these rules set up, the screening process began. Starting with the 48 contiguous states, areas were eliminated by a series of map overlays. Application of climatic restrictions to the base map resulted in the disqualification of large areas (Fig. 1b). Here, snowfall, snow cover and temperature were the criteria used.

Since the Federal government owns large areas of land—parks, monuments, forests and military installations—which would not be available, this also was a restrictive factor. The elimination of such properties by the screening process is shown in Fig. 1c. This does not show the location of government-owned grazing lands, whose availability for the intended use at the time of the study could not be firmly established. Fig. 1c also shows those areas eliminated because of distance from a large support city.

By using topographic maps (1:250,000 scale), the logical remaining areas of the country were checked for terrain and cultural features. Farmland values also entered into consideration. Ultimately, 14 logical areas were pinpointed. It was possible to complete this first screening process without even going into the field to see the prospective site areas.

Once the optimum areas are defined, the screening of communities and sites can begin. Here again, the established requirements continue to govern the procedure. Such requirements as acceptable community size, large water demands, the necessity of water transportation, and the absence of a competitor are typical of the restrictions frequently imposed. These restrictions reduce the possible locations to a manageable number prior to the start of detailed field investigation.

The Field Screening

The plant location task has reached the point where it is now feasible to begin visiting locations. The screening becomes finer, since "X" number of locations have now been narrowed down to not more than a couple of dozen potential sites most of which should come reasonably close to satisfying the requirements.

Even with pre-screening, some communities can be disqualified when they are visited in the field, and need not be investigated further. Some of the more frequent reasons why we have eliminated locations include:

• Prevalence of unusually high wage rates that our client could not meet and still remain competitive in his industry.

• Announcement of a new industry that would soak up a good portion of the available work force.

• Local resistance to new industry (most often found in college communities).

• Inadequate or marginal water supply.

• Labor shortages reported by local manufacturers.

• Inadequate or marginal municipal power source, with alternative suppliers excluded from the area.

• Absence of a workable site (this is more likely to be true in the case of process or heavy operations than for light manufacturing).

Applying Economic Factors

Screening for the remote test site mentioned previously did not permit much application of operating-cost requirements in the first phase of the study. The result was that even after 14 widely scattered areas had been selected, we were not able to establish the most favorable region of the country. This could not be ascertained until specific sites within the 14 areas were checked out.

More typical, perhaps, is the location study for an operation with a high dependence on freight, labor and utility factors. A consideration of these basic requirements often quickly defines the optimum-cost region.

Although an early determination of the most favorable area is always comforting, it is not always possible. Witness the case of a flat-glass manufacturer, where opposing cost factors kept in suspense not only the exact site but also the region of the U.S. until near the end of the selection process. Gas and power costs, of major importance, were low in the middle South. The market center was in the Midwest. The result was a standoff between the two areas on combined utility and outbound freight costs. For a high-tonnage product like glass, we would expect that proximity to raw materials would then make the difference, since labor rates for the industry apply nationally and would not influence the cost picture.

The principal raw materials of glass (mainly sand, dolomite, limestone, soda ash and salt cake), plus packaging materials, were then checked as to availability and suitability within the survey area. For sand, the largest-tonnage material, locations of sources of supply were quickly established. Most of the sand suppliers that could be considered were in the central Appalachian Mountain region. Further investigation soon revealed a number of underdeveloped sources, also within the favorable survey area. These included

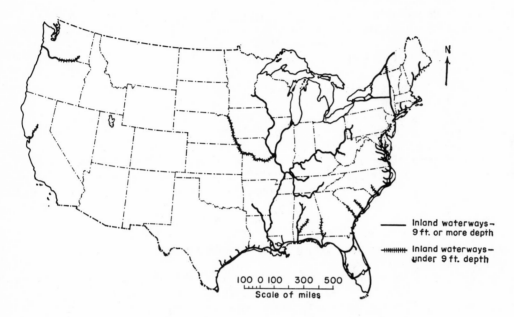

First step: major waterways are pinpointed—Fig. 1a

— Inland waterways— 9 ft. or more depth

‖‖‖‖‖ Inland waterways— under 9 ft. depth

100 0 100 300 500
Scale of miles

Then, cold climate eliminates some areas from consideration—Fig. 1b

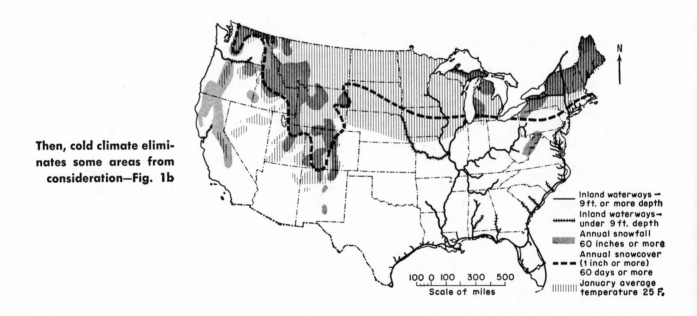

— Inland waterways— 9 ft. or more depth

‖‖‖‖‖ Inland waterways— under 9 ft. depth

Annual snowfall 60 inches or more

- - - Annual snowcover (1 inch or more) 60 days or more

‖‖‖ January average temperature 25 F.

100 0 100 300 500
Scale of miles

not only the Appalachian sandstones but midwestern sandstones, the unconsolidated sand deposits of central Tennessee and beach sands of the Gulf of Mexico. (We climbed several mountains and even rented a boat in the process of securing sand samples for testing.) Once a source was judged to be satisfactory, a cost for developing the supply and delivering it to the site also had to be established to make the picture complete.

Pinning down some of the other raw materials—particularly dolomite and limestone—proved to be nearly as difficult as sand. Here, we even considered the substitution of oyster shells for limestone in the Gulf area.

Finally, elimination of parks and remote areas further narrows the search area—Fig. 1c

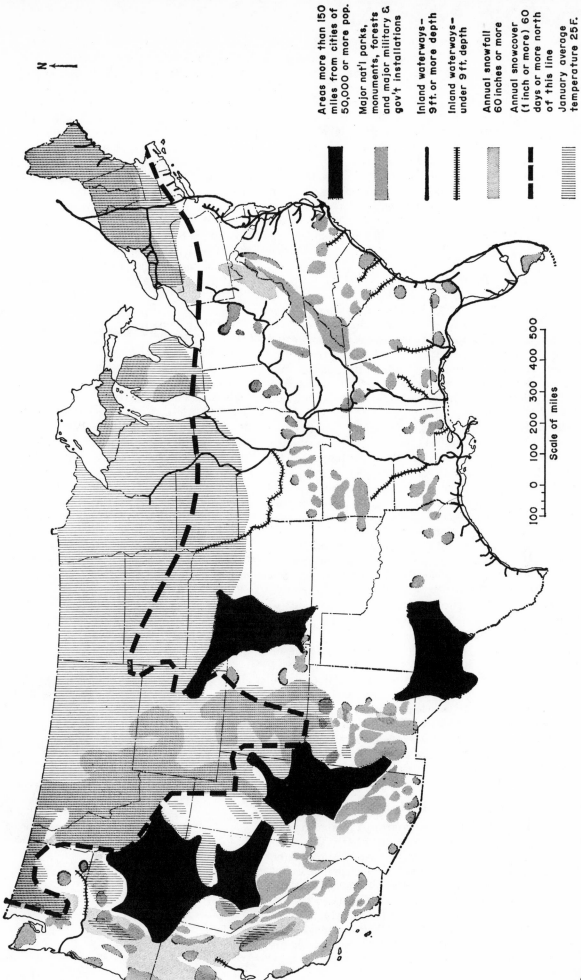

Areas more than 150 miles from cities of 50,000 or more pop.

Major nat'l parks, monuments, forests and major military & gov't installations

Inland waterways— 9 ft. or more depth

Inland waterways— under 9 ft. depth

Annual snowfall 60 inches or more

Annual snowcover (1 inch or more) 60 days or more north of this line

January average temperature 25 F.

Scale of miles

100   0   100 200 300 400 500

As with many process industries, availability of a large site was of major importance; the absence of such a site near a sand source was sufficient to disqualify some areas. If the location were to be in the Gulf Coast region and use beach sand and oyster shells, the site had to be accessible via barge transportation. For inland locations, situation on a navigable waterway also was desirable, since it offered some potential saving on soda ash transportation costs.

Adding all of the significant cost figures together showed that three of the potential locations, each in a different region, were nearly equal. Any one would have been a good location on the basis of meeting the tangible cost requirements established to guide the survey.

Applying the Intangibles

The glass-plant survey illustrates how the screening process continues to reduce the number of logical locations by application of cost criteria. At some point near the end of the screening, an impasse frequently occurs as the measurable cost differentials of the locations still in contention becomes less. This means that the intangibles will make the difference. It is time, then, for the final fine screening.

Some of the intangible criteria may reflect the corporate philosophy. Our experience is that the lighter types of industry are more likely to be influenced by the intangibles than the heavier ones because the easily measurable items of freight and utilities are relatively less important. One of the most extreme cases in our files is that of the space-age science company that requested that we compare an East Coast, a West Coast and a Gulf Coast location, to determine the cost of doing business in each. Our findings indicated a substantial advantage for the East Coast—with the West Coast running third. The company chose the West Coast location, despite the cost penalty, so it could be near the university that would contribute most to its technology.

The laboratory location problem is the epitome of the application of intangibles. Hardly anything about such a project can be assigned a dollar value. Yet the location of a laboratory should contribute to recruiting and retaining of the scientific personnel needed to ensure the operation's success. The presence of a major university (including access to the library) and "good living conditions" are widely proclaimed as essentials, but actually there are many cases where the absence of a university does not seem to be an insuperable handicap.

This is illustrated by a study we recently made for a chemical manufacturer headquartered in a moderate-sized city with no nearby major university. The problem was to determine whether another location might be more suitable for the laboratory—a move that would separate the laboratory from other company functions in the area. Investigation showed that qualified laboratory personnel were not recruited to the existing location as easily as they might have been in some of the major cities, but nevertheless could be attracted in adequate numbers. Once there, however, scientists were less likely to leave than would be probable in a larger metropolitan area. The company was overcoming the handicap of not having a major university by sponsoring special courses and cultural events, and by maintaining its own complete technical library—compensations that could be sustained indefinitely.

When all of the facts were evaluated, the company decided to retain its laboratories at the existing location. Had the laboratory been a newly conceived entity, it is likely that the important intangibles would have ruled out consideration of the community in which it was actually located. Superimposition of the advantages of remaining (also mostly intangibles, including easy intra-company liaison, general satisfaction of employees with community, reluctance of some to move, and a history of excellent labor relations) more than offset the probable advantages of another location in this case.

The Final Touches

We have seen how the location study begins with a large area, and by progressive screening, is narrowed down to several possibilities that are in close contention on a cost basis.

We are close to awarding the blue ribbon. Now is the time to be sure the school system has a high standard. Now is the time to meet local business leaders, to check further on labor conditions, to visit the country club, to talk with city officials, and to form a general impression of the community.

Somewhere along the way, one of the locations will check out a little better than the others. Then is the time to option the best available site, negotiate a tax assessment, seek utility extensions, analyze soil conditions and do anything else necessary to bring the project to a successful conclusion. If all goes well, and everything checks out as hoped, the number one choice will get the new plant. If not, the final investigation can be transferred easily to an alternate location, secured by the knowledge that screening has provided a logical backup site.

Meet the Author

Robert A. Will is manager of plant location surveys for The Austin Co., Cleveland. He joined Austin's plant location staff in 1955, having received a B.S. degree in geography from the University of Maryland in 1950, an M.S. from the University of Wisconsin in 1951, and served in the Pentagon as an intelligence analyst on geographical projects. Will is also a registered P.E. in Ohio.

Key Concepts for This Article

Active (8)	Passive (9)	Means/Methods (10)
Locating	Plants	Surveys
Selecting	Sites	
	Facilities	
	Installations	

(Words in bold are role indicators; numbers correspond to AIChE information retrieval system. Indexing details are described in Chem. Eng., Jan. 7, 1963.)

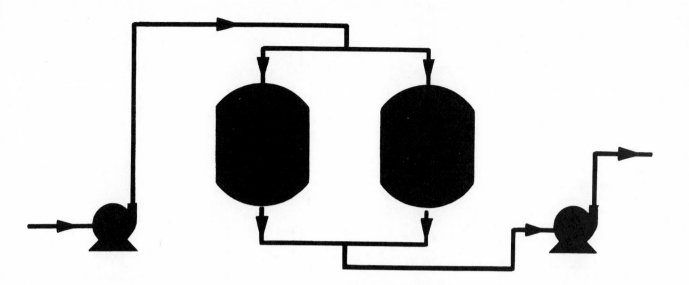

Minimize Batch Equipment Cost

Mathematical method calculates optimum equipment
size for batch processes by developing total cost
equation for each process.

S. E. KETNER, American Cyanamid Co., New York *

When designing a batch process, a designer must choose his equipment sizes to give minimum capital investment. This is an easy matter for a continuous process where equipment size depends upon the desired production rate. In a batch process, however, many combinations of equipment size will give the same amount of product. A typical example is a reactor with feed and discharge pumps. The production capacity depends not only on the reactor size but also on the pumping time required for feed and discharge. With very small pumps, a large reactor is required, and vice versa, for a given productive capacity.

With the cost equations derived in this article, the designer can calculate an estimate of the sizes for minimum investment. He should remember, however, that additional factors such as standard equipment sizes available, as well as safety and maintenance costs must still be considered in the final design.

We will assume the production capacity (W, gal./hr.) for a batch process to be fixed by market estimates. The problem, then, is to size the reaction and auxiliary equipment to minimize initial capital investment. If some pieces of equipment wear out faster than others, depreciation could be minimized instead, but we will consider only the first case. The time reactants are held at reaction temperatures, and other properties of the system

must be known as well as the effect of equipment capacity on installed cost. The first step is to determine expressions for the installed cost of each piece of equipment in the process. We have developed several typical examples of this type of expression, as shown below and on the following page.

• Reactor or tank cost $= C_1 V$ where

C_1 = Cost of reactor, \$/gal.
V = Vol. of reactants, gal./batch

• Pump cost $= \dfrac{C_2 V}{t_2}$ where

C_2 = Cost of pump, \$/(gal./hr.)
t_2 = Pumping time, hr.

• Heat exchanger cost $= C_3 A$ where

C_3 = Cost, \$/sq.ft. of area.

* At present with W. R. Grace & Co., Clarksville, Md.

445

A = Transfer area, sq.ft.

and, $A = \dfrac{QV}{U\,\Delta T_m t_3}$ where

Q = Total heat load, Btu./gal.
U = Heat transfer coefficient, Btu./(hr.)(sq.ft.)(°F.)
ΔT_m = Mean temperature diff., °F.
t_3 = Heat-up time, hr.

• Filter cost $= \dfrac{C_4 F V}{t_4}$ where

C_4 = Cost of filter, \$/sq.ft.
F = Filter area, sq.ft./(gal./hr.)
t_4 = Filtration time, hr.

Regardless of the equipment involved, the total cost equation will involve items such as those above. If the process has a piece of equipment that is used more than once in the cycle, then appropriate adjustments in the reactant volume processed and definition of the time involved must be made.

The total cost equation will always contain volume of reactants per batch (V) as a common factor. There are two types of batch equipment—true batch items such as tanks; and intermittent-continuous equipment such as pumps. The intermittent-continuous type will always contain a time-of-use factor while the true batch type will not. For the general case:

Cost of true batch equipment = $C_j V$

Nomenclature

A Heat transfer area, sq. ft.
C_i Cost of intermittent-continuous type equipment, \$/capacity rate.
C_j Cost of jth piece of true batch equipment, \$/gal. of reactants.
F Filter area required, sq. ft./(gal.) (hr. filtered).
I Total equipment investment, \$.
m Number of true batch pieces.
n Number of intermittent-continuous pieces.
Q Total heat load, Btu./gal.
t_i Time intermittent-continuous equipment is in use, hr.
t_j Time batch equipment is in use, hr.
Δt_m Mean temperature difference, °F.
U Heat transfer coefficient, Btu./ (hr.) (sq. ft.) (°F.).
V Volume of reactants, gal./batch.
W Production capacity, gal./hr.
y Investment per unit capacity, \$/(gal./hr.).
a_i Conversion factor characteristic of processed material.
     Subscripts 1, 2, . . . 6 refer to individual pieces of equipment.

Cost of intermittent-continuous batch equipment

$= C_i \alpha_i V / t_i$

The cycle time per batch is the sum of the appropriate processing times, taken here as:

$$\sum_1^m t_j + \sum_1^n t_i \tag{1}$$

The production capacity W, gal./hr., is:

$$W = \frac{V}{\Sigma t_j - \Sigma t_i} \tag{2}$$

The total equipment investment is the sum of the individual pieces, as follows:

$$I = V\Sigma C_j \Big]_1^m + V\Sigma (C_i \alpha_i / t_i) \Big]_1^n \tag{3}$$

The investment per unit capacity y is:

$$y = I/W = [\Sigma C_j + \Sigma(C_i \alpha_i / t_i)] \times [\Sigma t_j + \Sigma t_i] \tag{4}$$

To find the minimum investment per unit capacity with the corresponding cycle times, we take the partial derivatives of y with respect to the intermittent times and equate them to zero. The true batch equipment has a fixed time determined by reaction kinetics or other outside factors and the capacity is merely the reactant volume. The equations to be solved simultaneously, then, are:

$$\frac{\partial y}{\partial t_1} = \frac{\partial y}{\partial t_2} = \cdots \cdot \frac{\partial y}{dt_n} = 0$$

For the kth stage, processing time:

$$\frac{\partial y}{\partial t_k} = \left(\Sigma C_j + \Sigma \frac{C_i \alpha_i}{t_i} \right) - \frac{C_k \alpha_k}{t_k^2} (\Sigma t_j + \Sigma t_i) = 0 \tag{5}$$

and:

$$\frac{C_k \alpha_k}{t_k^2} = \left(\Sigma C_j + \Sigma \frac{C_i \alpha_i}{t_i} \right) \Big/ (\Sigma t_j + \Sigma t_i) \tag{6}$$

Since $C_k a_k / t_k^2$ is equal to the above summation for all i items, then:

$$C_1 \alpha_1 / t_1^2 = C_2 \alpha_2 / t_2^2 = \cdots C_n \alpha_n / t_n^2 \tag{7}$$

therefore:

$$t_i = t_1 \sqrt{C_i \alpha_i / C_1 \alpha_1} \tag{8}$$

and:

$$\Sigma t_i \Big]_1^n = \frac{t_1}{\sqrt{C_1 \alpha_1}} \Sigma \sqrt{C_i \alpha_i} \Big]_1^n \tag{9}$$

also, from Eq. (8):

$$\Sigma \frac{C_i \alpha_i}{t_i} \Big]_1^n = \frac{\sqrt{C_1 \alpha_1}}{t_1} \Sigma \sqrt{C_i \alpha_i} \Big]_1^n \tag{10}$$

Substituting Eq. (9) and (10)

in (6) and solving for t_1 gives:

$$t_1^2 = C_1 \alpha_1 \Sigma t_j \Big]_1^m \Big/ \Sigma C_j \Big]_1^m \tag{11}$$

Substituting Eq. (11) in (7) allows calculation of values for all t_i from:

$$t_i^2 = C_i \alpha_i \Sigma t_j \Big]_1^m \Big/ \Sigma C_j \Big]_1^m \tag{12}$$

In the above derivation, C_i was assumed independent of t_i. This assumption is not true, of course, but the derivatives of C_i with respect to t_i are sufficiently small in the area of the optimum to allow close convergence in a trial and error solution. That is, an initial C_i is determined, based on an assumed t_i. A new estimate of t_i is calculated from Eq. (12) and the procedure repeated until t_i remains constant. This will be the operating time for the ith piece of equipment corresponding to a minimum investment per unit production rate. The number of iterations necessary to determine t_i depends, naturally, on how close the first guess is to the best time. About six iterations at the most are required, with three to four as an average.

Although equipment such as pumps is available only in certain sizes, this should be ignored until an optimum time cycle is determined. Then further calculations can be made considering available standard sizes in the range indicated by the optimum.

Plot Cost Data

It is important to have continuous curves of the cost data plotted in the proper units of installed cost per unit working capacity versus total capacity. For instance, where tanks and reactors are concerned, the costs should be in terms of volume of reactants so that average is a part of the cost figure.

In the general solution that we have worked out, the problems of multi-units and overlapping in the time cycle were not considered, for the sake of simplicity. It turns out, however, that in these cases the solution is of the same form as Eq. (12) and convergence is just as fast as in the general case. In a very complicated batch processing scheme, the derivations may have to be repeated to obtain the proper equations.

Examples Illustrate Method

The following three batch processes are outlined to illustrate the method. In all cases, the cost data are from Aires and Newton, "Chemical Engineering Cost Estimation," McGraw-Hill (1955). The data have been adjusted to be in terms of installed cost per unit operating capacity.

Example I—Consider a simple batch process with a pump feeding a pressure reactor from a large storage tank. After reaction is complete, pressure developed during the cycle blows the batch out of the reactor. Reaction and discharge require 1 hr. operating time. Sales estimates indicate a production capacity W of 1,750 gal./hr.

$$W = \frac{V}{t_1 - t_2} = 1,750$$

where:

t_1 = Pumping time, hr.
t_2 = Reaction and discharge time, hr. or, 1 hr.

The total installed equipment cost is:

$$I = \frac{C_1 V}{t_1} + C_2 V$$

where:

C_1 = Hourly pumping cost, \$/gal.
C_2 = Reactor cost, \$/gal.

The installed cost per unit capacity is:

$$y = \frac{I}{W} = \left(\frac{C_1}{t_1} + C_2\right)(t_1 + t_2).$$

From the general solution, Eq. 12, the optimum pumping time is:

$t_1 = \sqrt{C_1 t_2 / C_2}$

An initial estimate of the time cycle must be made to estimate C_1 and C_2. Assume a pumping time of 30 min. to make a batch cycle of 1.5 hr.

First Estimate: Values of C_1 and C_2 are taken from plots of cost per unit capacity versus capacity for the following batch volume:

$V = W (t_1 + t_2) = 1,750 \times 1.50$
$= 2,625$ gal.

C_2 is \$16.0/gal. and C_1 is \$0.154/ (gal./hr.). The new estimate of pumping time is:

$t_1 = \sqrt{0.154 \times 1/16.0} = 0.098$ hr.

Second Estimate: With 0.098 hr. pumping time, the batch cycle time is 1.098 hr. The batch volume is now 1,922 gal. and the pumping rate is 17,860 gal./hr. From these data, C_1 and C_2 are 0.076 and 18.4 respectively. The second pumping time estimate is:

$t_1 = \sqrt{0.076 \times 1/18.4} = 0.064$ hr.

Third Estimate: Now the reactor volume is 1,862 gal. and the pumping rate is 29,094 gal./hr. The new C_1 and C_2 are 0.066 and 18.8, respectively. The pumping time now turns out to be 0.0592 hr.

Fourth Estimate: The reactor working volume is 1,853 gal. and the pumping rate is 31,400 gal./hr. The corresponding C_1 and C_2 are 0.0645 and 18.8, respectively. The new pumping time is 0.0588 hr.

Final Estimate: With pumping time of 0.0588 hr. or 3.5 min., the batch volume is again 1,853 gal. C_1 and C_2 are the same as in the fourth estimate. The final process, then, would have a 1,853 gal. reactor and a 31,500 gal./hr. pump. The pumping time is 3.5 min. and the reaction time 1.0 hr. to give 1,750 gal./hr. production capacity. The difference in installed cost between the first and final estimate is about 14%. Table I shows this cost comparison.

The next step would be to find

First Example Estimate Comparison—Table I

	First Estimate	Final Estimate
Pumping time, hr.	0.500	0.0588
Reactor working volume, gal.	2,625	1,853
Pump cost, \$	809	2,032
Reactor cost, \$	42,000	34,836
Total installed cost, \$	42,809	36,868

Flowsheet for Example II

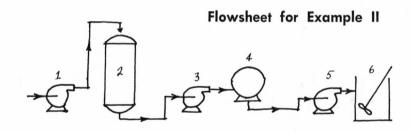

Second Example Estimate Comparison—Table II

Batch Cycle	First Estimate	Final Estimate (Optimum)
$t_1 + t_2 + t_3$, hr.	3.0	2.931
t_1, hr.	0.5	0.072
t_2, hr.	2.0	2.000
t_3	0.5	0.859
Reactor, gal.	1,500	1,466
Charging rate, gal./hr.	3,000	20,360
Filtering rate, gal./hr.	3,000	1,710
Production rate, gal./hr.	500	500
Installed Cost, \$		
Pump (1)	585	1,466
Reactor (2)	31,050	30,786
Pump (3)	585	495
Filter (4)	22,950	16,558
Pump (5)	585	495
Tank (6)	10,650	10,555
Total installed cost, \$	66,405	60,355

what equipment sizes are available in the vicinity of the calculated optimum, and then pick the combination giving the desired 1,750 gal./hr. at minimum investment.

With One Reactor

Example II—Here, six pieces of equipment are involved, as shown above. Material is pumped from a large, existing storage vessel by pump (1) to reactor (2). The reaction requires 2 hr. and the product goes out by pump (3) through a rotary drum filter (4) and is pumped (5) to a hold tank (6). The desired production rate W is 500 gal./hr. The time cycle is the sum of the charging time t_1, reaction time t_2, and filtering (discharging) time t_3. The reactor working volume is:

$$V = W(t_1 + t_2 + t_3)$$

and the installed cost in $/(gal./hr.) capacity is:

$$y = \left(\frac{C_1}{t_1} + C_2 + \frac{C_3}{t_3} + \frac{C_4 F}{t_3} + \frac{C_5}{t_3} + C_6 \right) \times (t_1 + t_2 + t_3)$$

where: C_1, C_3 and C_5 are installed pump costs, $/(gal./hr.) pumping rate; C_2 and C_6 are installed costs for the reactor and the agitated holding tank, $/gal.; C_4 is filter cost, $/sq. ft.; and F is filter area required, sq. ft./(gal./hr.). We will assume this last to be 0.05.

Rearranging the above equation with $C_3 = C_5$, we have:

$$y = \left[\frac{C_1}{t_1} + (C_2 + C_6) + \frac{2C_3 + C_4 F}{t_3} \right] \times [t_1 + t_2 + t_3]$$

From Equation 12, the best pumping and filtering times are:

$$t_1 = \sqrt{\frac{C_1 t_2}{C_2 + C_6}}$$

$$t_3 = \sqrt{\frac{(2C_3 + C_4 F) t_2}{C_2 + C_6}}$$

Table II shows the initial and

final calculated estimates of installed cost.

With Two Reactors

Example III—The rotary filter in Example II is the second most expensive piece of equipment. Two small reactors which would alternately feed the filtration unit would allow purchase of a smaller filter and possibly reduce the total investment. Also, the hold tank size could be cut in half. Consider, then, the system shown at left.

The initial time cycle is estimated so that the filtration unit will be in continuous operation—that is, $t_1 + t_2 = t_3$. In the following equations the subscripts are those of Example II. Remember, though, that the volume V is of one of the two reactors.

The time cycle is:
$$(t_1 + t_2 + t_3)/2$$
Reactor volume is:
$$V = W (t_1 + t_2 + t_3)/2$$
Cost/unit capacity is:

$$y = \left[\frac{C_1}{t_1} + (2C_2 + C_6) + \frac{2C_3 + C_4 F}{t_3} \right] \times [t_1 + t_2 + t_3]$$

The optimum operating times are:

$$t_1 = \sqrt{\frac{C_1 t_2}{2C_2 + C_6}}$$

$$t_3 = \sqrt{\frac{(2C_3 + C_4 F) t_2}{2C_2 + C_6}}$$

Table III shows the initial and final design cycles calculated from these equations. It is evident that the single-reactor system of Example II is cheaper than the two-reactor system. It is also evident that sizing to allow continuous operation of the filtration unit is not economical.

About the Author

S. E. KETNER graduated from Virginia Polytechnic Institute in 1950 with a B.S. in Chemical Engineering and obtain his M.S. in 1951 from the same school. From that time to 1956 he worked with the Virginia Smelting Co. on process improvement and new-product research. In 1956 he joined American Cyanamid Co. to work in production analysis, applying mathematics and statistics to process improvement. This year, Ketner joined W. R. Grace & Co. He is a licensed P.E. and a member of ACS, AIChE and the American Society for Quality Control.

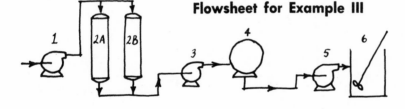

Flowsheet for Example III

Third Example Estimate Comparison—Table III

Batch Cycle	First Estimate	Final Estimate (Optimum)
$t_1 + t_2 + t_3$, hr.	5.0	2.623
t_1, hr.	0.5	0.0474
t_2, hr.	2.0	2.0
t_3, hr.	2.5	0.576
Reactor, gal. (each)	1,250	656
Charging rate, gal./hr.	2,500	13,960
Filtering rate, gal./hr.	500	1,135
Production rate, gal./hr.	500	500
Installed Cost, $		
Pump (1)	558	1,173
Reactor (2a)	28,375	20,992
Reactor (2b)	28,375	20,992
Pump (3)	346	443
Filter (4)	8,350	13,166
Pump (5)	346	443
Tank (6)	9,750	7,019
Total installed cost, $	76,100	64,228

Costs and installation labor for air-cooled heat exchangers—Fig. 1

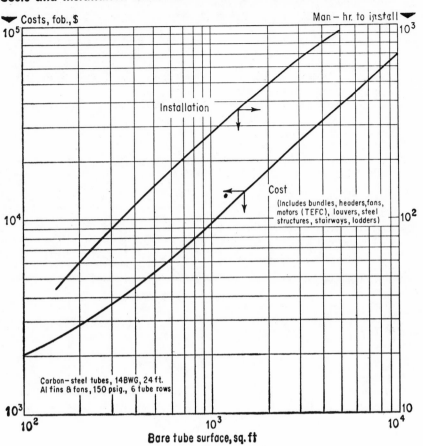

Is air cooling more expensive than water cooling? And what about cooling with seawater or brackish water? This analysis will show you how to make a preliminary economic analysis of cooling alternatives.

How to Compare . . .

Costs of Air vs. Water Cooling

JACKSON CLERK, *Holland, Mich.*

Many conflicting claims have been made in the literature about the question of air cooling vs. water cooling. One of the difficulties in making a cost comparison of alternate cooling methods is that while data can't be developed for the "general" case, many applications are unique and must be studied in detail for a meaningful comparison.

This Cost File attempts to provide the basis for detailed comparisons. Costs of air coolers are shown in Fig. 1. Data are for carbon-steel tubes and aluminum fins.* Costs of water coolers (shell-and-tube) are shown in Fig. 2. (All costs based on 1963-64 data.)

For other construction materials and design factors than are shown in Fig. 1, use the correction factors in Table I. For additional data on shell-and-tube coolers, see *Chem. Eng.*, Aug. 11, 1958, p. 151; Nov. 17, 1958, p. 166; Dec. 1, 1958, p. 123; Aug. 8, 1960, p. 152; Mar. 16, 1964, p. 137.

The uninstalled cost of a typical air cooler can be

separated into the following five major components:

Bundle	45%
Headers with plugs	12
Fans and motors (TEFC)	19
Louvers	12
Steel structures, stairways, ladders	12
	100%

Air coolers are often used in conjunction with trim coolers—these are usually low-fin tubes, double-pipe water coolers. Costs of one section of such a trim cooler having 180 sq. ft. of heat transfer surface is about $450 fob. plus approximately 25% of this cost for manifolds.

The data in Table II comparing air and water cooling are based on information from several sources.

Key Concepts for This Article

Active (8)	Passive (9)	Passive (cont.)	Means/Methods (10)
Comparing	Heat exchangers	Air	Charts
Estimating	Costs	Water	Curves
	Costs, installed	Coolers	Tables

(Words in bold are role indicators; numbers correspond to AIChE information retrieval system. Indexing details are described in *Chem. Eng.*, Jan. 7, 1963.)

---

* Extended-tube surface is about 17 times the bare-tube surface.

449

Note that the costs do not include operating costs. In most situations, operating costs will be lower for air cooling. In one application involving 11 refinery units, operating cost of air coolers was estimated at about 26% of the operating cost (including maintenance) of water coolers necessary to handle the same duty.

Correction factors for conversion of data in Fig. 1— Table I

Effect of tube wall thickness
(O.D. = 1 in.)

Thickness, In.	Correction Factor
0.109 (12 B.W.G.)....	1.1
0.083 (14 B.W.G.)....	1.0
0.065 (16 B.W.G.)....	0.9
0.049 (18 B.W.G.)....	0.8

Effect of pressure

Pressure, Psig.	Correction Factor
150.............	1.00
250.............	1.05
500.............	1.10
1,000.............	1.15
2,000.............	1.20
5,000.............	1.35

Effect of tube length

Length, Ft.	Correction Factor
15..............	1.10
20..............	1.05
24..............	1.00
30..............	0.95

Effect of tube rows

No. of Rows	Correction Factor
4................	1.15
5................	1.05
6................	1.00
8................	0.95
10................	0.90

Effect of tube materials
(aluminum fins)

Material	Correction Factor
Carbon steel........	1.00
Aluminum 3S........	1.30
Aluminum-Brass.......	1.50
Type 304 stainless....	2.20
Type 321 stainless....	2.50
Type 316 stainless....	3.00
Monel..............	3.20

Costs and installation labor for water-cooled heat exchangers—Fig. 2

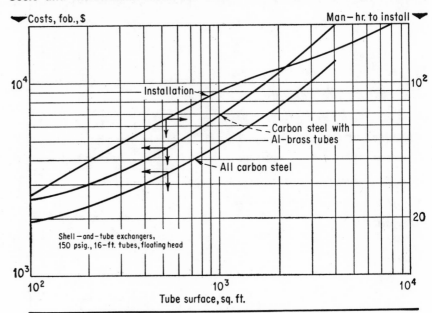

Air vs. water cooling: comparing total relative costs—Table II

	River Water Cooling	Brackish or Seawater Cooling	Air Cooling
Cooling unit, 150 psig., fob. costs			
All carbon steel......................	100		
Carbon-steel with Al-brass tubes....................		150	
Steel tubes with Al fins and fans, including trim coolers and steam coils.............			300
Ancillary equipment as addition to fob. costs of cooling unit			
Instruments and controls...............	10	7	3.5
Paving and foundations................	5	3.5	3.5
Steel structures......................	15	10	(Incl.)
Fireproofing........................	1	1	...
Electrical (switchgear, starters, cables)....	...	...	12
Piping			
Process.............................	15	10	5
Water..............................	30	30	...
Trim coolers and steam coils..........	...	...	2
Misc..............................	5	3.5	2
Total ancillary...................	81	65	28
Total fob. costs..................	181	215	328
Freight costs.....................	1	1	1
Erection costs (32%, 26%, 20% respectively of total fob. costs)............	58	56	65
Total installed costs..................	240	272	394
Utilities and general facilities			
Cooling tower, including pump and all related facilities....................	260		
Brackish or seawater facilities, pumps, reinforced-concrete pipelines to process unit, separator, return line...................		358	
Cooling facilities for trim coolers and additional facilities for steam coils for air coolers.........................			56
Total Relative Costs.................	500	630	450

Costs are based on several cases: cooling 20 to 100 million Btu./hr.; inlet temperature of liquid being cooled = 120-250 F. and outlet = 100-200 F.; ambient air, 80-105 F.

By spending more on your exchangers, you can actually cut down the over-all investment for your cooling-water system. How? By engineering the arrangement of your exchangers. Here is a design technique that shows how to arrive at the optimum cost.

Costlier Cooling Towers Require a New Approach to Water-Systems Design

P. M. PAIGE, *Consultant*

Cooling towers are becoming virtually mandatory for chemical plants. Sources of abundant cooling water are becoming scarce for new plants, and existing plants are facing rising restrictions on how much heat they can reject to public waters.

For a particular heat load—usually fixed by the process—the investment for piping, pumps and cooling tower relates directly to the gpm. of water circulated. However, boosting the circulation rate—for the same heat load—reduces exchanger investment, because lowering the temperature of the returning water improves the log-mean-temperature difference in the exchangers, which lessens cooling surface requirements.

Typically, exchangers represent a far larger investment than pumps and piping. Consequently, where cooling water is abundant, custom decrees that each exchanger be designed for a reasonable approach between the outlet water temperature and the process temperature. Water quantities for all the exchangers are subsequently summed up, and pumps and piping designated for the total. In effect, this procedure presupposes that individual exchangers optimumly designed produce an over-all optimum—a valid enough

assumption considering the preponderant size of the exchanger investment.

How a Cooling Tower Changes the Picture

Although exchangers still represent the largest investment in such a cooling system, the cost of the tower (plus that for piping and pumps) may now run as high as 70% of exchanger costs, and even higher. Obviously, when this is the case, approximating the optimum over-all economics of a cooling system through the summation of optimumly designed individual exchangers would not be good practice.

Four factors establish cooling-tower cost: (1) the total *heat load* rejected to the water by the process, Q in Btu./hr.; (2) the cold-water temperature required —actually this temperature minus the wet-bulb design temperature, which difference is called the *approach*, A in °F.; (3) the *wet-bulb temperature*, T_{WB} in °F.; and (4) the *range* in temperature between the hot water to the tower and the cold water leaving, R in °F.

Data published recently (1967) on tower costs (which do not include that for installation) for 29 combinations of conditions prompted the development of the following empirical equation, which has a max-

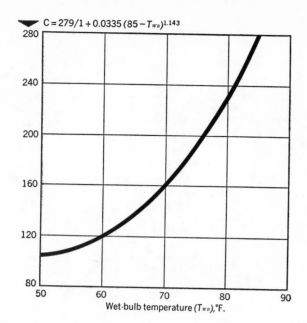

$$C = 279/1 + 0.0335\ (85 - T_{WB})^{1.143}$$

GRAPH solves the value of C in Eq. (1)—Fig. 1

imum error of less than 5% and a mean deviation of 2.5% :*

$$\text{Tower cost, } \$ = Q/(C \times A + 39.2R - 586) \quad (1)$$
With $C = 279/[1 + 0.0335\ (85 - T_{WB})^{1.143}]$

Fig. 1 presents a graphical solution for C.

The design of the process obviously establishes the heat load. Weighing the economic benefits of operating with colder cooling water against that of a costlier tower makes the approach a process design function as well. Plant site determines the design wet bulb. The remaining variable is the range.

For the same total cooling, 20,000 gpm. with a range of 90 F. to 120 F. would have to become 40,000 gpm. if the water were warmed from 90 F. to only 105 F. In other words, for a fixed Btu. load, range is inversely proportional to gpm. Maximizing the range will not only minimize tower cost, but also reduce the gpm. circulated, cutting pump and piping cost as well.

How Cooling With "Hot" Water Can Cut Costs

The wet bulb plus the approach fixes the cold-water temperature. Adding the range to this temperature sets the hot-water temperature. Range, then, is at its maximum when the water is heated as hot as possible.

Two limits hold down the hot-water temperature: an absolute limit based on the allowable scaling of exchanger tubes, and a relative limit set by process requirements. The former is typically 120 F. or 125 F., depending on scaling properties and blowdown rate. The process limit has its basis in the fact that any particular exchanger—depending on its service—can heat the water only so hot—not necessarily up to the scaling limit.

*Maze, R. W., "Practical Tips on Cooling Tower Sizing," *Hydrocarbon Processing*, Feb. 1967.

For example, ammonia at 220 psig. condenses at 106.6 F., requiring the removal of 470.5 Btu./lb. of ammonia condensed. The theoretical limit for the water would thus be 106.6 F. (with, of course, an infinitely large exchanger); the practical process limit might, in this case, be 100 F. If an appreciable part of the total process heat load limits the water temperature in this manner, the temperature of the total flow returning to the tower would be well below the scaling limit. Unless the water is used again—heated to higher temperatures a second or third time in the process unit —before it returns to the tower, the cost of the tower could be very high.

The principle of the reuse of water depends on having it first cool those services that require the coldest water, then on routing it through those services that can accept warmer water, so that when it returns to the tower, its temperature will be close or equal to the scaling limit.

Steps Toward the Optimum Balance

Of course, warmer water in some exchangers will reduce their log-mean-temperature differences, raising surface requirements and, therefore, cost—which must be balanced against the savings on the cooling-tower investment.

A procedure for rapidly arriving at an optimum arrangement would include the following steps:

(A) List all the cooling-water exchangers, their hot-side inlet and outlet temperatures, their Btu. duties, and finally total the duties.

(B) Classify all the exchangers into five groups: (1) those whose loads are negligible (a few percent of the total load); (2) those that require the coldest water and that can heat the water to the scaling limit (for example, process streams from 170 F. to 100 F., where the cold-water temperature is 90 F.); (3) those that need the coldest water but that cannot heat it to the maximum temperature (for instance, vapors condensing at 105 F.); (4) those that do not require the coldest water but that cannot heat it to the maximum (say, a surface condenser condensing steam at 120 F.); and finally, (5) those that do not need the coldest water and that can heat it to the maximum (process streams from 220 F. to 150 F.).

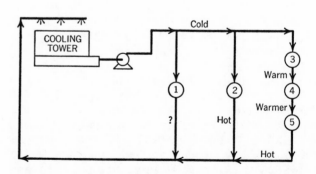

COOLERS arranged tentatively according to step C—Fig. 2

Exchanger data for the once-through cooling system of the hypothetical plant

Exchanger Number	Process Temperatures, °F.		Duty, 10^6 Btu.	Outlet Water Temperatures, °F.	Water Flow, Gpm.	ΔT_M, °F.	Cost, $	Class
	T_1	T_2						
101-C	115	100	3.5	105	467	10	10,500	1
102-C	150	100	40.0	120	2,670	18.2	25,000	2
103-C	105	105	100.0	100	20,000	9.1	110,000	3
104-C	120	120	120.0	110	12,000	18.2	80,000	4
105-C	220	150	130.0	120	8,660	78.0	66,500	5
106-C	150	120	20.0	120	1,330	30.0	37,000	5
			413.5		45,127			

Classes 2, 3, 4 and 5 cover every possible combination of inlet and outlet conditions:

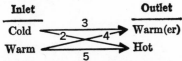

(C) A tentative first optimum balance might be set up as follows: Classes 1 and 2 in parallel, with a stream in series consisting of classes 3, 4 and 5, in that order (Fig. 2).

(D) Adding the duties for classes 3, 4 and 5 and postulating the maximum range for this stream, calculate the gpm. of flow through this class:

$$\text{Gpm.}_{(3,4,5)} = \text{Btu.}_{(3+4+5)}/(500 \times R_{max})$$

and similarly calculate the water quantities for classes 1 and 2. (For class 1, the range would be whatever is required—not R_{max}, of course.)

(E) Calculate how much the temperature of the water would rise across classes 3 and 4—i.e., the temperatures between 3 and 4 and between 4 and 5.

$$\Delta T_3 = \text{Btu.}_3/(500 \times \text{gpm.}_{(3,4,5)})$$

and

$$\Delta T_4 = \text{Btu.}_4/(500 \times \text{gpm.}_{(3,4,5)})$$

(F) With the water temperatures now known, calculate the countercurrent ΔT_M for each class 3, 4 and 5 exchanger:

$$\Delta T_M = (\Delta T_1 - \Delta T_2)/(\ln \Delta T_1/\Delta T_2)$$

The two values of ΔT that are required are: (1) the hotter process temperature minus the hotter water temperature, and (2) the cooler process temperature minus the cooler water temperature. Numerically, ΔT_1 will be the larger, ΔT_2 the smaller.

(G) Assuming that preliminary designs have already established the cost of each exchanger, as well as the ΔT_M (on a once-through water basis), and further, assuming that for a particular service, exchanger cost is directly proportional to exchanger surface (although not precisely true, adequate enough for present purposes), calculate the revised cost for exchangers on the basis of the tentative optimum balance:

Revised Cost, $ = (original cost, $) (revised area/original area) = (original cost, $) (original ΔT_M/revised ΔT_M).

(H) Compare the increase in exchanger costs to the savings on the cost of the cooling tower. By means of Eq. (1), calculate the cost for a once-through system (in which the water never passes through more than one exchanger before returning to the tower) and for a water reusing system (in which some, perhaps even a large quantity, of the water passes through two or more exchangers before returning to the tower).

(I) Investigating chances to improve savings, consider lowering the classification of exchangers 3, 4 and 5, especially that of those exchangers whose performance suffered the most because of the change in ΔT_M.

(J) With classes 3 and 4—if there is more than one exchanger in each class—try splitting the gpm.$_{(3, 4, 5)}$ differently, having more water going to those that are more expensive per Btu. duty and less water to cheaper ones.

(K) Finally, try increasing the total flow of water to classes 3, 4 and 5 to find a less than maximum range that results in over-all optimum costs.

Note that steps I through D are iterative, theoretically. In practice, the effects of changes in steps H through K can usually be evaluated almost intuitively—with a little experience.

Applying the Procedure to a Hypothetical Plant

Water-cooled exchangers, as designed for a once-through system (water passing once through an exchanger) for a hypothetical chemical plant (90 F. water from the tower), are shown in the table.

Exchanger classifications (following step B procedure) are tabulated in the last column. Compared to the total duty, that of 101-C is certainly negligible. At a 100 F. outlet, 102-C needs the coldest water but can easily heat it to 120 F. With a large condensing duty at 105 F., 103-C almost surely requires 90 F. water. Exchanger 104-C is a steam surface condenser. The next, 105-C, can use water at any temperature and

heat it to 120 F. Similar to 105-C is 106-C, although its small duty might place it in class 1. Its cost (a stainless-steel tube bundle) compared to its duty is relatively high. Temporarily, at least, leave it in class 5.

The tentative diagram is shown in Fig. 3, which also gives gpm. flows and intermediate temperatures calculated according to steps D and E.

Revised mean temperatures, following step F, come out as: 103-C, 10.4 F.; 104-C, 16.6 F.; 105-C, 67.0 F.; and 106-C, 19.8 F.

Design A: First Revision

Next, revise exchanger costs, multiplying (original cost) by (original ΔT_M/revised ΔT_M), according to step G:

Exchanger	Multiplication		Revised cost	Change
103-C	($110,000)	(9.1/10.4)	$96,000	−$14,000
104-C	($80,000)	(18.2/16.6)	$87,000	+ $7,800
105-C	($66,500)	(78.0/67.0)	$77,400	+$10,900
106-C	($37,000)	(30.0/19.8)	$56,000	+$19,000

Net change.... +$23,700

For once-through water flow at 45,127 gpm.: Range = $(413.5 \times 10^6)/(500 \times 45,127) = 18.3$ F.

For an 83 F. wet bulb and a 7 F. approach, the cost of the tower for the once-through system becomes: Cost, \$ = $(413.5 \times 10^6)/(257 \times 7 + 39.2 \times 18.3 - 586) = \$212,000$.

For the water reuse arrangement (some of the water through more than one exchanger), the water temperature will rise from 90 F. to 119.8 F.—i.e., R becomes 29.8 F.: Cost, \$ = $(413.5 \times 10^6)/(257 \times 7 + 39.2 \times 29.8 - 586) = \$174,000$.

By spending an additional $23,000 on exchangers, $38,000 ($212,000 − $174,000) will be saved.

Design B: Reclassifying 106-C to Class 4

For further savings, try reclassifying some of the exchangers (according to step I). An obvious target is 106-C. Shifting it to class 4 (parallel to 104-C) changes the ΔT across this classification to 11.3 F., altering the outlet temperature to 109.4 F.

Calculating revised ΔT_M's, costs are changed as follows (again, according to step 7):

Exchanger	Multiplication		Revised cost	Change
103-C	————not changed————			−$14,000
104-C	($80,000)	(18.2/15.6)	$93,200	+$13,200
105-C	($66,500)	(78.0/66.0)	$78,600	+$12,100
106-C	($37,000)	(30.0/30.4)	$36,500	−$ 500

Net change..... +$10,800

Design C: 106-C in Class 3

This is such an improvement that it would be interesting to try putting 106-C in class 3 (parallel to 103-C). Intermediate temperatures become revised to 99.7 F. and 109.4 F. (from 98.1 F. and 107.8 F., respectively).

Calculating again according to step G, the costs become as follows:

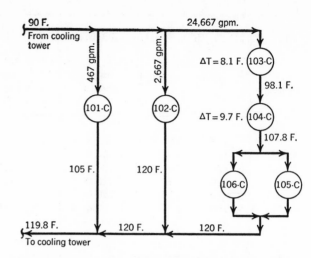

DESIGN A arrangement, 106-C placed in class 5—Fig. 3

Exchanger	Multiplication		Revised cost	Change
103-C	($110,000)	(9.1/ 9.3)	$107,800	− $2,200
104-C	($80,000)	(18.2/14.9)	$97,600	+$17,600
105-C	($66,500)	(78.0/66.0)	$78,600	+$12,100
106-C	($37,000)	(30.0/39.2)	$28,300	− $8,700

Net change..... +$18,800

Because shifting 106-C into class 3 would boost exchanger costs without concomitantly reducing tower cost (there being no change in water flow rate), 106-C should remain in class 4. (A hybrid arrangement in which 106-C is placed in parallel with the series flow of classes 103-C and 104-C would not be as economical —something the reader may want to calculate for himself.)

Design B: Splitting the Water Flow

According to step J, an examination should now be made of how to split the water between 104-C and 106-C. The flow to both is a total of 24,667 gpm. For an identical rise in temperature across each, the water flow to each should be in proportion to its duty (compared to the total duty of 140 million Btu.).

To 106-C, the water flow would be 3,520 gpm. (24,667 × 20/140). A 20% increase in this flow rate would bring it up to 4,220 gpm., and lower the ΔT to 9.5 F.:

$$\Delta T_{(106-C)} = (20 \times 10^6)/(500 \times 4,220) = 9.5$$

This would lower the water flow to 104-C to 20,447 gpm. (24,667 − 4,220), and raise its ΔT to 11.7 F.:

$$\Delta T_{(104-C)} = (120 \times 10^6)/(500 \times 20,447) = 11.7$$

Calculate the revised mean-temperature differences for these conditions, then—using Design B as the base —determine the revised costs for 104-C and 106-C:

Exchanger	Multiplication		Revised cost	Change
104-C	($80,000)	(15.6/15.4)	$81,000	+$1,000
106-C	($37,000)	(30.4/31.1)	$36,900	− $900

This is so close to a standoff that obviously the water split is practically an optimum for Design B. (Such is usually not the case, because, typically, the exchanger with the lower mean-temperature difference or the one

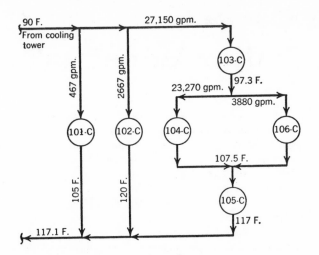

DESIGN D, final optimum balance, 106-C in class 4—Fig. 4

that is more costly on a Btu. duty basis will require more water. In this instance, it just happens that one of these factors applies to each of the exchangers—not both factors to one—and with equal strength.)

Design D: Zeroing in on the Optimum Range

Finally, it should be examined whether—even with the Design B arrangement—returning the water to the tower at 119.8 F. represents the best possible range—i.e., whether the range should be lowered (see step K).

Try narrowing the over-all range a small amount, perhaps about 2 F., and check over-all costs:

$$\text{Tower cost} = (413.5 \times 10^6)/(257 \times 7 + 39.2 \times 27.8 - 586)$$
$$= \$179,500$$

Of course, narrowing the range will raise the water flow:

$$\text{Water flow} = (413.5 \times 10^6)/(500 \times 27.8) = 29,700 \text{ gpm.}$$

Raising the water flow by 1,900 gpm., then, would increase the cost of the tower $5,500. The increment in tower cost for each gpm. increase in water flow would be about $3 ($5,500/1,900 gpm.).

Next, examine each parallel stream individually. Ignoring 101-C because of the small flow through it, try adding a small increment to the flow of 102-C (340 gpm.) and calculate the temperature rise across it:

$$\Delta T = 30 \text{ F. } (2,660/3,000) = 26.7 \text{ F.}$$

which range would give it a new ΔT_M of 19.4 F. Lastly, calculate the revised cost for 102-C:

$$\text{Cost, \$} = \$25,000 \ (18.2/19.4) = \$23,400$$

which cost constitutes a saving of $1,600. However, because an additional water flow of 340 gpm. would add another $1,000 to the cost of the cooling tower (and possibly as much as $1,500 over-all), it is apparent that nothing would be gained by lowering the range across 102-C.

Now try increasing the water flow by 11% through classes 103-C, 104-C and 106-C (which, of course, are now in parallel), and 105-C; in other words, reduce the

temperature rises across them by 10%. Outlet temperatures from these classes now become 97.3 F., 107.5 F. and 117 F., respectively. Finally, calculate new mean-temperature differences and how much exchanger costs would differ from Design B costs.

Exchanger	Multiplication	Revised cost	Change
103-C	($96,000) (10.4/10.95)	$91,200	−$4,800
104-C	($93,200) (15.6/17.1)	$85,000	−$8,200
105-C	($78,600) (66.0/68.5)	$75,800	−$2,800
106-C	($36,500) (30.4/31.6)	$35,100	−$1,400

Net change.....−$17,200

Raising the water flow 11% would increase the bare tower cost by $8,150 (11% × $3 × 24,667 gpm.), or a total of possibly $12,000, which would still leave a savings of $5,200.

Design E: Raising Water Flow 14% Instead of 11%

Using Design D as the base, try upping the water flow 14% rather than 11%, and recalculate costs:

Exchanger	Multiplication	Revised cost	Change
103-C	($91,200) (10.95/11.4)	$87,500	−$3,700
104-C	($85,000) (17.1/18.6)	$78,100	−$6,900
105-C	($75,800) (68.5/71)	$73,100	−$2,700
106-C	($35,100) (31.6/32.9)	$33,700	−$1,400

Net change.....−$14,700

Boosting the water flow 14%, however, would hike the bare tower cost by $10,400 (14% × $3 × 24,667 gpm.), or a total of possibly $15,000.

Design D, then, appears about optimum. The cost for all the exchangers would be $322,600 and the tower cost, for the 27.1 F. range, $181,000. This is approximately a 7% reduction from the once-through system (water flowing parallel through all the exchangers), which would cost $329,000 for the exchangers and $212,000 for the tower. Fig. 4 illustrates the final optimum design.

Meet the Author

Peter M. Paige (201 East 25th St., New York, N.Y. 10010) was a piping hydraulics systems engineer with M. W. Kellogg Co. for 12 years. Before this, he also was the head of the piping analytical section. His experience has been mainly in heat-transfer and pressure-vessel design, plant layout and refinery operations. He holds a B.S. in mechanical engineering from Columbia University and is a licensed engineer in the State of New York. A previous article of his, "Shortcuts to Optimum-Size Compressor Piping," appeared in the March 13, 1967 issue of *Chemical Engineering*.

Key Concepts for This Article

Active (8)	Passive (9)	Purpose of (4)
Design*	Cooling*	Minimum
Optimization	Water*	Cost*
	Towers*	Investment*
	Exchangers*	
	Condensers*	
	Coolers*	

(Words in bold are role indicators; numbers correspond to EJC-AIChE information retrieval system. Asterisks mark key concepts suggested for indexing. Others are added to improve reading as an abstract. Indexing is described in *Chem. Eng.*, Oct. 11, 1965, p. 187.)

Computers for Engineering Calculations:

How Do You Figure Dollar Benefits?

Does use of a computer for engineering calculations really pay off? That depends on how it's used and for what. Here is one way to analyze expenses and possible savings of a computer once it has been installed.

A good way to start an argument among a group of chemical engineers is to bring up the question of dollar benefits from computer applications. Everybody usually agrees that computers are useful in many areas of engineering work, including process simulation, process control, and engineering applications.

But how do you figure the dollar savings of a computer? Many of these savings come under the category of "intangibles"—time savings can be measured accurately, but how do you evaluate the benefits of a more refined and more accurate answer to a problem, compared to hand calculations?

The table (right) is one company's* attempt at evaluating the dollar benefits of a computer used in engineering calculations. The computer was an IBM 7090 applied for such problems as tray-to-tray distillation calculations, process-flow analyses, heat and material balances, unit operations calculations, and sizing and costing operations.

It should be noted that this is not a payout analysis. The large capital investment necessary for purchase of the computer is completely ignored. Nevertheless the approach is a useful one for getting order-of-magnitude data on dollar advantages of a computer.

Expenses and Savings

The expenses in the table include computer time costs, program-application manpower costs and program-development manpower costs.

Returns or savings are of three types. The first is the cost of manpower that would have been required to provide basic engineering services by hand calculations using hand methods, if the computer was not providing this service. The second is the cost of specialized outside services (primarily rented computer time) that would have been required for specialized studies, if in-company computer services had not been available.

The third type and the most significant is the dollar return from improvements in design and operations

through computer solutions, over and above those that could be expected from hand calculations.

A formal survey of expenses and returns for applications in 1962 is the base from which the evaluation was developed. Information gathered and derived from this survey included:

- Separate monthly costs for development and application time on the computer.
- Manpower costs (engineering and computer-communications departments) for program development.
- Application manpower costs for processing computer solutions.
- An estimate of the cost (savings) of outside services that would have been required (mainly for rented computer time), in the absence of in-company

Benefits from computer applications in engineering work

Expenses	Relative Dollar Amounts
Computer time for applications and developments[a]	0.4
Manpower for program development[b]	0.3
Manpower for application of computer[b]	0.3
Total computer services	1.0
Estimated savings	
Minimum alternate hand-calculation manpower for basic engineering problems[c,d]	1.0
Alternate outside services for special jobs	0.2
Improvements in design and operations due to sophisticated computer solutions[c,e]	1.3
Total savings	2.5
Estimated benefits	
Net savings	1.5
Ratio of net savings to expenses	1.5

---

[a] Hourly cost comparable to commercial service charge.

[b] All manpower costs include overhead charges.

[c] No account taken for use of programs by personnel outside the Engineering Dept.

[d] With hand methods, in absence of computer.

[e] Improvements and solutions over and above those that could be expected from hand calculations under normal circumstances.

[f] Absolute value is dependent on scaling factor used for relative dollars.

---

* "Computer Applications Payoff," by Edward Sorf of Socony Mobil Oil Co., New York, N. Y., at National Petroleum Refiner's Assn. Computer Conference, Pittsburgh, Pa., Oct. 15-17, 1963.

computer services, for vitally necessary calculations for specialized projects.

• An estimate of the cost (savings) of alternate hand-calculation manpower that would have been required in the absence of a computer. This is the minimum necessary engineering design and development service with hand methods for the jobs processed by the computer.

Alternate hand-calculation manpower for basic engineering calculations was estimated by those responsible for applications within each engineering function. First they determined the total number of case calculations made on the computer with a given program. Second they estimated the manpower that would have been required to cover these cases by hand calculations with hand methods commonly used. Finally they estimated the fraction of this total equivalent manpower that would have actually been committed for the basic engineering calculations in the absence of a computer.

• An estimate of savings in design, development and operating costs due to the ability of the computer to effect such improvements through solutions over and above those that could be expected from hand calculations.

These design savings were evaluated by averaging

two different estimates: one a direct but rough estimate; the other an indirect but relatively more precise and conservative estimate. The direct estimate was made by those involved with applications and consisted principally of savings in capital costs, product revenues or operating expenses of the designed plant.

The indirect estimate was derived from total equivalent hand-calculation manpower (corresponding to all cases run on the computer) as supplied by applications personnel. It was taken as that part of total equivalent hand-calculation manpower left over after accounting for manpower that would have been actually committed, in the absence of a computer, for basic engineering calculations.

A summary of the evaluation of dollar benefits, expressed as relative dollar amounts, from applications in 1962 is shown on the table. Dollar benefits were about 150% greater than the cost of computer services.

Manpower Benefits

Several secondary conclusions may also be drawn from other comparisons of the relative dollar amounts. For example:

• Total professional manpower (engineers, programmers, etc.) used for computer services was about 0.6 of the alternate engineering manpower that would have been needed for hand-method calculations: compare the sum of the relative dollar amounts for program development and application under expenses .(0.3 + 0.3 = 0.6) with the amount (1.0 relative dollar) for alternate hand calculations under estimated savings.

About three times as many engineers would have been needed for alternate hand-method calculations, without a computer, than were used for computer applications exclusive of program development. This is indicated by comparison of the 1.0 relative dollar amount for alternate hand calculations under estimated savings with the 0.3 relative dollar amount for program application under expenses.

Computer services costs in years prior to 1962 were available from computer time records and from known manpower assignments. Savings were evaluated for other years by apportioning savings in 1962 according to application hours on the computer in 1962 and in other years. The resulting trends are shown on the curve (left).

A breakeven point was reached in about three years after the start of operations. And accumulated dollar benefits through 1962 exceeded corresponding expenses by 60%.

Computer benefits: savings equaled expenses after three years

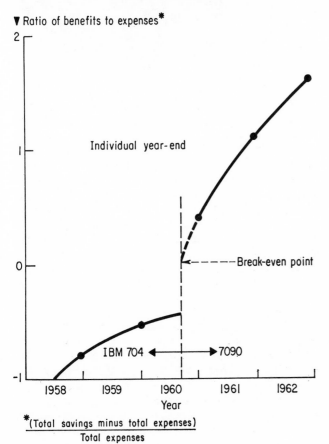

▼ Ratio of benefits to expenses*

Individual year-end

Break-even point

IBM 704 ◄——————► 7090

Year

*(Total savings minus total expenses)
Total expenses

Key Concepts for This Article

For indexing details, see *Chem. Eng.*, Jan. 7, 1963, p. 73 (Reprint No. 222). Words in bold are role indicators; numbers correspond to AIChE system.

Active (8)	Passive (9)	Means/Methods (10)
Comparing	Economics	Charts
Evaluating	Benefits	Curves
Analyzing	Computers	
Estimating	Engineering	
	Calculation	

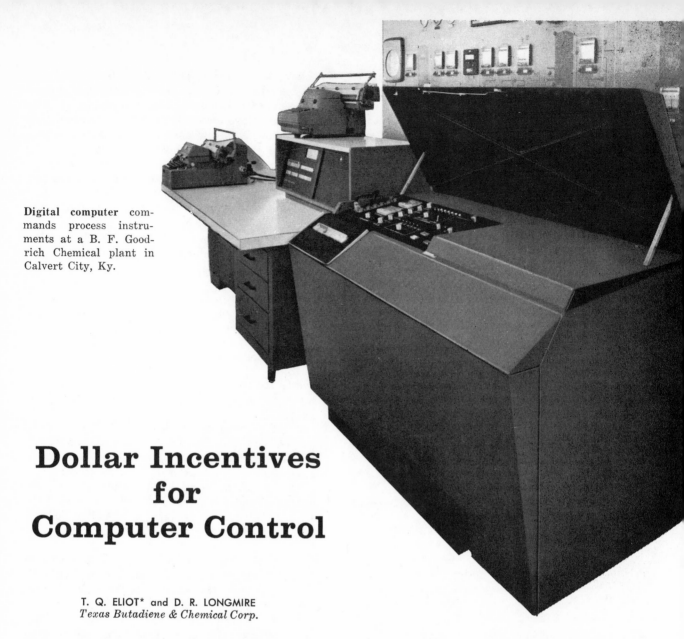

Digital computer commands process instruments at a B. F. Goodrich Chemical plant in Calvert City, Ky.

Dollar Incentives
for
Computer Control

T. Q. ELIOT* and D. R. LONGMIRE
Texas Butadiene & Chemical Corp.

Our purpose here is to provide some financial guides for preliminary evaluation of computer control used in process improvement. Many of our statements confirm an excellent general treatment that appeared last year.[1] We are concerned primarily with the individual unit process, operating on a large scale, with one or two feeds, and producing not more than two or three products.

A simple, flexible economic method is required for incentive analysis. An applicable method[8] is explained here and used to evaluate several example processes for computer control. Incentives for computer control of various processes are substantially different, and the difference between market-limited and production-limited situations is all-important, as is the size of the process being considered.

In the chemical engineering literature of recent years, discussion of methods for economic evaluation has been profuse. The system to be used here is that employed at Texas Butadiene & Chemical Corp., origi-

nated by Stokes[8] after exhaustive study of procedures reported by others. This system for economic evaluation is simple to apply and has a nomenclature consistent with commonly used financial-statement terms.

Corporate financial statements place capital or fixed-investment items on the asset side of a balance sheet. Total investment includes various forms of working capital and prepaid items, as well as investment in physical facilities (plants, warehouse, office buildings, etc.). (Though operations research and systems engineering[10] are often directed toward inventory and raw-material optimization, optimization of storage requirements and minimization of cash and receivable requirements, our concern is processing optimization.)

Possible interaction of working-capital items with the optimum design and operation of the various processes is ignored. Fixed investment FI is focused on and includes the particular processes, their share of off-site installations, and control computers.

The accounting of profit-and-loss items in economic evaluations is handled through an estimate of manufacturing cost of sales. This MCS includes cost of raw materials RM, conversion costs CC and depreciation D. Subtracting the sum of MCS and STA (sales, tech-

* T. Q. Eliot, formerly manager of the process development department of Texas Butadiene & Chemical, is now assistant to vice-president, sales, Scientific Design Co., New York. D. R. Longmire is associate professional engineer with his firm in Houston. Article is based on a paper originally presented at the Second Joint Automatic Control Conference, June 1961.

nical and administrative expenses) from net sales NS gives profit-before-taxes PBT. Conversion costs encompass such items as shown in Table I.

If we assume a corporate federal income tax rate of 52%, then profit-after-taxes PAT can be expressed in equation form as

$$PAT = 0.48[NS - STA - (RM + CC + D)] \qquad (1)$$

Cash accumulation is then $D + PAT$, and payout PO of the fixed investment FI after taxes becomes

$$PO = \frac{FI}{D + 0.48[NS - STA - (RM + CC + D)]} \qquad (2)$$

In the examples that follow, the payout of a computer installation is defined in terms of Eq. (2). Working-capital items have not been included in FI.

Note that we are talking here in terms of "fixed value" of dollars. Many published papers point out the importance of the "time value of money," and develop methods such as discounted-cash-flow, rate of return and profitability index.[12, 13] The system of economic factors used here can incorporate "time value of money" analysis by listing all costs for a given (usually yearly) period, and summing the resulting cash accumulations on a properly discounted basis. For screening economics of the type with which we are concerned, however, a simple calculation of payout after taxes suffices, since we are looking for relatively rapid payouts.

All costs for such projects can be reduced to the above elements. Product quality, for example, is expressed in sales price or, if the product is unsatisfactory, FI and CC must be increased to allow for re-running. In the worst case, off-spec products must be discarded, increasing FI, RM and CC. Even safety reduces to a simple cost item, since it must usually be paid for by additional fixed investment and conversion costs (monitoring instruments, special design, etc.).

At a given sales volume, the net sales price is determined by Eq. (3).

$$NS = (PBT + D) + CC + RM + STA \qquad (3)$$
$$NS = \Delta SP + CC + RM + STA \qquad (4)$$

The profitability criterion determines the quantity ΔSP which must be added to the sum of conversion costs, raw-material costs, and selling, technical and administrative expenses, so that the invested capital is adequately liquidated and made available for other profitable ventures. Stokes[9] has referred to ΔSP as the selling-price increment.

Note that ΔSP is a fixed percentage of FI for a given payout and method of depreciation. For example, the following figures are for 12-year life and double-declining-balance depreciation with change to straight-line after seven years—a system that is advantageous to businesses experiencing fast payouts on investments.

For payouts of 1-5 years respectively, ΔSP (% of FI) is 190, 87.7, 54.2, 38.1 and 28.6. Use of this ΔSP concept permits rapid determination of the economics of a proposed investment.

In a market-limited case, when a plant is operating at less than capacity, financial incentives for computer control must come from lower conversion and raw-material costs. This is the situation that faces most companies. Raw-material costs can be lowered by yield improvement or, in the case of multiple product grades, by quality improvement. Included in the conversion costs that lend themselves to cost reduction are utilities UTL, operating supplies and chemicals OSC, and various wage and salary items.

In line with current experience, we assume that no significant reduction in operating or maintenance labor can be credited to a computer-control system. Maintenance costs may indeed be increased by the needs of a computer system. Our effort will be concentrated on cost reduction for raw materials, utilities, and operating supplies and chemicals (including catalyst). The depreciation of the computer and allied instrumentation is added cost. Note that financial improvements under market-limited operation accrue to full-capacity operation as well.

When a plant is operated at capacity, the first limit on the profit potential is plant size itself. By closer control than usual, it is sometimes possible for a process-control computer to increase significantly the capacity of a chemical process plant.[8] Therefore, in considering computer control, one should also be interested in this production-limited market condition.

To determine the payout that might be obtained in increasing production, we must consider the additional raw-material and conversion costs. The latter can be separated into fixed conversion costs FCC and variable conversion costs VCC. By fixed, we mean constant-per-year for whatever total production is realized. A variable conversion cost is defined as an annual cost directly proportional to the attained production. Because of its importance, raw-material cost RM—itself a variable cost—is segregated from VCC here.

We use capital letters to signify annual over-all plant costs and sales, lower-case letters to denote the unit costs and unit sales values for products. For instance, ns is sales value per unit of product and pat is profit-after-taxes per unit of product. Then, if output is limited by plant capacity, incremental profit-after-taxes for an increase in capacity ΔP is

$$\Delta P(pat) = 0.48[ns\,\Delta P - (rm + vcc)\,\Delta P - \Delta D] \qquad (5)$$

in which ΔD is the increase in depreciation due to the computer and its associated equipment, and $(rm + vcc)$ is the variable cost associated with the additional production. We assume that the plant won't be run near throughputs where yields or product recoveries fall off rapidly, or where conversion costs (e.g., utilities and chemicals) rise quicker than usual.

Larger cost areas should be those investigated for potential improvements, since here is where the incentives should be most easily located. The cost areas that make up the sales dollar have been defined by Eq. (3). They are—in the order given by the equation—capital costs, conversion costs, raw-material costs, and selling, technical and administrative expenses. With some processes, only one of these items dominates, while

with others, there may be several areas of incentive for computer control.

STA costs are not under the influence of computer process control and will not be discussed further. In analyzing costs for computer-control incentives, we must allow for the normal rate of improvement in process technology, which is itself responsible for a certain enhancement of process efficiency.[11]

In the chemical process industries, cost of raw materials is usually an important factor. Raw-material cost can constitute as much as two-thirds of the MCS, but there are cases—e.g., SBR-rubber manufacture—in which ultimate raw-material use approaches 100% by repetitive recycling. In such cases, yield improvements can reduce the costs (both *FI* and *VCC*) of processing the recycle material. In most chemical processes, such as butadiene manufacture[18] and many polymer and resin processes, raw material use is considerably below 100%, and the incentive for improvement is, therefore, substantial.

Among conversion costs, those that are most subject to computer control are utilities, operating supplies and chemicals. Again, one is faced with evaluating computer-control improvements vs. improvement due to normal enhancement of process knowledge.

The cost of catalyst and chemicals can be substantial. Through closer control, consumption of chemicals such as expensive solvents can be minimized. In many catalytic processes, there is considerable incentive in more-efficient use of catalysts. This may be achieved by optimization of operating conditions, to obtain maximum yield or the best balance of yield, selectivity and catalyst life.

The capital cost portion of the product sales price is

$$\Delta SP = PBT + D$$

This relationship defines the profitability criterion for any process. It gives the desired profitability of operating and sales performance as a yardstick that is independent of the capital structure of the process or company involved. Capital cost is indeed a cost of doing business, and we shall see examples in which such a cost dominates the makeup of the sales price. This is quite common in the chemical process industries, and is due to

1. The large investments required per unit of product.

2. The relatively short payout time required for facilities because of rapid technological advances.

Several announcements about chemical processes operated by computers in closed loops have furnished practical benchmarks for use as economic criteria.[2, 3, 7] Some of the factors associated with the three installations described in these articles are as follows:

Company	Process	Installed Cost	Criteria
Texaco	Cat poly unit	$400,000	*PO* < 2 yr.
Goodrich	Vinyl chloride; Acrylonitrile	$225,000	*PO* < 3–4 yr. for vinyl chloride
Monsanto	Ammonia	$250,000	*PO* < 1 yr.

The Monsanto application has been given extensive

coverage[5, 6, 7] and that company is to be commended for allowing substantial details on its closed-loop ammonia system to be made public. The ammonia plant is a fairly complicated chemical process (reforming of natural gas with steam, followed by synthesis). Thus, the cost of the computer and accompanying instrumentation is probably representative of what the uninitiated might expect to invest when considering computer control for a new or existing chemical process.* To obtain the speculated payout at Monsanto in the production-limited case (plant operated at capacity), the computer was to increase ammonia plant capacity by 3.5-5%. In the examples for production-limited cases here, we have used 3.5% increase as a reasonable objective.

Less is known about market-limited cost savings for the existing installations. These must come from raw materials, utilities, chemicals and catalysts. In our own study of computer-control incentives at Texas Butadiene and Chemical, we have found that these incentives are an order-of-magnitude less than for the production-limiting situation. For the example processes, taken here to be operating at 80% capacity, we optimistically give a 10% improvement in the variable costs of utilities, chemicals and catalysts.

Depending upon the process, a nominal improvement in raw-material use is assumed. For chemicals, variable costs are taken to be 50% of total costs per year at 100% operating capacity. For utilities, variable costs are assumed to be 35% of total costs per year at 100% capacity. On this basis, and without on-the-spot knowledge of process operating details, computer payouts for the market-limited cases have been constructed.

A minimum standard can be specified for consideration of a process-control computer that with its necessary instrumentation and control would cost $250,000 to install. By stipulating a two-year payout as the profitability criterion, we have the following:

$$\Delta FI = \$250,000$$
$$PO = 2 \text{ years}$$
$$\Delta PBT + \Delta D = \Delta SP = \$219,000/\text{year}$$

Thus, to be an attractive investment, the computer should be responsible for savings of $219,000/year (12-year life and double-declining-balance depreciation with change to straight-line after seven years).

The Monsanto ammonia plant at Luling, La., with a capacity of about 525 tons/day,[7] is one of the plants now operating in closed-loop computer control. As an example in which the cost of capital is the largest element of the sales dollar, a 525-ton/day ammonia plant is considered here. The process is steam reforming of natural gas, and the economics have been developed from an estimate for a 200-ton/day plant, which was prepared six years ago.[14]

Fixed investment for the process and off-sites has been estimated by first multiplying the 1955 fixed investment for a 200-ton/day plant by a Nelson[15] inflation index of 1.28 for refinery construction. Then the new figure for a 200-ton/day plant is scaled proportionally

---

* R & D costs were not assigned to the project.

to 525 tons/day, since investment of this type usually consists of multiple units (reformers, compressors and synthesis reactors) in parallel. The fixed investment, then, for a 525-ton/day grass-roots ammonia plant is

$$(\$6,800,000)\ (1.28)\ (525/200) = \$22,900,000$$

The published article on a 200-ton/day plant[14] gives 1955 raw-material and conversion costs. The costs for materials and utilities appear to be in line with present-day Gulf Coast numbers. However, salary, wage and overhead items have been brought up-to-date using the Nelson factor[15] of 1.28 developed for operating labor. We have assumed that the operating manpower requirements for a 525-ton/day unit are about twice those for a 200-ton/day unit. Unit costs for utilities and chemicals at a capacity of 525 ton/day are taken as equal to the unit costs for the 200-ton/day plant operating at capacity. The various raw-material and conversion costs are shown in Table I. Selling, technical and administrative expenses are assumed to be $117,000/year for the case in which the product is sold as an intermediate.

From Eq. (4),

$$\begin{aligned}
\Delta SP &= \$8,730,000 \\
CC &= 3,807,000 \\
RM &= 1,220,000 \\
STA &= \underline{117,000} \\
NS &= \$13,874,000/\text{year} \\
ns &= \frac{\$13,874,000/\text{year}}{(525\ \text{ton/day})(350\ \text{day/year})}
\end{aligned}$$

It is evident that the principal cost in the ammonia sales dollar is capital cost, $(PBT + D)$ or ΔSP, which amounts to 63% of the total cost. Although large incentives will be found in the production-limited condition, the market-limited incentive will first be estimated.

If the process is operated properly, the raw material (methane) is completely and usefully converted in the primary and secondary reforming steps. Therefore, the search for an incentive must be concentrated on variable conversion costs. A 10% saving in variable chemicals OSC and variable utilities UTL at 80% of capacity amounts to $47,500/year, before income taxes and before taking depreciation on the computer. The payout of a $250,000 computer installation would take about seven years, as illustrated by the trial-and-error calculation below. For a seven-year payout, the depreciation factor is 10.3%. Therefore,

$$\begin{aligned}
\Delta FI &= \$250,000 \\
\Delta CC &= \$\ 47,500/\text{year} \\
\Delta D &= (0.1030)(250,000) = \underline{25,800} \\
\Delta PBT &= \$\ 21,700 \\
\Delta PAT &= 10,400 \\
\Delta(PAT+D) &= \$\ 36,200/\text{year} \\
PO &= \Delta FI/\Delta(PAT+D) = \$250,000/\$36,200 = 6.9\ \text{years}
\end{aligned}$$

Therefore, even with an optimistic saving of 10% in variable chemical and utility costs, along with the assumption that the computer would be responsible for generating the savings, the computer can in no sense be justified in the market-limited case.

The production-limited case is a different story. At capacity, we assume that the computer increases production by 3.5%. This incremental production amounts

Production costs for ammonia—Table I

| | Annual Dollars | |
	Fixed	Variable
Raw materials		$1,220,000
Conversion		
Operating salaries and wages*	$570,000	
Maintenance salaries and wages*	550,000	
Technical salaries and wages*	95,000	
Plant administrative payroll*	126,000	
Maintenance materials	370,000	
Operating supplies and chemicals	85,000	85,000
Purchased utilities	957,000	509,000
Royalties	....	
Local taxes	230,000	
Insurance	230,000	
	$3,213,000	$1,814,000

*Includes payroll overhead.

to 6,400 tons/year of ammonia. The payout has been determined by using Eq. (5). The unit values ns, rm and vcc have been developed from the information in Table I.

$$\begin{aligned}
ns &= & \$75.00/\text{ton} \\
vcc &= 594,000/(525\ \text{tons/day})(350\ \text{days/year}) = \$\ 3.23/\text{ton} \\
rm &= 1,220,000/(525\ \text{tons/day})(350\ \text{days/year}) = \$\ 6.65/\text{ton}
\end{aligned}$$

Therefore, from Eq. (5), for a 1.1-year payout, we obtain

$$\begin{aligned}
ns\Delta P &= (6,400\ \text{ton/yr.})(\$75.00/\text{ton}) = & \$480,000/\text{yr.} \\
(rm+vcc)\Delta P &= (6,400\ \text{ton/yr.})(\$9.88/\text{ton}) = \$63,000 \\
\Delta D &= (\$250,000)(0.1653) = 41,000 \\
& & \$104,000/\text{yr.} \\
\Delta PBT &= & \$376,000/\text{yr.} \\
\Delta PAT &= (\$376,000/\text{yr.})(0.48) = & \$180,000/\text{yr.} \\
\Delta D &= & \$\ 41,000/\text{yr.} \\
\Delta(PAT+D) &= & \$221,000/\text{yr.} \\
PO &= \frac{\$250,000}{\$221,000/\text{yr.}} = 1.1\ \text{yr.}
\end{aligned}$$

If capacity operation of such a plant is imminent in the extended future, installation of a $250,000 computer system should improve profitability, since without this increase in capacity the plant is paying itself out in four years. We assume, of course, that the computer itself, not merely improved process know-how and better control systems subsidiary to the computer, is responsible for the capacity improvement. These peripheral improvements certainly can be a significant portion of the total computer-system investment.

The actual Monsanto case was reported to involve an ammonia sales price of $92.00/ton.[7] If the cost figures for our example are representative, the Monsanto computer plus auxiliaries would be paid out in less than one year at full capacity plus 3.5%. If the computer increases plant capacity by 5%, the installation would be paid out in full in a still shorter period. The stimulus for computer control in this example arises from the

large capacity of the plant and the relative insignificance of the cost of the computer-system installation to that of fixed investment in the plant itself.

Besides the ammonia case, we have considered the economics of five other processes. The product capacity of these processes (ethylene, acetylene, butadiene, ethylbenzene and hydrogen cyanide) is believed to be typical for the petrochemical industry. Economic data for them have been taken from the literature and modified with appropriate factors to give investment and operating costs that exist today.

For all six processes, the capacity economics and computer-control incentives are presented in Table II. No claim is made for the absolute accuracy of these figures. Surely they do not agree with current results that are constantly changing. If your process is mentioned in this table, however, the figures given here should be good for order-of-magnitude evaluation.

The upper portion of Table II summarizes over-all economics for each process. The makeup of sales price is tabulated for 100% of capacity and for capital payouts that give net sales prices currently characteristic for the respective products. Too, this portion of the table indicates where the largest costs lie and thus where one might look for improvements to be implemented—perhaps by process-control computers.

In the lower portion of the table, payouts for a $250,000 computer-instrumentation system are tabulated for each process in both the market-limited and production-limited cases. In general, the production-limited situation gives the most favorable economic

picture for payout of incremental investment in process improvements—including computer control.

Among our examples, the most favorable payouts in addition to ammonia are for ethylene, acetylene and butadiene. These processes are the largest of those selected. Naturally, a fixed increase in capacity for an arbitrary process will pay out a particular computer (or process improvement in general) at faster rates for larger initial capacities, an observation illustrated graphically by Williams.[1] Large-capacity plants have a high fixed investment, and the computer along with associated instruments forms a smaller percentage of that total investment.

For the market-limited cases of the six processes examined, only two—ethylene and butadiene—show a reasonable payout. Therefore, when markets are limited the process must be large enough to allow a saving in utility, chemical and raw-material costs that will pay out the computer in at most two or three years. From this analysis, a more detailed investigation of ethylene and butadiene appears warranted.

Given a process of sufficient size to support a computer installation, in what way can the process be improved by sophisticated control techniques? Some general areas that might lend themselves to improvement through computer control are listed below, along with actual processes to which they might be adapted. For control purposes, these areas can require relationships with many variables and complex interactions, possibly impractical with standard control methods. Many of the ideas have been cited by others.[23,24]

Here are how economics of computer control prove out in six cases—Table II

	Ammonia (Chemico)	Ethylene[16] (Kellogg)	Acetylene[7] (Sachsse)	Butadiene[18] (Houdry)	Hydrogen Cyanide[21,22] (Girdler)	Ethyl-benzene[20] (UOP)
Production	525 t./d.	300 MM lb.	41 MM lb.	100 MM lb.	23 MM lb.	60 MM lb.
Fixed capital investment	$22,900 M	$12,000 M	$9,000 M	$18,000 M	$1,830 M	$1,040 M
Payout	4 years	2.5 years	5.0 years	5.0 years	5.0 years	3.0 years
Raw material cost	$1,220 M	$1,890 M	$352 M	$2,160 M	$932 M	$2,135 M
Conversion costs	3,807 M	3,607 M	1,660 M	5,654 M	453 M	516 M
Selling, technical and administrative expenses	117 M	300 M	50 M	220 M	25 M	....
Selling price increment	8,730 M	8,260 M	2,570 M	5,170 M	525 M	565 M
Net sales price	$13,874 M	$14,057 M	$4,632 M	$13,204 M	$1,935 M	$3,216 M
Sales value/unit of product	$75/ton	$0.047/lb.	$0.113/lb.	$0.132/lb.	$0.084/lb.	$0.054/lb.
Market-limited case						
Incremental fixed capital	$250,000	$250,000	$250,000	$250,000	$250,000	$250,000
Savings	47,500	236,000	23,700	202,000	24,000	8,700
Incremental depreciation	25,800	38,200	19,700	37,800	19,700	18,500
Incremental profit after taxes	10,400	95,000	1,920	79,000	3,310	−4,700
Payout	7 years	1.9 years	11.5 years	2.1 years	11 years	18 years
Production-limited case						
Incremental production	18.4 t./d.	10.5 MM lb.	1.44 MM lb.	3.5 MM lb.	0.8 MM lb.	2.1 MM lb.
Savings	$376.000	$694,700	$171,600	$590,000	$70,800	$40,100
Incremental depreciation	41,000	41,600	36,600	41,600	29,900	24,900
Incremental profit after taxes	180,000	313,000	64,800	263,000	20,000	7,300
Payout	1.1 years	0.7 years	2.5 years	0.8 years	5 years	8 years

1. *Multiple Units in Parallel.*—These might be reactors, feed-preparation units, compressors, or product-purification units. This application might apply to a number of types of chemical processing, including furnaces for manufacture of acetylene, ethylene, propylene and oxygenated chemicals. Parallel fixed-bed catalytic reactors are characteristic of butadiene processes, ethylene oxide plants, catalytic polymerization units, etc. Multiple-stage and parallel compression systems are common in ammonia plants, acetylene processes, ethylene monomer units, butadiene plants and large natural-gasoline plants.

2. *Feed Variations.*—Planned or uncontrolled feed transients might involve fluctuations in composition or feed rate. In refinery reforming units, feed composition is said to be the most important uncontrolled variable.[34] Peripheral processes such as ethylene-plant butadiene recovery, catalytic polymerization and the UOP Alkar processes for ethylbenzene and cumene might be subject to significant feed-rate variations, due to small operating changes in the heart of the refinery or central chemical plant. In such cases, what is the incentive for computer optimization of controllable variables along with measured variation in feed composition or rate?

3. *Variations in Catalyst Quality.*—In parallel catalytic units, differences in catalyst activity can cause variations in reactor output. This problem might arise in ethylene oxide plants or butane isomerization units, among many other processes. Also, activity variation for units in series, such as in refinery reformers, could present opportunities for computer control and optimization.

4. *Frequent Changes in Product Mix or Specifications.*—Scheduling can require changes in desired product mix and quality. This is a problem in polyolefin plants, where changes in polymer grade or type might be controlled by changes in polymerization pressure, catalyst concentration or some other variable. Optimum change from one operating condition to another might present computer-control incentives to eliminate useless intervening processing changes and off-grade products.

5. *Frequent Upsets.*—Upsets can occur due to human errors and inadequately designed control systems. External factors can contribute to the improper functioning of control systems, too—e.g., Monsanto took advantage of climatic conditions in its ammonia applications. Also, conventional feedback concepts might be inadequate while addition of feedforward control based on computers working with dynamic process relationships might bring favorable economic results.[25]

6. *Unrecognized Relationships of Variables.*—For some chemical processes, the technological fundamentals have not been developed to the extent that they have for many petrochemical processes. When theoretical concepts are difficult to apply, techniques such as evolutionary optimization might be possible with computer control. Perhaps, a combination of process theory and model adjustment by evolution is a possibility.

7. *Safety Considerations.* For the future, computer control of routine startup and shutdown or during emergency shutdown has often been mentioned as a possibility. This problem involves control of transient conditions vs. steady-state operation, and appears to be more difficult than the computer-control applications reported thus far.

Although preliminary screening might show substantial potential for savings, one should beware of processes in their twilight of economic operation. For instance, a very high percentage of the sales dollar for viscose rayon is taken up by conversion costs, which in turn involve labor, utilities and chemicals in substantial proportions.[26] Although the economics for computer control might appear favorable, should the necessary time and money be spent on a process that is on the way out? Might not the effort be better applied to a newer fiber process? This is particularly important with major processes for which systems study, computer specification and manufacture, and instrument installation can take two years or longer.

For any process, computer control is more attractive in installations having high capacities (the percentage of capital taken up by computer is lower). Attention should be focused on cost elements that dominate the sales dollar. When conversion costs or raw-material costs (with substantial unselective conversion) are higher than capital costs, incentives for computer control might be found under market-limited conditions—the position in which most businesses find themselves today. If capital costs dominate, attractive computer-control incentives probably exist in the production-limited case only—the situation that is generally the most favorable.

It appears to these authors, as it has been pointed out elsewhere,[1] that the small size of many petrochemical processes continues to limit the adaptation of chemical-plant computer control, at least as it has been attempted in the last several years.

References

1. Williams, T. J., *ISA Journal* **8**, No. 1, 50 (1961).
2. Schall, W. C., *Chem. Eng.* **66**, No. 21, 102 (1959).
3. Madigan, J. M., *Chem. Eng. Progr.*, **55**, No. 5, 63 (1960).
4. Schall, W. C., *Chem. Eng.* **67**, No. 6, 78 (1960).
5. Eisenhardt, R. D., Williams, T. J., *Control Eng.* **7**, No. 11, 103 (1960).
6. Bozeman, H. C., *Oil & Gas J.* **58**, No. 46, 148 (1960).
7. Schall, W. C., *Chem. Eng.* **67**, No. 23, 110 (1960).
8. Stokes, C. A., *Chem. Eng. Progr.*, **55**, No. 5, 60 (1959).
9. Stokes, C. A., unpublished notes, 1957.
10. Rio, J., Shorkey, A. F., *Chem. Eng. Progr.*, **53**, No. 1, 17-J (1957).
11. Hall, C. R., *Chem. Eng. Progr.*, **56**, No. 2, 97 (1960).
12. Baldwin, R. H., *Harvard Business Review* **37**, No. 3, 98 (1959).
13. Ravenscroft, E. A., *Ibid.* **38**, No. 2, 97 (1960).
14. Eickmeyer, A. J., Marshall, W. H., Jr., *Chem. Eng. Progr.* **51**, 418 (1955).
15. Nelson, W. L., *Oil & Gas J.* **59**, No. 1, 73 (1961).
16. "Better Yields, Higher Purity from Ethylene Plants by M. W. Kellogg," M. W. Kellogg Co., New York.
17. Forbath, T. P., Gaffney, B. J., *Pet. Refiner* **33**, No. 12, 160 (1954).
18. Hornaday, G. F., *Pet. Refiner* **33**, No. 12, 173 (1954).
19. Grote, H. W., Gerald, C. F., *Chem. Eng. Progr.* **56**, No. 1, 60 (1960).
20. Jones, E. K., *Oil & Gas J.* **58**, No. 9, 80 (1960).
21. *Pet. Refiner* **38**, No. 11, 261 (1959).
22. Updegraff, N., *Pet. Refiner* **32**, No. 9, 197 (1953).
23. Hall, C. R., *Chem. Eng.* **66**, No. 6, 153 (1960).
24. Lane, J. W., Johnson, R. C., *Oil & Gas J.* **58**, No. 26, 116 (1960).
25. Williams, T. J., *Chem. Eng.* **68**, No. 3, 107 (1961).
26. American Viscose Corp., annual report 1959, p. 22.

How to Find Optimum Economic Size For Multistage Heat Exchangers

This second of two articles explains a graphical method to enable evaluation of the optimum economic design by expressing financial return as a function of stage effectiveness, number of transfer units, and number of exchangers.

HAROLD S. MICKLEY and **ERNEST I. KORCHAK**
Massachusetts Institute of Technology

The size and number of heat exchangers to be used in a multistage heat-exchange system will be determined by the profitability of the system. A frequently occurring example is the recovery of heat from process streams. The rate of increase in effectiveness will become smaller with increasing exchanger area. When a certain area is exceeded, the rate of increase of cost will grow faster than that of the effectiveness, so that an optimum can be achieved.

Bosnjakovic and others[1] have used the *E-NTU* charts successfully for optimization in the case of a single heat exchanger. To explain the use of a graphical construction, Bosnjakovic's work will be discussed first. Then we will extend this principle to multistage heat-exchanger systems.

Heat Recovery With a Single Heat Exchanger

The fuel consumption per pound of steam produced by a boiler will be reduced by preheating the air used for combustion with the heat in the waste-gas stream. Each unit of enthaply, supplied to the air in the preheater, will save an amount of fuel equivalent to that unit of enthaply. Thus the saving in fuel costs will be K_f, \$/Btu. If Q, Btu./hr. are transferred, and the boiler operates during a fraction x of the 8,760 hr./yr., the annual value of the heat transferred will be:

$$\text{Value of heat} = 8,760xQK_f \tag{1}$$

The installed cost of the exchanger will be K_a, \$/sq. ft. of heat transfer area. Depreciation, maintenance, etc. will be aK_a, \$/(yr.)(sq. ft.). Operating costs, such as pumping power to overcome the pressure drop across the heat exchanger, will amount to K_o, \$/(hr.)(sq. ft.).

We will assume that operating costs K_o are proportional to the heat-transfer area of the exchanger. The validity of this assumption should be checked for any particular application. If the assumption is invalid,

then K_o must be expressed as a function of the area A. Total annual expense for a heat exchanger with A, sq. ft. of surface area will be:

$$\text{Annual expense} = (aK_a + 8,760xK_o)A \tag{2}$$

The annual savings S will be:

$$S = 8,760xQK_f - (aK_a + 8,760xK_o)A \tag{3}$$

To achieve maximum annual savings, an optimum value A_o for the heat-exchanger area must be determined.

The exchanger effectiveness E can be introduced by expressing the heat transferred Q as:

$$Q = EC_s(T_0 - t_2) \tag{4}$$

where C_s ($=c_sW_s$) is the smaller of the heat-capacity rates, T_0 is the temperature at which the hot gas enters the exchanger, and t_2 is the temperature of the entering air.

Substituting Eq. (4) into Eq. (3) gives:

$$S = 8,760xK_f C_s (T_0 - t_2) E - A(aK_a + 8,760xK_o)$$
$$= 8,760xK_f C_s (T_0 - t_2) E - KA(U/C_S) \tag{5}$$

where

$$K = \frac{K_o + (aK_a/8,760x)}{U(T_0 - t_2)K_f} \tag{6}$$

The cost ratio K is a function of installation, operating and fuel costs; the actual time the heat exchanger is in use; the difference between the temperatures of the entering streams; and the over-all heat-transfer coefficient. For given conditions, K will be constant, but uncertainties as to the accuracy of the values of U and x must be kept in mind.

How to Find Optimum Exchanger Area

The value of S in Eq. (5) will be a maximum if b is maximal in the equation:

$$b = E - K(UA/C_S) = E - K(NTU) \tag{7}$$

To find the maximum value of b, the *E-NTU* diagram (Fig. 1) should be consulted for the type of exchanger under consideration. If a line with slope K is drawn through the origin, then the ordinate will be K for NTU equal to one. A tangent to the *E-NTU* curve for the appropriate C_s/C_L value parallel to this

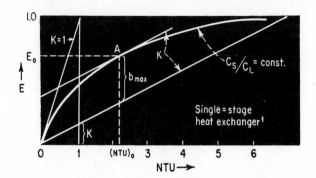

How to determine optimum conditions from the E-NTU diagram—Fig. 1

K-line will then give $b_{max.}$ from which the optimum values for E_o and $(NTU)_o$ can be found. The latter will have the optimum area A_o from:

$$A_o = C_S (NTU)_o/U \qquad (8)$$

Increasing the effectiveness beyond E_o by increasing the heat-transfer area would result in a less economical operation because above A_o, b would be less than $b_{max.}$.

It is also evident from Fig. 1 that the larger the value of K, the smaller $(NTU)_o$ will be. A lower value of NTU means that a smaller heat exchanger is economically justifiable. If K were to become equal to, or greater than one, the installation of a heat exchanger would not be justified.

High operating and equipment costs, and particularly small temperature differences $(T_o - t_2)$ will lead to high values of the cost ratio K. We can then state that the use of a heat exchanger may be justified if:

$$(T_o - t_2) K_f > \frac{K_o + (aK_a/8,760x)}{U} \qquad (9)$$

The value of the left-hand side of Eq. (9) will be determined by external conditions, but the right-hand side of the equation depends only on the choice of heat exchanger.

When choosing the most economical heat exchanger, not only its size but also the type should be considered such as a crossflow, countercurrent flow, or 1-2 parallel-counterflow exchanger. For this purpose Eq. (5) and Eq. (7) can be combined in the expression:

$$\frac{S}{8,760xK_f C_S (T_o - t_2)} = E - K (NTU) = b \qquad (10)$$

In the denominator of the left side of Eq. (10) are collected the variables imposed externally on the exchanger, which are independent of the type of exchanger. On the right-hand side, the externally imposed conditions and the variables determined by the choice of exchanger are expressed in terms of E and K.

For different types of heat exchanger, $b_{max.}$ must be determined. For example, Bosnjakovic[1] showed $b_{max.}$ as a function of K for cocurrent, countercurrent, and crossflow type exchangers. Fig. 2 indicates that a pure countercurrent heat exchanger would be the most economical type for any value of K. This was based on the assumption of equal installed cost per unit area K_a, operating cost per unit area K_o, and over-all heat-transfer coefficient U for the three types of exchangers. However, if any or several of these variables were to differ, according to the type of exchanger, then it would be possible for the countercurrent exchanger to become less economical than some other types.

It should be stressed that the previous argument is based on economic conditions as expressed by Eq. (5). Different processes may require a different treatment, although the same approach would apply.

Optimization of Multistage Systems

The example of preheating of a cold feed to a column will be used to illustrate a typical application of the optimization of a multistage exchange system. Assume that a hot process-stream is available to preheat the column feed. Any heat supplied to the feed will mean a saving of fuel in the reboiler.

In the first place, we will treat the system as a single heat exchanger by using the inlet temperature of the cold stream t_i, the inlet temperature of the hot stream T_o, and the heat-capacity rate ratio C_S/C_L. The cost ratio K is:

$$K = \frac{K_o + (aK_a/8,760x)}{U (T_o - t_i) K_f} \qquad (11)$$

Although at this stage, U cannot be calculated accurately, a reasonable estimate can be made from a knowledge of the physical properties of both streams and the mass flow rates. Inlet temperatures and all the cost factors may be assumed known. Having made a choice of the most economical heat-exchanger type, the optimum size of this exchanger is to be determined.

With the use of the E-NTU curve for the type of exchanger and heat-capacity rate ratio under consideration, the optimum effectiveness E_o and the number of transfer units $(NTU)_o$ can be determined graphically as shown for the single-stage case.

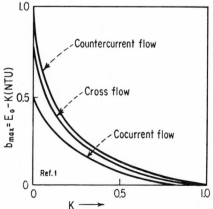

Each type of exchanger has optimum design— Fig. 2

If the optimum area A_o, as found from $(NTU)_o$ is fairly small, it would be improbable that a multistage heat-exchanger unit would be most economical. However, if a large heat-transfer area were found to give optimum conditions, the feasibility of a multistage system should be considered.

Initially, the heat-transfer coefficients and both heat-capacity rates will be assumed known. Let the cost of each unit of enthalpy transferred to the stream be K_f, $/Btu. The value of the heat transferred at the rate of Q, Btu./hr., from the hot process stream to the column feed will then be $8,760xQK_f$, $/yr. where x is the fraction of time the unit is in operation, referred to the total hours per year.

The installed cost of several heat exchangers cannot be based on a cost per unit area of heat-transfer area alone. This would imply that the installed cost of a heat exchanger of 1,000 sq. ft. is the same as that of two exchangers of 500 sq. ft. each. It is better to put the cost of an exchanger of A sq. ft. equal to $(K_e + AK_a)$. Depreciation and maintenance costs will then be: $a(K_e + AK_a)$, $/yr., or $[aK_a + (a/A)K_e]$, $/(yr.) (sq. ft.)$.

It will be assumed that operating costs, such as power for pumps, are proportional to the heat-transfer area, and amount to K_o, $/(yr.) (sq. ft.)$. This assumption should be checked for each application, and if not valid, K_o should be expressed as a function of area. The total annual expense for a heat exchanger with a surface area of A, sq. ft., will then be:

Annual expense $= aK_e + (aK_a + 8,760xK_o) A$ (12)

For exchanger n, in a series of N exchangers (Fig. 3), the heat transferred Q_n may be written as:

$$Q_n = E_n C_S (T_{n-1} - t_{n+1}) \tag{13}$$

The annual savings S_n due to the nth exchanger will be:

$$S_n = 8,760xQK_f - aK_e - A_n (aK_a + 8,760xK_o) \tag{14}$$

Substituting Eq. (13) into Eq. (14) yields:

$$S_n = 8,760xK_f C_S (T_{n-1} - t_{n+1}) E_n - aK_e - (aK_a + 8,760xK_o) A_n$$

$$S_n = 8,760xK_f C_S \left[\frac{C_t}{C_S} (t_n - t_{n+1}) - \frac{aK_e}{8,760xK_f C_S} - \right.$$

$$\left. \frac{\left(K_o + \dfrac{aK_a}{8,760x} \right)_n A_n U_n}{U_n K_f C_S} \right]$$

For all n exchangers of the multistage system:

$$S = 8,760xK_f C_S \left[\frac{C_t}{C_S} (t_1 - t_i) - \frac{a \sum\limits_{i}^{N} (K_e)_n}{8,760xK_f C_S} - \right.$$

$$\left. \sum_{n=1}^{N} \left(\frac{K_o + \dfrac{aK_a}{8,760x}}{U K_f} \right)_n \frac{(UA)_n}{C_S} \right] \tag{15}$$

But since in terms of the over-all effectiveness $\overline{E}$,

$$C_t (t_1 - t_i)/C_S = \overline{E} (T_0 - t_i)$$

we may write:

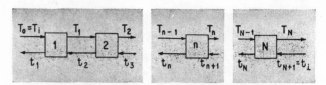

Multistage arrangement of a series of heat exchangers—Fig. 3

$$S = g \left[\overline{E} - \frac{a}{g} \sum_{n=1}^{N} (K_e)_n - \sum_{n=1}^{N} K_n (NTU)_n \right] \tag{16}$$

where $g = 8,760xK_f C_S (T_0 - t_i)$ (17)

The value of S will be a maximum if b is maximal in the equation:

$$b = \overline{E} - \left[\frac{a}{g} \sum_{n=1}^{N} (K_e)_n + \sum_{n=1}^{N} K_n (NTU)_n \right] \tag{18}$$

To determine an optimal system, it will be necessary to find the over-all effectiveness $\overline{E}$, the number of heat exchangers N, and the number of transfer units NTU.

If several exchangers with known effectiveness E, number of transfer units NTU, and cost factors K and K_e are available, then it is possible to calculate b for a number of arrangements. By plotting b vs. NTU, the optimum arrangement for b_{max} can be obtained.

Another method that will give a better visual insight into the performance of the system is to plot the over-all effectiveness E, and the function

$$\left[(a/g) \sum_{n=1}^{N} (K_e)_n + \sum_{n=1}^{N} K_n (NTU)_n \right]$$

for different numbers of transfer units NTU. Since the system consists of discrete units, this type of plot will give points rather than a continuous line as for the single-stage unit.

Note that the number of transfer units is additive, but that the over-all effectiveness for two or more units may be found by the graphical method.[2]

Example Illustrates Method

A multistage heat-exchanger array contains three separate exchangers that are externally connected in counterflow as shown.

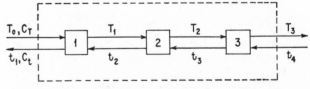

Will it be economical to reduce the number of exchangers from three to two or one? The graphical construction to determine T_3 and t_1 was developed in Example 2 of Part 1.[3] Data for this example are shown in an accompanying table.

Use These Data in the Example

$E_1 = 0.45$	$T_0 = 700$ F.
$E_2 = 0.50$	$t_i = 80$ F.
$E_3 = 0.40$	$C_T = 10,800$ Btu./(hr.) (°F.)
$(NTU)_1 = 0.8$	$C_t = 6,000$ Btu./(hr.) (°F.)
$(NTU)_2 = 0.9$	$U = 15$ Btu./(hr.) (sq. ft.) (°F.)
$(NTU)_3 = 0.65$	$K_a = \$3$/sq. ft.
$C_T/C_t = 1.8$	$K_e = \$2,000$
$a = 0.23$	$K_f = \$0.4 \times 10^{-6}$/Btu.
$x = 0.95$	$K_0 = \$22.8 \times 10^{-6}$/Btu.

From these data, we will compute the following constants for substitution into Eq. (18):

$$g = 8,760 x K_f C_s (T_0 - t_i) = 12,400$$
$$a/g = 0.23/12,400 = 1.853 \times 10^{-5}$$

$$K = \frac{K_0 + \dfrac{aK_a}{8,760x}}{U (T_0 - t_i) K_f} = 0.0284$$

We can now calculate the maximum value of b in Eq. (18) for each of the heat-exchanger arrangements.

Single Heat-Exchanger—If exchanger number 1 is used alone, then: $aK_e/g) + K(NTU) =$

$$(1.853) (10^{-5}) (2,000) + (0.0284) (0.80) = 0.0598$$

Since $E_1 = 0.45$, substituting the necessary values in Eq. (18) yields: $b_1 = 0.45 - 0.0598 = 0.3902$.

Two Heat-Exchangers—If exchangers number 1 and 2 are used, we must first find the over-all stage effectiveness $\overline{E}_2$. The over-all stage efficiency $\overline{\eta}_2$ is found to be 0.686 as worked out for the graphical construction of Example 2 in Part 1.[2] Since $C_s = C_t$, the over-all stage effectiveness $\overline{E}_2 = \overline{\eta}_2 C_t/C_s = 0.686$.

Next, we calculate the exit and entrance temperatures of the streams as follows:

$\overline{E}_2 = (t_1 - t_i)/(T_0 - t_i)$
$0.686 = (t_1 - 80)/(700 - 80)$, or $t_1 = 505$ F.

$E_1 = \eta_t = (t_1 - t_2)/(T_0 - t_2)$
$0.45 = (505 - t_2)/(700 - t_2)$, or $t_2 = 354$ F.

$C_t (t_1 - t_2) = C_T (T_0 - T_1)$
$0.555 (505 - 354) = (700 - T_1)$, or $T_1 = 616$ F.

$C_t (t_2 - t_i) = C_T (T_1 - T_2)$
$0.555 (354 - 80) = (616 - T_2)$, or $T_2 = 464$ F.

Values of K for exchangers 1 and 2 are now required for substitution in Eq. (18). These values are calculated by substituting in Eq. (11) the appropriate temperatures for each exchanger. Hence:

$$K_1 = 17.6/(T_0 - t_2) = 17.6/346$$
$$K_2 = 17.6/(T_1 - t_i) = 17.6/536$$

Therefore, the value for $[K(NTU)]_1 = 0.0510$ and for $[K(NTU)]_2 = 0.328$. We now evaluate the last two terms of Eq. (18) as follows:

$$(a/g) \sum_{n=1}^{2} K_e + \sum_{n=1}^{2} K (NTU) = 0.07412 + 0.0838 = 0.1579$$

Since $\overline{E}_2 = 0.686$, substituting in Eq. (18) yields: $b_2 = 0.686 - 0.1579 = 0.5281$.

Three Heat-Exchangers—If exchangers 1, 2 and 3 are used, we proceed as before to determine the exit

and entrance temperatures of the hot and cold streams. Using the solution of $t_1 = 561$ F., found in Example 2, Part 1,[2] we calculate the temperatures as follows:

$E_1 = \eta_t = (t_1 - t_2) (T_0 - t_2)$
$0.45 = (561 - t_2)/(700 - t_2)$, or $t_2 = 447$ F.

$C_t (t_1 - t_2) = C_T (T_0 - T_1)$
$0.555 (561 - 447) = (700 - T_1)$, or $T_1 = 637$ F.

$E_2 = (t_2 - t_3)/(T_1 - t_3)$
$0.50 = (447 - t_3)/(637 - t_3)$, or $t_3 = 258$ F.

$C_t (t_2 - t_3) = C_T (T_1 - T_2)$
$0.555 (447 - 258) = (637 - T_2)$, or $T_2 = 532$ F.

$K_1 = 17.6/(T_0 - t_2) = 0.0695$

$K_2 = 17.6/(T_1 - t_3) = 0.0465$

$K_3 = 17.6/(T_2 - t_i) = 0.0391$

The last two terms of Eq. (18) now yield:

$$(a/g) \sum_{n=1}^{3} K_e + K (NTU) = (1.853) (10^{-5}) (3) (2,000) + (0.0555 + 0.0418 + 0.0254) = 0.2339$$

The over-all stage efficiency $\overline{\eta}_3$ is 0.7775 as found in Example 2, Part 1.[2] Since $C_s = C_t$, the over-all stage effectiveness $\overline{E}_s = \overline{\eta}_3 C_t/C_s = 0.7775$. Substituting the values into Eq. (18) then gives: $b_3 = 0.7775 - 0.2339 = 0.5436$.

An examination of the values for b shows that $b_3 > b_2 > b_1$. Hence the three exchanger system is the most economical. By using Eq. (16), we can calculate the annual savings as: $S = (12,400) (0.5436) = \$6,741$/yr.

Nomenclature

a	Depreciation constant.
A	Area, sq. ft.
C	Heat-capacity rate, Btu./(hr.)(°F.).
E	Stage effectiveness.
E_n	Over-all stage effectiveness of n stages.
K	Cost factor.
K_a	Installed cost, \$/sq. ft.
K_e	Cost of exchanger, \$.
K_f	Fuel cost, \$/Btu.
K_0	Operating cost, \$/(hr.)(sq. ft.).
Q	Heat flow, Btu./hr.
S	Annual savings, \$/yr.
t	Temperature of cold stream, °F.
T	Temperature of hot stream, °F.
U	Over-all heat-transfer coefficient, Btu./(hr.)(sq. ft.) (°F.).
η	Efficiency.
Subscripts	
i	Inlet.
L	Stream with the greater heat-capacity rate.
S	Stream with the smaller heat-capacity rate.
T	Hot stream.
t	Cold stream.

References

1. Bosnjakovic, F., Vilicic, M. and Slipcevic, B., "VDI-Forschungsheft 432," Ausgabe B, Band 17 (1951).
2. Mickley, H. S. and Korchak, E. I., *Chem. Eng.*, Oct. 15, 1962, pp. 181-188.

Selecting Heat-Transfer Media by Cost Comparison

Here is a quick method for comparing and estimating
the heat-exchange costs of various heat-transfer fluids.

STANLEY KASPER, Blaw-Knox Co.

Every processing plant entails some sort of heat
transfer. And, of course, the most popular heat-
transfer medium is water: it is cheap, readily available in quantity, nontoxic, nonflammable and stable;
it provides good heat-transfer coefficients, has the
highest sensible latent-heat capacities, and its technology is well understood. Its utility, however, is
limited at the low-temperature end by the freezing
point, and at the high-temperature end by high vapor
pressure. Heat-transfer media other than water, therefore, are considered only when these limits must be
exceeded—but even then, the other media only supplement the cooling water-steam systems. Common
examples are flue gases, cooling air, electric heaters,
and refrigerants utilizing phase changes.

The selection of a suitable heat-transfer medium
for unusual temperature ranges can be influenced by
many factors such as fluid properties, type of application, climate, or even personal prejudices. The
primary requisite, of course, is that the medium be
a fluid in the temperature range under consideration.
It is preferable that the material remain fluid under
the lowest ambient temperature encountered to eliminate freezing problems. It is also desirable that the
vapor pressure of the material be close to one atmosphere as the highest temperature encountered
in the system to eliminate the need for equipment
capable of withstanding high pressures.

Some of the properties that are of interest are
freezing point, vapor pressure, thermal stability, flash
point, explosive limits, fouling characteristics, cor-
rosiveness and toxicity. Other factors are initial cost,
unpleasant odor, difficulty of retention, reactivity with
process materials, nuclear cross section, or other
nuclear properties. One or more of these factors may
be of prime importance and dictate the choice of a
particular fluid in spite of minor disadvantages. More
often, some factor is given a large and perhaps undue
measure of importance because of the preferences
and previous experience of the engineering personnel
participating in the choice of a medium or because
some justification must usually be presented for the
choice.

Cost Calculation

An engineering decision should be based on economics, although it is often difficult to attach a correct monetary value to the factors considered. In
selecting a heat-transfer medium, the factors and
properties mentioned above may be of minor economic
importance. The direct cost of installing and operating
a heat-transfer system then becomes much more significant. A partial guide to comparison of operating
costs of heat-transfer media may be found in Carberry's study of power requirements per unit of heat
transferred.[2] Methods for optimum sizing of heat
exchangers have also been established,[9] and it appeared that one of these could be utilized in an
economic evaluation of heat-transfer media. The scope
of this evaluation is limited to comparison of liquid
media because of mathematical simplicity and because,
in the case of phase change, other factors such as ease
of temperature control and maintenance of equipment
come into consideration.

The total annual cost of exchanging heat can be
written as a function of exchanger area and power
required for pumping heat-transfer media.

$$C = \frac{e}{100} A R_a + P R_p \theta \tag{1}$$

Power requirements are related to flow and pressure
drop through the exchanger as follows:

$$P = V\Delta p \times \frac{100}{E} \times \frac{0.746}{550} = 0.1357 \frac{V\Delta p}{E} \tag{2}$$

Assuming that the heat-transfer medium is flowing
through the tubeside, the pressure drop can be represented by the Fanning equation.

$$\Delta p = \frac{2fLv^2}{gD} = \frac{fLv^2 \rho}{16.2 D} \tag{3}$$

Flow in terms of velocity for a single-tube exchanger is:

$$V = \frac{\pi D^2 v}{4} \tag{4}$$

Assuming turbulent flow, the friction factor can be written as a function of the Reynolds Number.[10]

$$f = \frac{0.046}{N_{Re}^{0.2}} = \frac{0.046\, \mu^{0.2}}{\rho^{0.2}\, v^{0.2}\, D^{0.2}} \tag{5}$$

Also,

$$L = \frac{A}{\pi D} \tag{6}$$

Substituting Eq. (3), (4), (5), (6) in Eq. (2) gives

$$P = \frac{K_1 v^{2.8}\, \rho^{0.8}\, A \mu^{0.2}}{E\, D^{0.2}} \tag{7}$$

Since:

$$A = \frac{q}{U \Delta T_{lm}} \tag{8}$$

substitution of (7) and (8) into (1) gives:

$$C = \frac{e\, R_a\, q}{100 U\, \Delta T_{lm}} + \frac{K_1 v^{2.8}\, \rho^{0.8}\, \mu^{0.2}\, q\, R_p\, \theta}{E\, D^{0.2}\, U\, \Delta T_{lm}} \tag{9}$$

For the case where the tubeside film coefficient is controlling, $U = h_i$.

The heat-transfer film coefficient can be represented by the Colburn equation[8]:

$$\frac{h_i D}{k} = 0.023 \left(\frac{D\, G}{\mu} \right)^{0.8} \left(\frac{C_p\, \mu}{k} \right)^{0.333} \tag{10}$$

where $G = v\rho$
then:

$$h_i = \frac{0.023\, k^{0.667}\, v^{0.8}\, \rho^{0.8}\, C_p^{0.333}}{D^{0.2}\, \mu^{0.467}} \tag{11}$$

Substituting for U in Eq. (9) gives:

$$C = \frac{43.5\, R_a\, q\, D^{0.2}\, \mu^{0.467}\, e}{\Delta T_{lm}\, k^{0.667}\, v^{0.8} \rho^{0.8}\, C_p^{0.333}} + \frac{K_2\, v^2\, q\, \mu^{0.667}\, R_p\, \theta}{\Delta T_{lm}\, E\, k^{0.667}\, C_p^{0.333}} \tag{12}$$

Defining the optimum velocity as the velocity that will give the minimum annual cost per unit (Btu./hr.) of heat transferred, we may set:

$$\frac{d\,(C/q)}{d\,v} = 0 \tag{13}$$

Differentiating and solving for v, we get:

$$v = \frac{K_3\, D^{0.0714}}{\mu^{0.0714}\, \rho^{0.286}} \left(\frac{R_a\, e\, E}{R_p\, \theta} \right)^{0.357} \tag{14}$$

Substituting the optimum velocity back in Eq. (12), we get:

$$\frac{C}{q_{min.}} = K(R_a\, e)^{0.714}\, \frac{D^{0.1428}}{E^{0.286}\, \Delta T_{lm}} \times$$
$$\frac{\mu^{0.525}}{k^{0.667}\, C_p^{0.333}\, \rho^{0.572}}\, (R_p\, \theta)^{0.286} \tag{15}$$

When comparing different heat-transfer media, it is evident that e, R_p, and θ are constant. Also, the comparison will be made on the basis of constant tube diameter. The cost of heat-exchange surface per unit area decreases with total area. However, if the total area is large and the areas being compared are not greatly different, it is reasonable to assume a constant value for R_a. It is assumed that for a given application, the material of construction will be the same for all heat-transfer media considered. It is also assumed that for a given application, pump efficiency can be considered constant.

The log mean temperature difference will obviously differ from one medium to another unless, by coincidence, the $m\, C_p$ products for two media are the same for a particular application. The temperature difference at the end where the media enter will be constant since the media compared will be assumed to enter at the same temperature. If terminal differences are large, the effect of varying temperature at the exit end of the media will be minimized. In order to simplify the mathematics, it will be assumed that this condition exists and ΔT_{lm} will be assumed constant.

Eq. (15) may now be written with only physical properties as independent variables.

$$C/q_{min.} = K\, \frac{\mu^{0.525}}{k^{0.667}\, C_p^{0.333}\, \rho^{0.572}} \tag{16}$$

Nomenclature

A	Surface area, sq.ft.
A_s	Cross-section area of tube, sq.ft.
C	Total annual cost, $/yr.
C_a	Annual fixed charges, $/yr.
C_o	Annual operating cost, $/yr.
C_p	Specific heat, Btu./lb.°F.
D	Tube diameter (I.D.), ft.
E	Over-all pump efficiency
e	Fixed charges as % of capital cost
f	Friction factor
g	Gravitational constant
h_i	Inside film coefficient, Btu./(hr.) (sq.ft.) (F.)
h_o	Outside film coefficient, Btu./(hr.) (sq.ft.) (F.)
k	Thermal conductivity, Btu./(hr.)(ft.)(F.)
K, K_1	Constants
L	Tube length, ft.
m	Mass flow, lb./hr.
N_{Re}	Reynolds number
P	Power consumption, kw.
Δp	Pressure drop, lb./sq.ft.
q	Heat transferred, Btu./hr.
R_a	Exchanger cost, $/sq.ft.
R_p	Power cost, $/kwh.
S	Sum of wall and dirt resistances, Btu./(hr.) (sq.ft.) (F.)
ΔT_{lm}	Log mean temperature difference
ΔT	Temperature change in a stream, F.
U	Over-all heat transfer coefficient, Btu./(hr.) (sq.ft.) (F.)
V	Volume flow, cu.ft./sec.
v	Velocity, ft./sec.
v_o	Optimum velocity, ft./sec.
θ	Operating time, hr./yr.
ρ	Density, lb./cu.ft.
μ	Viscosity, lb./ft./sec.

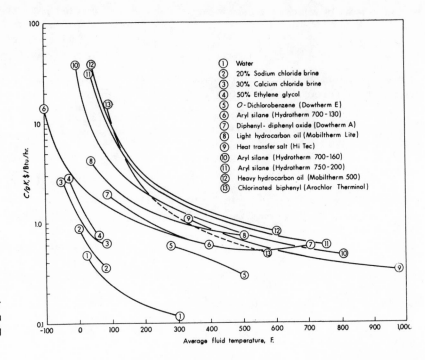

RELATIVE COST for heat transfer using various fluid media (media coefficients being the controlling factors)—Fig. 1

Legend:
1. Water
2. 20% Sodium chloride brine
3. 30% Calcium chloride brine
4. 50% Ethylene glycol
5. O-Dichlorobenzene (Dowtherm E)
6. Aryl silane (Hydrotherm 700-130)
7. Diphenyl- diphenyl oxide (Dowtherm A)
8. Light hydrocarbon oil (Mobiltherm Lite)
9. Heat transfer salt (Hi Tec)
10. Aryl silane (Hydrotherm 700-160)
11. Aryl silane (Hydrotherm 750-200)
12. Heavy hydrocarbon oil (Mobiltherm 500)
13. Chlorinated biphenyl (Arochlor Therminol)

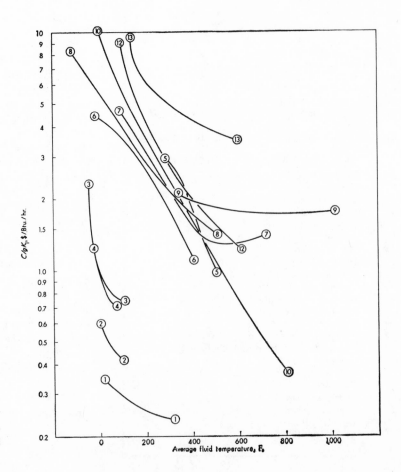

RELATIVE COST for heat transfer using various fluid media (process coefficients being the controlling factors)—Fig. 2

Annual Costs at a Glance

In Fig. 1, C/qK_{min} has been plotted as a function of temperature for a number of liquid heat-transfer media that have been employed in industrial plants. Each curve extends over the recommended or useful operating range. It is apparent from Fig. 1 that water would be a very economical heat-transfer medium on the basis of its physical properties alone. Physical properties were obtained from published data[1,4,5,7,8,11,12,13] except for thermal conductivities of Arochlor and HTS. These were calculated by the methods of Gambill[5] over part of the temperature range.

It will be noted that liquid metals are absent from Fig. 1. Heat-transfer characteristics of these materials are partially dependent on their ability to wet the tube wall. Consequently, ordinary film coefficient equations such as Eq. (10) are not applicable, and these materials could not be included in the comparison. A spot check on mercury at 260 F. and NaK at 212 F. indicated that they would rate as very economical heat-transfer media on the basis of the method of analysis used in this article.

For the case where the shellside coefficient is controlling, the heat-transfer area is not a function of the type of heat-transfer medium (again neglecting small changes in terminal temperature difference). Therefore, the total annual cost is dependent only on power requirements. Modifying Eq. (9), we get:

$$C = \frac{K_1 \, v^{2.8} \, \rho^{0.8} \, \mu^{0.2} \, q \, R_p \, \theta}{E \, D^{0.2} \, U \, \Delta T_{lm}} \qquad (17)$$

Heat transferred to or from the heat-transfer medium is:

$$q = m \, C_p \, \Delta T \qquad (18)$$

where:

$$m = \rho V = \rho v A$$

therefore: $q = \rho v A C_p \Delta T$ or $v = \dfrac{q}{C_p \, \rho A \, \Delta T}$

substituting for v in Eq. (17), we get:

$$C = \frac{K_1 \, q^{3.8} \, \mu^{0.2} \, R_p \, \theta}{C_p^{2.8} \, E D^{0.2} \, U \, \Delta T_{lm} \, \rho^2 \, A^{2.8} \, (\Delta T)^{2.8}} \quad \text{or} \qquad (19)$$

$$C = \frac{K_1 \, q^{2.8} \, \mu^{0.2} \, R_p \, \theta}{C_p^{2.8} \, E D^{0.2} \, \rho^2 \, A^{1.8} \, (\Delta T)^{2.8}} \qquad (19a)$$

In comparing various heat-transfer media, we may say that in accordance with our assumptions, all factors are constant except viscosity, density, and specific heat of the fluid, and the cost per unit of heat transferred per hour is expressed as:

$$\frac{C}{K_1 q} = \frac{\mu^{0.2}}{C_p^{2.8} \, \rho^2} \qquad (20)$$

In Fig. 2, $C/K_1 q$ has been plotted as a function of bulk fluid temperature for several heat-transfer media covering the same ranges as shown in Fig. 1. Fig. 2 indicates that there is not nearly as much difference in relative cost of transferring heat by the various media in this case. Again, water is seen to be a

superior medium. From Eq. (20), it is evident that high density and high specific heat liquids would tend to be more economical. This is borne out by the fact that the heat-transfer salt shows up as an economical fluid. Spot checks on mercury and NaK (not plotted) indicate that their high density is outweighed by low specific heat, and the economics of these liquid metals are unfavorable.

Fig. 1 or 2 can be used to make a rough comparison of the economics of heat-transfer media by estimating the average operating temperature of the medium and noting the relative positions of the curves at this temperature. The lowest curve will indicate the most economical medium. If a more detailed comparison is desired, actual temperature ranges based on optimum velocities can be computed and average C/Kq values may then be obtained graphically for the specific ranges.

In a given plant, there may be numerous users of a heat-transfer fluid. In some applications, process fluid coefficients may be controlling, in others the media coefficients may be controlling, and in many cases neither coefficient may control. Furthermore, several different temperature levels could be involved. Thus, a detailed economic study for selection of a heat-transfer medium could be a formidable and expensive undertaking.

An experienced engineer, however, should be able to estimate where the bulk of the costs for heat exchange will be and to use Fig. 1 and 2 as a guide in selection of an economic heat-transfer medium.

References

1. American Hydrotherm Corp., Bulletins "Hydrotherm 700-10A, 700-130, 700-160, and 750-200."
2. Carberry, J.J., "Media for Heat Transport," *Chem. Eng.*, June, 1953, p. 225.
3. Colburn, A.P., *Trans A.I.Ch.E.*, 29, 1933, pp. 174-200.
4. Dow Chemical Co., Bulletin "Dowtherm Heat Transfer Fluids."
5. Gambill, W.R., "Fused Salt Thermal Conductivity," *Chem. Eng.*, Aug. 10, 1959, p. 129.
6. Geiringer, P.L., "Heat With Liquids," *Chem. Eng.*, Oct., 1950, p. 136.
7. Keenan, J.H., and F.G. Keyes, "Thermodynamic Properties of Steam."
8. Kirst, W.E., Nagle, W.M., and J.B. Castner, "A New Heat Transfer Medium for High Temperature," *Trans. A.I.Ch.E.*, 36, 1940, pp. 371-94.
9. Lewis, W.K., Ward, J.T., and E. Voss, *Ind. Eng. Chem.*, 16, 1924, pp. 467-468.
10. McAdams, W.H., "Heat Transmission," 3rd ed., p. 155.
11. McArdle, M., Garrett, L.C., and P.G. Benignus, "An Indirect Aroclor Heater for Unit Chemical Operations," *Ind. Eng. Chem.*, 41, July 1949, p. 1341.
12. Shields, C.T., "Brine Film Resistance to Heat Flow of Calcium Chloride, Sodium Chloride, and Ethylene Glycol," *Refrig. Eng.*, Sept., 1951, p. 880.
13. Socony Mobil Oil Co., Bulletin, "Aromatic Heat Transfer Oils."

Meet the Author

Stanley Kasper, who now heads the process design section at Blaw-Knox Co., Chemical Plants Div. (300 Sixth Ave., Pittsburgh, Pa. 15222), has nearly 20 years of experience in the process and project engineering of a wide variety of plants that run the gamut from food processing and petrochemicals to hydrometallurgy and atomic energy. A registered professional engineer in Pennsylvania, Mr. Kasper obtained his B.S. in chemical engineering from MIT, and is a member of AIChE and Alpha Chi Sigma.

Calculating Minimum-Cost
Ion-Exchange Units

*A new equation permits you to calculate minimum
operating costs for ion-exchange units; it also yields
optimum regeneration levels and resin volumes.*

D. G. DOWNING, *Rohm & Haas Co.*

Many variables have to be considered in the design
and sizing of ion-exchange units. There is usually
some combination of these variables (i.e., regeneration
level, resin volume, resin life, etc.) that will produce
the lowest over-all operating cost. We have developed
a simple equation for determining the size of an ion-
exchange unit that will yield the most economical operation for a specific influent. This equation can also
be used to calculate optimum regeneration levels and
resin volumes.

The equation is as follows:

$$O_c = \frac{I_v I_k}{46.84 A_x} \left[R_r R_1 + \frac{L_t L_r}{A_v} \right] + \frac{A_v A_c}{A_r} + \frac{E_c}{E_r}$$

Here, operating cost, O_c, is in units of \$/yr.

Deriving the Equation

The four major cost items that were considered in
developing the equation are (a) regeneration cost , R_c,

(b) resin-depreciation cost, A_d, (c) equipment-de-
preciation cost, E_d, and (d) labor cost, L_c. The summation of these four factors yields the major over-all
operating cost:

$$O_c = R_c + A_d + E_d + L_c \qquad (1)$$

In designing a plant for industrial water treatment,
two factors are usually known—the quantity of water
to be treated, I_v, and the influent concentration, I_k.
These two factors, with resin capacity, A_x, and volume,
A_v, allow us to compute the volume of water treated
per cycle, I_n, and the number of cycles per year, N.
Eqs. (2), (3) and (4) demonstrate these relationships:

$$I_n = A_x, \ \frac{\text{Kgr.}}{(\text{cu. ft.}) (\text{cycle})} \times \frac{17.1 \ (\text{ppm.}) (\text{gal.})}{\text{gr.}} \times$$

$$\frac{1,000 \ \text{gr.}}{\text{Kgr.}} \times \frac{A_v, \text{cu. ft.}}{I_k, \text{ppm.}}$$

$$I_n = 17,100 \ A_x A_v / I_k \qquad (2)$$

$$N = I_v, \ \frac{\text{gal.}}{\text{day}} \times 365 \ \frac{\text{days}}{\text{yr.}} \times \frac{1, \text{cycle}}{I_n, \text{gal.}} = 365 \ I_v / I_n \qquad (3)$$

$$N = \frac{I_v I_k}{46.84 A_x A_v} \qquad (4)$$

The regenerant requirements per year, R_p, can be obtained from substituting Eq. (4) into Eq. (5).

$$R_p = R_1, \ \frac{\text{lb. reg.}}{(\text{cu. ft.}) \ (\text{cycle})} \times A_v, \ \text{cu. ft.} \times N, \ \frac{\text{cycles}}{\text{yr.}}$$

$$= R_1 A_v N \qquad (5)$$

$$R_p = \frac{R_1 I_v I_k}{46.84 A_x} \qquad (6)$$

The cost of the regenerant per year, R_c, one of the major cost components, is obtained from Eqs. (6) and (7).

$$R_c = R_p, \ \frac{\text{lb.}}{\text{yr.}} \times R_r, \ \frac{\$}{\text{lb.}} = R_p R_r \qquad (7)$$

$$R_c = \frac{R_r R_1 I_v I_k}{46.84 A_x} \qquad (8)$$

Depreciation and Labor

Both resin and equipment depreciation costs per year, A_d and E_d, are calculated from Eqs. (9) and (10). In both cases, the resin and equipment depreciation times, A_r and E_r, were assumed to follow a straight-line rate of depreciation. Normally, equipment is depreciated over a 10-yr. period, while resin depreciation would range from 3 to 5 yr. Anion-exchange resins usually have shorter lives than cation-exchange resins.

$$A_d = \frac{A_v, \ \text{cu. ft.} \times A_c, \ \$/\text{cu. ft.}}{A_r, \ \text{yr.}} = A_v A_c / A_r \qquad (9)$$

$$E_d = E_c, \$/E_r, \text{yr.} = E_c / E_r \qquad (10)$$

The cost of labor, L_c, is based upon the number of cycles, N; operator charges, L_r; and the time charged per cycle, L_t. This factor can be reduced to the main variables by substituting Eq. (4) into Eq. (11).

$$L_c = N, \ \frac{\text{cycles}}{\text{yr.}} \times L_t, \ \frac{\text{hr.}}{\text{cycle}} \times L_r, \ \frac{\$}{\text{hr.}} = N L_t L_r \qquad (11)$$

$$L_c = \frac{I_v I_k L_t L_r}{46.84 A_x A_v} \qquad (12)$$

Thus, the final Eq. (13) representing the total operating cost, O_c, is obtained by substituting Eqs. (8), (9), (10) and (12) into Eq. (1).

$$O_c = \frac{I_v I_k}{46.84 A_x} \left[R_r R_1 + \frac{L_t L_r}{A_v} \right] + \frac{A_v A_c}{A_r} + \frac{E_c}{E_r} \qquad (13)$$

In the case of an operation where one wishes to compare the operating cost of one resin to another in an identical piece of equipment, Eq. (13) can be rewritten as Eq. (14).

$$O_c = \frac{I_v I_k}{46.84 A_x} \left[R_r R_1 + \frac{L_t L_r}{A_v} \right] + \frac{A_v A_c}{A_r} \qquad (14)$$

Calculation Technique

In using Eq. (13) for a specific design, the known operating factors, I_v and I_k, and the three factors that are either known or estimated with relative accuracy —R_r, L_t and L_r—are all constant. Total operating cost, O_c, can be calculated for each of several values of the following factors:

1. Regeneration level, R_r.
2. Resin volume, A_v, and appropriate resin cost, A_c.
3. Assumed life of the resin, A_r.
4. Average resin capacity, A_x, which is a function of regeneration level and assumed life of the resin.

The calculated total operating costs, O_c, are then plotted as a function of resin volume with regeneration level as the parameter. This graph will yield a minimum point equivalent to the minimum total operating cost.

Two or Three-Bed System?

Let us assume we are building a deionization plant to treat 500 gpm. (720,000 gal./day) of water that has the analysis given in Table I. The problem is to determine whether a two-bed (cation and strong-base anion exchangers) or three-bed system (cation and weak- and strong-base anion exchangers) is the most economical. In both cases, a degasifier will be used. In the three-bed system, the degasifier will follow the weak-base unit. Since the size of the cation-exchange unit is common to both systems, the main economic difference will be between the cost of the anion-exchange units for both systems.

As shown in Table II, three regeneration levels (the common levels used industrially) were selected, and corrected resin capacities were computed based on the water analysis (Table I). To determine the effect of resin operating life, the optimum system for both 3 and 5 years was calculated. For the 3-yr. life, it was assumed that there was an average capacity loss of 20%/yr., or for the 3-yr. period that the average capacity was 70% of the original charge (1st year, 90%; 2nd year, 70%; and 3rd year, 50%). For the 5-yr. life, it was assumed that a drop in capacity of 10%/yr. for an average capacity of 75% of the original would be obtained. Thus, the average capacities, A_x, to be substituted into Eq. (13) are in Table II.

The cost of equipment, including costs for automatic controls and installation was estimated from published

---

Nomenclature for Eq.(13)

A_c	Resin cost, \$/cu.ft.
A_r	Resin depreciation time, yr.
A_v	Resin volume, cu.ft.
A_x	Resin capacity, average over-all life of resin in kilograins, (Kgr,)/(cu.ft.) (cycle)
E_c	Equipment cost, \$
E_r	Equipment depreciation time, yr.
I_k	Influent conc., ppm.
I_v	Effluent requirements, gal./day
L_r	Labor rate, \$/hr.
L_t	Labor time, hr./cycle
R_1	Regeneration level, lb./(cu.ft) (cycle)
R_r	Regenerant cost, \$/lb.
O_c	Total operating cost, \$/yr.

literature.* Fig. 1 illustrates the equipment cost for rubber-lined units as a function of bed volume.

To account for the increase in equipment cost from 1956 to 1965, the values obtained from Fig. 1 were multiplied by 1.25. Equipment was depreciated over a 10-yr. period and on a straight-line basis.

Labor time required for an operator on a given unit was based on requiring one operator for two hours during regeneration and one hour during exhaustion (for analytical purposes). Thus, $L_t = 3$ hr./cycle.

The minimum resin volume required was based on an exhaustant flow rate of 2 gpm./cu. ft. Thus, for a 500-gpm. plant, the minimum resin volume would be 250 cu. ft. To determine the effect of resin volume on the cost, volumes of 300, 350, 400, 450, and 500 cu. ft. were chosen.

For the strong-base unit in the two-bed system, the appropriate values were substituted into Eq. (13),

Chem. Eng. Prog. Jan. 1957, p. 37.

tabulated in Table III, and plotted as in Fig. 2.

For the three-bed system, separate calculations were made for both the strong-base and weak-base units. Because of different water compositions entering the strong base unit, its capacities were somewhat different than for the strong-base unit in the two-bed system (see Table II). The cost figures, O_c, for this system are compiled in Table IV.

By inspecting Fig. 2, the cost graphs for the two-bed system, it can be seen that the minimum cost for the strong-base unit is at a 4-lb. regeneration level; a resin volume of 250 cu. ft. for a 3-yr. life, and 300 cu. ft. for a 5-yr. resin life. For the three-bed system, a 4-lb. regeneration level and a resin volume of 250 cu. ft. seems appropriate.

A summary of optimum costs for the two systems is tabulated in Table V. Note that for the three-bed system, consideration has been given to the additional capital required for the weak-base unit. The cost required for the weak-base unit in the three-bed system

(Text Cont'd p. 476)

Typical water analysis for ion-exchange calculations—Table I

| | | Analysis (ppm. as $CaCO_3$) | | |
| | | Two-Bed System | Three-Bed System | |
	Raw Water	Strong-Base Influent	Weak-Base Influent	Strong-Base Influent
Alkalinity	70	5	70	5
Chloride	60	60	60	..
Sulfate	40	40	40	..
Silica	7	7	7	7
Total	177	112	177	12
Exchangeable anions		112	100	12

Calculations for resin capacity—A_x—Table II

Two-Bed System, Strong-Base Resin (IRA-402)

| Reg. Level, Lb. NaOH/Cu.Ft. at 120 F. | SiO_2-Cl Capacity, Kgr./Cu.Ft. | SiO_2-SO_4 Capacity, Kgr./Cu.Ft. | SiO_2-HCO_3 Capacity, Kgr./Cu.Ft. | SiO_2 Capacity, Kgr./Cu. Ft. | Capacity Corrected for Component Fraction, Kgr./Cu.Ft. | | | | | Resin Life, Yr. | Avg. Cap. After Loss, % | A_x, Kgr./Cu.Ft. |
					SiO_2-Cl	SiO_2-SO_4	SiO_2-HCO_3	SiO_2	Total			
4	12.0	13.8	20.5	14.3	6.4	4.9	0.9	0.9	13.1	3	70	9.2
6	14.0	15.8	22.7	16.8	7.5	5.6	1.0	1.1	15.2	3	70	10.6
8	15.5	17.4	23.9	18.3	8.3	6.2	1.1	1.1	16.7	3	70	11.7
4	12.0	13.8	20.5	14.3	6.4	4.9	0.9	0.9	13.1	5	75	9.8
6	14.0	15.8	22.7	16.8	7.5	5.6	1.0	1.1	15.2	5	75	11.4
8	15.5	17.4	23.9	18.3	8.3	6.2	1.1	1.1	16.7	5	75	12.5

Three-Bed System, Strong-Base Resin (IRA-402)

| Reg. Level Lb. NaOH/Cu.Ft. at 120 F. | SiO_2-HCO_3 Capacity Kgr./Cu.Ft. | SiO_2 Capacity, Kgr./Cu. Ft. | Corrected Capacity, Kgr./Cu.Ft. | | | Resin Life, Yr. | Avg. Cap. After Loss, % | A_x, Kgr./Cu.Ft. |
			SiO_2-HCO_3	SiO_2	Total			
4	17.0	14.3	7.1	8.3	15.4	3	70	10.8
6	19.5	16.8	8.1	9.8	17.9	3	70	12.5
8	20.8	18.3	8.7	10.7	19.4	3	70	13.6
4	17.0	14.3	7.1	8.3	15.4	5	75	11.5
6	19.5	16.8	8.1	9.8	17.9	5	75	13.4
8	20.8	18.3	8.7	19.7	19.4	5	75	14.5

Weak-Base Resin (IR-45)

Reg. Level Lb. NaOH/CuFt.	Orig. Cap., Kgr./Cu.Ft.	Resin Life, Yr.	Avg. Cap. After Loss, %	A_x Kgr./Cu.Ft.
4	28	3	70	19.6
4	28	5	75	21.0

Calculations for strong-base resin unit—Table III*

A_z	R_1	A_v	A_c	A_r	E_c	O_c
9.2	4	250	62.45	3	24,400	36,836
		300	62.45		27,000	37,014
		350	62.45		29,400	37,493
		400	60.00		31,700	37,836
		450	60.00		34,000	38,598
		500	60.00		36,000	39,423
10.6	6	250	62.45		24,400	42,725
		300	62.45		27,000	43,052
		350	62.45		29,400	43,636
		400	60.00		31,700	44,059
		450	60.00		34,000	44,883
		500	60.00		36,000	45,758
11.7	8	250	62.45		24,400	48,256
		300	62.45		27,000	48,674
		350	62.45		29,400	49,324
		400	60.00		31,700	49,795
		450	60.00		34,000	50,657
		500	60.00		36,000	51,563
9.8	4	250	62.45	5	24,400	32,967
		300	62.45		27,000	32,797
		350	62.45		29,400	32,909
		400	60.00		31,700	33,003
		450	60.00		34,000	33,394
		500	60.00		36,000	33,842
11.4	6	250	62.45		24,400	38,182
		300	62.45		27,000	38,160
		350	62.45		29,400	38,377
		400	60.00		31,700	38,550
		450	60.00		34,000	39,003
		500	60.00		36,000	39,501
12.5	8	250	62.45		24,400	43,575
		300	62.45		27,000	43,633
		350	62.45		29,400	43,907
		400	60.00		31,700	44,123
		450	60.00		34,000	44,609
		500	60.00		36,000	45,133

* Using Eq. (13). $I_v = 720,000$; $I_k = 112$; $R_r = 0.03$; $L_t = 3$; $L_r = 3$, $E_r = 10$.

Calculations for three-bed ion-exchange unit—Table IV

Strong-Base Unit*

A_z	R_1	A_v	A_c	A_r	E_c	O_c
10.8	4	250	62.45	3	24,400	10,309
		300	62.45		27,000	11,509
		350	62.45		29,400	12,714
		400	60.00		31,700	13,604
		450	60.00		34,000	14,791
		500	60.00		36,000	15,956
12.5	6	250	62.45		24,400	10,831
		300	62.45		27,000	12,043
		350	62.45		29,400	13,261
		400	60.00		31,700	14,158
		450	60.00		34,000	15,351
		500	60.00		36,000	16,521
13.6	8	250	62.45		24,400	11,387
		300	62.45		27,000	12,607
		350	62.45		29,400	13,829
		400	60.00		31,700	14,730
		450	60.00		34,000	15,926
		500	60.00		36,000	17,099
11.5	4	250	62.45	5	24,400	8,064
		300	62.45		27,000	8,852
		350	62.45		29,400	9,648
		400	60.00		31,700	10,255
		450	60.00		34,000	11,045
		500	60.00		36,000	11,813
13.4	6	250	62.45		24,400	8,535
		300	62.45		27,000	9,337
		350	62.45		29,400	10,143
		400	60.00		31,700	10,757
		450	60.00		34,000	11,553
		500	60.00		36,000	12,325
14.5	8	250	62.45		24,400	9,073
		300	62.45		27,000	9,881
		350	62.45		29,400	10,691
		400	60.00		31,700	11,309
		450	60.00		34,000	12,107
		500	60.00		36,000	12,882

* Using Eq. (13). $I_v = 720,000$; $I_k = 12$; $R_r = 0.03$; $L_t = 3$; $L_r = 3$, $E_r = 10$.

Weak-Base Unit†

A_z		A_v	A_c	A_r	E_c	O_c
19.6		250	62.45	3	24,400	19,878
		300	62.45		27,000	20,708
		350	62.45		29,400	21,653
		400	60.00		31,700	22,345
		450	60.00		34,000	23,379
		500	60.00		36,000	24,422
21.0		250	62.45	5	24,400	16,981
		300	62.45		27,000	17,426
		350	62.45		29,400	17,977
		400	60.00		31,700	18,400
		450	60.00		34,000	19,047
		500	60.00		36,000	19,701

$I_v = 720,000$; $I_k = 100$; $R_r = 0.03$; $R_1 = 4$; $L_t = 3$; $L_r = 3$; $E_r = 10$.

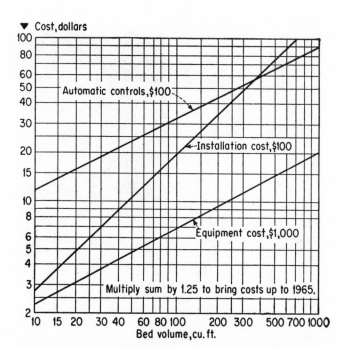

▼ Cost, dollars

Automatic controls, $100

Installation cost, $100

Equipment cost, $1,000

Multiply sum by 1.25 to bring costs up to 1965.

Bed volume, cu. ft.

Costs for ion-exchange units; equipment, installation, controls—Fig. 1

Summary of cost calculations—Table V

	Two-Bed		Three-Bed	
Resin life, A_r, years	3	5	3	5
Strong-base unit				
Optimum bed volume, cu.ft	250	300	250	250
Optimum reg. level, lb./cu.ft	4	4	4	4
Minimum cost, $/yr	36,836	32,797	10,309	8,064
Weak-base unit				
Optimum bed volume, cu.ft	......	......	250	250
Minimum cost, $/yr	......	......	19,878	16,981
Interest lost, $/yr*	......	......	1,193	1,019
Total cost, $/yr	36,836	32,797	31,380	26,064

* Because of the additional capital investment required for the weak-base unit in the three-bed system, it was assumed that this money could be invested at a rate of 6%/yr.

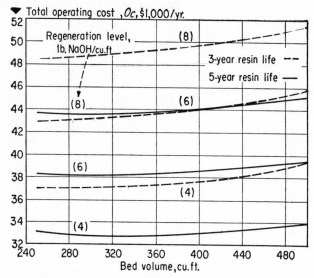

Operating costs for two-bed system—Fig. 2

Calculations for silica-removal problem—Table VI*

A_x	R_l	A_v	E_c	O_c
10.7	4	3.5	3,400	921
		5.0	3,400	786
		10.0	4,400	769
		20.0	5,400	905
12.6	6	3.5	3,400	850
		5.0	3,400	738
		10.0	4,400	748
		20.0	5,400	898
13.7	8	3.5	3,400	822
		5.0	3,400	720
		10.0	4,400	743
		20.0	5,400	899

* Using Eq. (13). $I_v = 1 \times 10^4$; $I_k = 10$; $R_r = 0.03$; $L_t = 3$; $L_r = 3$; $A_c = 62.95$; $A_r = 5$; $E_r = 10$.

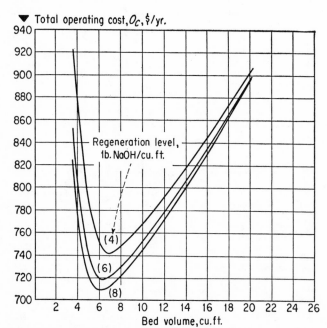

Optimum costs for SiO₂ removal—Fig. 3

could be invested at an interest rate of 6%/yr. if the two-bed system were used.

From the data in Table V, it can be seen that the optimum plant for this design, regardless of resin life, is the three-bed system.

Removing Silica

Assume it is desired to design the optimum size strong-base ion-exchange unit to treat 10,000 gal./day of a water supply that contains 10 ppm. SiO₂. For such a plant, let us apply the following factors:

I_v = 10,000 gal./day
I_k = 10 ppm. SiO₂ (as CaCO₃)
L_t = 3 hr./cycle
L_r = \$3/hr.
A_r = 5 years
A_v = 3.5, 5.0, 10.0 and 20 cu.ft.
R_r = \$0.03/lb.
A_c = \$62.95/cu. ft.
E_r = 10 years

Average over-all resin capacity, A_x, is computed as follows for the three regeneration levels:

Reg. Level Lb. NaOH/Cu. Ft. at 120 F.	Orig. Capacity, Kgr./Cu. Ft.	Avg. Cap. Loss, %	Resin Cap., A_{x3} Kgr./Cu. Ft.
4	14.3	0.25	10.7
6	16.8	0.25	12.6
8	18.3	0.25	13.7

Equipment costs, E_c, are obtained from Fig. 1. The above values are substituted in Eq. (13), as shown in Table VI, and operating costs, O_c, are calculated. From the results plotted in Fig. 3 it can be seen that the optimum volume for the strong-base ion-exchange unit is 6 cu. ft., regenerated at 8 lb./cu. ft.

Meet the Author

Donald G. Downing is a Group Leader in the Ion Exchange Applications Laboratory of the Research Dept., Rohm & Haas, Philadelphia, Pa. 19105. Since joining the company in 1948, he has specialized in the engineering and technical-service problems of ion-exchange resin applications. Mr. Downing is a graduate of the U. of Pennsylvania (B.S., Ch.E.) and holds an M.S. in chemical engineering from Villanova. He is a member of A.I.Ch.E.

Key Concepts for This Article

Active (8)	**Passive (9)**	**Means/Methods (10)**
Optimization*	Costs*	Equations
	Costs, operating*	Graphs
	Ion exchange	
	Ion exchangers*	
Dep. Var. (7)	**Ind. Var. (6)**	**Ind. Var. (6)**
Costs, operating	Labor Equipment	Regeneration Resins, ion exchange

(Words in bold are role indicators; numbers correspond to EJC-AIChE information retrieval system. Asterisks mark key concepts suggested for indexing. Others are added to improve reading as an abstract. Indexing is described in *Chem. Eng.,* Oct. 11, 1965, p. 187.)

Optimum Insulation Thickness– By Computer

Calculating optimum thicknesses for insulating pipes and vessels is time-consuming, tedious and costly. Here's how one company developed a computer program to do the job.

WESLEY J. DODGE, *Socony Mobil Oil Co., Inc.*

Simple in principle but time-consuming in practice, the selection of economical thicknesses for insulation is ideally suited to computer solution. True, there are many mathematical formulas and graphical methods (e.g., Stoever[1]) available for the problem (and these are still useful where only a few line sizes are being considered). But when the engineer needs insulation thicknesses for a large construction job involving many different pipe and vessel sizes with varying design wall temperatures, a more rapid and simpler method is almost a must.

Prior to adopting a computer solution, the Socony Mobil Engineering Dept. used hot and cold surface temperatures as the basis for insulation specification. While this was easy, it was recognized that if the most economical thicknesses were used it was by coincidence only. More often, we found ourselves in the somewhat illogical position of using the same insulation in locations where fuel costs were 20¢/MM.Btu. as in locations where fuel cost 65¢/MM.Btu.—a threefold difference. At the same time, we were also using the same thickness in tropical areas as in our most northerly refineries.

These and other drawbacks eventually led to a new set of objectives, around which our new thickness specification is built:

• It must permit competitive evaluation of any material (from any supplier) for use in any suitable service.

• It must be easy to understand and use.

• It must be adaptable for obtaining lump sum or unit price contracts on jobs of any size.

• It must provide an economical and practical selection of composition and thickness for each application, considering both capital and operating costs.

The basic approach we adopted in meeting these objectives involves adding the annual cost of capital investment in insulation to the annual cost of heat loss; the selected insulation, for any given case, is that which provides the lowest possible total annual cost.

The computer does not tell us which *type* of insulation to use—only the optimum thickness for the type chosen. Normally we run the program for several possible materials, and the results are included in the plant specifications. The contractor then selects the least expensive material on the basis of subcontractors' bids.

A Twelve-Variable Problem

The first step in developing the program was to list the twelve major variables involved, and then adapt them for use in machine calculations.

1. *Hot Surface Temperature:* The program is designed to handle only those temperatures between ambient and 1,200 F.
2. *Ambient Temperature and Wind Velocity:* Average annual values are used for the site being considered.

These twelve variables define the program:

1. Hot Surface Temperature
2. Ambient Temperature and Wind Velocity
3. Shape of Surface
4. Operating Hours
5. Types and Thicknesses of Insulation
6. Number of Layers
7. Thermal Conductivity
8. Installed Cost
9. Fuel Cost
10. Incremental Heater/Boiler Cost
11. Rate of Return on Investment
12. Maintenance Cost

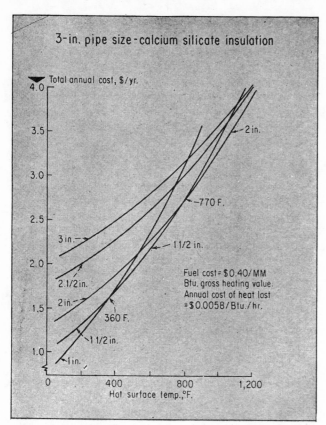

3-in. pipe size - calcium silicate insulation

Total annual cost, $/yr.

Fuel cost = $0.40/MM Btu. gross heating value. Annual cost of heat lost = $0.0058/Btu./hr.

Hot surface temp.,°F.

Optimum thickness for each pipe size is related to hot surface temperature. These curves are never plotted—instead, the computer locates the intersections—Fig. 1.

3. *Shape of Surface Covered:* The program caters to 18 pipe sizes and a flat surface. For simplicity, external metal temperatures are assumed the same as those of the contained liquid.

4. *Operating Hours:* Since it was apparent that we could afford to provide more insulation on steam than on process lines, separate cases were set up for each. Operating time for process lines (allowing for shutdowns): 8,000 hr. Operating time for steam lines: the full 8,760 hr.

5. *Insulation types and thicknesses:* Any suitable insulating material in commercial thicknesses for a particular job can be processed.

6. *Single or Double Layer:* 4-in. pipes and under (at all temperatures) use a single layer. Larger pipe sizes at temperatures over 800 F. must have a double layer.

7. *"K" Values:* The program uses a rationalized curve based on data provided by the Mellon Institute of Industrial Research and the insulation manufacturers.

8. *Installed Cost:* In conjunction with two insulation manufacturers, an estimate was made of installed costs of each thickness of each type of insulation commonly available for each of the pipe sizes, and for the flat surface included in the program. These costs include: material, freight, insurance, labor, fringe benefits, handling, overhead, and profit.

9. *Fuel Cost:* Current fuel cost at the site being considered is used; unless future changes can be definitely quantified in advance, they are not normally considered.

10. *Incremental Heater/Boiler Cost and Efficiency:* The greater the heat loss, the larger must be the heaters and boilers. Incremental cost of heater capacity is assumed as $1,800/MM.Btu. with an efficiency of 70% of gross heating value. Incremental boiler costs (including cost of water treating and distribution facilities) is assumed to be $5,000/MM.Btu. at 77% efficiency. These costs are put on an equivalent annual cost/Btu. basis and considered as part of the fuel cost.

11. *Rate of Return on Investment:* The "payout" method was adopted since savings that vary from year to year need not be considered. In selecting the payout period, however, lack of risk must be considered since savings are assured so long as the plant operates.

12. *Maintenance Cost:* This is ignored in the calculations because, for undisturbed, suitably specified insulation, it is small.

Program Formulation

The first stage of programming gives the computer some method of determining heat loss. This is a trial-and-error calculation based on two formulas for heat loss through insulation:

$$Q_I = \frac{T_a - T_b}{t/K}$$

where Q_I = heat loss through insulation
T_a = pipe wall temperature
T_b = insulation surface temperature
t = effective insulation thickness
K = thermal conductivity of insulation

$$Q_F = \frac{T_b - T_c}{1/C}$$

where Q_F = heat loss through surface air film
T_c = ambient temperature
C = air film coefficient

Since the quantity of heat passing through the insulation must equal that passing through the film, Q_I must equal Q_F.

In executing the program, the computer arrives at a value of T_b (which is within ± 1 F. of that which makes $Q_I = Q_F$) for each pipe size, temperature increment and commercial insulation thickness. Knowing the area/linear ft., the computer then calculates the heat loss/linear ft., and stores this information for later use in solving the economic problem.

In order to calculate heat losses in a given situation, both K and C must be formulated so that the computer can use them. Fortunately, both these values can be represented by linear equations:

$$K = A + B(T_a + T_b)/2$$
$$C = E + G(T_b - T_c)$$

Values of A and B for all common insulation materials, based on their K values, were given in a recent article by Chapman and Holland[2]; E and G are numbers based on the air film coefficient for the assumed wind velocity.

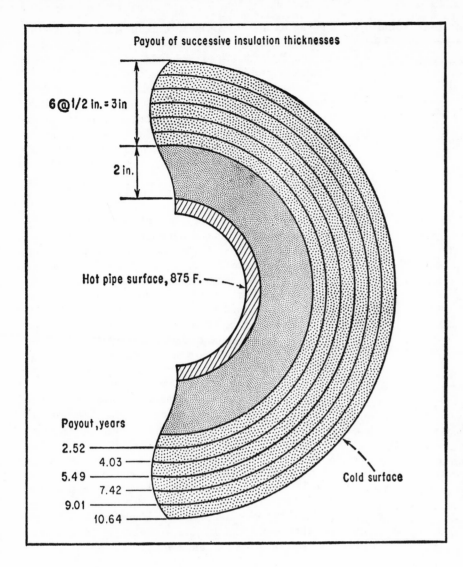

Payout of successive insulation thicknesses

6 @ 1/2 in. = 3 in

2 in.

Hot pipe surface, 875 F.

Cold surface

Payout, years

2.52
4.03
5.49
7.42
9.01
10.64

The economic effect of successively adding ½-in. thicknesses of insulation. Users satisfied with 10-yr. payouts should choose the 5-in insulation; those wanting their money back in 3 yr. must choose 2 in.—Fig. 2.

To calculate total annual cost, the program uses the relationship:

$$C_{ta} = ZC_u + QA_LC_F$$

where C_{ta} = total annual cost
Z = adjusted writeoff factor
C_u = unit cost of insulation
Q = heat loss/sq.ft.
A_L = outside area/ft. length
C_F = annual cost per Btu./hr. of heat.

ZC_u represents the annual cost of insulation; QA_LC_F is the annual cost of heat losses. Values for Z, the adjusted writeoff factor, are derived from:

$$Z = yW$$

where $W = \dfrac{1}{D} + \dfrac{1/P - 1/D}{1 - R}$

and y = correction factor—adjustment to over-all investment costs that includes contractor's overhead, area differentials, cost index changes, etc.
W = writeoff factor—the percent of investment cost needed annually to cover payout
D = straight-line depreciation rate (yr.)
P = desired payout after taxes
R = income tax rate

Annual cost of fuel to provide 1 Btu./hr. (adjusted to include cost of incremental heater or boiler for converting fuel into useful Btu.'s) is as follows:

$$C_F = \frac{\text{fuel cost/MM.Btu.} \times \text{hr./yr}}{\text{heater efficiency} \times 10^6} + \frac{\text{heater cost} \times Z}{10^6}$$

which becomes for:

Process lines $C_F = \dfrac{\text{fuel cost} \times 8,000}{0.7 \times 10^6} + \dfrac{1,800\ Z}{10^6}$

Steam lines $C_F = \dfrac{\text{fuel cost} \times 8,760}{0.77 \times 10^6} + \dfrac{5,000\ Z}{10^6}$

Selecting Optimum Thickness

Once the above formulas are written into the program, the computer is in a position to calculate total annual cost of any insulation on any pipe under any of the stated conditions. The next step is to select particular insulation thickness with the least total annual cost.

At this point in the program, the computer contains an array of numbers—total annual cost at various temperatures—for a specific set of conditions using several insulation thicknesses. If they were printed out and plotted, these numbers would give a set of curves such as those shown in Fig. 1. By in-

Typical thermal insulation data sheet—Table 1

Nominal Pipe Size In.	Nominal Insulation Thickness, In.				
	1	1½	2	2½	3
	Application Range Of Hot Surface Temp., F.				
½	156–787	788–1,200			
¾	141–650	651–1,050	1,051–1,200		
1	126–766	767–970	971–1,200		
1½	110–583	584–632	632–1,200		
2	102–457	458–772	772–1,095	1,095–1,200	
2½	96–334	335–872	872–1,180	1,180–1,200	
3	91–345	345–642	642–960	960–1,200	
4	88–187	187–625	625–910	910–965	965–1,200

spection, it can be seen that the lowest total annual cost is obtained when we use 1-in. insulation for temperatures up to 360 F.; similarly, 1½ in. should be used for higher temperatures up to 770 F., and 2-in. insulation for all temperatures above 770F.

In the actual program, rather than print out or plot these curves, the computer is simply instructed to locate the points of intersection. This is done by determining total annual cost at fixed intervals of temperature, noting whenever there is a reversal between two insulation thicknesses, and finally (by trial and error) locating the intersection to within ± 1 F.

What Kind of Input Is Needed

A design engineer using our program must complete the input forms to supply six kinds of information:

• Ambient condition factors—temperature and wind velocity constants E and G.

• Annual cost of 1 Btu./hr. of heat—data for both process lines and steam lines must be given.

• K-value constant A and B—types of insulation being considered.

• Z-factor—payout and corrections.

• Pipe diameters to be examined.

• Single or double-layer insulation.

Given this information, the computer will perform all the necessary calculations and print out the temperatures for each point of intersection where insulation must increase as the hot wall temperature increases. The relevant part of this information then becomes part of the job specifications.

A typical battery-limits job involves only about two to six data sheets for insulation specifications when this approach is used. One reason for this small number of sheets is that one tabulation can be used for more than one type of insulation. Also, some types of

insulation, for instance calcium silicate, asbestos fiber, and 85% magnesia are closely competitive. Values of A and B are assumed to be the same, and the same thickness requirements are used for each.

On large jobs, the program allows us to distinguish between the long straight runs of pipe outside battery limits and the complicated piping systems within the plant. For the latter, larger Z-factors can be used to cover the extra cost/linear ft. caused by the large number of fittings.

Payout of Successive Thicknesses

Fig. 2 illustrates the effect of selecting insulation thicknesses on the basis of economics. This cross-section of an insulated surface has a surface temperature of 875 F. and starts with a 2-in. thickness for a two-year payout. Adding ½ in. more insulation increases the payout to 2.52 yr., the next ½ in. gives a 4.03-yr. payout and the last ½-in. increment has a payout of 10.64 yr. If the company is satisfied with 10-yr. payouts, then 5-in. insulation should be used: if, however, this money can be invested elsewhere for a better return, the capital spent for the extra thickness is not a good investment even though it does show a return.

As with all studies leading to the development of a computer program, many man-hours were required to reach a successful conclusion. Nevertheless, using the program over the past few years has convinced us that the time was well spent. Not only has it reduced over-all insulation costs, but it has enabled us to invest our insulation dollars where they will do the most good, i.e. optimizing operating costs over the life of the unit to which the insulation is applied. This is true whether or not the line has to be heated from a separate source to insure pumpability.

References

1. Stoever, H. J., "Applied Heat Transmission," p. 204, Mc-Graw-Hill, New York, 1941.
2. Chapman, F. S., Holland, F. A., "Keeping Piping Hot," *Chem. Eng.*, Dec. 20, 1965, p. 79.

Meet the Author

Wesley J. Dodge is manager of the cost engineering section of Socony Mobil Oil Co.'s engineering dept. (150 E. 42 St., New York City). He has been with Socony for 24 yr. A civil engineering graduate of the University of Michigan, he is a licensed professional engineer in New York. This article was adapted from a paper given to the National Assn. of Cost Engineers, which body he has served as president (1964), vice-president (1963), and treasurer (1962).

Key Concepts for This Article

Active (8)	Passive (9)	Means/Methods (10)
Estimating	Insulating (Process)*	Computer*
Computing	Thickness*	
Optimization*	Costs*	

(Words in bold are role indicators; numbers correspond to EJC-AIChE information retrieval system. Asterisks mark key concepts suggested for indexing. Others are added to improve reading as an abstract. Indexing is described in *Chem. Eng.*, Oct. 11, 1965, p. 187.)

Some properties are only available on a lease basis; other times, the engineer must evaluate leasing vs. outright purchase. Methods outlined here provide the basis for making sound decisions.

When Should You Lease?

HERBERT E. KROEGER, *Huntington Station, N. Y.*

When a company wants to acquire new fixed assets, it frequently must make a choice between outright purchase or long-term leasing. At other times, desired land, buildings and some types of equipment (e.g., crude desalters and can fillers) can only be obtained on a lease basis.

The analyzing of leasing arrangements in the normal profitability study can create serious problems. If an engineer attempts to calculate a discounted-cash-flow (DCF) rate of return for a project and includes the rentals for leased components as operating expenses in the cash flows, he will seriously overstate the indicated profitability.

This article will set forth methods for evaluating leasing decisions on a basis comparable to the discounted-cash-flow rate of return that is normally calculated for purchased properties.

Leasing vs. Debt Financing

Leasing means acquiring the use of property for a period of time at an agreed rental without the benefit of ownership. If the property is depreciable and used up during the lease term, the rental consists of principal and interest. If it is nondepreciable and remains intact during the lease term, the rental consists solely of interest or economic rent.

Table I shows comparisons of leasing vs. debt financing (via bank loans) of a nondepreciable fixed asset. It was assumed in these illustrations that: (1) the depreciable and the nondepreciable fixed assets cost $100,000 each; (2) the useful life of the depreciable fixed asset is ten years, and the lease term for the two assets is for the same period; (3) the lessor wants to earn 6% on his money. Therefore, the lessor has to charge an annual rental of $13,587 for the depreciable

fixed asset and $6,000 for the nondepreciable one; (4) the company's current direct borrowing rate is 5%. Consequently the annual bank payment is $12,951 if the assets are debt-financed.

Obviously, in the case of depreciable properties, leasing is similar to debt financing because both the $13,587 annual rent payment and the $12,951 annual bank payments consist of repayment of principal and interest. Therefore, the total funds committed under the two financing methods are identical except for the $6,360 ($35,870 minus $29,510) higher interest expense of leasing compared with debt financing.

On the other hand, in the case of nondepreciable properties, leasing differs from debt financing because the annual rent payments consist of interest or economic rent only. In Table I, the annual rental is only $6,000 compared with the $12,951 annual bank payment. Consequently, the total amount committed for this nondepreciable property is $69,510, or 53.7% lower under leasing than debt financing. Despite this difference, leasing nondepreciable properties is nevertheless a financing device since it involves the use of someone else's money for which interest is paid in the form of lease rentals.

Comparative Costs: Leasing vs. Debt Financing

Under normal circumstances, depreciable fixed assets are financed more economically via bank borrowing than leasing because (a) the interest rate in lease rentals is generally higher than companies' current direct borrowing rate, (b) the timing of the tax savings is more advantageous under owning and debt-financing than leasing, and (c) no residual value for the assets accrues to the companies under leasing. Generally speaking, therefore, the profit indicator

(DCF rate of return, for example) for depreciable properties should be lower when they are leased than when they are owned and debt-financed.

Conversely, nondepreciable fixed assets such as land are normally financed more economically through leasing than debt-financing. The reasons are that the total funds committed for such properties are generally substantially lower under leasing than debt financing and that no tax benefits accrue to companies on nondepreciable fixed assets under ownership.

When to Lease

Although it is generally more economic to lease than to own nondepreciable properties such as land, companies nevertheless frequently favor purchasing assets primarily in order to:

• Continue operations at the location beyond the normal life of the facilities erected on the site.

• Avoid escalator clauses for upward adjustments of lease rentals.

• Profit from future appreciation of the land's market value.

Frequently, depending on local tax laws and business customs, companies can enjoy both the benefits that accrue to them as a result of ownership and those that come about as a result of leasing, by financing the land through some form of purchase, sale and leaseback arrangement.

On the other hand, while it is generally more costly to lease than to own and debt-finance depreciable fixed assets, leasing may be the recommended procedure when special circumstances produce benefits that outweigh the higher lease-financing cost. Such special circumstances could arise when the assets:

• Are subject to high obsolescence rates and, therefore, liable to rapid declines in market or resale value.

• Are expected to require substantial repairs by the manufacturer during their useful lives, and it is believed that more-efficient service will be obtained when the assets are leased instead of owned.

• Have useful lives that exceed the economic life of the over-all new project in which they will be used, and when their market or resale values will decline rapidly.

The above situations assume that buy versus lease alternatives exist. This isn't always the case, because frequently properties can only be leased. This situation creates problems in the project evaluation that will be discussed next.

Evaluating "Must Lease" Investments

Reasons why owners may be unwilling to sell properties include:

1. Land and other real properties frequently are considered good hedges against inflation.

2. Owners expect to profit from market appreciation that may result from population increases or shifts.

3. Local tax laws or the owners' tax positions favor leasing over selling.

Comparing cost of leasing vs. debt financing—Table I

LEASE FINANCING

1. Depreciable Fixed Asset

	(1)	(2)	(3)	(4)
	Lessor's Investment Balance	Rent Payments		
		Total	Principal Repayment	Interest at 6%
Method	(1) - (3)	$100,000 ÷ 7.3601	(2) - (4)	(1) X 6%
Year				
1	$100,000	$ 13,587	$ 7,587	$ 6,000
2	92,413	13,587	8,042	5,545
3	84,371	13,587	8,525	5,062
4	75,846	13,587	9,036	4,551
5	66,810	13,587	9,578	4,009
6	57,232	13,587	10,153	3,434
7	47,079	13,587	10,762	2,825
8	36,317	13,587	11,408	2,179
9	24,909	13,587	12,092	1,495
10	12,817	13,587	12,817	770
Total	0	$135,870	$100,000	$35,870

2. Non-Depreciable Fixed Asset

	(5)	(6)	(7)	(8)
	Lessor's Investment Balance	Rent Payments		
		Total	Principal Recovery	Interest at 6%
Method			(6) - (8)	(5) X 6%
Year				
1	$100,000	$ 6,000	—	$ 6,000
2	100,000	6,000	—	6,000
3	100,000	6,000	—	6,000
4	100,000	6,000	—	6,000
5	100,000	6,000	—	6,000
6	100,000	6,000	—	6,000
7	100,000	6,000	—	6,000
8	100,000	6,000	—	6,000
9	100,000	6,000	—	6,000
10	100,000	6,000	—	6,000
Total	$100,000	$60,000	—	$60,000

DEBT FINANCING

3. Depreciable or Non-Depreciable Fixed Asset

	(9)	(10)	(11)	(12)
	Bank Loan Balance	Bank Payments		
		Total	Principal Repayment	Interest at 5%
Method	(9) - (11)	$100,000 ÷ 7.7216	(10) - (12)	(9) X 5%
Year				
1	$100,000	$ 12,951	$ 7,951	$ 5,000
2	92,049	12,951	8,349	4,602
3	83,700	12,951	8,766	4,185
4	74,934	12,951	9,204	3,747
5	65,730	12,951	9,664	3,287
6	56,066	12,951	10,148	2,803
7	45,918	12,951	10,655	2,296
8	35,263	12,951	11,188	1,763
9	24,075	12,951	11,747	1,204
10	12,328	12,951	12,328	623
Total	0	$129,510	$100,000	$29,510

4. Small businessmen and others who are not entitled to pensions often acquire properties for retirement income.

5. Building contractors acquire properties to develop areas in which component service establishments are offered for lease only.

To determine the profitability and liquidity of proposed "must lease" investments, four factors enter into the rate of return and payout computations: (1) project cost, (2) project cash income, (3) project life and (4) timing of the cash flows.

The source of the funds (whether equity, debt or lease money) as well as the interest expense (if outside capital is used) are not factors to be considered in the economic evaluation of proposed new investments for two reasons:

Computing the investment equivalent of leasing a depreciable fixed asset—Table II

	1. Rate of Return Computed on the Purchase Price			2. Rate of Return Computed on Lease Rental Netted in the Project Cash Flow			3. Computing the Investment Equivalent		
	(1)	(2)	(3)	(4)	(5)	(6)	(7)	(8)	(9)
	Purchase Price and Tax Savings on Depreciation	Operating Income After Tax	Project Cash Flow	Rent Payments After Tax	Operating Income After Tax	Project Cash Flow	Rent Payments Before Tax	Discount Factor at 5%	Present Value
Method	Footnote a.	(3) - (1)	(1) + (2)	(7) X 50%	(2)	(5) + (4)	Table 1, Col. 2	Interest Table	(7) X (8)
Year									
0	$(100,000)	–	$(100,000)		–	0	–	–	–
1	10,000	$ 6,275	16,275	$ (6,794)	$ 6,275	$ (519)	$ 13,587	.9524	$ 12,940
2	8,000	8,275	16,275	(6,794)	8,275	1,481	13,587	.9070	12,323
3	6,400	9,875	16,275	(6,794)	9,875	3,081	13,587	.8638	11,736
4	5,120	11,155	16,275	(6,794)	11,155	4,361	13,587	.8227	11,178
5	4,096	12,179	16,275	(6,794)	12,179	5,385	13,587	.7835	10,645
6	3,277	12,998	16,275	(6,793)	12,998	6,205	13,587	.7462	10,139
7	3,277	12,998	16,275	(6,793)	12,998	6,205	13,587	.7107	9,656
8	3,277	12,998	16,275	(6,793)	12,998	6,205	13,587	.6768	9,196
9	3,277	12,998	16,275	(6,793)	12,998	6,205	13,587	.6446	8,758
10	3,276	12,999	16,275	(6,793)	12,999	6,206	13,587	.6139	8,341
Total	$(50,000)	$112,750	$ 62,750	$(67,935)	$112,750	$44,815	$135,870	–	$104,912
DCF Rate			10.0%			∞	–	–	–

a. Computed on $100,000 by Double Declining Method

	4. Determining the Amortization Amounts of the Investment Equivalent				5. Rate of Return Computed on the Investment Equivalent		
	(10)	(11)	(12)	(13)	(14)	(15)	(16)
	Investment Equivalent Balance Start of Year	Total	Amortization Amounts	Interest at 5%	Investment Equivalent & Tax Savings on Amortization	Operating Income After Tax	Project Cash Flow
Method	(10) - (12)	(7)	(11) - (13)	(10) X 5%	(12) X 50%	(2)	(14) + (15)
Year							
0					$(104,912)	–	$(104,912)
1	$104,912	$ 13,587	$ 8,341	$ 5,246	4,171	$ 6,275	10,446
2	96,571	13,587	8,758	4,829	4,379	8,275	12,654
3	87,813	13,587	9,196	4,391	4,598	9,875	14,473
4	78,617	13,587	9,656	3,931	4,828	11,155	15,983
5	68,961	13,139	10,139	3,448	5,069	12,179	17,248
6	58,822	13,587	10,645	2,942	5,322	12,998	18,320
7	48,177	13,587	11,178	2,409	5,589	12,998	18,587
8	36,999	13,587	11,736	1,851	5,868	12,998	18,866
9	25,263	13,587	12,323	1,264	6,162	12,998	19,160
10	12,940	13,587	12,940	647	6,470	12,999	19,469
Total	–	$135,870	$104,912	$30,958	$ (52,456)	$112,750	$ 60,294
DCF Rate	–	–	–	–			8.3%

1. The same principle that applies in measuring the productivity of total assets (in which interest expense is customarily added back to net operating income) applies also in assessing the profitability of proposed new investments. This means that in order to determine a new project's earning power, interest expense must be excluded from the cash flows.

2. The bulk of new investments is generally financed with equity money. Therefore, new investments are normally evaluated on an ex-financing (interest) cost basis in order to preclude prejudicing those that will be financed directly with outside capital.

The problem with calculating rates of return for "must lease" investments is twofold: (1) the capital cost in terms of lease rentals is spread over the entire investment life—thus, there is no lump-sum investment cost, as in purchased properties, on which to compute a rate of return; (2) the lease rentals include an interest factor, whereas the capital cost for purchased properties excludes interest expense. Consequently, to make possible the computation of rates of return for "must lease" investments comparable to those for purchased properties, it is necessary to convert the stream of future rent payments to a lump-sum investment that excludes interest.

This is accomplished by discounting, which has the effect of providing a lump-sum investment as well as extracting from the lease rentals interest expense that is equal to the rate at which the rentals are being discounted. The resultant capitalized rental or investment equivalent may then be used to compute a rate of return for the "must lease" property comparable to those for purchased properties. If the lease rentals include items incidental to ownership, such as taxes, maintenance and repair, etc., these must be excluded from the rentals before they are discounted. Such items should then be included in the new project's cash flow as operating expenses.

Determining the interest factor included in lease payments for a depreciable fixed asset—Table III

	(1)	(2)	(3)	(4)	(5)
	Rent Payments Before Tax	Discount Factor at 5%	Present Value	Discount Factor at 6%	Present Value
Method	Table 1 Col. 2	Interest Table	(1) X (2)	Interest Table	(1) X (4)
Year					
1	$ 13,587	.9524	$ 12,940	.9434	$ 12,817
2	13,587	.9070	12,323	.8900	12,092
3	13,587	.8638	11,736	.8396	11,408
4	13,587	.8227	11,178	.7921	10,762
5	13,587	.7835	10,645	.7472	10,152
6	13,587	.7462	10,139	.7050	9,579
7	13,587	.7107	9,656	.6651	9,037
8	13,587	.6768	9,196	.6274	8,524
9	13,587	.6446	8,758	.5919	8,042
10	13,587	.6139	8,341	.5584	7,587
TOTAL	$135,870	–	$104,912	–	$100,000

This procedure is illustrated in Table II, in which it is assumed that the depreciable fixed asset: (a) costs $100,000 to purchase, or $13,587 annually to lease for ten years, (b) has a ten-year life and (c) generates operating income after taxes as shown in Column 2. It is assumed further that the company's current prime rate for direct borrowing is 5%, compared with the 6% interest factor included in the $13,587 annual lease rental. As shown in Schedule 1, if this fixed asset is purchased for $100,000, the total profits are $62,750 and the DCF rate of return for this project is 10%.

On the other hand, Schedule 2 shows that if this fixed asset is leased for ten years, and the $13,587 annual rent payments are netted in the project cash flow, the total profits are reduced to $44,815 but the DCF rate of return for this project increases to infinity. Obviously this rate of return is fallacious, and another method must be found for calculating it when this fixed asset is leased, which is comparable to the 10% DCF rate of return in Schedule 1.

To enable us to do this, an investment equivalent must be computed, i.e., a lump sum investment amount excluding interest expense in lieu of the annual rent payments, on which this rate may then be calculated.

As shown in Schedule 3, this is accomplished by discounting the $13,587 annual rent payments at the company's current 5% direct borrowing rate. The resultant $104,912 total present value is the investment equivalent that may be defined as the amount of a bank loan the company could obtain at its current 5% direct borrowing rate if it were committed to make annual bank payments equal to the $13,587 annual rent payments. The difference between the $135,870 total rent commitment and the $104,912 investment equivalent represents the interest expense extracted from the future rent payments equal to the company's current 5% direct borrowing rate.

Schedule 5 shows that when the $104,912 investment equivalent and the tax savings on the amortization charges computed in Schedule 4 (Column 12) are substituted in Column 14 for the $13,587 annual rent payments, the DCF rate for this depreciable fixed asset when it is leased is 8.3%. This rate is comparable to the 10.0% rate in Schedule 1; it is 1.7% lower because of: (a) the $4,912 premium the company is paying for acquiring the depreciable fixed asset via the more expensive lease-financing method, and (b) the slower return of tax benefits under this method. The $4,912 premium represents the present value of the 1% difference in the interest expense between the 6% interest factor included in the $13,587 annual rent payments and the company's current 5% direct borrowing rate.

If the annual lease rentals are constant level amounts, separate computations of the annual amortization amounts as shown in Schedule 4 are unnecessary. The annual present value amounts in their reverse order in Schedule 3, Column 9, are identical with the annual amortization in Schedule 4.

Schedule 4 is included here only to show the prin-

ciple involved in amortizing the investment equivalent and the method to be used for calculating the annual amortization amounts when the annual lease rentals are uneven over the lease term. For the purpose of simplicity, this illustration assumes that items incidental to ownership are not included in the lease rentals, and that this fixed asset has no residual value at the end of its ten-year life.

The Rate to Use for Discounting

As mentioned previously, leasing is a financing device, and the interest factor in lease rentals is generally higher than companies' current direct borrowing rate. With new depreciable properties, for which buy vs. lease alternatives exist, a company's objective in normal situations is to acquire them via the lowest financing cost. The same principle applies to depreciable properties that are available for lease only.

The incremental lease-financing cost, therefore, must be reflected in the investment equivalents and, consequently, in the rates of return for "must lease" investments, because the latter compete for funds with investments that are equity- or debt-financed. As shown in Table II, this is accomplished by discounting the lease rentals for "must lease" investments at the company's current direct borrowing rate.

If parent companies, located in the U. S., operate on a worldwide basis through foreign affiliated companies, the question arises whether the U. S. borrowing rate or the foreign borrowing rate should be used for discounting. The latter rate is recommended because:

1. The bulk of all new projects including "must lease" investments normally is approved locally. To avoid confusion, therefore, all new investments, whether approved locally or by the parent companies in the U. S., should be discounted at the local rate.

2. Lease rentals reflect the local borrowing rate. Therefore, discounting at the foreign borrowing rate will result in investment equivalents that will be closer to the depreciable fixed assets' purchase prices than if a worldwide interest rate were used.

3. The earnings rates for various types of local investments will be geared to the local borrowing rate. Therefore, the rates of return calculated on investment equivalents that are arrived at by discounting at the local rate will be much closer to the earnings rates that companies are looking for in similar investments in the particular foreign country.

Buy vs. Lease Evaluations

The first step in evaluating proposed investments in which buy vs. lease alternatives exist is to determine whether they generate sufficient profits or savings to yield acceptable rates of return commensurate with the risks involved. These investment decisions are generally based on the rate of return and payout criteria calculated on the properties' purchase prices.

The second step is to determine whether the new properties should be leased instead of owned, providing the company is interested in some form of outside (lease or debt) financing.

If outside financing is desirable, the only analysis needed in most instances is to determine the interest factor included in the lease rentals. This is accomplished by discounting the lease rentals at various interest rates until the one is found that discounts the rentals down to a total present value equal to the pur-

Calculating the after-tax cost of leasing as opposed to owning—Table IV

	(1)	(2)	(3)	(4)	(5)	(6)	(7)	(8)
	Rent Payments		Purchase Price and (Tax Savings) on Depreciation	Incremental After Tax Cost (Savings) Leasing over Purchasing	Discount Factor at 3%	DCF Rate Present Value	Computation Discount Factor at 4%	Present Value
	Before Tax	After Tax						
Method	Table I, Col. 2	(1) X 50%	Table II, Col. 1	(2) - (3)	Interest Table	(4) X (5)	Interest Table	(4) X (7)
Year 0	—	—	$100,000	$(100,000)	1.000	$(100,000)	1.000	$(100,000)
1	$ 13,587	$ 6,794	$(10,000)	$ 16,794	.9709	$ 16,305	.9615	$ 16,147)
2	13,587	6,794	(8,000)	14,794	.9426	13,945	.9246	13,679
3	13,587	6,794	(6,400)	13,194	.9151	12,074	.8890	11,729
4	13,587	6,794	(5,120)	11,914	.8885	10,586	.8548	10,184
5	13,587	6,794	(4,096)	10,890	.8626	9,394	.8219	8,950
6	13,587	6,793	(3,277)	10,070	.8375	8,434	.7903	7,958
7	13,587	6,793	(3,277)	10,070	.8131	8,188	.7599	7,652
8	13,587	6,793	(3,277)	10,070	.7894	7,949	.7307	7,358
9	13,587	6,793	(3,277)	10,070	.7664	7,718	.7026	7,075
10	13,587	6,793	(3,276)	10,069	.7441	7,492	.6756	6,803
Total	$135,870	$67,935	$50,000	$ 17,935	—	$ 2,085	—	$ (2,465)
After-tax cost of leasing				3.46%				

chase price. The rate that does this is the interest factor included in the lease rental.

This is illustrated in Table III, in which it is assumed that the company has the alternative of buying a depreciable fixed asset for $100,000 or leasing it at an annual rental of $13,587 for ten years. When the annual rent payments are discounted at 5%, the total present value is $104,912 or $4,912 greater than the purchase price. This indicates that the interest factor in the lease rental exceeds 5%. On the other hand, when the rent payments are discounted at 6%, the total present value is equal to the $100,000 purchase price. This means that the interest factor in the lease rental is 6% and, assuming the company's current direct-borrowing rate is 5%, that this asset should be purchased instead of leased, providing depreciation is allowed as a tax deduction.

Even when the interest factor in the lease rental is identical with the company's borrowing rate, owning is more economical than leasing due to the tax advantage owning has over leasing, assuming that depreciation is recognized for tax purposes.

Consequently, no after-tax leasing-cost analysis is required unless: (a) the interest factor in the lease rental is lower than the company's current direct-borrowing rate (which occurs rarely), or (b) special circumstances prevail (as discussed earlier) which, if lease financing isn't too much higher than debt financing, could favor leasing.

The method for determining after-tax leasing cost is illustrated in Table IV, in which all assumptions are identical with those in Table III. The incremental after-tax cost of leasing this depreciable fixed asset for ten years at an annual rental of $13,587, as opposed to buying it for $100,000, is $17,935, or 3.46%. Assuming that the company's current direct-borrowing rate is 5% before taxes, or 2.5% after taxes, the depreciable fixed asset should be owned, unless the benefits that will accrue to the company as a result of leasing will exceed the additional after-tax cost.

The rate developed in Table IV is commonly misinterpreted to mean that the fixed asset should be leased because the company stands to earn only 3.46% if it were owned, compared with the 10% it might earn if the $100,000 were invested in another project.

This interpretation is incorrect because it confuses financing cost with rate of return. The 3.46% is an after-tax lease financing rate that should be compared to the company's after-tax direct-borrowing rate to determine whether the fixed asset should be leased or debt-financed if outside financing is desirable. The 3.46% is not an earnings rate for this fixed asset because the latter is computed on cash flows (Table II) that are based on ownership investment and operating income and exclude cash flows that are purely financial.

To avoid confusion with regard to making buy vs. lease decisions, it is necessary to adhere to this basic principle: leasing is merely another financing method, and under normal conditions new acquisitions should be financed at the lowest cost. For example, why should the company pay 6% interest, or $13,587 annually for ten years, to lease the depreciable fixed asset in this example (Table IV) when it can borrow money at 5%, buy the fixed asset and debt-finance it at annual payments of only $12,951 for ten years (Table I) and in addition enjoy the benefits of higher tax savings in the early years and the residual value of the asset at the end of the project's ten-year life.

Illusory Benefits of Leasing

Normally, the after-tax cost of leasing is higher than companies' current borrowing rate. Leasing depreciable fixed assets nevertheless is frequently believed to be more attractive for these reasons:

• To keep the long-term rent obligations and their related property rights off company books. This is deemed desirable to: (a) improve the company's over-all earnings rates based on total assets employed, total invested capital, etc., and to (b) enhance the company's general credit standings.

• To have the lease rentals for the fixed assets appear in the project cash flows as operating expenses instead of as integral parts of the project cost. This yields higher rates of return for prospective projects.

• To keep the commitments of funds for new fixed assets out of company's annual capital expenditure budgets, which are generally confined to purchased properties. This keeps the capital expenditure budget low, and obtains management's approval more easily.

It is my opinion that these so-called leasing advantages are nonexistent because lending institutions and investors in industrial stocks and bonds have long been accustomed to making adjustments in their analyses and evaluations to compensate for undisclosed financial facts before making loan or investment decisions. Therefore the practice of not showing leased fixed assets and their related long-term rent obligations in the body of published reports serves no useful purpose. Likewise, showing lease rentals in project cash flows as operating expenses, instead of capitalizing them and including them in the investment base of new projects, could conceivably lead to many wrong decisions.

Key Concepts for This Article

(Words in bold are role indicators; numbers correspond to AIChE information retrieval system. Indexing details are described in *Chem. Eng.*, Jan. 7, 1963.)

Meet the Author

Herbert E. Kroeger is a financial analyst in the budgeting department of a major oil company. His work involves the preparation and analysis of financial data for management reports and studies. Prior to his current assignment, Kroeger was a methods analyst. He holds a B.S. degree in accounting from New York University. Kroeger lives at 3 Arizona Place, Huntington Station, N. Y.

Economics of Long- vs. Short-Life Materials

When is it economical to select relatively inexpensive, short-life materials of construction? When should you choose more-expensive materials that will last longer? Two charts help in making these decisions.

J. R. BRAUWEILER, *Research and Development Dept., American Oil Co., Whiting, Ind.*

One of the most vexing problems that usually must be faced in the design and construction of process plants centers around the economic choice of materials of construction.

A plant may be built of corrosion-resistant (usually more expensive) materials to decrease or eliminate product contamination and maintenance expense. These require a larger capital investment. Alternatively, less costly (usually less corrosion resistant) materials can be used, involving a more frequent maintenance expense that is immediately deductible for tax purposes.

All chemicals plants are designed to be shut down periodically for inspection and rehabilitation. Production lost during shutdown is part of the cost of doing business. Of course, materials failure may result in an unplanned shutdown or an extension of a planned shutdown. If so, the added cost of this loss of production, which is properly charged against the materials failure, frequently overshadows the savings resulting from using lower cost materials in the first place.

Balanced against this is a possibility of something unexpected happening to cause serious deterioration or failure in the more resistant materials. This is particularly true in any new process about which not too much is known.

Currently, the capitalized cost of an original installation is depreciated over its life at a rate set by the Government. However, costs of subsequent replacements, as long as they do not upgrade the equipment, are looked upon as repairs to help the equipment reach that life, and are expended (i.e., are deductible for tax purposes in the year incurred), even though they last more than a year. Thus, current tax laws and rates can profoundly influence materials selection.

Taking these factors into account, we have developed a method that will assist in decisions involving selection between long-life and short-life components (see charts, p. 130).

With these curves, if any three of the four variables —(1) earnings rate, (2) life of long-life alternative, (3) life of short-life alternative, (4) ratio of the costs of the two alternatives—are known, the fourth can be read directly. Thus when the life of the second alternative is unknown, as usually is the case, the two charts can be used to determine what the life of this alternative will have to be to break even.

For example, a given piece of equipment can be built either of carbon steel with a relatively high corrosion rate so it has an expected life of two years or of a low-alloy steel that costs twice as much. If the required earnings rate is 15%, the chart indicates that the low-alloy steel will have to last more than six years to be economical. However, for an earnings rate of 25%, the low-alloy steel would have to last considerably more than 15 years, on the same basis.

Basis for Charts

As an aid to correct use of the curves, we will give a brief summary of how the charts were constructed.

In each alternative, cost of the first installation, I, is considered to be capitalized and to be depreciated over d years. This gives a total depreciation credit of $I \times t$ in terms of net cash flow if the tax rate is t.

The present worth of this total depreciation credit, compounding the interest continuously,[1] is $I \times t \times D_d$ where D_d is the present worth factor for declining cash flows of duration d years, given by

$$\frac{2}{Rd}\left[1 - \frac{(1 - e^{-Rd})}{Rd}\right] = D_d$$

where R is the earnings rate, as a decimal, and d is the depreciable life of the project in years.

Each replacement or succeeding installation then costs $(I + E)$ where E is the added expense of replacement, such as removal of existing worn-out components, cost of downtime in excess of that normally scheduled, etc. These succeeding installations are expensed for an immediate tax credit $(I + E)\,t$. Thus, the net after-tax cost of the first and later replacements is: $I - (I\,t\,D_d) + (I + E)(1 - t)[e^{-XR} + e^{-2XR} + e^{-3XR} + \cdots \cdots e^{-(n-1)XR}]$ where e^{-XR} is the present worth factor for X years at a rate R; X is the expected life of each installation; and n is the total number of installations.

This can be written as:

$$(I + E)[(1 - t)(1 + e^{-XR} + e^{-2XR} + \cdots \cdots) - (tD_d) + t] - E(1 - tD_d)$$

To compare two alternatives, say one of long-life (subscript L) and one short-life (subscript S), the total present worths of the two programs are equated. To simplify matters somewhat, the reasonable assumption is made that the added expense of replacement, E, is the same for the two alternatives. Then,

$$(I_L + E)[(1-t)(1 + e^{-LR} + \cdots \cdots) - (tD_d) + t] - E(1 - tD_d) = (I_S + E)[(1-t)(1 + e^{-SR} + \cdots \cdots) - (tD_d) + t] - E(1 - tD_d)$$

The last term on each side cancels. It is then con-

Charts help in selecting long-life or short-life alternative

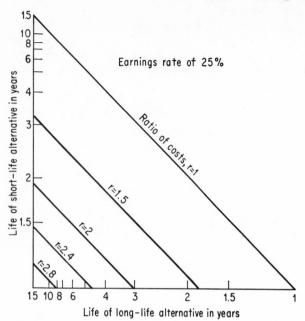

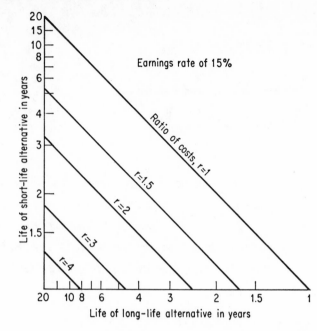

venient to write this equation in terms of the cost ratio, r, of the two alternatives and to evaluate the series term

$$\sum_{i=0}^{n-1} e^{-iXR} = \frac{1 - e^{-nXR}}{1 - e^{-XR}}$$

For a large number of installations:

$1 - e^{-nXR} \cong 1$, so: $r = \dfrac{I_L + E}{I_S + E} = \dfrac{(1-t)[1/(1-e^{-SR})] + t(1-D_d)}{(1-t)[1/(1-e^{-LR})] + t(1-D_d)}$

This ratio is easier to evaluate if the formula is rearranged to use the discount factor, C, for cash flows occurring continuously over a period of years after the zero or reference point.[2] Such factors are given by the expression $C = (1 - e^{-RT})/RT$ where R is the earnings rate, as a decimal, and T is the time in years over which the cash flows. Therefore $1/(1 - e^{-RS}) = 1/(C_s R S)$ and $1/(1 - e^{-LR}) = 1/(C_L R L)$. Substituting,

$$r = \frac{I_L + E}{I_S + E} = \frac{[(1 - t)/(C_s S)] + Rt(1 - D_d)}{[(1 - t)/(C_L L)] + Rt(1 - D_d)}$$

The relationship between the cost ratio of the two alternatives and the expected life of the long-life and the short-life alternatives is shown on the charts for typical earnings rates of 15% and 25% and a tax rate of 54%.

These earnings rates are only intended as guides because decisions regarding acceptable values for each project must be decided by management either as a general policy or individually. And, of course, the curves do not apply if some unexpected breakdown occurs other than that considered in constructing the charts.

Nomenclature

C	Discount factors for uniform cash flows for X yr.
E	Added expense of replacement, $
I	Cost of first installation, $
R	Earnings rate
T	Time for cash flow, yr.
X	Expected life of each installation, yr.
D_d	Discount factors for declining cash flows for d yr.
d	Depreciable life of equipment or component, yr.
n	Total number of installations
r	Ratio of cost, long-life component/short-life component
t	Tax rate
S	Short-life component or material, yr. of life.
L	Long-life component or material, yr. of life.

References

1. Gregory, J.G, "Interest Tables for Determining Rates of Return," The Atlantic Refining Co., Philadelphia, Pa., 1946.
2. Weaver, J.B., Reilly, R.J., *Chem. Eng. Prog.*, Oct. 1956 (for values of C and D factors).

Key Concepts for This Article
For indexing details, see Chem. Eng., Jan. 7, 1963, p. 73. (Reprint No. 222).
Words in bold are role indicators; numbers correspond to AIChE system.

Active Concept (8)
Selection
Comparison

Passive Concept (9)
Materials of construction
Costs
Alternatives
Return on investment

Means/Methods (10)
Economic evaluation
Investment analysis
Present worth

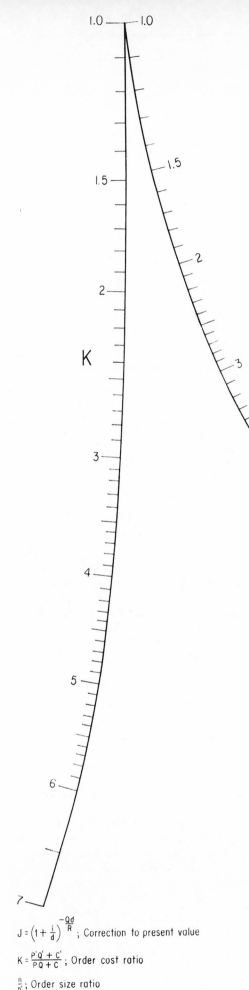

$$J = \left(1 + \frac{i}{d}\right)^{-\frac{Qd}{R}} \text{; Correction to present value}$$

$$K = \frac{P'q' + c'}{PQ + c} \text{; Order cost ratio}$$

$$\frac{n}{n'} \text{; Order size ratio}$$

Evaluating Quantity Discounts

Should you take
a quantity discount?
Here's how to
calculate
the cost of money
tied up in
larger inventories.

GEORGE E. MAPSTONE
Dermacult S.A. (Pty.) Ltd.
Johannesburg, South Africa

It is seldom simple for a purchaser to decide whether he is justified in placing an order for a large quantity of a commodity, so that he can take advantage of a larger discount or lower unit price. While he can save on his actual cash outlay, per unit purchased, by taking advantage of the discount, he has to balance this saving against the disadvantage of having the necessary working capital tied up in extra stocks. The need of arriving at the correct decision becomes more important if this additional capital has to be borrowed at high interest rates.

The decision regarding the size of the order is usually based on custom, i.e., size of the previous order for the same materials, or the size of an order for similar materials. Other factors that may be of importance are the persuasiveness of the salesman, and the current cash situation. For example, orders are frequently curtailed near the end of the financial year or when tax demands or other large cash commitments are to be met.

Obviously, there is an optimum size of order that gives the lowest unit cost when the price discount offered is balanced against the interest costs involved on the additional capital outlay. This article presents an impartial method for determining the savings that may arise from the quantity discount.

Cost of Order

The cost of money borrowed to pay for an order is the interest charged on the loan. If the capital is on hand, the cost is the interest that could be obtained

by employing the cash in making investments or loans.

In order to eliminate variations due to the interest rate, we can base our calculations on the *present value* of the order; this is the amount of money that would have to be invested at the current rate of interest to equal the purchase price at the time of purchase.

Cost of Placing an Order

The cost of placing and paying for an order can be a significant portion of the cost of the material purchased, amounting to $8 per order in many cases. Although this amount may be insignificant with large or infrequent orders, it should not be overlooked with small or frequent orders.

This cost is seldom if ever isolated and is usually absorbed in general office overhead. It is, nevertheless, a cost that is incurred and, as such, it reduces the profits of the company. In a few offices, one may hear the comment, "If they didn't have these orders to handle they would have nothing to do." Such offices are undoubtedly overstaffed, and savings in the form of reduced costs are not welcome. This article is not for this minority of firms.

The composition and amount of the cost of placing and handling each order will not be considered further but will be assumed to be C dollars, incurred effectively on the date of payment of the order.

Additional Costs of Larger Orders

Normally, the main factor in deciding whether to place a larger order is the reduced price. However, the larger orders sometimes involve additional warehousing or storage charges. When incurred, these additional charges should be included in C, the cost of placing and handling the order.

The present value of n successive orders for Q units each is as follows (assuming constant stock usage):

Order	Total Cost†	
1st	$PQ + C$	
2nd	$(PQ + C)\left(1 + \dfrac{i}{d}\right)^{-Qd/R}$	$= (PQ + C)J$
3rd	$(PQ + C)\left(1 + \dfrac{i}{d}\right)^{-2Qd/R}$	$= (PQ + C)J^2$
nth	$(PQ + C)\left(1 + \dfrac{i}{d}\right)^{(1-n)Qd/R}$	$= (PQ + C)J^{(n-1)}$

Summing gives the total present cost of the nQ units so purchased:

$$T = (PQ + C)\frac{(1 - J^n)}{(1 - J)} \qquad (1)$$

Now, if Q units can be purchased at a price of $P/$unit while Q' units can be purchased at a price of $P'/$unit, any saving will be the difference between the present costs of ordering the same total quantity in units of Q and Q', or where:

$$nQ = n'Q' \qquad (2)$$
$$(n \text{ and } n' \text{ are always integers})$$

This saving will be given by:

$$S = (PQ + C)\frac{(1 - J^n)}{(1 - J)} - (P'Q' + C')\frac{(1 - J'^{n'})}{(1 - J')} \qquad (3)$$

Now $J = \left(1 + \dfrac{i}{d}\right)^{-Qd/R}$

and $J' = \left(1 + \dfrac{i}{d}\right)^{-Q'd/R}$

From Eq. (2)
$$Q' = Q^{n/n'}$$

whence

$$J' = \left(1 + \frac{i}{d}\right)^{-(Qd/R)(n/n')}$$

i.e., $J' = J^{n/n'}$ \qquad (4)

Eliminating J' between Eq. (3) and (4) gives

$$S = \left(\frac{PQ + C}{1 - J} - \frac{P'Q' + C'}{1 - J^{n/n'}}\right)(1 - J^n) \qquad (5)$$

If S is positive, the savings will be made by purchasing the quantity Q' instead of Q on each order. On the other hand, if S is negative, the quantity Q is the more economic order level. In either case, the saving will be the absolute value of S.

If $S = 0$, neither order level shows any advantage over the other. This condition means that the interest rate is such that the interest charges on the additional capital just balance the quantity discount. In this case, Eq. (5) reduces to:

$$\frac{PQ + C}{1 - J} = \frac{P'Q' + C'}{1 - J^{n/n'}} \qquad (6)$$

whence

$$\frac{P'Q' + C'}{PQ + C} = \frac{1 - J^{n/n'}}{1 - J} = K \qquad (7)$$

Solution of Eq. (7) for J, and hence for i, will then give the effective interest rate that will just balance the discount offered on the larger quantity.

The number of orders, n, for the smaller quantity Q will be larger than the number of orders, n', for the larger quantity Q'. Also n and n' will be integral and will usually bear a simple relationship to one another. By writing:

$$J = F^{n'} \qquad (8)$$

Eq. (7) becomes:

$$\frac{1 - F}{1 - F^{n'}} = K \qquad (9)$$

which can be expanded in descending powers of F to give:

$$F^{(n-1)} + F^{(n-2)} + \ldots + F^{n'} + (1 - K)F^{(n'-1)} + (1 - K)F^{(n'-2)}$$
$$+ \ldots + (1 - K)F^2 + (1 - K)F + (1 - K) = 0 \qquad (10)$$

Eq. (10) can be solved for F but the method to be adopted depends on the values of n and n'. In general, the equation is best solved for values of n above three by using Horner's method* to find the root between zero and one. However, since n and n' are integral quantities and usually in simple ratio to one another, the effect of substituting common ratios in Eq. (9) and (10) gives the following special cases:

† See nomenclature table on last page.
* An explanation of Horner's method for solving polynomial equations may be found in many texts, e.g., Sokolnikoff, I. S., and E. S., "Higher Mathematics for Engineers and Physicists," 2nd edition, McGraw-Hill Book Co., 1941.

n	n'	Equation	Eq. No.
2	1	$F = K - 1$	(11)
3	1	$F = -\frac{1}{2} + \frac{1}{2}\sqrt{4K - 3}$	(12)
3	2	$F = \frac{1}{2}(K - 1) + \frac{1}{2}\sqrt{K^2 + 2K - 3}$	(13)
4	1	$F^3 + F^2 + F + (1 - K) = 0$	(14)

Eq. 14 can be solved by the general methods applicable to cubic equations, or by Horner's method.

Example 1

Should an order be placed for 1,000 gross of oil cans at $12.50/gross or for 500 gross at $14.50/gross, when the annual requirement is 1,500 gross? The larger order will not affect warehouse costs. Money is worth 8%/yr., compounded monthly.

Then: $Q = 500$; $Q' = 1,000$; $P = 14.50$; $P' = 12.50$; $i = 0.08$; $d = 12$, i.e., compounded 12 times a year; and $R = 1,500$. Since no data are given for the costs of placing and handling the orders, C and C' are neglected.

We can now calculate the savings that arise from buying 1,000 gross at a time, rather than 500 gross.

From Eq. (2): $500n = 1,000n'$
Since n and n' should have the smallest possible integral values in order to simplify the calculations, put $n' = 1$; then $n = 2$. Then:

$$J = \left(1 + \frac{i}{d}\right)^{-Qd/R}$$
$$= (1 + 0.08/12)^{-500(12/1,500)}$$
$$= 0.973769$$

From Eq. (5), we have:

$$S = \frac{14.5 \times 500}{1 - 0.973769} - \frac{12.5 \times 1,000}{1 - 0.973769^2}(1 - 0.973769^2)$$
$$= \$1,826.10$$

That is, the purchase every eight months of 1,000 gross cans, instead of 500 gross each four months at the higher price, will effect a saving of $1,826.10 each eight months.

In order to calculate the effective interest rate on the extra financial outlay equivalent to this saving, solve Eq. (7) for J and then for i. Since $n = 2$ and $n' = 1$, Eq. (11) can be applied.

$$F = J = K - 1$$
$$K = \frac{12.5 \times 1,000}{14.5 \times 500}$$
$$= 1.724$$
$$J = 0.724$$
$$= \left(1 + \frac{i}{12}\right)^{-4}$$
$$i = 1.044 \text{ or } 104.4\%$$

The saving possible from placing an order for the larger quantity, therefore, involves a saving equivalent to an annual interest of 104.4% on the additional outlay. In other words, if the cash must be borrowed, any interest rate less than 104.4%, compounded monthly, is an economic proposition in this case.

The above example is a relatively simple though common one and can be checked as follows: expenditure on 1,000 gross at $12.50/gross is $12,500; ex-

penditure on 500 gross at $14.50/gross is $7,250; additional expenditure on larger order is $5,250. The present value of the $7,250 that will need to be spent in four months time, if the smaller order is placed with money worth 8% compounded monthly, is:

$$\$7,250\left(1 + \frac{0.08}{12}\right)^{-4} = \$7,076.10$$

but the additional amount actually spent on the larger order is $5,250, hence the saving of $1,826.10.

This saving is due to the additional investment made at the time of the first purchase, and four months prior to the time the second purchase would have had

Solution of Eq. (7) for Example 3

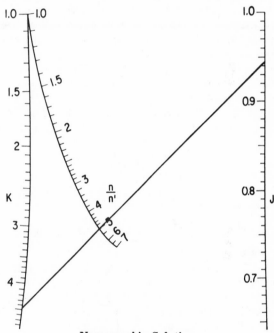

Nomographic Solution

In order to reduce the labor involved in solving Eq. (7) by Horner's method, the accompanying nomograph was designed. Since the design required some simplifying assumptions, it is not quite so accurate for the higher values of K and n/n' as the mathematical method, but is sufficiently accurate for most purposes. This is shown by the value of J calculated in the worked examples as compared with the values obtained from the chart:

Example No.	K	n/n'	J (Calc.)	J (Chart)
1	1.724	2.0	0.724	0.724
2	4.76	5.0	0.985	0.980

Attempts to design an extension to the chart to allow the direct solution of i from J and Qd/R were unsuccessful. However, since the methods of calculating present values are well known and straightforward, it is felt that the single chart presented on the first page of this article should be adequate for most purposes.

to have been made. The effective interest rate can then be computed as follows:

$$7,250 = 5,250\left(1 + \frac{i}{12}\right)^4$$
$$i = 1.044 \text{ or } 104.4\%$$

This check confirms the applicability of the method in the frequently occurring simple case where the order size was doubled to take advantage of the quantity discount.

Example 2

An additive is used at the rate of 350 gal./mo. and is purchased in drums containing 45 gal. The cost is $5.50/gal., in lots of 10 to 49 drums and $5.25/gal., in lots of 50 or more drums. The cost of placing, handling and paying for an order is $8, and money is worth $6\frac{3}{4}\%$, compounded monthly. Storage is adequate to handle a 50-drum order without incurring extra expense. Is there any advantage in purchasing 50 rather than 10 drums and, if so, what is the extent of the advantage?

Here: $C = 8$; $d = 12$; $i = 0.0675$; $P = 5.50$; $P' = 5.25$; $Q = 450$; $Q' = 5 \times 450 = 2,250$; and $R = 350 \times 12 = 4,200$.

Since the alternative orders are for 10 or 50 drums of additive, we can write $n = 5$ and $n' = 1$. Then:

$$J = \left(1 + \frac{0.0675}{12}\right)^{-450(12/2,250)}$$
$$= 0.986622$$

The saving that can be realized is then given by Eq. (5) as:

$$S = \frac{5.50 \times 450 + 8}{1 - 0.986622} - \frac{5.28 \times 2,250 + 8}{1 - 0.986622^5}(1 - 0.986622^5)$$
$$= \$26.75$$

In this case, the order for the larger quantity is favored by the narrow margin of only $26.75 on an expenditure of $11,812.50. It is therefore an advantage to know the effective interest rate of this saving. This will be obtained by the solution of Eq. (10) after substituting the values $n = 5$ and $n' = 1$ to give:

$$J^4 + J^3 + J^2 + J + (1 - K) = 0$$
$$K + \frac{5.25 \times 2,250 + 8}{5.50 \times 450 + 8}$$
$$= 4.76057$$
$$\text{i.e., } J^4 + J^3 + J^2 + J - 3.76057 = 0$$

Solution by Horner's method gives the root between zero and one as $J = 0.98502$, whence $i = 0.07568$, or approximately 7.6%. Consequently, should the interest rate increase from $6\frac{3}{4}\%$ to slightly over $7\frac{1}{2}\%$, it would be advisable to order the smaller lots of 10 drums each.

Example 3

A chemical intermediate is purchased in 200-lb. lots at $15.90/lb. Orders for 1,000 lb. cost $14.30/lb. What annual consumption would justify the larger order? The cost of placing and handling an order is

estimated to be $9 and the cost of money is 12%, compounded monthly. Then: $P = 1,000$; $P' = 200$; $C = C' = 9$; $i = 12$; $Q = 14.30$; $Q' = 15.90$; $n/n' = 5$; and $d = 12$

$$K = \frac{14.30 \times 1,000 + 9}{15.90 \times 200 + 9} = 4.487$$

Using the nomograph with this value of K, and $n/n' = 5$, gives $J = 0.944$. Substituting the various values in the expression for J yields:

$$0.944 = 1.01^{-12,000/R}$$
$$R = 2,064$$

Therefore, it is more economical to order 1,000 lb. at a time, if the annual consumption exceeds 2,064 lb.

Nomenclature

C	Cost of placing and handling an order.
d	Number of interest determinations per year.
F'	$(J)^{1/n'}$ [see Eq. (8)].
i	Interest rate as a decimal (6% = 0.06).
J	$(1 + i/d)^{-Qd/R}$, correction from future to present value.
K	$\dfrac{P'Q' + C'}{PQ + C}$, ratio of costs of orders.
m	Number of periods of interest determinations between consecutive orders.
n	Number of orders.
P	Price per unit purchased.
Q	Number of units per order.
R	Rate of usage, units/yr.
S	Savings.
T	Total cost of a series of orders.

Units with primes (') refer to alternative orders.

Meet the Author

GEORGE E. MAPSTONE *is chief chemist and factory manager of Dermacult S.A. (Pty.) Ltd., Johannesburg, South Africa. He was born and educated in New Zealand, receiving his B.Sc. and M.Sc. at the University of Canterbury. He later obtained his Ph.D. from the University of South Africa where he recently received the degree of Bachelor of Commerce.*

Although he is now in the cosmetics field, most of his previous experience has been in the manufacture of gasoline from shale oil, both in Australia and South Africa.

Dr. Mapstone is a Fellow of the Royal Economic Society and a member of many societies concerned with chemistry, chemical engineering, petroleum and fuels.

Investing in Major Spare Parts

Buying a major spare part is as much an investment as purchasing bonds or new plant facilities. Using investment concepts such as cash outlay, future value, rate of return and risk can make major parts expenditures more profitable.

H. ELGEE, *Lignosol Chemicals Ltd.*

When deciding whether or not to buy a major spare part, an engineer should ask and answer these basic questions:

• How much will it cost?
• Will it ever be used?
• How much will be saved if it is used?

If money is not available, perhaps because of other capital commitments or a business recession, the part may not be bought, regardless of how profitable buying it appears. If money is available, two possibilities that arise from the second question must be carefully considered.

A spare part might never be used because the item that it is supposed to replace might not fail. This could occur because of a tendency to play it safe (ordering spare parts for items not really showing signs of failure), or because the obsolescence of the machine or the process for which the part is intended was not anticipated, or because of other miscalculations.

Or, a spare part may be needed that was never acquired. The failure to have expensive spare parts on hand is usually due to an unwillingness to invest large sums of money in parts that are not obviously necessary.

In either case, money is lost. Besides the loss of the investment in the unused spare part, expenses may also be incurred in purchasing, storing, inventorying and in giving protective maintenance. On the other hand, not having the part to begin with can cause sales losses and costly emergency repairs.

Hedging Against Risks

J. C. Enevoldsen[1] suggests that the risks in spare part investment decisions may be lessened by spreading the availability and cost of spare parts throughout several companies, or throughout divisions of the same company.

He proposes setting up a mutually financed, geographically proximate, parts depot to stock spare parts commonly used by the member companies.

Capital is better employed by avoiding investments in duplicate spare parts by each of the compaines. More-efficient operation is also possible because spare parts that the members individually would not stock would be available.

The difficulty of finding an acceptable economic yard-

stick to measure the value of such a capital investment might be a stumbling block to the establishment of this type of service organization, Enevoldsen points out.

However, the problem may not require such a solution. It is the purpose of this article to suggest a method for evaluating the desirability of stocking major spare parts.

There are two types of situations involving spare parts. In one, the part is acquired, kept intact and used intermittently to provide uninterrupted process operation when equipment fails. In this case, the part may be used many times and, from an accounting point of view, is considered as capital equipment. The situation can be handled with the normal techniques of evaluating capital projects. Standby pumps, jets and generators are examples of this type of capital spare part.

In the other situation, the part is carried in stock and on spare-parts inventory accounts until such time as it is used to replace a broken or worn out part. At this time, the value of the spare part is transferred from inventory and charged to plant operation as a maintenance expense. Major items like cylinders or pistons for reciprocating compressors, fan runners, or large drive-gear sets are examples of this type of spare part, and it is with this type of parts situation that this article deals.

Money is invested in spare parts with the intention of recovering not only the amount invested but something extra. Spare parts buying should, therefore, be considered an investment opportunity having all of the elements of a true investment.

These elements are: cash outlay, future value, time, rate of return, risk and intangibles.

How these elements are related to investments in spare parts is now described:

Cash Outlay—The cash investment is, of course, the delivered cost of the spare part. The storage and handling charges for the part and the interest charges on the investment money should not be counted as cash outlay but as an expense that reduces future savings.

Future Value—The future value of the investment is the savings that accrue through having the spare part available when it is needed. When calculating savings, the potential loss to profit and overhead of lost sales, as well as the direct cost of idle and non-productive machine time, should be included. As noted, these savings will be reduced by storage and handling charges, as well as by interest charges.

Time—The time, or term, of an investment is the period between the time of payment for the part and the time the part is used. Term has a pronounced effect on the calculated interest rate, which in turn strongly influences the decision of acquiring a spare part. If an inspection and data-processing system enables good forecasting of when the spare will be used, calculating the interest rate may be precise. If the time of failure could be accurately forecast, it would, of course, only be necessary to acquire a part just before it were needed. Since such forecasting is not likely, spare parts

must be ordered well in advance of the anticipated date of use.

Rate of Return—The rate of return on an investment in a spare part is the excess of future savings over the capital cost during the term of the investment. In comparing the value of investing in alternative projects, rate of return is an important criterion, and several methods of comparing it have been described in the literature.[2, 3, 4]

To calculate the rate of return on a spare part investment, the familiar present-worth type of formula may be used:

$$C = \frac{S}{(1+i)^n}$$

where C = delivered cost of the spare part; S = future net savings; n = time between the acquisition and the use of the spare part; i = interest rate necessary to make future savings equal to the present delivered cost of the spare part.

A set of interest tables makes the formula easier to use.

The cost and potential savings of acquiring spare parts are usually known, or can be calculated. Management usually establishes the rate of interest above which capital investments will be considered and below which they will be deferred. Most doubt exists about when the spare will be required. But, if the acceptable rate of interest, the capital cost and the potential savings are known, the value of n necessary to get the rate of return can be calculated. The engineer must now determine if the part will be required within the calculated time.

For example, a main drive shaft may have failed three times in five years. If the failures were at the end of the first year, the fourth year and the fifth year, an engineer would find it difficult to determine accurately when a spare shaft would be required. If, knowing the cost and the potential savings, he calculated that the acceptable interest rate would be earned if the spare shaft were used at any time up to the sixth year, he must now find out if there were a good chance of using the part within six years. If he would be justified in believing that the part would be used, it could be bought with confidence that the investment would earn at least the acceptable rate of return.

When interest charges decrease future savings, the present worth formula can be modified to enable direct calculation of n:

$$n = \frac{\log S/C}{\log (1+i_1)(1+i_2)}$$

where i_1 = cutoff rate of interest acceptable to management; i_2 = interest rate charged to the funds invested in the part, including storage and handling charges.

Risk—In evaluating investments in capital projects, risk is a difficult factor to assess. Hawkins and Martin[5] propose a method of capital evaluation that takes into account the vagaries of estimates that plague evaluations. The method is lengthy and requires many calculations. Ross *et al*,[6] suggest that capital projects

be rated simply as low, medium or high risks. But whether evaluating the risk is complex or simple, the results reflect that some doubt exists about the return on the investment, even about whether the investment will be recovered at all.

The risk involved in losing an investment in a spare part, however, can be evaluated. It depends upon whether or not the part is likely to be used, which in turn depends upon the probability of failure. Risk varies inversely with this probability.

Combining modern inspection techniques with a data storage and retrieval system takes a lot of the guess work not only out of deciding whether a part will be required but also when it will be required. Good inspection and data storage and retrieval systems make investing in spare parts about the least risky of industrial investments. Since a lessening of risk carries with it a reduction in the rate of return required to make the investment attractive, the acceptable interest rate for an investment in a spare part should be lower than for other investments.

But, no matter how accurate is the forecast of failure, it will be of little value if the process for which the part is intended becomes obsolete before the failure occurs. Rates of wear, corrosion and failure data should be compared with rate of obsolescence before the forecast is made.

Intangibles—Intangible factors such as customer service and public image should not be overlooked when considering a spare part investment. A breakdown that causes shipping delays might give rise to customer dissatisfaction and a permanent loss of business. Or the failure of a critical component in a waste disposal system, while not essential to direct operation of the plant, might result in atmospheric or water pollution, and the possibilities of law suits for property damage or public nuisance charges. Although these factors are difficult to assess, they should be considered, since they may be the governing ones.

Example Problem

A plant for a new product is designed around a novel processing system. A critical part of the system is a 15,000-cu.ft./min., 10-psig., centrifugal compressor. Because the plant is new and the markets for its product are undeveloped, stocking major spare parts is not considered necessary.

At the end of the first year of operation, the compressor is inspected, and a small, but measurable, amount of corrosion is seen on the compressor rotor. The rate of corrosion indicates the rotor might have a further service life of ten years.

Because a new rotor will cost $30,000, the old one is not badly corroded, and the plant has ample excess capacity for existing sales, no immediate action is taken.

When the next annual overhaul rolls around and the compressor is opened again, a distinct etching is found on the rotor, and the estimate is that it will not last more than five years. While a replacement will clearly be required eventually, the time seems distant,

so a decision is deferred until the next annual inspection.

Another year passes, and the appearance of the rotor is not much different. The only sign of change is that a dynamic balancing of the unit is required and some metal has to be ground off the hub. Since some change in the rotor has taken place to cause an imbalance and since there is no sign of a buildup of extraneous deposit, the conclusion is that the imbalance was caused by a corrosive loss of material.

During the three-year interval, markets for the product have expanded, so that the plant is operating at 90% of rated capacity. Growth patterns indicate that sales will exceed plant capacity sometime during the next year. An engineering study indicates that plant capacity can be increased by approximately 15% if the process is operated at a slightly higher pressure. To achieve this, however, the operating speed of the compressor will have to be increased. But higher compressor speeds will accelerate the corrosion rate.

The dilemma is whether to buy a spare compressor rotor now or to wait until the next annual inspection.

The following data are assembled and an estimate is made of the potential profitability of stocking a rotor:

Capital Cost—The delivered cost of a rotor has now increased to $35,000, and the best delivery is given as three months. The capital investment in the rotor is therefore taken as $35,000.

Term—Finding the term of this investment presents a problem because it is difficult to predict exactly when the spare will be required. If the corrosion rate accelerates, the rotor might be needed within a year. If the rate stays the same, however, the rotor should not be required for another four years.

Future Savings—If there is a failure and no spare is available, plant output will be reduced by about 75% for the three months it will take to get delivery of a new rotor. The loss in contribution to profit and overhead is calculated using the normal profit on the amount of sales that will be made if no breakdown happens. This amount is calculated at $82,000. To this is added $30,000 in processing cost for the extra work and the emergency measures that will be required to maintain plant ouput at 25% of its rated level. Storage and handling charges are calculated at 1½% per annum. Interest charges on the investments are compounded at 6½%. These interest charges are included as an expense because, if the $35,000 is not invested in the spare, this amount will be deducted from the company's current bank loan on which interest is being charged at 6½% per annum.

Cost of breakdown without spare rotor:

Loss in contribution to profit and overhead due to loss in sales	=	$82,000
Increase in emergency manufacturing costs	=	30,000
Total cash loss if no spare available	=	$112,000

Cost of stocking rotor:

| Cost of storing and handling | = 1½%/yr. |
| Cost of interest on investment | = 6½%/yr. |

Total cost of rotor = 8% of purchase cost/yr.

Risk—Because the corrosion is apparently continu-

ing, the rotor will eventually have to be changed. Also, since the intended operation at a higher running speed will probably increase the corrosion rate, the rotor will likely be required within four years. Since the estimate is that the process cannot become obsolete for at least eight years, the risk of losing the investment is small.

Time—The rate of return being used by the company at this time as the cutoff rate for capital investments is 18%.

Using the formula:

$$n = \frac{\log S/C}{\log (1 + i_1)(1 + i_2)}$$

and substituting the following values: $S = 112,000$, $C = 35,000$, i_1 = cutoff rate = 0.18, i_2 = storage and interest rate being charged = 0.08, n is calculated as 4.8 years.

It seems likely that the rotor will be required within four years and that the investment will therefore earn something in excess of the minimum rate of return required by the company for its other investments.

Intangibles—In evaluating the intangibles, this investment is compared with the investment required to expand the production facilities. The following data are compiled:

Capital Requirement—$60,000 is required for a new compressor drive, a new starting gear, and additional processing equipment.

Cash Generated—If the productive capacity is increased 15%, and if all of this extra product can be sold, net over-all contribution to profit and overhead will be $20,000 per year.

Term—The life of this venture is taken as eight years, not because the venture will cease to exist in eight years but because competitive products and potential substitute make profitability obscure so far ahead.

Risk—Sales have shown a steady growth and it is reasonably certain that growth will continue at least at the projected rate. But the intrusion of competitive products of lower quality and price are threatening to cause some loss of sales. Otherwise, the investment on expanded production facilities is considered reasonably safe.

Rate of Interest—To calculate the effective interest rate, a method outlined by Edge[2], using the capital recovery formula, is applied:

$$C = R \div \frac{i(1+i)^n}{(1+i)^n - 1}$$

where C = capital invested; R = annual cash generated; n = life of product; i = return on investment. For this case,

$$\$60,000 = 20,000 \div \frac{i(1+i)^8}{(1+i)^8 - 1}$$

and solving, $i = 29\%$.

Superficially it appears profitable to invest in expanded production facilities. However, the uncertainties of the competitive situation, and the possibility that projected sales (and consequent profits) will not expand at the assumed rate, also indicate that expan-

sion might not be as safe a course as buying a spare rotor.

Moreover, if the rotor is found to be structurally weak on the next inspection, and the unit has to be shut down for lack of a replacement, there may be a permanent loss of sales because of dissatisfied customers. If the investment in the expanded facilities is to be protected, a spare rotor is all the more necessary.

Based on these considerations, the decision is made to acquire a spare rotor and to inspect the compressor at more frequent intervals.

Conclusion

Spare parts are necessary to minimize cash outflow, and earning investments are used to bring in cash. Any business uses both type of investments.

The choice between spare parts and other investments cannot always be based wholly upon economic considerations. But if cash outflows, such as spare parts investments, are treated similarly to investments in bonds or production facilities, a step will be taken toward more profitable investment decisions.

Acknowledgments

The author wishes to thank W. G. Stanley for his assistance and for suggesting the use of investment term at acceptable rates of interest to evaluate this type of situation, and P. Ryan for his help in arranging the mathematics.

References

1. Enevoldsen, J. C., Stocking of Capital Spares, *Pulp and Paper Magazine of Canada*, 66, 1, Jan. 1965, pp. 76-77.
2. Edge, C. G., 'The Appraisal of Capital Expenditure," The Society of Industrial and Cost Accountants of Canada, Hamilton, Ont.
3. Weaver, J. B., "Venture Analysis," *Chem. Eng. Progr.* Technical Manual, AIChE, 1961.
4. Ruel, R. I., *TAPPI*, 43, 5, May 1960, p. 40A.
5. Hawkins, H. M., Martin, O. E., *Chem. Eng. Progr.*, 60, 12, Dec. 1964, pp. 58-63.
6. Ross, J., others, *Chem. Eng.*, Aug. 17, 1963, pp. 145-149.

Meet the Author

H. Elgee became production manager of Lignosol Chemicals Ltd., Quebec, Que., in 1960, after eight years as the plant superintendent. He was a research chemist for the Anglo-Canadian Pulp and Paper Mills, and the supervisor of a spent pulping recovery pilot plant for the Gaspesia Paper Co. He graduated with a B. Sc. in chemical engineering from Queen's University in 1947 and is a member of the AIChE, the Chemical Institute of Canada, the Canadian Pulp and Paper Assn. and the Corporation of Professional Engineers of Quebec.

Key Concepts for This Article

Active (8)	Passive (9)	Means/Methods (10)
Purchasing	Parts	Equations
Evaluating	Spares	Analysis
Comparing	Investments	

Dep. Var. (7)		Ind. Var. (6)
Rate of return		Costs
Savings		Obsolescence
		Risks (cost)

(Words in bold are role indicators; numbers correspond to AIChE information retrieval system. Indexing details are described in *Chem. Eng.*, Jan. 7, 1963.)

When is Electricity Cheaper Than Steam for Pipe Tracing?

Electricity is challenging steam's dominance in pipe tracing. Lower rates, precise thermostats and better insulations have added to its past advantage of low-cost maintenance. Here are methods for finding which will be cheaper.

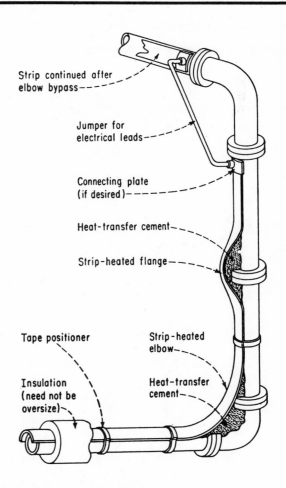

Strip continued after elbow bypass

Jumper for electrical leads

Connecting plate (if desired)

Heat-transfer cement

Strip-heated flange

Tape positioner

Insulation (need not be oversize)

Strip-heated elbow

Heat-transfer cement

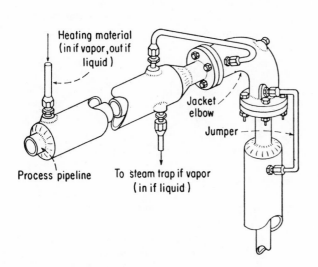

Heating material (in if vapor, out if liquid)

Jacket elbow

Jumper

Process pipeline

To steam trap if vapor (in if liquid)

C. H. BUTZ, *E. I. du Pont de Nemours & Co.*

Tracing pipes with steam has for economic reasons generally been chosen over tracing with electricity, except for those specialized processes that require critical temperature control. Now, however, the variables that once favored steam for ordinary processes have turned about and are giving economic impetus to the growth of electrical systems.

The cost of electrical power has continued to tumble; accurate temperature controllers and thermostats have become readily available; and better electrical insulation materials have made the components of pipe-tracing systems more reliable.

Electrical systems have always had attractive main-

For a survey of the different types of pipe tracing, see: Keeping Piping Hot, *Chem. Eng.,* Dec. 20, 1965, p. 79 and Jan. 17, 1966, p. 133

tenance advantages. It is the improved engineering and lower electrical energy costs of recent years that have brought electrical pipe tracing to the point where, for pipe temperatures of less than 140 F., the energy costs for electrical tracing will usually be lower than for steam.

This conclusion can be demonstrated by first defining what the energy costs for heating process pipe electrically are, and secondly by comparing these costs to those of steam-heated systems—by means of either graphs or formulas into which the pertinent data for particular circumstances may be put.

That it costs two to four times more to produce a Btu. with electricity than with steam is generally accepted. It is much less widely known, however, that thermostatic control, along with other innovations, has often made it possible to heat pipes more cheaply with electrical power, despite the disparity in cost between the two ways of producing a Btu.*

Graphical Method of Finding Costs

Graphs such as Fig. 1 can be used to either (1) determine whether the cost of electrical tracing is competitive with that of steam when the cost of producing a Btu. by both methods is known, or (2) find out what

---

*The procedures and formulas in this article were derived from W. H. Holstein's "What It Costs to Steam and Electrically Trace Pipelines," *Chem. Eng. Prog.,* March 1966, p. 107.

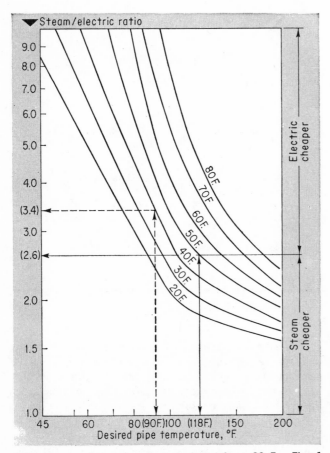

With flow in pipe and system designed for −20 F.—Fig. 1

the actual operating cost is to heat a foot of pipe by each of these systems.†

These graphs (Fig. 1 and those for other pipe diameters) were based on the following assumptions: (1) pipes of less than 8 in. dia. have 1-in.-thick thermal insulation whose thermal conductivity is 0.35; (2) pipes of 8-in. and larger diameter are insulated with 1½-in. material having a thermal conductivity of 0.35; (3) both systems are designed to operate satisfactorily at temperatures as low as −20 F.; (4) the steam system operates at 267 F. (25-psig. steam) and has 15% excess capacity, which enables it to handle sudden drops in temperature below −20 F.; (5) the electrical system is also overdesigned by 15%, but this power is only consumed when it is called for by the thermostat; finally, (6) process fluid is flowing through the pipe.

Examples of Graphical Method

The following example illustrates how energy costs can be calculated from these graphs.

Take the cost of steam to be 80¢/1,000 lb., and electricity 0.75¢/kwh. The heat of vaporization of 25-psig. steam is 930 Btu./lb. Therefore, 930,000 Btu./hr. generated by steam will cost 80¢. Similarly, 1 kwh. produces 3,413 Btu., so the cost to turn out 930,000 Btu./hr. electrically will be $2.05. Producing a Btu. with electricity thus costs 2.6 times more than with steam.

However, because energy consumption will be closely regulated, thermostatic control will substantially reduce the cost of generating heat with electricity.

Taking the case for the 3-in. pipe of Fig. 1, the line across the graph at the 2.6 steam-electric ratio divides those applications in which pipes could be heated more cheaply with electricity (above the line) from those that could be heated more economically with steam (below the line).

If the average ambient temperature in the plant were 50 F., the crossover point between steam and electrical tracing for a 3-in. pipe through which process fluid is flowing would be 118 F. (Fig. 1). In other words, it would be cheaper to depend on electrical tracing when the process fluid in the pipe were to be maintained at some temperature below 118 F.

How much it will cost to heat a pipe of a specified diameter to a certain temperature can also be determined from Fig. 1, as is shown in this second example.

Assume that the process fluid in a 3-in.-dia. pipe must be kept at 90 F., and that the average temperature when the tracing will be in service is 40 F.

Locate 90 F. on Fig. 1 and continue upward in a line until the 40 F. parameter is touched. At this juncture, the steam/electric ratio is 3.4—which in this case means that the heat generated by the steam will be 3.4 times more than necessary.

How many Btu.'s will be consumed can be found by means of Fig. 2. With the process at 90 F. and the ambient temperature at 40 F., the temperature dif-

---

† Because of space limitations, only one graph—that for the 3-in.-dia. pipe—is shown. For similar graphs of 1, 1½, 2, 4, 6, 8, 10 and 12-in. pipes, address requests to: W. P. Bithell, Fluorocarbons Div., Plastics Dept., Du Pont Co., Wilmington, Del. 19898.

498

Steps in the Calculation of Annual Costs for Pipe Tracing

Steam Tracing:

1. Calculate the thermal conductance required of the steam tracing at the coldest ambient temperature.

$$C_{t(min.)} = 1.15C_i \frac{(t_p - t_{a(min.)})}{t_s - t_p}$$

2. Find C_t, the actual thermal conductance of the steam tracing, by selecting the equal or the next larger value to the $C_{t(min.)}$ in Table III.

3. Calculate the annual energy cost for steam tracing from the following formula:

$$\text{Annual Cost} = C_t SLT(t_s - t_p)/1,000h$$

Electrical Tracing:

1. From the following formula, calculate the annual energy cost:

$$\text{Annual Cost} = C_i ELT(t_p - t_a)/3,413$$

---

ferential will, of course, be 50 F. From Fig. 2, the heat output that will be required from the electrical tracing will be 20 Btu./hr. per ft. of this 3-in. pipe. With steam costing 80¢/930,000 Btu., and electricity $2.05/930,000 Btu., the cost to heat 1 ft. of this 3-in. pipe for 1 hr. will be 0.0047¢ with steam, and 0.0044¢ with electricity.

If this difference seems negligible, consider what it would cost to keep 10,000 ft. of this pipe at 90 F. for a whole year. The savings for one year could be as much as $259.20 (24 hr. × 360 days × 10,000 ft. × 0.0003¢), not including the money that would be saved because there would be no worn steam traps to replace or leaks to repair.

Mathematical Methods

Steam and electrical energy tracing costs can also be calculated by means of the formulas in the box and the data in Tables I, II and III.

The water that will flow through 12,000 ft. of 3-in.

Nomenclature

C_i	Thermal conductance of pipe insulation, Btu./hr./ft.
D	Pipe diameter, in.
E	Cost of electricity, ¢/kwh.
h	Heat of vaporization of steam, Btu./lb.
L	Length of pipe being traced, ft.
S	Cost of steam, ¢/1,000 lb.
T	Time per year that tracing will be in service, hr.
t_a	Average ambient temperature, °F.
$t_{a(min.)}$	Minimum ambient temperature, °F.
t_p	Process temperature, °F.
t_s	Steam temperature, °F.

pipe is to be kept from freezing by having its temperature maintained at a minimum of 50 F. for twenty-four hours a day, five months of the year. While the average ambient temperature during this period is expected to be about 28 F., the temperature could drop to as low as −10 F.

Steam will be available at 25 psig. (267 F.) for 80¢/1,000 lb. The local rate for electricity is ¾¢/kwh. The pipe is to be insulated thermally with 1-in.-thick asbestos.

What would be the annual energy costs for electrical and steam tracing? Follow the steps of this example:

$D = 3$-in.	$E = 0.75$¢/kwh.
$t_p = 50$ F.	$S = 80$¢/1,000 lb.
$t_a = 28$ F.	$h = 930$ Btu./lb.
$t_{a(min.)} = -10$ F.	$L = 1,200$ ft.
$t_s = 267$ F.	$T = 3,600$ hr.

Thermal conductivity of insulation materials—Table I

Type of Insulation	Thermal Conductivity, k Btu./(hr.)(sq.ft.)(°F/in.)
Asbestos (corrugated, 4 plies/in.)......	0.49
Calcium silicate (hydrous)...........	0.37
Diatomaceous earth (with asbestos)...	0.66
Glass (cellular).....................	0.39
Glass felts (fiber)...................	0.25
Magnesia (85%)....................	0.41
Polyurethane (expanded, cellular).....	0.17
Vegetable cork....................	0.30

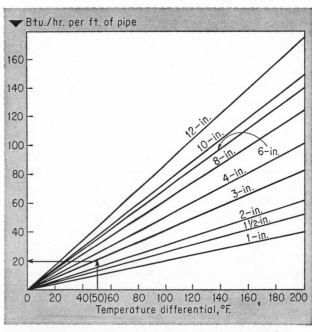

Differential is process temperature less ambient—Fig. 2

The thermal conductance of pipe insulation—Table II

$$C_i = (C_i/k)k$$

Pipe Size, In.	Insulation Thickness, In.	C_i/k
1	1.0	0.566
1½	1.0	0.728
2	1.0	0.856
3	1.0	1.158
4	1.0	1.424
6	1.0	1.752
8	1.5	1.984
10	1.5	2.126
12	1.5	2.478

The thermal conductance of steam tracing—Table III

Tubing Size, In.	Thermal Conductance, C_t Btu./Hr./Ft. Without Heat-Transfer Cement	With Heat-Transfer Cement
1⅜	0.393	3.44
1½	0.524	4.58
1⅝	0.654	5.73
2⅜	0.786	6.87
2½	1.047	9.16
2⅝	1.309	11.46
3½	1.571	13.75
4⅝	2.62	22.9
5⅝	3.27	28.6
6⅝	3.93	24.4

Taking data from Tables I and II, calculate the thermal conductance of the pipe insulation:

$$C_i = 1.158 \times 0.47 = 0.568 \text{ Btu./(hr.)(ft.)}$$

Electrical tracing: annual energy cost = (0.568) (50 − 28)(0.0075)(1,200)(3,600)/(3,413) = $119.

Steam tracing: $C_{t(min.)}$ = (1.15)(0.428)[50—(−10)]/(267 − 50) = 0.139 Btu./hr.-ft. From Table III, C_t = 0.393. The annual energy cost for steam tracing will be: (0.393)(267 — 50)(0.80)(1,200)(3,600)/(1,000)(930) = $317.

For a second example of the mathematical method, assume that 500 ft. of 10-in. pipe must be heated to 210 F. for fifty weeks of the year. Take the average ambient temperature for the year at 70 F. and the minimum at 15 F. calcium silicate, 1½ in. thick, will be used to thermally insulate the pipe.

Plant steam will be furnished at 25 psig. for 85¢/1,000 lb. Electricity will cost 0.70¢/kwh. Process liquid will flow continuously through the pipe.

The annual energy costs for electrical and steam tracing can be calculated as the following:

D = 10 in.	E = 0.70¢/kwh.
t_p = 210 F.	S = 85¢/1,000 lb.
t_a = 70 F.	h = 930 Btu./hr.
$t_{a(min.)}$ = 15 F.	L = 500 ft.
t_s = 267 F.	T = 8,400 hr.

From the thermal conductivity value in Table I:

$$C_i = 2.126 \times 0.37 = 0.786 \text{ Btu./(hr.)(ft.)}$$

Electrical tracing: annual energy cost = (0.786) (210 — 70)(0.007)(500)(8,400)/(3,413) = $950.

Steam tracing: $C_{t(min.)}$ = (1.15)(0.786)(210 − 15)/(267 − 210) = 3.10. From Table III: C_t = 3.27. The annual energy cost for steam: (3.27)(267 — 210)(0.85)(500)(8,400)/(1,000)(930) = $715.

These examples, like the graphical ones, show that energy costs can frequently favor electrical tracing. While steam tracing has the advantage of lower-cost Btu.'s, electrical tracing consumes less Btu.'s. Because steam tracing cannot be controlled thermostatically, it must be overdesigned.

Of course, energy cost is not the only basis on which steam and electrical tracing can be compared. There are, in addition, initial costs, which include design, materials and installation, and maintenance.

Reducing the estimation of these initial and maintenance costs to simple formulas is difficult because the factors on which such estimates depend vary widely from job to job and plant to plant. Generally, however, the initial cost for electrical tracing is competitive with that for steam tracing. Maintenance costs almost always favor electrical tracing by a substantial margin.

The past advantages of electrical tracing—precise temperature control, cleanliness, efficient operation and the elimination of such costly maintenance problems as steam leaks and the downtime caused by them—added to the more-recent attraction of lower electrical energy costs pose for engineers and managers a new choice between steam and electrical tracing for pipes.

Meet the Author

Charles H. Butz has been with E.I. du Pont de Nemours & Co. for the four years since he received his B.S. degree in chemistry from Bethany College. Currently, he is a marketing representative for Du Pont's Fluorocarbons Div. in the southeastern U. S. Before this, he served in assignments at Du Pont's experimental station, technical services laboratory and marketing section of the Fluorocarbons Div.

Key Concepts for This Article

Active (8)	Passive (9)
Comparing	Pipes*
Calculating	Tracing*
	Heating*
	Electricity*
	Steam*
	Costs*

(Words in bold are role indicators; numbers correspond to EJC-AIChE information retrieval system. Asterisks mark key concepts suggested for indexing. Others are added to improve reading as an abstract. Indexing is described in *Chem. Eng.*, Oct. 11, 1965, p. 187.)

8

Selling-Price and Sales-Volume Estimates

Selling Price vs. Raw-Material Cost

A rough rule of thumb here undergoes a significant refinement, permitting the price of bulk organic chemicals to be quickly estimated from their raw-material costs.

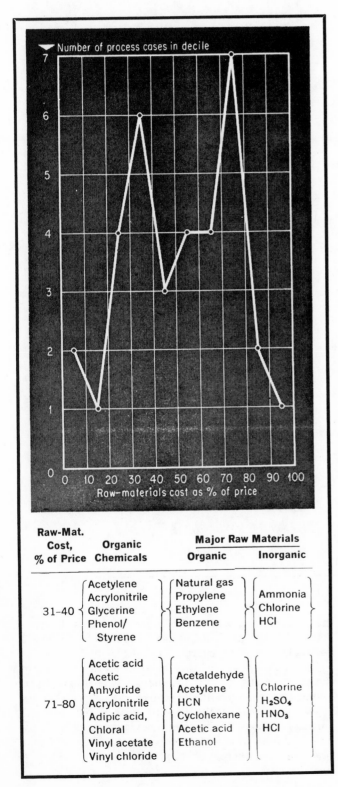

Raw-Mat. Cost, % of Price	Organic Chemicals	Major Raw Materials	
		Organic	Inorganic
31-40	Acetylene Acrylonitrile Glycerine Phenol/ Styrene	Natural gas Propylene Ethylene Benzene	Ammonia Chlorine HCl
71-80	Acetic acid Acetic Anhydride Acrylonitrile Adipic acid, Chloral Vinyl acetate Vinyl chloride	Acetaldehyde Acetylene HCN Cyclohexane Acetic acid Ethanol	Chlorine H_2SO_4 HNO_3 HCl

E. L. GRUMER, *Celanese Chemical Co.*

In evaluating the cursory economics of a chemical process, the usual approach is to gather and digest pertinent data, draw a flowsheet, calculate a material balance, do a heat balance, size equipment, estimate capital costs, and apply appropriate factors for final presentation. Since each of these steps seldom requires less than a day, an estimated sales price becomes available after seven working days at best. More-reliable results can be obtained by spending two to three weeks on the project, depending on the process novelty and complexity, and on the degree to which computer techniques can be used.

Attempts to bypass the equipment sizing and to eliminate detailed material and heat balances usually produce such a low degree of accuracy that the estimate is nearly worthless. A really meaningful estimate without equipment sizing does not appear to be feasible.

Occasions do arise, however, when time does not allow a detailed economic evaluation. One must then settle for broad deviations from the correct answer. To make such a very quick estimate, particularly for a new chemical, a direct proportion is frequently proposed between raw-material costs and selling price. As a rule of thumb, raw-material costs are considered to constitute 50% of the selling price; the remaining 50% includes operating costs and return on capital.[1] (This criterion assumes that a commercial process has evolved, usually with considerable development work.) The ratio will naturally vary with the relative costs of raw materials, equipment and labor.

Basis for the Study

Selecting raw-material costs as a basis has the advantage of taking into account both stoichiometry and yield. The term "raw material," however, immediately raises a question of definition. Obviously, nylon made from adipic acid and hexamethylenediamine would seem far more costly than nylon made from benzene and ammonia. This illustrates that processed raw materials must be treated differently from those substances that nature provides in bulk

TWIN PEAKS are shown in this frequency distribution of raw-material cost as percent of product price for 34 process cases. Constituents of the two peaks are shown in the table below the graph. This binodal relationship leads to a useful price estimating method.

Relationship Between Raw-Material Cost and Product Price

Organic Chemical	Raw Material	Raw-Material Cost, $/Lb. Product	Organic Chemical Price, $/Lb.	Raw-Material Cost, % of Price
1. Acetaldehyde	Acetylene	0.0397	0.09	44
	Ethylene, O_2, HCl, Cat.	0.0392	0.09	43
2. Acetic Acid	Acetaldehyde, Cat.	0.0708	0.09	78
	Butane	0.0075	0.09	8
3. Acetic Anhydride	Acetaldehyde, $CoAc_2$	0.1090	0.14	78
	Acetic Acid	0.1215	0.14	87
4. Acetylene	Natural gas, DMF	0.0218	0.064	34
	Natural gas, DMF, O_2	0.0370	0.064	58
5. Acrylonitrile	Propylene, Ammonia	0.0462	0.145	32
	Acetylene, HCN	0.1074	0.145	74
6. Adipic Acid	Cyclohexane	0.0257	0.29	9
	Cyclohexane, HNO_3, Cat.	0.2054	0.29	71
7. Benzoic Acid	Phthalic Anhydride	0.1050	0.1502	70
8. Butadiene	Butylenes	0.0726	0.1175	62
	Butane	0.0200	0.1175	17
9. Chloral	Ethanol, Cl_2, H_2SO_4	0.1672	0.21	80
10. Crotonaldehyde	Acetaldehyde	0.1460	0.215	68
11. Ethyl Acetate	Ethanol, Acetic Acid	0.1114	0.13	86
12. Ethylene Glycol	Ethylene, Silver	0.0360	0.135	27
	Ethylene, Cl_2, Lime, NaOH	0.1262	0.135	93
13. Ethylene Oxide	Ethylene, Silver	0.0440	0.155	28
	Ethylene, Cl_2, Lime	0.1024	0.155	66
14. Formaldehyde	Methanol	0.0191	0.035	55
15. Glycerine	Propylene, Cl_2, NaOH, Lime	0.0886	0.2375	37
16. Isopropanol	Propylene, Sulfuric Acid	0.0188	0.0733	26
17. MIBK	Acetone, Hydrogen	0.0743	0.13	57
18. Nitrobenzene	Benzene, H_2SO_4, HNO_3	0.0513	0.095	54
19. Phenol	Benzene, HCl	0.0434	0.1125	39
20. Phthalic Anhydride	Naphthalene	0.0469	0.105	45
	Orthoxylene	0.0292	0.105	28
21. Styrene	Benzene, Ethylene, $AlCl_3$	0.0451	0.1246	36
	Benzene, Ethylene	0.0471	0.1246	38
22. Vinyl Acetate	Acetylene, Acetic Acid	0.0847	0.115	74
23. Vinyl Chloride	Acetylene, HCl, $HgCl_2$	0.0638	0.08	80

without undergoing much chemical reaction by man. Hereafter I shall differentiate raw materials already subjected to chemical processing by the term "intermediates."

As was done in the previous search for a rule of thumb,[1] raw-materials costs were examined for the organic chemicals prominent in the American economy. To provide a broad and consistent source, the book "Industrial Chemicals"[2] was used for raw-materials consumption of 30 major compounds. We selected the *Oil, Paint and Drug Reporter*[3] to provide raw-material and product prices, supplemented where necessary by other current literature quotations.

The table shown on this page includes 23 organic chemicals made by 34 processes. The other seven compounds were omitted because raw-material costs appeared to exceed the product cost. This apparent anomaly reflects the widely known fact that many intermediates produced largely for captive consumption are transferred on the company books at less

than market price. Most of these cases, for instance, involve alcohols. Even without these seven cases, however, one is somewhat startled to note that raw materials range from 8% to 93% of the final price. This indicates the wide deviations one must expect from the unrefined approach.

Refining the Rule of Thumb

The arithmetic average raw-material cost from the "Industrial Chemicals" survey is slightly above 50% of the product price. If a single figure is desired, this supports the common rule of thumb that product price is twice the raw-material cost.

But if we go a step further and subdivide the 34 process cases into deciles, we arrive at the interesting distribution plotted in the chart on the first page. The binodal nature is explained in the table below the chart. This shows that a lower ratio of raw-material cost to price applies to those processes using basic hydrocarbons, as opposed to those using oxidized or other intermediates and a greater proportion of inorganic acids.

Averaging the cases falling in the three deciles from 21 to 50% gives us a calculated raw-material cost to price relation of 35%. This is less than half the 75% average value for cases in the three deciles from 61 to 90%.

For comparison purposes, a further investigation was made involving the production costs of certain chemicals that undergo multiple processing and purification steps before the final stage. Production costs were arbitrarily augmented by calculating a net transfer value for a 30% return before taxes on total fixed and working capital. The latter was estimated at two month's production cost, excluding depreciation.

The results of this investigation give an average raw-material percentage of product value as 23%. It is probably more than coincidental that the over-all value one would anticipate from the "Industrial Chemicals" survey is very similar (i.e., 26%), because:

(Basic Raw-Material Percentage of Intermediates)
$\times$ (Intermediates Percentage of Final Product)

= Basic Raw-Material Percentage of Final Product; or
$$0.35 \times 0.75 = 0.26$$

This consistency is a particularly impressive check because the sources are of differing natures. It leads us to the following refined rule of thumb for bulk organic chemicals:

• Basic hydrocarbon raw materials constitute 35% of the prices for chemical intermediates.

• Organic intermediates, such as oxidation products, and considerable inorganic acid raw materials account for 75% of their final product prices.

This rule is equally applicable to open market quotations and internal transfer prices, assuming one employs a consistent source for raw-material prices. Of course, the rule is not intended as a substitute for a detailed economic evaluation when time permits.

References

1. Schuman, S..C., How Plant Size Affects Unit Costs, *Chem. Eng.*, May 1955, p. 173.
2. Faith, W. L., Keyes, D. B., Clark, R. L., "Industrial Chemicals," 3rd Ed., Wiley, 1965.
3. *Oil, Paint and Drug Reporter*, July 25, 1966.

Meet the Author

Eugene L. Grumer is a research engineer at Celanese Chemical Co.'s Technical Center at Corpus Christi, Tex. As a member of the economic evaluation group in the Chemical Engineering Section, he does both preliminary cost estimating for new processes, and development of methods. He has S.M. and Ch.E. degrees from M.I.T., following a B.S. degree in fuel technology and chemical engineering from Pennsylvania State University.

Key Concepts for This Article

Active (8)	Passive (9)	Means/Methods (10)
Estimating*	Pricing*	Raw materials*
	Chemicals, organic*	Costs*
		Relationships

(Words in bold are role indicators; numbers correspond to EJC-AIChE information retrieval system. Asterisks mark key concepts suggested for indexing. Others are added to improve reading as an abstract. Indexing is described in *Chem. Eng.*, Oct. 11, 1965, p. 187.)

How to Price New Products

This theoretical approach to product pricing is based upon variable costs and market potential for the new product. It helps decide whether or not to enter the market, and gives fast indication of product acceptance once the market has been entered.

L. SEGLIN, *FMC Corp.*

In this era of rapid development of new products, it is appropriate to seek the best way, at least theoretically, to price new materials. This is not to suggest that the answer can be determined by theory alone. Such an analysis can, however, establish a logical starting point towards a sound pricing strategy.

This article develops and presents a rational approach to the subject. It points out the information necessary for the analysis, and how to estimate the theoretical optimum price from this information. Further, it may be used as a guide for deciding whether or not to enter into the expensive commercial product-development phases of a new product.

The problem faced by the new-product entrepreneur is defined here as getting the product fully accepted by the market with the maximum of profits (or the minimum of losses). In the past, this problem has been considered as one of gaining market acceptance, while at the same time holding losses, if any, to some reasonable (but undefined) level during the critical developmental period. Losses are intuitively much to be avoided in any way possible. Hence, the prevalent approach has been to set prices on a cost-plus basis.[8]

Under this system, the price is based on an estimated cost plus some "fair" profit percentage. This completely ignores demand. It frequently includes costs that are not relevant to the pricing decision, for example, annual fixed costs. Cost-plus pricing rarely serves the seller's best interest.[7]

Optimum pricing relationships have been developed in the classical econometric theory.[1, 4, 9] However, how to apply that theory to new-product pricing is not at all obvious because the theory applies only to static, nontransient conditions. It is based on a demand-price relationship that does not change with time.[8] By definition, the demand for a new product is a continually changing phenomenon, at least until that product has reached maturity. This presents a seemingly insurmountable barrier to solving the new-product-pricing problem because, while defining a demand-price relationship for a new product at any one time is difficult enough, extending these prognostications for all times during the growth phase is nigh impossible.

The approach to be developed here is an extension of the classical treatment to this non-equilibrium, transient problem. It introduces the concept of potential demand; and this, together with reasonable assumptions usually met with in practice, makes the problem soluble. The potential demand, as its name implies, is the demand for the product as a function of price when the product has reached market maturity. After maturity, the demand for the product will either decrease because of obsolescence (replacement by other new products) or will increase or decrease slowly due to changes in the total economy.

The development of the potential demand-price relationship for a new product is achieved through market research.[10] Some of the methods used by market research analysts are described by Dean and others.[3, 5] The importance of such market research is obvious, not only when approaching the pricing problem quantitatively as in this article but also when dealing qualitatively with such vital factors as market development and financial planning.

Basic Assumptions and Conclusions

The mathematical development of the proposed optimum pricing strategy, which is not included because of space limitations, is based on these assumptions:

1. Price of a product has a definable effect on the size of the potential market for that product. This relationship is usually developed as a result of market research and analysis.[3, 5]

2. Growth of the market for the new product can be expressed by some function of time with an upper limit equal to the market potential at any given price.

3. Recommended sales price during any phase of the market development is the price that maximizes the profits during that phase. Each phase is defined by a production facility or facilities having constant variable costs.

4. Cost of producing the product is the sum of (a) fixed costs per calendar period (e.g., overhead, taxes, insurance, depreciation), and (b) variable costs that vary directly with production (e.g., raw materials, utilities).

5. Effect of this selling price upon the competitors'

$\blacktriangledown \alpha' - \alpha_0$

Values for $\alpha' - \alpha_0$, given R and N:

R	$N=8$	$N=10$	$N=11$	$N=12$	$N=15$	$N=16$	$N=20$
5	0.1567	0.2157	0.2452	0.2734	0.3498	0.3729	0.4543
10	0.2415	0.3206	0.3557	0.3873	0.4655	0.4866	0.5533
15	0.2876	0.3636	0.3947	0.4214	0.4813	0.4960	0.5369
20	0.3079	0.3726	0.3971	0.4172	0.4581	0.4671	0.4896
25	0.3125	0.3642	0.3824	0.3965	0.4228	0.4280	0.4396

Data for both table and graph:

Market growth: 20% of capacity the first year to 100% capacity the ninth year, in equal annual increments of 10%.
$X_1 = 0.2$; $X_2 = 0.3$; $X_3 = 0.4$; $X_4 = 0.5$; $X_5 = 0.6$; $X_6 = 0.7$; $X_7 = 0.8$; $X_8 = 0.9$; $X_9 = X_N = 1.0$

capacity per annual pound is no higher than the investment used to estimate the LREP; replace part or all of existing plant facilities if they have failed or are expected to fail soon.

Pricing—For either new or existing products, LREP estimates can help determine what price should be charged and whether the present or proposed price invites competition by being substantially above the LREP.

Research Appraisal—Research management can use LREP estimates, even if very crude, to help decide whether to speed up, continue, slow down or drop a program. The key to this decision is how LREP compares with the market price of competitive or substitute materials.

Competitive Position—This is probably the hardest to evaluate since data are often incomplete and/or incorrect. But management can make sounder decisions on how to cope with competitors if it knows even the approximate LREP's of competing products. If a competitor's LREP is 2¢/lb. lower than yours, he could lower the price by 2¢/lb. and still obtain a return on his investment equal to your return at the higher price. This knowledge and an estimate of the risk of a price war can provide some valuable insight concerning your abilities to stand up to competitive actions.

services should be charged at their estimated LRTV's. Steam should be valued at the cost to produce plus a minimum rate of return on the steam plant. Water should be valued at pumping and purification costs plus a profit on the water facilities, etc.

In summary, LRTV is the long-range value of a product or service at the plant gate or within battery limits. LREP covers costs and investment needed to deliver the product or service to customers and to remain competitive. Table I lists main cost and investment components.

How to Use LREP and LRTV

As already mentioned, the most important uses for both LREP and LRTV are to evaluate capital investment proposals, guide in pricing decisions, appraise research projects, and evaluate competitive positions. But the LRTV can also be used to help decide on forward and backward integration (self-sufficiency), to determine values of utilities and services, and to guide in interdivisional pricing practices.

Investment Decisions—The basic rule here is that if a product's LREP is less than or about the same as the forecast price, it would be profitable to: build a plant to make a new product; expand an existing plant for a current product if the investment for added

Operating flows for three points during time of project life—Table II

	Construction Period	1st Year	nth Year
Fixed investment......	$-F$		
Sales income..........		X_1AP	X_nAP
Construction expense..	$-Y_1F$		
Startup expense.......		$-Y_2F$	
Fixed costs associated with fixed investment		$-Y_3F$	$-Y_3F$
Other fixed costs......		$-C_f$	$-C_f$
Variable costs.........		$-X_1AC_v$	$-X_nAC_v$
Depreciation..........		$-D_1$	$-D_n$
Corporate overhead....		$-Y_4X_1AP$	$-Y_4X_nAP$

Back-Integration—If the LRTV of a raw material is lower than the present or expected purchase price, it would be more profitable to back-integrate. Conversely, if the LRTV for a raw material exceeds forecast purchase price, it would be preferable to buy. (This argument, of course, excludes intangibles such as assurance of supply.)

Value of Utilities and Services—LRTV is a good tool to determine utility and service values for plant decisions. For instance, assume you can build a steam plant and produce steam for an LRTV of 80¢/1,000 lb. If a nearby utility will sell steam over-the-fence for a long-term price of 70¢/1,000 lb., the better choice is to buy. Similarly, if steam from an existing plant costs 50¢/1,000 lb. out-of-pocket, and has a LRTV of 80¢/1,000 lb., a 70¢/1,000 lb. contract price would be preferred when major equipment must be replaced or capacity expanded.

Interdivisional Pricing—LRTV as a transfer price allows each division to make a minimum profit. If policy dictates that products be transferred between divisions at cost, LRTV's can still be used to measure the extent to which one division is benefiting from profits that really belong to another.

Interaction of LREP and Market Price

The following prices or values are pertinent to the determination of the price of a product:

Published Price—This is the price given by manufacturers in their literature and published in periodicals.

Actual Price—The price the customer must actually pay for a product (this is very often never known by competitors).

Use Value—The maximum amount a customer would be willing to pay for a product in filling his needs, usually determined by other products available for the same applications.

The relationship of LREP to the foregoing prices and values, and some of the factors determining these values, is as follows:

1. Since LREP is simply a calculated price required to achieve a given long-term rate of return on investment, it is neither a ceiling nor floor on price over any given period of time. It will seldom equal actual product price.

2. The larger, more efficient, more back-integrated a producer, the lower will be his LREP.

3. Use value determines the upper limit on a product's actual price. Especially for new products, use value is often considerably higher than LREP.

4. The higher the actual price of a product compared with its LREP, the greater is the incentive for competitors to make the same or a similar product.

5. The lower the actual price of a product below its use value, the faster should be the rate at which the market accepts it.

6. Actual price often drops below LREP when capacity exceeds demand. Excess capacity leads producers to reduce actual price to large users below the published price, in the hope of increasing their share of the market and the total market size. If this practice becomes widespread, the published price also necessarily drops.

7. How long the actual price stays below the LREP depends on the period of overcapacity. The depressed price can speed growth in the total market. And growth of demand, with no corresponding capacity increase, is the main force to bring prices back to economic levels.

8. If a new product substitutes for an existing product, producers of the existing product can be expected to fight to keep their share of the market. This special form of excess capacity also can cause actual prices

---

Nomenclature

A	Plant capacity, units/yr.
C_f	Fixed costs not proportional to fixed investment, dollars/yr.
C_v	Variable cost, dollars/yr.
D_n	Depreciation charged during the nth year (n=any year), dollars
F	Fixed investment, dollars
$f_{i,N}$	Discount function at any given instant* for cash flow at the end of the Nth year
$f_{u,n}$	Uniform discount function* for cash flow during the nth year
$f_{u,1}$	Uniform discount function for cash flow during the first year
$f_{u,-m}$	Uniform discount function for cash flow during the year preceding time zero
I	1 − income tax rate
N	Economic life, years
P	LREP, dollars/unit
Q	α/N for straight-line depreciation and α'' for sum-of-the-years digits depreciation
R	Raw-material cost, \$/unit
X_n	Fraction of capacity produced in the nth year
Y_1	Fraction of F to cover construction expense
Y_2	Fraction of F for startup expense
Y_3	Fraction of F incurred as a fixed cost
Y_4	Fraction of sales to cover corporate overhead
Y_5	Fraction of sales for cash requirements
Y_6	Fraction of sales as accounts receivable
Y_7	Fraction of annual raw-material needs as accounts payable
Y_8	Fraction of annual raw-material needs inventoried
Y_9	Fraction of F maintained as spare parts
Y_{10}	Fraction of annual production in inventory
α	$\sum_1^N f_{u,n}$
α'	$\sum_1^N X_n f_{u,n}$
α''	$\sum_1^N \dfrac{2}{N}\left[\dfrac{N+1-n}{N+1}\right] f_{u,n}$
α_0	$\sum_2^N X_{n-1} f_{u,n} + X_N f_{i,N}$
β	$f_{u,1}$
γ	$f_{i,N}$
θ	$f_{u,-m}$

*Discount function at an instant is defined as e^{rt}, and uniform discount function over the individual years as $(e^{rt}-1)/rt$. Here e = the mathematical factor 2.71828 r = annual rate of interest (expressed as a decimal), and t = time in years.

Elements of working capital—Table III

Working capital	1st Year	nth Year	End of Economic Life
Cash	$Y_5 X_1 A P$	$Y_5 X_n A P$	
Accounts receivable	$Y_6 X_1 A P$	$Y_6 X_n A P$	
Accounts payable	$Y_7 X_1 A R$	$Y_7 X_n A R$	
Raw-material inventory	$Y_8 X_1 A R$	$Y_8 X_n A R$	
Spare parts	$Y_9 F$	$Y_9 F$	
Product inventory	$Y_{10}(Y_3 F + C_f + X_1 A C_v)$	$Y_{10}(Y_3 F + C_f + X_n A C_v)$	
Change in working capital	$X_1 A R (Y_7 - Y_8) -$ $P(Y_5 + Y_6) - Y_{10} C_v -$ $F(Y_9 - Y_{10} Y_3) - Y_{10} C_f$	$A(X_n - X_{n-1}) R (Y_7 - Y_8) -$ $P(Y_5 + Y_6) - Y_{10} C_v)$	$X_n A R (Y_8 - Y_7) +$ $P(Y_5 + Y_6) + Y_{10} C_v +$ $F(Y_9 + Y_{10} Y_3) + Y_{10} C_f$

for the existing product to fall below its LREP, since investment in facilities is "sunk." Such price competition decreases the advantage for the new product and slows its rate of penetration.

9. In a highly competitive situation, actual prices may trend toward the breakeven point of the most efficient producers. The worst competitive situation would be where large excess capacity exists for a new product, and that product threatens the markets for well-established older products.

Calculation of LREP and LRTV

To simplify the work of calculating LREP's longhand, we have derived a general equation and calculated its constants, using the interest rate of return method. The LREP equation would be simpler but, we feel, less versatile if the rate of return were not based on the time value of money.

The basic form of the LREP equation is:

$$P = f_1(F/A) + f_2(C_f/A) + f_3(R) + f_4(C_v)$$

For given assumptions of the f variables, LREP and LRTV change only with the investment per unit of capacity, the fixed cost per unit of capacity, and the variable cost (which includes raw materials).

Table II shows elements of LRTV and LREP in algebraic form for three points in time over the project life. The first column indicates the construction period when a net cash outflow results from purchase of equipment and plant erection. During the first year, startup expenses are expected that will usually not continue later in the life of the project. The third column shows any year after the first. Thus, the operating flows consist of the incoming flow of sales revenue and the outgoing flow of startup expense in the first year—and, in all years of the project, fixed costs and variable costs. Since fixed costs are often shown as a fraction of the fixed investment plus other fixed costs, the same distinction has been made.

Table III shows the parts of working capital that may change with time. Because working capital is recovered at the end of the economic life, a positive net cash flow develops.

To get the main LREP or LRTV equation that follows below: sum up the profit after tax, the change in capital requirements, and depreciation; apply appropriate discount factors; express depreciation in terms of F; and rearrange terms:

$$
\begin{aligned}
P = {}& \frac{F}{A}\left[\frac{\theta - Q(1 - I) + Y_2 I\beta + Y_3 I\alpha}{(1 - Y_4) I\alpha' - (Y_5 + Y_6)(\alpha' - \alpha_o)}\right] + \\
& \frac{F}{A}\left[\frac{(Y_9 + Y_{10} Y_3)(\beta - \gamma) + Y_1 I\theta}{(1 - Y_4) I\alpha' - (Y_5 + Y_6)(\alpha' - \alpha_o)}\right] + \\
& \frac{C_f}{A}\left[\frac{I\alpha + Y_{10}(\beta - \gamma)}{(1 - Y_4) I\alpha' - (Y_5 + Y_6)(\alpha' - \alpha_o)}\right] - \\
& R\left[\frac{(Y_7 - Y_8)(\alpha' - \alpha_o)}{(1 - Y_4) I\alpha' - (Y_5 + Y_6)(\alpha' - \alpha_o)}\right] + \\
& C_v\left[\frac{I\alpha' + Y_{10}(\alpha' - \alpha_o)}{(1 - Y_4) I\alpha' - (Y_5 + Y_6)(\alpha' - \alpha_o)}\right]
\end{aligned}
$$

Note that this form is merely an enlargement of the basic equation. Here f_1, f_2, f_3, f_4 have become the bracketed modifiers divided by a common denominator. Sensitivity analysis is greatly simplified. Once the values of the modifiers have been calculated, minor algebraic computation will show the economic effect of changes in investments or costs. Also, the area of greatest significance to the economic evaluation can be quickly identified.

Depreciation enters the equation only in the second term within the brackets modifying F/A. In the equation for straight-line depreciation, this is the term $(a/N)(1 - I)$, and for sum-of-the-years digits depreciation it is the term $a''(1 - I)$. If the depreciable life is different from the economic life, choose the value of a or a'' at N equal to the depreciable life. All other terms in the equation would be at N equal to the economic life.

Use of Tables and Graphs

The use of the equation can be simplified through tables and graphs. Because the terms F, A, I, C_f, R, C_v, and the subscripted Y's must remain as variables, all that remain are the Greek-letter terms. The only difference between straight-line depreciation and sum-of-the-years-digits depreciation is the use of a'' instead of a/N for the value of Q.

Since the Greek-letter terms are only discount functions dependent on economic and depreciable life, rate of return, and rate of market growth, these variables can be presented in table or graph form. Atlas Chemical Industries, Inc., has prepared a pamphlet of tables and graphs of these factors for a wide range of economic lives.* Each variable is shown in both forms

---

* Copies of this pamphlet are available to readers upon request to the authors.

as illustrated above by the typical table and graph for $(\alpha' - \alpha_0)$. Here N refers to the economic life and R is the desired interest rate of return.

In short, LREP and LRTV can be used to evaluate investments, new research products, your position in established products, and to provide a pricing guideline. LRTV, on the other hand, helps to price products between divisions, evaluate integration, and in addition determine the value of services and in-plant transfers.

With the help of a general equation such as presented here, some of whose terms can be found by the use of tables and graphs, the LREP and LRTV can be quickly calculated. The sensitivity to variations in costs or investments can be likewise found for many assumptions.

Example of Use of LREP Equation

Problem: Product X has been developed in the laboratory, and Engineering has prepared an estimate for a 5-million-lb./yr. plant based on the results of a market research study. Discussions with Production have indicated the operating requirements. What price would be needed to give a 15% per year rate of return on the project? The expected growth rate is from 30% to 100% of capacity in eight years.

Data:

A Plant capacity . . . . 5 million lb./yr.

C_f Fixed operating costs not dependent on investment value . . . . $210,000/yr.

C_v Variable operating costs including raw materials . . . . $0.11/lb.

F Fixed investment . . . . $2 million

I 1 − income tax rate . . . . 0.5 (50% tax rate)

N Economic and depreciable life . . . . 11 yr.

R Raw-material cost . . . . $0.10/lb.

Y_1 Fraction of F to cover $60,000 construction expense 0.03

Y_2 Fraction of F for $100,000 start-up expense 0.05

Y_3 Fraction of F incurred as a fixed cost . . . . 0.05 (5% of F/yr.)

Y_4 Fraction of sales to cover corporate overhead . . . . 0.15 (15% of sales)

Y_5 Fraction of sales for cash requirements . . . . 0.01 (1% of sales)

Y_6 Fraction of sales as accounts receivable . . . . 0.083 (1 month of sales)

Y_7 Fraction of annual raw-material needs as accounts payable . . . . 0.083 (1 month material needs)

Y_8 Fraction of annual raw-material needs inventoried . . . . 0.083 (1 month material inventory)

Y_9 Fraction of F maintained as spare parts 0.0

Y_{10} Fraction of annual production in inventory . . . . 0.083 (1 month product inventory)

From prepared table* or by calculation:

$$\alpha = \sum_{1}^{N} f_{u,n} \dots\dots\dots\dots\dots 5.3862$$

$$\alpha' = \sum_{1}^{N} X_n f_{u,n} \dots\dots\dots\dots\dots 3.3975$$

$$\alpha'' = \sum_{1}^{N} \frac{2}{N}\left[\frac{N+1-n}{N+1}\right] f_{u,n} \dots\dots\dots 0.6068$$

$$\alpha' - \alpha_o \quad \left(\text{where } \alpha_o = \sum_{2}^{N} X_{n-1} f_{u,n} + X_N f_{i,N}\right) \dots 0.4596$$

$$\beta = f_{u,1} \dots\dots\dots\dots\dots\dots 0.9286$$

$$\beta - \gamma \quad (\text{where } \gamma = f_{i,N}) \dots\dots\dots\dots 0.7366$$

$$\theta = f_{u,-m} \dots\dots\dots\dots\dots\dots 1.0789$$

Substituting in the terms of the equation for sum-of-the-years digits depreciation ($Q = \alpha''$):

$F/A = 0.40$	$Y_{10}(\beta - \gamma) = 0.0611$
$\alpha''(1 - I) = 0.3034$	$R = 0.10$
$Y_2 I \beta = 0.0232$	$(Y_7 - Y_8)(\alpha' - \alpha_0) = 0.0$
$Y_3 I \alpha = 0.1347$	$C_v = 0.11$
$(Y_9 + Y_{10}Y_3)(\beta - \gamma) = 0.0031$	$I\alpha' = 1.6988$
$Y_1 I \theta = 0.0162$	$Y_{10}(\alpha' - \alpha_0) = 0.03815$
$C_f/A = 0.042$	$(1 - Y_4)I\alpha' = 1.4439$
$I\alpha = 2.6931$	$(Y_5 + Y_6)(\alpha' - \alpha_0) = 0.0427$

Answer: P = $0.49/lb., which is the average price for product X that must be received over the next 11 years to yield a 15% per year return on all associated invested capital.

* Values for $f_{i,N}$, $f_{u,n}$, $f_{u,1}$ and $f_{u,-m}$ obtained from tables prepared by J. C. Gregory, The Atlantic Refining Co., Philadelphia (1946).

Key Concepts for This Article

Active (8)	Passive (9)	Purpose of (4)	Means/Methods (10)
Calculating*	Long-range*	Covering	Formulas*
	Prices*	Expenses	Equations
		Rate of return	

(Words in bold are role indicators; numbers correspond to EJC-AIChE information retrieval system. Asterisks mark key concepts suggested for indexing. Others are added to improve reading as an abstract. Indexing is described in *Chem. Eng.*, Oct. 11, 1965, p. 187.)

Meet the Authors

Richard T. Cheslow has been long-range forecasting engineer of Atlas Chemical Industries, Inc., Wilmington, Del., since the end of 1962. Prior to this, he worked as a chemical engineer and in economic development for Hooker Chemical Corp., after first having been laboratory shift supervisor for U.S. Rubber Co. He holds a B.S.Ch.E. degree from the Massachusetts Institute of Technology and an M.B.A. degree from the University of Buffalo. He is a member of AACE, AIChE and also of the National Assn. of Business Economists.

Alan G. Bates has been manager of economic evaluation of Atlas Chemical Industries, Inc., Wilmington, Del., since 1960. He joined Atlas' economic evaluation group in 1958 after working as process design supervisor for Celanese Corp. of America and, prior to this, as process engineer for Union Carbide Nuclear Corp. He holds both B.S. in Ch.E. and M.S.Ch.E. degrees from the Massachusetts Institute of Technology. A member of AIChE, ACS, AACE and the Institute of Management Sciences, he has numerous publications to his credit.

Supply and Demand Curves In Profitability Analysis

This graphical technique is useful for investigating the profitability of a venture when the demand curve for the product can be defined—e.g., for identifying price-volume combinations at which the discounted rate of return for the venture is maximized, or at which profitability exceeds preestablished minimum acceptable rates of return.

CALVIN S. MOORE, Ethyl Corp.

The relationship between supply and demand curves for a product determines how the profitability varies with production or sales volume. An optimum volume, or range of volumes, may exist at which profitability (discounted rate of return) is maximized. Similarly, a minimum rate of return may occur at one or more production levels. In other possible cases,

no maximum or minimum occurs, and the profitability varies either directly or inversely with volume.

Occurrence of these various situations can be explained by the relative positions of the supply and demand curves. Also, if a minimum acceptable rate of return can be established for a project, the relationship of the supply curve (corresponding to the minimum rate of return) and the demand curve, indicates whether satisfying any portion of the demand curve will result in an acceptable venture.

Curve Characteristics

For all products, some relationship exists between the price of the product and the size of the potential market. This relationship is commonly called a demand curve. If the slope of the demand curve is zero

SUPPLY AND DEMAND RELATIONSHIP showing a maximum return at a single intermediate volume—Fig. 1

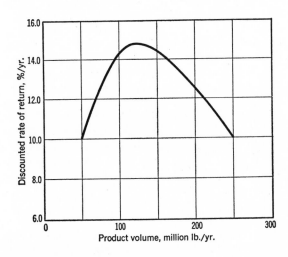

MAXIMUM RETURN of 14.8% is obtained at the single intermediate volume of 125 million lb./yr.—Fig. 2

at all points, demand is perfectly elastic, and a single price exists at which all volumes are satisfied (i.e., demand dries up completely with any increase in price). If the demand curve has an infinite slope at all points, it is perfectly inelastic, and wide changes in price produce no change in the volume demand. Generally, however, the demand curve will have a negative slope, indicating that volume demand increases as price decreases. The examples used in this discussion are based on the last case, although the principles involved are applicable to the special cases of perfectly elastic and inelastic demand curves.

A relationship also exists between the price of a product and the volume that manufacturers are willing to supply; this is described by a supply curve. The conventional supply curve discussed in economics has a positive slope and correctly indicates that industry becomes willing to supply greater quantities as the price increases. However, a different kind of supply curve can be established that reflects industry's capability of supplying a product and returning a profit on the required investment. Because potential profitability is the key criterion for industrial investment decisions, this discussion concerns the second type of supply curve. Like the demand curve, the supply curve can be elastic or inelastic depending on slope; the general principles covered are applicable to both cases.

For most industrial processes, unit production costs decrease as product volume increases for at least two reasons:

1. Fixed costs (independent of production volume) are distributed over a larger volume base, and

2. Increases in capacity usually do not require corresponding percentage changes in investment.

For the latter reason, also, the unit profit required to yield a fixed rate of return on investment decreases with increasing volume. In turn, the sales price required to yield a constant return on investment also decreases. Thus, for any specified return on investment, a supply curve with a negative slope can be established, indicating that as volume becomes larger, industry can offer lower prices, and will be willing to do this if satisfactory profitability can be achieved. This does not contradict the concept of the conventional supply curve, because if industry can supply larger volumes at lower prices, the incentive to supply larger volumes at higher prices is certainly even greater. Presumably, industry will supply according to any supply schedule for which the rate of return on investment is greater than the cost of capital.

The equilibrium level of supply and demand occurs at the intersection(s) of the supply and demand curves. The relative positions of these two curves determine the nature of their intersection(s) and are important in properly analyzing an investment project.

Analyzing the Curves: Typical Cases

The following examples illustrate various supply and demand relationships. These possibilities are not all that exist, and many extensions and combinations will be obvious.

It is essential that any analysis of this type be conducted over the entire range for which valid data are available. Failure to do this may prevent discovery of an important point. Also, any attempts to extrapolate beyond the region for which valid data are available may be erroneous, particularly since the demand curve is necessarily based on wholly empirical data. Thus, any conclusions reached must be restricted to the region studied. What appears to be one case over a narrow range may be a small part of an entirely different case viewed over a wider range.

CASE 1—Maximum Investment Return at a Single Intermediate Volume. Consider an evaluation of a project for the manufacture of a new product requiring new capacity and investment. Production, and sales and distribution costs can be divided into four categories, as shown below:

1. Unit costs independent of volume (i.e., raw materials, utilities and freight) = $0.04/lb.

2. Unit costs inversely proportional to volume (i.e., labor, supervision, overhead and other fixed costs) = $4.0-million/yr.

3. Unit costs variable with investment (i.e., insurance, maintenance and taxes) = 30% of investment annually.

4. Unit costs dependent on sales price (i.e., royalty and sales expenses) = 10% of sales price.

Although depreciation is an investment-related cost, it is not included because it can be more readily handled in the following calculations as a separate item. Total costs (except depreciation) can be expressed by:

$$C_T = 0.04V + 4.0 + 0.3I + 0.10PV$$

where: C_T = Total cost, million \$/yr.
V = Volume, million lb./yr.
I = Investment, million \$.
P = Sales price, \$/lb.

After-tax profits before depreciation will then be 1.0 minus the tax rate, times revenue (sales), less costs or: $0.52 (PV - C_T)$.

Working capital is a function of volume described by the following equation:

$$W_c = 0.668 + 0.00167V$$

where W_c = Working capital in million \$.

Of the working capital requirement, 50% is distributed over the year prior to plant startup, and the remaining 50% will be required at the end of the first year of operation. The investment in working capital will be recovered taxfree during the last year of the project life. These considerations are important for subsequent discounting calculations

A \$10-million capital investment is estimated for a plant with a capacity of 200 million lb./yr., and the investment varies with ratio of capacities to the 0.6 power. Simplified, this relationship is:

$$I = 0.415 \, V^{0.6}$$

In this example, production volume, sales volume, and plant capacity are assumed to be equal. Although

this is a simplifying assumption to facilitate calculations, it also corresponds to the actual anticipated long-term situation in most cases. Moreover, the method employed is not limited by this assumption.

The initial investment will be required as follows: 30% equally distributed over the three years prior to startup, 40% equally distributed over the two years immediately prior to startup, and the remaining 30% in the year immediately preceding startup. Using this distribution for compounding requires only a single calculation for each period, taking factors from J. C. Gregory's tables of continuous discounting.[1] The above distribution is equivalent to a distribution of 10% of the investment during the third year prior to startup, 30% during the second year, and the remaining 60% during the year immediately before startup.

Depreciation is based on the sum-of-the-years-digits method for a ten-year depreciable life, corresponding to the assumed project life. Calculations are based on a 48% federal income tax rate, and do not include any surtax.

This information is sufficient to calculate the supply curve for a given profitability criterion, which in this case is the discounted rate of return. For given production volumes, therefore, we must calculate the sale price that yields a specified discounted rate of return. This approach is analogous to that proposed for determining the long-range economic price,[2] and involves finding the price at which the sum of the cash flows, discounted to present value at the specified return rate, equals zero. At that price, the sum of the discounted revenues (including depreciation) equals the sum of the compounded investments.

Table I shows the annual cash flows during the life of the undertaking that constitutes our illustrative example.

To obtain the total cash flow for the project, the sum of the cash flows in Table I must be modified to include recovery of the remaining depreciation (52% is already included, because no depreciation was deducted in calculating after-tax profits). Use of the proper discounting factor allows this correction to be made by adding a single term, $(0.48)\,0.415V^{0.6}$, or 48% of the depreciable investment. The equation

for the sum of the discounted cash flows will then be:

$$\Sigma = -\,(0.3)0.415V^{0.6}(f_{3-\text{to }0}) - (0.4)0.415V^{0.6}(f_{-2\text{ to }0}) -$$
$$(0.3)0.415V^{0.6}(f_{-1\text{ to }0}) - 0.5(0.668 + 0.00167V)(f_{-1\text{ to }0}) -$$
$$0.5(0.668 + 0.00167V)(f_1) +$$
$$\Sigma\,\{PV - [0.04V + 4.0 + (0.3)0.415V^{0.6} + 0.10PV]\}0.52f_n +$$
$$(0.668 + 0.00167V)f_{10} + (0.48)0.415V^{0.6}f_d \quad (1)$$

where f is the appropriate discounting factor, and n ranges from 1 to 10.

Substituting factors for a 10% discounted rate of return and setting the sum of the discounted cash flows equal to zero gives:

$$0 = -\,(0.3)0.415V^{0.6}(1.166) - (0.4)0.415V^{0.6}(1.107) -$$
$$(0.3)0.415V^{0.6}(1.052) - 0.5(0.668 + 0.00167V)(1.052) -$$
$$0.5(0.668 + 0.00167V)(0.905) +$$
$$\{PV - [0.04V + 4.0 + (0.3)0.415V^{0.6} + 0.10PV]\}0.52(6.32) +$$
$$(0.668 + 0.00167V)(0.387) + (0.48)0.415V^{0.6}(0.736) \quad (2)$$

Simplifying and solving this expression for P gives the equation of the supply curve for a 10% discounted rate of return:

$$P = 0.245/V^{0.4} + 4.59/V + 0.0449 \quad (3)$$

The comparable equation for a 25% discounted rate of return is:

$$P = 0.395/V^{0.4} + 4.62/V + 0.0452 \quad (4)$$

Fig. 1 shows these two supply curves and an arbitrarily established demand curve. The supply curve for a 10% return on investment and the demand curve intersect twice, once at $V = 50$ million lb./yr., and again at $V = 250$ million lb./yr. Thus the potential market could be supplied at either of these volumes and their corresponding prices, while obtaining a 10% return on investment.

The 25% return-on-investment curve does not intersect the demand curve anywhere in the region investigated. Therefore, there is no point in this range in which the market can be satisfied while obtaining a 25% discounted rate of return. This does not eliminate the possibility that a solution may exist somewhere outside this range.

Even when mathematical expressions exist for both the supply and demand curves, care must be taken to

Annual cash flows for sample project — Table I

Year	Cash Flow	Explanation
−3 to 0	$-(0.3)0.415V^{0.6}$	30% of investment
−2 to 0	$-(0.4)0.415V^{0.6}$	40% of investment
−1 to 0	$-(0.3)0.415V^{0.6} - 0.5(0.668 + 0.00167V)$	30% of investment plus 50% of working capital
1	$\{PV - [0.04V + 4.0 + (0.3)0.415V^{0.6} + 0.10PV]\}0.52 - 0.5(0.668 + 0.00167V)$	After-tax profit on operations (after depreciation) plus 50% of working capital
2 to 9	$\{PV - [0.04V + 4.0 + (0.3)0.415V^{0.6} + 0.10PV]\}0.52$	After-tax profit on operations (after depreciation)
10	$\{PV - [0.04V + 4.0 + (0.3)0.415V^{0.6} + 0.10PV]\}0.52 + (0.668 + 0.00167V)$	After-tax profit on operations (after depreciation) plus working capital recovery

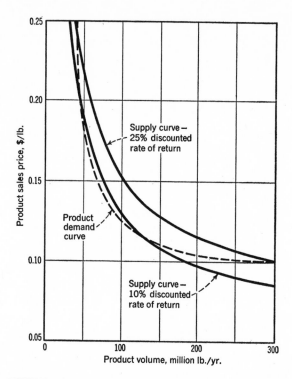

SUPPLY AND DEMAND RELATIONSHIP for situation showing a minimum investment return at a single intermediate volume—Fig. 3

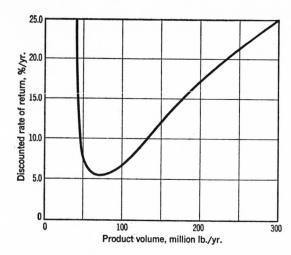

MINIMUM RETURN for this situation is 5.5%—Fig. 4

ensure that any result, although mathematically correct, has a valid physical meaning. For example, an intersection may be calculated for the supply and demand curve at a volume far exceeding the maximum potential market. This situation obviously has no meaning. A similar situation may be one yielding a solution at a negative volume.

In Fig. 1, the demand curve lies between the supply curves for 10% and 25% return at volumes between 50 million and 250 million lb./yr. Therefore, there appear to be intermediate volumes at which returns on investment between 10% and 25% are possible. This area is investigated by using the equation for the summation of cash flows to determine the discounted rate of return at various volumes and prices corresponding to the demand curve. This is a trial-and-error procedure illustrated in the following sample calculation.

At a volume of 100 million lb./yr., a sale price of $0.135/lb. is required to satisfy the demand curve. Since Fig. 1 shows that the solution at this volume will be at a discounted rate of return greater than 10%, we use the compounding and discounting factors for a 10% rate of return in the first trial. Thus, in Eq. 2, if we substitute $P = 0.135$, and $V = 100$, we find that $\Sigma = +1.730$.

Using factors for a 15% return in the second trial gives $\Sigma = -0.368$. Because the summation of cash flows has changed from positive to negative, the solution rate of return ($\Sigma = 0$) must lie between 10%

and 15%. Over this narrow range, the actual solution can be found by interpolation:

$$\text{Percent return} = 10 + 5 \left[\frac{1.730}{1.730 + 0.368} \right] = 14.1$$

Therefore, the demand curve intersects the supply curve for a 14.1% return at 100 million lb./yr. Following this procedure for other points along the demand curve and plotting the percent return versus the corresponding volume gives us Fig. 2. This figure shows that a maximum return on investment of 14.8% can be obtained at a volume of 125 million lb./yr. Thus, this is the optimum volume over the volume range considered. A single optimum solution occurs when the curvatures of the supply curves are greater than the curvature of the demand curve except at a single point of tangency, where the curvatures must be the same.

CASE 2—Minimum Investment Return at a Single Intermediate Volume. Fig. 3 illustrates the case in which the curvature of the demand curve is greater than those of the supply curves except at a single intersection. Here, a volume exists at which a minimum rate of return is obtained. An analysis similar to that for Case 1 shows that a minimum return is 5.5%, and occurs at a volume of 75 million lb./yr. Again, the data are insufficient to permit extrapolating the curve in Fig. 4. Therefore, it cannot be correctly concluded that an infinite rate of return can be obtained at very small or very large volumes. Likewise, we cannot determine whether a maximum return occurs somewhere outside of this region.

CASES 3 and 4—Maximum and Minimum Investment Return Plateaus. Cases 1 and 2 involve situations in which the supply and demand curves are tangent at a single common point, and result in volumes producing either a single maximum or minimum rate of return. Whether the result is a maximum or minimum depends on the relative curvature of the supply and demand curves. An analogous situation can exist for supply and demand curves whose intersection occurs

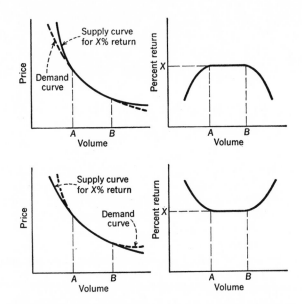

MAXIMUM AND MINIMUM investment-return plateaus are shown by these supply and demand curves—Fig. 5

NEITHER MAXIMUMS NOR MINIMUMS are represented by the investment return plateaus shown above—Fig. 6

at more than a single common point. The result is a maximum or minimum plateau. In this case, a range of volume may yield the same maximum or minimum rate of return. Fig. 5 shows curves for such cases. For the plateau, as for a single point, whether a maximum or minimum occurs depends on the relative curvature of the supply and demand curves at other than the points of intersection.

CASES 5 and 6—Investment-Return Plateaus That Are Neither Maximums Nor Minimums. Cases 3 and 4 can be combined to give plateaus that are neither maximums nor minimums. Here, the relative curvature of the supply and demand curves changes at the intersection. The resulting curves are shown in Fig. 6.

SPECIAL CASES—If the supply and demand curves have exactly the same equation, the result is a single rate of return that will be achieved at any volume. Or, if there is only a single intersection between the demand curve and any rate-of-return curve, the result is either a constantly increasing or decreasing rate of return with changes in volume. These are very special cases, and have not been plotted.

The Graphical Procedure

When the significance of the possible relationships between supply and demand curves is recognized, we do not need to actually graph the curves in analyzing every investment venture. In general, graphing the rate of return versus volume for those points that satisfy the demand curve will be adequate to determine whether a maximum or minimum exists, and these are probably the most important possibilities. However, if a maximum or minimum is indicated, enough points must be chosen to distinguish between single-point intersections and plateaus. If there is any

doubt, drawings of the supply and demand curves may be useful. Also, if a minimum acceptable rate of return is established for a project, drawing the supply line corresponding to this rate will indicate whether satisfying any portion of the demand curve will result in an acceptable venture.

All of the concepts treated graphically in this discussion can be handled mathematically. However, this usually requires fitting an equation to empirical data points for the demand curve, and the subsequent simultaneous solution with the supply curve is generally complicated and difficult. Similarly, in looking for maximum and minimum rates of return, the cash flow must be expressed as a function of the discounting and compounding factors, differentiated, set equal to zero, and solved for volume. This procedure can be treated graphically much more easily.

References

1. Gregory, J. C., Atlantic-Richfield Refining Co. These tables are reproduced in Nemmers, E. S., "Managerial Economics," Wiley, 1962.
2. Cheslow, R. T., Bates, A. G., "Long-Range Economic Price," *Chem. Eng.*, Nov. 22, 1965, p. 161.

Meet the Author

Calvin S. Moore is in the Economic Evaluation Section of Ethyl Corp.'s R&D Dept. (P.O. Box 341, Baton Rouge, La.). He joined Ethyl in 1961, and was staff economist at the firm's laboratories in Detroit for four years before taking up his present assignment last June. Mr. Moore has a B.Ch.E. degree from Georgia Institute of Technology, and has done graduate work in business administration and chemical engineering at Louisiana State University. He belongs to AIChE and AACE.

Chemical Marketing Research

What is the place of marketing research in the chemical industry? How and why is it done? What is the place of the chemical engineer in it?

ROBERT B. STOBAUGH, *Consultant*

Marketing is the business of moving goods from the producer to the consumer. It is not synonymous with selling, although selling is the heart of it, because it also includes such functions as product development, testing, pricing, distribution and advertising.

Marketing research is the systematic collecting and analyzing of marketing information, out of which come the forecasts and reports that, if accurate, aid management in making profitable decisions.

Because of the diversity of the chemical industry, its marketing research is not as yet a clearly defined activity. Sometimes it is done by professional marketing researchers, who usually belong to the Chemical Marketing Research Assn. Frequently it is carried out by such part-time researchers as field salesmen, marketing managers, market development specialists and executives.

Marketing research projects range from getting a single fact about one customer to extensive surveying and interviewing that lasts for months or years. They may explore possible markets for chemicals that do not yet exist, or probe markets for chemicals as old as sulfuric acid.

The Place of Marketing Research

When a company is only a few years old, any marketing research is likely to be handled by the president, who is usually the chief salesman as well. Later, when the company is marketing more than one product and its sales are several million dollars annually, the president, now finding his functions too complex, adds a trained marketing researcher to his staff. Finally, when the company has grown into a large, multidivisional corporation, it has a corporate marketing research group that concentrates solely on long-range planning and on problems not connected with any specific division. Each division has one or more of its own marketing researchers.

Because a large part of the marketing research of a growing company deals with new products, marketing researchers in such a company either work closely with or are in the group that develops products commercially. When being commercialized, a new chemical passes through four stages of development, and at each of these stages, the marketing researcher contributes important information and advice.

When a chemical product is nothing more than an idea, a decision must be made on whether or not money should be spent for research. At the idea stage, marketing researchers serve as consultants about the general economic prospects for the new product or conduct an exploratory survey that entails some field work. The exploratory survey may take days or weeks to complete, depending on the complexity of the problems.

At the second stage, when the chemical is being researched and a manufacturing process is being developed, whether or not to fully commercialize the chemical must be decided. This decision is generally preceded by a formal market survey that requires three to four months of fact finding and cross-checking of information.

At the third, semicommercial, stage, a decision must be made about whether or not to go ahead with large expenditures for a commercial plant. Now, the marketing researcher, working closely with the market development group, conducts a formal research project that must be as complete and reliable as possible. His field work, which must be extensive, may take six months to finish. This is when the marketing researcher's contribution is most critical, because how a company spends its capital in the present determines what its position in the industry will be in the future.

After the new product has been successfully commercialized—the fourth stage—the work of the marketing researcher is still not finished. New markets for the chemical may be found; activities of competitors must be checked; decisions must be made if markets for the chemical continue to grow about whether to expand the old plant or build a new one; salesmen must be trained to think about the needs of customers if the company is to be "market oriented"; the cheapest and fastest transportation must be selected; the operations of the marketing department must be made as efficient as possible; and, finally, when sales of the chemical are in the declining phase, the decision of

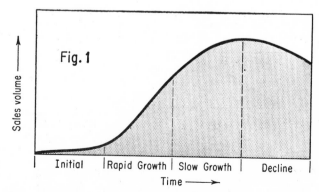

Four stages in the sales cycle of a chemical

whether or not to continue manufacturing the chemical must be faced.

When a product is in the initial phase of its sales cycle (Fig. 1), marketing research is mostly concerned with its possible uses and with the economics of producing and selling it. During the period when sales of the product are growing rapidly, the research, while still dealing with economics, is primarily directed at estimating the extent of market growth. When sales are growing slowly or are beginning to decline, marketing research mostly works at luring customers away from competitors, not by resorting to price cutting but by studying customer's needs so that they can

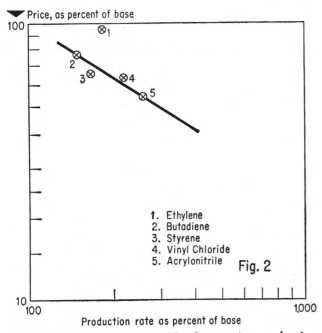

Price change correlated with change in production rate, 1962-63 figures compared to 1956-57 base.

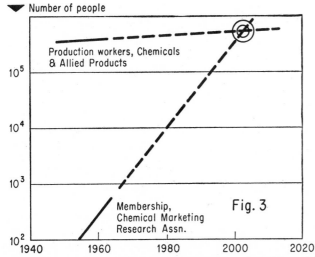

How data can be misused: CMRA membership will equal number of production workers in 35 years.

be satisfied better. So that price cutting by competitors can be met, ways of lowering costs, such as distribution expenses, are continually sought. Finally, marketing research helps decide whether or not to stop producing the chemical.

First of Four Steps

There are four steps in a marketing research study: defining the problem, getting the information, analyzing the information, and reporting the results.

Defining the problem is the key step in marketing research, for unless the essential problem is isolated and studied, a marketing effort that is not showing results cannot be corrected.

The problem must also be defined specifically enough so that it can be broken down into manageable proportions. If, for example, sales are falling in a particular territory, the obvious question is: Why are sales falling? Although the obvious problem is how can the sales decline be stopped, it must be attacked by determining the reasons behind the decline. This downtrend may have been triggered by the renewed activity of competitors, by product deficiencies, or by ineffectual advertising. Whatever the essential cause of the sales decline, pinpointing it is the vital first step in remedying the situation.

Part of the defining process consists of a "situation analysis." When he makes a "situation analysis," the marketing researcher talks with the people of his own company who might be knowledgeable about the subject, searches the literature, reviews the pertinent records of his company and, possibly, contacts several informed people outside the company. By such means, he becomes thoroughly familiar with the problem he is handling.

When he has finished his informal investigation, the researcher may have solved the problem. If he has not solved it, he at least has a thorough knowledge of it and now knows best how to attack it. He then prepares detailed plans for his research project, which includes an estimate of the time and money that must be spent to complete it.

Two Sources of Information

A marketing researcher's information comes mainly from two sources: records and publications, and personal contacts. Some important sources of written information are the following: the files and records of the researcher's own department, those of other departments (such as manufacturing, accounting, or traffic), trade journals, government publications (both state and federal), and publications of private research organizations.

When he starts a project, the researcher usually checks the files and records of his own department first. But he can always find valuable data elsewhere in his company if he knows where to look.

Most professionals keep extensive files of clippings of news items about new plants and products; or, if they do not file clippings, they file index cards upon

Examples of divergent estimates—Table I

Chemical	Estimate of:	Differing Estimates	
		A[16, 18, 20]	B[17, 19]
Benzene	Capacity in 1965, million gal./yr. by:		
	Ashland Oil Co.........	11	30
	Gulf Oil Co...........	44	59
	Hess Oil Co...........	35	30
	Suntide Oil Co........	18	25
Benzene	Production in 1965, million gal./yr........	800	842
Ethylene	Production in 1970, billion lb./yr..........	12	13–15

which are noted the sources of news items that appear in trade journals.

The statistics that the U.S. government publishes about the chemical industry are particularly reliable sources of information because companies are not apt to deliberately send false statements to the government. Foremost among government publications is the annual report put out by the U.S. Tariff Commission: *Synthetic Organic Chemicals, U.S. Production and Sales.* This report, which lists the companies that produce most of the organic chemicals made in the U.S., gives the following information about each organic chemical: how much is produced, how much is sold (excluding the quantity consumed captively by companies), the value of the total quantity sold, and the unit value of total sales (the average price per pound). Data about a chemical that is not produced by at least three manufacturers are not generally reported because publishing them would reveal confidential information.

There are two drawbacks to this report: it is not issued until nine months after the end of the year, and the exact composition of some chemicals is not stated. A report of monthly production figures of major chemicals, also published by the Tariff Commission, partially offsets the first drawback. The second problem can only be overcome by the marketing researcher's knowledge or by his investigations.

The U.S. Dept. of Commerce's *Chemical Statistics Directory*[1] lists the government publications that give information about particular chemicals.

Private organizations engaged in marketing research put out regular reports about specific subjects, make special studies that are sold to clients, and prepare studies for clients on specification.

Marketing researchers frequently contact people personally, either by telephone or interviews. If a short answer is all that he wants and if he has a large number of acquaintances (as he must if he is going to be effective), the researcher has in the telephone an inexpensive, convenient way of getting information. This is especially true when what he wants is not confidential.

If his problem is complex, the researcher usually depends on the personal interview. What people he interviews depends, of course, upon his project. He may talk with purchasing agents, research scientists, marketing specialists, development engineers, or other marketing researchers. Because the people that he may want to see are usually scattered geographically the researcher must plan his itinerary carefully, both with respect to the persons on whom he calls and to the subject that he intends to discuss. He frequently finds that people tell him more about other companies than about their own. A good researcher pays particular attention to such information.

Chemical marketing researchers spend a lot of their time interviewing people. Traveling takes up as much as a quarter to a half of his time. However, the researcher should never overlook the information that he can get within his own company because he can get this more easily. Fellow employees are usually more cooperative sources of information than outsiders.

How to Analyze Market Data

What method a researcher uses to analyze data depends, of course, upon the type of problem he is analyzing. For example, evaluating the effectiveness of a sales force calls for a different type of analysis from that used to project a product's future price and volume. Since price and volume are the two most important variables affecting the profitability of a plant,[2] the methods by which marketing researchers estimate them are important.

Estimating Price—No method has yet been published that claims to predict short-term price changes; sales managers generally have the best idea of what might be expected of prices for the immediate future. For predicting prices of the intermediate and distant future, two methods are frequently used, although neither is very accurate.

In one method, the marketing researcher calculates how much the lowest-cost competitor spends to produce the product. To this figure he adds a reasonable return on investment. He then assumes that all prices for the product will tend toward this figure.

The weakness of this method is apparent: it is, of course, often difficult to determine a competitor's costs, especially when his operation is integrated and he charges the product as a raw material to the second step of the process on an arbitrary basis.

In the other method, the researcher estimates how much greater the volume of the product in the U.S. will be six years later. If he expects the volume to become two-to-three times larger, he assumes that the future price of the product will be approximately two-thirds of the present price. On a long-term basis, substantial increases in the volume of a product is in most cases accompanied by a substantial decrease in price.[3,4,5] Fig. 2 shows this relationship for five chemicals.

The weakness of this method is that the results are only approximate and must be used with caution. Circumstances particular to a chemical must always be carefully considered.

Estimating Volume—Marketing researchers have

several methods of predicting future consumption of a chemical. If there is time, more than one method is used so that results can be checked. Of the methods to be discussed, the first is qualitative, the last four are quantitative.

1. Sample of opinions: Because this is one of the simplest methods, it tends to be popular. The researcher contacts a number of industry experts with whom he discusses future demand for a product and the factors that might influence it. His contacts are people who are especially knowledgeable about the product or in a position to influence its future course. The researcher then forms his forecast from a composite of the estimate of others, exercising judgment to skillfully derive a consensus out of conflicting opinions.

Marketing researchers who do not like this technique point out that averaging good and bad estimates does not necessarily lead to accurate forecasting. Researchers who are more thorough do not sample opinions until they have made their own study independently. They then use the opinions of others to insure that they have not overlooked key factors. A researcher who has first made his own study is also in a better position because he can discuss the future of a product more intelligently with experts.

2. Time series analysis and projection: Basically, the researcher projects past trends into the future with this technique. After drawing a straight line through the data he has plotted, he simply extends the line, assuming that the past trend will continue without change into the future. This is the simplest way of projecting. Fig. 3 shows, however, that such an approach can obviously lead to ridiculous results. The plot, based upon this method, indicates that by the end of the century that there will be more members in the Chemical Marketing Research Assn. than production workers in the chemical industry!

Obviously more-complex methods must be employed: Gompertz, logistic, exponential, or other growth curves might be fitted to historic data. Plotting exponentially weighted moving averages is a much better way of curve-fitting than the graphical methods by which chemical sales are ordinarily forecast. Marketing researchers generally consider the exponential method to be more accurate than the standard growth curves for intermediate-term estimates (up to five years), and they will probably employ it more extensively in the future.[6]

The time-series analysis and projection has two advantages: it is objective and it can be finished quickly. But it should never be used by itself because of the unreliability of the assumption upon which it is based: that marketing and technological conditions will not change and that the future is simply an extension of the past. Another disadvantage of the method is that it is of little value to the researcher when sufficient data are not available, which is generally the situation with new products.

3. Consumption pattern analysis: With this method, the researcher breaks down the total market for a particular chemical into its individual markets. He

A box score of chemical-consumption prediction techniques—Table II

Phenol (forecasts made in 1956)

Forecasting Method	Four-Year Forecast, Million Lb./Yr.	Deviation From 1960 Actual	Six-Year Forecast, Million Lb./Yr.	Deviation From 1962 Actual
1. Time series analysis and projection	730	−5.5%	830	−0.5%
2. Correlation analysis[1]	766	−0.9%	908	+10.0%
3. Consumption pattern analysis	735	−4.9%	854	+3.4%
Average	744	−3.7%	864	+4.6%
Actual consumption[2]	773	....	826	....

Polyethylene (forecasts made in 1960)

Forecasting Method	One-Year Forecast Million Lb./Yr.	Deviation From 1961 Actual	Five-Year Forecast Million Lb./Yr.	Deviation From 1965 Average
1. Historical analogy[3]	71	+9.2%	305	−4.4%[4]
2. Consumption pattern analysis	62	−4.6%	333	+4.4%[4]
Average of (1) and (2)	67	+3.1%	318	....
3. Sample of industry opinion	89	+37%	458	+25.4%[5]
Average of (1), (2) and (3)	74	+13.8%	365	....
Actual[6]	65	....	?[7]	....

1. Consumption correlated with adjusted FRB index of industrial production. 2. Based on U. S. Tariff Commission data. 3. Compared with linear polyethylene. 4. Deviation from the average of Methods 1 and 2 in polyethylene forecasts. 5. Deviation from the average of all three methods in polyethylene forecasts. 6. Based on current market appraisal. 7. Current forecast indicates range of 320-350 million lb./yr.

first forecasts how much of the chemical will be consumed in each of its end-uses, then he combines all of these forecasts to predict total consumption.

This technique is especially useful for forecasting sales of new products, when little data are available. When data are lacking, estimates of market penetration must be subjective.

Consumption pattern analysis, when used for established products, is often more accurate than methods that base estimates wholly upon the total market. Its disadvantage is that it takes a lot of work to estimate the market for each of a products end-uses.

4. Historical analogy: With this method, the researcher relates the growth of a new product to the growth of a similar, established product at a comparable period in the development. The method is highly subjective but, in conjunction with others, it is a handy device for predicting the growth of new products. Fig. 4 illustrates the technique: the market for polyester fibers is forecast by comparing its growth with that of nylon fibers at comparable stages in their development.[7]

5. Correlation analysis: Here, the researcher correlates the demand for a chemical with one or more independent variables. The procedure is objective. Its drawback is that the researcher must be able to forecast accurately the action of the independent variable. He must also relate demand for the chemical to the action of the variable.

Indicators such as the Federal Reserve Board Index of Industrial Production and the Gross National Product are most often employed, because forecasts of these indexes are easily obtained. And chemicals that are marketed widely are most easily correlated with them. Phenol, for example, is such a versatile chemical (phenolic resins, epoxy resins, oil additives, plastics, fibers, plasticizers, pertoleum refining and salicylates) that its U. S. consumption can be correlated with an adjusted FRB index of industrial productions, Fig. 5. (The FRB index was adjusted by taking only half of the points contributed by military production.)[7]

Fourth Step: Reporting Results

Although preparing a report takes only a small part of his time on a project, the researcher must remember that it is the primary means of communicating his results to other people. While he may often be called upon to make personal presentations to groups of executives, his report will be referred to long after they have forgotten the details of his oral presentation. Also, the summary of his report may be featured in a request for funds for a new plant.

The difficulties that generally trouble marketing researchers can be broadly classified: defining the essential problem; deciding when to stop the project and report the results; planning the project and judging data. A special problem occurs when the data in literature are widely divergent in reporting such things as capacities, future consumption and future prices.

Defining the Problem—This is, as has been mentioned, the key to successful researching. How much

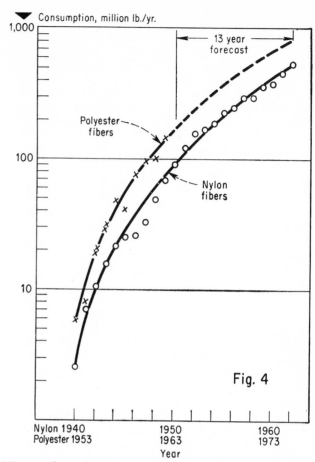

Historical analogy is sometimes an accurate guide.

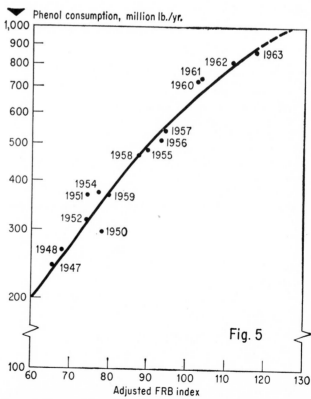

Correlation with other forecasts is another method.

work the researcher must do and how accurate his results will be depend wholly upon how well he has defined the problem at the outset. The inexperienced researcher, who frequently spends too little time defining the problem, often comes to realize half way through the project that his objective is in error.

When to Stop and Report—The researcher must guard against the natural tendency to make "just one more check" before he issues his report. Often, although he has defined the problem accurately and designed his project well, he is tempted to take additional time getting a little more data. What sometimes happens when he does this is that management goes ahead with decisions about a project without the information that he has worked so hard to collect and report. A timely report that contains a statement of its limitations is much better than a masterpiece issued six months too late.

Planning the Projects and Judging Data

When he plans his project, the researcher must decide: what method of market forecasting is most suited to the project, whom to call on to discuss the subject, and how much data must be obtained.

Because it is usually difficult to apply statistical methods in chemical marketing research, the researcher must use sound judgment when sampling opinions and analyzing data. This is especially true when the information comes from personal interviews.

To arrive at a meaningful answer to a question, the researcher must often read between the lines.

When attempting to analyze what a competitor might do in the future, the researcher must anticipate the competitor's actions by properly weighing the variables that the competitor faces. The researcher must be imaginative, analytical and experienced to be able to take a few facts, combine them with a few guesses, and realistically estimate a competitor's manufacturing costs.

Divergent Estimates in Literature—The researcher often finds that the plant capacities reported in literature vary considerably. Because chemical companies are secretive, he must expect such discrepancies (see Table 1) and should generally be skeptical about the price forecasts that are published, especially when they are made by people who might be biased.

In a typical example of such price predictions, one publication reported: "The opinion of most is that butadiene will hold at 14½¢ in the foreseeable future."[8] By the end of the following year, the price was down to 12¾¢, a drop of 12%. Another said, "It would appear that ethylene consumers, by '65, will be paying at least an average of 5.5¢/lb. for regular, and perhaps 5.7¢/lb. for high-purity material."[9] The price of ethylene at the beginning of 1965 was 4¢/lb., or 30% less than predicted. A company that had built a new plant because of such predictions would have been rudely shocked!

Despite the hazards of making such speculations, marketing researchers occasionally indulge in them.[10]

The estimates of future phenol consumption that were made by the California Chemical Co. (now Chevron Chemical Co.) vary from −0.9% to +10.0%, compared with what phenol consumption actually was four to six years after the estimates were made (Table II). Note also that in the polypropylene table the forecasts based on objective methods were far more accurate than those made from sampling opinions.

Many of the published estimates of future volume of products tend to err on the low side (Table III).

Generalizing from these meager data, it appears that predictions from the private files of companies are on the whole more accurate than many that are usually published.

The Future of Chemical Marketing Research

Chemical marketing research will continue to grow as long as the market for established chemicals keeps

Comparison of forecasts with actual U. S. production of selected chemicals—Table III

Chemical	Units	Forecast Made	Forecast For	Amount Forecast	Amount Actual	Error
Benzene from petroleum[21]	M gal./yr.	1961	1965	425	680*	+60%
Styrene[22]	MM lb./yr.	1959	1965	1,900	2,870*	+51%
Ethylene[23]	MM lb./yr.	1959	1965	6,500	9,700*	+49%
Acrylonitrile[24]	MM lb./yr.	1962	1964	415	594	+43%
Butadiene[25]	MM lb./yr.	1959	1962	1,900	2,145	+13%
Cyclohexane[26]	M gal./yr.	early 1964	1963	150	166	+11%

* Estimated by author from latest production data available from U. S. Tariff Commission.

getting more competitive, the needs of customers continually change, and research departments put out more new products. Marketing research is changing, and the following are some trends.

Marketing Research, Not Market Research—In the future, there will be more emphasis on the whole marketing process rather than on markets, because companies are adopting a "market concept," which focuses primarily upon the needs of customers. This trend will result in closer cooperation between the marketing research and the line marketing functions, because much of future marketing research will be devoted to improving the operations of line marketing.

The Chemical Marketing Research Assn. indicated its awareness of this trend when, in 1965, it amended the "market" in its name to "marketing."

Integration Toward Consumers—To insure itself a strong marketing position, the chemical industry must, many executives feel, integrate towards the consumer. As this movement occurs, the chemical marketing researcher will find himself facing problems considerably different from those encountered with large-volume industrial markets.

For example, he will have to become more knowledgeable about sampling and questionnaires, and about such special devices as the "semantic differential" (with which the meaning of an object to a consumer is measured).[11] Many of these newer techniques will require a thorough knowledge of mathematics.

New Tools—The computer has given rise to a greater interest in mathematics as a marketing tool. Besides its use to predict customer behavior, mathematics will also be used increasingly to systematically explore alternative marketing policies.[12,13]

Because more emphasis will be placed on anticipating the actions of competitors, statistical decision theory will be employed more frequently.[14] Mathematics will also be applied in dynamic simulation models.[15]

Long-Range Planning—Chemical companies have become more concerned with long-range planning during the last few years, and this trend can be expected to continue. Chemical marketing researchers have been among the leaders of this movement, and today many corporate planning groups are managed by men who were originally marketing researchers. Those who have remained in marketing research are now spending more of their time making long-range forecasts for management and corporate planners.

Marketing Research Personnel—Most chemical marketing researchers hold one or more degrees in chemical engineering or chemistry, and many have M.B.A. degrees additionally. Although many of the future marketing researchers will possess degrees in either chemical engineering or chemistry, in the future a greater percentage of them will have M.B.A. degrees as well. Because companies are becoming more oriented toward the consumer, new methods are being evolved for solving business problems, and profits rather than products are being emphasized, training in business administration will become more important to chemical marketing researchers.

References

1. U.S. Dept. of Commerce, *Chemical Statistics Directory*, Government Printing Office, Washington, D.C.
2. "Why Profitability Estimates Go Wrong," *Chem. Eng.*, Oct. 28, 1963, p. 154.
3. Stobaugh, R. B., "Can Chemical Prices be Predicted by Empirical Methods?" Paper presented at meeting of AIChE, Las Vegas, Nev., Sept. 21, 1964.
4. Stobaugh, R. B., "The Relationship Between the Prices of Certain Chemicals and Three Variable Conditions," *U. of Houston Business Review*, Fall of 1965.
5. Stobaugh, R. B., "Why Do Prices Drop?" *Chem. Eng. Progr.*, Dec. 1964, p. 13.
6. Winters, P. R., "Forecasting Sales by Exponentially Weighted Moving Averages," *Management Science*, Apr. 1960.
7. Van Arnum, K. J., "Key Factors Involved in Measuring and Forecasting Markets in the Chemical Industry." Paper presented at meeting of AIChE, Las Vegas, Nev., Sept. 21, 1964.
8. *Oil, Paint and Drug Reporter*, Mar. 23, 1959, p. 38.
9. "Ethylene Chemicals Outlook: Surge to '65," *Chem. Week*, May 9, 1959, p. 92.
10. Stobaugh, R. B., "Benzene: How, Where, Who and Future," *Hydrocarbon Proc. & Pet. Ref.*, Sept. 1965, p. 209.
11. Selltiz, C., others, "Research Methods in Social Relations," p. 380, Holt, Rinehart & Winston, New York, 1959.
12. Buzzell, R. D., "New Tools for Marketing Management in the Chemical Industries." Paper presented at meeting of AIChE, Boston, Dec. 6, 1964.
13. Buzzell, R. D., A Basic Bibliography on Mathematical Models in Marketing, AMA Bibliography Series No. 7 American Marketing Assn., Chicago, 1962.
14. Meal, H. C., "Statistical Decision Theory in Marketing Decisions." Paper presented at meeting of AIChE, Boston, Dec. 6, 1964.
15. Hegeman, G. B., "Dynamic Simulation for Market Planning," *Chem. Eng. News*, Jan. 4, 1965, p. 64.
16. "Benzene's New Look," *Chem. Week*, Mar. 6, 1965, p. 50.
17. *Oil, Paint and Drug Reporter*, June 14, 1965, p. 36.
18. Stobaugh, R. B., "Benzene: How, Where, Who and Future," *Hydrocarbon Proc. & Pet. Ref.*, Sept. 1965, p. 209.
19. Sweeney, R. C., "No Slow-Up in Petrochemicals," *The Oil and Gas Journal*, Sept. 6, 1965, p. 110.
20. Laibe, J. W., "Observations on Olefins," *Chem. Eng. Progr.*, Apr. 1965, p. 29.
21. Labine, R. A., "Petrochemicals: What's Ahead," *Chem. Eng.*, Sept. 4, 1961, p. 118. (Quotes an estimate of Cities Service Co.)
22. "Styrene to Break Records," *Chem. Eng. News*, Nov. 30, 1959, p. 26.
23. "Ethylene Chemicals Outlook: Surge to '65," *Chem. Week*, May 9, 1959, p. 83.
24. "Export Markets Put Zip in Acrylonitrile Output," *Chem. Eng. News*, Dec. 31, 1962, p. 11.
25. *Oil, Paint and Drug Reporter*, Mar. 23, 1959, p. 3.
26. *Oil, Paint and Drug Reporter*, Feb. 24, 1964, p. 52.
27. Backman, J., *Chemicals in the National Economy*, Manufacturing Chemists' Assn., Washington, D.C., 1964, p. 53. (Estimate of Chemical Marketing Research Assn. membership by author).

Meet the Author

Robert B. Stobaugh, a consultant in Newtonville, Mass., is currently studying for a doctoral degree at the Harvard Business School. Formerly the manager of economic evaluation for the worldwide activities of Monsanto Co.'s Hydrocarbons Div., he was responsible for profit analysis, economic studies, and appraisal of major ventures.

He received his B.S. degree in chemical engineering at Louisiana State University and has taken graduate work in business administration and economics at the University of Houston. He is a registered engineer (Texas) and a member of the Chemical Marketing Research Assn.

Key Concepts for This Article

Active (8)	Passive (9)	Means/Methods (10)
Marketing Research*	Costs	Analyzing
	Customers	Collection
	Markets*	Estimates*
	Products	Forecasting*
	Production	Interviews

Ind. Variable (6)	Dep. Variable (7)
Costs	Prices
Volume	

(Words in bold are role indicators; numbers correspond to EJC-AIChE information retrieval system. Asterisks mark key concepts suggested for indexing. Others are added to improve reading as an abstract. Indexing is described in *Chem. Eng.*, Oct. 11, 1965, p. 187.)

Market Simulation Makes a Science Out of Forecasting

It is now possible to arrive at more-accurate long-range forecasts by simulating in mathematical models the interplay of market forces. Such simulation models are unique in that they exploit the untapped reservoir of management experience and insight.

TERRY W. ROTHERMEL, Arthur D. Little, Inc.

Planning, in the broadest sense, entails preparation of a program to cope with, or to initiate, future changes in one's environment.

When the environment happens to be "the market"—be it the stock market, the housing market, the plastics market, or what have you—the most wanted information is: "How will the market behave?" which really means: "What will the price be?"

Hence, planning depends on forecasting. If the forecast is accurate, in the sense that it objectively projects into the future the resultant of all significant market forces, then even a hastily prepared planning program has a good chance of success. But there's a catch: How does one evaluate objectively "all the significant market forces"?

Identifying the Problem

Gathering marketing data, difficult as it may often be, is not the problem; many companies already have more marketing information than they know what to do with. Developing sophisticated mathematical and statistical tools is not the problem either since these tools are readily available.

The real problem consists of how best to capitalize on the wide spectrum of qualitative and quantitative information that a corporation has at its disposal, and how to combine management experience with a quantitative technique so as to better grasp the complex relationships that exist within the market.

The solution is achieved by establishing a simulation model that approximates the market conditions. By viewing the market as a system where all the important inputs, relationships and feedback mechanisms are reproduced and analyzed continuously, one can then greatly reduce the uncertainty and risk of projecting the course of events over time.

Why Simulation?

Historically, market forecasting has been the end-result of the efforts spent by market researchers in integrating the experience of a few individual managers with basic industry data. Though these estimates utilize management experience, they can be incomplete because most market researchers grasp only a few of the factors governing future developments. And statistical techniques, such as correlation analysis and exponential smoothing used for short-term forecasting, are merely extrapolative tools

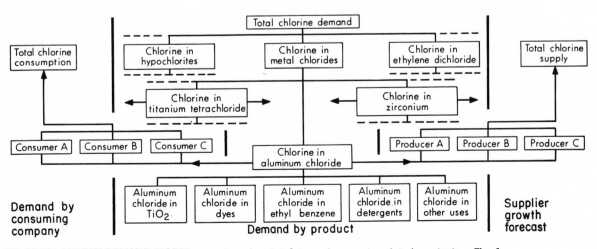

CHLORINE SUPPLY-DEMAND MODEL can also pinpoint future changes in related products—Fig. 1

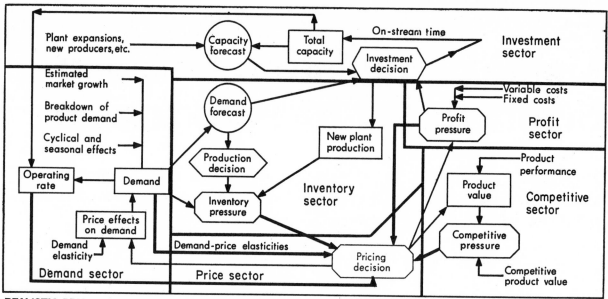

REALISTIC PRICE FORECASTING is possible with a model that integrates all key market forces—Fig. 2

for historical data; since they break down when confronted with changing technology and highly competitive industrial and consumer markets, they are of little value for the long-term type of forecasting required by corporate planning.

Dynamic simulation models, on the other hand, can be part of a continuing planning program since they can be used for both long-term and short-term forecasting. When put in the form of a mathematical model, a market simulation offers the completeness and rigor of a quantitative technique, while permitting a broader utilization of management's experience and insight, which statistical techniques ignore. Moreover, it can often be expanded, with minor changes, to forecast competitive or related product markets.

The development of a demand forecast, for example, is a major function in planning, and depends on certain assumptions of price and competition—assumptions that a management is uniquely qualified to make. Usually, however, the typical demand forecast winds up in a report, which is seldom revised and eventually loses whatever accuracy it might have had. But by building a network-demand forecast, one has a mechanism to prevent estimates from getting stale. Arthur D. Little Inc., for instance, recently developed a market model for a multi-client chlor-alkali study to show the impact of the growth of all end-uses on the demand for each of the key products (chlorine, caustic soda, soda ash and hydrochloric acid). One sector of this model was geared to chlorine demand (Fig. 1) and made it possible to establish the anticipated supply-demand balances for the related products. It also made possible development of both customer-demand and supplier-growth forecasts.*

* Hegeman, G. B., "The Computer and Forecasting of Market Demand and Prices," Advances in Chemistry series No. 88, "Chemical Marketing: The Challenges of the 70's," Copyright 1968, American Chemical Society.

A network-demand forecast of this type works as well with statistical projections as with subjective estimates of end-use demands so often required for those products where no basic market data exist.

The primary raw material for designing these models is management insight, as will be shown shortly. Hence as a first step toward model development, planning personnel get together with management to determine a preliminary flow diagram of the major relationships that exist in the marketplace and that are expected to continue to do so in the future.

It might appear that this dependence on management's views amounts to no more than a translation of managerial biases into a quantitative format. But as far as model simulation of the market is concerned, this so-called bias is management's most valuable and untapped resource.

Since a simulation model is primarily a logical network of how market forces are related, it need not be dependent upon complex mathematical functions except where such functions aid in producing the desired effect.

Ultimately, the forecasting model will consist of a network of various sectors, each representing specific factors—demand, investment, inventory, competition, profit, and price (Fig. 2).

The following will describe the typical workings of a model that might be used to forecast the price of an industrial chemical.

Model Design

Let us begin with the demand sector (Fig. 2).

To arrive at a reasonable demand forecast, the major inputs into this sector consist of a breakdown of current product-demand and of a quantitized estimate of its growth; these inputs form a basic

demand curve that can be altered by the model as additional information becomes available. It may also be appropriate to include other determinants (e.g., seasonal cycles) that affect market behavior.

But this is not enough. The model is missing the crucial relationship between price and demand. Although the theories of price elasticity and supply–demand curves are well known, in practice there are virtually no data available to feed into the model for a particular product. It is at this point that a marketing manager's insight becomes invaluable because, in many cases, he is very sensitive as to how much he will gain or lose from a change in price relative to a competing product—and this is often sufficient for defining a crude demand-elasticity function that can serve as input to the model.

Investment Sector. In industries where price and profit cycles are evident, the heart of a simulation lies in the logic of an investment sector. It is in capacity decisions that most of the cyclic behavior of industrial markets begins.

It is not very difficult to arrive at a reliable forecast of industrial capacity for one or two years on the basis of announced plant expansions and new plants expected to go onstream. But what about predicting the capacity cycles further in the future?

Here again, management's experience and insight of an industry's investment patterns are invaluable. A product manager, for instance, is aware that whenever projected profits rise above existing levels, an "optimism factor" comes into play—i.e., the higher profits will attract some new producer who, on entering the market, will upset everybody else's plans. This optimism factor has to be introduced into the investment sector to force the model into simulating the real behavior that the product manager knows all too well.

Other factors that must also be included are provisions for a margin of error in demand-and-capacity forecasts, and the simulated entry of those new producers who operate *without* the benefit of good market information.

It is also necessary to include an obvious, though often overlooked, factor: when building further capacity, the appropriate onstream time governs the delay during which the plant becomes a producing unit.

Profit Sector. A sector that calculates costs, depreciation and profits will show the changing cost-profit pressures on the investment and pricing decisions in an industry. Variable costs are multiplied by total production or sales to arrive at total variable costs. Depreciation costs may be calculated on the basis of an appropriate depreciation system. Other fixed costs are added to give a total-cost picture. Revenue is calculated on the basis of price and sales. Total profits may then be calculated.

Inventory Sector. The direct result of overcapacity is usually overproduction, or at least a tendency towards it. A typical production decision may be simulated based on a typical scheme for forecasting short-term demand, as well as on a plant's operating rate at or above breakeven levels. In addition, it is usually realistic to include the periodic excess production due to new plans coming onstream, which usually contribute significantly to an oversupply.

Competitive Sector. Since most industrial products compete against substitute or similar materials, a simulation model should have as input the effects of alternative products on the market. The objective is to keep a continuing record of the value of the product under study relative to competitive products.

Price Sector. This sector integrates the interim result of the other sectors into the major output of a price-forecasting model, which attempts to simulate the workings of the competitive process. Thus, one can see how the profit level can weigh against price falling below cost, or how overproduction created in the inventory sector can cause precipitous effects on price, or how an imbalance of product value with competitive products will affect prices, or how trends in operating rates can influence price trends.

Computer Implementation

The programming and computerization of the model need not begin until after the various sectors have been jointly formulated by management and the planning staff. While programming the model and collecting information, a number of opportunities for improvement will appear either in the form

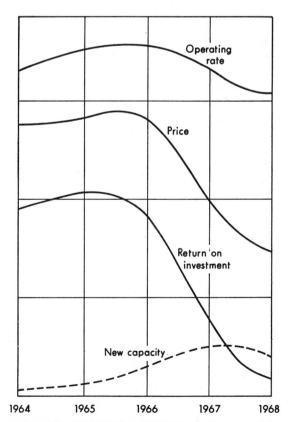

FERTILIZER INDUSTRY'S HISTORY was simulated by a model that can forecast future developments—Fig. 3

of tying together the loose ends of the formulation or better relating the model to available data.

In addition to FORTRAN, a number of other programming languages can be used in simulation. The choice of a programming language will depend on the computer system capabilities, input and output characteristics, and particularly on the ease of understanding it offers to management.

The cooperation of management and the planning staff becomes important once again, as the computer program is "tuned" to a final and satisfactory operating form. This tuning is directed not at a fixed result but at a believable model behavior that meets the expectations of its designers.

The great advantage of simulation models is their ability to answer such questions as: "What if something unexpected comes up? What if there's a shift in the market? What happens to the forecast? What if three new competing products enter the market? What do we do then?" By testing the implications of information monitored in the marketplace, a computerized simulation model can recalculate hundreds of times a whole gamut of assumptions about the future course of events. Thus, passive forecasting is converted into an aggressive program of anticipation, and continual monitoring of competition.

One byproduct of a simulation model is the indication it gives of the relative importance of different types of information. Priorities can be placed on the collection of various data, depending on the results of sensitivity analyses. It is quite possible that in some instances a forecasting model will be most sensitive to an item for which little or no information is available. In these cases, the forecasts themselves will be impaired, but management will have been warned to fill this particular information gap.

These models also hold implications for investment strategy. Without explicit enlargement into a planning and strategy model, the forecasting results may be combined with appropriate financial analyses (e.g., discounted cash flow) to provide, say, a measure of new-plant requirements and a better basis for a crucial aspect of investment decisions: timing.

The Simulation of an Industry

An illustration of how the logical structure of a model provides a framework for future strategy becomes very revealing, particularly when the lessons of recent marketing history have been as dramatic and costly as in the fertilizer industry.

Many changes shook this industry in the 1950's. High-analysis materials such as triple superphosphate, diammonium phosphate, ammonium nitrate, and urea began to replace the traditional products (such as normal superphosphate, ammonium sulfate, and others). Also, the form of the product changed from the usual pulverized form to the now preferred granular and liquid materials. And as the type of materials changed, so did the methods of distribution: retail outlets serving much smaller geo-graphical areas began to specialize in high-analysis liquid and bulk materials. Finally, economies of scale followed in the wake of the development of large potash deposits and new technology for ammonia manufacturers.

All these changes occurred while large corporations went into an acquisition binge of both large and small fertilizer manufacturers. The reasons for such acquisitions included the glamorous growth image that had been attributed to the fertilizer-food industry, the limited capital resources of the independent companies facing larger capital requirements, and the general search by large corporations for diversification. This surge of merger activity in the late 1950's and early 1960's resulted in an excess of investment in fertilizer manufacturing. By 1966, the oversupply caused major reductions—as much as 30-40% of previous levels—in prices and profits. The price history of the past few years has been reflected in a simulation model.

As demand grew from 1964 to 1966, the simulated price of fertilizers (Fig. 3) increased—and as demand was growing faster with the entry of new capacity, the industry operating rate kept increasing as well. This combination of rising prices and increased operating rates resulted in higher returns on investment during those years.

In 1966, the entry of new capacity was simulated by the model just as it happened in the marketplace: new capacity resulted in overproduction, which in turn caused competitive price-cutting; and as operating rates began to fall, return on investment deteriorated. The irrevocable decisions to boost capacity accelerated the consequences in 1967, and simulated prices fell to even lower levels much as they did in the real market, while operating rates and return on investment kept on falling through the year.

So much for past history. How useful is this model now? By feeding data inputs for 1968 instead of 1964, the simulation model can forecast future developments in price, operating rates, new capacity, and return on investment. Of particular interest is the timing and level at which price deterioration begins to bottom out.

As can be seen from this example, a simulation model that often begins as a forecasting tool for prices and markets can also predict overall industry dynamics, thus providing a more sound foundation for corporate planning.

Meet the Author

Terry W. Rothermel is a member of the Corporate and Public Management Div., Arthur D. Little, Inc. (Acorn Park, Cambridge, Mass. 02140), where he has worked on a number of simulations for price forecasting, demand forecasting, and venture strategy models. He received his B.E. and B.S. degrees in chemical engineering and industrial administration at Yale, and his M.S. from the Arthur P. Sloan School of Management at the Massachusetts Institute of Technology.

Author Index

Subject Index